APPLIED NUMERICAL METHODS USING MATLAB®

APPLIED NUMERICAL METHODS USING MATLAB®

SECOND EDITION

Won Y. Yang

Wenwu Cao

Jaekwon Kim

Kyung W. Park

Ho-Hyun Park

Jingon Joung

Jong-Suk Ro

Han L. Lee

Cheol-Ho Hong

Taeho Im

WILEY

This edition first published 2020
© 2020 John Wiley & Sons, Inc

Edition History
Wiley-Interscience (1e 2005)

All rights reserved. No part of this publication may be reproduced, stored in a retrieval system, or transmitted, in any form or by any means, electronic, mechanical, photocopying, recording or otherwise, except as permitted by law. Advice on how to obtain permission to reuse material from this title is available at http://www.wiley.com/go/permissions.

The right of Won Y. Yang, Wenwu Cao, Jaekwon Kim, Kyung W. Park, Ho-Hyun Park, Jingon Joung, Jong-Suk Ro, Han L. Lee, Cheol-Ho Hong, Taeho Im to be identified as the authors of this work has been asserted in accordance with law.

Registered Office
John Wiley & Sons, Inc., 111 River Street, Hoboken, NJ 07030, USA

Editorial Office
111 River Street, Hoboken, NJ 07030, USA

For details of our global editorial offices, customer services, and more information about Wiley products visit us at www.wiley.com.

Wiley also publishes its books in a variety of electronic formats and by print-on-demand. Some content that appears in standard print versions of this book may not be available in other formats.

Limit of Liability/Disclaimer of Warranty

MATLAB® is a trademark of The MathWorks, Inc. and is used with permission. The MathWorks does not warrant the accuracy of the text or exercises in this book. This work's use or discussion of MATLAB® software or related products does not constitute endorsement or sponsorship by The MathWorks of a particular pedagogical approach or particular use of the MATLAB® software. While the publisher and authors have used their best efforts in preparing this work, they make no representations or warranties with respect to the accuracy or completeness of the contents of this work and specifically disclaim all warranties, including without limitation any implied warranties of merchantability or fitness for a particular purpose. No warranty may be created or extended by sales representatives, written sales materials or promotional statements for this work. The fact that an organization, website, or product is referred to in this work as a citation and/or potential source of further information does not mean that the publisher and authors endorse the information or services the organization, website, or product may provide or recommendations it may make. This work is sold with the understanding that the publisher is not engaged in rendering professional services. The advice and strategies contained herein may not be suitable for your situation. You should consult with a specialist where appropriate. Further, readers should be aware that websites listed in this work may have changed or disappeared between when this work was written and when it is read. Neither the publisher nor authors shall be liable for any loss of profit or any other commercial damages, including but not limited to special, incidental, consequential, or other damages.

Library of Congress Cataloging-in-Publication Data

Names: Yang, Won Y., 1953- author. | Cao, Wenwu, author. | Kim,
 Jaekwon, 1972- author. | Park, Kyung W., 1976- author. | Park, Ho Hyun,
 1964- author. | Joung, Jingon, 1974- author. | Ro, Jong Suk, 1975-
 author. | Lee, Han L., 1983- author. | Hong, Cheol Ho, 1977- author. |
 Im, Taeho, 1979- author.
Title: Applied numerical methods using MATLAB® / Won Y. Yang, Wenwu Cao,
 Jaekwon Kim, Kyung W. Park, Ho Hyun Park, Jingon Joung, Jong Suk Ro, Han
 L. Lee, Cheol Ho Hong, Taeho Im.
Description: Second edition. | Hoboken, NJ : Wiley, 2020. | Includes
 bibliographical references and index.
Identifiers: LCCN 2019030074 (print) | LCCN 2019030075 (ebook) | ISBN
 9781119626800 (hardback) | ISBN 9781119626718 (adobe pdf) | ISBN
 9781119626824 (epub)
Subjects: LCSH: MATLAB. | Numerical analysis–Data processing.
Classification: LCC QA297 .A685 2020 (print) | LCC QA297 (ebook) | DDC
 518–dc23
LC record available at https://lccn.loc.gov/2019030074
LC ebook record available at https://lccn.loc.gov/2019030075

Cover Design: Wiley
Cover Image: © Yurchanka Siarhei/Shutterstock

Set in 10.25/12pt, TimesLTStd by SPi Global, Chennai, India

10 9 8 7 6 5 4 3 2 1

*To our parents and families
who love and support us
and
to our teachers and students
who enriched our knowledge*

CONTENTS

Preface xv

Acknowledgments xvii

About the Companion Website xix

1 MATLAB Usage and Computational Errors 1

    1.1   Basic Operations of MATLAB / 2
- 1.1.1 Input/Output of Data from MATLAB Command Window / 3
- 1.1.2 Input/Output of Data Through Files / 3
- 1.1.3 Input/Output of Data Using Keyboard / 5
- 1.1.4 Two-Dimensional (2D) Graphic Input/Output / 6
- 1.1.5 Three Dimensional (3D) Graphic Output / 12
- 1.1.6 Mathematical Functions / 13
- 1.1.7 Operations on Vectors and Matrices / 16
- 1.1.8 Random Number Generators / 25
- 1.1.9 Flow Control / 27

    1.2   Computer Errors vs. Human Mistakes / 31
- 1.2.1 IEEE 64-bit Floating-Point Number Representation / 31
- 1.2.2 Various Kinds of Computing Errors / 35
- 1.2.3 Absolute/Relative Computing Errors / 37
- 1.2.4 Error Propagation / 38
- 1.2.5 Tips for Avoiding Large Errors / 39

    1.3   Toward Good Program / 42
- 1.3.1 Nested Computing for Computational Efficiency / 42
- 1.3.2 Vector Operation vs. Loop Iteration / 43
- 1.3.3 Iterative Routine vs. Recursive Routine / 45
- 1.3.4 To Avoid Runtime Error / 45

1.3.5 Parameter Sharing via GLOBAL Variables / 49
1.3.6 Parameter Passing Through VARARGIN / 50
1.3.7 Adaptive Input Argument List / 51
Problems / 52

2 System of Linear Equations — 77

2.1 Solution for a System of Linear Equations / 78
    2.1.1 The Nonsingular Case ($M = N$) / 78
    2.1.2 The Underdetermined Case ($M < N$): Minimum-norm Solution / 79
    2.1.3 The Overdetermined Case ($M > N$): Least-squares Error Solution / 82
    2.1.4 Recursive Least-Squares Estimation (RLSE) / 83
2.2 Solving a System of Linear Equations / 86
    2.2.1 Gauss(ian) Elimination / 86
    2.2.2 Partial Pivoting / 88
    2.2.3 Gauss-Jordan Elimination / 97
2.3 Inverse Matrix / 100
2.4 Decomposition (Factorization) / 100
    2.4.1 LU Decomposition (Factorization) – Triangularization / 100
    2.4.2 Other Decomposition (Factorization) – Cholesky, QR and SVD / 105
2.5 Iterative Methods to Solve Equations / 108
    2.5.1 Jacobi Iteration / 108
    2.5.2 Gauss-Seidel Iteration / 111
    2.5.3 The Convergence of Jacobi and Gauss-Seidel Iterations / 115
Problems / 117

3 Interpolation and Curve Fitting — 129

3.1 Interpolation by Lagrange Polynomial / 130
3.2 Interpolation by Newton Polynomial / 132
3.3 Approximation by Chebyshev Polynomial / 137
3.4 Pade Approximation by Rational Function / 142
3.5 Interpolation by Cubic Spline / 146
3.6 Hermite Interpolating Polynomial / 153

3.7 Two-Dimensional Interpolation / 155
3.8 Curve Fitting / 158
   3.8.1 Straight-Line Fit – A Polynomial Function of Degree 1 / 158
   3.8.2 Polynomial Curve Fit – A Polynomial Function of Higher Degree / 160
   3.8.3 Exponential Curve Fit and Other Functions / 165
3.9 Fourier Transform / 166
   3.9.1 FFT vs. DFT / 167
   3.9.2 Physical Meaning of DFT / 169
   3.9.3 Interpolation by Using DFS / 172
   Problems / 175

4 Nonlinear Equations   197

4.1 Iterative Method toward Fixed Point / 197
4.2 Bisection Method / 201
4.3 False Position or Regula Falsi Method / 203
4.4 Newton(-Raphson) Method / 205
4.5 Secant Method / 208
4.6 Newton Method for a System of Nonlinear Equations / 209
4.7 Bairstow's Method for a Polynomial Equation / 212
4.8 Symbolic Solution for Equations / 215
4.9 Real-World Problems / 216
    Problems / 223

5 Numerical Differentiation/Integration   245

5.1 Difference Approximation for the First Derivative / 246
5.2 Approximation Error of the First Derivative / 248
5.3 Difference Approximation for Second and Higher Derivative / 253
5.4 Interpolating Polynomial and Numerical Differential / 258
5.5 Numerical Integration and Quadrature / 259
5.6 Trapezoidal Method and Simpson Method / 263
5.7 Recursive Rule and Romberg Integration / 265
5.8 Adaptive Quadrature / 268
5.9 Gauss Quadrature / 272

- 5.9.1 Gauss-Legendre Integration / 272
- 5.9.2 Gauss-Hermite Integration / 275
- 5.9.3 Gauss-Laguerre Integration / 277
- 5.9.4 Gauss-Chebyshev Integration / 277
- 5.10 Double Integral / 278
- 5.11 Integration Involving PWL Function / 281
- Problems / 285

6 Ordinary Differential Equations 305

- 6.1 Euler's Method / 306
- 6.2 Heun's Method – Trapezoidal Method / 309
- 6.3 Runge-Kutta Method / 310
- 6.4 Predictor-Corrector Method / 312
  - 6.4.1 Adams-Bashforth-Moulton Method / 312
  - 6.4.2 Hamming Method / 316
  - 6.4.3 Comparison of Methods / 317
- 6.5 Vector Differential Equations / 320
  - 6.5.1 State Equation / 320
  - 6.5.2 Discretization of LTI State Equation / 324
  - 6.5.3 High-order Differential Equation to State Equation / 327
  - 6.5.4 Stiff Equation / 328
- 6.6 Boundary Value Problem (BVP) / 333
  - 6.6.1 Shooting Method / 333
  - 6.6.2 Finite Difference Method / 336
- Problems / 341

7 Optimization 375

- 7.1 Unconstrained Optimization / 376
  - 7.1.1 Golden Search Method / 376
  - 7.1.2 Quadratic Approximation Method / 378
  - 7.1.3 Nelder-Mead Method / 380
  - 7.1.4 Steepest Descent Method / 383
  - 7.1.5 Newton Method / 385
  - 7.1.6 Conjugate Gradient Method / 387
  - 7.1.7 Simulated Annealing / 389
  - 7.1.8 Genetic Algorithm / 393

7.2 Constrained Optimization / 399
    7.2.1 Lagrange Multiplier Method / 399
    7.2.2 Penalty Function Method / 406
7.3 MATLAB Built-In Functions for Optimization / 409
    7.3.1 Unconstrained Optimization / 409
    7.3.2 Constrained Optimization / 413
    7.3.3 Linear Programming (LP) / 416
    7.3.4 Mixed Integer Linear Programming (MILP) / 423
7.4 Neural Network[K-1] / 433
7.5 Adaptive Filter[Y-3] / 439
7.6 Recursive Least Square Estimation (RLSE)[Y-3] / 443
    Problems / 448

8 Matrices and Eigenvalues   467

8.1 Eigenvalues and Eigenvectors / 468
8.2 Similarity Transformation and Diagonalization / 469
8.3 Power Method / 475
    8.3.1 Scaled Power Method / 475
    8.3.2 Inverse Power Method / 476
    8.3.3 Shifted Inverse Power Method / 477
8.4 Jacobi Method / 478
8.5 Gram-Schmidt Orthonormalization and QR Decomposition / 481
8.6 Physical Meaning of Eigenvalues/Eigenvectors / 485
8.7 Differential Equations with Eigenvectors / 489
8.8 DoA Estimation with Eigenvectors[Y-3] / 493
    Problems / 499

9 Partial Differential Equations   509

9.1 Elliptic PDE / 510
9.2 Parabolic PDE / 515
    9.2.1 The Explicit Forward Euler Method / 515
    9.2.2 The Implicit Backward Euler Method / 516
    9.2.3 The Crank-Nicholson Method / 518
    9.2.4 Using the MATLAB function 'pdepe()' / 520
    9.2.5 Two-Dimensional Parabolic PDEs / 523

xii CONTENTS

    9.3   Hyperbolic PDES / 526
           9.3.1 The Explicit Central Difference Method / 526
           9.3.2 Two-Dimensional Hyperbolic PDEs / 529
    9.4   Finite Element Method (FEM) for Solving PDE / 532
    9.5   GUI of MATLAB for Solving PDES – PDE tool / 543
           9.5.1 Basic PDEs Solvable by PDEtool / 543
           9.5.2 The Usage of PDEtool / 545
           9.5.3 Examples of Using PDEtool to Solve PDEs / 549
    Problems / 559

Appendix A Mean Value Theorem   575

Appendix B Matrix Operations/Properties   577

    B.1   Addition and Subtraction / 578
    B.2   Multiplication / 578
    B.3   Determinant / 578
    B.4   Eigenvalues and Eigenvectors of a Matrix / 579
    B.5   Inverse Matrix / 580
    B.6   Symmetric/Hermitian Matrix / 580
    B.7   Orthogonal/Unitary Matrix / 581
    B.8   Permutation Matrix / 581
    B.9   Rank / 581
    B.10 Row Space and Null Space / 581
    B.11 Row Echelon Form / 582
    B.12 Positive Definiteness / 582
    B.13 Scalar (Dot) Product and Vector (Cross) Product / 583
    B.14 Matrix Inversion Lemma / 584

Appendix C Differentiation W.R.T. A Vector   585

Appendix D Laplace Transform   587

Appendix E Fourier Transform   589

Appendix F Useful Formulas   591

Appendix G Symbolic Computation   595

    G.1   How to Declare Symbolic Variables and Handle Symbolic Expressions / 595
    G.2   Calculus / 597
           G.2.1 Symbolic Summation / 597

G.2.2 Limits / 597
G.2.3 Differentiation / 598
G.2.4 Integration / 598
G.2.5 Taylor Series Expansion / 599
G.3 Linear Algebra / 600
G.4 Solving Algebraic Equations / 601
G.5 Solving Differential Equations / 601

Appendix H Sparse Matrices   603

Appendix I MATLAB   605

References   611

Index   613

Index for MATLAB Functions   619

Index for Tables   629

PREFACE

This book introduces applied numerical methods for engineering and science students in sophomore to senior levels; it targets the students of today who do not like and/or do not have time to derive and prove mathematical results. It can also serve as a reference to MATLAB applications for professional engineers and scientists, since many of the MATLAB codes presented after introducing each algorithm's basic ideas can easily be modified to solve similar problems even by those who do not know what is going on inside the MATLAB routines and the algorithms they use. Just as most drivers have to know only where to go and how to drive a car to get to their destinations, most users have to know only how to formulate their problems that they want to solve using MATLAB and how to use the corresponding routines for solving them. We never deny that detailed knowledge about the algorithm (engine) of the program (car) is helpful for getting safely to the solution (destination); we only imply that one-time users of any MATLAB program or routine may use this book as well as the readers who want to understand the underlying principle/equations of each algorithm.

This book mainly focuses on helping readers understand the fundamental mathematical concepts and practice problem-solving skills using MATLAB-based numerical methods, skipping some tedious derivations/proofs. Obviously, basic concepts must be taught so that readers can properly formulate the mathematics problems. Afterward, readers can directly use the MATLAB codes to solve practical problems. Almost every algorithm introduced in this book is followed by example MATLAB code with a friendly interface so that students can easily modify the code to solve their own problems. The selection of exercises follows the same philosophy of making the learning easy and practical. Readers should be able to solve similar problems immediately after reading the materials and codes listed in this book. For most students – and particularly non-math majors – understanding how to use numerical tools correctly in solving their problems of interest is more important than doing lengthy proofs and derivations.

MATLAB is one of the most developed software packages available today. It provides many numerical methods and it is very easy to use, even for those having no programming technique or experience. We have supplemented MATLAB's built-in functions with over 100 small MATLAB routines. Readers should find these routines handy and useful. Some of these routines give better results for some problems than the built-in functions. Readers are encouraged to develop their own routines following the examples.

Compared with the first edition, Bairstow's method (Section 4.7), Integration Involving PWL Function (Section 5.11), Mixed Integer Linear Programming (Section 7.3.4), Neural Network (Section 7.4), Adaptive Filter (Section 7.5), Recursive Least-Squares Estimation (Section 7.6), and DoA Estimation (Section 8.8) have been added to the second edition.

Program files can be downloaded from <https://wyyang53.wixsite.com/mysite/publications>. Any questions, comments, and suggestions regarding this book are welcome and they should be mailed to wyyang53@hanmail.net.

March 2020   Won Y. Yang et al.

ACKNOWLEDGMENTS

The knowledge in this book is derived from the work of many eminent scientists, scholars, researchers, and MATLAB developers, all of whom we thank. We thank our colleagues, students, relatives, and friends for their support and encouragement. We thank the reviewers, whose comments were so helpful in tuning this book. We gratefully acknowledge the editorial, Brett Kurzman and production staff of John Wiley & Sons, Inc. including Project Editor Antony Sami and Production Editor Gayathree Sekar for their kind, efficient, and encouraging guide.

ABOUT THE COMPANION WEBSITE

Don't forget to visit the companion website for this book:

www.wiley.com/go/yang/appliednumericalmethods

Scan this QR code to visit the companion website:

There you will find valuable material designed to enhance your learning, including:

- Learning Outcomes for all chapters
- Exercises for all chapters
- References for all chapters
- Further reading for all chapters
- Figures for chapters 16, 22 and 30

1

MATLAB USAGE AND COMPUTATIONAL ERRORS

CHAPTER OUTLINE

1.1 Basic Operations of MATLAB	2
1.1.1 Input/Output of Data from MATLAB Command Window	3
1.1.2 Input/Output of Data Through Files	3
1.1.3 Input/Output of Data Using Keyboard	5
1.1.4 Two-Dimensional (2D) Graphic Input/Output	6
1.1.5 Three Dimensional (3D) Graphic Output	12
1.1.6 Mathematical Functions	13
1.1.7 Operations on Vectors and Matrices	16
1.1.8 Random Number Generators	25
1.1.9 Flow Control	27
1.2 Computer Errors vs. Human Mistakes	31
1.2.1 IEEE 64-bit Floating-Point Number Representation	31
1.2.2 Various Kinds of Computing Errors	35
1.2.3 Absolute/Relative Computing Errors	37
1.2.4 Error Propagation	38
1.2.5 Tips for Avoiding Large Errors	39
1.3 Toward Good Program	42

Applied Numerical Methods Using MATLAB®, Second Edition. Won Y. Yang, Jaekwon Kim, Kyung W. Park, Donghyun Baek, Sungjoon Lim, Jingon Joung, Suhyun Park, Han L. Lee, Woo June Choi, and Taeho Im.
© 2020 John Wiley & Sons, Inc. Published 2020 by John Wiley & Sons, Inc.
Companion website: www.wiley.com/go/yang/appliednumericalmethods

1.3.1 Nested Computing for Computational Efficiency — 42
1.3.2 Vector Operation vs. Loop Iteration — 43
1.3.3 Iterative Routine vs. Recursive Routine — 45
1.3.4 To Avoid Runtime Error — 45
1.3.5 Parameter Sharing via GLOBAL Variables — 49
1.3.6 Parameter Passing Through VARARGIN — 50
1.3.7 Adaptive Input Argument List — 51
Problems — 52

1.1 BASIC OPERATIONS OF MATLAB

MATLAB is a high-level software package with many built-in functions that make the learning of numerical methods much easier and more interesting. In this section, we will introduce some basic operations that will enable you to learn the software and build your own programs for problem solving. In the workstation environment, you type "matlab" to start the program, while in the PC environment, you simply double-click the MATLAB icon.

Once you start the MATLAB program, a Command window will open with the MATLAB prompt ≫. On the command line, you can type MATLAB commands, functions together with their input/output arguments, the names of script files containing a block of statements to be executed at a time or functions defined by users. The MATLAB program files must have the extension name ***.m to be executed in the MATLAB environment. If you want to create a new M-file or edit an existing file, you click File/New/M-file or File/Open in the top left corner of the main menu, find/select/load the file by double-clicking it, and then begin editing it in the Editor window. If the path of the file you want to run is not listed in the MATLAB search path, the file name will not be recognized by MATLAB. In such cases, you need to add the path to the MATLAB-path list by clicking the menu 'Set_Path' in the Command window, clicking the 'Add_Folder' button, browsing/clicking the folder name and finally clicking the SAVE button and the Close button. The *lookfor* command is available to help you find the MATLAB commands/functions that are related with a job you want to be done. The *help* command helps you know the usage of a particular command/function. You may type directly in the Command window

```
»lookfor repeat           or           »help for
```

to find the MATLAB commands in connection with 'repeat' or to find information about the 'for loop'

1.1.1 Input/Output of Data from MATLAB Command Window

MATLAB remembers all input data in a session (anything entered through direct keyboard input or running a script file) until the command 'clear()' is given or you exit MATLAB.

One of the many features of MATLAB is that it enables us to deal with the vectors/matrices in the same way as scalars. For instance, to input the matrices/vectors,

$$A = \begin{bmatrix} 1 & 2 & 3 \\ 4 & 5 & 6 \end{bmatrix}, \quad B = \begin{bmatrix} 3 \\ -2 \\ 1 \end{bmatrix}, \quad C = [1 \quad -2 \quad 3 \quad -4]$$

type the following statements in the MATLAB Command window:

```
»A=[1 2 3;4 5 6]
 A= 1    2    3
    4    5    6
»B=[3;-2;1]; %put the semicolon at the end of the statement to
suppress the result printout onto the screen
»C=[1 -2  3 -4]
```

At the end of the statement, press <Enter> key if you want to check the result of executing the statement immediately. Otherwise, type a semicolon ';' before pressing <Enter> key so that the Command window will not be overloaded by a long display of results.

1.1.2 Input/Output of Data Through Files

MATLAB can handle two types of data files. One is the binary format mat-files named ***.mat. This kind of files can preserve the values of more than one variable, but will be handled only in the MATLAB environment and cannot be shared with other programming environments. The other is the ASCII dat-files named ***.txt, which can be shared with other programming environments, but preserve the values of only one variable.

Beneath are a few sample statements for storing some data into a mat-file in the current directory and reading the data back from the mat-file.

```
»save ABC A B C %store the values of A,B,C into the file 'ABC.mat'
»clear A C %clear the memory of MATLAB about A,C
»A %what is the value of A?
   Undefined function or variable 'A'
»load ABC A C %read the values of A,C from the file 'ABC.mat'
»A % the value of A
   A= 1    2    3
      4    5    6
```

MATLAB USAGE AND COMPUTATIONAL ERRORS

If you want to store the data into an ASCII dat-file (in the current directory), make the filename the same as the name of the data and type '-ascii' at the end of the *save* statement.

```
»save B.txt B -ascii
```

However, with the save/load commands into/from a dat-file, the value of only one variable having the lowercase name can be saved/loaded, a scalar or a vector/matrix. Besides, nonnumeric data cannot be handled by using a dat-file. If you save a string data into a dat-file, its ASCII code will be saved. If a dat-file is constructed to have a data matrix in other environments than MATLAB, every line (row) of the file must have the same number of columns. If you want to read the data from the dat-file in MATLAB, just type the (lowercase) filename ***.txt after 'load', which will also be recognized as the name of the data contained in the dat-file.

```
»load b.txt %read the value of variable b from the ascii file 'b.txt'
```

At the MATLAB prompt, you can type 'nm112' (the filename excluding the extension name part ".m") and <Enter> key to run the following M-file "nm112.m" consisting of several file input(save)/output(load) statements. Then you will see the effects of the individual statements from the running results appearing on the screen.

```
%nm112.m
clear
A=[1 2 3;4 5 6]
B=[3;-2;1];
C(2)=2; C(4)=4
disp('Press any key to see the input/output through Files')
save ABC A B C %save A,B & C as a MAT-file named 'ABC.mat'
clear('A','C') %remove the memory about A and C
load ABC A C %read MAT-file to recollect the memory about A and C
save B.txt B -ascii %save B as an ASCII-file file named 'b.txt'
clear B
load b.txt %read ASCII-file to recollect the memory about b
b
x=input('Enter x:')
format short e
x
format rat, x
format long, x
format short, x
```

1.1.3 Input/Output of Data Using Keyboard

The command 'input' enables the user to input some data via the keyboard. For example,

```
»x=input('Enter x: ')
  Enter x: 1/3
  x= 0.3333
```

Note that the fraction 1/3 is a nonterminating decimal number, but only four digits after the decimal point is displayed as the result of executing the above command. This is a choice of formatting in MATLAB. One may choose to display more decimal places by using the command 'format', which can make a fraction show up as a fraction, as a decimal number with more digits, or even in an exponential form of a normalized number times 10 to the power of some integer. For instance:

```
»format rat % as a rational number
»x
  x= 1/3
»format long % as a decimal number with 14 digits
»x
  x= 0.33333333333333
»format long e % as a long exponential form
»x
  x= 3.333333333333333e-001
»format hex % as a hexadecimal form as represented/stored in memory
»x
  x= 3fd5555555555555
»format short e % as a short exponential form
»x
  x= 3.3333e-001
»format short % back to a short form(default)
»x
  x= 0.3333
```

Note that the number of displayed digits is not the actual number of significant digits of the value stored in computer memory. This point will be made clear in Section 1.2.1.

There are other ways of displaying the value of a variable and a string on the screen than typing the name of the variable. Two useful commands are 'disp()' and 'fprintf()'. The former displays the value of a variable or a string without 'x=' or 'ans='; the latter displays the values of several variables in a specified format and with explanatory/cosmetic strings. For example:

```
»disp('The value of x='), disp(x)
  The value of x= 0.3333
```

Table 1.1 summarizes the type specifiers and special characters that are used in 'fprintf()' statements.

6 MATLAB USAGE AND COMPUTATIONAL ERRORS

Table 1.1 Conversion type specifiers and special characters used in fprintf() statements.

Type specifier	Printing form: fprintf('**format string**', variables_to_be_printed,..)	Special character	Meaning
%c	character type	\n	new line
%s	string type	\t	tab
%d	decimal integer number type	\b	backspace
%f	floating point number type	\r	CR return
%e	decimal exponential type	\f	form feed
%x	hexadecimal integer number	%%	%
%bx	floating number in 16 hexadecimal digits(64 bits)	''	'

Beneath is a script named "nm113.m", which uses the command 'input' so that the user could input some data via the keyboard. If we run the script, it gets a value of the temperature in Fahrenheit (F) via the keyboard from the user, converts it into the temperature in Centigrade (°C) and then prints the results with some remarks both onto the screen and into a data file named 'nm113.txt'.

```
%nm113.m
f=input('Input the temperature in Fahrenheit[F]:');
c=5/9*(f-32);
fprintf('%5.2f(in Fahrenheit) is %5.2f(in Centigrade).\n',f,c)
fid=fopen('nm113.txt','w')
fprintf(fid,'%5.2f(Fahrenheit) is %5.2f(Centigrade).\n',f,c)
fclose(fid)
```

In case you want the keyboard input to be recognized as a string, you should add the character 's' as the second input argument.

```
»ans=input('Answer <yes> or <no>: ','s')
```

1.1.4 Two-Dimensional (2D) Graphic Input/Output

How do we plot the value(s) of a vector or an array? Suppose data reflecting the highest/lowest temperatures for five days are stored as a 5×2 array in an ASCII file named 'temp.txt'.

The job of the MATLAB script "nm01f01.m" is to plot these data. Running the script yields the graph shown in Figure 1.1a. Note that the first two lines are comments about the name and the functional objective of the program (file), and the fifth and sixth lines are auxiliary statements that designate the graph title and units of the vertical/horizontal axis; only the third and fourth lines are indispensable in drawing the colored graph. We need only a few MATLAB statements for this artwork, which shows the power of MATLAB.

BASIC OPERATIONS OF MATLAB 7

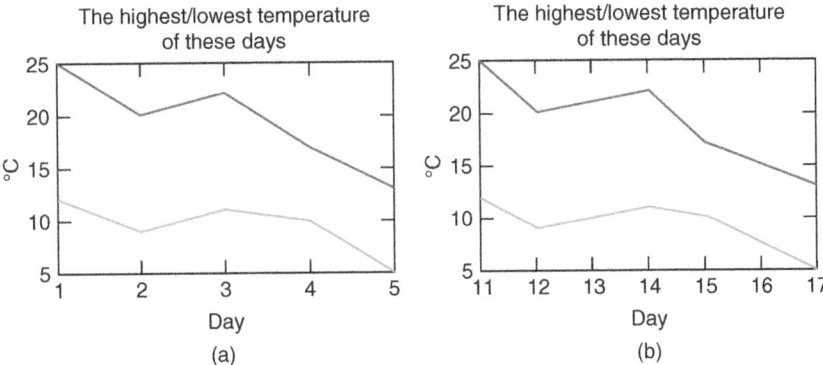

Figure 1.1 Plot of a 5×2 matrix data representing the variations of the highest/lowest temperature. Domain of the horizontal variable unspecified (a) and specified (b).

```
%nm01f01.m
% plot the data of a 5x2 array stored in "temp.txt"
load temp.txt    % load the data file "temp.txt"
clf, plot(temp)  % clear any existent figure and plot
title('The highest/lowest temperature of these days')
ylabel('degrees[C]'), xlabel('day')
```

Here are several things to keep in mind.

- The command 'plot()' reads along the columns of the 5×2 array data given as its input argument and recognizes each column as the value of a vector.
- MATLAB assumes the domain of the horizontal variable to be [1 2 ... 5] by default, where 5 equals the length of the vector to be plotted (see Figure 1.1a).
- The graph is constructed by connecting the data points with the straight lines and is piecewise-linear (PWL), while it looks like a curve as the data points are densely collected. Note that the graph can be plotted as points in various forms according to the optional input argument described in Table 1.2.

Table 1.2 Graphic line specifications used in the 'plot()' command.

Line type		Point type (marker symbol)					Color		
-	Solid line	.	Dot	+	Plus	*	Asterisk	r: Red	m: Magenta
:	Dotted line	^:	△	>:	>	o	Circle	g: Green	y: Yellow
--	Dashed line	p:	*	v:	▽	x:	x-Mark	b: Blue	c: Cyan (sky blue)
-.	Dash-dot	d:	◇	<:	<	s:□	Square	k: Black	

8 MATLAB USAGE AND COMPUTATIONAL ERRORS

(Q1) Suppose the data in the array named 'temp' are the highest/lowest temperatures measured on the 11th, 12th, 14th, 16th, and 17th days, respectively. How should we modify the above script to have the actual days shown on the horizontal axis?

(A1) Just make the day vector [11 12 14 16 17] and use it as the first input argument of the 'plot()' command.

```
»days=[11 12 14 16 17]; plot(days,temp)
```

Running these statements yields the graph in Figure 1.1b.

(Q2) How do we change the ranges of the horizontal/vertical axes into 10–20 and 0–30, respectively, and draw the grid on the graph?

(A2) »axis([10 20 0 30]), grid on

(Q3) How can we change the range of just the horizontal or vertical axis, separately?

(A3) »xlim([11 17]); ylim([0 30])

(Q4) How do we make the scales of the horizontal/vertical axes equal so that a circle appears round, not like an ellipse?

(A4) »axis('equal')

(Q5) How do we fix the tick values and their labels of the horizontal/vertical axes?

(A5) »set(gca,'xtick',[11:3:17],'xticklabel',{'11','14','17'})
»set(gca,'ytick',[5:5:15],'yticklabel',{'5','15','25'})

where gca is the current (figure) axis and gcf is the current figure handle.

(Q6) How do we have another graph overlapped onto an existing graph?

(A6) If you run the 'hold on' command after plotting the first graph, any following graphs in the same section will be overlapped onto the existing one(s) rather than plotted newly. For example:

```
»hold on, plot(days,temp(:,1),'b*', days,temp(:,2),'ro')
```

This will be good until you issue the command 'hold off' or clear all the graphs in the graphic window by using the 'clf' command.

Sometimes we need to see the inter-relationship between two variables. Suppose we want to plot the lowest/highest temperature, respectively, along the horizontal/vertical axis in order to grasp the relationship between them. Let us try using the following command:

```
»plot(temp(:,1),temp(:,2),'kx') % temp(:,2) vs. temp(:,1) in black 'x'
```

This will produce a point-wise graph like Figure 1.2a, which is fine. But if you replace the third input argument by 'b-' or just omit it to draw a PWL graph connecting the data points like Figure 1.2b, the graphic result looks clumsy,

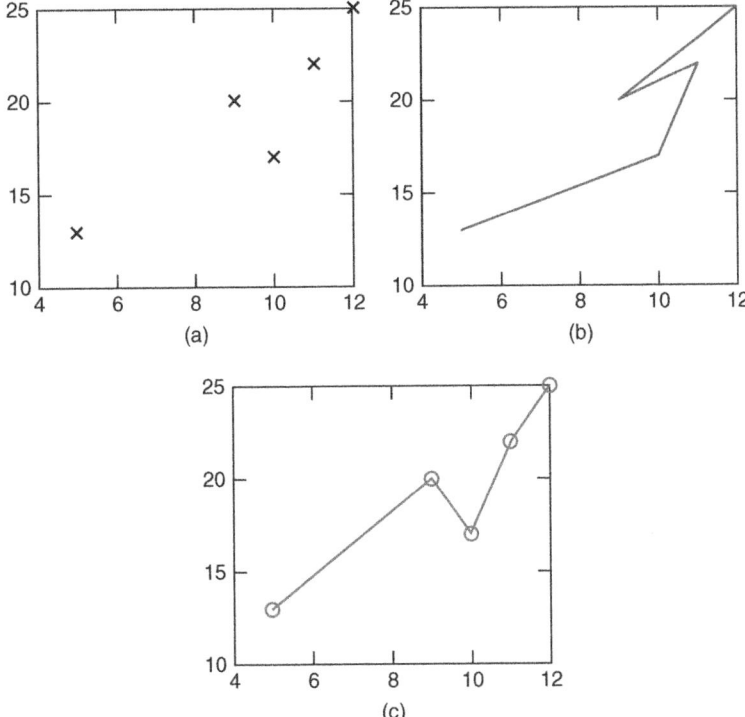

Figure 1.2 Examples of graphs obtained using the 'plot()' command. (a) Data not arranged – pointwise, (b) data not arranged – interpolated linearly, and (c) data arranged along the horizontal axis.

because the data on the horizontal axis are not arranged in ascending or descending order. The graph will look better if you sort the data on the horizontal axis and also the data on the vertical axis accordingly and then plot the relationship in the PWL style by running the following MATLAB statements:

```
»[temp1,I]=sort(temp(:,1)); temp2=temp(I,2);
»plot(temp1,temp2)
```

This will yield the graph like Figure 1.2c, which looks more informative than Figure 1.2b.

We can also use the 'plot()' command to draw a circle.

```
»r=1; th=[0:0.01:2]*pi; % [0:0.01:2] makes [0 0.01 0.02 .. 2]
»plot(r*cos(th),r*sin(th))
»plot(r*exp(j*th)) % Alternatively,
```

Note that the 'plot()' command with a sequence of complex numbers as its first input argument plots the real/imaginary parts along the horizontal/vertical axis.

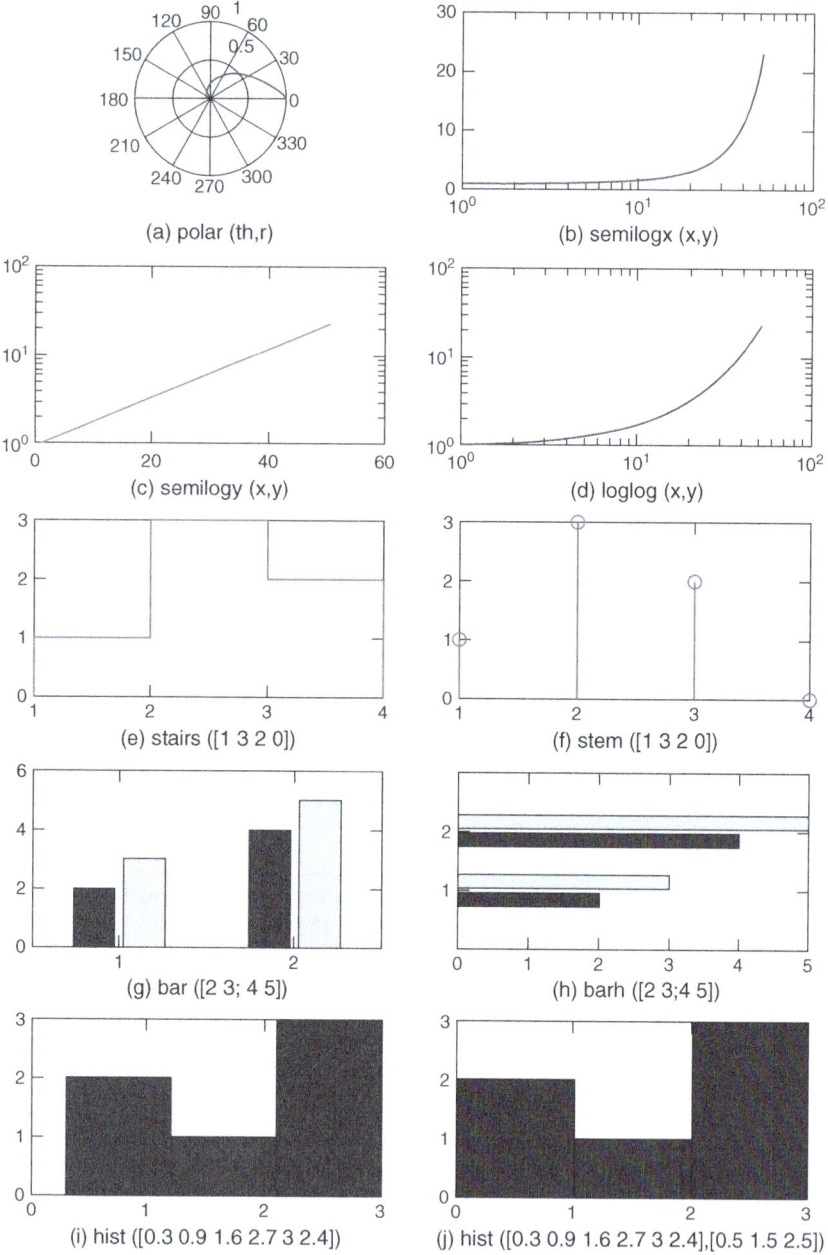

Figure 1.3 Graphs drawn by using various graphic commands.

The 'polar()' command plots the phase (in radians) and magnitude given as its first and second input arguments, respectively (see Figure 1.3a).

```
»polar(th,exp(-th)) % polar plot of a spiral
```

Several other plotting commands, such as 'semilogx()' (with a base 10 logarithmic scale for the x-axis), 'semilogy()' (with a base 10 logarithmic scale for the y-axis), 'loglog()' (with a base 10 logarithmic scale for both the x-axis and the y-axis), 'stairs()' (stairstep graph), 'stem()' (discrete graph), 'bar()'/'barh()' (vertical/horizontal bar graph), and 'hist()' (histogram), may be used to draw various graphs (shown in Figure 1.3). Readers may use the 'help' command to get the detailed usage of each one and try running the following MATLAB script "nm01f03.m" to draw various types of graphs.

```
%nm01f03.m: plot several types of graph
th=[0:0.02:1]*pi;
subplot(521), polar(th,exp(-th)) % polar graph
subplot(522), semilogx(exp(th)) % with a base 10 logarithmic scale
for x-axis
subplot(523), semilogy(exp(th)) % with a base 10 logarithmic scale
for y-axis
subplot(524), loglog(exp(th)) % with a logarithmic scale for
x-/y-axis
subplot(525), stairs([1 3 2 0]) % stairstep graph
subplot(526), stem([1 3 2 0]) % discrete graph
subplot(527), bar([2 3; 4 5]) % vertical bar graph
subplot(528), barh([2 3; 4 5]) % horizontal bar graph
y=[0.3 0.9 1.6 2.7 3 2.4];
subplot(529), hist(y,3) % histogram
subplot(5,2,10), hist(y,0.5+[0 1 2])
```

Moreover, the commands 'sprintf()', 'text()', and 'gtext()' are used for combining supplementary statements with the value(s) of one or more variables to construct a string and printing it at a certain location on the existing graph. For instance, let us run the following statements:

```
»f=1./[1:10]; plot(f)
»n=3; [s,errmsg]=sprintf('f(%1d)=%5.2f',n,f(n))
»text(3,f(3),s) %writes the text string at the point (3,f(3))
»gtext('f(x)=1/x') %writes the input string at point clicked by mouse
```

The command 'ginput()' allows you to obtain the coordinates of a point by clicking the mouse button on the existent graph. Let us try the following commands:

```
»[x,y,butkey]=ginput %get the x,y coordinates & # of the mouse button
        or ascii code of the key pressed till pressing the ENTER key
»[x,y,butkey]=ginput(n) %repeat the same job for up to n points clicked
```

1.1.5 Three Dimensional (3D) Graphic Output

MATLAB has several three-dimensional (3D) graphic plotting commands such as 'plot3()', 'mesh()', and 'contour()'. 'plot3()' plots a two-dimensional (2D) valued-function of a scalar-valued variable; 'mesh()'/'contour()' plots a scalar valued-function of a 2D variable in a mesh/contour-like style, respectively.

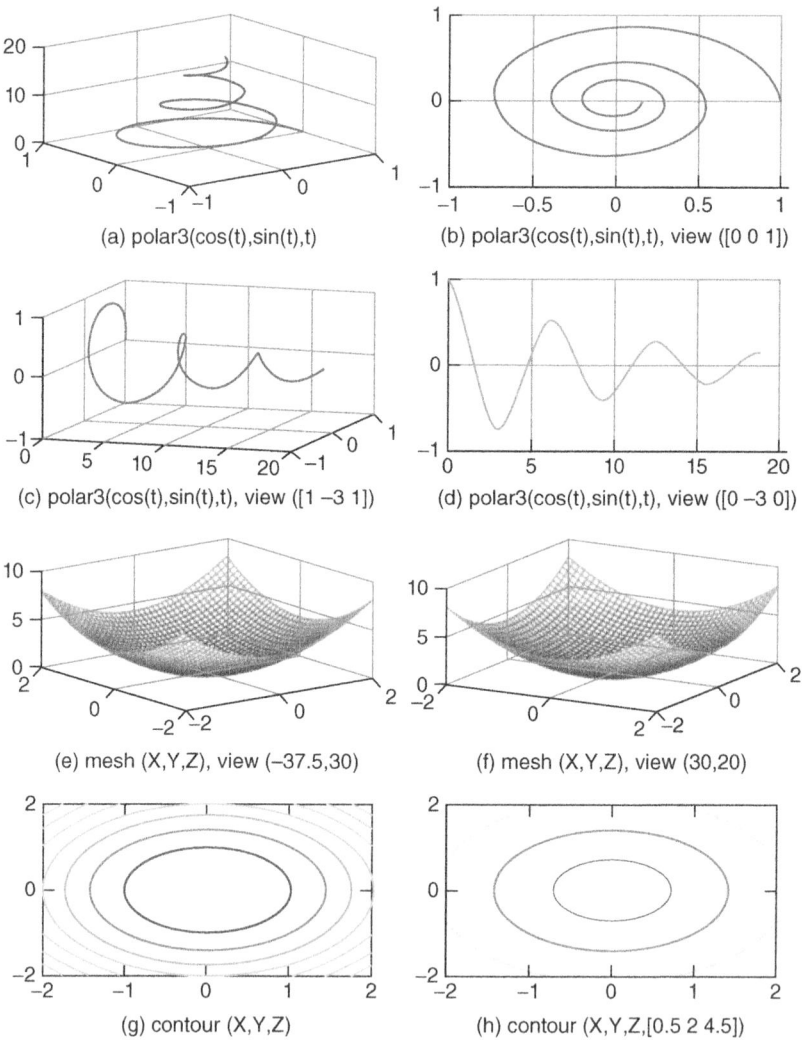

Figure 1.4 3D graphs drawn by using plot3(), mesh(), and contour().

Readers are recommended to use the 'help' command for detailed usage of each command. Try running the above MATLAB script "nm01f04.m" to see what figures will appear (Figure 1.4).

```
%nm01f04.m: to plot 3D graphs
t=0:pi/50:6*pi;
expt= exp(-0.1*t);
xt= expt.*cos(t); yt= expt.*sin(t);
% dividing the screen into 2x2 sections
clf
subplot(521), plot3(xt,yt,t), grid on %helix
subplot(522), plot3(xt,yt,t), grid on, view([0 0 1])
subplot(523), plot3(t,xt,yt), grid on, view([1 -3 1])
subplot(524), plot3(t,yt,xt), grid on, view([0 -3 0])
x=-2:.1:2;  y=-2:.1:2;
[X,Y] = meshgrid(x,y); Z =X.^2 + Y.^2;
subplot(525), mesh(X,Y,Z), grid on %[azimuth,elevation]=[-37.5,30]
subplot(526), mesh(X,Y,Z), view([0,20]), grid on
pause, view([30,30])
subplot(527), contour(X,Y,Z)
subplot(528), contour(X,Y,Z,[.5,2,4.5])
```

1.1.6 Mathematical Functions

Mathematical functions and special reserved constants/variables defined in MATLAB are listed in Table 1.3.

MATLAB also allows us to define our own function and store it in a file named after the function name so that it can be used as if it were a built-in function. For instance, we can define a scalar-valued function:

$$f_1(x) = 1/(1 + 8x^2)$$

and a vector-valued function

$$f_{49}(x) = \begin{bmatrix} f_1(x_1,x_2) \\ f_2(x_1,x_2) \end{bmatrix} = \begin{bmatrix} x_1^2 + 4x_2^2 - 5 \\ 2x_1^2 - 2x_1 - 3x_2 - 2.5 \end{bmatrix}$$

as follows:

function y=f1(x)	function y=f49(x)
y=1./(1+8*x.^2);	y(1)= x(1)^2+4*x(2)^2 -5; y(2)=2*x(1)^2-2*x(1)-3*x(2) -2.5;

Once we store these functions as M-files, each named "f1.m" and "f49.m" after the function names, respectively, we can call and use them as needed inside another M-file or in the MATLAB Command window.

Table 1.3 Functions and variables inside MATLAB.

Function	Remark	Function	Remark
cos(x)		exp(x)	Exponential function
sin(x)		log(x)	Natural logarithm
tan(x)		log10(x)	Common logarithm
acos(x)	$\cos^{-1}(x)$	abs(x)	Absolute value
asin(x)	$\sin^{-1}(x)$	angle(x)	Phase of a complex number (rad)
atan(x)	$-\pi/2 \leq \tan^{-1}(x) \leq \pi/2$	sqrt(x)	Square root
atan2(y,x)	$-\pi \leq \tan^{-1}(y/x) \leq \pi$	real(x)	Real part
cosh(x)	$(e^x + e^{-x})/2$	imag(x)	Imaginary part
sinh(x)	$(e^x - e^{-x})/2$	conj(x)	Complex conjugate
tanh(x)	$(e^x - e^{-x})/(e^x + e^{-x})$	round(x)	The nearest integer (round-off)
acosh(x)	$\cosh^{-1}(x)$	fix(x)	The nearest integer toward 0
asinh(x)	$\sinh^{-1}(x)$	floor(x)	The greatest integer $\leq x$
atanh(x)	$\tanh^{-1}(x)$	ceil(x)	The smallest integer $\geq x$
max	Maximum and its index	sign(x)	1 (positive)/0/−1 (negative)
min	Minimum and its index	mod(y,x)	Remainder of y/x
sum	Sum	rem(y,x)	Remainder of y/x
prod	Product	eval(f)	Evaluate an expression
norm	Norm	feval(f,a)	Function evaluation
sort	Sort in the ascending order	polyval	Value of a polynomial function
clock	Present time	poly	Polynomial with given roots
find	Index of element(s) satisfying given condition	roots	Roots of polynomial
flops(0)	Reset the flops count to zero	tic	Start a stopwatch timer
flops	Cumulative # of floating point *operations* (no longer available)	toc	Read the stopwatch timer (elapsed time from tic)
date	Present date	magic	Magic square

BASIC OPERATIONS OF MATLAB **15**

Table 1.3 (Continued)

Function	Remark	Function	Remark
Reserved variables with special meaning			
i, j	$\sqrt{-1}$	pi	π
eps	Machine epsilon	Inf, inf	Largest number (∞)
realmax, realmin	Largest/smallest positive number	NaN	Not_a_Number (undetermined)
end	The end of for-loop or if, while, case statement or an array index	break	Exit while/for loop
nargin	# of input arguments	nargout	# of output arguments
varargin	Variable input argument list	varargout	Variable output argument list

```
»f1([0 1])  % several values of a scalar function of a scalar variable
   ans= 1.0000   0.1111
»f49([0 1])  % a value of a 2D vector function of a vector variable
   ans= -1.0000  -5.5000
»feval('f1',[0 1]), feval('f49',[0 1])  % equivalently, yields the same
   ans= 1.0000   0.1111
   ans= -1.0000  -5.5000
```

(Q7) With the function $f_1(x)$ defined as a scalar function of a scalar variable, we enter a vector as its input argument to obtain a seemingly vector-valued output. What's going on?

(A7) It is just a set of function values $[f_1(x_1) \quad f_1(x_2) \ldots]$ obtained at a time for several values $[x_1 \quad x_2 \ldots]$ of x. In expectation of one-shot multioperation, it is a good practice to put a dot (.) just before the arithmetic operators *(multiplication), /(division), and ^(power) in the function definition so that the element-by-element (element-wise or term-wise) operation can be done any time.

Note that we can define a simple function not only in an independent M-file but also inside a script by using the function handle operator @, the 'inline()' command, or just in a form of literal expression that can be evaluated by the command 'eval()'.

```
»f1h=@(x)1./(1+8*x.^2);  % Using function handle
»f1i=inline('1./(1+8*x.^2)','x');  % Usng inline()
»f1h([0 1]), feval(f1h,[0 1]), f1i([0 1]), feval(f1i,[0 1])
   ans= 1.0000   0.1111
   ans= 1.0000   0.1111
»f1='1./(1+8*x.^2)'; x=[0 1]; eval(f1)
   ans= 1.0000   0.1111
```

As far as a polynomial function is concerned, it can simply be defined as its coefficient vector arranged in descending order. It may be called to yield its value for certain value(s) of its independent variable by using the command 'polyval()'.

```
»p=[1 0 -3 2];  % Polynomial function p(x) = x^3-3x+2
»polyval(p,[0 1])  % Polynomial function values at x=0 and 1
    ans=  2.0000   0.0000
```

The multiplication of two polynomials can be performed by taking the convolution of their coefficient vectors representing the polynomials in MATLAB, since

$$(a_N x^N + \cdots + a_1 x + a_0)(b_N x^N + \cdots + b_1 x + b_0) = c_{2N} x^{2N} + \cdots + c_1 x + c_0$$

where

$$c_k = \sum_{m=\max(0,k-N)}^{\min(k,N)} a_{k-m} b_m \text{ for } k = 2N, 2N-1, \ldots, 1, 0$$

This operation can be done by using the MATLAB built-in command 'conv()' as illustrated beneath:

```
»a=[1 -1]; b=[1 1 1]; c=conv(a,b)
    c=  1   0   0  -1  % meaning that (x-1)(x^2+x+1) = x^3+0·x^2+0·x-1
```

But, in case you want to multiply a polynomial by only x^n, you can simply append n zeros to the right end of the polynomial coefficient vector to extend its dimension.

```
»a=[1 2 3]; c=[a 0 0] % equivalently, c=conv(a,[1 0 0])
    c=  1   2   3   0   0  % meaning that (x^2+2x+3)x^2 = x^4+2x^3+3x^2+0·x+0
```

1.1.7 Operations on Vectors and Matrices

We can define a new scalar/vector/matrix or redefine any existing ones in terms of the existent ones or irrespective of them. In the MATLAB Command window, let us define A and B as

$$A = \begin{bmatrix} 1 & 2 & 3 \\ 4 & 5 & 6 \end{bmatrix}, \quad B = \begin{bmatrix} 3 \\ -2 \\ 1 \end{bmatrix}$$

by running

```
»A=[1 2 3;4 5 6], B=[3;-2;1]
```

We can modify them or take a portion of them. For example:

```
»A=[A;7 8 9]
    A=  1   2   3
        4   5   6
        7   8   9
```

```
»B=[B [1 0 -1]']
  B=  3    1
     -2    0
      1   -1
```

Here, the apostrophe (prime) operator (') takes the complex conjugate transpose and functions virtually as a transpose operator for real-valued matrices. If you want to take just the transpose of a complex-valued matrix, you should put a dot (.) before ', i.e. '.''.

When extending an existing matrix or defining another one based on it, the compatibility of dimensions should be observed. For instance, if you try to annex a 4×1 matrix into the 3×1 matrix B, MATLAB will reject it squarely, giving you an error message.

```
»B=[B ones(4,1)]
   Error using horzcat
   Dimensions of matrices being concatenated are not consistent.
```

We can modify or refer to a portion of a given matrix.

```
»A(3,3)=0
  A= 1   2   3
     4   5   6
     7   8   0
»A(2:3,1:2) % from 2nd row to 3rd row, from 1st column to 2nd column
   ans= 4   5
        7   8
»A(2,:) % 2nd row, all columns
   ans= 4   5   6
```

The colon (:) is used for defining an arithmetic (equal difference) sequence without the bracket [] as

```
»t=0:0.1:2
```

which makes

```
t=[0.0 0.1 0.2 ... 1.9 2.0]
```

(Q8) What if we omit the increment between the left/right boundary numbers?
(A8) By default, the increment is 1.

```
»t=0:2
   t= 0 1 2
```

(Q9) What if the right boundary number is smaller/greater than the left boundary number with a positive/negative increment?
(A9) It yields an empty matrix, which is useless.

```
»t=0:-2
   t= Empty matrix: 1-by-0
```

(Q10) If we define just some elements of a vector not fully, but sporadically, will we have a row vector or a column vector and how will it be filled in between?

(A10) We will have a row vector filled with zeros between the defined elements.

```
»D(2)=2; D(4)=3
  D= 0  2  0  3
```

(Q11) How do we make a column vector in the same style?

(A11) We must initialize it as a (zero-filled) row vector, prior to giving it a value.

```
»D=zeros(4,1); D(2)=2; D(4)=3
  D= 0
     2
     0
     3
```

(Q12) What happens if the specified element index of an array exceeds the defined range?

(A12) It is rejected. MATLAB does not accept nonpositive or noninteger indices.

```
»D(5)
  Index exceeds matrix dimensions.
»D(0)=1;
  Subscript indices must either be real positive integers or logicals.
»D(1.2)
  Subscript indices must either be real positive integers or logicals.
```

(Q13) How do we know the size (the numbers of rows/columns) of an already-defined array?

(A13) Use the 'length()', 'size()', and 'numel()' commands as indicated as follows:

```
»length(D)
  ans = 4
»[M,N]=size(A)
  M = 3, N = 3
»Number_of_elements=numel(A)
  Number_of_elements = 9
```

MATLAB enables us to handle vector/matrix operations in almost the same way as scalar operations. However, we must make sure of the dimensional compatibility between vectors/matrices, and put a dot (.) in front of the operator for elementwise (element-by-element) operations. The addition of a matrix and a scalar adds the scalar to every element of the matrix. The multiplication of a matrix by a scalar multiplies every element of the matrix by the scalar.

There are several things to know about the matrix division and inversion.

Remark 1.1 Rules of Vector/Matrix Operation.

(1) For a matrix to be invertible, it must be square and nonsingular, i.e. the numbers of its rows and columns must be equal and its determinant must not be zero.

(2) The MATLAB command 'pinv(A)' provides us with a matrix X of the same dimension as A^T such that $AXA = A$ and $XAX = X$. We can use this command to get the right/left pseudo (generalized) inverse $A^T[AA^T]^{-1}/[A^TA]^{-1}A^T$ for a matrix A given as its input argument, depending on whether the number (M) of rows is smaller or greater than the number (N) of columns, so long as the matrix is of full rank, i.e. rank(A) = min(M,N) [K-2, Section 6.4]. Note that $A^T[AA^T]^{-1}/[A^TA]^{-1}A^T$ is called the right/left inverse since it is multiplied onto the right/left side of A to yield an identity matrix.

(3) You should be careful when using the 'pinv(A)' command for a rank-deficient matrix, because its output is no longer the right/left inverse, which does not even exist for rank-deficient matrices.

(4) The value of a scalar function having an array value as its argument is also an array with the same dimension.

Suppose we have defined vectors a_1, a_2, b_1, b_2, and matrices A_1, A_2, B as follows:

```
»a1=[-1 2 3]; a2=[4 5 2]; b1=[1 -3]'; b2=[-2 0];
```
$$a_1 = \begin{bmatrix} -1 & 2 & 3 \end{bmatrix}, \quad a_2 = \begin{bmatrix} 4 & 5 & 2 \end{bmatrix}, \quad b_1 = \begin{bmatrix} 1 \\ -3 \end{bmatrix}, \quad b_2 = \begin{bmatrix} -2 & 0 \end{bmatrix}$$

```
»A1=[a1;a2], A2=[a1;[b2 1]], B=[b1 b2']
```
$$A_1 = \begin{bmatrix} -1 & 2 & 3 \\ 4 & 5 & 2 \end{bmatrix}, \quad A_2 = \begin{bmatrix} -1 & 2 & 3 \\ -2 & 0 & 1 \end{bmatrix}, \quad B = \begin{bmatrix} 1 & -2 \\ -3 & 0 \end{bmatrix}$$

The results of various operations on these vectors/matrices are as follows (pay attention to the error message):

```
»A3=A1+A2, A4=A1-A2, 1+A1  % matrix/scalar addition/subtraction
   A3= -2  4  6    A4= 0  0  0    ans= 0  3  4
        2  5  3         6  5  1          5  6  3
»AB=A1*B  %AB(m,n) = ΣA₁(m,k)B(k,n)  matrix multiplication?
                      k
   Error using  *
   Inner matrix dimensions must agree.
»BA1=B*A1 % regular matrix multiplication
   BA1=-9  -8  -1
        3  -6  -9
»AA=A1.*A2 % element-wise (term-wise) multiplication
   AA=  1   4   9
       -8   0   2
```

20 MATLAB USAGE AND COMPUTATIONAL ERRORS

```
»AB=A1.*B  %AB(m, n) = A₁(m, n)B(m, n) element-wise multiplication
  Error using .*
  Matrix dimensions must agree.
»A1_1=pinv(A1),A1'*(A1*A1')^-1,eye(size(A1,2))/A1 %A₁ᵀ[A₁A₁ᵀ]⁻¹
  A1_1= -0.1914    0.1399   %right inverse of a 2x3 matrix A1
         0.0617    0.0947
         0.2284   -0.0165
»A1*A1_1 %A1/A1=I implies the validity of A5_1 as the right inverse
  ans= 1.0000    0.0000
       0.0000    1.0000
»A5=A1'; % a 3x2 matrix
»A5_1=pinv(A5),(A5'*A5)^-1*A5',A5\eye(size(A5,1)) % [A₅ᵀA₅]⁻¹A₅ᵀ
  A5_1= -0.1914    0.0617    0.2284  %left inverse of a 3x2 matrix A5
         0.1399    0.0947   -0.0165
»A5_1*A5 %A5\A5=I implies the validity of A5_1 as the left inverse
  ans= 1.0000   -0.0000
      -0.0000    1.0000
»A1_li=(A1'*A1)^-1*A1' %the left inverse of matrix A1 with M<N?
  Warning: Matrix is close to singular or badly scaled.
            Results may be inaccurate. RCOND = 6.211163e-18.
  A1_li = -0.2500         0
           0.2500   -0.5000
           0.5000    0.2500
```

(Q14) Does the left inverse of a matrix having rows fewer than columns exist?

(A14) No. There is no $N \times M$ matrix that is premultiplied on the left of an $M \times N$ matrix with $M < N$ to yield a nonsingular matrix, far from an identity matrix. In this context, MATLAB should have rejected the above case on the ground that $[A_1^T A_1]$ is singular and so its inverse does not exist. But, because the round-off errors make a very small number appear to be a zero or a real zero appear to be a very small number (as will be mentioned in Remark 2.3), it is not easy for MATLAB to tell a near-singularity from a real singularity. That is why MATLAB dares not to declare the singularity case and instead issues just a warning message to remind you to check the validity of the result so that it will not be blamed for a delusion. Therefore, you must be alert for the condition mentioned in Remark 1.1(2), which says that, in order for the left inverse to exist, the number of rows must not be less than the number of columns.

```
»A1_li*A1 %No identity matrix, since A1_li isn't the left inverse
  ans =  1.2500    0.7500   -0.2500
        -0.2500    0.5000    0.7500
         1.5000    3.5000    2.5000
»det(A1'*A1) %A1 is not left-invertible for A1'*A1 is singular
  ans = 0
```

(cf) Let us be nice to MATLAB as it is to us. From the standpoint of promoting mutual understanding between us and MATLAB, we acknowledge that MATLAB tries to show us apparently good results to please us like always, sometimes even pretending not to be obsessed by the demon of 'ill-condition' in order not to make us feel uneasy. How kind MATLAB is! But, we should be always careful not to be spoiled by its benevolence and not to accept the computing results every inch as it is. In this case, even though the matrix [A1'*A1] is singular and so not invertible, MATLAB tried to invert it and that's all. MATLAB must have felt something abnormal as can be seen from the ominous warning message prior to the computing result. Who would blame MATLAB for being so thoughtful and loyal to us? We might well be rather touched by its sincerity and smartness.

In the aformentioned statements, we see the slash(/)/backslash(\) operators. These operators are used for right/left division, respectively; B/A is the same as B*inv(A) and A\B is the same as inv(A)*B when A is invertible and the dimensions of A and B are compatible. Noting that B/A is equivalent to (A'\B')', let us take a close look at the function of the backslash(\) operator.

```
»X=A1\A1 % an identity matrix?
     X=    1.0000       0        -0.8462
              0       1.0000      1.0769
              0          0           0
```

(Q13) It seems that A1\A1 should have been an identity matrix, but it is not, contrary to our expectation. Why?

(A13) We should know more about the various functions of the backslash(\), which can be seen by typing 'help slash' into the MATLAB Command window. Let Remark 1.2 answer this question in cooperation with the next case.

```
»A1*X-A1 %zero if X is the solution to A1*X=A1?
   ans= 1.0e-015 *  0         0        0
                    0         0     -0.4441
```

Remark 1.2 The Function of Backslash(\) Operator.

Overall, for the command 'A\B', MATLAB finds a solution to the equation A*X=B. Let us denote the row/column dimension of the matrix A by M and N.

(1) If matrix A is square and upper/lower-triangular in the sense that all of its elements below/above the diagonal are zero, then MATLAB finds the solution by applying backward/ forward substitution method (Section 2.2.1).

(2) If matrix A is square, symmetric (Hermitian), and positive definite, then MATLAB finds the solution by using Cholesky factorization (Section 2.4.2).

(3) If matrix A is square and has no special feature, then MATLAB finds the solution by using lower-upper (LU) decomposition (Section 2.4.1).

MATLAB USAGE AND COMPUTATIONAL ERRORS

(4) If matrix A is rectangular, then MATLAB finds a solution by using QR factorization (Section 2.4.2). In case A is rectangular and of full rank with rank(A) = min(M,N), it will be the least-squares (LSs) solution (Eq. (2.1.10)) for M>N (over-determined case) and one of the many solutions that is not always the same as the minimum-norm solution (Eq. (2.1.7)) for M<N (under-determined case). But for the case where A is rectangular and has rank deficiency, what MATLAB gives us may be useless. Therefore, you must pay attention to the warning message about rank deficiency, which might tell you not to count on the dead-end solution made by the backslash(\) operator. To find an alternative in the case of rank deficiency, you had better resort to the singular value decomposition (SVD) (see Problem 2.8 for details).

For the moment, let us continue to try more operations on matrices.

```
»A1./A2  % termwise right division
   ans=  1    1    1
        -2   Inf   2
»A1.\A2  % termwise left division
   ans=  1    1    1
        -0.5  0   0.5
»format rat, B^-1  %represent the numbers (of B⁻¹) in fractional form
   ans=  0   -1/3
        -1/2 -1/6
»inv(B)  % inverse matrix, equivalently
   ans=  0   -1/3
        -1/2 -1/6
»B.^-1   % element-wise inversion(reciprocal of each element)
   ans=  1   -1/2
        -1/3  Inf
»B^2     % square of B, i.e., B²=B*B
   ans=  7   -2
        -3    6
»B.^2    % element-wise square(square of each element)
   ans=  1 ($b_{11}^2$)  4 ($b_{12}^2$)
         9 ($b_{21}^2$)  0 ($b_{22}^2$)
»2.^B    % 2 to the power of each number in B
   ans=  2 ($2^{b_{11}}$)   1/4 ($2^{b_{12}}$)
         1/8 ($2^{b_{21}}$) 1 ($2^{b_{22}}$)
»A1.^A2  % element of A1 to the power of each element in A2
   ans=  -1 ($A_1(1,1)^{A_2(1,1)}$)  4 ($A_1(1,2)^{A_2(1,2)}$)  27 ($A_1(1,3)^{A_2(1,3)}$)
         1/16 ($A_1(2,1)^{A_2(2,1)}$) 1 ($A_1(2,2)^{A_2(2,2)}$) 2 ($A_1(2,3)^{A_2(2,3)}$)
»format short, exp(B)  %elements of $e^B$ with 4 digits below the dp
   ans=  2.7183 ($e^{b_{11}}$)  0.1353 ($e^{b_{12}}$)
         0.0498 ($e^{b_{21}}$)  1.0000 ($e^{b_{22}}$)
```

There are more useful MATLAB commands worthwhile to learn by heart.

Remark 1.3 More Useful Commands for Vector/Matrix Operations.

(1) We can use the commands 'zeros()', 'ones()', and eye() to construct a matrix of specified size or the same size as an existing matrix that has only zeros, only ones, or only ones/zeros on/off its diagonal.
```
»Z=zeros(2,3) % or zeros(size(A1)) yielding a 2x3 zero matrix
    Z= 0   0   0
       0   0   0
»E=ones(size(B)) % or ones(3,2) yielding a 3x2 one matrix
    E= 1   1
       1   1
       1   1
»I=eye(2) % yielding a 2x2 identity matrix
    I= 1   0
       0   1
```

(2) We can use the 'diag()' command to make a column vector composed of the diagonal elements of a matrix or to make a diagonal matrix with on-diagonal elements taken from a vector given as the input argument.
```
»A1=[-1  2  3; 4  5  2]
    A1=  -1   2   3
          4   5   2
»a=diag(A1) % The column vector consisting of diagonal elements
    a= -1
        5
»diag(a) % The column vector consisting of diagonal elements
    ans= -1   0
          0   5
```

(3) We can use the commands 'sum()'/'prod()' to get the sum/product of elements in a vector or a matrix, column-wisely first (along the first non-singleton dimension).
```
»sa1=sum(a1) % sum of all the elements in vector a₁
    sa1= 4  %∑a₁(n) = -1+2+3 = 4
»sA1=sum(A1) % sum of all the elements in each column of matrix A₁
    sA1= 3   7   5   %sA1(n) = ∑_{m=1}^{M} A₁(m,n) = |-1+4  2+5  3+2|
»SA1=sum(sum(A1)) % sum of all elements in matrix A₁
    SA1= 15   %SA1 = ∑_{n=1}^{N} ∑_{m=1}^{M} A₁(m,n) = 3+7+5 = 15
»pa1=prod(a1) % product of all the elements in vector a₁
    pa1= -6   %∏a₁(n) = (-1)×2×3 = -6
»pA1=prod(A1) % product of all the elements in each column of matrix A₁
    pA1= -4   10   6   %pA1(n) = ∏_{m=1}^{M} A₁(m,n) = |-1×4  2×5  3×2|
»PA1=prod(prod(A1)) % product of all the elements of matrix A₁
    PA1= -240  %PA1 = ∏_{n=1}^{N} ∏_{m=1}^{M} A₁(m,n) = (-4)×10×6 = -240
```

(4) We can use the commands 'max()'/'min()' to find the first maximum/minimum number and its index in a vector or in a matrix given as the input argument.

```
»[aM,iM]=max(a2)
  aM= 5, iM= 2  %means that the max. element of vector a2 is a2(2)=5
»[AM,IM]=max(A1)
  AM= 4   5   3
  IM= 2   2   1
  %means that the max. elements of each column of A1 are
     A1(2,1)=4, A1(2,2)=5, A1(1,3)=3
»[AMx,J]=max(AM)
  AMx= 5, J= 2
  %implies that the max. element of A1 is A1(IM(J),J)=A1(2,2)=5
```

(5) We can use the commands 'rot90()'/'fliplr()'/'flipud()' to rotate a matrix by an integer multiple of 90° and to flip it left–right/up–down.

```
»A1, A3=rot90(A1), A4=rot90(A1,-2)
  A1= -1    2    3
       4    5    2
  A3=  3    2         %90° rotation
       2    5
      -1    4
  A4=  2    5    4    %90 °x(-2) rotation
       3    2   -1
»A5=fliplr(A1)  %flip left-right
  A5=  3    2   -1
       2    5    4
»A6=flipud(A1)  %flip up-down
  A6=  4    5    2
      -1    2    3
```

(6) We can use the 'reshape()' command to change the row–column size of a matrix with its elements preserved (column-wisely first).

```
»A7=reshape(A1,3,2)
  A7= -1    5
       4    3
       2    2
»A8=reshape(A1,6,1), A8=A1(:)  %makes super-column vector
  A8= -1
       4
       2
       5
       3
       2
```

(7) We can use the 'repmat(A,M,N)' command to repeat a matrix A to make a large matrix consisting of an M×N tiling of copies of A.

```
»A9=repmat(A1,2,3)
  A9= -1    2    3   -1    2    3   -1    2    3
       4    5    2    4    5    2    4    5    2
      -1    2    3   -1    2    3   -1    2    3
       4    5    2    4    5    2    4    5    2
```

BASIC OPERATIONS OF MATLAB   25

(8) We can use the 'find' command to find the row/column indices of the entries satisfying the condition specified as its input argument.

```
»[ir,ic]=find(2<A1&A1<5)
    ir = 2           ic = 1
         1                3
```

This means that the matrix A1 has two entries, A1(2,1) and A1(1,3), satisfying the two inequalities, 'greater than 2' and 'smaller than 5'.

1.1.8 Random Number Generators

MATLAB has the built-in functions, 'rand()'/'randn()', to generate random numbers having uniform/normal(Gaussian) distributions, respectively [K-2, Chapter 22].

```
rand(M,N): Generates an MxN matrix consisting of uniformly distributed
random numbers
randn(M,N): Generates an MxN matrix consisting of normally distributed
random numbers
```

1.1.8.1 Random Number Having Uniform Distribution

The numbers in a matrix generated by the MATLAB function 'rand(M,N)' have uniform probability distribution over the interval [0,1], as described by $U(0,1)$. The random number x generated by 'rand()' has the probability density function:

$$f_X(x) = u_s(x) - u_s(x-1) \left(u_s(x) = \begin{cases} 1 & \forall x \geq 0 \\ 0 & \forall x < 0 \end{cases} : \text{the unit step function} \right) \quad (1.1.1)$$

whose value is 1 over [0,1] and 0 elsewhere. The average of this standard uniform number x is

$$m_X = \int_{-\infty}^{\infty} x f_X(x)\, dx = \int_0^1 x\, dx = \left.\frac{x^2}{2}\right|_0^1 = \frac{1}{2} \quad (1.1.2)$$

and its variance or deviation is

$$\sigma_X^2 = \int_{-\infty}^{\infty} (x - m_X)^2 f_X(x)\, dx = \int_0^1 \left(x - \frac{1}{2}\right)^2 dx = \frac{1}{3}\left(x - \frac{1}{2}\right)^3 \Big|_0^1 = \frac{1}{12} \quad (1.1.3)$$

If you want another random number y with uniform distribution $U(a,b)$, transform the standard uniform number x as follows:

$$y = a + (b - a)x \quad (1.1.4)$$

For practice, we make a vector consisting of 1000 standard uniform numbers, transform it to make a vector of numbers with uniform distribution $U(-1,+1)$, and then draw the histograms showing the shape of the distribution for the two uniform number vectors (Figure 1.5a1,a2) by running the following block of MATLAB statements.

```
%nm01f05a.m
N=1e4; x=rand(N,1); % An Nx1 noise vector with U(0,1)
mx=mean(x); sgm2=sum((x-mx).^2)/(N-1), var(x)
subplot(221), M=20; hist(x,M) % Histogram having M=20 bins
a=-1; b=1; y=(b-a)*x+a; %Eq.(1.1.4): 1000x1 noise vector with U(-1,1)
subplot(222), hist(y,M) % Histogram
```

1.1.8.2 Random Number with Normal (Gaussian) Distribution

The numbers in a matrix generated by the MATLAB function 'randn(M,N)' have normal (Gaussian) distribution with average $m = 0$ and variance $\sigma^2 = 1$, as described by $N(0,1)$. The random number x generated by 'rand()' has the probability density function

$$f_X(x) = \frac{1}{\sqrt{2\pi}} e^{-x^2/2} \quad (1.1.5)$$

(a1) Uniform distribution U[0, 1]

(a2) Uniform distribution U[−1, 1]

(b1) Gaussian distribution N(0, 1)

(b2) Gaussian distribution N(0, 1/2²)

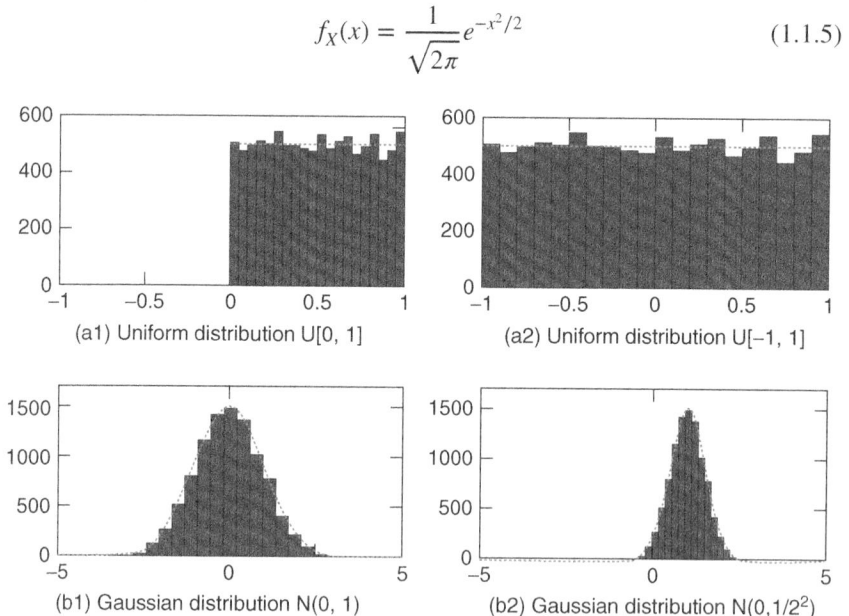

Figure 1.5 Distribution (histogram) of noise generated by the `rand()`/`randn()` command.

If you want another Gaussian number y with a general normal distribution $N(m,\sigma^2)$, transform the standard Gaussian number x as follows:

$$y = \sigma x + m \qquad (1.1.6)$$

The probability density function of the new Gaussian number generated by this transformation is obtained by substituting $x = (y - m)/\sigma$ into Eq. (1.1.5) and dividing the result by the scale factor σ (which can be seen in $dx = dy/\sigma$) so that the integral of the density function over the whole interval $(-\infty,+\infty)$ amounts to 1.

$$f_Y(y) = \frac{1}{\sqrt{2\pi}\sigma} e^{-(y-m)^2/2\sigma^2} \qquad (1.1.7)$$

```
%nm01f05b.m
N=1e4; x=randn(N,1); % An Nx1 noise vector with N(0,1)
subplot(223), M=20; hist(x,M) % Histogram having M=20 bins
f=@(x,m,sgm)exp(-(x-m).^2/2/sgm^2)/sqrt(2*pi)/sgm; %Gaussian density ftn
[Ns,Cs]=hist(x,20); dx=Cs(2)-Cs(1); % Bin width
x_=[-5:0.01:5]; fx=f(x_,0,1); hold on, plot(x_,fx*dx*N,'r:')
sgm=1/2; m=1; y=sgm*x+m; % An Nx1 noise vector with N(m,sgm^2)
subplot(224), hist(y,M) % Histogram
[Ns,Cs]=hist(y,M); dy=Cs(2)-Cs(1); % Bin width
y_=[-5:0.01:5]; fy=f(y_,m,sgm);
hold on, plot(y_,fy*dy*N,'r:')
```

For practice, we make a vector consisting of 1000 standard Gaussian numbers, transform it to make a vector of numbers having normal distribution $N(1,1/4)$, with mean $m = 1$ and variance $\sigma^2 = 1/4$, and then draw the histograms for the two Gaussian number vectors (Figure 1.5b1,b2) by running the above block of MATLAB statements.

1.1.9 Flow Control

1.1.9.1 *if-end* and *switch-case-end* **Statements**

An `if-end` block basically consists of an `if` statement, a sequel part and an `end` statement categorizing the block. An `if` statement, having a condition usually based on the relational/logical operator (Table 1.4), is used to control the program flow, i.e. to adjust the order in which statements are executed according to whether or not the condition is met, mostly depending on unpredictable situations. The sequel part consisting of one or more statements may contain `else` or `elseif` statements, possibly in a nested structure containing another `if` statement inside it. The `switch-case-end` block might replace a multiple `if-elseif-...-end` statement in a neat manner.

Let us see the following examples:

(Ex. 1) A Simple `if-else-end` Block

```
%nm119_1.m: example of if-end block
t=0;
if t>0
   sgnt= 1;
  else
    sgnt= -1;
end
```

(Ex. 2) *A Simple* `if-elseif-end` *Block*

```
%nm119_2.m: example of if-elseif-end block
if t>0
   sgnt= 1
  elseif t<0
    sgnt= -1
end
```

(Ex. 3) *An* `if-elseif-else-end` *Block*

```
%nm119_3.m: example of if-elseif-else-end block
if t>0, sgnt= 1
  elseif t<0, sgnt= -1
  else    sgnt= 0
end
```

Table 1.4 Relational operators and logical operators.

Relational operator	Remark
<	Less than
<=	Less than or equal to
==	Equal
>	Greater than
>=	Greater than or equal to
~=	Not equal ($\neq$)
&	and
|	or
~	not

(Ex. 4) *An* `if-elseif-elseif-...-else-end` *Block*

```
%nm119_4.m: example of if-elseif-elseif-else-end block
point= 85;
if point>=90, grade= 'A'
  elseif point>=80, grade= 'B'
  elseif point>=70, grade= 'C'
  elseif point>=60, grade= 'D'
  else             grade= 'F'
end
```

(Ex. 5) *A* `switch-case-end` *Block*

```
%nm119_5.m: example of switch-case-end block
point= 85;
switch floor(point/10) %floor(x): integer less than or equal to x
  case  9, grade= 'A'
  case  8, grade= 'B'
  case  7, grade= 'C'
  case  6, grade= 'D'
  otherwise grade= 'F'
end
```

1.1.9.2 `for index=i_0:increment:i_last-end` Loop

A `for` loop makes a block of statements executed repeatedly for a specified number of times, with its loop index increasing from `i_0` to a number not greater than `i_last` by a specified step (`increment`) or by 1 if not specified. The loop iteration normally ends when the loop index reaches `i_last`, but it can be stopped by a `break` statement inside the `for` loop. The `for` loop with a positive/negative increment will never be iterated if the last value (`i_last`) of the index is smaller/greater than the starting value (`i_0`).

(Ex. 6) *A* `for` *Loop*

```
%nm119_6.m: example of for loop
point= [76 85 91 65 87];
for n=1:length(point)
   if point(n)>=80,  pf(n,:)= 'pass';
      elseif point(n)>=0, pf(n,:)= 'fail';
      else   %if point(n)<0
      pf(n,:)= '????';
      fprintf('\n\a Something wrong with the data??\n');
      break;
   end
end
pf
```

30 MATLAB USAGE AND COMPUTATIONAL ERRORS

1.1.9.3 `while` Loop

A `while` loop will be iterated as long as its predefined condition is satisfied and a `break` statement is not encountered inside the loop.

(Ex. 7) A `while` Loop

```
%nm119_7.m: example of while loop
r=1;
while r<10
   r= input('\nType radius (or nonpositive number to stop):');
   if r<=0, break, end %isempty(r)|r<=0, break, end
   v= 4/3*pi*r^3;
   fprintf('The volume of a sphere with radius %3.1f =%8.2f\n',r,v);
end
```

(Ex. 8) `while` *Loops to find the minimum/maximum positive numbers represented in MATLAB*

```
%nm119_8.m: example of while loops
x=1; k1=0;
% To repeat division-by-2 to find the minimum positive number
while x/2>0
   x=x/2; k1=k1+1;
end
k1, x_min=x;
fprintf('x_min is %20.18e\n',x_min)

% To repeat multiplation-by-2 to find the maximum posi-
tive number
x=1; k2=0;
while 2*x<inf
   x=x*2; k2=k2+1;
end
k2, x_max0=x;

tmp=1; k3=0;
while x_max0*(2-tmp/2)<inf
   tmp=tmp/2; k3=k3+1;
end
k3, x_max=x_max0*(2-tmp);
fprintf('x_max is %20.18e\n',x_max)

format long e, x_min,-x_min,x_max,-x_max
format hex, x_min,-x_min,x_max,-x_max
format short
```

The following script "nm119_8.m" contains three `while` loops. In the first one, `x=1` continues to be divided by 2 till just before reaching zero and it will hopefully end up with the smallest positive number that can be represented in MATLAB. In the second one, `x=1` continues to be multiplied by 2 till just before reaching `inf` (the infinity defined in MATLAB) and seemingly it will get the largest positive number (`x_max0`) that can be represented in MATLAB. But while this number reaches or may exceed `inf` if multiplied by 2 once more, it still is not the largest number in MATLAB (slightly less than `inf`) that we want to find. How about multiplying `x_max0` by $(2-1/2^n)$? In the third `while` loop, the temporary variable `tmp` starting with the initial value of 1 continues to be divided by 2 till just before `x_max0*(2-tmp)` reaches `inf` and apparently it will end up with the largest positive number (`x_max`) that can be represented in MATLAB.

1.2 COMPUTER ERRORS VS. HUMAN MISTAKES

Digital systems like calculators and computers hardly make a mistake, since they follow the programmed order faithfully. Nonetheless, we often encounter some numerical errors in the computing results made by digital systems, mostly coming from representing the numbers in finite bits, which is an intrinsic limitation of digital world. If you let the computer compute something without considering what is called the finite-word-length effect, you might come across a weird answer. In that case, it is not the computer, but yourself as the user or the programmer, who is to blame for the wrong result. In this context, we should always be careful not to let the computer produce a farfetched output. In this section, we will see how the computer represents and stores the numbers. Then we think about the cause and the propagation effect of computational error in order not to be deceived by unintentional mistake of the computer and hopefully, to be able to take some measures against them.

1.2.1 IEEE 64-bit Floating-Point Number Representation

MATLAB uses the IEEE 64-bit floating-point number system to represent all numbers. It has a word structure consisting of the sign bit, the exponent field, and the mantissa field as follows:

63	62		52	51		0
S	Exponent			Mantissa		

Each of these fields expresses S, E, and M of a number f in the way described as follows:

32 MATLAB USAGE AND COMPUTATIONAL ERRORS

- Sign bit

$$S = b_{63} = \begin{cases} 0 & \text{for positive numbers} \\ 1 & \text{for negative numbers} \end{cases}$$

- Exponent field $(b_{62}b_{61}b_{60} \cdots b_{52})$: adopting the excess 1023 code

$$E = \text{Exp} - 1023 = \{0, 1, \ldots, 2^{11} - 1 = 2047\} - 1023$$
$$= \{-1023, -1022, \ldots, +1023, +1024\}$$
$$= \begin{cases} -1023 + 1 & \text{for } |f| < 2^{-1022} (\text{Exp} = 0000000000\ 0) \\ -1022 \sim +1023 & \text{for } 2^{-1022} \le |f| < 2^{1024} (\text{normailzed ranges}) \\ +1024 & \text{for } \pm\infty \end{cases}$$

- Mantissa field $(b_{51}b_{50} \cdots b_1b_0)$:
  In the un-normalized range where the numbers are so small that they can be represented only with the value of hidden bit 0, the number represented by the mantissa is

$$M = 0.b_{51}b_{50} \ldots b_1b_0 = [b_{51}b_{50} \ldots b_1b_0] \times 2^{-52} \qquad (1.2.1)$$

You might think that the value of the hidden bit is added to the exponent, instead of to the mantissa.

In the normalized range, the number represented by the mantissa together with the value of hidden bit $b_h = 1$ is

$$M = 1.b_{51}b_{50}, \ldots, b_1b_0 = 1 + [b_{51}b_{50}, \ldots, b_1b_0] \times 2^{-52}$$
$$= 1 + b_{51} \times 2^{-1} + b_{50} \times 2^{-2} + \cdots + b_1 \times 2^{-51} + b_0 \times 2^{-52}$$
$$= \{1, 1 + 2^{-52}, 1 + 2 \times 2^{-52}, \ldots, 1 + (2^{52} - 1) \times 2^{-52}\}$$
$$= \{1, 1 + 2^{-52}, 1 + 2 \times 2^{-52}, \ldots, (2 - 2^{-52})\}$$
$$= \{1, 1 + \Delta, 1 + 2\Delta, \ldots, 1 + (2^{52} - 1)\Delta = 2 - \Delta\} \ (\Delta = 2^{-52}) \quad (1.2.2)$$

The set of numbers S, E, and M, each represented by the sign bit S, the exponent field Exp and the mantissa field M, represents a number as a whole

$$f = \pm M \cdot 2^E \qquad (1.2.3)$$

We classify the range of numbers depending on the value (E) of the exponent and denote it as

$$R_E = [2^E, 2^{E+1}) \ with -1022 \le E \le +1023 \qquad (1.2.4)$$

In each range, the least unit, i.e. the value of least significant bit (LSB) or the difference between two consecutive numbers represented by the mantissa of 52 bits is

$$\Delta_E = \Delta \times 2^E = 2^{-52} \times 2^E = 2^{E-52} \qquad (1.2.5)$$

Let us take a closer look at the bit-wise representation of numbers belonging to each range:

(0) 0 (zero)

63	62　　　　52	51　　　　　　　　　　0
S	000 ⋯ 0000	0000 0000 ⋯ 0000 0000

(1) Un-normalized range (with the value of hidden bit $b_h = 0$)

$$R_{-1023} = [2^{-1074}, 2^{-1022}) \text{ with Exp} = 0, E = \text{Exp} - 1023 + 1 = -1022 \qquad (1.2.6.1a)$$

1 bit	Exp:11 bits	M : 52 bits	Hidden bit b_h
S	000 ... 0000	0000 0000 0000 ... 0000 0001	$(0+1\times2^{-52})\times2^{-1022} = 1\times2^{-1074}$
...	... ...	... ...	
S	000 ... 0000	1111 1111 ... 1111 1111	$\{0+(2^{52}-1)2^{-52}=2-2^{-52}\}\times2^{-1022}$

The value of LSB : $\Delta_{-1023} = \Delta_{-1022} = 2^{-1022-52} = 2^{-1074}$ (1.2.6.1b)

(2) The smallest normalized range (with the value of hidden bit $b_h = 1$)

$$R_{-1022} = [2^{-1022}, 2^{-1021}) \text{ with Exp} = 1, E = \text{Exp} - 1023 = -1022 \qquad (1.2.6.2a)$$

1 bit	Exp:11 bits	M : 52 bits	Hidden bit b_h
S	000 ... 0001	0000 0000 ... 0000 0000	$(1+0)\times2^{1-1023} = 1\times2^{-1022}$: realmin
S	000 ... 0001	0000 0000 ... 0000 0001	$(1+1\times2^{-52})\times2^{-1022}$
...	... ...	... ...	
S	000 ... 0001	1111 1111 ... 1111 1111	$\{1+(2^{52}-1)2^{-52}=2-2^{-52}\}\times2^{-1022}$

The value of LSB : $\Delta_{-1022} = 2^{-1022-52} = 2^{-1074}$ (1.2.6.2b)

(3) Basic normalized range (with the value of hidden bit $b_h = 1$)

$$R_0 = [2^0, 2^1) \text{ with Exp} = 2^{10} - 1 = 1023, E = \text{Exp} - 1023 = 0 \qquad (1.2.6.3a)$$

34 MATLAB USAGE AND COMPUTATIONAL ERRORS

1 bit	Exp:11 bits	M : 52 bits	Hidden bit	
S	011 ... 1111	0000 0000 ... 0000 0000	$(1+0) \times 2^{1023-1023}$	$= 1 \times 2^0 = 1$
S	000 ... 0001	0000 0000 ... 0000 0001	$(1+1 \times 2^{-52}) \times 2^{1023-1023}$	$= 1 + 2^{-52}$
...	... ...	... ...		
S	000 ... 0001	1111 1111 ... 1111 1111	$\{1+(2^{52}-1)2^{-52} = 2-2^{-52}\} \times 2^0$	

$$\text{The value of LSB}: \Delta_0 = 2^{-52} \qquad (1.2.6.3\text{b})$$

(4) The largest normalized range (with the value of hidden bit $b_h = 1$)

$$R_{1023} = [2^{1023}, 2^{1024}) \text{ with Exp} = 2^{11} - 2 = 2046,$$

$$E = \text{Exp} - 1023 = 1023 \qquad (1.2.6.4\text{a})$$

1 bit	Exp:11 bits	M : 52 bits	Hidden bit	
S	111 ... 1110	0000 0000 ... 0000 0000	$(1+0) \times 2^E = 1 \times 2^{1023}$	
S	000 ... 0001	0000 0000 ... 0000 0001	$(1+2^{-52}) \times 2^{1023}$	
...	... ...	... ...		
S	000 ... 0001	1111 1111 ... 1111 1111	$\{1+(2^{52}-1)2^{-52} = 2-2^{-52}\} \times 2^{1023}$	: realmax

$$\text{The value of LSB}: \Delta_{1023} = 2^{1023-52} = 2^{971} \qquad (1.2.6.4\text{b})$$

(5) $\pm\infty(\text{inf})$ with Exp $= 2^{11} - 1 = 2047$, $E = \text{Exp} - 1023 = 1024$ (meaningless)

1 bit	Exp:11 bits	M : 52 bits	Hidden bit	
0	111 ... 1111	0000 0000 ... 0000 0000	$+(1+0) \times 2^{1024} \sim +\infty$	
1	111 ... 1111	0000 0000 ... 0000 0000	$-(1+0) \times 2^{1024} \sim -\infty$	
1	111 ... 1111	0000 0000 ... 0000 0001	Invalid (not used)	
...	... ...	... ...		
S	111 ... 1111	1111 1111 ... 1111 1111	Invalid (not used)	

From what has been mentioned earlier, we know that the minimum and maximum positive numbers are, respectively,

$$f_{\min} = (0 + 2^{-52}) \times 2^{-1022} = 2^{-1074} \approx 4.9406564584124654 \times 10^{-324} \qquad (1.2.7\text{a})$$

$$f_{\max} = (2 - 2^{-52}) \times 2^{1023} \approx 1.7976931348623157 \times 10^{308} \qquad (1.2.7\text{b})$$

where the three MATLAB constants, i.e. eps, realmin, and realmax, represent 2^{-52}, 2^{-1022}, and $(2 - 2^{-52}) \times 2^{1023}$, respectively. This can be checked by running the script "nm109.m" in Section 1.I..

Now, in order to gain some idea about the arithmetic computational mechanism, let us see how the addition of two numbers, 3 and 14, represented in the IEEE 64-bit floating number system, is performed.

Digital-to-binary conversion → Normalization → 64-bit Representation
Decimal-to-binary conversion

$3_{10} \to 11_2 \to 1.1_2 \times 2^1 = \boxed{1}.1_2 \times 2^{1024-1023}$
$14_{10} \to 1110_2 \to 1.11_2 \times 2^3 = \boxed{1}.11_2 \times 2^{10264-1023}$

$2\underline{)3 \;...\; 1}$
$3_{10} \to 11_2$

$2\underline{)14 \;...\; 0}$
$2\underline{)7 \;...\; 1}$
$2\underline{)3 \;...\; 1}$
$14_{10} \to 1110_2$

64-bit Representation

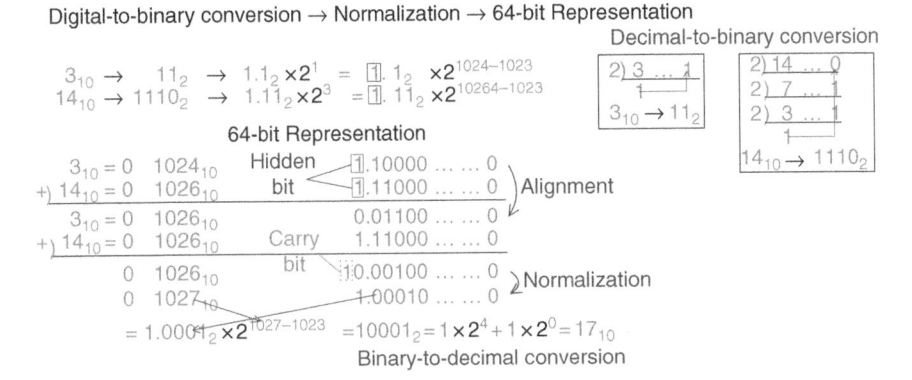

$3_{10} = 0 \quad 1024_{10} \quad$ Hidden $\quad \boxed{1}.10000 \;...\; ... \;0$
$+)\; 14_{10} = 0 \quad 1026_{10} \quad$ bit $\quad \boxed{1}.11000 \;...\; ... \;0$ } Alignment
$3_{10} = 0 \quad 1026_{10} \quad\quad\quad\quad 0.01100 \;...\; ... \;0$
$+)\; 14_{10} = 0 \quad 1026_{10} \quad$ Carry $\quad 1.11000 \;...\; ... \;0$
$0 \quad 1026_{10} \quad$ bit $\quad 10.00100 \;...\; ... \;0$ } Normalization
$0 \quad 1027_{10} \quad\quad\quad\quad 1.00010 \;...\; ... \;0$
$= 1.00010_2 \times 2^{1027-1023} \quad = 10001_2 = 1 \times 2^4 + 1 \times 2^0 = 17_{10}$

Binary-to-decimal conversion

Figure 1.6 Process of adding two numbers, 3 and 14, in MATLAB.

In the process of adding the two numbers illustrated in Figure 1.6, an alignment is made so that the two exponents in their 64-bit representations equal each other; and it will kick out the part smaller by more than 52 bits, causing some numerical error. For example, adding 2^{-23} to 2^{30} does not make any difference, while adding 2^{-22} to 2^{30} does, as we can see by typing the following statements into the MATLAB Command window.

```
»x=2^30; x+2^-22==x, x+2^-23==x
      ans= 0(false)     ans= 1(true)
```

(cf) Each range has a different minimum unit (LSB value) described by Eq. (1.2.5). It implies that the numbers are uniformly distributed within each range. The closer the range is to 0, the denser the numbers in the range are. Such a number representation makes the absolute quantization error large/small for large/small numbers, decreasing the possibility of large relative quantization error.

1.2.2 Various Kinds of Computing Errors

There are various kinds of errors that we encounter when using a computer for computation.

- *Truncation error*: Caused by adding up to a finite number of terms, while we should add infinitely many terms to get the exact answer in theory.
- *Round-off error*: Caused by representing/storing numeric data in finite bits.
- *Overflow/underflow*: Caused by too large or too small numbers to be represented/stored properly in finite-bits, more specifically, the numbers having absolute values larger/smaller than the maximum (f_{max})/minimum (f_{min}) number that can be represented in MATLAB.
- *Negligible addition*: Caused by adding two numbers of magnitudes differing by over 52 bits, as can be seen in the last section.

36 MATLAB USAGE AND COMPUTATIONAL ERRORS

- *Loss of significance*: Caused by a 'bad subtraction', which means a subtraction of a number from another one that is almost equal in value.
- *Error magnification*: Caused and magnified/propagated by multiplying/dividing a number containing a small error with a large/small number.
- Errors depending on the numerical algorithms, step size, and so on.

For all that we cannot be free from these kinds of inevitable errors in some degree, it is not computers, but we, human beings, who must be responsible the computing errors. While our computer may insist on its innocence for an unintended lie, we programmers and users cannot escape from the responsibility of taking measures against the errors and would have to pay for being careless enough to be deceived by a machine. We should, therefore, try to decrease the errors and minimize their impact on the final results. In order to do so, we must know the sources of computing errors and also grasp the computational properties of numerical algorithms.

For instance, consider the following two formulas:

$$f_1(x) = \sqrt{x}(\sqrt{x+1} - \sqrt{x}),\ f_2(x) = \frac{\sqrt{x}}{\sqrt{x+1} + \sqrt{x}} \qquad (1.2.8)$$

These are theoretically equivalent, whence we expect them to give exactly the same value. However, running the following MATLAB script "nm122.m" to compute the values of the two formulas, we see a surprising result that, as x increases, the step of $f_1(x)$ incoherently moves hither and thither, while $f_2(x)$ approaches 1/2 at a steady pace. We might feel betrayed by the computer and have a doubt about its reliability. Why does such a flustering thing happen with $f_1(x)$? It is because the number of significant bits abruptly decreases when the subtraction ($\sqrt{x+1} - \sqrt{x}$) is performed for large values of x, which is called '*loss of significance*'. In order to take a close look at this phenomenon, let $x = 10^{15}$. Then we have

$$\sqrt{x+1} = 3.162\ 277\ 660\ 168\ 381 \times 10^7 = 31622776.601\ 683\ 81$$

$$\sqrt{x} = 3.162277660168379 \times 10^7 = 31622776.60168379$$

These two numbers have 52 significant bits, or equivalently 16 significant digits ($2^{52} \approx 10^{52 \times 3/10} \approx 10^{15}$) so that their significant digits range from 10^8 to 10^{-8}. Accordingly, the least significant digit (LSD) of their sum and difference is also the eighth digit after the decimal point (10^{-8}).

$$\sqrt{x+1} + \sqrt{x} = 63245553.20336761$$

$$\sqrt{x+1} - \sqrt{x} = 0.0000000186264514923 0957 \approx 0.00000002$$

Note that the number of significant digits of the difference decreased to 1 from 16. Could you imagine that a single subtraction may kill most of the significant digits? This is the very 'loss of significance', which is often called 'catastrophic cancellation'.

```
%nm122.m
f1=@(x) sqrt(x)*(sqrt(x+1)-sqrt(x));
f2=@(x) sqrt(x)./(sqrt(x+1)+sqrt(x));
x=1;
format long e
for k=1:15
   fprintf('At x=%15.0f, f1(x)=%20.18f, f2(x)=%20.18f', x,f1(x),f2(x));
   x= 10*x;
end
sx1=sqrt(x+1);   sx=sqrt(x);
d=sx1-sx;   s=sx1+sx;
fprintf('sqrt(x+1)=%25.13f, sqrt(x)=%25.13f ',sx1,sx);
fprintf('   diff=%25.23f, sum=%25.23f ',d,s);
```

```
»nm122
At x=                    1, f1(x)=0.414213562373095150, f2(x)=0.414213562373095090
At x=                   10, f1(x)=0.488088481701514750, f2(x)=0.488088481701515480
At x=                  100, f1(x)=0.498756211208899460, f2(x)=0.498756211208902730
At x=                 1000, f1(x)=0.499875062461021870, f2(x)=0.499875062460964860
At x=                10000, f1(x)=0.499987500624854420, f2(x)=0.499987500624960890
At x=               100000, f1(x)=0.499998750005928860, f2(x)=0.499998750006249940
At x=              1000000, f1(x)=0.499999875046341910, f2(x)=0.499999875000062490
At x=             10000000, f1(x)=0.499999987401150920, f2(x)=0.499999987500000580
At x=            100000000, f1(x)=0.500000005558831620, f2(x)=0.499999998749999950
At x=           1000000000, f1(x)=0.500000077997506340, f2(x)=0.499999999874999990
At x=          10000000000, f1(x)=0.499999441672116520, f2(x)=0.499999999987500050
At x=         100000000000, f1(x)=0.500004449631168080, f2(x)=0.499999999998750000
At x=        1000000000000, f1(x)=0.500003807246685030, f2(x)=0.499999999999874990
At x=       10000000000000, f1(x)=0.499194546973835970, f2(x)=0.499999999999987510
At x=      100000000000000, f1(x)=0.502914190292358400, f2(x)=0.499999999999998720
At x=     1000000000000000, f1(x)=0.589020114423405180, f2(x)=0.499999999999999830
sqrt(x+1)=    100000000.0000000000000, sqrt(x)=    100000000.0000000000000
   diff=0000000000000000, sum=41a7d78400000000
```

1.2.3 Absolute/Relative Computing Errors

The absolute/relative error of an approximate value x to the true value X of a real-valued variable is defined as follows:

$$\varepsilon_x = X(\text{true value}) - x(\text{approximate value}) \quad (1.2.9)$$

$$\rho_x = \frac{\varepsilon_x}{X} = \frac{X-x}{X} \quad (1.2.10)$$

If the LSD is the dth digit after the decimal point, then the magnitude of the absolute error is not greater than half the value of LSD.

$$|\varepsilon_x| = |X - x| \leq \frac{1}{2}10^{-d} \quad (1.2.11)$$

If the number of significant digits is s, then the magnitude of the relative error is not greater than half the relative value of LSD over most significant digit (MSD).

$$|\rho_x| = \frac{|\varepsilon_x|}{|X|} = \frac{|X-x|}{|X|} \leq \frac{1}{2}10^{-s} \qquad (1.2.12)$$

1.2.4 Error Propagation

In this section, we will see how the errors of two numbers, x and y, are propagated with the four arithmetic operations. *Error propagation* means that the errors in the input numbers of a process or an operation cause the errors in the output numbers.

Let their absolute errors be ε_x and ε_y, respectively. Then the magnitudes of the absolute/relative errors in the sum and difference are

$$\varepsilon_{x\pm y} = (X \pm Y) - (x \pm y) = (X - x) \pm (Y - y) = \varepsilon_x \pm \varepsilon_y;$$

$$|\varepsilon_{x\pm y}| \leq |\varepsilon_x| + |\varepsilon_y|; \qquad (1.2.13)$$

$$|\rho_{x\pm y}| = \frac{|\varepsilon_{x\pm y}|}{|X \pm Y|} \leq \frac{|X||\varepsilon_x/X| + |Y||\varepsilon_y/Y|}{|X \pm Y|} = \frac{|X||\rho_x| + |Y||\rho_y|}{|X \pm Y|} \qquad (1.2.14)$$

From this, we can see why the relative error is magnified to cause the '*loss of significance*' in the case of subtraction when the two numbers X and Y are almost equal so that $|X - Y| \approx 0$.

The magnitudes of the absolute and relative errors in the multiplication/division are

$$|\varepsilon_{xy}| = |XY - xy| = |XY - (X + \varepsilon_x)(Y + \varepsilon_y)| \approx |X\varepsilon_y \pm Y\varepsilon_x|;$$

$$|\varepsilon_{xy}| \leq |X||\varepsilon_y| + |Y||\varepsilon_x|; \qquad (1.2.15)$$

$$|\rho_{xy}| = \frac{|\varepsilon_{xy}|}{|XY|} \leq \frac{|\varepsilon_y|}{|Y|} + \frac{|\varepsilon_x|}{|X|} = |\rho_x| + |\rho_y| \qquad (1.2.16)$$

$$|\varepsilon_{x/y}| = \left|\frac{X}{Y} - \frac{x}{y}\right| = \left|\frac{X}{Y} - \frac{X+\varepsilon_x}{Y+\varepsilon_y}\right| \approx \frac{|X\varepsilon_y - Y\varepsilon_x|}{Y^2};$$

$$|\varepsilon_{x/y}| \leq \frac{|X||\varepsilon_y| + |Y||\varepsilon_x|}{Y^2}; \qquad (1.2.17)$$

$$|\rho_{x/y}| = \frac{|\varepsilon_{x/y}|}{|X/Y|} \leq \frac{|\varepsilon_x|}{|X|} + \frac{|\varepsilon_y|}{|Y|} = |\rho_x| + |\rho_y| \qquad (1.2.18)$$

This implies that, in the worst-case, the relative error in multiplication/division may be as large as the sum of the relative errors of the two numbers.

1.2.5 Tips for Avoiding Large Errors

In this section, we will look over several tips to reduce the chance of large errors occurring in calculations.

First, to decrease the magnitude of round-off errors and to lower the possibility of overflow/underflow errors, make the intermediate result as close to 1 (one) as possible in consecutive multiplication/division processes. According to this rule, when computing xy/z, we program the formula as

- $(xy)/z$ when x and y in the multiplication are very different in magnitude,
- $x(y/z)$ when y and z in the division are close in magnitude, and
- $(x/z)y$ when x and z in the division are close in magnitude.

```
%nm125_1.m
x=36; y=1e16;
for n=[-20 -19 19 20]
    fprintf('y^%2d/e^%2dx=%25.15e\n',n,n,y^n/exp(n*x));
    fprintf(' (y/e^x)^%2d=%25.15e\n',n, (y/exp(x))^n);
end
```

For instance, when computing y^n/e^{nx} with $x \gg 1$ and $y \gg 1$, we would program it as $(y/e^x)^n$ rather than as y^n/e^{nx}, so that overflow/underflow can be avoided. You may verify this by running the above MATLAB script "nm125_1.m".

```
»nm125_1
y^-20/e^-20x=  0.000000000000000e+000
 (y/e^x)^-20=  4.920700930263814e-008
y^-19/e^-19x=  1.141367814854768e-007
 (y/e^x)^-19=  1.141367814854769e-007
y^19/e^19x=    8.761417546430845e+006
 (y/e^x)^19=   8.761417546430843e+006
y^20/e^20x=                       NaN
 (y/e^x)^20=   2.032230802424294e+007
```

Second, in order to prevent '*loss of significance*', it is important to avoid a '*bad subtraction*' (Section 1.B.), i.e. a subtraction of a number from another number having almost equal value. Let us consider a simple problem of finding the roots of a second-degree equation $ax^2 + bx + c = 0$ by using the quadratic formula

$$x_1 = \frac{-b + \sqrt{b^2 - 4ac}}{2a}, \quad x_2 = \frac{-b - \sqrt{b^2 - 4ac}}{2a} \qquad (1.2.19)$$

Let $|4ac| \ll b^2$. Then, depending on the sign of b, a 'bad subtraction' may be encountered when we try to find x_1 or x_2, which is a smaller one of the two roots. This implies that it is safe from the 'loss of significance' to compute the root having the larger absolute value first and then obtain the other root by using the relation (between the roots and the coefficients) $x_1 x_2 = c/a$ (see Problem 1.20(b)).

40 MATLAB USAGE AND COMPUTATIONAL ERRORS

```
%nm125_2.m: roundoff error test
f1=@(x)(1-cos(x))/x/x;           % Eq.(1.2.20-1)
f2=@(x)sin(x)*sin(x)/x/x/(1+cos(x));  % Eq.(1.2.20-2)
for k=0:1
   x=k*pi;   tmp= 1;
   for k1=1:8
      tmp=tmp*0.1; x1= x+tmp;
      fprintf('At x=%10.8f, ', x1)
      fprintf('f1(x)=%18.12e; f2(x)=%18.12e', f1(x1),f2(x1));
   end
end
```

For another instance, we consider the following two formulas, which are analytically the same, but numerically different.

$$f_1(x) = \frac{1-\cos x}{x^2}, \quad f_2(x) = \frac{\sin^2 x}{x^2(1+\cos x)} \qquad (1.2.20)$$

It is safe to use $f_1(x)$ for $x \approx \pi$ since the term $(1 + \cos x)$ in $f_2(x)$ is a 'bad subtraction', while it is safe to use $f_2(x)$ for $x \approx 0$ since the term $(1 - \cos x)$ in $f_1(x)$ is a 'bad subtraction'. Let us run the above MATLAB script "nm125_2.m" to confirm this. Under is the running result. This implies that we might use some formulas to avoid a 'bad subtraction'.

```
»nm125_2
  At x=0.10000000, f1(x)=4.995834721974e-01; f2(x)=4.995834721974e-01
  At x=0.01000000, f1(x)=4.999958333474e-01; f2(x)=4.999958333472e-01
  At x=0.00100000, f1(x)=4.999999583255e-01; f2(x)=4.999999583333e-01
  At x=0.00010000, f1(x)=4.999999969613e-01; f2(x)=4.999999995833e-01
  At x=0.00001000, f1(x)=5.000000413702e-01; f2(x)=4.999999999958e-01
  At x=0.00000100, f1(x)=5.000444502912e-01; f2(x)=5.000000000000e-01
  At x=0.00000010, f1(x)=4.996003610813e-01; f2(x)=5.000000000000e-01
  At x=0.00000001, f1(x)=0.000000000000e+00; f2(x)=5.000000000000e-01
  At x=3.24159265, f1(x)=1.898571371550e-01; f2(x)=1.898571371550e-01
  At x=3.15159265, f1(x)=2.013534055392e-01; f2(x)=2.013534055391e-01
  At x=3.14259265, f1(x)=2.025133720884e-01; f2(x)=2.025133720914e-01
  At x=3.14169265, f1(x)=2.026294667803e-01; f2(x)=2.026294678432e-01
  At x=3.14160265, f1(x)=2.026410772244e-01; f2(x)=2.026410604538e-01
  At x=3.14159365, f1(x)=2.026422382785e-01; f2(x)=2.026242248740e-01
  At x=3.14159275, f1(x)=2.026423543841e-01; f2(x)=2.028044503269e-01
  At x=3.14159266, f1(x)=2.026423659946e-01; f2(x)=               Inf
```

It may be helpful for avoiding a 'bad subtraction' to use the Taylor series expansion [W-5] rather than using the exponential function directly for the computation of e^x. For example, suppose we want to find

$$f_3(x) = \frac{e^x - 1}{x} \text{ at } x = 0 \qquad (1.2.21)$$

COMPUTER ERRORS VS. HUMAN MISTAKES   41

We can use the Taylor series expansion up to just the fourth-order of e^x about $x = 0$:

$$g(x) = e^x \approx g(0) + g'(0)x + \frac{g''(0)}{2!}x^2 + \frac{g^{(3)}(0)}{3!}x^3 + \frac{g^{(4)}(0)}{4!}x^4$$

$$= 1 + x + \frac{1}{2!}x^2 + \frac{1}{3!}x^3 + \frac{1}{4!}x^4$$

to approximate the above function (1.2.19) as

$$f_3(x) = \frac{e^x - 1}{x} \approx 1 + \frac{1}{2!}x + \frac{1}{3!}x^2 + \frac{1}{4!}x^3 = f_4(x) \quad (1.2.22)$$

Noting that the true value of (1.2.21) is computed to be 1 by using the L'Hopital's rule [W-7], we run the MATLAB script "nm125_3.m" to find which one of the two formulas $f_3(x)$ and $f_4(x)$ is better for finding the value of the (1.2.21) at $x = 0$. Would you compare them based on the running result shown below? How can the approximate formula $f_4(x)$ outrun the true one $f_3(x)$ for the numerical purpose, though not usual? It is because the zero factors in the numerator/denominator of $f_3(x)$ are cancelled to set $f_4(x)$ free from the terror of a 'bad subtraction'.

```
»nm125_3
   At x=0.100000000000,  f3(x)=1.051709180756e+00;  f4(x)=1.084166666667e+00
   At x=0.010000000000,  f3(x)=1.005016708417e+00;  f4(x)=1.008341666667e+00
   At x=0.001000000000,  f3(x)=1.000500166708e+00;  f4(x)=1.000833416667e+00
   At x=0.000100000000,  f3(x)=1.000050001667e+00;  f4(x)=1.000083334167e+00
   At x=0.000010000000,  f3(x)=1.000005000007e+00;  f4(x)=1.000008333342e+00
   At x=0.000001000000,  f3(x)=1.000000499962e+00;  f4(x)=1.000000833333e+00
   At x=0.000000100000,  f3(x)=1.000000049434e+00;  f4(x)=1.000000083333e+00
   At x=0.000000010000,  f3(x)=9.999999939225e-01;  f4(x)=1.000000008333e+00
   At x=0.000000001000,  f3(x)=1.000000082740e+00;  f4(x)=1.000000000833e+00
   At x=0.000000000100,  f3(x)=1.000000082740e+00;  f4(x)=1.000000000083e+00
   At x=0.000000000010,  f3(x)=1.000000082740e+00;  f4(x)=1.000000000008e+00
   At x=0.000000000001,  f3(x)=1.000088900582e+00;  f4(x)=1.000000000001e+00
```

```
%nm125_3.m: reduce the roundoff error using Taylor series
f3=@(x)(exp(x)-1)/x;   % LHS of Eq.(1.2.22)
f4=@(x)((x/4+1)*x/3)+x/2+1;  % RHS of Eq.(1.2.22)
x=0;
tmp=1;
for k1=1:12
   tmp=tmp*0.1; x1=x+tmp;
   fprintf('At x=%14.12f, ', x1)
   fprintf('f3(x)=%18.12e; f4(x)=%18.12e', f3(x1),f4(x1));
end
```

1.3 TOWARD GOOD PROGRAM

Among the various criteria about the quality of a general program, the most important one is how robust its performance is against the change of the problem properties and the initial values. A good program guides the program users who do not know much about the program and at least give them a warning message without runtime error for their minor mistake. There are many other features that need to be considered, such as user friendliness, compactness and elegance, readability. But, as far as the numerical methods are concerned, the accuracy of solution, execution speed (time efficiency), and memory utilization (space efficiency) are of utmost concern. Since some tips to achieve the accuracy or at least to avoid large errors (including overflow/underflow) are given in the last section, we will look over the issues of execution speed and memory utilization.

1.3.1 Nested Computing for Computational Efficiency

The execution speed of a program for a numerical solution depends mostly on the number of function (subroutine) calls and arithmetic operations performed in the program. Therefore, we like the algorithm requiring fewer function calls and arithmetic operations. For instance, suppose we want to evaluate the value of a polynomial

$$p_4(x) = a_1 x^4 + a_2 x^3 + a_3 x^2 + a_4 x + a_5 \qquad (1.3.1)$$

It is better to use the *nested* (*computing*) structure (as follows) than to use the above form as it is

$$p_{4n}(x) = (((a_1 x + a_2)x + a_3)x + a_4)x + a_5 \qquad (1.3.2)$$

Note that the numbers of multiplications needed in Eqs. (1.3.2) and (1.3.1) are 4 and $(4+3+2+1=9)$, respectively. This point is illustrated by the script "nm131_1.m", where a polynomial $\sum_{i=0}^{N-1} a_i x^i$ of degree $N = 10^6$ for a certain value of x is computed by using the three methods, i.e. Eq. (1.3.1), Eq. (1.3.2), and the MATLAB built-in function 'polyval()'. Interested readers could run this script to see that the nested multiplication (Eq. (1.3.2)) and 'polyval()' (fabricated in a nested structure) take less time than the plain method (Eq. (1.3.1)), while 'polyval()' may take longer time because of some overhead time for being called.

```
%nm131_1.m: nested multiplication vs. plain multiple multiplication
N=1000000+1; a=[1:N]; x=1;
tic % initialize the timer
p=sum(a.*x.^[N-1:-1:0]); % Plain multiplication
p
time_plain=toc % Operation time for the sum of multiplications
tic, pn=a(1);
for i=2:N % Nested multiplication
   pn = pn*x +a(i);
end
pn, time_nested=toc % Operation time for the nested multiplications
tic, polyval(a,x)
time_polyval=toc % Operation time for using polyval()
```

Programming in a nested structure is not only recommended for time-efficient computation, but also may be critical to the solution. For instance, consider a problem of finding the value

$$S(K) = \sum_{k=0}^{K} \frac{\lambda^k}{k!} e^{-\lambda} \text{ for } \lambda = 100 \text{ and } K = 155 \qquad (1.3.3)$$

%nm131_2_1.m: nested structure	%nm131_2_2.m: not nested structure
lam=100; K=155;	lam=100; K=155;
p=exp(-lam);	tic
tic	S=0;
S=0;	for k=1:K
for k=1:K	p=lam^k/factorial(k);
p=p*lam/k; S=S+p;	S=S+p;
end	end
S	S*exp(-lam)
time_nested=toc	time_not_nested=toc

The above two scripts are made for computing Eq. (1.3.3). Noting that this sum of Poisson probability distribution [W-4] is close to 1 for such a large K, we can run them to find that one works fine, while the other gives a quite wrong result. Could you tell which one is better?

1.3.2 Vector Operation vs. Loop Iteration

It is time-efficient to use vector operations rather than loop iterations to perform a repetitive job for an array of data. The following script "nm132_1.m" compares a *loop iteration* and a *vector operation* (for computing the sum of 10^5 numbers) in terms of the execution speed. Could you tell which one is faster?

44 MATLAB USAGE AND COMPUTATIONAL ERRORS

```
%nm132_1.m: Vector operation vs. Loop iteration
N=1e5; th=[0:N-1]/50000*pi;
tic
s1=sin(th(1));
for i=2:N, s1= s1+sin(th(i)); end   % Loop iteration
time_loop=toc, s1
tic
s2=sum(sin(th));  % Vector operation
time_vector=toc, s2
```

As a more practical example, let us consider a problem of finding the discrete-time Fourier transform (DtFT) [W-2 of a given sequence $x[n]$:

$$X(\Omega) = \sum_{n=0}^{N-1} x[n]\, e^{-j\Omega n} \text{ for } \Omega = [-100 : 100]\frac{\pi}{100} \qquad (1.3.4)$$

```
%nm132_2.m: Vector operation vs. Loop iteration
N=1000; x=rand(1,N);
kk=[-100:100]; W=kk*pi/100; % Frequency range
% for for loop
tic
for k =1:length(W)
   X_for1(k)=0; %zeros(size(W));
   for n=1:N, X_for1(k) = X_for1(k) +x(n)*exp(-j*W(k)*(n-1)); end
end
time_loop_loop=toc
% for vector loop
tic
X_for2 =0 ; %zeros(size(W));
for n=1:N
   X_for2 = X_for2 +x(n)*exp(-j*W*(n-1));
end
time_vector_loop=toc
% Vector operation
tic
nn=[1:N].'; X_vec = x*exp(-j*(nn-1)*W);
time_vector_vector=toc
discrepancy1= norm(X_for1-X_vec)
discrepancy2= norm(X_for2-X_vec)
```

The above script "nm132_2.m" compares a vector operation vs. a loop iteration for computing the DtFT in terms of the execution speed. Could you tell which one is faster?

1.3.3 Iterative Routine vs. Recursive Routine

In this section, we compare an iterative routine and a *recursive routine* performing the same job. Consider the following two functions 'fctrl1(n)'/'fctrl2(n)', whose common objectives is to get the factorial of a given nonnegative integer k.

$$k! = k(k-1) \cdots 2 \cdot 1 \qquad (1.3.5)$$

They differ in their structure. While 'fctrl1()' uses a for loop structure, 'fctrl2()' uses the *recursive* (*self-calling*) structure that a program uses itself as a subroutine to perform a sub-job. Compared with 'fctrl1()', 'fctrl2()' is easier to program as well as to read, but is subject to runtime error that is caused by the excessive use of stack memory as the number of recursive calls increases with large n. Another disadvantage of 'fctrl2()' is that it is time-inefficient for the number of function calls, which increases with the input argument (n). In this case, a professional programmer would consider the standpoint of users to determine the programming style. Some algorithms like the adaptive integration (Section 5.8), however, may fit the recursive structure perfectly.

```
function m=fctrl1(n)
m=1;
for k=2:n, m=m*k; end
```

```
function m=fctrl2(n)
if n<=1, m=1;
 else   m=n*fctrl2(n-1);
end
```

1.3.4 To Avoid Runtime Error

A good program guides the program users who do not know much about the program and at least give them a warning message without runtime error for their minor mistake. If you do not know what runtime error is, you can experience one by taking the following steps:

```
function m=fctrl(n)
if n<0, error('The factorial of negative number ??');
 else   m=1; for k=2:n, m=m*k; end
end
```

1. Make and save the following routine 'fctrl()' in an M-file named "fctrl.m" in a directory listed in the MATLAB search path.
2. Type fctrl(-1) into the MATLAB Command window. Then you will see

    ```
    »fctrl1(-1)
    ans = 1
    ```

46 MATLAB USAGE AND COMPUTATIONAL ERRORS

This seems to imply that $(-1)! = 1$, which is not true. It is caused by the mistake of the user who tries to find $(-1)!$ without knowing that it is not defined. This kind of runtime error seems to be minor because it does not halt the process. But it needs special attention for it may not be easy to detect. If you are a good programmer, you will insert some error handling statements in the function 'fctrl()' as above. Then, when someone happens to execute fctrl(-1) in the Command window or through an M-file, the execution stops and he will see the error message in the Command window as

```
Error using fctrl (line 2)
The factorial of negative number ??
```

This shows the error message (given as the input argument of the 'error()' routine) together with the name of the routine in which the accidental 'error' happens, which is helpful for the user to avoid the error.

Most common runtime errors are caused by an 'out of domain' index of array and the violation of matrix dimension compatibility, as illustrated in Section 1.G.. For example, consider the 'Gauss(A,B)' routine in Section 2.2.2, whose job is to solve a system of linear equations Ax=B for x. To appreciate the role of the fifth line handling the dimension compatibility error in the routine, remove the line (by putting the comment mark % before the line in the M-file defining 'Gauss()') and type the following statements in the Command window:

```
»A=rand(3,3); B=rand(2,1); x=Gauss(A,B)
 Index exceeds matrix dimensions.
 Error in Gauss (line 10)
 AB=[A(1:NA,1:NA) B(1:NA,1:NB)]; % Augmented matrix
```

Then MATLAB gives you an error message together with the suspicious statement line and the routine name. But it is hard to figure out what caused the runtime error and you may get nervous lest the routine should have some bug. Now, restore the fifth line in the routine and type the same statements in the Command window:

```
»x=Gauss(A,B)
 Error using Gauss (line 8)
 A and B must have compatible dimension
```

This error message (provided by the programmer of the routine) helps you to realize that the source of the runtime error is the incompatible matrices/vectors A and B given as the input arguments to the 'Gauss()' routine. Very like this, a good program has a scenario for possible user mistakes and fires the 'error' routine for each abnormal condition to show the user the corresponding error message.

Many users often give more/fewer input arguments than supposed to be given to the MATLAB functions/routines and sometimes give wrong types/formats of data. To experience this type of error, let us try using the MATLAB function 'sinc1(t,D)', that is defined in an M-file named "sinc1.m" as

```
function x=sinc1(t,D)
if nargin<2, D=1; end
%t(find(t==0))=eps;
x=sin(pi*t/D)./(pi*t/D);
```

to plot the graph of a sinc function

$$\text{sinc}(t/D) = \frac{\sin(\pi t/D)}{\pi t/D} \quad \text{with} \quad D = 0.5 \quad \text{and} \quad t = \begin{bmatrix} -2 & 2 \end{bmatrix} \quad (1.3.6)$$

With this purpose, type the following statements in the Command window:

```
D=0.5; b1=-2; b2=2; t=b1+[0:100]/100*(b2-b1);
subplot(221), plot(t,sinc1(t,D)), axis([b1 b2 -0.4 1.2])
hold on, plot(t,sinc1(t),'r:')
```

The two plotting commands 'sinc1(t,D)' and 'sinc1(t)' yield the solid and dotted line graphs, respectively, as shown in Figure 1.7a where it should be noted that 'sinc1()' does not bother us and works fine even without the second input argument D. We owe the second line in the function 'sinc1()' for the nice error-handling service:

```
    if nargin<2, D=1; end
```

This line takes care of the case where the number of input arguments (nargin) is <2, by assuming that the second input argument is D=1 by default. This programming technique is the key to making the MATLAB functions adaptive to different number/type of input arguments, which is very useful for adding the user-convenience into the MATLAB functions. To appreciate its role, we remove the second line or deactivate it by putting the comment mark % in the beginning of the line in the M-file defining 'sinc1()', and type the same statement in the Command window, trying to use 'sinc1()' without the second input argument.

```
»plot(t,sinc1(t),'k:')
  Not enough input arguments.
  Error in sinc1 (line 4)
  x=sin(pi*t/D)./(pi*t/D);
```

This time we get a serious error message with no graphic result. It is implied that the MATLAB function without appropriate error-handling capability does not allow the user's default or carelessness.

48 MATLAB USAGE AND COMPUTATIONAL ERRORS

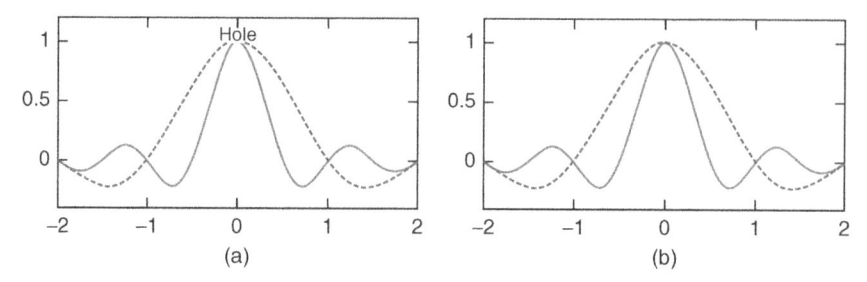

Figure 1.7 Graphs of sinc functions. Using 'sinc1()' without division-by-zero handling (a) and with division-by-zero handling (b).

However, you may feel annoyed by the hole at $t = 0$ in the graphs shown in Figure 1.7a. What caused such an 'accident of hole'? It happened (with no warning message about 'division-by-zero') because not only the denominator but also the numerator of Eq. (1.3.6) becomes for $t = 0$ to make the value of Eq. (1.3.6) NaN (not a number: undetermined) while the true value of the sinc function at $t = 0$ is sinc(0) = 1 as obtained by applying the L'Hopital's rule [W-7]. To avoid the 'accident of hole' due to the NaN problem, activate the third line (in 'sinc1()') by deleting the comment mark % in the beginning of the line where the third line is another error-handling statement:

```
t(find(t==0))=eps;
   or equivalently, for i=1:length(t), if t(i)==0, t(i)=eps; end, end
```

This statement changes any zero entry in the t vector into eps (2.2204e − 16). What is the real purpose of this statement? It is actually to remove the possibility of division-by-zero in the next statement, which is a mathematical expression having t in the denominator:

```
x=sin(pi*t/D)./(pi*t/D);
```

Using the modified function 'sinc1()' with the third line activated, you will get the graphs with no hole like Figure 1.7b.

Lastly, consider of the fourth line in 'sinc1()', which is only one essential statement performing the main job:

```
x=sin(pi*t/D)./(pi*t/D);
```

What is the dot (.) before division operator (/) for? Regarding this, authors gave you a piece of advice that you had better put a dot (.) just before the arithmetic operators *(multiplication), /(division), and ^(power) in the function definition so that the element-by-element (elementwise or termwise) operation can be done any time (Section 1.1.6 (A7)). To appreciate the existence of the dot (.), we remove it from the M-file defining 'sinc1()', and run the following statements:

```
»clf, plot(t,sinc1(t,D)), sinc1(t,D), sin(pi*t/D)/(pi*t/D)
   ans =   -0.0183
```

What do you see in the graphic window on the screen? To our surprise, nothing appears and that with no warning message! What is more surprising, the value of `sinc1(t,D)` or `sin(pi*t/D)/(pi*t/D)` shows up as a scalar. It is hoped that this accident will help you realize how important it is for right term-by-term operations to put dot (.) before the arithmetic operators *, /, and ^. By the way, aren't you curious about how MATLAB deals with a vector division without dot (.)? If so, try with the following statements:

```
»A=[1:10]; B=2*A; A/B, A*B'*(B*B')^-1, A*pinv(B)
  ans =    0.5
```

To understand this response of MATLAB, you can see Section 1.1.7 or 2.1.2.

In this section, we looked over several sources of runtime error, hoping that it aroused reader's attention to runtime errors.

1.3.5 Parameter Sharing via GLOBAL Variables

When we discuss the runtime error that may be caused by user's default in passing some parameter as input argument to the corresponding function, you might feel that the parameter passing job is troublesome. OK, it is understandable as a beginner in MATLAB. How about declaring the parameters as global so that they can be accessed/shared from anywhere in the MATLAB world as far as the declaration is valid? If you want to, you can declare any variable(s) by inserting the following statement in both the main program and all the functions using the variables.

```
global Gravity_Constant Dielectric_Constant
```

```
%plot_sinc.m
clear, clf
global D
D=0.5; b1=-2; b2=2;
t=b1+[0:100]/100*(b2-b1);
%passing the parameter(s) through arguments of the function
subplot(221), plot(t, sinc1(t,D)), axis([b1 b2 -0.4 1.2])
%passing the parameter(s) through global variables
subplot(222), plot(t, sinc2(t)), axis([b1 b2 -0.4 1.2])
```

function x=**sinc1**(t,D)	function x=**sinc2**(t)
if nargin<2, D=1; end	global D
t(find(t==0))=eps;	t(find(t==0))=eps;
x=sin(pi*t/D)./(pi*t/D);	x=sin(pi*t/D)./(pi*t/D);

Then, how convenient it would be, since you do not have to bother about passing the parameters. But as you get proficient in programming and handle many functions/routines that are involved with various sets of parameters, you might find that the global variable is not always convenient, because of the following reasons:

- Once a variable is declared as global, its value can be changed in any of the MATLAB functions having declared it as global, without being noticed by other related functions. Therefore, it is usual to declare only the constants as global and use long names (with all capital letters) as their names for easy identification.
- If some variables are declared as global and modified by several functions/routines, it is not easy to see the relationship and the interaction among the related functions in terms of the global variable. In other words, the program readability gets worse as the number of global variables and related functions increases.

For example, let us look over the above script "plot_sinc.m" and the function 'sinc2()'. They both have a declaration of D as global and consequently, 'sinc2()' does not need the second input argument for getting the parameter D. Running the script, you will see that both 'sinc1()' (accepting D=0.5 as an input argument) and 'sinc2()' (knowing D=0.5 as a global variable) result in the same graphic result as the solid-line graph shown in Figure 1.7b.

1.3.6 Parameter Passing Through VARARGIN

In this section, we see two kinds of routines that get a function name (string) with its parameters as its input argument and play with the function.

First, let us look over the routine 'ez_plot1()', which gets a function name (ftn), its parameters (p), and the lower/upper bounds (bounds=[b1 b2]) of horizontal axis as its first, third, and second input arguments, respectively, and plots the graph of the given function over the interval set by the bounds. Since the given function may or may not have its parameter, the two cases are determined and processed by the number of input arguments (nargin) in the if-else-end block.

```
%plot_sinc1.m
D=1; b1=-2; b2=2;
t=b1+[0:100]/100*(b2-b1);
bounds=[b1 b2];
subplot(223), ez_plot1('sinc1',bounds,D)
axis([b1 b2 -0.4 1.2])
subplot(224), ez_plot('sinc1',bounds,D)
axis([b1 b2 -0.4 1.2])
```

``function ez_plot1(ftn,bounds,p)
b1=bounds(1); b2=bounds(2);
t=b1+[0:100]/100*(b2-b1);
if nargin<=2, x=feval(ftn,t);
  else x=feval(ftn,t,p);
end
plot(t,x)```
	``function ez_plot(ftn,bounds,varargin)
	b1=bounds(1); b2=bounds(2);
	t=b1+[0:100]/100*(b2-b1);
	x=feval(ftn,t,varargin{:});
	plot(t,x)```

Now, let us see the routine 'ez_plot()', which does the same plotting job as 'ez_plot1()'. Note that it has a MATLAB keyword varargin (*variable length argument list*) as its last input argument and passes it into the MATLAB built-in function 'feval()' as its last input argument. Since varargin can represent comma-separated multiple parameters including expression/strings, it paves the high-way for passing the parameters in relays. As the number of parameters increases, it becomes much more convenient to use varargin for passing the parameters than to deal with the parameters one-by-one as in 'ez_plot1()'. This technique will be widely used later in Chapter 4 (Nonlinear Equations), Chapter 5 (on numerical differentiation/integration), Chapter 6 (Ordinary Differential Equations), and Chapter 7 (Optimization).

(cf) Note that MATLAB has a built-in graphic function 'ezplot()', which is much more powerful and convenient to use than 'ez_plot()'. You can type 'help ezplot' to see its function and usage.

1.3.7 Adaptive Input Argument List

A MATLAB function/routine is said to be 'adaptive' to users in terms of input arguments if it accepts different number/type of input arguments and makes a reasonable interpretation. For example, let us see the nonlinear equation solver 'Newton()' in Section 4.4. Its input argument list is

```
(f,df,x0,TolX,MaxIter)
```

where f, df, x0, tol, and MaxIter denote the name of function (to be solved), the name of the derivative function, the initial guess (for solution), the error tolerance, and the maximum number of iterations, respectively. Suppose the user, not having the derivative function, tries to use the routine with just four input argument as follows:

```
»Newton(f,x0,tol,MaxIter)
```

At first, these four input arguments will be accepted as f, df, x0, and tol, respectively. But when the second line of the program body

```
if nargin==4&isnumeric(df), MaxIter=TolX; TolX=x0; x0=df; end
```

is executed, the routine will notice something wrong from that df is not any function name but a number, and then interpret the input arguments as f, x0, tol, and kmax to the idea of the user. This allows the user to use the routine in two ways, depending on whether he is going to supply the routine with the derivative function or not. This scheme is conceptually quite similar to function overloading of C++, but C++ requires us to have several functions having the same name, with different argument list.

PROBLEMS

1.1 Creating a Data File and Retrieving/Plotting Data Saved in a Data File

(a) Using the MATLAB editor, make a script "nm01p01a.m" that lets its user input data pairs of heights (ft) and weights (lb) of as many persons as he wants till he presses <Enter> and save the whole data in the form of an $N \times 2$ matrix into an ASCII data file '$_{***}$.txt' named by the user. If you have no idea how to compose such a script, you can permutate the statements in the box beneath to make your script. Store the script in the file named "nm01p01a.m" and run it to save the following data into the data file named 'hw.txt':

$$5.5 \quad 162$$
$$6.1 \quad 185$$
$$5.7 \quad 170$$
$$6.5 \quad 195$$
$$6.2 \quad 191$$

```
%nm01p01a.m: input data pairs and save them into an ASCII data file
k=0;
while 1
end
k=k+1;
x(k,1)=h;
h=input('Enter height:')
x(k,2)=input('Enter weight:')
if isempty(h), break; end
cd('c:\MATLAB\nma') % to change current working directory
filename=input('Enter filename(.txt):','s');
filename=[filename '.txt']; % String concatenation
save(filename,'x','-ascii')
```

(b) Make a MATLAB script "nm01p01b.m" that reads (loads) the data file 'hw.txt' made in (a), plots the data as in Figure 1.1a in the upper-left region of the screen divided into four regions like Figure 1.5 and plots the data in the form of PWL graph describing the relationship between the height

```
%nm01p01b.m: to read the data file and plot the data
cd('c:\MATLAB\nma') % to change current working directory
weight=hw(I,2);
load hw.txt
clf
subplot(221)
plot(hw)
subplot(222)
axis([5 7 160 200])
plot(height,weight,'-+')
[height,I]=sort(hw(:,1));
```

and the weight in the upper-right region of the screen. Let each data pair be denoted by the symbol '+' on the graph. Also let the ranges of height and weight be [5,7] and [160,200], respectively. If you have no idea, you can permutate the statements in the below box. Additionally, run the script to check if it works fine.

1.2 Text Printout of Alphanumeric Data

Make a MATLAB function 'max_array(A)' that uses the 'max()' command to find one of the maximum elements of a matrix A given as its input argument and uses the 'fprintf()' command to print it onto the screen together with its row/column indices in the following format.

```
'\n Max(A) is A(%2d,%2d)=%5.2f\n',row_index,col_index,maxA
```

Additionally, try it to have the maximum element of an arbitrary matrix (generated by the following two consecutive commands) printed in this format onto the screen.

```
»rand('state',sum(100*clock)), rand(3)
```

1.3 Plotting the Mesh Graph of a Two-dimensional Function

Consider the MATLAB script "nm01p03a.m", whose objective is to draw a cone.

(a) The statement on the sixth line seems to be dispensable. Run the script with and without this line and see what happens.

(b) If you want to plot the function 'fcone(x,y)' defined in another M-file "fcone.m", as an inline function, or as a function handle, how will you modify this script?

(c) If you replace the fifth line by 'Z=1-abs(X)-abs(Y);', what difference does it make?

```
%nm01p03a.m: to plot a cone
x=-1:0.02:1; y=-1:0.02:1;
[X,Y]=meshgrid(x,y);
Z=1-sqrt(X.^2+Y.^2);
Z=max(Z,zeros(size(Z)));
mesh(X,Y,Z)
```
```
function z=fcone(x,y)
z=1-sqrt(x.^2+y.^2);
```

1.4 Plotting the Mesh Graph of Stratigraphic Structure

Consider the incomplete MATLAB script "nm01p04.m", whose objective is to draw a stratigraphic structure of the area around Pennsylvania State University from the several perspective point of view. The data about the depth of the rock layer at 5×5 sites are listed in Table P1.4. Supplement the incomplete parts of the script so that it serves the purpose and run the script to answer the following questions. If you complete it properly and run it, MATLAB will show you the

Table P1.4 The depth of the rock layer.

y-Coordinate	x-Coordinate				
	0.1	1.2	2.5	3.6	4.8
0.5	410	390	380	420	450
1.4	395	375	410	435	455
2.2	365	405	430	455	470
3.5	370	400	420	445	435
4.6	385	395	410	395	410

four similar graphs at the four corners of the screen and be waiting for you to press any key.

(a) At what value of k does MATLAB show you the mesh/surface-type graphs that are the most similar to the first graphs? From this result, what do you guess are the default values of the azimuth or horizontal rotation angle and the vertical elevation angle (in degrees) of the perspective view point?

(b) As the first input argument Az of the command 'view(Az,El)' decreases, in which direction does the perspective view point revolve round the z-axis, clockwise or counterclockwise (seen from the above)?

(c) As the second input argument El of the command 'view(Az,El)' increases, does the perspective view point move up or down along the z-axis?

(d) What is the difference between the plotting commands 'mesh()' and 'meshc()'?

(e) What is the difference between the usages of the command 'view()' with two input arguments Az, El, and with a three-dimensional vector argument [x,y,z]?

```
%nm01p04.m: to plot a stratigraphic structure
clear, clf
x=[0.1 .. .. . ];
y=[0.5 .. .. . ];
Z=[410 390 .. .. .. .. ];
[X,Y]=meshgrid(x,y);
subplot(221), mesh(X,Y,500-Z)
subplot(222), surf(X,Y,500-Z)
subplot(223), meshc(X,Y,500-Z)
subplot(224), meshz(X,Y,500-Z)
pause
for k=0:7
   Az=-12.5*k; El=10*k; Azr=Az*pi/180; Elr=El*pi/180;
   subplot(221), view(Az,El)
   subplot(222),
   k, view([sin(Azr),-cos(Azr),tan(Elr)]), pause %pause(1)
end
```

PROBLEMS

1.5 Plotting the Graph of a Function over an Interval Containing Its Singular Point
Noting that the tangent function $f(x) = \tan(x)$ is singular at $x = \pi/2, 3\pi/2, \ldots$, let us plot its graph over $[0, 2\pi]$ as follows:

(a) Define the domain vector x consisting of sufficiently many intermediate point x_is along the x-axis and the corresponding vector y consisting of the function values at x_is and plot the vector y over the vector x. You may use the following statements.

```
»x=[0:0.01:2*pi]; y=tan(x);
»subplot(221), plot(x,y); xlim([0 2*pi])
```

Which one is the most similar to what you have got, among the graphs shown in Figure P1.5? Is it far from your expectation?

(b) Expecting to get the better graph, we scale it up along the y-axis by using the following command.

```
»axis([0 2*pi -10 10])
```

Which one is the most similar to what you have got, among the graphs shown in Figure P1.5? Is it closer to your expectation than what you got in (a)?

(c) Most probably, you must be nervous about the straight-lines at the singular points $x = \pi/2$ and $3\pi/2$. The severer you get disturbed by the lines that must not be there, the better you are at the numerical stuffs. As an alternative to avoid such a singular happening, you can try dividing the interval into three sections excluding the two singular points as follows:

```
x1=[0:0.01:pi/2-0.01]; x2=[pi/2+0.01:0.01:3*pi/2-0.01];
x3=[3*pi/2+0.01:0.01:2*pi];
y1=tan(x1); y2=tan(x2); y3=tan(x3);
subplot(222), plot(x1,y1,x2,y2,x3,y3), axis([0 2*pi -10 10])
```

(d) How about trying the easy plotting command 'ezplot()' or 'fplot()'? Do they work properly?

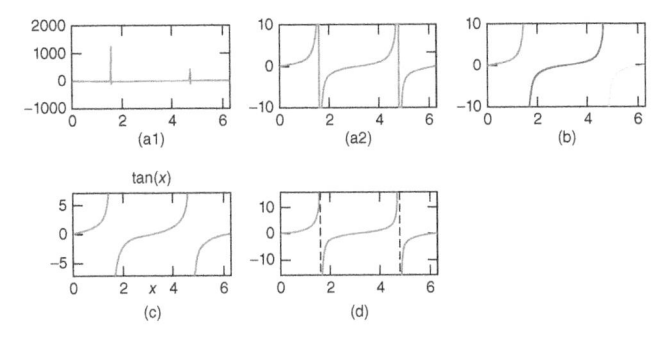

Figure P1.5 Plotting the graph of $f(x) = \tan x$.

```
»ezplot('tan(x)',[0 2*pi])
»fplot(@(x)tan(x),[0 2*pi])  % plot an anonymous function
```

1.6 Plotting the Graph of a Sinc Function
The sinc function is defined as

$$f(x) = \frac{\sin x}{x} \qquad (P1.6.1)$$

whose value at $x = 0$ is

$$f(0) = \lim_{x \to 0} \frac{\sin x}{x} = \left.\frac{(\sin x)'}{x'}\right|_{x=0} = \left.\frac{\cos x}{1}\right|_{x=0} = 1 \qquad (P1.6.2)$$

We are going to plot the graph of this function over $[-4\pi, 4\pi]$.

(a) Casually, you may try as follows:

```
»x=[-100:100]*pi/25; y=sin(x)./x;
»plot(x,y), axis([-15 15 -0.4 1.2])
```

Even if no warning message shows up, is there anything odd about the graph?

(b) How about trying with a different domain vector?

```
»x=[-4*pi:0.1:+4*pi]; y=sin(x)./x;
»plot(x,y), axis([-15 15 -0.4 1.2])
```

Surprisingly, MATLAB gives us the function values without any complaint and presents a nice graph of the sinc function. What is the difference between (a) and (b)?

(cf) Actually, we would have no problem if we used the MATLAB built-in function 'sinc()'.

1.7 Term-Wise (Element-by-Element) Operation in MATLAB User Functions

(a) Let the function $f_1(x)$ be defined without one or both of the dot (.) operators in Section 1.1.6. Could we still get the output vector consisting of the function values for the several values in the input vector? You can run the following statements and see the results.

```
»f1=@(x)1./(1+8*x^2);   f1([0 1])
»f1=@(x)1/(1+8*x.^2);   f1([0 1])
```

(b) Let the function $f_1(x)$ be defined with both of the dot (.) operators as in Section 1.1.6. What would we get by running the following statements?

```
»f1=@(x)1./(1+8*x.^2);  f1([0 1]')
```

1.8 In-line Function and M-file Function with the Integral Function 'quad()'
As will be seen in Section 5.8, one of the MATLAB built-in functions for computing numerical integrals is 'quad()', the usual usage of which is

$$\text{quad}(f,a,b,\text{tol},\text{trace},p1,p2,..) \text{ for } \int_a^b f(x,p1,p2,\ldots)dx \quad \text{(P1.8.1)}$$

where

f: the name of the integrand function (M-file name should be categorized by ' ')

a,b: the lower/upper bound of the integration interval

tol: the error tolerance (10^{-6} by default []), trace: 1(on)/0(off) (0 by default [])

p1,p2, ..: additional parameters to be passed directly to function f

Let us use this 'quad()' function with an in-line function and an M-file function to obtain

$$\int_{m-10}^{m+10} (x-x_0)f(x)dx \quad \text{(P1.8.2a)}$$

and

$$\int_{m-10}^{m+10} (x-x_0)^2 f(x)dx \quad \text{(P1.8.2b)}$$

where

$$x_0 = 1, f(x) = \frac{1}{\sqrt{2\pi}\sigma} e^{-(x-m)^2/2\sigma^2} \text{ with } m=1, \sigma=2 \quad \text{(P1.8.3)}$$

Complete and run the following script "nm01p08.m" to compute the two integrals (P1.8.2a and P1.8.2b).

```
%nm01p08.m
m=1; sigma=2; x0=1;
Gpdf='exp(-(x-?).^2/2/?????^2)/sqrt(2*pi)/sigma';
xGpdf=inline(['(x-x0).*' Gpdf],'x','m','?????','x0');
x2Gpdf=inline(['(x-x0).^2.*' Gpdf],'x','?','sigma','x0');
int_xGpdf=quad(xGpdf,m-10,m+10,[],0,?,sigma,x0)
int_x2Gpdf=quad(x2Gpdf,m-10,m+10,[],0,m,?????,x0)
```

1.9 μ-Law Function Defined in an M-File

The so-called μ-law function and μ^{-1}-law function used for uniform quantization is defined as

$$y = g_\mu(x) = |y|_{\max} \frac{\ln(1+\mu|x|/|x|_{\max})}{\ln(1+\mu)} \text{sign}(x) \quad \text{(P1.9.1a)}$$

$$x = g_\mu^{-1}(y) = |x|_{\max} \frac{(1+\mu)^{|y|/|y|_{\max}} - 1}{\mu} \text{sign}(y) \quad \text{(P1.9.1b)}$$

58 MATLAB USAGE AND COMPUTATIONAL ERRORS

Below are the incomplete μ-law function 'mulaw()', μ^{-1}-law function 'mulaw_inv()', and a script "nm01p09.m" that are all supposed to be saved as M-files. Complete them and run the script to do the following jobs with $\mu = 10$, 50, and 255:

- Find the values y of the μ-law function for x=[-1:0.005:1] and plot the graph of y vs. x.
- Find the values x0 of the μ^{-1}-law function for y.
- Compute the discrepancy between x and x0.

```
function [y,xmax]=mulaw(x,mu,ymax)
if nargin<3, ymax=1; end
xmax=max(abs(x));
y=ymax*log(1+mu*???(x/xmax))./log(1+mu).*????(x); % Eq.(P1.9a)
```

```
function x=mulaw_inv(y,mu,xmax)
% Inverse of mu-law
if nargin<3, xmax=1; ymax=1;
 else ymax=max(abs(y));
end
if nargin<2, mu=255; end
x=xmax.*(((1+??).^(abs(?)/y???)-1)/??).*sign(y); % Eq.(P1.9b)
```

```
%nm01p09.m: to plot the mulaw curve
x=[-1:.005:1];   mu=[10 50 255];
for i=1:3
   [y,xmax]=mulaw(x,mu(i),1);
   x0=mulaw_inv(y,mu(i),xmax);
   discrepancy=norm(x-x0)
   plot(x,y,'b-', x,x0,'r-'), hold on
end
```

1.10 Analog-digital converter (ADC)

Below are two ADC routines 'adc1(a,b,c)' and 'adc2(a,b,c)', which assign the corresponding digital value c(i) to each one of the analog data belonging to the quantization interval [b(i),b(i+1)]. Let the boundary vector and the centroid (level) vector be, respectively,

    b=[-3 -2 -1 0 1 2 3];   c=[-2.5 -1.5 -0.5 0.5 1.5 2.5];

Note that the digital value corresponding to an analog data smaller than b(1) (the smallest boundary) should be c(1) (the smallest centroid) and that corresponding to an analog data larger than b(N) (the largest boundary) should be c(N) (the largest centroid) where the boundary and centroid vectors are assumed to be arranged in ascending order and N is the number of the centroids (quantization levels) or quantization intervals.

(a) Explain the jobs of lines 8, 9, and 10 of the following ADC function 'adc1()'.
(b) Explain the jobs of lines 3, 5, and 7 of the following ADC function 'adc2()'.

PROBLEMS 59

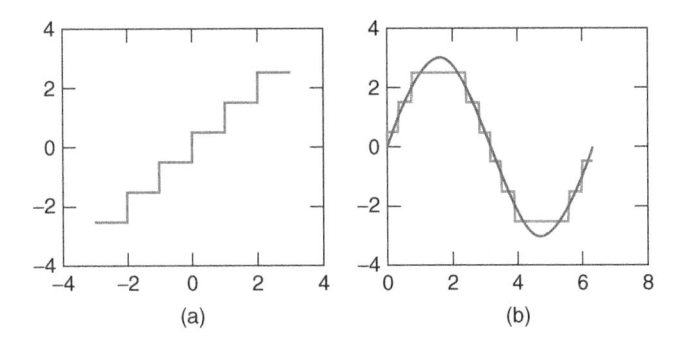

Figure P1.10 The characteristic of an analog-to-digital converter (ADC).

(c) Run the script "nm01p10.m" (see below box) to get two graphs as shown in Figure P1.10. Explain about the graphs.

```
function d=adc1(a,b,c)
%Analog-to-Digital Converter
%Input    a=analog signal, b(1:N+1)=boundary vector
%         c(1:N)=centroid vector
%Output:  d=digital samples
N=length(c);
for n=1:length(a)
    I=find(a(n)<b(2:N));
    if ~isempty(I), d(n)=c(I(1));
    else            d(n)=c(N);
    end
end
```

```
function d=adc2(a,b,c)
N=length(c);
d(find(a<b(2)))=c(1);
for i=2:N-1
    index=find(b(i)<=a&a<b(i+1));   d(index)=c(i);
end
d(find(b(N)<=a))=c(N);
```

```
%nm01p10.m
b=[-3 -2 -1 0 1 2 3]; % Boundary vector
c=[-2.5 -1.5 -0.5 0.5 1.5 2.5]; % Centroid vector
% Plot the input-output relationship
xa=[-300:300]/100; % Analog data in the range [-3~+3]
xd1=adc1(xa,b,c); xd2=adc2(xa,b,c);
subplot(221)
plot(xa,xd1, xa,xd2,'r:')
% Output of ADC to a sinusoidal input
t=[0:200]/100*pi; xa=3*sin(t);
xd1=adc1(xa,b,c); %xd2=adc2(xa,b,c);
subplot(222)
plot(t,xa, t,xd1,'r')
```

1.11 Decimal-to-Binary/Octal Conversions

(a) Referring to the decimal-to-binary conversion process illustrated in Figure 1.6, complete the following function 'dec2bin_my()' so that it can perform the conversion process. Then, using the function, convert two decimal numbers 3 and 14 to their corresponding binary numbers and compare the results with those obtained using the MATLAB built-in function 'dec2bin()'. Can you modify the function so that the binary numbers corresponding to each of x can be placed columnwise, that is, from top to bottom, rather than from left to right when a vector consisting of multiple decimal numbers is given as the first input argument x?

(b) Complete the following function 'bin2dec_my()' so that it can perform the binary-to-digital conversion process:

```
function y=dec2bin_my(x)
% converts given decimal numbers into binary numbers of N bits
y=[]; xmax=max(x); N=1;
while xmax>1
  xmax=floor(xmax/2); N=N+1;
end
for n=1:length(x)
  xn=x(n);
  for i=?:-1:1;
    xn_1= floor(xn/?);
    yn(i)= xn-?*xn_1; xn=xn_1;
  end
  y= [y yn];
end
```

```
function y=bin2dec_my(x)
[M,N]=size(x); % Number of binary numbers and Number of bits
for m=1:M
  y(m,:)=x(m,:)*?.^[?-1:-1:0]';
end
```

Then, using the function, convert two binary numbers [0 0 1 1] and [1 1 1 0] to their corresponding digital numbers and compare the results with those obtained using the MATLAB built-in function 'bin2dec()', by typing the following statements at MATLAB prompt:

```
bin2dec_my([0 0 1 1; 1 1 1 0])
bin2dec(['0011'; '1110'])
```

(c) Modify the decimal-to-binary converting function 'dec2bin_my()' into another function 'dec2oct_my()' so that it can perform the decimal-to-octa conversion process. Then, using the function, convert two decimal numbers 14 and 3 to their corresponding octal numbers and compare the results with those obtained using the MATLAB built-in function 'deci2oct()'.

1.12 Truth Table for a Logical (Boolean) Expression

Noting that the logical AND, OR, and NOT operators are &, |, and ~ in MATLAB, complete and run the following script "make_truth_table.m" to construct the truth table for the following logical (Boolean) expression:

$$(\overline{A}BC) + (B\overline{C}) \tag{P1.12.1}$$

that lists the (output) value of the expression for all possible values of the input variables.

```
%make_truth_table.m
% makes the truth table for a logical (Boolean) expression
N=3; % Number of Boolean variables
Inputs=dec2bin_my([0:2^N-1]); %All possible values of input variables
A=Inputs(:,1); B=Inputs(:,2); C=Inputs(:,3);
truth_table=[Inputs (?A&?&C)?(B?~C)]
```

1.13 Playing with Polynomials

(a) Polynomial evaluation: polyval()

Write a MATLAB statement to compute

$$p(x) = x^8 - 1 \text{ for } x = 1 \tag{P1.13.1}$$

(b) Polynomial addition/subtraction by using compatible vector addition/subtraction

Write a MATLAB statement to add the following two polynomial coefficient vectors:

$$p_1(x) = x^4 + 1, p_2(x) = x^3 - 2x^2 + 1 \tag{P1.13.2}$$

(c) Polynomial multiplication: conv()

Write a MATLAB statement to get the following product of polynomials:

$$p(x) = (x^4 + 1)(x^2 + 1)(x + 1)(x - 1) \tag{P1.13.3}$$

(d) Polynomial division – deconv()

Write a MATLAB statement to get the quotient and remainder of the following polynomial division:

$$p(x) = x^8/(x^2 - 1) \tag{P1.13.4}$$

(e) Routines for differentiation/integration of a polynomial

What you see in the below box is the function 'poly_der(p)', which gets a polynomial coefficient vector p (in the descending order) and outputs the coefficient vector pd of its derivative polynomial. Likewise, compose a function 'poly_int(p)', which outputs the coefficient vector of the integral polynomial for a given polynomial coefficient vector.

(cf) MATLAB has the built-in functions 'polyder()'/'polyint()' that can be used to find the derivative/integral of a polynomial.

```
function pd=poly_der(p)
%p: Vector of polynomial coefficients in descending order
N=length(p);
if N<=1, pd=0; % constant
   else
      for i=1:N-1, pd(i)=p(i)*(N-i); end
end
```

```
function pi=poly_int(p)
% p: Vector of polynomial coefficients in descending order
N=length(p); pi(N+1)=0;
for i=1:N, pi(i)=p(?)/(N-?+1); end
```

(f) Roots of a polynomial equation: roots()
Write a MATLAB statement to get the roots of the following polynomial equation:
$$p(x) = x^8 - 1 = 0 \quad (P1.13.5)$$

You can check if the result is right, by using the MATLAB command 'poly()', which generates a polynomial having a given set of roots.

(g) Partial fraction expansion (PFE) of a rational polynomial function — residue()/residuez()

(i) The MATLAB function '[r,p,k]=residue(B,A)' finds the PFE for a ratio of given polynomials $B(s)/A(s)$ as

$$\frac{B(s)}{A(s)} = \frac{b_1 s^{M-1} + b_2 s^{M-2} + \cdots + b_M}{a_1 s^{N-1} + a_2 s^{N-2} + \cdots + a_N} = k(s) + \sum_i \frac{r(i)}{s - p(i)} \quad (P1.13.6a)$$

which is good for taking the inverse Laplace transform. Use the function to find the PFE for

$$X(s) = \frac{4s + 2}{s^3 + 6s^2 + 11s + 6} = \frac{}{s+} + \frac{}{s+} + \frac{}{s+} \quad (P1.13.7a)$$

(ii) The MATLAB function '[r,p,k]=residuez(B,A)' finds the PFE for a ratio of given polynomials $B(z)/A(z)$ as

$$\frac{B(z)}{A(z)} = \frac{b_1 + b_2 z^{-1} + \cdots + b_M z^{-(M-1)}}{a_1 + a_2 z^{-1} + \cdots + a_N z^{-(N-1)}} = k(z^{-1}) + \sum_i \frac{r(i) z}{z - p(i)} \quad (P1.13.6b)$$

which is good for taking the inverse z-transform. Use the function to find the PFE for

$$X(z) = \frac{4 + 2z^{-1}}{1 + 6z^{-1} + 11z^{-2} + 6z^{-3}} = \frac{1}{1 + z^{-1}} + \frac{-12}{1 + 2z^{-1}} + \frac{15}{1 + 3z^{-1}}$$

$$= \frac{z}{z+1} + \frac{-12z}{z+2} + \frac{15z}{z+3} \qquad \text{(P1.13.7b)}$$

(h) Piecewise polynomial − mkpp()/ppval()
Suppose we have an $M \times N$ matrix P, the rows of which denote M (piecewise) polynomials of degree $(N-1)$ for different (nonoverlapping) intervals with $(M+1)$ boundary points bb=[b(1) .. b(M+1)], where the polynomial coefficients in each row are supposed to be generated with the interval starting from $x = 0$. Then we can use the MATLAB command 'pp=mkpp(bb,P)' to construct a structure of piecewise polynomials, which can be evaluated by using 'ppval(pp)'.

Figure P1.13 shows a set of piecewise polynomials $\{p_1(x+3), p_2(x+1), p_3(x-2)\}$ for the intervals [-3, -1], [-1, 2], and [2, 4], respectively, where

$$p_1(x) = x^2,\, p_2(x) = -(x-1)^2 \text{ and } p_3(x) = x^2 - 2 \qquad \text{(P1.13.8)}$$

(i) Complete and run the upper part of the above MATLAB script "nm01p13h.m" to use 'mkpp()'/'ppval()' for making this graph.
(ii) Complete and run the lower part of the above MATLAB script "nm01p13h.m" to use 'polyval()' for making this graph.
(cf) You can type 'help mkpp' to see a couple of examples showing the usage of 'mkpp()'.

```
%nm01p13h.m
% to plot the graph of a piecewise polynomial
bb=[-3 -? 2 4]; % Boundary point vector
% Matrix having three polynomial coefficient vectors
%   in its rows
P=[1 0 0;-1 2 -?;1 0 -2];
pp=mkpp(??,P)
xx=bb(1)+[0:1000]/1000*(bb(end)-bb(1)); % Whole range
plot(xx,ppval(??,xx)), hold on

% Alternative without using ppval()
%   to plot the polynomial curves one by one
for i=1:size(P,1)
    xx=[0:100]/100*(bb(i+1)-bb(i));
    plot(xx+??(i),polyval(?(i,:),xx),'r:')
end
```

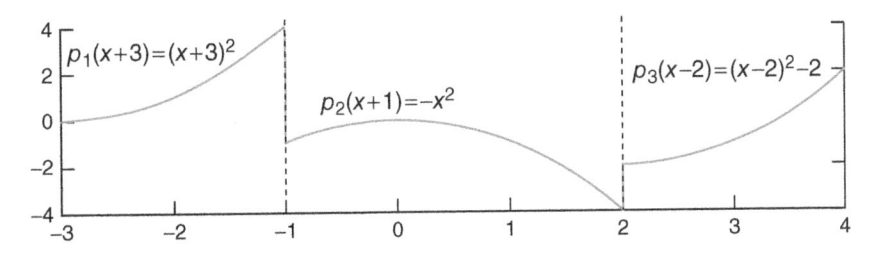

Figure P1.13 The graph of piece-wise polynomial functions.

64 MATLAB USAGE AND COMPUTATIONAL ERRORS

1.14 Routine for Matrix Multiplication

Assuming that MATLAB cannot perform direct multiplication on vectors/matrices, supplement the following incomplete function 1.1 '`multiply_matrix(A,B)`' so that it can multiply two matrices given as its input arguments only if their dimensions are compatible, but display an error message if their dimensions are not compatible. Try it to get the product of two arbitrary 3×3 matrices generated by the command '`rand(3)`' and compare the result with that obtained by using the direct multiplicative operator *. Note that the matrix multiplication can be described as

$$C(m,n) = \sum_{k=1}^{K} A(m.k)B(k,n) \qquad \text{(P1.14.1)}$$

```
function C=multiply_matrix(A,B)
[M,K]=size(A);   [K1,N]=size(B);
if K1~=K
  error('The # of columns of A is not equal to the # of rows of B')
else
  for m=1:?
     for n=1:?
        C(m,n)=A(m,1)*B(1,n);
        for k=2:?
           C(m,n)=C(m,n)+A(m,k)*B(k,n);
        end
     end
  end
end
```

1.15 Function for Finding Vector Norm

Assuming that MATLAB does not have the '`norm()`' command finding us the norm of a given vector/matrix, make a routine '`norm_vector(v,p)`' which computes the norm of a given vector as

$$\|\mathbf{v}\|_p = \sqrt[p]{\sum_{n=1}^{N} |v_n|^p} \qquad \text{(P1.15.1)}$$

for any positive integer `p`, finds the maximum absolute value of the elements for `p=inf`, and computes the norm as if `p=2`, even if the second input argument `p` is not given. If you have no idea, permutate the statements in the below box and save it in a file named "norm_vector.m". Additionally, try it to get the norm with p=1,2,∞ (`inf`) for an arbitray vector generated by the command '`rand(2,1)`'. Compare the result with that obtained by using the '`norm()`' command.

```
function nv=norm_vector(v,p)
if nargin<2, p=2; end
nv= sum(abs(v).^p)^(1/p);
nv= max(abs(v));
if p>0&p~=inf
  elseif p==inf
end
```

1.16 Backslash (\) Operator

Let us play with the backslash (\) operator.

(a) Use the backslash (\) command, the minimum-norm solution (2.1.7), and the 'pinv()' command to solve the following equations, find the residual error $\|A_i \mathbf{x} - \mathbf{b}_i\|$'s and the rank of the coefficient matrix A_i, and fill in Table 1.6 with the results.

$$\text{(i)} \ A_1 \mathbf{x} = \begin{bmatrix} 1 & 2 & 3 \\ 4 & 5 & 6 \end{bmatrix} \begin{bmatrix} x_1 \\ x_2 \\ x_3 \end{bmatrix} = \begin{bmatrix} 6 \\ 15 \end{bmatrix} = \mathbf{b}_1 \quad \text{(P1.16.1)}$$

$$\text{(ii)} \ A_2 \mathbf{x} = \begin{bmatrix} 1 & 2 & 3 \\ 2 & 4 & 6 \end{bmatrix} \begin{bmatrix} x_1 \\ x_2 \\ x_3 \end{bmatrix} = \begin{bmatrix} 6 \\ 8 \end{bmatrix} = \mathbf{b}_2 \quad \text{(P1.16.2)}$$

$$\text{(iii)} \ A_3 \mathbf{x} = \begin{bmatrix} 1 & 2 & 3 \\ 2 & 4 & 6 \end{bmatrix} \begin{bmatrix} x_1 \\ x_2 \\ x_3 \end{bmatrix} = \begin{bmatrix} 6 \\ 12 \end{bmatrix} = \mathbf{b}_3 \quad \text{(P1.16.3)}$$

(b) Use the backslash(\) command, the least-squares (LSs) solution (2.1.10), and the 'pinv()' command to solve the following equations and find the residual error $\|A_i \mathbf{x} - \mathbf{b}_i\|$'s and the rank of the coefficient matrix A_i, and fill in Table P1.16 with the results.

$$\text{(i)} \ A_4 \mathbf{x} = \begin{bmatrix} 1 & 2 \\ 2 & 3 \\ 3 & 4 \end{bmatrix} \begin{bmatrix} x_1 \\ x_2 \end{bmatrix} = \begin{bmatrix} 2 \\ 6 \\ 7 \end{bmatrix} = \mathbf{b}_4 \quad \text{(P1.16.4)}$$

$$\text{(ii)} \ A_5 \mathbf{x} = \begin{bmatrix} 1 & 2 \\ 2 & 4 \\ 3 & 6 \end{bmatrix} \begin{bmatrix} x_1 \\ x_2 \end{bmatrix} = \begin{bmatrix} 1 \\ 5 \\ 8 \end{bmatrix} = \mathbf{b}_5 \quad \text{(P1.16.5)}$$

$$\text{(iii)} \ A_6 \mathbf{x} = \begin{bmatrix} 1 & 2 \\ 2 & 4 \\ 3 & 6 \end{bmatrix} \begin{bmatrix} x_1 \\ x_2 \end{bmatrix} = \begin{bmatrix} 3 \\ 6 \\ 9 \end{bmatrix} = \mathbf{b}_6 \quad \text{(P1.16.6)}$$

(cf) If some or all of the rows of the coefficient matrix A in a system of linear equations can be expressed as a linear combination of other row(s), the corresponding equations are dependent, which can be revealed by the rank deficiency, i.e. rank(A) < min(M,N) where M and N are the row and column dimensions, respectively. If some equations are dependent, they may have either inconsistency (no exact solution) or redundancy (infinitely many solutions), which can be distinguished by checking if augmenting the RHS vector $\mathbf{b}$ to the coefficient matrix A increases the rank or not, that is, rank([A $\mathbf{b}$]) > rank(A) or not [M-2].

(c) Based on the results obtained in (a) and (b) and listed in Table P1.16, answer the following questions:

  (i) Based on the results obtained in (a-(i)), which one yielded the nonminimum-norm solution among the three methods, i.e. the backslash (\) operator, the minimum-norm solution (2.1.7), and the 'pinv()' command? Note that the minimum-norm solution means the solution whose norm (||**x**||) is the minimum over the many solutions.

  (ii) Based on the results obtained in (a), which one is most reliable as a means of finding the minimum-norm solution among the three methods?

  (iii) Based on the results obtained in (b), choose two reliable methods as a means of finding the LS solution among the three methods, i.e. the backslash (\) operator, the LS solution (2.1.10), and the 'pinv()' command. Note that the LS solution means the solution for which the residual error (||A**x**−**b**||) is the minimum over the many solutions.

Table P1.16 Results of operations with backslash(\) operator and 'pinv()' command.

	backslash (\)		Minimum-norm or least-squares (LS)		pinv()		Remark
	x	$\|A_i\mathbf{x} - \mathbf{b}_i\|$	x	$\|A_i\mathbf{x} - \mathbf{b}_i\|$	x	$\|A_i\mathbf{x} - \mathbf{b}_i\|$	rank(A_i) redundant/ inconsistent
$A_1\mathbf{x} - \mathbf{b}_1$	1.5000 0 1.5000	6.4047 × 10^{-15}					
$A_2\mathbf{x} - \mathbf{b}_2$					0.3143 0.6286 0.9429	1.7889	
$A_3\mathbf{x} - \mathbf{b}_3$			∞ ∞ ∞	∞			
$A_4\mathbf{x} - \mathbf{b}_4$			2.5000 0.0000	1.2247			
$A_5\mathbf{x} - \mathbf{b}_5$							
$A_6\mathbf{x} - \mathbf{b}_6$							

1.17 Operations on Vectors

(a) Find the mathematical expression for the computation to be done by the following MATLAB statements:

```
»n=0:100; S=sum(2.^-n)
```

(b) Write a MATLAB statement which performs the following computation:

$$\left(\sum_{n=0}^{10000} \frac{1}{(2n+1)^2}\right) - \frac{\pi^2}{8}$$

(c) Write a MATLAB statement that uses the commands 'prod()' and 'sum()' to compute the product of the sums of each row of a 3×3 random matrix.

(d) How does the following function 'repetition(x,M,N)' convert a scalar or a matrix given as the first input argument x to make a new sequence y? How does it compare with the MATLAB built-in function 'repmat()'?

```
function y=repetition(x,M,N)
y1=x; for n=2:N, y1 = [y1 x]; end
y=y1; for m=2:M, y = [y; y1]; end
```

(e) Complete the following function 'zero_insertion(x,M,m)' so that it can insert m zeros just after every Mth element of a given row vector sequence x to make a new sequence. Write a MATLAB statement to use the function for inserting two zeros just after every third element of $x = [1\ 3\ 7\ 2\ 4\ 9]$ to get $y = [1\ 3\ 7\ 0\ 0\ 2\ 4\ 9\ 0\ 0]$.

```
function y=zero_insertion(x,M,m)
Nx=length(x); N=floor(Nx/M);
y=[reshape(x(1:M*N),?,N); ????s(m,N)];
y=[y(:)' x(M*N+1:Nx)];
```

(f) How does the following function 'zeroing(x,M,m)' convert a given row vector sequence x to make a new sequence y?

```
function y=zeroing(x,M,m)
m=mod(m,M); Nx=length(x); N=floor(Nx/M);
y=x; y(M*[1:N]-m)=0;
```

(g) Complete the following function 'sampling(x,M,m)' so that it can sample every $(kM - m)$th element of a given row vector sequence x to make a new sequence y. Write a MATLAB statement to use the function for sampling every $(3k - 2)$th element of $x = [1\ 3\ 7\ 2\ 4\ 9]$ to get

$$y = [1\quad 2]$$

```
function y=sampling(x,M,m)
m=mod(m,M); Nx=numel(x); N=floor(Nx/M);
y=x([1:N]*?-?);
if Nx-N*M>=M-m, y=[y x(N*M+M-m)]; end
```

(h) Complete the following function 'rotate_r(x,M)' so that it can rotate a given row vector sequence x right by M samples to make a new sequence y. Write a MATLAB statement to use the function for rotating $x = [1\ 2\ 3\ 4\ 5]$ to get
$$y = [3\ 4\ 5\ 1\ 2]$$

```
function x1=rotate_r(x,M)
N=size(x,2); M=mod(M,N);
x1=[x(:,end-?+1:end) x(:,1:end-?)];
```

1.18 Distribution of a Random Variable – Histogram
Complete the following function 'randu(N,a,b)' that uses the MATLAB function 'rand()' to generate an N-dimensional random vector having the uniform distribution over [a,b] and draws the histogram (with 20 bins) for the distribution of the elements of the generated vector as Figure 1.5. Then, find the average height of the histogram that you can get by running the following statement:

```
»randu(1000,-2,2)
```

```
function x=randu(N,a,b)
% generates an N-dimensional random vector with U(a,b)
x=a+(?-?)*rand(1,N);
if nargout==0, hist(x(:),20); end
```

1.19 Number Representation
In Section 1.2.1, we looked over how a number is represented in 64 bits. For example, the IEEE 64-bit floating-point number system represents the number $3(2^1 \leq 3 < 2^2)$ belonging to the range $R_1 = [2^1, 2^2)$ with $E = 1$ as

0	100 0000 0000	1000 0000 0000	....	0000 0000 0000 0000 0000
4	0 0 8	0 0	....	0 0 0 0 0

where the exponent and the mantissa are

$$\text{Exp} = E + 1023 = 1 + 1023 = 1024 = 2^{10} = 100\ 0000\ 0000$$
$$M = (3 \times 2^{-E} - 1) \times 2^{52} = 2^{51}$$
$$= 1000\ 0000\ 0000\ \ldots\ .\ 0000\ 0000\ 0000\ 0000\ 0000$$

This can be confirmed by typing the following statement into MATLAB command window:

```
»fprintf('3=%bx\n',3)    or    »format hex, 3, format short
```

which will print out onto the screen

```
4008000000000000
```

This is exactly the hexadecimal representation of the number 3 as we expected. Find the IEEE 64-bit floating-point number representation of the number 14 and use the command 'fprintf()' to check if the result is right.

(cf) Since the INTEL system stores numbers in little-endian (byte order) format with more significant bytes in the memory of higher address number, you might see

```
0000000000000840
```

in the old versions of MATLAB, which is reversely ordered in the unit of byte (8 bits = 2 hexadecimal digits) so that the number is represented with the most/least significant byte on the right/left side.

1.20 Resolution of Number Representation and Quantization Error

In Section 1.2.1, we have seen that adding 2^{-22} to 2^{30} makes some difference, while adding 2^{-23} to 2^{30} makes no difference due to the bit shift by over 52 bits for alignment before addition. How about subtracting 2^{-23} from 2^{30}? In contrast with the addition of 2^{-23} to 2^{30}, it makes difference as you can see by running the following MATLAB statement:

```
»x=2^30; x+2^-23==x, x-2^-23==x
```

which will give you the logical answer 1 (true) and 0 (false). Justify this result based on the difference of resolution of two ranges $[2^{30}, 2^{31})$ and $[2^{29}, 2^{30})$ to which the true values of computational results $(2^{30} + 2^{-23})$ and $(2^{30} - 2^{-23})$ belong, respectively. Note from Eq. (1.2.5) that the resolutions, i.e. the maximum quantization errors are $\Delta_E = 2^{E-52} = 2^{-52+30} = 2^{-22}$ and $2^{-52+29} = 2^{-23}$, respectively. For details, refer to Figure P1.20, which illustrates the process of addition/subtraction with 4 mantissa bits, 1 hidden bit, and 1 guard bit.

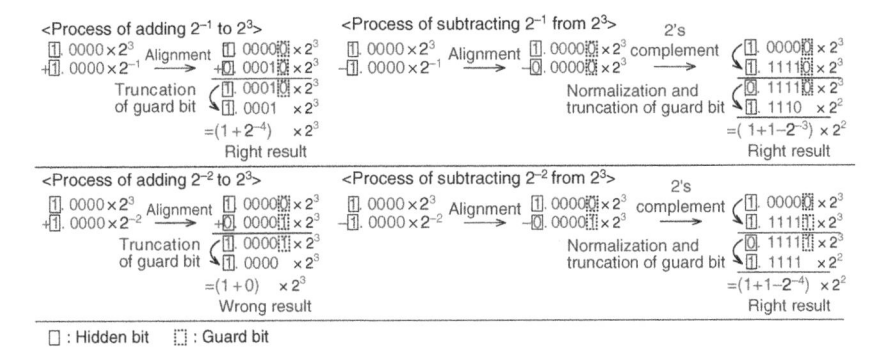

Figure P1.20 Process of addition/subtraction with four mantissa bits.

1.21 Resolution of Number Representation and Quantization Error

(a) What is the result of running the following statement?

```
»7/100*100-7
```

How do you compare the absolute value of this answer with the resolution Δ of the range to which 7 belongs?

(b) Find how many numbers are susceptible to this kind of quantization error caused by division/ multiplication by 100, among the numbers from 1 to 31.

(c) What will be the result of running the following script? Why?

```
%nm01p21.m: Quantization Error
x=2-2^-50;
for n=1:2^3
    x=x+2^-52; fprintf('%20.18E\n',x)
end
```

1.22 Avoiding Large Errors/Overflow/Underflow

(a) For $x = 9.8^{201}$ and $y = 10.2^{199}$, evaluate the following two expressions that are mathematically equivalent and tell which is better in terms of the power of resisting the overflow.

$$z = \sqrt{x^2 + y^2} \qquad \text{(P1.22.1a)}$$

$$z = y\sqrt{(x/y)^2 + 1} \qquad \text{(P1.22.1b)}$$

Also for $x = 9.8^{-201}$ and $y = 10.2^{-199}$, evaluate the above two expressions and tell which is better in terms of the power of resisting the underflow.

(b) With $a = c = 1$ and for 100 values of b over the interval $[10^{7.4}, 10^{8.5}]$ generated by the MATLAB command 'logspace(7.4,8.5,100)', evaluate the following two formulas (for the roots of a quadratic equation) that are mathematically equivalent and plot the values of the second root of each pair. Noting that the true values are not available and so the shape of solution graph is only one practical basis on which we can assess the quality of numerical solutions, tell which is better in resisting the loss of significance.

(i) $\left[x_1 = \dfrac{1}{2a}(-b - \text{sign}(b)\sqrt{b^2 - 4ac}), \right.$

$\left. x_2 = \dfrac{1}{2a}(-b + \text{sign}(b)\sqrt{b^2 - 4ac}) \right]$ (P1.22.2a)

(ii) $\left[x_1 = \dfrac{1}{2a}(-b - \text{sign}(b)\sqrt{b^2 - 4ac}), \; x_2 = \dfrac{c/a}{x_1} \right]$ (P1.22.2b)

(c) For 100 values of x over the interval $[10^{-9}, 10^{-7.4}]$, evaluate the following two expressions that are mathematically equivalent, plot them, and based on the graphs, tell which is better in terms of resisting the loss of significance.

$$(i)\ y = \sqrt{2x^2 + 1} - 1 \qquad \text{(P1.22.3a)}$$

$$(ii)\ y = \frac{2x^2}{\sqrt{2x^2 + 1} + 1} \qquad \text{(P1.22.3b)}$$

(d) For 100 values of x over the interval $[10^{14}, 10^{16}]$, evaluate the following two expressions that are mathematically equivalent, plot them, and based on the graphs, tell which is better in terms of resisting the loss of significance.

$$(i)\ y = \sqrt{x+4} - \sqrt{x+3} \qquad \text{(P1.22.4a)}$$

$$(ii)\ y = \frac{1}{\sqrt{x+4} + \sqrt{x+3}} \qquad \text{(P1.22.4b)}$$

(e) On purpose to find the value of $(300^{125}/125!)e^{-300}$, run the following statement:

```
»300^125/prod([1:125])*exp(-300)
```

What is the result? Is it of any help to change the order of multiplication/division? As an alternative, referring to the script "nm131_2_1.m" (Section 1.3.1), complete the following function `Poisson_pdf()` so that it can compute

$$p(k) = \frac{\lambda^k}{k!} e^{-\lambda} \text{ for } \lambda = 300 \text{ and an integer } k \qquad \text{(P1.22.5)}$$

with nested computing (in a recursive way), say, like $p(k+1) = p(k) * \lambda/k$ and then, use the function to find the value of $(300^{125}/125!)e^{-300}$.

```
function p=Poisson_pdf(K,lambda)
p=exp(-??????);
for k=1:?, p= ?*lambda/?; end
```

(f) Make a function that computes the sum

$$S(K) = \sum_{k=0}^{K} \frac{\lambda^k}{k!} e^{-\lambda} \text{ for } \lambda = 100 \text{ and an integer } K \qquad \text{(P1.22.6)}$$

and then uses the function to find the value of $S(155)$.

```
function S=Poisson_pdf_sum(K,lambda)
p=exp(-lambda); S=0;
for k=1:K,  p= ?*lambda/k; S=?+p;   end
```

1.23 Nested Computing for Computational Efficiency

(a) The Hermite polynomial [K-2, W-3]

Consider the Hermite polynomial defined as

$$H_0(x) = 1 \; ; \quad H_N(x) = (-1)^N e^{x^2} \frac{d^N}{dx^N} e^{-x^2} \qquad (P1.23.1)$$

(i) Show that the derivative of this polynomial function can be written as

$$H'_N(x) = (-1)^N 2x \, e^{x^2} \frac{d^N}{dx^N} e^{-x^2} + (-1)^N e^{x^2} \frac{d^{N+1}}{dx^{N+1}} e^{-x^2}$$

$$= 2x \, H_N(x) - H_{N+1}(x) \qquad (P1.23.2)$$

whence the $(N+1)$th-degree Hermite polynomial can be obtained recursively from the Nth-degree Hermite polynomial as

$$H_{N+1}(x) = 2x \, H_N(x) - H'_N(x) \qquad (P1.23.3)$$

(ii) Complete the following MATLAB function 'Hermitp(N)' so that it can use Eq. (P1.23.3) to generate the Nth-degree Hermite polynomial $H_N(x)$.

```
function p=Hermitp(N)
% Hn+1(x)=2xHn(x)-Hn'(x)
if N<=0,   p=1;
  else     p=[2 ?];
           for n=2:N, p= 2*[p ?]-[0 ? polyder(p)]; end
end
```

(b) The Bessel function of the first kind [K-2, W-1]

Consider the Bessel function of the first kind of order k defined as

$$J_k(\beta) = \frac{1}{\pi} \int_0^\pi \cos(k\delta - \beta \sin \delta) \, d\delta \qquad (P1.23.4a)$$

$$= \left(\frac{\beta}{2}\right)^k \sum_{m=0}^\infty \frac{(-1)^m \beta^{2m}}{4^m m!(m+k)!} \equiv (-1)^k J_{-k}(\beta) \qquad (P1.23.4b)$$

```
function [J,JJ]=Jkb(K,beta) %first kind of kth-order Bessel ftn
tmpk= ones(size(beta));
for k=0:K
   tmp= tmpk; JJ(k+1,:)=tmp;
   for m=1:100
      tmp= -tmp.*????.^2/4/m/(m+?);
      JJ(k+1,:)= JJ(k+1,:)?tmp;
      if norm(tmp)<.001, break; end
   end
   tmpk=tmpk.*beta/2/(k+1);
end
J=JJ(K+1,:);
```

```
function y=Bessel_integrand(x,beta,k)
y=cos(k*x-beta*sin(x));
```

```
%nm01p23b.m: Bessel_ftn
beta=0:.05:15; K=15;
tic
for i=1:length(beta) % Integration
   J15_qd(i)=quad('Bessel_integrand',0,pi,[],0,beta(i),K)/pi;
end
time_qd=toc
tic, J15_Jkb=Jkb(K,beta); time_Jkb=toc/K   % Nested Computing
tic, J15_bj=besselj(K,beta); time_bj=toc
discrepancy1 = norm(J15_qd-J15_bj)
discrepancy2 = norm(J15_Jkb-J15_bj)
```

(i) Define the integrand of (P1.23.4a) in the name of 'Bessel_integrand (x,beta,k)' and save it in an M-file named "Bessel_integrand.m".

(ii) Complete the above function 'Jkb(K,beta)' so that it can use (P1.23.4b) in a recursive way to compute $J_k(\beta)$ of order $k = 1:K$ for given K and β (beta).

(iii) Run the above script "nm01p21b3.m", which computes $J_{15}(\beta)$ for $\beta = 0 : 0.05 : 15$ in three ways, that is, using Eq. (P1.23.4a), Eq. (P1.23.4b) (cast into 'Jkb()'), and the MATLAB built-in function 'besselj()'. Do they conform with each other?

(cf) Note that 'Jkb(K,beta)' computes $J_k(\beta)$ of order $k = 1:K$, while the integration does for only $k = K$.

(cf) Note also that the MATLAB built-in function 'besselj(k,beta)' computes the value of the Bessel function of the first kind, $J_k(\beta)$, of order k even when k is not an integer and β is a complex number.

1.24 Find the four functions in Chapters 5 and 7 that are fabricated in a recursive (self-calling) structure.

(cf) Does not those algorithms, which are the souls of the functions, seem to have been born to be in a nested structure?

1.25 Avoiding Runtime Error in Case of Deficient/Non-admissible Input Arguments

(a) Consider the MATLAB function 'rotate_r(x,M)', that you made in Problem 1.17(h). Does it work somehow when the user gives a negative integer as the second input argument M? If not, modify it so that it can perform the rotation left by –M samples for M<0, say, making

```
rotate_r([1 2 3 4 5],-2)=[3 4 5 1 2]
```

(b) Consider the function 'trpzds(f,a,b,N)' in Section 5.6, which computes the integral of function f over [a, b] by dividing the integration interval into N sections and applying the trapezoidal rule. If the user tries to use it without the fourth input argument N, will it work? If not, make it work with N = 1000 by default even without the fourth input argument N.

```
function INTf=trpzds(f,a,b,N)
%integral of f(x) over [a,b] by trapezoidal rule with N segments
if abs(b-a)<eps|N<=0, INTf=0; return; end
h=(b-a)/N; x=a+[0:N]*h;
fx=feval(f,x); values of f for all nodes
INTf= h*((fx(1)+fx(N+1))/2+sum(fx(2:N))); % Eq.(5.6.1)
```

1.26 Parameter Passing Through `varargin`

Consider the integration function 'trpzds(f,a,b,N)' in Section 5.6. Can you use it to compute the integral of a function with some parameter(s), like the 'Bessel_integrand(x,beta,k)' that you defined in Problem 1.23? If not, modify it so that it works for a function with some parameter(s) (see Section 1.3.6) and save it in the M-file named "trpzds_par.m". Then replace the 'quad()' statement in the script "nm01p23b.m" (presented in Problem 1.23) by an appropriate 'trpzds_par()' statement (with N=1000) and run the script. What is the discrepancy between the integration results obtained by this function and the nested computing based on Eq. (P1.23.4b)? Is it comparable with that obtained with 'quad()'? How do you compare the running time of this function with that of 'quad()'? Why do you think it takes so much time to execute the 'quad()' function?

1.27 Adaptive Input Argument to Avoid Runtime Error in Case of Different Input Arguments

Consider the integration function 'trpzds(f,a,b,N)' in Section 5.6. If some user tries to use this function with the following statement, will it work?

        trpzds(f,[a b],N)   or   trpzds(f,[a b])

```
function INTf=trpzds_bnd(f,a,b,N)
if numel(a)==2
  if nargin>2, N=b; else N=1000; end
  b=a(?); a=a(?);
 else
  if nargin<4, N=1000; end
end
% . . . . . . . . . . . . . . . . . . .
```

If not, modify it so that it works for such a usage (with a bound vector as the second input argument) as well as for the standard usage and save it in the M-file named "trpzds_bnd.m". Then try it to find the integral of e^{-t} for [0,100] by running the following statements. What did you get?

1.28 Continuous-time Fourier transform (CtFT) of a Signal

Consider the following definitions of CtFT and inverse continuous-time Fourier transform (ICtFT) [W-8].

$$X(\omega) = \mathcal{F}\{x(t)\} = \int_{-\infty}^{\infty} x(t)e^{-j\omega t}dt \; : \text{CtFT} \qquad (\text{P1.28.1a})$$

$$x(t) = \mathcal{F}^{-1}\{X(\omega)\} = \frac{1}{2\pi}\int_{-\infty}^{\infty} X(\omega)e^{j\omega t}d\omega \; : \text{ICtFT} \qquad (\text{P1.28.1b})$$

(a) Complete the following two MATLAB functions, 'CtFT1(x,Dt,w)' computing the CtFT (P1.28.1a) of x(t) over [-Dt,Dt] for w and 'ICtFT1(X,Bw,t)' computing the ICtFT (P1.28.1b) of X(w) over [-Bw, Bw] for t. You can choose whatever integral function including 'trpzds_par()' (Problem 1.25) and 'quad()', considering the running time.

(b) The following script "nm01p28.m" finds the CtFT of a rectangular pulse (with duration [-1,1]) defined by 'rDt()' for $\omega = [-6\pi,+6\pi]$ and the ICtFT of a sinc spectrum (with bandwidth 2π) defined by 'sincBw()' for $t = [-5,+5]$. After having saved the functions and script into M-files with the appropriate names, run the script to see the rectangular pulse, its CtFT spectrum, a sinc spectrum, and its ICtFT as shown in Figure P1.28. If it does not work, modify/supplement the functions so that you can rerun it to see the signals and their CtFT spectra.

```
function Xw=CtFT1(x,Dt,w)
% CtFT (Continuous-time Fourier Transform)
x_ejkwt=inline(['x '(t).*exp(-j*w*t)'],'t','w');
Xw=trpzds_par(x_ejkwt,-Dt,??,1000,?);
```

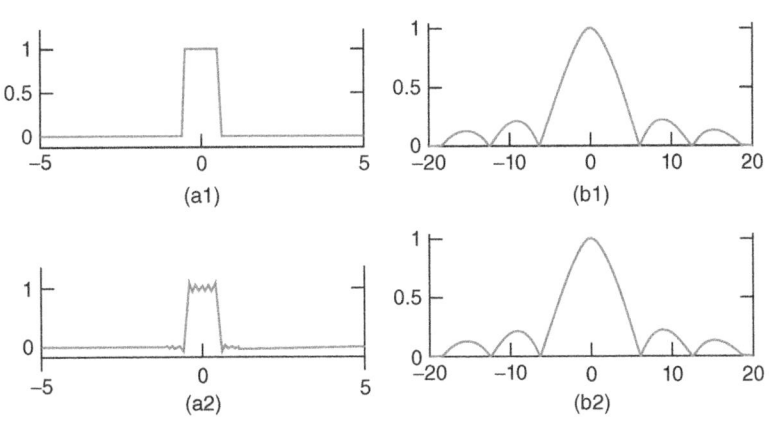

Figure P1.28 Graphs for Problem 1.28. (a1) A rectangular pulse function $r_D(t)$; (a2) the ICtFT of $X_B(\omega)$; (b1) the CtFT spectrum of $r_D(t)$; and (b2) a sinc spectrum $X_B(\omega)$.

76 MATLAB USAGE AND COMPUTATIONAL ERRORS

```
function xt=ICtFT1(X,Bw,t)
% ICtFT (Inverse Continuous-time Fourier Transform)
Xejkwt=inline([X '(w).*exp(j*w*t)'],'w','t');
xt=trpzds_par(Xejkwt,-??,Bw,1000,?)/2/pi;
```

```
%nm01p28.m : CtFT and ICtFT
global B D
% CtFT of A Rectangular Pulse Function
t=[-50:50]/10; % Time vector
w=[-60:60]/10*pi; % Frequency vector
D=1; % Duration of a rectangular pulse rD(t)
for k=1:length(w)
  Xw(k)=CtFT1('rDt',D*5,w(k));
end
subplot(221), plot(t,rDt(t)); subplot(222), plot(w,abs(Xw))
% ICtFT of a Sinc Spectrum
B=2*pi; % Bandwidth of a sinc spectrum sncB(w)
for n=1:length(t)
  xt(n)=ICtFT1('sincBw',B*5,t(n));
end
subplot(223), plot(t,real(xt)); subplot(224), plot(w,sincBw(w))
```

```
function x=rDt(t)
% Rectangular pulse function
global D
x=(-D/2<=t?t<=D/2);
```

```
function X=sincBw(w)
% Sinc function
global B
X=2*pi/?*sinc(?/B);
```

2

SYSTEM OF LINEAR EQUATIONS

CHAPTER OUTLINE

2.1 Solution for a System of Linear Equations	78
2.1.1 The Nonsingular Case ($M = N$)	78
2.1.2 The Underdetermined Case ($M < N$): Minimum-norm Solution	79
2.1.3 The Overdetermined Case ($M > N$): Least-squares Error Solution	82
2.1.4 Recursive Least-Squares Estimation (RLSE)	83
2.2 Solving a System of Linear Equations	86
2.2.1 Gauss(ian) Elimination	86
2.2.2 Partial Pivoting	88
2.2.3 Gauss-Jordan Elimination	97
2.3 Inverse Matrix	100
2.4 Decomposition (Factorization)	100
2.4.1 *LU* Decomposition (Factorization) – Triangularization	100
2.4.2 Other Decomposition (Factorization) – Cholesky, *QR* and SVD	105
2.5 Iterative Methods to Solve Equations	108
2.5.1 Jacobi Iteration	108
2.5.2 Gauss-Seidel Iteration	111
2.5.3 The Convergence of Jacobi and Gauss-Seidel Iterations	115
Problems	117

Applied Numerical Methods Using MATLAB®, Second Edition. Won Y. Yang, Jaekwon Kim, Kyung W. Park, Donghyun Baek, Sungjoon Lim, Jingon Joung, Suhyun Park, Han L. Lee, Woo June Choi, and Taeho Im.
© 2020 John Wiley & Sons, Inc. Published 2020 by John Wiley & Sons, Inc.
Companion website: www.wiley.com/go/yang/appliednumericalmethods

78 SYSTEM OF LINEAR EQUATIONS

In this chapter, we discuss several numerical schemes for solving a system of equations

$$a_{11}x_1 + a_{12}x_2 + \cdots + a_{1N}x_N = b_1$$
$$a_{21}x_1 + a_{22}x_2 + \cdots + a_{2N}x_N = b_2$$
$$\vdots \qquad = \vdots$$
$$a_{M1}x_1 + a_{M2}x_2 + \cdots + a_{MN}x_N = b_M \qquad (2.0.1a)$$

which can be written in a compact form by using a matrix-vector notation as

$$A_{M \times N}\mathbf{x} = \mathbf{b} \qquad (2.0.1b)$$

where

$$A_{M \times N} = \begin{bmatrix} a_{11} & a_{12} & \cdots & a_{1N} \\ a_{21} & a_{22} & \cdots & a_{2N} \\ \vdots & \vdots & \cdots & \vdots \\ a_{M1} & a_{M1} & \cdots & a_{MN} \end{bmatrix}, \quad \mathbf{x} = \begin{bmatrix} x_1 \\ x_2 \\ \cdot \\ x_N \end{bmatrix}, \quad \mathbf{b} = \begin{bmatrix} b_1 \\ b_2 \\ \cdot \\ b_M \end{bmatrix}$$

We will deal with the three cases:

(i) The case where the number (M) of equations and the number (N) of unknowns are equal ($M = N$) so that the coefficient matrix $A_{M \times N}$ is square.
(ii) The case where the number (M) of equations is smaller than the number (N) of unknowns ($M < N$) so that we might have to find the minimum-norm solution among infinitely many solutions.
(iii) The case where the number of equations is greater than the number of unknowns ($M > N$) so that there might exist no exact solution, and we must find a solution based on a global error minimization, like the least squares error (LSE) solution.

2.1 SOLUTION FOR A SYSTEM OF LINEAR EQUATIONS

2.1.1 The Nonsingular Case ($M = N$)

If the number (M) of equations and the number (N) of unknowns are equal ($M = N$), then the coefficient matrix A is square so that the solution can be written as

$$\mathbf{x} = A^{-1}\mathbf{b} \qquad (2.1.1)$$

so long as the matrix A is not singular. Here are the MATLAB statements for this job:

```
»A=[1 2;3 4]; b=[-1;-1];
»x=A^-1*b  % or, x=inv(A)*b
  x =   1.0000
       -1.0000
```

SOLUTION FOR A SYSTEM OF LINEAR EQUATIONS   79

What if A is square, but singular?

```
»A=[1 2;2 4]; b=[-1;-1];
»x=A^-1*b
  Warning: Matrix is singular to working precision.
  x = -Inf
      -Inf
```

This is the case where some or all of the rows of the coefficient matrix A are dependent on other rows and so the rank of A is deficient, which implies that there are some equations equivalent (*redundant*) to or *inconsistent* with other equations. If we remove the dependent rows until all the (remaining) rows are independent of each other so that A is of full rank (equal to M), it leads to the case of $M < N$, which will be dealt with in the next section.

2.1.2 The Underdetermined Case ($M < N$): Minimum-norm Solution

If the number (M) of equations is less than the number (N) of unknowns, the solution is not unique, but numerous. Suppose the M rows of the coefficient matrix A are independent. Then, any N-dimensional vector can be decomposed into two components

$$\mathbf{x} = \mathbf{x}^+ + \mathbf{x}^- \qquad (2.1.2)$$

where the one is in the row space $\mathcal{R}(A)$ of A that can be expressed as a linear combination of the M row vectors

$$\mathbf{x}^+ = A^T \boldsymbol{\alpha} \qquad (2.1.3)$$

and the other is in the null space $\mathcal{N}(A)$ orthogonal (perpendicular) to the row space so that

$$A\mathbf{x}^- = 0 \qquad (2.1.4)$$

Substituting the arbitrary N-dimensional vector representation (2.1.2) into Eq. (2.0.1) yields

$$A_{M \times N}\mathbf{x} \stackrel{(2.0.1)}{=} \mathbf{b} \stackrel{(2.1.2)}{\rightarrow} A(\mathbf{x}^+ + \mathbf{x}^-) \stackrel{(2.1.3)}{=} AA^T\boldsymbol{\alpha} + A\mathbf{x}^- \stackrel{(2.1.4)}{=} AA^T\boldsymbol{\alpha} = \mathbf{b} \qquad (2.1.5)$$

Since AA^T is supposedly a nonsingular $M \times M$ matrix resulting from multiplying an $M \times N$ matrix by an $N \times M$ matrix, we can solve this equation for α to get

$$\boldsymbol{\alpha}^o = [AA^T]^{-1}\mathbf{b} \qquad (2.1.6)$$

Then, substituting Eq. (2.1.6) into Eq. (2.1.3) yields

$$\mathbf{x}^{o+} \stackrel{(2.1.3)}{=} A^T \boldsymbol{\alpha}^o \stackrel{(2.1.6)}{=} A^T [AA^T]^{-1}\mathbf{b} \qquad (2.1.7)$$

This satisfies Eq. (2.0.1) and thus qualifies as its solution. However, it is far from being unique because the sum of $\mathbf{x}^{o+}$ and any vector $\mathbf{x}^-$ belonging to the null space satisfying Eq. (2.1.4) still satisfies Eq. (2.0.1) (as seen from Eq. (2.1.5)), yielding infinitely many solutions.

Based on the principle that any one of the two perpendicular legs is shorter than the hypotenuse in a right-angled triangle, Eq. (2.1.7) is believed to represent the *minimum-norm solution*. Note that the matrix $A^T[AA^T]^{-1}$ is called the *right pseudo (generalized) inverse* of A (see Remark 1.1(2)).

The MATLAB function 'pinv()' or Eq. (2.1.7) can be used to find the minimum-norm solution (2.1.7) to the system of linear equations (Eq. (2.0.1)).

```
»A=[1 2]; b=3;
»x=pinv(A)*b %x=A'*(A*A')^-1*b or eye(size(A,2))/A*b
  x = 0.6000
      1.2000
```

Remark 2.1 Projection Operator and Minimum-norm Solution.

(1) The solution (2.1.7) can be viewed as the projection of an arbitrary solution $\mathbf{x}^o$ onto the row space $\mathscr{R}(A)$ of the coefficient matrix A spanned by the row vectors. The remaining component of the solution $\mathbf{x}^o$:

$$\mathbf{x}^{o-} = \mathbf{x}^o - \mathbf{x}^{o+} = \mathbf{x}^o - A^T[AA^T]^{-1}\mathbf{b} = \mathbf{x}^o - A^T[AA^T]^{-1}A\,\mathbf{x}^o$$
$$= [I - A^T[AA^T]^{-1}A]\,\mathbf{x}^o$$

is in the null space $\mathscr{N}(A)$, since it satisfies Eq. (2.1.4). Note that

$$P_A = I - A^T[AA^T]^{-1}A$$

is called the *projection operator*.

(2) The solution (2.1.7) can be obtained by applying the Lagrange multiplier method (Section 7.2.1) to the constrained optimization problem in which we must find a vector $\mathbf{x}$ minimizing the (squared) norm $\|\mathbf{x}\|^2$ subject to the equality constraint $A\mathbf{x} = \mathbf{b}$.

$$\text{Min } J(\mathbf{x}, \lambda) \stackrel{\text{Eq.(7.2.2)}}{=} \frac{1}{2}\|\mathbf{x}\|^2 - \lambda^T(A\mathbf{x} - \mathbf{b}) = \frac{1}{2}\mathbf{x}^T\mathbf{x} - \lambda^T(A\mathbf{x} - \mathbf{b})$$

By using Eq. (7.2.3), we get

$$\frac{\partial}{\partial \mathbf{x}}J = \mathbf{x} - A^T\lambda = 0; \quad \mathbf{x} = A^T\lambda = A^T[AA^T]^{-1}\mathbf{b}$$
$$\frac{\partial}{\partial \lambda}J = A\mathbf{x} - \mathbf{b} = 0; \quad AA^T\lambda = \mathbf{b}; \quad \lambda = [AA^T]^{-1}\mathbf{b}$$

Example 2.1. Minimum-norm Solution.
Consider the problem of solving the equation

$$[1 \ 2]\begin{bmatrix}x_1\\x_2\end{bmatrix} = 3; \quad A\mathbf{x} = \mathbf{b} \text{ where } A = [1 \ 2], \quad \mathbf{b} = 3 \quad \text{(E2.1.1)}$$

This has infinitely many solutions and any $\mathbf{x} = [x_1 \ x_2]^T$ satisfying this equation, or equivalently,

$$x_1 + 2x_2 = 3; \quad x_2 = -\frac{1}{2}x_1 + \frac{3}{2} \quad \text{(E2.1.2)}$$

is a qualified solution. Equation (E2.1.2) describes the solution space as depicted in Figure 2.1.

On the other hand, any vector in the row space of the coefficient matrix A can be expressed by Eq. (2.1.3) as

$$\mathbf{x}^+ = A^T\alpha = \begin{bmatrix}1\\2\end{bmatrix}\alpha \quad (\alpha \text{ is a scalar, since } M = 1) \quad \text{(E2.1.3)}$$

and any vector in the null space of A can be expressed by Eq. (2.1.4) as

$$A\mathbf{x}^- = [1 \ 2]\begin{bmatrix}x_1^-\\x_2^-\end{bmatrix} = 0; \quad x_2^- = -\frac{1}{2}x_1^- \quad \text{(E2.1.4)}$$

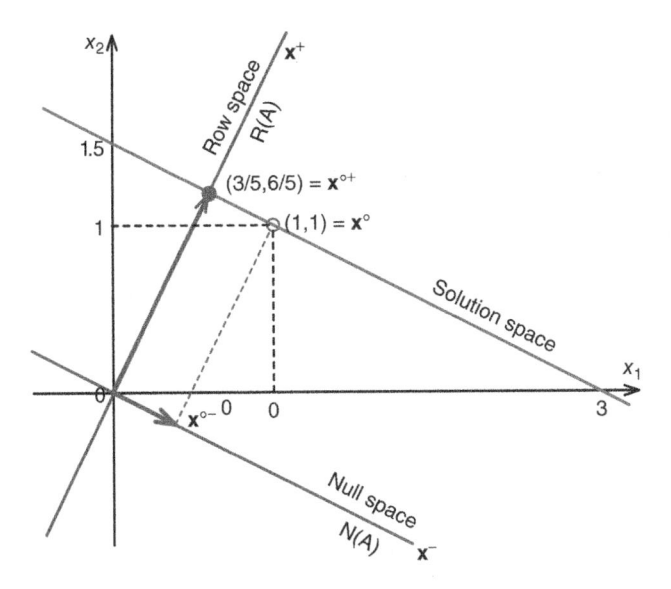

Figure 2.1 A minimum-norm solution.

We use Eq. (2.1.7) to obtain the minimum-norm solution

$$\mathbf{x}^{o+} = A^{\mathrm{T}}[AA^{\mathrm{T}}]^{-1}\mathbf{b} = \begin{bmatrix}1\\2\end{bmatrix}\left(\begin{bmatrix}1 & 2\end{bmatrix}\begin{bmatrix}1\\2\end{bmatrix}\right)^{-1} 3 = \frac{3}{5}\begin{bmatrix}1\\2\end{bmatrix} = \begin{bmatrix}0.6\\1.2\end{bmatrix} \qquad (E2.1.5)$$

Note from Figure 2.1 that the minimum-norm solution $\mathbf{x}^{o+}$ is the intersection of the solution space and the row space and is the closest to the origin among the vectors in the solution space.

2.1.3 The Overdetermined Case (M > N): Least-squares Error Solution

If the number (M) of (independent) equations is greater than the number (N) of unknowns, there exists no solution satisfying all the equations strictly. Thus, we try to find the *LS* (*least squares*) or *LSE* (*least squares error*) solution minimizing the norm of the (inevitable) error vector

$$\mathbf{e} = A\mathbf{x} - \mathbf{b} \qquad (2.1.8)$$

Then, our problem is to minimize the objective function

$$J = \frac{1}{2}\|\mathbf{e}\|^2 = \frac{1}{2}\|A\mathbf{x} - \mathbf{b}\|^2 = \frac{1}{2}[A\mathbf{x} - \mathbf{b}]^{\mathrm{T}}[A\mathbf{x} - \mathbf{b}] \qquad (2.1.9)$$

whose solution can be obtained by setting the derivative of this function (2.1.9) w.r.t. $\mathbf{x}$ to zero.

$$\frac{\partial}{\partial \mathbf{x}}J = A^{\mathrm{T}}[A\mathbf{x} - \mathbf{b}] = 0; \quad \mathbf{x}^o = [A^{\mathrm{T}}A]^{-1}A^{\mathrm{T}}\mathbf{b} \qquad (2.1.10)$$

This is called the LS or LSE solution. Note that the matrix A having more rows than columns ($M > N$) does not have its inverse, but has its *left pseudo (generalized) inverse* $[A^{\mathrm{T}} A]^{-1}A^{\mathrm{T}}$ as long as A is not rank-deficient (full rank), i.e., all of its columns are independent of each other (see Remark 1.1(2)). The left pseudo inverse matrix can be found by using the MATLAB command 'pinv()'.

The LSE solution (2.1.10) can be obtained by using the pinv() command or the backslash(\) operator.

```
»A=[1; 2]; b=[2.1; 3.9];
»x=pinv(A)*b  % A\b or x=(A'*A)^-1*A'*b
  x = 1.9800
```

The following MATLAB function 'lin_eq()' is designed to solve a given set of equations, covering all of the three cases in Sections 2.1.1-2.1.3.

SOLUTION FOR A SYSTEM OF LINEAR EQUATIONS

```
function x=lin_eq(A,B)
%This function finds the solution to Ax=B
[M,N] =size(A);
if size(B,1)~=M
  error('Incompatible dimension of A and B in lin_eq()!')
end
if M==N, x=A^-1*B; %x=inv(A)*B or Gaussj(A,B); %Eq. (2.1.1)
  elseif M<N % Minimum-norm solution (2.1.7)
    x=pinv(A)*B; %A'*(A*A')^-1*B; or eye(size(A,2))/A*B
  else % LSE solution (2.1.10) for M>N
    x=pinv(A)*B; %(A'*A)^-1*A'*B or x=A\B
end
```

(cf) The power of the 'pinv()' command is beyond our imagination as you might have felt in Problem 1.14. Even in case of $M < N$, it finds us an LS solution if the equations are *inconsistent*. Even in the case of $M > N$, it finds us a minimum-norm solution if the equations are *redundant*. Actually, the three cases can be dealt with by a single 'pinv()' command in the following function.

2.1.4 Recursive Least-Squares Estimation (RLSE)

In this section, we will see the so-called recursive least squares estimation (RLSE) algorithm, which is a recursive method to compute the LSE solution. Suppose we know the theoretical relationship between the temperature t (°C) and the resistance r (Ω) of a resistor as

$$c_1 t + c_2 = r$$

where we have many pairs of experimental data $\{(t_1, r_1), (t_2, r_2), \ldots, (t_k, r_k)\}$ collected up to time k. Since the above equation cannot be satisfied for all the data with any value of the parameters c_1 and c_2, we should try to get the parameter estimates that are optimal in some sense. This corresponds to the overdetermined case dealt with in the previous section and can be formulated as an LSE problem that we must solve a set of linear equations

$$A_k \mathbf{x}_k \approx \mathbf{b}_k, \quad \text{where } A_k = \begin{bmatrix} t_1 & 1 \\ t_2 & 1 \\ \cdot & \cdot \\ t_k & 1 \end{bmatrix}, \quad \mathbf{x}_k = \begin{bmatrix} c_{1,k} \\ c_{2,k} \end{bmatrix}, \quad \text{and} \quad \mathbf{b}_k = \begin{bmatrix} r_1 \\ r_2 \\ \cdot \\ r_k \end{bmatrix}$$

for which we can apply Eq. (2.1.10) to get the solution as

$$\mathbf{x}_k \stackrel{(2.1.10)}{=} [A_k^T A_k]^{-1} A_k^T \mathbf{b}_k \qquad (2.1.11)$$

Now, we are given an additional pair of data (t_{k+1}, r_{k+1}) and must find the new parameter estimate

84 SYSTEM OF LINEAR EQUATIONS

$$\mathbf{x}_{k+1} \stackrel{(2.1.10)}{=} [A_{k+1}^T A_{k+1}]^{-1} A_{k+1}^T \mathbf{b}_{k+1} \qquad (2.1.12)$$

with

$$A_{k+1} = \begin{bmatrix} t_1 & 1 \\ \vdots & \vdots \\ t_k & 1 \\ t_{k+1} & 1 \end{bmatrix}, \quad \mathbf{x}_{k+1} = \begin{bmatrix} c_{1,k+1} \\ c_{2,k+1} \end{bmatrix}, \quad \text{and} \quad \mathbf{b}_{k+1} = \begin{bmatrix} r_1 \\ \vdots \\ r_k \\ r_{k+1} \end{bmatrix}$$

How do we compute this? If we discard the previous estimate $\mathbf{x}_k$ and make direct use of Eq. (2.1.12) to compute the next estimate $\mathbf{x}_{k+1}$ every time a new data pair is available, the size of matrix A will get bigger and bigger as the data pile up, eventually defying any powerful computer in this world.

How about updating the previous estimate by just adding a correction term based on the new data to get a new estimate? This is the basic idea of the RLSE algorithm, which we are going to study. To get the idea, let us define the following notations

$$A_{k+1} = \begin{bmatrix} A_k \\ \mathbf{a}_{k+1}^T \end{bmatrix}, \quad \mathbf{a}_{k+1} = \begin{bmatrix} t_{k+1} \\ 1 \end{bmatrix}, \quad \mathbf{b}_{k+1} = \begin{bmatrix} \mathbf{b}_k \\ R_{k+1} \end{bmatrix}, \quad \text{and} \quad P_k = [A_k^T A_k]^{-1} \qquad (2.1.13)$$

and see how the inverse matrix P_k is to be updated on the arrival of new data (t_{k+1}, R_{k+1}):

$$P_{k+1} = [A_{k+1}^T A_{k+1}]^{-1} = \left[[A_k^T \ \mathbf{a}_{k+1}] \begin{bmatrix} A_k \\ \mathbf{a}_{k+1}^T \end{bmatrix} \right]^{-1} = [A_k^T A_k \ \mathbf{a}_{k+1} \mathbf{a}_{k+1}^T]^{-1}$$

$$= [P_k^{-1} \ \mathbf{a}_{k+1} \mathbf{a}_{k+1}^T]^{-1} \qquad (2.1.14)$$

(Matrix Inversion Lemma (B.14) in Section B.14.)

$$P_{k+1} = P_k - P_k \mathbf{a}_{k+1} [\mathbf{a}_{k+1}^T P_k \mathbf{a}_{k+1} + 1]^{-1} \mathbf{a}_{k+1}^T P_k \qquad (2.1.15)$$

It is interesting that $[\mathbf{a}_{k+1}^T P_k \mathbf{a}_{k+1} + 1]$ is nothing but a scalar and so we do not need to compute the matrix inverse thanks to the Matrix Inversion Lemma (Appendix B). It is much better in the computational aspect to use this recursive formula (2.1.15) than to compute $P_{k+1} = [A_{k+1}^T A_{k+1}]^{-1}$ directly. We can also write Eq. (2.1.12) in recursive form as

$$\mathbf{x}_{k+1} \stackrel{(2.1.12, 2.1.14)}{=} P_{k+1} A_{k+1}^T \mathbf{b}_{k+1} \stackrel{(2.1.13)}{=} P_{k+1} [A_k^T \ \mathbf{a}_{k+1}] \begin{bmatrix} \mathbf{b}_k \\ r_{k+1} \end{bmatrix}$$

$$= P_{k+1} [A_k^T \mathbf{b}_k + \mathbf{a}_{k+1} r_{k+1}]$$

$$\stackrel{(2.1.13)}{=} P_{k+1} \left[(A_{k+1}^T A_{k+1} - \mathbf{a}_{k+1} \mathbf{a}_{k+1}^T) \mathbf{x}_k + \mathbf{a}_{k+1} r_{k+1} \right]$$

$$\stackrel{(2.1.13)}{=} P_{k+1} \left[P_{k+1}^{-1} \mathbf{x}_k - \mathbf{a}_{k+1} \mathbf{a}_{k+1}^T \mathbf{x}_k + \mathbf{a}_{k+1} r_{k+1} \right];$$

SOLUTION FOR A SYSTEM OF LINEAR EQUATIONS

$$\mathbf{x}_{k+1} = \mathbf{x}_k + P_{k+1}\mathbf{a}_{k+1}\left(r_{k+1} - \mathbf{a}_{k+1}^T \mathbf{x}_k\right) \quad (2.1.16)$$

Equation (2.1.15) can be used to rewrite the gain matrix $P_{k+1}\mathbf{a}_{k+1}$ premultiplied by the 'error' $(r_{k+1} - \mathbf{a}_{k+1}^T \mathbf{x}_k)$ to make the correction term on the RHS (right-hand side) of Eq. (2.1.16) as

$$K_{k+1} = P_{k+1}\mathbf{a}_{k+1} \stackrel{(2.1.15)}{=} \left[P_k - P_k \mathbf{a}_{k+1} [\mathbf{a}_{k+1}^T P_k \mathbf{a}_{k+1} + 1]^{-1} \mathbf{a}_{k+1}^T P_k \right] \mathbf{a}_{k+1}$$

$$= P_k \mathbf{a}_{k+1} \left[I - [\mathbf{a}_{k+1}^T P_k \mathbf{a}_{k+1} + 1]^{-1} \mathbf{a}_{k+1}^T P_k \mathbf{a}_{k+1} \right]$$

$$= P_k \mathbf{a}_{k+1} [\mathbf{a}_{k+1}^T P_k \mathbf{a}_{k+1} + 1]^{-1} \left\{ [\mathbf{a}_{k+1}^T P_k \mathbf{a}_{k+1} + 1] - \mathbf{a}_{k+1}^T P_k \mathbf{a}_{k+1} \right\};$$

$$K_{k+1} = P_k \mathbf{a}_{k+1} [\mathbf{a}_{k+1}^T P_k \mathbf{a}_{k+1} + 1]^{-1} \quad (2.1.17)$$

and substitute this back into Eq. (2.1.15) to write it as

$$P_{k+1} = P_k - K_{k+1}\mathbf{a}_{k+1}^T P_k \quad (2.1.18)$$

The following MATLAB function 'rlse_online()' implements this RLSE algorithm that updates the parameter estimates by using Eqs. (2.1.17), (2.1.16) and (2.1.18). The MATLAB script "do_rlse.m" updates the parameter estimates every time new data arrives and compares the results of the online processing with those obtained by the offline (batch job) processing, i.e. by using Eq. (2.1.12) directly. Noting that

- the matrix $[A_k^T A_k]$ as well as $\mathbf{b}_k$ consists of information and is a nonnegative matrix, and
- as valuable information data accumulate, $[A_k^T A_k]$ will get larger, or equivalently, $P_k = [A_k^T A_k]^{-1}$ will get smaller and consequently, the gain matrix K_k will get smaller,

one could understand that P_k is initialized to a very large identity matrix, since no information is available in the beginning. Since a large/small P_k makes the correction term on the RHS of Eq. (2.1.16) large/small, the RLSE algorithm becomes more conservative and reluctant to learn from new data as the data pile up, while it is willing to make use of new data for updating the estimates when it is hungry for information in the beginning.

```
function [x,K,P]=rlse_online(a_k,b_k,x,P)
% One step of RLSE (Recursive Least Squares Estimation) algorithm
a_k=a_k(:); b_k=b_k(:); % To make sure them column vectors
K = P*a_k1/(a_k'*P*a_k+1); % Eq.(2.1.17)
x = x +K*(b_k-a_k'*x); % Eq.(2.1.16)
P = P -K*a_k'*P; % Eq.(2.1.18)
```

```
%do_rlse.m
clear
xo=[2  1]'; %The true value of unknown coefficient vector
NA=length(xo);
x =zeros(NA,1); P =100*eye(NA,NA);
for k=1:100
   A(k,:)=[k*0.01 1];
   b(k,:)=A(k,:)*xo +0.2*rand;
   [x,K,P] =rlse_online(A(k,:),b(k,:),x,P);
end
x    % the final parameter estimate
A\b % for comparison with the off-line processing (batch job)
```

2.2 SOLVING A SYSTEM OF LINEAR EQUATIONS

2.2.1 Gauss(ian) Elimination

For simplicity, we assume that the coefficient matrix A in Eq. (2.0.1) is a nonsingular 3×3 matrix with $M = N = 3$. Then, we can write the equation as

$$a_{11}x_1 + a_{12}x_2 + a_{13}x_3 = b_1 \tag{2.2.1a}$$

$$a_{21}x_1 + a_{22}x_2 + a_{23}x_3 = b_2 \tag{2.2.1b}$$

$$a_{31}x_1 + a_{32}x_2 + a_{33}x_3 = b_3 \tag{2.2.1c}$$

First, to remove the x_1 terms from equations (2.2.1-m) (with $m \neq 1$), we subtract (2.2.1a) $\times a_{m1}/a_{11}$ from each of them to get

$$a_{11}^{(0)}x_1 + a_{12}^{(0)}x_2 + a_{13}^{(0)}x_3 = b_1^{(0)} \tag{2.2.2a}$$

$$a_{22}^{(1)}x_2 + a_{23}^{(1)}x_3 = b_2^{(1)} \tag{2.2.2b}$$

$$a_{32}^{(1)}x_2 + a_{33}^{(1)}x_3 = b_3^{(1)} \tag{2.2.2c}$$

with

$$a_{mn}^{(0)} = a_{mn}, \quad b_m^{(0)} = b_m \quad \text{for } m, n = 1, 2, 3 \tag{2.2.3a}$$

$$a_{mn}^{(1)} = a_{mn}^{(0)} - \frac{a_{m1}^{(0)}}{a_{11}^{(0)}} a_{1n}^{(0)}, \quad b_m^{(1)} = b_m^{(0)} - \frac{a_{m1}^{(0)}}{a_{11}^{(0)}} b_1^{(0)} = b_m \quad \text{for } m, n = 2, 3 \tag{2.2.3b}$$

We call this work 'doing *ERO*s (*elementary row operations*) at a_{11}' and the center element a_{11} a 'pivot (element)'.

Next, to remove the x_2 term from Eq. (2.2.2-m) (with $m \neq 1, 2$), we subtract (2.2.2b) $\times a_{m2}^{(1)}/a_{22}^{(1)}$ ($m = 3$) from it to get

$$a_{11}^{(0)}x_1 + a_{12}^{(0)}x_2 + a_{13}^{(0)}x_3 = b_1^{(0)} \tag{2.2.4a}$$

$$a_{22}^{(1)}x_2 + a_{23}^{(1)}x_3 = b_2^{(1)} \tag{2.2.4b}$$

$$a_{33}^{(2)}x_3 = b_3^{(2)} \tag{2.2.4c}$$

with

$$a_{mn}^{(2)} = a_{mn}^{(1)} - \frac{a_{m2}^{(1)}}{a_{22}^{(1)}} a_{2n}^{(1)} \tag{2.2.5a}$$

$$b_m^{(2)} = b_m^{(1)} - \frac{a_{m2}^{(1)}}{a_{22}^{(1)}} b_2^{(1)} \tag{2.2.5b}$$

(for $m, n = 3$). We call this procedure '*Gauss (forward) elimination*' or '*Gaussian (forward) elimination*' and can generalize the updating formula (2.2.3)/(2.2.5) as

$$a_{mn}^{(k)} = a_{mn}^{(k-1)} - \frac{a_{mk}^{(k-1)}}{a_{kk}^{(k-1)}} a_{kn}^{(k-1)} \quad \text{for } m, n = k+1, k+2, \ldots, M \tag{2.2.6a}$$

$$b_m^{(k)} = b_m^{(k-1)} - \frac{a_{mk}^{(k-1)}}{a_{kk}^{(k-1)}} b_k^{(k-1)} \quad \text{for } m = k+1, k+2, \ldots, M \tag{2.2.6b}$$

After having the triangular matrix-vector equation like Eq. (2.2.4), we can solve the last equation (2.2.4c) first to get

$$x_3 = \frac{b_3^{(2)}}{a_{33}^{(2)}} \tag{2.2.7a}$$

and then substitute this result into Eq. (2.2.4b) to get

$$x_2 = \frac{b_2^{(1)} - a_{23}^{(1)}x_3}{a_{22}^{(1)}} \tag{2.2.7b}$$

Successively, we substitute Eqs. (2.2.7a) and (2.2.7b) into Eq. (2.2.4a) to get

$$x_1 = \frac{b_1^{(0)} - \sum_{n=2}^{3} a_{1n}^{(0)} x_n}{a_{11}^{(0)}} \tag{2.2.7c}$$

We call this procedure '*backward substitution*' and can generalize the solution formula (2.2.7) as

$$x_m = \frac{b_m^{(m-1)} - \sum_{n=m+1}^{M} a_{mn}^{(m-1)} x_n}{a_{mm}^{(m-1)}} \quad \text{for } m = M, M-1, \ldots, 1 \tag{2.2.8}$$

88 SYSTEM OF LINEAR EQUATIONS

In this way, the Gauss elimination procedure consists of two steps, i.e. forward elimination and backward substitution. Noting that

(1) this procedure has nothing to do with the specific values of the unknown variable x_ms and involves only the coefficients, and
(2) the formulas (2.2.6a) on the coefficient matrix A and (2.2.6b) on the RHS vector **b** conform with each other,

we will augment A with **b** and put formulas (2.2.6a) and (2.2.6b) together into one framework when programming the Gauss forward elimination procedure.

2.2.2 Partial Pivoting

The core formula (2.2.6) used for Gauss elimination requires division by $a_{kk}^{(k-1)}$ at the kth stage, where $a_{kk}^{(k-1)}$ is the diagonal element in the kth row. What if $a_{kk}^{(k-1)} = 0$? In such a case, it is customary to switch the kth row and another row below it having the element of the largest absolute value in the kth column. This procedure, called 'partial pivoting', is recommended for reducing the round-off error even in the case where the kth pivot $a_{kk}^{(k-1)}$ is not zero.

Let us consider the following example:

$$\begin{bmatrix} 0 & 1 & 1 \\ 2 & -1 & -1 \\ 1 & 1 & -1 \end{bmatrix} \begin{bmatrix} x_1 \\ x_2 \\ x_3 \end{bmatrix} = \begin{bmatrix} b_1 = 2 \\ b_2 = 0 \\ b_3 = 1 \end{bmatrix} \qquad (2.2.9)$$

We construct an augmented matrix by combining the coefficient matrix and the RHS vector as

$$\begin{bmatrix} a_{11} & a_{12} & a_{13} & b_1 \\ a_{21} & a_{22} & a_{23} & b_2 \\ a_{31} & a_{32} & a_{33} & b_3 \end{bmatrix} = \begin{bmatrix} 0 & 1 & 1 & 2 \\ 2 & -1 & -1 & 0 \\ 1 & 1 & -1 & 1 \end{bmatrix} \begin{matrix} : r_1 \\ : r_2 \\ : r_3 \end{matrix} \qquad (2.2.10)$$

and apply the Gauss elimination procedure.

In the stage of forward elimination, we want to do EROs at a_{11}, but $a_{11} = 0$ cannot be used as the pivot element because it is zero. So we switch the first row and the second row having the element of the largest absolute value in the first column:

$$\begin{bmatrix} a_{11}^{(1)} & a_{12}^{(1)} & a_{13}^{(1)} & b_1^{(1)} \\ a_{21}^{(1)} & a_{22}^{(1)} & a_{23}^{(1)} & b_2^{(1)} \\ a_{31}^{(1)} & a_{32}^{(1)} & a_{33}^{(1)} & b_3^{(1)} \end{bmatrix} = \begin{bmatrix} 2 & -1 & -1 & 0 \\ 0 & 1 & 1 & 2 \\ 1 & 1 & -1 & 1 \end{bmatrix} \begin{matrix} : r_1^{(1)} \\ : r_2^{(1)} \\ : r_3^{(1)} \end{matrix} \qquad (2.2.11a)$$

Then we do EROs at $a_{11}^{(1)} = 2$ by applying Eq. (2.2.3) to write

$$
\begin{array}{c} r_1^{(1)} \to \\ r_2^{(1)} - a_{21}^{(1)}/a_{11}^{(1)} \times r_1^{(1)} \to \\ r_3^{(1)} - a_{31}^{(1)}/a_{11}^{(1)} \times r_1^{(1)} \to \end{array}
\begin{bmatrix} a_{11}^{(2)} & a_{12}^{(2)} & a_{13}^{(2)} & b_1^{(2)} \\ a_{21}^{(2)} & a_{22}^{(2)} & a_{23}^{(2)} & b_2^{(2)} \\ a_{31}^{(2)} & a_{32}^{(2)} & a_{33}^{(2)} & b_3^{(2)} \end{bmatrix}
= \begin{bmatrix} 2 & -1 & -1 & 0 \\ 0 & 1 & 1 & 2 \\ 0 & 3/2 & -1/2 & 1 \end{bmatrix}
\begin{array}{l} : r_1^{(2)} \\ : r_2^{(2)} \\ : r_3^{(2)} \end{array}
$$
(2.2.11b)

Here, instead of doing EROs at $a_{22}^{(2)}$, we switch the second row and the third row having the element of the largest absolute value among the elements not above $a_{22}^{(2)}$ in the second column to write

$$
\begin{bmatrix} a_{11}^{(3)} & a_{12}^{(3)} & a_{13}^{(3)} & b_1^{(3)} \\ a_{21}^{(3)} & a_{22}^{(3)} & a_{23}^{(3)} & b_2^{(3)} \\ a_{31}^{(3)} & a_{32}^{(3)} & a_{33}^{(3)} & b_3^{(3)} \end{bmatrix}
= \begin{bmatrix} 2 & -1 & -1 & 0 \\ 0 & 3/2 & -1/2 & 1 \\ 0 & 1 & 1 & 2 \end{bmatrix}
\begin{array}{l} : r_1^{(3)} \\ : r_2^{(3)} \\ : r_3^{(3)} \end{array} \qquad (2.2.11c)
$$

Then we do EROs at $a_{22}^{(3)}$ by applying Eq. (2.2.5), more generally, Eq. (2.2.6) to get the upper-triangularized form:

$$
\begin{array}{c} r_1^{(3)} \to \\ r_2^{(3)} \to \\ r_3^{(3)} - a_{32}^{(3)}/a_{22}^{(3)} \times r_2^{(3)} \to \end{array}
\begin{bmatrix} a_{11}^{(4)} & a_{12}^{(4)} & a_{13}^{(4)} & b_1^{(4)} \\ a_{21}^{(4)} & a_{22}^{(4)} & a_{23}^{(4)} & b_2^{(4)} \\ a_{31}^{(4)} & a_{32}^{(4)} & a_{33}^{(4)} & b_3^{(4)} \end{bmatrix}
= \begin{bmatrix} 2 & -1 & -1 & 0 \\ 0 & 3/2 & -1/2 & 1 \\ 0 & 0 & 4/3 & 4/3 \end{bmatrix}
\begin{array}{l} : r_1^{(4)} \\ : r_2^{(4)} \\ : r_3^{(4)} \end{array}
$$
(2.2.11d)

Now, in the stage of backward substitution, we apply Eq. (2.2.7), more generally, Eq. (2.2.8) to get the final solution as

$$
\left. \begin{array}{l} x_3 = \dfrac{b_3^{(4)}}{a_{33}^{(4)}} = \dfrac{4/3}{4/3} = 1 \\[2mm] x_2 = \dfrac{b_2^{(4)} - a_{23}^{(4)} x_3}{a_{22}^{(4)}} = \dfrac{1 - (-1/2) \times 1}{3/2} = 1 \\[2mm] x_1 = \dfrac{b_1^{(4)} - \sum_{n=2}^{3} a_{1n}^{(4)} x_n}{a_{11}^{(4)}} = \dfrac{0 - (-1) \times 1 - (-1) \times 1}{2} = 1 \end{array} \right\} \to \begin{bmatrix} x_3 \\ x_2 \\ x_1 \end{bmatrix} = \begin{bmatrix} 1 \\ 1 \\ 1 \end{bmatrix}
$$
(2.2.12)

90 SYSTEM OF LINEAR EQUATIONS

Let us consider another system of equations:

$$\begin{bmatrix} 1 & 0 & 1 \\ 1 & 1 & 1 \\ 1 & -1 & 1 \end{bmatrix} \begin{bmatrix} x_1 \\ x_2 \\ x_3 \end{bmatrix} = \begin{bmatrix} b_1 = 2 \\ b_2 = 3 \\ b_3 = 1 \end{bmatrix} \qquad (2.2.13)$$

We construct an augmented matrix by combining the coefficient matrix and the RHS vector as

$$\begin{bmatrix} a_{11} & a_{12} & a_{13} & b_1 \\ a_{21} & a_{22} & a_{23} & b_2 \\ a_{31} & a_{32} & a_{33} & b_3 \end{bmatrix} = \begin{bmatrix} 1 & 0 & 1 & 2 \\ 1 & 1 & 1 & 3 \\ 1 & -1 & 1 & 1 \end{bmatrix} \begin{matrix} : r_1 \\ : r_2 \\ : r_3 \end{matrix} \qquad (2.2.14)$$

and apply the Gauss elimination procedure.

First, noting that all the elements in the first column have the same absolute value and so we do not need to switch the rows, we do EROs at a_{11}.

$$\begin{bmatrix} a_{11}^{(1)} & a_{12}^{(1)} & a_{13}^{(1)} & b_1^{(1)} \\ a_{21}^{(1)} & a_{22}^{(1)} & a_{23}^{(1)} & b_2^{(1)} \\ a_{31}^{(1)} & a_{32}^{(1)} & a_{33}^{(1)} & b_3^{(1)} \end{bmatrix} = \begin{bmatrix} 1 & 0 & 1 & 2 \\ 0 & 1 & 0 & 1 \\ 0 & -1 & 0 & -1 \end{bmatrix} \begin{matrix} : r_1^{(1)} \\ : r_2^{(1)} \\ : r_3^{(1)} \end{matrix} \qquad (2.2.15a)$$

Second, without having to switch the rows, we do EROs at $a_{11}^{(1)}$ to write

$$\begin{matrix} r_1^{(1)} \to \\ r_2^{(1)} \to \\ r_3^{(1)} - a_{32}^{(1)}/a_{22}^{(1)} \times r_2^{(1)} \to \end{matrix} \begin{bmatrix} a_{11}^{(2)} & a_{12}^{(2)} & a_{13}^{(2)} & b_1^{(2)} \\ a_{21}^{(2)} & a_{22}^{(2)} & a_{23}^{(2)} & b_2^{(2)} \\ a_{31}^{(2)} & a_{32}^{(2)} & a_{33}^{(2)} & b_3^{(2)} \end{bmatrix} = \begin{bmatrix} 1 & 0 & 1 & 2 \\ 0 & 1 & 0 & 1 \\ 0 & 0 & 0 & 0 \end{bmatrix} \begin{matrix} : r_1^{(2)} \\ : r_2^{(2)} \\ : r_3^{(2)} \end{matrix} \qquad (2.2.15b)$$

Now, we are at the stage of backward substitution, but $a_{33}^{(2)}$, which is supposed to be the denominator in Eq. (2.2.8), is zero. We may face such a weird situation of zero-division even during the forward elimination process where the pivot is zero and besides, we cannot find any (nonzero) element below it in the same column and on its right in the same row except the RHS element. In this case, we cannot go further. This implies that some or all rows of coefficient matrix A are dependent on others, corresponding to the case of redundancy (infinitely many solutions) or inconsistency (no exact solution). Noting that the RHS element of the zero row in Eq. (2.2.15b) is also zero, we should declare the case of redundancy and may have to be satisfied with one of the infinitely many solutions being the RHS vector as

$$[x_1 \; x_2 \; x_3] = [b_1^{(2)} \; b_2^{(2)} \; b_3^{(2)}] = [2 \; 1 \; 0] \qquad (2.2.16)$$

Furthermore, if we remove the all-zero row(s), the problem can be treated as an underdetermined case handled in Section 2.1.2. Note that, if the RHS element were not zero, we would have to declare the case of inconsistency, as will be illustrated.

Suppose that $b_1 = 1$ in Eq. (2.2.13). Then, the Gauss elimination would have proceeded as follows:

$$\begin{bmatrix} 1 & 0 & 1 & 1 \\ 1 & 1 & 1 & 3 \\ 1 & -1 & 1 & 1 \end{bmatrix} \xrightarrow[r_3-r_1]{r_2-r_1} \begin{matrix} r_1 \\ \\ \end{matrix} \begin{bmatrix} 1 & 0 & 1 & 1 \\ 0 & 1 & 0 & 2 \\ 0 & -1 & 0 & 0 \end{bmatrix} \rightarrow \begin{matrix} r_1 \\ r_2 \\ r_3+r_2 \end{matrix} \begin{bmatrix} 1 & 0 & 1 & 1 \\ 0 & 1 & 0 & 2 \\ 0 & 0 & 0 & 2 \end{bmatrix} \quad (2.2.17)$$

This ended up with an all-zero row except the nonzero RHS element, corresponding to the case of inconsistency. Thus, we must declare the case of 'no exact solution' for this problem.

The following MATLAB function 'Gauss()' implements the Gauss elimination algorithm and the script "do_Gauss.m" is designed to solve Eq. (2.2.9) by using 'Gauss()'. Note that at every operation in the function 'Gauss()', the pivot row is divided by the pivot element so that every diagonal element becomes one and that we do not need to perform any computation for the kth column at the kth stage, since the column is supposed to be all zeros but the kth element $a_{kk}^{(k)} = 1$.

```
%do_Gauss.m
A=[0 1 1;2 -1 -1;1 1 -1]; b=[2 0 1]'; % Eq.(2.2.9)
x=Gauss(A,b)
x1=A\b %for comparison with the result of backslash operation
```

(cf) The number of floating-point multiplications required in this function 'Gauss()' is

$$\sum_{k=1}^{N_A}\{\overset{(3)}{(N_A-k+1)}\overset{(2)}{(N_A+N_B-k)} + \overset{(1)}{N_A-k+1}\} + \sum_{k=1}^{N_A-1}\overset{(4)}{(N_A-k)N_B}$$

$$\overset{N_A-k+1\rightarrow k}{=} \sum_{k=1}^{N_A} k(k+N_B) - N_B\sum_{k=1}^{N_A} k + \sum_{k=1}^{N_A} N_A \cdot N_B = \sum_{k=1}^{N_A} k^2 + \sum_{k=1}^{N_A} N_A \cdot N_B$$

$$= \frac{1}{6}(N_A+1)N_A(2N_A+1) + N_A^2 N_B \approx \frac{1}{3}N_A^3 \quad \text{for } N_A \gg N_B \quad (2.2.18)$$

where N_A: the size of the matrix A, N_B: the column dimension of the RHS matrix B.

92 SYSTEM OF LINEAR EQUATIONS

```
function x=Gauss(A,B)
%This function solves Ax=B by Gauss elimination algorithm.
%The sizes of matrices A,B are supposed to be NAxNA and NAxNB.
NA=size(A,2); [NB1,NB]=size(B); N=NA+NB;
if NB1~=NA, error('The coefficient matrix A must be square'); end
AB=[A(1:NA,1:NA) B(1:NA,1:NB)]; % Augmented matrix
epss=eps*ones(NA,1);
for k=1:NA
   %Scaled Partial Pivoting at AB(k,k) by Eq.(2.2.20)
   [akx,kx]=max(abs(AB(k:NA,k))./ ...
     max(abs([AB(k:NA,k+1:NA) epss(1:NA-k+1)]'))'); %(1) in Eq.(2.2.18)
   if akx<eps, error('Singular matrix');
     x=pinv(A)*B;
     if abs(AB(k,NA+1))>eps
       fprintf('Inconsistent: A SVD-based LS solution')
      else
       fprintf('Indeterminant: A SVD-based minimum-norm solution')
     end
     return;
   end
   if kx>1 % Row change if necessary
     mx=k+kx-1; % the index of row to be exchanged with row k
     tmp_row =AB(k,k:N); AB(k,k:N)=AB(mx,k:N); AB(mx,k:N)=tmp_row;
   end
   % Gauss forward elimination
   AB(k,k+1:N)=AB(k,k+1:N)/AB(k,k); %(2) in Eq.(2.2.18)
   AB(k,k)=1; % make each diagonal element one
   for m=k+1:NA %(3) in Eq.(2.2.18)
      AB(m,k+1:N)=AB(m,k+1:N)-AB(m,k)*AB(k,k+1:N); % Eq.(2.2.6)
      AB(m,k)=0;
   end
end
%backward substitution for a upper-triangular matrix equation
% having all the diagonal elements equal to one
x(NA,:)= AB(NA,NA+1:N);
for m=NA-1:-1:1  %(4) in Eq.(2.2.18)
   x(m,:)= AB(m,NA+1:N)-AB(m,m+1:NA)*x(m+1:NA,:); % Eq.(2.2.8)
end
```

Here are several things to note:

Remark 2.2 Partial Pivoting and Undetermined/Inconsistent Case.

(1) In Gauss or Gauss-Jordan elimination, some row switching is performed to avoid the zero division. Even without that purpose, it may be helpful for reducing the round-off error to fix

$$\text{Max}\{|a_{mk}|, \quad k \leq m \leq M\} \qquad (2.2.19)$$

as the pivot element in the kth iteration through some row switching, which is called 'partial pivoting'. Actually, it might be better off to fix

$$\text{Max}\left\{ \frac{|a_{mk}|}{\text{Max}\{|a_{mn}|, \ k \le n \le M\}}, \ k \le m \le M \right\} \quad (2.2.20)$$

as the pivot element in the kth iteration, which is called '*scaled partial pivoting*' or to do column switching as well as row switching for choosing the best (largest) pivot element, which is called 'full (complete) pivoting'. Note that if the columns are switched, the order of the unknown variables should be interchanged accordingly.

(2) What if some diagonal element a_{kk} and all the elements beneath it in the same column are zero and besides, all the elements in the row including a_{kk} are also zero except the RHS element? It implies that some or all rows of the coefficient matrix A are dependent on others, corresponding to the case of redundancy (infinitely many solutions) or inconsistency (no exact solution). If even the RHS element is zero, it should be declared to be the case of redundancy. In this case, we can get rid of the all-zero row(s) and then treat the problem as the underdetermined case handled in Section 2.1.2. If the RHS element is only one nonzero in the row, it should be declared to be the case of inconsistency.

Example 2.2. Delicacy of Partial Pivoting.
To get an actual feeling about the delicacy of partial pivoting, consider the following systems of linear equations, which apparently have $x^o = \begin{bmatrix} 1 & 1 \end{bmatrix}^T$ as their solutions.

(a) $\quad A_1 x = b_1 \quad$ with $\quad A_1 = \begin{bmatrix} 10^{-15} & 1 \\ 1 & 10^{11} \end{bmatrix}, \quad b_1 = \begin{bmatrix} 1 + 10^{-15} \\ 10^{11} + 1 \end{bmatrix} \quad$ (E2.2.1)

Without any row switching, the Gauss elimination procedure will find us the true solution only if there is no quantization error.

$$[A_1 \ b_1] = \begin{bmatrix} 10^{-15} & 1 & 1 + 10^{-15} \\ 1 & 10^{11} & 10^{11} + 1 \end{bmatrix}$$

$$\xrightarrow{\text{forward elimination}} \begin{bmatrix} 1 & 10^{15} & 10^{15} + 1 \\ 0 & 10^{11} - 10^{15} & 10^{11} - 10^{15} \end{bmatrix} \xrightarrow{\text{backward substitution}} x = \begin{bmatrix} 1 \\ 1 \end{bmatrix}$$

But because of the round-off error, it will deviate from the true solution.

$$\xrightarrow{\text{forward elimination}} \begin{bmatrix} 1 & 1/10^{-15} & 10^{15}+1 \\ & = 9.9999999999\ldots\text{e}+014 & = 1.000000000\ldots\text{1e}+015 \\ 0 & 10^{11}-1/10^{-15} & 10^{11}+1-(1+10^{-15})/10^{-15} \\ & = -9.9989999999\ldots\text{e}+014 & = -9.99900\ldots\ldots\ldots\text{e}+014 \end{bmatrix}$$

$$\xrightarrow{\text{backward substitution}} \mathbf{x} = \begin{bmatrix} 8.7500000000000000\text{e}-001 \\ 1.0000000000000000\text{e}+000 \end{bmatrix}$$

If we enforce the strategy of partial pivoting or scaled partial pivoting, the Gauss elimination procedure will give us much better result as follows:

$$[A_1\ \mathbf{b}_1] \xrightarrow{\text{row swap}} \begin{bmatrix} 1 & 10^{11} & 10^{11}+1 \\ 10^{-15} & 1 & 1+10^{-15} \end{bmatrix}$$

$$\xrightarrow{\text{forward elimination}} \begin{bmatrix} 1 & 10^{11} = 1.000\text{e}+011 & 10^{11}+1 = 1.0000000000100000\text{e}+011 \\ 0 & 1-10^{-4} = 9.999\text{e}-001 & 9.999000000000001\text{e}-001 \end{bmatrix}$$

$$\xrightarrow{\text{backward substitution}} \mathbf{x} = \begin{bmatrix} 9.999847412109375\text{e}-001 \\ 1.0000000000000000\text{e}+000 \end{bmatrix}$$

(b) $$A_2 \mathbf{x} = \mathbf{b}_2 \text{ with } A_2 = \begin{bmatrix} 10^{-14.6} & 1 \\ 1 & 10^{15} \end{bmatrix}, \quad \mathbf{b}_2 = \begin{bmatrix} 1+10^{-14.6} \\ 10^{15}+1 \end{bmatrix} \quad (E2.2.2)$$

Without partial pivoting, the Gauss elimination procedure will give us a quite good result.

$$[A_1\ \mathbf{b}_1] = \begin{bmatrix} 1 & 10^{14.6} = 3.981071705534969\text{e}+014 & 10^{14.6}+1 = 3.981071705534979\text{e}+014 \\ 0 & 6.018928294465030\text{e}+014 & 6.018928294465030\text{e}+014 \end{bmatrix}$$

$$\rightarrow \begin{bmatrix} 1 & 3.981071705534969\text{e}+014 & 3.981071705534979\text{e}+014 \\ 0 & 1 & 1 \end{bmatrix} \xrightarrow{\text{backward substitution}} \mathbf{x} = \begin{bmatrix} 1 \\ 1 \end{bmatrix}$$

But if we exchange the first row with the second row having the larger element in the first column according to the strategy of partial pivoting, the Gauss elimination procedure will give us a rather surprisingly bad result as follows:

$$\xrightarrow[\text{forward elimination}]{\text{row swapping}} \begin{bmatrix} 1 & 10^{15} = 1.0000000000000000\text{e}+015 & 10^{15}+1 = 1.000000000000001\text{e}+015 \\ 0 & 1-10^{15}\cdot 10^{-14.6} & 1+10^{-14.6}-(1+10^{15})\cdot 10^{-14.6} \\ & = -1.5118864315095819 & = -1.5118864315095821 \end{bmatrix}$$

$$\xrightarrow{\text{backward substitution}} \mathbf{x} = \begin{bmatrix} 0.7500000000000000 \\ 1.0000000000000002 \end{bmatrix}$$

One might be happy to have the scaled partial pivoting scheme (Eq. (2.2.21)), which does not switch the rows in this case, since the relative magnitude (dominancy) of a_{11} in the first row is greater than that of a_{21} in the second row, that is, $10^{-14.6}/1 > 1/10^{15}$.

(c) $A_3 \mathbf{x} = \mathbf{b}_3$ with $A_3 = \begin{bmatrix} 10^{15} & 1 \\ 1 & 10^{-14.6} \end{bmatrix}$, $\mathbf{b}_3 = \begin{bmatrix} 10^{15} + 1 \\ 1 + 10^{-14.6} \end{bmatrix}$ (E2.2.3)

With any pivoting scheme, we do not need to switch the rows, since the relative magnitude as well as the absolute magnitude of a_{11} in the first row is greater than those of a_{21} in the second row. Thus, the Gauss elimination procedure will go as follows:

$$\xrightarrow{\text{forward elimination}} \begin{bmatrix} 1 & 1.000000000000000e-015 & 1.000000000000001e+000 \\ 0 & 1.511886431509582e-015 & 1.332267629550188e-015 \end{bmatrix}$$

$$\xrightarrow{\text{backward substitution}} \mathbf{x} = \begin{bmatrix} 1.000000000000000 \\ 0.811955724875121 \end{bmatrix}$$

(cf) Note that the coefficient matrix A_3 is the same as would be obtained by applying the full pivoting scheme for A_2 to have the largest pivot element. This example implies that the Gauss elimination with full pivoting scheme may produce a worse result than would be obtained with scaled partial pivoting scheme. As a matter of factor, we cannot say that some pivoting scheme always yields better solution than other pivoting schemes, because the result depends on the random round-off error as well as the pivoting scheme (see Problem 2.2). But, in most cases, the scaled partial pivoting shows a reasonably good performance and that is why we adopt it in our function 'Gauss()'.

Remark 2.3 Computing Error, Singularity, and Ill-condition.

(1) As the size of the matrix grows, the round-off errors are apt to accumulate and propagated in matrix operations to such a degree that zero may appear to be an absolutely small number, or a nonzero number very close to zero may appear to be zero. Therefore, it is not a simple task to determine whether a zero or a number very close to zero is a real zero or not.

(2) It is desirable, but not so easy, for us to discern the case of singularity from the case of ill-condition, and distinguish the case of redundancy from the case of inconsistency. In order to be able to give such a qualitative judgment in the right way based on some quantitative analysis, we should be equipped with theoretical knowledge as well as practical experience.

(3) There are several criteria by which we judge the degree of ill-condition, such as how discrepant AA^{-1} is with the identity matrix, and how far $\det\{A\}\det\{A^{-1}\}$ stays away from one (1), etc.

$$AA^{-1} \stackrel{?}{=} I, \quad [A^{-1}]^{-1} \stackrel{?}{=} A, \quad \det(A)\det(A^{-1}) \stackrel{?}{=} 1 \quad (2.2.21)$$

96 SYSTEM OF LINEAR EQUATIONS

The MATLAB command 'cond()' tells us the degree of ill-condition for a given matrix by the size of the condition number, which is defined as

$$\text{cond}(A) = \|A\|\|A^{-1}\| \quad \text{with} \quad \|A\| = \text{largest eigenvalue of } A^T A,$$

i.e. largest singular value of A

Example 2.3. Hilbert Matrix – Ill-conditioned or Near-singular (Close to Singularity).
The Hilbert matrix defined by

$$A = [a_{mn}] = \left[\frac{1}{m+n-1}\right] \tag{E2.3.1}$$

is notorious for its ill-condition.

We increase the dimension of the Hilbert matrix from $N = 9$ to 12 and make use of the MATLAB commands 'cond()' and 'det()' to compute the condition number and $\det(A)\det(A^{-1})$ in the MATLAB script "do_condition.m". Especially for $N = 8$, we will see the degree of discrepancy between AA^{-1} and the identity matrix. Note that the number RCOND following the warning message about near-singularity or ill-condition given by MATLAB is a reciprocal condition number, which can be computed by the 'rcond()' command and is supposed to get close to 1/0 for a well/badly conditioned matrix.

```
»do_condition
  AAI =
    1.0000   -0.0000   -0.0005   -0.0160   -0.0999   -0.1455   -0.0392   -0.0000
   -0.0000    1.0000   -0.0004   -0.0124   -0.0827   -0.1257   -0.0348   -0.0000
   -0.0000   -0.0000    0.9997   -0.0101   -0.0704   -0.1105   -0.0313   -0.0000
   -0.0000   -0.0000   -0.0002    0.9915   -0.0613   -0.0986   -0.0285   -0.0000
   -0.0000   -0.0000   -0.0002   -0.0073    0.9458   -0.0890   -0.0261   -0.0000
   -0.0000   -0.0000   -0.0002   -0.0064   -0.0486    0.9190   -0.0241   -0.0000
   -0.0000   -0.0000   -0.0001   -0.0057   -0.0440   -0.0744    0.9776    0.0000
   -0.0000   -0.0000   -0.0001   -0.0052   -0.0402   -0.0687   -0.0209    1.0000
  N= 8: cond(A)=1.525758e+10,  det(A)det(A^-1)=0.833858
  N= 9: cond(A)=4.931537e+11,  det(A)det(A^-1)=-6.789373
  N=10: cond(A)=1.602520e+13,  det(A)det(A^-1)=-124287068.968264
  N=11: cond(A)=5.229057e+14,  det(A)det(A^-1)=74092582851351.766000
     Warning: Matrix is close to singular or badly scaled.
              Results may be inaccurate. RCOND = 2.635296e-17.
     > In do_condition.m at line 12
  N=12: cond(A)=1.633018e+16,
        det(A)det(A^-1)=-308335072702678510000000000000000.000000
```

These results show that the condition number increases as the numerical values of AA^{-1} and $\det(A)\det(A^{-1})$ deviate from their theoretical values that are identity matrices of the same size and 1, respectively, for any square matrix A.

```
%do_condition.m
for m=1:8
   for n=1:8, A(m,n)=1/(m+n-1); end % A=hilb(6), Eq.(E2.3)
end
AAI=A*A^-1
c=cond(A);  d=det(A)*det(A^-1);
fprintf('N=%2d: cond(A)=%e,  det(A)det(A^-1)=%8.6f\n', m,c,d);
for N=9:12
   for m=1:N,    A(m,N) =1/(m+N-1);  end
   for n=1:N-1,  A(N,n) =1/(N+n-1);  end
   c =cond(A);  d=det(A)*det(A^-1);
   fprintf('N=%2d: cond(A)=%e,  det(A)det(A^-1)=%8.6f\n', N,c,d);
end
```

2.2.3 Gauss-Jordan Elimination

While Gauss elimination consists of forward elimination and backward substitution as explained in Section 2.2.1, Gauss-Jordan elimination consists of forward/backward elimination, which makes the coefficient matrix A an identity matrix so that the resulting RHS vector will appear as the solution.

For simplicity, we start from the triangular matrix-vector equation (2.2.4) obtained by applying the forward elimination:

$$\begin{bmatrix} a_{11}^{(0)} & a_{12}^{(0)} & a_{13}^{(0)} & b_1^{(0)} \\ 0 & a_{22}^{(1)} & a_{23}^{(1)} & b_2^{(1)} \\ 0 & 0 & a_{33}^{(2)} & b_3^{(2)} \end{bmatrix} \quad (2.2.22)$$

First, we divide the last row by $a_{33}^{(2)}$

$$\begin{bmatrix} a_{11}^{(0)} & a_{12}^{(0)} & a_{13}^{(0)} & b_1^{(0)} \\ 0 & a_{22}^{(1)} & a_{23}^{(1)} & b_2^{(1)} \\ 0 & 0 & a_{33}^{[1]}=1 & b_3^{[1]}=b_3^{(2)}/a_{33}^{(2)} \end{bmatrix} \quad (2.2.23)$$

and subtract (the third row $\times a_{m3}^{(m-1)}$, $m=1,2$) from the above two rows to get

$$\begin{bmatrix} a_{11}^{(0)} & a_{12}^{(0)} & a_{13}^{[1]}=0 & b_1^{[1]}=b_1^{(0)}-a_{13}^{(0)}b_3^{[1]} \\ 0 & a_{22}^{(1)} & a_{23}^{[1]}=0 & b_2^{[1]}=b_2^{(1)}-a_{23}^{(1)}b_3^{[1]} \\ 0 & 0 & a_{33}^{[1]}=1 & b_3^{[1]} \end{bmatrix} \quad (2.2.24)$$

98 SYSTEM OF LINEAR EQUATIONS

Now, we divide the second row by $a_{22}^{(1)}$

$$\begin{bmatrix} a_{11}^{(0)} & a_{12}^{(0)} & 0 & b_1^{[1]} \\ 0 & a_{22}^{[2]} = 1 & 0 & b_2^{[2]} = b_2^{[1]}/a_{22}^{(1)} \\ 0 & 0 & a_{33}^{[1]} = 1 & b_3^{[1]} \end{bmatrix} \qquad (2.2.25)$$

and subtract (the second row $\times a_{m2}^{(m-1)}, m = 1$) from the above first row to get

$$\begin{bmatrix} a_{11}^{(0)} & 0 & 0 & b_1^{[2]} = b_1^{[1]} - a_{12}^{(0)} b_2^{[2]} \\ 0 & 1 & 0 & b_2^{[2]} \\ 0 & 0 & 1 & b_3^{[1]} \end{bmatrix} \qquad (2.2.26)$$

Lastly, we divide the first row by $a_{11}^{(0)}$ to get

$$\begin{bmatrix} 1 & 0 & 0 & b_1^{[3]} = b_1^{[2]}/a_{11}^{(0)} \\ 0 & 1 & 0 & b_2^{[2]} \\ 0 & 0 & 1 & b_3^{[1]} \end{bmatrix} \qquad (2.2.27)$$

which denotes a system of linear equations having an identity matrix as the coefficient matrix and consequently, take the $(N+1)$th column vector (i.e. virtually the RHS vector) as the final solution.

$$\begin{bmatrix} x_1 & x_2 & x_3 \end{bmatrix} = \begin{bmatrix} b_1^{[3]} & b_2^{[2]} & b_3^{[1]} \end{bmatrix} \qquad (2.2.28)$$

Note that we do not have to distinguish the two steps, the forward/backward elimination. In other words, during the forward elimination, we do EROs in such a way that the pivot becomes one and other elements above/below the pivot in the same column become zeros.

Consider the following system of linear equations:

$$\begin{bmatrix} -1 & -2 & 2 \\ 1 & 1 & -1 \\ 1 & 2 & -1 \end{bmatrix} \begin{bmatrix} x_1 \\ x_2 \\ x_3 \end{bmatrix} = \begin{bmatrix} -1 \\ 1 \\ 2 \end{bmatrix} \qquad (2.2.29)$$

We construct the augmented matrix by combining the coefficient matrix and the RHS vector as

SOLVING A SYSTEM OF LINEAR EQUATIONS   99

$$\begin{bmatrix} a_{11} & a_{12} & a_{13} & b_1 \\ a_{21} & a_{22} & a_{23} & b_2 \\ a_{31} & a_{32} & a_{33} & b_3 \end{bmatrix} = \begin{bmatrix} -1 & -2 & 2 & -1 \\ 1 & 1 & -1 & 1 \\ 1 & 2 & -1 & 2 \end{bmatrix} \begin{matrix} : r_1 \\ : r_2 \\ : r_3 \end{matrix} \quad (2.2.30)$$

and apply Gauss-Jordan elimination procedure.

First, we divide the first row r_1 by $a_{11} = -1$ to make the new first row $r_1^{(1)}$ have the pivot $a_{11}^{(1)} = 1$ and subtract $r_1^{(1)} \times a_{m1}(m = 2, 3)$ from the second/third row r_2/r_3 to write

$$\begin{matrix} r_1 \div (-1) \to \\ r_2 - 1 \times r_1^{(1)} \to \\ r_3 - 1 \times r_1^{(1)} \to \end{matrix} \begin{bmatrix} a_{11}^{(1)} & a_{12}^{(1)} & a_{13}^{(1)} & b_1^{(1)} \\ a_{21}^{(1)} & a_{22}^{(1)} & a_{23}^{(1)} & b_2^{(1)} \\ a_{31}^{(1)} & a_{32}^{(1)} & a_{33}^{(1)} & b_3^{(1)} \end{bmatrix} = \begin{bmatrix} 1 & 2 & -2 & 1 \\ 0 & -1 & 1 & 0 \\ 0 & 0 & 1 & 1 \end{bmatrix} \begin{matrix} : r_1^{(1)} \\ : r_2^{(1)} \\ : r_3^{(1)} \end{matrix} \quad (2.2.31a)$$

Then, we divide the second row $r_2^{(1)}$ by $a_{22}^{(1)} = -1$ to make the new second row $r_2^{(2)}$ have the pivot $a_{22}^{(2)} = 1$ and subtract $r_2^{(2)} \times a_{m2}^{(1)}(m = 1, 3)$ from the first/third row $r_1^{(1)}/r_3^{(1)}$ to get

$$\begin{matrix} r_1^{(1)} - 2 \times r_2^{(2)} \to \\ r_2^{(1)} \div (-1) \to \\ r_3^{(1)} - 0 \times r_2^{(2)} \to \end{matrix} \begin{bmatrix} a_{11}^{(2)} & a_{12}^{(2)} & a_{13}^{(2)} & b_1^{(2)} \\ a_{21}^{(2)} & a_{22}^{(2)} & a_{23}^{(2)} & b_2^{(2)} \\ a_{31}^{(2)} & a_{32}^{(2)} & a_{33}^{(2)} & b_3^{(2)} \end{bmatrix} = \begin{bmatrix} 1 & 0 & 0 & 1 \\ 0 & 1 & -1 & 0 \\ 0 & 0 & 1 & 1 \end{bmatrix} \begin{matrix} : r_1^{(2)} \\ : r_2^{(2)} \\ : r_3^{(2)} \end{matrix} \quad (2.2.31b)$$

Lastly, we divide the third row $r_3^{(2)}$ by $a_{33}^{(2)} = 1$ to make the new third row $r_3^{(3)}$ have the pivot $a_{33}^{(3)} = 1$ and subtract $r_3^{(3)} \times a_{m3}^{(2)}(m = 1, 2)$ from the first/second row $r_1^{(2)}/r_2^{(2)}$ to get

$$\begin{matrix} r_1^{(2)} - 0 \times r_3^{(3)} \\ r_2^{(2)} - (-1) \times r_3^{(3)} \\ r_3^{(2)} \end{matrix} \to \begin{bmatrix} a_{11}^{(3)} & a_{12}^{(3)} & a_{13}^{(3)} & b_1^{(3)} \\ a_{21}^{(3)} & a_{22}^{(3)} & a_{23}^{(3)} & b_2^{(3)} \\ a_{31}^{(3)} & a_{32}^{(3)} & a_{33}^{(3)} & b_3^{(3)} \end{bmatrix} = \begin{bmatrix} 1 & 0 & 0 & 1 = x_1 \\ 0 & 1 & 0 & 1 = x_2 \\ 0 & 1 & 1 & 1 = x_3 \end{bmatrix} \begin{matrix} : r_1^{(3)} \\ : r_2^{(3)} \\ : r_3^{(3)} \end{matrix}$$
(2.2.31c)

After having the identity matrix-vector form like this, we take the $(N + 1)$th column vector (i.e. virtually the RHS vector) as the solution.

The general formula applicable for Gauss-Jordan elimination is the same as Eq. (2.2.6), except that the index set is $m \neq k$, i.e. all the numbers from $m = 1$ to $m = M$ except $m = k$. Interested readers are recommended to make their own functions to implement this algorithm (see Problem 2.3).

2.3 INVERSE MATRIX

In the previous section, we looked over some algorithms to solve a system of linear equations. We can use such algorithms to solve several systems of linear equations having the same coefficient matrix

$$A\mathbf{x}_1 = \mathbf{b}_1, \quad A\mathbf{x}_2 = \mathbf{b}_2, \quad \ldots, \quad A\mathbf{x}_{N_B} = \mathbf{b}_{N_B}$$

by putting different RHS vectors into one RHS matrix as

$$A\begin{bmatrix}\mathbf{x}_1 & \mathbf{x}_2 & \cdots & \mathbf{x}_{N_B}\end{bmatrix} = \begin{bmatrix}\mathbf{b}_1 & \mathbf{b}_2 & \cdots & \mathbf{b}_{N_B}\end{bmatrix}; \quad AX = B;$$

$$X = A^{-1}B \tag{2.3.1}$$

If we substitute an identity matrix I for B into this equation, we will get the matrix inverse $X = A^{-1}I = A^{-1}$. We, however, usually use the MATLAB command 'inv(A)' or 'A^-1' to compute the inverse of a matrix A.

2.4 DECOMPOSITION (FACTORIZATION)

2.4.1 *LU* Decomposition (Factorization) – Triangularization

LU decomposition (factorization) of a nonsingular (square) matrix A means expressing the matrix as the multiplication of a lower triangular matrix L and an upper triangular matrix U, where a lower/upper triangular matrix is a matrix having no nonzero elements above/below the diagonal. For the case where some row switching operation is needed like in the Gauss elimination, we include a permutation matrix P representing the necessary row switching operation(s) to write the *LU* decomposition as

$$PA = LU \tag{2.4.1}$$

The usage of a permutation matrix is exemplified by

$$PA = \begin{bmatrix}0 & 0 & 1\\1 & 0 & 0\\0 & 1 & 0\end{bmatrix}\begin{bmatrix}a_{11} & a_{12} & a_{13}\\a_{21} & a_{22} & a_{23}\\a_{31} & a_{32} & a_{33}\end{bmatrix} = \begin{bmatrix}a_{31} & a_{32} & a_{33}\\a_{11} & a_{12} & a_{13}\\a_{21} & a_{22} & a_{23}\end{bmatrix} \tag{2.4.2}$$

which denotes switching the first and third rows followed by switching the second and third rows. An interesting and useful property of the permutation matrix is that its transpose agrees with its inverse.

$$P^{\mathrm{T}}P = I; \quad P^{\mathrm{T}} = P^{-1} \tag{2.4.3}$$

DECOMPOSITION (FACTORIZATION)

To take a close look at the LU decomposition, we consider a 3×3 nonsingular matrix

$$\begin{bmatrix} a_{11} & a_{12} & a_{13} \\ a_{21} & a_{22} & a_{23} \\ a_{31} & a_{32} & a_{33} \end{bmatrix} = \begin{bmatrix} 1 & 0 & 0 \\ l_{21} & 1 & 0 \\ l_{31} & l_{32} & 1 \end{bmatrix} \begin{bmatrix} u_{11} & u_{12} & u_{13} \\ 0 & u_{22} & u_{23} \\ 0 & 0 & u_{33} \end{bmatrix};$$

$$\begin{bmatrix} a_{11} & a_{12} & a_{13} \\ a_{21} & a_{22} & a_{23} \\ a_{31} & a_{32} & a_{33} \end{bmatrix} = \begin{bmatrix} u_{11} & u_{12} & u_{13} \\ l_{21}u_{11} & l_{21}u_{12} + u_{22} & l_{21}u_{13} + u_{23} \\ l_{31}u_{11} & l_{31}u_{12} + l_{32}u_{22} & l_{31}u_{13} + l_{32}u_{23} + u_{33} \end{bmatrix} \quad (2.4.4)$$

First, equating the first rows of both sides yields

$$u_{1n} = a_{1n}, \quad n = 1, 2, 3 \quad (2.4.5a)$$

Then, equating the second rows of both sides yields

$$a_{21} = l_{21}u_{11}, \quad a_{22} = l_{21}u_{12} + u_{22}, \quad a_{23} = l_{21}u_{13} + u_{23}$$

from which we can get

$$l_{21} = a_{21}/u_{11}, \quad u_{22} = a_{21} - l_{21}u_{12}, \quad u_{23} = a_{23} - l_{21}u_{13} \quad (2.4.5b)$$

Now, equating the third rows of both sides yields

$$a_{31} = l_{31}u_{11}, \quad a_{32} = l_{31}u_{12} + l_{32}u_{22}, \quad a_{33} = l_{31}u_{13} + u_{32}u_{23} + u_{33}$$

from which we can get

$$l_{31} = a_{31}/u_{11}, \quad l_{32} = (a_{32} - l_{31}u_{12})/u_{22}, \quad u_{33} = (a_{33} - l_{31}u_{13}) - l_{32}u_{23} \quad (2.4.5c)$$

In order to put these formulas in one framework to generalize them for matrices having dimension greater than 3, we split this procedure into two steps and write the intermediate lower/upper triangular matrices into one matrix for compactness as

$$\text{Step 1}: \begin{bmatrix} a_{11} & a_{12} & a_{13} \\ a_{21} & a_{22} & a_{23} \\ a_{31} & a_{32} & a_{33} \end{bmatrix} \rightarrow \begin{bmatrix} u_{11} = a_{11} & u_{12} = a_{12} & u_{13} = a_{13} \\ l_{21} = a_{21}/u_{11} & a_{22}^{(1)} = a_{22} - l_{21}u_{12} & a_{23}^{(1)} = a_{23} - l_{21}u_{13} \\ l_{31} = a_{31}/u_{11} & a_{32}^{(1)} = a_{32} - l_{31}u_{12} & a_{33}^{(1)} = a_{33} - l_{31}u_{13} \end{bmatrix}$$

$$(2.4.6a)$$

$$\text{Step 2}: \rightarrow \begin{bmatrix} u_{11} & u_{12} & u_{13} \\ l_{21} & u_{22} = a_{22}^{(1)} & u_{23} = a_{23}^{(1)} \\ l_{31} & l_{32} = a_{32}^{(1)}/u_{22} & a_{33}^{(2)} = a_{33}^{(1)} - l_{32}u_{23} \end{bmatrix} \quad (2.4.6b)$$

102 SYSTEM OF LINEAR EQUATIONS

This leads to an *LU* decomposition algorithm generalized for an $NA \times NA$ nonsingular matrix as described in the following box. The MATLAB function '`lu_dcmp()`' (listed below) implements this algorithm to find not only the lower/upper triangular matrix L and U but also the permutation matrix P. We run it for a 3×3 matrix to get L, U, and P and then reconstruct the matrix $P^{-1}LU = A$ from L, U, and P to ascertain whether the result is right.

(cf) The number of floating-number multiplications required in this function '`lu_dcmp()`' is

$$\sum_{k=1}^{N_A-1}(N_A - k)(N_A - k + 1) = \sum_{k=1}^{N_A-1}\{N_A(N_A + 1) - (2N_A + 1)k + k^2\}$$

$$= (N_A - 1)N_A(N_A + 1) - \frac{1}{2}(2N_A + 1)(N_A - 1)N_A + \frac{1}{6}(N_A - 1)N_A(2N_A - 1)$$

$$= \frac{1}{3}(N_A - 1)N_A(N_A + 1) \approx \frac{1}{3}N_A^3 \quad \text{with } N_A: \text{the size of matrix } A \qquad (2.4.7)$$

```
»A=[1 2 5;0.2 1.6 7.4; 0.5 4 8.5];
»[L,U,P]=lu_dcmp(A)  %LU decomposition
   L = 1.0   0    0     U = 1   2   5     P = 1   0   0
       0.5   1.0  0         0   3   6         0   0   1
       0.2   0.4  1.0       0   0   4         0   1   0
»P'*L*U-A  %check the validity of the result (P'=P^-1)
   ans =  0   0   0
          0   0   0
          0   0   0
»[L,U,P]=lu(A) %for comparison with the MATLAB built-in function
```

---

0. Initialize $A^{(0)} = A$, or equivalently, $a_{mn}^{(0)} = a_{mn}$ for $m = 1, \ldots, N_A$ and for $n = 1, \ldots, N_A$
1. Let $k = 1$.
2. If $a_{kk}^{(k-1)} = 0$, do an appropriate row switching operation so that $a_{kk}^{(k-1)} \neq 0$. When it is not possible, then declare the case of singularity and stop.
3. $a_{kn}^{(k)} = a_{kn}^{(k-1)} = u_{kn}$ for $n = k, \ldots, N_A$ (Just leave the kth row as it is.)

   (2.4.8a)

   $a_{mk}^{(k)} = a_{mk}^{(k-1)}/a_{kk}^{(k-1)} = l_{mk}$ for $m = k+1, \ldots, N_A$ (2.4.8b)

4. $a_{mn}^{(k)} = a_{mn}^{(k-1)} - a_{mk}^{(k)}a_{kn}^{(k)}$ for $m = k+1, \ldots, N_A$ and $n = k+1, \ldots, N_A$

   (2.4.9)

5. Increment k by 1 and if $k < N_A - 1$, go to step 1; otherwise, go to step 6.
6. Set the part of the matrix $A^{(N_A-1)}$ below the diagonal to L (lower triangular matrix with the diagonal of 1's) and the part on and above the diagonal to U (upper triangular matrix).

What is the *LU* decomposition for? It can be used for solving a system of linear equations as

$$A\mathbf{x} = \mathbf{b} \qquad (2.4.10)$$

Once we have the *LU* decomposition of the coefficient matrix $A = P^T LU$, it is more efficient to use the lower/upper triangular matrices for solving Eq. (2.4.10) than to apply the Gauss elimination method. The procedure is as follows:

$$P^T LU\mathbf{x} = \mathbf{b}; \quad LU\mathbf{x} = P\mathbf{b}; \quad U\mathbf{x} = L^{-1}P\mathbf{b}; \quad \mathbf{x} = U^{-1}L^{-1}P\mathbf{b} \qquad (2.4.11)$$

```
function [L,U,P]=lu_dcmp(A)
%This gives LU decomposition of A with the permutation matrix P
%   denoting the row switch(exchange) during factorization
NA=size(A,1);
AP=[A eye(NA)]; %augment with the permutation matrix.
for k=1:NA-1
  %Partial Pivoting at AP(k,k)
  [akx,kx] =max(abs(AP(k:NA,k)));
  if akx<eps
    error('Singular matrix and No LU decomposition')
  end
  mx =k+kx-1;
  if kx>1 % Row change if necessary
    tmp_row =AP(k,:);
    AP(k,:) =AP(mx,:);
    AP(mx,:) =tmp_row;
  end
  % LU decomposition
  for m=k+1: NA
    AP(m,k) =AP(m,k)/AP(k,k); % Eq.(2.4.8b)
    AP(m,k+1:NA) =AP(m,k+1:NA)-AP(m,k)*AP(k,k+1:NA); % Eq.(2.4.9)
  end
end
P =AP(1:NA, NA+1:NA+NA); % Permutation matrix
for m=1:NA
  for n=1:NA
    if m==n,  L(m,m)=1.;   U(m,m)=AP(m,m);
      elseif m>n, L(m,n)=AP(m,n);  U(m,n) =0.;
      else L(m,n)=0.; U(m,n)=AP(m,n);
    end
  end
end
if nargout==0
  disp('L*U=P*A with'); L,U,P,
end
%You can check if P'*L*U=A?
```

104 SYSTEM OF LINEAR EQUATIONS

Table 2.1 Residual error and the number of floating-point operations of various solutions.

	tmp=forsubst(L,P*b) backsubst(U,tmp)	Gauss(A,b)	A\b	A^-1*b
$\|Ax_i - b\|$	1.3597e−016	5.5511e−017	1.7554e−016	3.0935e−012
Number of flops	123	224	155	50

The numbers of flops for the *LU* decomposition and the inverse of the matrix *A* are not counted. Note that the command 'flops' to count the number of floating-point operations is no longer available in MATLAB 6.x and higher versions.

Note that the premultiplication of a vector by L^{-1} and U^{-1} can be performed by the forward and backward substitutions, respectively. The following script "do_lu_dcmp.m" applies the *LU* decomposition method, Gauss elimination algorithm as well as the MATLAB operators '\' and 'inv' or '^-1' to solve Eq. (2.4.10), where *A* is the five-dimensional Hilbert matrix (introduced in Example 2.3) and $b = Ax^o$ with $x^o = \begin{bmatrix} 1 & 1 & 1 & 1 & 1 \end{bmatrix}^T$. The residual error $\|Ax_i - b\|$ of the solutions obtained by the four methods and the numbers of floating-point operations required for carrying out them are listed in Table 2.1. The table shows that, once the inverse matrix A^{-1} is available, the inverse matrix method requiring only N^2 multiplications/additions (*N*: the dimension of the coefficient matrix or the number of unknown variables) is the most efficient in computation, but the worst in accuracy. Therefore, if we need to continually solve the system of linear equations with the same coefficient matrix *A* for different RHS vectors, it is a reasonable choice in terms of computation time and accuracy to save the *LU* decomposition of the coefficient matrix *A* and apply the forward/backward substitution process.

```
%do_lu_dcmp.m
% Use LU decomposition, Gauss elimination to solve Ax=b
A=hilb(5);
[L,U,P]=lu_dcmp(A); %LU decomposition
x=[1 -2 3 -4 5 -6 7 -8 9 -10].';
b=A*x(1:size(A,1));
x_lu=backsubst(U,forsubst(L,P*b)); % Eq.(2.4.11)
x_Gauss=Gauss(A,b);
x_bs=A\b;
AI=A^-1; x_iv=AI*b;
% assuming that we have already got the inverse matrix
disp('       x_lu         x_Gauss        x_bs         x_iv')
format short e
solutions=[x_lu x_Gauss x_bs x_iv]
errs=[norm(A*x_lu-b) norm(A*x_Gauss-b) norm(A*x_bs-b) norm(A*x_iv-b)]
format short
```

```
function x=forsubst(L,B)
%forward substitution for a lower-triangular matrix equation Lx=B
N= min(size(L));
x(1,:)= B(1,:)/L(1,1);
for m=2:N
   x(m,:)= (B(m,:)-L(m,1:m-1)*x(1:m-1,:))/L(m,m);
end
```

```
function x=backsubst(U,B)
%backward substitution for a upper-triangular matrix equation Ux=B
N= min(size(U));
x(N,:)= B(N,:)/U(N,N);
for m=N-1: -1:1
   x(m,:)= (B(m,:)-U(m,m+1:N)*x(m+1:N,:))/U(m,m);
end
```

2.4.2 Other Decomposition (Factorization) – Cholesky, QR and SVD

There are several other matrix decompositions such as Cholesky decomposition, QR decomposition and singular value decomposition (SVD). Instead of looking into the details of these algorithms, we will simply survey the MATLAB built-in functions implementing these decompositions.

Cholesky decomposition factors a positive definite symmetric/Hermitian matrix into an upper triangular matrix premultiplied by its transpose as

$$A = U^H U = U^{*T} U \quad (U: \text{an upper triangular matrix}) \quad (2.4.12)$$

and is implemented by the MATLAB built-in function 'chol()'.

- (cf) If a (complex-valued) matrix A satisfies $A^H \triangleq A^{*T} = A$, i.e. the conjugate transpose of a matrix equals itself, it is said to be *Hermitian*. It is said to be just symmetric in the case of a real-valued matrix with $A^T = A$.
- (cf) If a square matrix A satisfies $\mathbf{x}^{*T} A \mathbf{x} > 0 \ \forall \mathbf{x} \neq \mathbf{0}$, the matrix is said to be positive definite (see Appendix B).

```
»A=[2 3 4;3 5 6;4 6 9]; % a positive definite symmetric matrix
»U=chol(A) %Cholesky decomposition
    U= 1.4142    2.1213    2.8284
       0         0.7071    0.0000
       0         0         1.0000
»U'*U-A % to check if the result is right
```

QR decomposition is to express a square or a rectangular matrix as the product of an orthogonal (unitary) matrix Q and an upper triangular matrix R as

$$A = QR \quad (2.4.13)$$

where $Q^T Q = I$ ($Q^{*T} Q = I$). This is implemented by the MATLAB built-in function 'qr()'.

(cf) If all the columns of a (complex-valued) matrix A are orthonormal to each other so that $A^{*T}A = I$; $A^{*T} = A^{-1}$, it is said to be unitary. It is said to be orthogonal for a real-valued matrix with $A^T = A^{-1}$.

SVD is to express an $M \times N$ matrix A in the following form

$$A = USV^H \qquad (2.4.14)$$

where U is an orthogonal (unitary) $M \times M$ matrix, V an orthogonal (unitary) $N \times N$ matrix, and S a real diagonal $M \times N$ matrix having the singular values (s_{mm},s) of A (the square roots of the eigenvalues of $A^H A$) in decreasing order on its diagonal. Note that we have

$$A^H AV = (USV^H)^H USV^H V = VS^H U^H US = V(S^H S) = V[s^2_{mm}], \qquad (2.4.15a)$$

$$AA^H U = USV^H (USV^H)^H U = USV^H VS^H = U(SS^H) = U[s^2_{mm}], \qquad (2.4.15b)$$

which implies that s^2_{mm}s are the common eigenvalues of $A^H A$ and AA^H and that the columns of V and U are the eigenvectors of $A^H A$ and AA^H, respectively. This is implemented by the MATLAB built-in function 'svd()'.

```
»A=[1 2;2 3;3 5]; %a rectangular matrix
»[U,S,V]=svd(A) %Singular Value Decomposition
U= -0.3092   0.7557  -0.5774    S= 7.2071   0         V= 0.5184  -0.8552
   -0.4998  -0.6456  -0.5774       0        0.2403       0.8552   0.5184
   -0.8090   0.1100   0.5774       0        0
»err=U*S*V'-A %to check if the result is right
  err= 1.0e-015*  -0.2220  -0.2220
                   0        0
                   0.4441   0
```

Like the LU decomposition, the QR factorization, and SVD can also be used for solving a system of linear equations as

$$Ax = b \qquad (2.4.16)$$

With the QR (unitary-triangular) decomposition of the coefficient matrix $A = QRP^T$, we can solve this equation as follows:

$$QRP^T x = b; \quad RP^T x = Q^H b; \quad P^T x = R^{-1} Q^H b; \quad x = PR^{-1} Q^H b \qquad (2.4.17)$$

where P is the permutation matrix (required, say, to make the diagonal elements of R arranged in a descending order) and the premultiplication of R^{-1} by a vector can be performed by the backward substitution. Note that the LSE solution for an overdetermined set of linear equations is easy to get with the QR decomposition since

DECOMPOSITION (FACTORIZATION)

$$\mathbf{x} \stackrel{(2.1.10)}{=} (A^H A)^{-1} A^H \mathbf{b} = ((QRP^T)^H QRP^T)^{-1} (QRP^T)^H \mathbf{b}$$
$$= (PR^H Q^H QRP^T)^{-1} PR^H Q^H \mathbf{b}$$
$$= P(R^H R)^{-1} R^H Q^H \mathbf{b} \stackrel{R^H R \text{ is of full rank}}{=} PR^{-1} Q^H \mathbf{b} \quad (2.4.18)$$

With the SVD of the coefficient matrix $A = U S V^T$, we can solve the equation as follows:

$$U S V^H \mathbf{x} = \mathbf{b}; \quad S V^H \mathbf{x} = U^H \mathbf{b}; \quad V^H \mathbf{x} = S^{-1} U^H \mathbf{b}; \quad \mathbf{x} = V S^{-1} U^H \mathbf{b} \quad (2.4.19)$$

where the inverse matrix S^{-1} of the diagonal matrix $S = [s_{mm}]$ is $S^{-1} = [s_{mm}^{-1}]$, which can be obtained by just taking the reciprocal of each diagonal element if only it is not a (virtual) zero. Note that the SVD can be used to find a minimum-norm solution to a (possibly) rank-deficient least squares problem with rank$(A) < \min(M, N)$ (M,N: the numbers of rows/columns of A), which minimizes both $\|\mathbf{x}\|$ and $\|A\mathbf{x}-\mathbf{b}\|$. The minimum-norm least squares solution can be found as

$$\mathbf{x} = V(:, 1:K) S(1:K, 1:K)^{-1} U(:, 1:K)^H \mathbf{b} \quad \text{with } K = \text{rank}(A) \quad (2.4.20)$$

Let us see the following example:

Example 2.4. QR Decomposition for Solving a Set of Overdetermined Linear Equations.
Consider the following sets of linear equations:

(a) $\begin{bmatrix} 2 & -1 \\ 1 & -5 \\ -3 & 6 \end{bmatrix} \begin{bmatrix} x_1 \\ x_2 \end{bmatrix} = \begin{bmatrix} 5 \\ 10 \\ -14 \end{bmatrix}$ (b) $\begin{bmatrix} 2 & -1 & 1 \\ 1 & -5 & 2 \\ -3 & 6 & -3 \end{bmatrix} \begin{bmatrix} x_1 \\ x_2 \\ x_3 \end{bmatrix} = \begin{bmatrix} 7 \\ 17 \\ -24 \end{bmatrix}$ (c) $\begin{bmatrix} 2 & 1 & -3 \\ -1 & -5 & 1 \end{bmatrix} \begin{bmatrix} x_1 \\ x_2 \\ x_3 \end{bmatrix} = \begin{bmatrix} -9 \\ 12 \end{bmatrix}$

rank$(A) = 2 = \min(M, N)$ rank$(A) = 2 < 3 = \min(M, N)$ rank$(A) = 2 = \min(M, N)$

(E2.4.1)

Among the aforementioned three sets of equations, the coefficient matrices of Eq. (E2.4.1b) is of deficient rank, while those of Eqs. (E2.4.1a,c) are of full rank. Equation (E2.4.1a) with rank$([Ab])$ = rank$(A) + 1$ is inconsistent and we should find the LSE solution for it. Equation (E2.4.1b) with rank$([Ab])$ = rank(A) = 2 is redundant and it has infinitely many solutions. Equation (E2.4.1c) has also infinitely many solutions since the number of unknowns is greater than that of equations. The script "nm02e04.m" solves each of these sets of equations by using the QR method, the SVD method, and the backslash operator, yielding the solutions listed in the following table. This result shows the robustness of the SVD for finding the LSE or minimum-norm solution irrespective of the rank deficiency of A.

```
%nm02e04.m
EPS=1e-12; format short e
for k=1:3
   if k==1, A=[2 -1; 1 -5; -3 6]; b =[5; 10; -14]; % Eq.(E2.4.1a)
   elseif k==2, A=[2 -1 1;1 -5 2; -3 6 -3]; b=[7; 17;-24]; %Eq.(E2.4.1b)
   else A= [2 1 -3; -1 -5 1];  b= [-9; 12]; % Eq.(E2.4.1c)
   end
   [Q,R,P] = qr(A); x_qr = backsubst(R,Q'*b); % permutated QR solution
   [M,N] = size(A);
   if N>M
     x_qr = [x_qr; zeros(N-M,1)];
   end % possibly augmented
   x_qr = P*x_qr; % depermutated QR solution (2.4.17)
   [U,S,V] = svd(A); eig2 = diag(S); % SVD decomposition
   id = find(eig2>=EPS); eig2 = eig2(id) ;
   K = length(eig2); % the rank of A possibly rank-deficient
   x_svd = V(:,1:K)*diag(1./eig2)*U(:,1:K)'*b; % SVD solution (2.4.20)
   x_bs = A\b; % solution using the backslash operator
   disp('      x_qr          x_svd         x_bs')
   solutions=[x_qr x_svd x_bs]
   norms= [norm(x_qr) norm(x_svd) norm(x_bs)]
   errs= [norm(A*x_qr-b) norm(A*x_svd-b) norm(A*x_bs-b)]
end
format short
```

Equation		(a)	(b)	(c)
QR solution, Eq. (2.4.17)	x	[1.5185 −1.6296]	[9.515 −10.52 −22.54]	[0 −1.929 2.357]
	‖x‖	2.2275	26.633	3.0456
	‖Ax − b‖	0.5774	3.6231×10^{-14}	8.8818×10^{-15}
SVD solution, Eq. (2.4.20)	x	[1.5185 −1.6296]	[1.546 −2.546 1.364]	[−1.166 −1.845 1.608]
	‖x‖	2.2275	3.2753	2.7109
	‖Ax − b‖	0.5774	6.4047×10^{-15}	3.5527×10^{-15}
Backslash solution, $x = A\backslash b$	x	[1.5185 −1.6296]	[2.000 −3.000 0]	[0 −1.929 2.357]
	‖x‖	2.2275	3.6056	3.0456
	‖Ax − b‖	0.5774	0	7.5364×10^{-15}

2.5 ITERATIVE METHODS TO SOLVE EQUATIONS

2.5.1 Jacobi Iteration

Let us consider the following equation:

$$3x + 1 = 0$$

which can be cast into an iterative scheme as

$$2x = -x - 1; \quad x = -\frac{x+1}{2} \rightarrow x_{k+1} = -\frac{1}{2}x_k - \frac{1}{2}$$

ITERATIVE METHODS TO SOLVE EQUATIONS   109

Starting from some initial value x_0 for $k = 0$, we can incrementally change k by 1 each time to proceed as beneath:

$$x_1 = -2^{-1} - 2^{-1}x_0$$
$$x_2 = -2^{-1} - 2^{-1}x_1 = -2^{-1} + 2^{-2} + 2^{-2}x_0$$
$$x_3 = -2^{-1} - 2^{-1}x_2 = -2^{-1} + 2^{-2} - 2^{-3} - 2^{-3}x_0$$
$$\vdots$$

Whatever the initial value x_0 is, this process will converge to the sum of a geometric series with the ratio of $(-1/2)$ as

$$x_k = \frac{a_0}{1-r} = \frac{-1/2}{1-(-1/2)} = -\frac{1}{3} = x^o \quad \text{as } k \to \infty$$

and what is better, the limit is the very true solution to the given equation. We are happy with this, but might feel uneasy, because we are afraid that this convergence to the true solution is just a coincidence. Will it always converge, no matter how we modify the equation so that only x remains on the LHS?

To answer this question, let us try another iterative scheme.

$$x = -2x - 1 \to x_{k+1} = -2x_k - 1;$$
$$x_1 = -1 - 2x_0$$
$$x_2 = -1 - 2x_1 = -1 - 2(-1 - 2x_0) = -1 + 2 + 2^2 x_0$$
$$x_3 = -1 - 2x_2 = -1 + 2 - 2^2 - 2^3 x_0$$
$$\vdots$$

This iteration will diverge regardless of the initial value x_0. But we are never disappointed, since we know that no one can be always lucky.

To understand the essential difference between these two cases, we should know the fixed-point theorem (Section 4.1). Apart from this, let us go into a system of equations.

$$\begin{bmatrix} 3 & 2 \\ 1 & 2 \end{bmatrix} \begin{bmatrix} x_1 \\ x_2 \end{bmatrix} = \begin{bmatrix} 1 \\ -1 \end{bmatrix}; \quad A\mathbf{x} = \mathbf{b}$$

Dividing the first equation by 3, transposing all term(s) other than x_1 to the RHS, dividing the second equation by 2, and transposing all term(s) other than x_2 to the RHS, we have

SYSTEM OF LINEAR EQUATIONS

$$\begin{bmatrix} x_{1,k+1} \\ x_{2,k+1} \end{bmatrix} = \begin{bmatrix} 0 & -2/3 \\ -1/2 & 0 \end{bmatrix} \begin{bmatrix} x_{1,k} \\ x_{2,k} \end{bmatrix} + \begin{bmatrix} 1/3 \\ -1/2 \end{bmatrix} \begin{bmatrix} x_{1,k+1} \\ x_{2,k+1} \end{bmatrix}$$

$$= \begin{bmatrix} 0 & -2/3 \\ -1/2 & 0 \end{bmatrix} \begin{bmatrix} x_{1,k} \\ x_{2,k} \end{bmatrix} + \begin{bmatrix} 1/3 \\ -1/2 \end{bmatrix};$$

$$\mathbf{x}_{k+1} = \widetilde{A}\,\mathbf{x}_k + \widetilde{\mathbf{b}} \qquad (2.5.1)$$

Assuming that this scheme works well, we set the initial value to zero ($\mathbf{x}_0 = 0$) and proceed as

$$\mathbf{x}_k \rightarrow [I + \widetilde{A} + \widetilde{A}^2 + \cdots]\widetilde{\mathbf{b}} = [I - A]^{-1}\widetilde{\mathbf{b}} = \begin{bmatrix} 1 & 2/3 \\ 1/2 & 1 \end{bmatrix}^{-1} \begin{bmatrix} 1/3 \\ -1/2 \end{bmatrix}$$

$$= \frac{1}{1 - 1/3} \begin{bmatrix} 1 & -2/3 \\ -1/2 & 1 \end{bmatrix} \begin{bmatrix} 1/3 \\ -1/2 \end{bmatrix} = \frac{1}{2/3} \begin{bmatrix} 2/3 \\ -2/3 \end{bmatrix} = \begin{bmatrix} 1 \\ -1 \end{bmatrix} = \mathbf{x}^o \qquad (2.5.2)$$

which will converge to the true solution $\mathbf{x}^o = \begin{bmatrix} 1 & -1 \end{bmatrix}^T$. This suggests another method of solving a system of equations, which is called Jacobi iteration. It can be generalized for an $N \times N$ matrix-vector equation as follows:

$$a_{m1}x_1 + a_{m2}x_2 + \cdots + a_{mm}x_m + \cdots + a_{mN}x_N = b_m;$$

$$x_m^{(k+1)} = -\sum_{n \neq m}^{N} \frac{a_{mn}}{a_{mm}} x_n^{(k)} + \frac{b_m}{a_{mm}} \qquad \text{for } m = 1, 2, \cdots, N \text{ and for each time stage } k;$$

$$\mathbf{x}_{k+1} = \widetilde{A}\mathbf{x}_k + \widetilde{\mathbf{b}} \qquad (2.5.3)$$

where

$$\widetilde{A}_{N \times N} = \begin{bmatrix} 0 & -a_{12}/a_{11} & \cdots & -a_{1N}/a_{11} \\ -a_{21}/a_{22} & 0 & \cdots & -a_{2N}/a_{22} \\ \vdots & \vdots & \cdots & \vdots \\ -a_{N1}/a_{NN} & -a_{N2}/a_{NN} & \cdots & 0 \end{bmatrix}, \quad \widetilde{\mathbf{b}} = \begin{bmatrix} b_1/a_{11} \\ b_2/a_{22} \\ \vdots \\ b_N/a_{NN} \end{bmatrix}$$

This scheme is implemented by the following MATLAB function 'jacobi()'. We can use it to solve the abovementioned equation by running the following statements:

```
»A=[3 2;1 2]; b=[1  -1]'; %the coefficient matrix and RHS vector
»x0=[0  0]'; %the initial value
»x=jacobi(A,b,x0,20) %to repeat 20 iterations starting from x0
    x= 1.0000
      -1.0000
»jacobi(A,b,x0,20) %omit output argument to see intermediate results
    X= 0.3333    0.6667    0.7778    0.8889    0.9259  ......
      -0.5000   -0.6667   -0.8333   -0.8889   -0.9444  ......
```

```
function X=jacobi(A,B,X0,kmax)
%This function finds a solution to Ax=B by Jacobi iteration.
if nargin<4, tol=1e-6; kmax=100; %called by jacobi(A,B,X0)
  elseif kmax<1, tol=max(kmax,1e-16); kmax=100; %jacobi(A,B,X0,tol)
  else tol=1e-6;  %jacobi(A,B,X0,kmax)
end
if nargin<3, X0=zeros(size(B)); end
NA=size(A,1); X=X0;  At=zeros(NA,NA);
for m=1:NA
   for n=1:NA
      if n~=m, At(m,n)=-A(m,n)/A(m,m); end
   end
   Bt(m,:)=B(m,:)/A(m,m);
end
for k=1:kmax
   X =At*X +Bt; % Eq.(2.5.3)
   if nargout==0, X, end % To see the intermediate results
   if norm(X-X0)/(norm(X0)+eps)<tol, break; end
   X0=X;
end
```

2.5.2 Gauss-Seidel Iteration

Let us take a close look at Eq. (2.5.1). Each iteration of Jacobi method updates the whole set of N variables at a time. However, so long as we do not use a multiprocessor computer capable of parallel processing, each one of N variables is updated sequentially one by one. Therefore, it is no wonder that we could speed up the convergence by using all the most recent values of variables for updating each variable even in the same iteration as follows:

$$x_{1,k+1} = -\frac{2}{3}x_{2,k} + \frac{1}{3}$$

$$x_{2,k+1} = -\frac{1}{2}x_{1,k+1} - \frac{1}{2}$$

This scheme is called Gauss-Seidel iteration, which can be generalized for an $N \times N$ matrix-vector equation as follows:

$$x_m^{(k+1)} = \frac{b_m - \sum_{n=1}^{m-1} a_{mn} x_n^{(k+1)} - \sum_{n=m+1}^{N} a_{mn} x_n^{(k)}}{a_{mm}} \qquad (2.5.4)$$

for $m = 1, \cdots, N$ and for each time stage k.

This computational scheme has been cast into the following MATLAB function 'Gauseid()', which we will use to solve the earlier mentioned equation as follows:

112 SYSTEM OF LINEAR EQUATIONS

```
function  X=Gauseid(A,B,X0,kmax)
%This function finds x=A^-1 B by Gauss-Seidel iteration.
if nargin<4, tol=1e-6; kmax=100;
  elseif kmax<1, tol=max(kmax,1e-16); kmax=1000; else tol=1e-6;
end
if nargin<3, X0=zeros(size(B)); end
NA=size(A,1); X =X0;
for k=1: kmax
   X(1,:) =(B(1,:)-A(1,2:NA)*X(2:NA,:))/A(1,1);
   for m=2:NA-1
      tmp =B(m,:)-A(m,1:m-1)*X(1:m-1,:)-A(m,m+1:NA)*X(m+1:NA,:);
      X(m,:) =tmp/A(m,m); %Eq. (2.5.4)
   end
   X(NA,:) =(B(NA,:)-A(NA,1:NA-1)*X(1:NA-1,:))/A(NA,NA);
   if nargout==0, X, end %To see the intermediate results
   if norm(X-X0)/(norm(X0)+eps)<tol, break; end
   X0=X;
end
```

```
»A=[3 2;1 2]; b=[1  -1]'; % the coefficient matrix and RHS vector
»x0=[0  0]'; % Initial value
»Gauseid(A,b,x0,10) % omit output argument to see intermediate results
  X= 0.3333    0.7778    0.9259    0.9753    0.9918   ......
     -0.6667   -0.8889   -0.9630   -0.9877   -0.9959  ......
```

As with the Jacobi iteration in the previous section, we can see this Gauss-Seidel iteration converging to the true solution $\mathbf{x}^o = [1 \ -1]^T$ and that with fewer iterations. But if we use a multiprocessor computer capable of parallel processing, the Jacobi iteration may be better in speed even with more iterations, since it can exploit the advantage of simultaneous parallel computation.

Note that the Jacobi and Gauss-Seidel iterative schemes seem unattractive and even unreasonable if we are given a standard form of linear equations as

$$A\mathbf{x} = \mathbf{b}$$

because the computational overhead for converting it into the form of Eq. (2.5.3) may be excessive. But it is not always the case, especially when the equations are given in the form of Eq. (2.5.3)/(2.5.4). In such a case, we simply repeat the iterations without having to use such ready-made functions as 'jacobi()' or 'Gauseid()'. Let us see the following example.

Example 2.5.   Jacobi or Gauss-Seidel Iterative Scheme.
Suppose the temperature of a metal rod of length 10 m has been measured to be 0 and 10 °C at each end, respectively. Find the temperatures x_1, x_2, x_3, and x_4 at the four points equally spaced with the interval of 2 m, assuming that the temperature at each point is the average of the temperatures of both neighboring points.

```
%nm02e05.m
N=4; % Number of unknown variables/equations
kmax=20; tol=1e-6;
At=[0 1 0 0; 1 0 1 0; 0 1 0 1; 0 0 1 0]/2;
x0=0; x5=10; % Boundary values
b=[x0/2 0 0 x5/2]'; % RHS vector
% initialize all the values to the average of boundary values
xp=ones(N,1)*(x0+x5)/2;
%Jacobi iteration
for k=1:kmax
    x= At*xp +b; % Eq.(2.5.3) with Eq.(E2.5)
    if norm(x-xp)/(norm(xp)+eps)<tol, break; end
    xp=x;
end
k, xj=x
%Gauss-Seidel iteration
xp=ones(N,1)*(x0+x5)/2; x=xp; %initial value
for k=1:kmax
    for n=1:N, x(n)= At(n,:)*x +b(n); end % Eq.(2.5.4) with Eq.(E2.5)
    if norm(x-xp)/(norm(xp)+eps)<tol, break; end
    xp=x;
end
k, xg=x
```

We can formulate this problem into a system of equations as

$$x_1 = \frac{x_0 + x_2}{2}, \quad x_2 = \frac{x_1 + x_3}{2}, \quad x_3 = \frac{x_2 + x_4}{2},$$

$$x_4 = \frac{x_3 + x_5}{2} \quad \text{with } x_0 = 0 \text{ and } x_5 = 10 \quad \text{(E2.5.1)}$$

This can easily be cast into Eq. (2.5.3) or (2.5.4) as programmed in the above script "nm02e05.m.".

The following example illustrates that the Jacobi iteration and the Gauss-Seidel iteration can also be used for solving a system of nonlinear equations, although there is no guarantee that it will work for every nonlinear equation.

Example 2.6. Gauss-Seidel Iterative Scheme for Solving a System of Nonlinear Equations.
We are going to use the Gauss-Seidel iteration to solve a set of nonlinear equations as

$$x_1^2 + 10x_1 + 2x_2^2 = 13$$
$$2x_1^3 - x_2^2 + 5x_2 = 6 \quad \text{(E2.6.1)}$$

In order to do so, we write these equations in the following form (each with only one variable on the LHS), which suits the Gauss-Seidel scheme.

114 SYSTEM OF LINEAR EQUATIONS

$$\begin{bmatrix} x_1 \\ x_2 \end{bmatrix} = \begin{bmatrix} (13 - x_1^2 - 2x_2^2)/10 \\ (6 - 2x_1^3 + x_2^2)/5 \end{bmatrix} \qquad (E2.6.2)$$

Then we make the following MATLAB script "nm02e06.m":

```
%nm02e06.m
% use Gauss-Seidel iteration to solve a set of nonlinear equations
kmax=100; tol=1e-5;
x=zeros(2,1); % Initial value
for k=1:kmax
    xp=x; % to remember the previous solution
    x(1)=(13-x(1)^2-2*x(2)^2)/10; % Eq.(E2.6.2a)
    x(2)=(6-2*x(1)^3+x(2)^2)/5;   % Eq.(E2.6.2b)
    if norm(x-xp)/(norm(xp)+eps)<tol, break; end
end
k, x
```

and run it to get

```
»nm02e06
   k =
        36
   x =
       1.0000    1.0000
```

This means that the simple iterative scheme reached the solution within the given tolerance (tol = 10^{-5}) of (apparent) error in 36 iterations. How marvelous it is to solve the set nonlinear equations without any special algorithm!

Have you any doubt about this solution? If so, run the following statements:

```
»eq=@(x)[x(1)^2+10*x(1)+2*x(2)^2-13; % Eq.(E2.6.1a)
         2*x(1)^3-x(2)^2+5*x(2)-6];  % Eq.(E2.6.1b)
 Residual_error=eq(x)
```

to get

```
         Residual_error =   1.0e-04 *
                               0.4111
                              -0.2581
```

Aren't these residual errors of each equation small enough to make you believe that **x** = [1.0000 1.0000] is the proper solution? Another way to check the validity of the solution is to plot the graphs for Eq. (E2.6.1) by running

```
»ezplot('x^2+10*x+2*y^2-13'); hold on;
 ezplot('2*x^3-y^2+5*y-6')
```

This yields the graphs in Figure 2.2, which implies that **x** = [1.0000 1.0000] is one of the two intersections between the two graphs where the two data points were made available by clicking on the Data Cursor button ( ) in the toolbar of the Figure window and then clicking on the intersections.

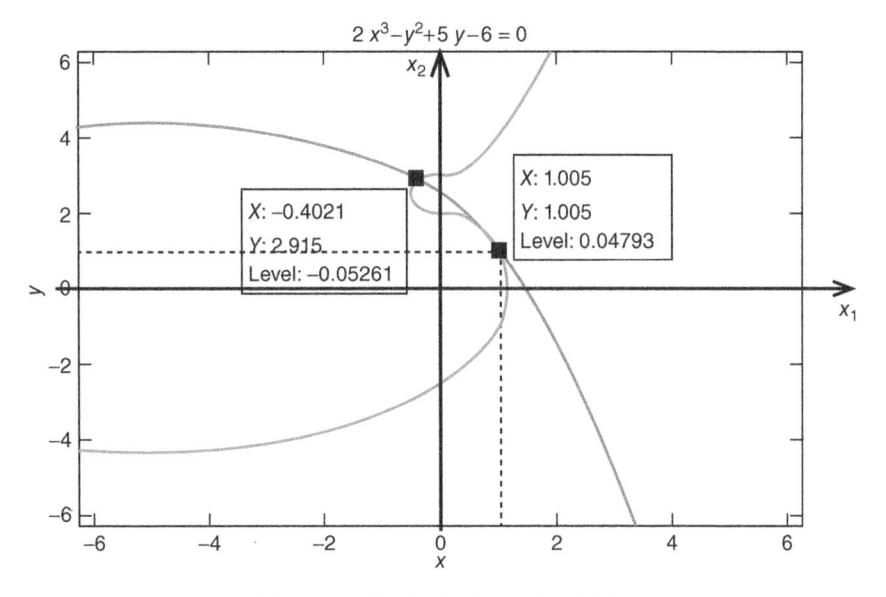

Figure 2.2 Graphs for Eqs. (E2.6.1a,b).

(cf) Due to its remarkable capability to deal with a system of nonlinear equations, the Gauss-Seidel iterative method plays an important role in solving partial differential equations (see Chapter 9).

2.5.3 The Convergence of Jacobi and Gauss-Seidel Iterations

Jacobi and Gauss-Seidel iterations have a very simple computational structure for they do not need any matrix inversion. So it may be of practical use, if only the convergence is guaranteed. However, everything cannot always be fine, as illustrated in Section 2.5.1. Then, what is the convergence condition? It is the diagonal dominancy of coefficient matrix A, which is stated as follows:

$$|a_{mm}| > \sum_{n \neq m}^{N} |a_{mn}| \quad \text{for } m = 1, 2, \ldots, N \quad (2.5.5)$$

This implies that the convergence of the iterative schemes is assured if, in each row of coefficient matrix A, the absolute value of the diagonal element is greater than the sum of the absolute values of the other elements. It should be noted, however, that this is sufficient, not a necessary, condition. In other words, the iterative scheme may work even if the above condition is not strictly satisfied.

SYSTEM OF LINEAR EQUATIONS

One thing to note is the *relaxation* technique, which may be helpful in accelerating the convergence of Gauss-Seidel iteration. It is a slight modification of Eq. (2.5.4) as

$$x_m^{(k+1)} = (1-\omega)x_m^{(k)} + \omega \frac{b_m - \sum_{n=1}^{m-1} a_{mn}x_n^{(k+1)} - \sum_{n=m+1}^{N} a_{mn}x_n^{(k)}}{a_{mm}} \quad \text{with } 0 < \omega < 2$$

(2.5.6)

and is called SOR p(successive overrelaxation) for the relaxation factor $1 < \omega < 2$ and successive underrelaxation for $0 < \omega < 1$. But regrettably, there is no general rule for selecting the optimal value of the relaxation factor.

PROBLEMS

2.1 Recursive Least Square Estimation (RLSE)

(a) Run the script "do_rlse.m" (in Section 2.1.4) with another value of the true parameter

```
xo=[1  2]'
```

What is the parameter estimate obtained from the RLS solution?

(b) Run the script "do_rlse.m" with a small matrix P generated by the following MATLAB statement:

```
P=0.01*eye(NA,NA);
```

What is the parameter estimate obtained from the RLS solution? Is it still close to the value of the true parameter?

(c) Insert the statements in the following box at appropriate places in the MATLAB script "do_rlse.m" that appear in Section 2.1.4. Remove the last two statements and run it to compare the times spent for using the RLS solution and the standard LS(E) solution to get the parameter estimates online.

```
%nm02p01.m
.. .. .. .. .. .. .. ..
time_on=0; time_off=0;
.. .. .. .. .. .. .. ..
tic
.. .. .. .. .. .. .. ..
time_on=time_on+toc;
tic
.. .. .. .. .. .. .. ..
time_off=time_off+toc;
.. .. .. .. .. .. .. ..
solutions=[xk_on xk_off]
discrepancy=norm(xk_on-xk_off)
times=[time_on time_off]
```

2.2 Delicacy of Scaled Partial Pivoting

As a complement to Example 2.2, we want to compare no pivoting, partial pivoting, scaled partial pivoting, and full pivoting in order to taste the delicacy of row switching strategy. To do it in a systematic way, add the third input argument (`pivoting`) to the Gauss elimination function 'Gauss()' and modify its contents by inserting the following statements into appropriate places so that the new function 'Gauss(A,b,pivoting)' implements the partial pivoting procedure optionally depending on the value of 'pivoting'. You can also remove any unnecessary parts.

```
-if nargin<3, pivoting=2; end %scaled partial pivoting by default
-switch pivoting
   case 2, [akx,kx]=max(abs(AB(k:NA,k))./...
              max(abs([AB(k:NA,k+1:NA) eps*ones(NA-k+1,1)]')))');
   otherwise, [akx,kx]=max(abs(AB(k:NA,k))); %partial pivoting
end
-&pivoting>0 %partial pivoting not to be done for pivot=1
```

(a) Use this function with `pivoting=0/1/2`, the '\' operator and the 'inv()' command to solve the systems of linear equations with the coefficient matrices and the RHS vectors shown below and fill in Table P2.2 with the residual error $\|A_i x - b_i\|$ to compare the results in terms of how well the solutions satisfy the equation, i.e. $\|A_i x - b_i\| \approx 0$.

(1) $A_1 = \begin{bmatrix} 10^{-15} & 1 \\ 1 & 10^{11} \end{bmatrix}$, $b_1 = \begin{bmatrix} 1 + 10^{-15} \\ 10^{11} + 1 \end{bmatrix}$

(2) $A_2 = \begin{bmatrix} 10^{-14.6} & 1 \\ 1 & 10^{15} \end{bmatrix}$, $b_2 = \begin{bmatrix} 1 + 10^{-14.6} \\ 10^{15} + 1 \end{bmatrix}$

(3) $A_3 = \begin{bmatrix} 10^{11} & 1 \\ 1 & 10^{-15} \end{bmatrix}$, $b_3 = \begin{bmatrix} 10^{11} + 1 \\ 1 + 10^{-15} \end{bmatrix}$

(4) $A_4 = \begin{bmatrix} 10^{14.6} & 1 \\ 1 & 10^{-15} \end{bmatrix}$, $b_4 = \begin{bmatrix} 10^{14.6} + 1 \\ 1 + 10^{-15} \end{bmatrix}$

(b) Which pivoting strategy yields the worst result for Problem 2.1 in (a)? Has the row swapping been done during the process of partial pivoting and scaled partial pivoting? If yes, did it work to our advantage? Did the '\' operator or the 'inv()' command give you any better result?

(c) Which pivoting strategy yields the worst result for Problem 2.2 in (a)? Has the row swapping been done during the process of partial pivoting and scaled partial pivoting? If yes, did it produce a positive effect for this case? Did the '\' operator or the 'inv()' command give you any better result?

(d) Which pivoting strategy yields the best result for Problem 2.3 in (a)? Has the row swapping been done during the process of partial pivoting and scaled partial pivoting? If yes, did it produce a positive effect for this case?

Table P2.2 Comparison of 'Gauss()' with different pivoting methods in terms of $\|Ax - b\|$.

	$A_1 x = b_1$	$A_2 x = b_2$	$A_3 x = b_3$	$A_4 x = b_4$
Gauss(A,b,0) (no pivoting)	1.2498e−001			
Gauss(A,b,1) (partial pivoting)		4.4409e−016		
Gauss(A,b,2) (scaled partial pivoting)			0	
A\b				6.2500e−002
A^-1*b		1.2500e−001		

(e) The coefficient matrix A_3 is the same as would be obtained by applying the full pivoting scheme for A_1 to have the largest pivot element. Does the full pivoting give better result than no pivoting or the (scaled) partial pivoting?

(f) Which pivoting strategy yields the best result for Problem 2.4 in (a)? Has the row swapping been done during the process of partial pivoting and scaled partial pivoting? If yes, did it produce a positive effect for this case? Did the '\' operator or the 'inv()' command give you any better result?

2.3 Gauss-Jordan Elimination Algorithm vs. Gauss Elimination Algorithm

Gauss-Jordan elimination algorithm mentioned in Section 2.2.3 is trimming the coefficient matrix A into an identity matrix and then takes the RHS vector/matrix as the solution, while Gauss elimination algorithm introduced with the corresponding function 'Gauss()' in Section 2.2.1 makes the matrix an upper-triangular one and performs backward substitution to get the solution. Since Gauss-Jordan elimination algorithm does not need backward substitution, it seems to be simpler than Gauss elimination algorithm.

(a) Modify the function 'Gauss()' into a function 'Gaussj()' that implements Gauss-Jordan elimination algorithm and count the number of multiplications consumed by the function, excluding those required for partial pivoting. Compare it with the number of multiplications consumed by 'Gauss()' (Eq. (2.2.19)). Does it support or betray our expectation that Gauss-Jordan elimination would take fewer computations than Gauss elimination?

(b) Use both of the functions, the '\' operator and the 'inv()' command or '^-1' to solve the system of linear equations

$$Ax = b \quad \quad (P2.3.1)$$

where A is the 10-dimensional Hilbert matrix (see Example 2.3) and $b = Ax^o$ with $x^o = \begin{bmatrix} 1 & 1 & 1 & 1 & 1 & 1 & 1 & 1 & 1 & 1 \end{bmatrix}^T$. Fill in Table P2.3 with the residual errors

$$\|A x_i - b\| \approx 0 \quad \quad (P2.3.2)$$

as a way of describing how well each solution satisfies the equation.

(cf) The numbers of floating-point operations required for carrying out the computations are listed in Table P2.3 so that readers can compare the computational loads of different approaches. Those data were obtained by using the MATLAB command 'flops()', which was available only in MATLAB of version below 6.0.

Table P2.3 Comparison of several methods for solving a set of linear equations.

	Gauss(A,b)	Gaussj(A,b)	A\b	A^-1*b
$\|Ax_i - b\|$	6.6613e−16		2.7195e−16	
Number of flops	1124	1744	785	7670

2.4 Tridiagonal System of Linear Equations

Consider the following system of linear equations

$$
\begin{aligned}
a_{11}x_1 + a_{12}x_2 &= b_1 \\
a_{21}x_1 + a_{22}x_2 + a_{23}x_3 &= b_2 \\
&\vdots \\
a_{N-1,N-2}x_{N-2} + a_{N-1,N-1}x_{N-1} + a_{N-1,N}x_N &= b_{N-1} \\
a_{N,N-1}x_{N-1} + a_{N,N}x_N &= b_N
\end{aligned}
\quad \text{(P2.4.1)}
$$

which can be written in a compact form by using a matrix-vector notation as

$$A_{N \times N} \mathbf{x} = \mathbf{b} \quad \text{(P2.4.2)}$$

where

$$
A_{N \times N} = \begin{bmatrix}
a_{11} & a_{12} & 0 & 0 & 0 \\
a_{21} & a_{22} & a_{23} & 0 & 0 \\
0 & \vdots & \vdots & \vdots & 0 \\
0 & 0 & a_{N-1,N-2} & a_{N-1,N-1} & a_{N-1,N} \\
0 & 0 & 0 & a_{N,N-1} & a_{NN}
\end{bmatrix}, \quad
\mathbf{x} = \begin{bmatrix} x_1 \\ x_2 \\ \vdots \\ x_{N-1} \\ x_N \end{bmatrix}, \quad
\mathbf{b} = \begin{bmatrix} b_1 \\ b_2 \\ \vdots \\ b_{N-1} \\ b_N \end{bmatrix}
$$

This is called a tridiagonal system of equations on account of that the coefficient matrix A has nonzero elements only on its main diagonal and super/subdiagonals.

(a) Modify the Gauss elimination function 'Gauss()' (Section 2.2.1) in such a way that this special structure can be exploited for reducing the computational burden. Give the name 'trid()' to the modified function and save it in an m-file named "trid.m" for future use.

(b) Modify the Gauss-Seidel iteration function 'Gauseid()' (Section 2.5.2) in such a way that this special structure can be exploited for reducing the computational burden. Let the name of the modified function be 'Gauseid1()'.

(c) Noting that Eq. (E2.5) in Example 2.5 can be trimmed into a tridiagonal structure as (P2.4.2), use the functions 'Gauss()', 'trid()', 'Gauseid()', 'Gauseid1()', and the backslash (\) operator to solve the problem.

(cf) The numbers of floating-point operations required for carrying out the computations are listed in Table P2.4 so that readers can compare the computational loads of the different approaches.

Table P2.4 The computational load of the methods to solve a tridiagonal system of equations.

	Gauss(A,b)	trid(A,b)	Gauseid()	Gauseid1()	A\b
Number of flops	141	50	2615	2082	94

2.5 *LU* Decomposition of a Tridiagonal Matrix

Modify the *LU* decomposition function 'lu_dcmp()' (Section 2.4.1) in such a way that the tridiagonal structure can be exploited for reducing the computational burden. Give the name 'lu_trid()' to the modified function and use it to get the *LU* decomposition of the tridiagonal matrix

$$A = \begin{bmatrix} 2 & -1 & 0 & 0 \\ -1 & 2 & -1 & 0 \\ 0 & -1 & 2 & -1 \\ 0 & 0 & -1 & 2 \end{bmatrix} \quad \text{(P2.5.1)}$$

You may run the following MATLAB statements:

```
»A=[2 -1 0 0; -1 2 -1 0; 0 -1 2 -1; 0 0 -1 2];
»[L,U]=lu_trid(A)
»L*U-A %=0(No error)?
```

2.6 Least Squares (LS) Solution by Backslash Operator and *QR* (Orthogonal–Upper Triangular) Decomposition

The backslash ('A\b') operator and the matrix left division ('mldivide(A,b)') function turn out to be the most efficient means for solving a system of linear equations as Eq. (P2.3.1). They are also capable of dealing with the under/overdetermined cases. Let us see how they handle the under/overdetermined cases.

(a) For a underdetermined system of linear equations

$$A_1 x = b_1; \quad \begin{bmatrix} 1 & 2 & 3 \\ 4 & 5 & 6 \end{bmatrix} \begin{bmatrix} x_1 \\ x_2 \\ x_3 \end{bmatrix} = \begin{bmatrix} 14 \\ 32 \end{bmatrix} \quad \text{(P2.6.1)}$$

find the minimum-norm solution (2.5.7) and the solutions that can be obtained by running the following MATLAB statements:

```
»A1=[1 2 3; 4 5 6]; b1=[14 32]';
»x_mn=A1'*(A1*A1')^-1*b1, x_pi=pinv(A1)*b1, x_bs=A1\b1
```

Are the three solutions the same?

(b) For another underdetermined system of linear equations

$$A_2 x = b_2; \quad \begin{bmatrix} 1 & 2 & 3 \\ 2 & 4 & 6 \end{bmatrix} \begin{bmatrix} x_1 \\ x_2 \\ x_3 \end{bmatrix} = \begin{bmatrix} 14 \\ 28 \end{bmatrix} \quad \text{(P2.6.2)}$$

find the solutions by using Eq. (2.5.7), 'pinv()', and backslash(\). If you are not pleased with the result obtained from Eq. (2.5.7), you can remove one of the two rows from the coefficient matrix A_2 and try again. Identify the minimum solution(s). Are the equations redundant or inconsistent?

122 SYSTEM OF LINEAR EQUATIONS

(c) For another underdetermined system of linear equations

$$A_2 \mathbf{x} = \mathbf{b}_3; \quad \begin{bmatrix} 1 & 2 & 3 \\ 2 & 4 & 6 \end{bmatrix} \begin{bmatrix} x_1 \\ x_2 \\ x_3 \end{bmatrix} = \begin{bmatrix} 21 \\ 21 \end{bmatrix} \qquad \text{(P2.6.3)}$$

find the solutions by using Eq. (2.5.7), 'pinv()', and backslash(\). Does any of them satisfy Eq. (P2.6.3) closely? Are the equations redundant or inconsistent?

(d) For an overdetermined system of linear equations

$$A_4 \mathbf{x} = \mathbf{b}_4; \quad \begin{bmatrix} 1 & 2 \\ 2 & 3 \\ 4 & -1 \end{bmatrix} \begin{bmatrix} x_1 \\ x_2 \end{bmatrix} = \begin{bmatrix} 5.2 \\ 7.8 \\ 2.2 \end{bmatrix} \qquad \text{(P2.6.4)}$$

find the LS or LSE (least square error) solution (2.5.10), which can be obtained by running the following statements. Fill in the corresponding blanks of Table P2.6.1 with the results.

```
»A4=[1 2; 2 3; 4 -1]; b4=[5.2 7.8 2.2]';
» x_ls=(A4'*A4)\A4'*b4, x_pi=pinv(A4)*b4, x_bs=A4\b4
```

(e) We can use QR decomposition to solve a system of linear equations as Eq. (P2.3.1), where the coefficient matrix A is square and nonsingular or rectangular with the row dimension greater than the column dimension. The procedure is explained as follows:

$$A\mathbf{x} = QR\mathbf{x} = \mathbf{b}; \quad R\mathbf{x} = Q^{-1}\mathbf{b} = Q'\mathbf{b}; \quad \mathbf{x} = R^{-1}Q'\mathbf{b} \qquad \text{(P2.6.5)}$$

Note that $Q'Q = I$; $Q' = Q^{-1}$ (orthogonality) and the premultiplication of R^{-1} can be performed by backward substitution, because R is an upper-triangular matrix. You are supposed not to count the number of floating-point operations needed for obtaining the LU and QR decompositions, assuming that they are available.

Table P2.6.1 Comparison of several methods for computing the LS solution.

	QR	LS: Eq.(2.5.10)	pinv(A)*b	A\b
$\|A\mathbf{x}_i - \mathbf{b}\|$	2.8788e−016		2.8788e−016	
Number of flops	25	89	196	92

Table P2.6.2 Comparison of several methods for solving a system of linear equations.

	LU	QR	Gauss(A,b)	A\b
$\|A\mathbf{x}_i - \mathbf{b}\|$		7.8505e−016		2.7195e−16
Number of flops	453	327	1124	785

(i) Apply the *QR* decomposition, the *LU* decomposition, Gauss elimination, and the backslash (\) operator to solve the system of linear equations whose coefficient matrix is the 10-dimensional Hilbert matrix (see Example 2.3) and fill in the corresponding blanks of Table P2.6.2 with the results.

(ii) Apply the *QR* decomposition to solve the system of linear equations given by Eq. (P2.6.4) and fill in the corresponding blanks of Table P2.6.2 with the results.

(cf) This problem illustrates that *QR* decomposition is quite useful for solving a system of linear equations, where the coefficient matrix *A* is square and nonsingular or rectangular with the row dimension greater than the column dimension and no rank deficiency.

2.7 Cholesky Factorization of a Symmetric Positive Definite Matrix
If a matrix *A* is symmetric and positive definite, we can find its *LU* decomposition such that the upper triangular matrix *U* is the transpose of the lower triangular matrix *L*, such that

$$LL^T = U^T U = A \quad \quad (P2.7.1)$$

which is called Cholesky factorization.
Consider the Cholesky factorization process for a 4×4 matrix:

$$\begin{bmatrix} a_{11} & a_{12} & a_{13} & a_{14} \\ a_{12} & a_{22} & a_{23} & a_{24} \\ a_{13} & a_{23} & a_{33} & a_{34} \\ a_{14} & a_{24} & a_{34} & a_{44} \end{bmatrix} = \begin{bmatrix} u_{11} & 0 & 0 & 0 \\ u_{12} & u_{22} & 0 & 0 \\ u_{13} & u_{23} & u_{33} & 0 \\ u_{14} & u_{24} & u_{34} & u_{44} \end{bmatrix} \begin{bmatrix} u_{11} & u_{12} & u_{13} & u_{14} \\ 0 & u_{22} & u_{23} & u_{24} \\ 0 & 0 & u_{33} & u_{34} \\ 0 & 0 & 0 & u_{44} \end{bmatrix}$$

$$= \begin{bmatrix} u_{11}^2 & u_{11}u_{12} & u_{11}u_{13} & u_{11}u_{14} \\ u_{12}u_{11} & u_{12}^2 + u_{22}^2 & u_{12}u_{13} + u_{22}u_{23} & u_{12}u_{14} + u_{22}u_{24} \\ u_{13}u_{11} & u_{13}u_{12} + u_{23}u_{22} & u_{13}^2 + u_{23}^2 + u_{33}^2 & u_{13}u_{14} + u_{23}u_{24} + u_{33}u_{34} \\ u_{14}u_{11} & u_{14}u_{12} + u_{24}u_{22} & u_{14}u_{13} + u_{24}u_{23} + u_{34}u_{33} & u_{14}^2 + u_{24}^2 + u_{34}^2 + u_{44}^2 \end{bmatrix}$$
$$(P2.7.2)$$

Equating every row of the matrices on both sides yields

$$u_{11} = \sqrt{a_{11}}, \quad u_{12} = a_{12}/u_{11}, \quad u_{13} = a_{13}/u_{11}, \quad u_{14} = a_{14}/u_{11} \quad (P2.7.3)$$

$$u_{22} = \sqrt{a_{22} - u_{12}^2}, \quad u_{23} = (a_{23} - u_{13}u_{12})/u_{22}, \quad u_{24} = (a_{24} - u_{14}u_{12})/u_{22}$$
$$(P2.7.4)$$

$$u_{33} = \sqrt{a_{33} - u_{23}^2 - u_{13}^2}, \quad u_{34} = (a_{43} - u_{24}u_{23} - u_{14}u_{13})/u_{33} \quad (P2.7.5)$$

$$u_{44} = \sqrt{a_{44} - u_{34}^2 - u_{24}^2 - u_{14}^2} \quad (P2.7.6)$$

```
function U=Cholesky(A)
% U'U decomposition of a symmetric positive definite matrix A
NA=size(A,1);    U=zeros(NA,NA);
for k=1: NA
    A_UU = A(k,?)-U(1:k-1,k)'*U(1:k-1,k);
    U(k,k)=sqrt(A_UU);
    for m=k+1:NA
        U(k,m) = (A(k,m)-U(1:k-?,m)'*U(1:k-1,k))/U(k,?);
    end
end
```

which can be combined into two formulas as

$$u_{kk} = \sqrt{a_{kk} - \sum_{i=1}^{k-1} u_{ik}^2} \quad \text{for } k = 1 : N \tag{P2.7.7}$$

$$u_{km} = \left(a_{km} - \sum_{i=1}^{k-1} u_{im} u_{ik}\right)/u_{kk} \quad \text{for } m = k+1 : N \text{ and } k = 1 : N \tag{P2.7.8}$$

(a) Complete the above MATLAB function 'Cholesky()' so that it can use Eqs. (P2.7.7) and (P2.7.8) to perform Cholesky factorization.

(b) Try your function 'Cholesky()' for the following matrix and check if $U^T U - A \approx O$ (U: the upper triangular matrix). Compare the result with that obtained by using the MATLAB built-in function 'chol()'.

$$A = \begin{bmatrix} 1 & 2 & 4 & 7 \\ 2 & 13 & 23 & 38 \\ 4 & 23 & 77 & 122 \\ 7 & 38 & 122 & 294 \end{bmatrix} \tag{P2.7.9}$$

(c) Use the function 'lu_dcmp()' to get the LU decomposition for the above matrix (P2.7.9) and check if $P^T LU - A \approx O$, where L and U are the lower/upper triangular matrix, respectively. Compare the result with that obtained by using 'lu()'.

2.8 Usage of Singular Value Decomposition (SVD)
What is SVD good for? Suppose we have the SVD of an $M \times N$ real-valued matrix A as

$$A = USV^T \tag{P2.8.1}$$

where U is an orthogonal $M \times M$ matrix, V an orthogonal $N \times N$ matrix, and S a real diagonal $M \times N$ matrix having the singular value σ_i's of A (the square roots of the eigenvalues of $A^T A$) in decreasing order on its diagonal. Then, it is possible to improvise the pseudo-inverse even in the case of rank-deficient matrices (with rank(A) < min(M,N)) for which the left/right pseudo-inverse cannot be found. The virtual pseudo-inverse can be written as

$$\hat{A}^{-1} = \hat{V}\hat{S}^{-1}\hat{U}^{T} \qquad (P2.8.2)$$

where $\hat{S}^{-1}$ is the diagonal matrix having $1/\sigma_i$ on its diagonal that is reconstructed by removing all-zero(like) rows/columns of the matrix S and substituting $1/\sigma_i$ for $\sigma_i \neq 0$ into the resulting matrix; $\hat{V}$ and $\hat{U}$ are reconstructed by removing the columns of V and U corresponding to the zero singular value(s). Consequently, SVD has a specialty in dealing with the singular cases. Let us take a closer look at this through the following problems.

(a) Consider the problem of solving

$$A_1 x = \begin{bmatrix} 1 & 2 & 3 \\ 2 & 4 & 6 \end{bmatrix} \begin{bmatrix} x_1 \\ x_2 \\ x_3 \end{bmatrix} = \begin{bmatrix} 6 \\ 12 \end{bmatrix} = \mathbf{b}_1 \qquad (P2.8.3)$$

Since this is an underdetermined case ($M = 2 < 3 = N$), it seems that we can use Eq. (2.5.7) to find the minimum-norm solution.

(i) Run the following statements at the MATLAB prompt:

```
»A1=[1 2 3;2 4 6]; b1=[6;12];
  x_MN=A1'*(A1*A1')^-1*b1 % Eq.(2.1.7)
```

Is the minimum-norm solution fine? If not, explain why it is so.

(ii) Run the following block of statements to find the SVD-based minimum-norm solution by Eq. (P2.8.2). What are the values of $\mathbf{x} = \hat{A}_1^{-1} \mathbf{b}_1 = \hat{V}\hat{S}^{-1}\hat{U}^T \mathbf{b}_1$ and $\|A_1 \mathbf{x} - \mathbf{b}_1\|$?

```
[U,S,V]=svd(A1); r=rank(A1); % Eq.(P2.8.1)
u=U(:,1:r); v=V(:,1:r); s=S(1:r,1:r);
AIp=v*diag(1./diag(s))*u'; % faked pseudo-inverse (P2.8.2)
x_svd=AIp*b1 % Minimum-norm solution for singular underdetermined
err=norm(A1*x_svd-b1) % Residual error
```

(iii) To see that the norm of this solution is less than that of any other solution that can be obtained by adding any vector in the null space of the coefficient matrix A_1, run the following statements. What is implied by the result?

```
nullA=null(A1); normx=norm(x_svd);
for n=1:1000
   if norm(x_svd+nullA*(rand(size(nullA,2),1)-0.5))<normx
      disp('What the hell smaller-norm sol: not
            minimum norm!');
   end
end
```

(b) For the problem

$$A_2 \mathbf{x} = \begin{bmatrix} 1 & 2 & 3 \\ 2 & 3 & 4 \end{bmatrix} \begin{bmatrix} x_1 \\ x_2 \\ x_3 \end{bmatrix} = \begin{bmatrix} 6 \\ 9 \end{bmatrix} = \mathbf{b}_2 \qquad (P2.8.4)$$

compare the minimum-norm solution based on SVD and that obtained by Eq. (2.5.7).

```
%nm02p08c.m
A3=[1 2 3; 4 5 9;7 11 18;-6 9 3]; b3=[1;2;3;4];
X_LS=(A3'*A?)^-1*A3'*?3  % Eq.(2.1.10)
r=rank(A3), [U,S,V]=svd(A3);
u=U(:,1:r); v=V(:,1:r); s=S(1:r,1:r);
AIp=?*diag(1./diag(?))*?';
x_svd=AIp*?3, err=norm(A3*x_svd-b3)
for n=1:1000
   if norm(A3*(x_svd+rand(size(x_svd))-0.5)-b3)<err
      disp('What the hell smaller error sol - not LSE?');
   end
end
```

(c) Consider the problem of solving

$$A_3 \mathbf{x} = \begin{bmatrix} 1 & 2 & 3 \\ 4 & 5 & 9 \\ 7 & 11 & 18 \\ -6 & 9 & 3 \end{bmatrix} \begin{bmatrix} x_1 \\ x_2 \\ x_3 \end{bmatrix} = \begin{bmatrix} 1 \\ 2 \\ 3 \\ 4 \end{bmatrix} = \mathbf{b}_3 \qquad (P2.8.5)$$

Since this is an overdetermined case ($M = 4 > 3 = N$), it seems that we can use Eq. (2.1.10) to find the LS or LSE solution. To check if this idea is right and find the SVD solution, complete and run the above block of statements, and answer the following questions:

(i) Is the LS solution (obtained using Eq. (2.1.10)) all right? If not, explain why it is so in connection with the rank of A_3.
(ii) Does the SVD-based least-squares solution seem to be an LSE solution?

(d) For the problem

$$A_4 \mathbf{x} = \begin{bmatrix} 1 & 2 & 3 \\ 4 & 5 & 9 \\ 7 & 11 & -1 \\ -2 & 3 & 1 \end{bmatrix} \begin{bmatrix} x_1 \\ x_2 \\ x_3 \end{bmatrix} = \begin{bmatrix} 1 \\ 2 \\ 3 \\ 4 \end{bmatrix} = \mathbf{b}_4 \qquad (P2.8.6)$$

compare the LSE solution based on SVD and that obtained by Eq. (2.1.10).

(cf) This problem illustrates that SVD can be used for making a universal solution of a set of linear equations, minimum-norm or LSE, for all the possible rank deficiency of the coefficient matrix A.

2.9 QR Decomposition, SVD, and Backslash Operator

Referring to Example 2.4, apply the QR decomposition, SVD, and backslash operator to solve the following sets of linear equations and fill the blanks in the following table with the appropriate answers. Which method presents the LSE/minimum-norm solutions irrespective of the rank deficiency?

$$\text{(a)} \begin{bmatrix} 2 & -1 & 1 \\ 1 & -5 & 2 \\ -3 & 6 & -3 \end{bmatrix} \begin{bmatrix} x_1 \\ x_2 \\ x_3 \end{bmatrix} = \begin{bmatrix} 8 \\ 16 \\ -23 \end{bmatrix} \quad \text{(b)} \begin{bmatrix} 2 & 1 & -3 \\ 4 & 2 & -6 \end{bmatrix} \begin{bmatrix} x_1 \\ x_2 \\ x_3 \end{bmatrix} = \begin{bmatrix} -9 \\ 12 \end{bmatrix}$$

$$\text{(c)} \begin{bmatrix} 2 & 1 & -3 \\ 4 & 2 & -6 \end{bmatrix} \begin{bmatrix} x_1 \\ x_2 \\ x_3 \end{bmatrix} = \begin{bmatrix} -9 \\ -18 \end{bmatrix} \quad \text{(P2.9.1)}$$

Equation		(a)	(b)	(c)
Rank deficient? Rank(A) < min(M,N)			Rank deficient 1 < 2	
Rank([A b]) < rank(A) + 1? Inconsistent/redundant?		3 = 2 + 1	Inconsistent	1 < 1 + 1
QR solution, Eq. (2.5.17)	x	$[-1.22\ 1.22\ 3.66] \times 10^{15}$		$[-15.27\ 0\ -7.18]$
	‖x‖		6.93×10^{16}	
	‖Ax − b‖	3.0000		0.0000
SVD solution, Eq. (2.5.20)	x		$[0.429\ 0.214\ -0.643]$	
	‖x‖	3.2937		2.4054
	‖Ax − b‖		13.416	
Backslash solution, x = A\b	x	$[-3.00\ 3.00\ 9.01] \times 10^{15}$		$[0\ 0\ 3.0000]$
	‖x‖		1.0000	
	‖Ax − b‖	1.0000		7.9441×10^{-15}

2.10 Gauss-Seidel Iterative Method with Relaxation Technique

(a) Try the relaxation technique (introduced in Section 2.5.3) with several values of the relaxation factor $\omega = 0.2, 0.4, \ldots, 1.8$ for the following problems. Find the best one among these values of the relaxation factor for each problem, together with the number of iterations required for satisfying the termination (stopping) criterion $\|\mathbf{x}_{k+1} - \mathbf{x}_k\|/\|\mathbf{x}_k\| < 10^{-6}$.

$$\text{(i)}\ A_1 \mathbf{x} = \begin{bmatrix} 5 & -4 \\ -9 & 10 \end{bmatrix} \begin{bmatrix} x_1 \\ x_2 \end{bmatrix} = \begin{bmatrix} 1 \\ 1 \end{bmatrix} = \mathbf{b}_1 \quad \text{(P2.10.1)}$$

$$\text{(ii)}\ A_2 \mathbf{x} = \begin{bmatrix} 2 & -1 \\ -1 & 4 \end{bmatrix} \begin{bmatrix} x_1 \\ x_2 \end{bmatrix} = \begin{bmatrix} 1 \\ 3 \end{bmatrix} = \mathbf{b}_2 \quad \text{(P2.10.2)}$$

(iii) The nonlinear equations (E2.6.1) are given in Example 2.6

(b) Which of the two matrices A_1 and A_2 has stronger diagonal dominance in the above equations? For which equation does Gauss-Seidel iteration converge faster, Eq. (P2.10.1) or (P2.10.2)? What would you conjecture about the relationship between the convergence speed of Gauss-Seidel iteration for a set of linear equations and the diagonal dominancy of the coefficient matrix A?

(c) Is the relaxation technique always helpful for improving the convergence speed of Gauss-Seidel iterative method regardless of the value of the relaxation factor ω?

2.11 Gauss-Seidel Iteration vs. Jacobi Iteration for a Set of Nonlinear Equations
Consider a set of nonlinear equations in $N-1$ unknown values of temperature $\{T(2), \ldots, T(N)\}$ at $x = x_i = x_0 + i\Delta x$ (with $\Delta x = (x_f - x_0)/N$) for $i = 1:N-1$ along a rod of length $L = x_f - x_0 = 10$ m:

$$\frac{T_{i-1} - 2T_i + T_{i+1}}{\Delta x^2} = h(T_i - T_a) + \sigma(T_i^4 - T_a^4) \quad \text{for } i = 1, \ldots, N-1$$
(P2.11.1)

where $h = 0.05$ m^{-2}, $\sigma = 2.7 \times 10^{-9}$ K^{-3} m^{-2}, $T_a = 200$ K, $T(1) = T_0 = 300$ K at $x = x_0 = 0$ m, and $T(N+1) = T_f = 400$ K at $x = x_f = 10$ m. This can be rearranged into the following form:

$$T_i = \frac{h\Delta x^2 T_a + \sigma \Delta x^2 (T_a^4 - T_i^4) + T_{i-1} + T_{i+1}}{2 + h\Delta x^2}$$
(P2.11.2)

With the length of the rod into $N = 5$ segments each of length $\Delta x = 10/5 = 2$ m, apply the Jacobi iteration and Gauss-Seidel iteration to solve Eq. (P6.13.2) and compare the two solutions in terms of the residual (mismatch) error. To do this job, complete and run the above script "nm02p11.m".

```
%nm02p11.m
N=5; x0=0; xf=10; dx=(xf-x0)/N; xx=x0+[0:N]*dx;
h=0.05; s=2.7e-9; K1=h*dx^2; K2=s*dx^2; Ta=200; T0=300; Tf=400;
er_T=@(T,N)norm((T(2:N)-(K1*Ta+K2*(Ta^4-T(1:N-1).^4)...
     +T(1:N-1)+T(3:N+1))/(2+K1))./T(2:N));
tol=1e-6*N/2; % Error tolerance for stopping criterion
disp('Jacobi method')
Ts=[T0 zeros(1,N-1) Tf].'; % Ts=[T0 ones(1,N-1)*(T0+Tf)/2 Tf].';
for k=1:1000
   Ts0=Ts;
   Ts(2:N) = (K1*Ta+K2*(Ta^4-Ts(?:N).^4)+Ts(1:N-?)+Ts(3:?+1))/(2+K1);
   if norm((Ts-Ts0)./Ts)<tol, break; end
end
k, T_J=Ts.'; er_J=er_T(T_J,N), if N<11, T_J, end
disp('Gauss-Seidel method')
Ts=[T0 zeros(1,N-1) Tf].'; % Ts=[T0 ones(1,N-1)*(T0+Tf)/2 Tf].';
for k=1:1000
   Ts0=Ts;
   for i=2:N
      Ts(i) = (K1*Ta+K2*(Ta^4-Ts(?).^4)+Ts(i-?)+Ts(?+1))/(2+K1);
   end
   if norm((Ts-Ts0)./Ts)<tol, break; end
end
k, T_GS=Ts.'; er_GS=er_T(T_GS,N)
if N<11, T_GS, end   % Compare with Example 24.7 of Chapra
plot(xx,T_J, xx,T_GS,'r')
```

3

INTERPOLATION AND CURVE FITTING

CHAPTER OUTLINE

3.1 Interpolation by Lagrange Polynomial	130
3.2 Interpolation by Newton Polynomial	132
3.3 Approximation by Chebyshev Polynomial	137
3.4 Pade Approximation by Rational Function	142
3.5 Interpolation by Cubic Spline	146
3.6 Hermite Interpolating Polynomial	153
3.7 Two-Dimensional Interpolation	155
3.8 Curve Fitting	158
3.8.1 Straight-Line Fit – A Polynomial Function of Degree 1	158
3.8.2 Polynomial Curve Fit – A Polynomial Function of Higher Degree	160
3.8.3 Exponential Curve Fit and Other Functions	165
3.9 Fourier Transform	166
3.9.1 FFT vs. DFT	167
3.9.2 Physical Meaning of DFT	169
3.9.3 Interpolation by Using DFS	172
Problems	175

Applied Numerical Methods Using MATLAB®, Second Edition. Won Y. Yang, Jaekwon Kim, Kyung W. Park, Donghyun Baek, Sungjoon Lim, Jingon Joung, Suhyun Park, Han L. Lee, Woo June Choi, and Taeho Im.
© 2020 John Wiley & Sons, Inc. Published 2020 by John Wiley & Sons, Inc.
Companion website: www.wiley.com/go/yang/appliednumericalmethods

130 INTERPOLATION AND CURVE FITTING

There are two topics to be dealt with in this chapter, interpolation[1] and curve fitting. Interpolation is to connect discrete data points in a plausible way so that one can get reasonable estimates of data points between the given points. The interpolation curve goes through all data points. Curve fitting, on the other hand, is to find a curve that could best indicate the trend of a given set of data. The curve does not have to go through the data points. In some cases, the data may have different accuracy/reliability/uncertainty and we need the weighted least squares (WLS) curve fitting to process such a data.

3.1 INTERPOLATION BY LAGRANGE POLYNOMIAL

For a given set of $N+1$ data points $\{(x_0, y_0), (x_1, y_1), \ldots, (x_N, y_N)\}$, we want to find the coefficients of an Nth-degree polynomial function to match them:

$$p_N(x) = a_0 + a_1 x + a_2 x^2 + \cdots + a_N x^N \qquad (3.1.1)$$

The coefficients can be obtained by solving the following system of linear equations.

$$\begin{aligned} a_0 + x_0 a_1 + x_0^2 a_2 + \cdots + x_0^N a_N &= y_0 \\ a_0 + x_1 a_1 + x_1^2 a_2 + \cdots + x_1^N a_N &= y_1 \\ &\vdots \\ a_0 + x_N a_1 + x_N^2 a_2 + \cdots + x_N^N a_N &= y_N \end{aligned} \qquad (3.1.2)$$

But, as the number of data points increases, so does the number of unknown variables and equations, consequently, it may be not so easy to solve. That is why we look for alternatives to get the coefficients $\{a_0, a_1, \ldots, a_N\}$.

One of the alternatives is to make use of the Lagrange polynomials

$$l_N(x) = y_0 \frac{(x-x_1)(x-x_2)\cdots(x-x_N)}{(x_0-x_1)(x_0-x_2)\cdots(x_0-x_N)} + y_1 \frac{(x-x_0)(x-x_2)\cdots(x-x_N)}{(x_1-x_0)(x_1-x_2)\cdots(x_1-x_N)}$$

$$+ \cdots + y_N \frac{(x-x_0)(x-x_1)\cdots(x-x_{N-1})}{(x_N-x_0)(x_N-x_1)\cdots(x_N-x_{N-1})};$$

$$l_N(x) = \sum_{m=0}^{N} y_m L_{N,m}(x) \quad \text{with} \quad L_{N,m}(x) = \frac{\prod_{k \neq m}^{N}(x-x_k)}{\prod_{k \neq m}^{N}(x_m-x_k)} = \prod_{k \neq m}^{N} \frac{x-x_k}{x_m-x_k} \qquad (3.1.3)$$

---
[1] If we estimate the values of the unknown function at the points that are inside/outside the range of collected data points, we call it the interpolation/extrapolation.

It can easily be shown that the graph of this function matches every data point

$$l_N(x_m) = y_m \quad \forall \ m = 0, 1, \ldots, N \tag{3.1.4}$$

since the Lagrange coefficient polynomial $L_{N,m}(x)$ is 1 only for $x = x_m$ and zero for all other data points $x = x_k$ ($k \neq m$). Note that the Nth-degree polynomial function matching the given $N+1$ points is unique and so Eq. (3.1.1) having the coefficients obtained from Eq. (3.1.2) must be the same as the Lagrange polynomial (3.1.3).

Now, we have the MATLAB function 'Lagranp()', which finds us the coefficients of Lagrange polynomial (3.1.3) together with each Lagrange coefficient polynomial $L_{N,m}(x)$. In order to understand this function, you should know that MATLAB deals with polynomials as their coefficient vectors arranged in descending order and the multiplication of two polynomials corresponds to the convolution of the coefficient vectors as mentioned in Section 1.1.6.

```
function [l,L]=Lagranp(x,y)
%Input : x=[x0 x1 ... xN], y=[y0 y1 ... yN]
%Output: l=Lagrange polynomial coefficients of degree N
%        L=Lagrange coefficient polynomial
N=length(x)-1; %the degree of polynomial
l=0;
for m=1:N+1
   P=1;
   for k=1:N+1
      if k~=m, P=conv(P,poly(x(k)))/(x(m)-x(k)); end
   end
   L(m,:)=P; % Lagrange coefficient polynomial
   l = l+y(m)*P; % Lagrange polynomial (3.1.3)
end
```

```
%do_Lagranp.m
x=[-2 -1 1 2]; y=[-6 0 0 6]; % given data points
l=Lagranp(x,y) % find the Lagrange polynomial
xx=[-2: 0.02 : 2]; yy=polyval(l,xx); % interpolate for [-2,2]
clf, plot(xx,yy,'b', x,y,'*') % plot the graph
```

We make the MATLAB script "do_Lagranp.m" to use the function 'Lagranp()' for finding the third-degree polynomial $l_3(x)$ that matches the four given points
$$\{(-2, -6), (-1, 0), (1, 0), (2, 6)\}$$

and to check if the graph of $l_3(x)$ really passes the four points. The results from running this script are depicted in Figure 3.1.

```
>>do_Lagranp
   l =   1    0   -1    0 % meaning l_3(x) = 1·x^3 + 0·x^2 - 1·x + 0
```

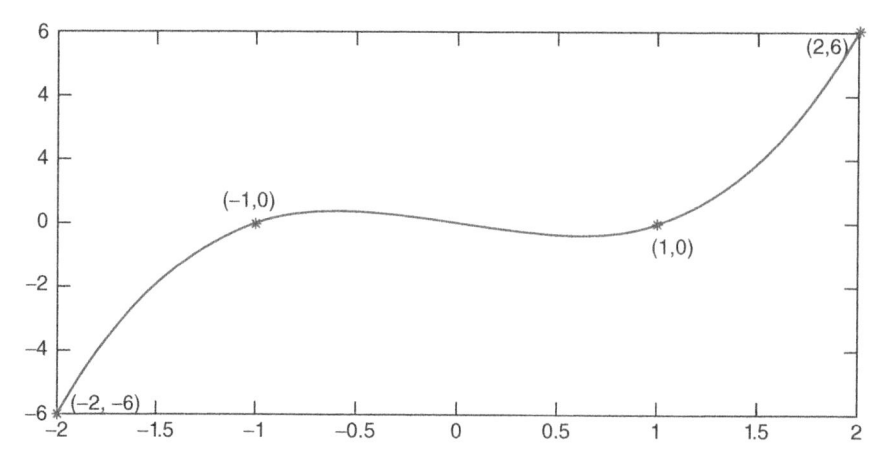

Figure 3.1 The graph of a third-degree Lagrange polynomial.

3.2 INTERPOLATION BY NEWTON POLYNOMIAL

Although the Lagrange polynomial works pretty well for interpolation irrespective of the interval widths between the data points along the x-axis, it requires restarting the whole computation with heavier burden as data points are appended. Unlike this, the Nth-degree Newton polynomial matching the $N + 1$ data points $\{(x_0, y_0), (x_1, y_1), \ldots, (x_N, y_N)\}$ can be recursively obtained as the sum of the $(N - 1)$th-degree Newton polynomial matching the N data points $\{(x_0, y_0), (x_1, y_1), \ldots, (x_{N-1}, y_{N-1})\}$ and one additional term.

$$n_N(x) = a_0 + a_1(x - x_0) + a_2(x - x_0)(x - x_1) + \cdots$$
$$= n_{N-1}(x) + a_N(x - x_0)(x - x_1) \cdots (x - x_{N-1}) \quad \text{with} \quad n_0(x) = a_0 \quad (3.2.1)$$

In order to derive a formula to find the successive coefficients $\{a_0, a_1, \ldots, a_N\}$ that make this equation accommodate the data points, we will determine a_0 and a_1 so that

$$n_1(x) = n_0(x) + a_1(x - x_0) \quad (3.2.2)$$

matches the first two data points (x_0, y_0) and (x_1, y_1). We need to solve the two equations

$$n_1(x_0) = a_0 + a_1(x_0 - x_0) = y_0$$
$$n_1(x_1) = a_0 + a_1(x_1 - x_0) = y_1$$

to get

$$a_0 = y_0, \qquad a_1 = \frac{y_1 - a_0}{x_1 - x_0} = \frac{y_1 - y_0}{x_1 - x_0} \equiv Df_0 \quad (3.2.3)$$

INTERPOLATION BY NEWTON POLYNOMIAL 133

Starting from this first-degree Newton polynomial, we can proceed to the second-degree Newton polynomial

$$n_2(x) = n_1(x) + a_2(x - x_0)(x - x_1) = a_0 + a_1(x - x_0) + a_2(x - x_0)(x - x_1)$$
(3.2.4)

which, with the same coefficients a_0 and a_1 as (3.2.3), still matches the first two data points (x_0, y_0) and (x_1, y_1), since the additional (third) term is zero at (x_0, y_0) and (x_1, y_1). This is to say that the additional polynomial term does not disturb the matching of previous existing data. Therefore, given the additional matching condition for the third data point (x_2, y_2), we only have to solve

$$n_2(x_2) = a_0 + a_1(x_2 - x_0) + a_2(x_2 - x_0)(x_2 - x_1) \equiv y_2$$

for only one more coefficient a_2 to get

$$a_2 = \frac{y_2 - a_0 - a_1(x_2 - x_0)}{(x_2 - x_0)(x_2 - x_1)} = \frac{y_2 - y_0 - \frac{y_1 - y_0}{x_1 - x_0}(x_2 - x_0)}{(x_2 - x_0)(x_2 - x_1)}$$

$$= \frac{y_2 - y_1 + y_1 - y_0 - \frac{y_1 - y_0}{x_1 - x_0}(x_2 - x_1 + x_1 - x_0)}{(x_2 - x_0)(x_2 - x_1)}$$

$$= \frac{\frac{y_2 - y_1}{x_2 - x_1} - \frac{y_1 - y_0}{x_1 - x_0}}{x_2 - x_0} = \frac{Df_1 - Df_0}{x_2 - x_0} \equiv D^2 f_0$$
(3.2.5)

Generalizing these results, Eqs. (3.2.3) and (3.2.5) yield the formula to get the Nth coefficient a_N of the Newton polynomial function (3.2.1) as

$$a_N = \frac{D^{N-1}f_1 - D^{N-1}f_0}{x_N - x_0} \equiv D^N f_0$$
(3.2.6)

This is the divided difference, which can be obtained successively from the second row of Table 3.1.

Table 3.1 Divided difference table.

x_k	y_k	Df_k	$D^2 f_k$	$D^3 f_k$	
x_0	y_0	$Df_0 = \dfrac{y_1 - y_0}{x_1 - x_0}$	$D^2 f_0 = \dfrac{Df_1 - Df_0}{x_2 - x_0}$	$D^3 f_0 = \dfrac{D^2 f_1 - D^2 f_0}{x_3 - x_0}$	—
x_1	y_1	$Df_1 = \dfrac{y_2 - y_1}{x_2 - x_1}$	$D^2 f_1 = \dfrac{Df_2 - Df_1}{x_3 - x_1}$	—	
x_2	y_2	$Df_2 = \dfrac{y_3 - y_2}{x_3 - x_2}$	—		
x_3	y_3	—			

134 INTERPOLATION AND CURVE FITTING

```
function [n,DD]=Newtonp(x,y)
%Input  : x=[x0 x1 ... xN]
%         y=[y0 y1 ... yN]
%Output: n=Newton polynomial coefficients of degree N
N= length(x)-1;
DD =zeros(N+1,N+1);
DD(1:N+1,1)=y';
for k=2:N+1
   for m=1: N+2-k %Divided Difference Table
      DD(m,k)=(DD(m+1,k-1)-DD(m,k-1))/(x(m+k-1)-x(m));
   end
end
a=DD(1,:); % Eq.(3.2.6)
n=a(N+1); % Begin with Eq.(3.2.7)
for k=N:-1:1 %Eq.(3.2.7)
   n=conv(n,[1 -x(k)])+[zeros(size(n)) a(k)]; %n(x)*(x-x(k-1))+a_k-1
end
```

Note that, as mentioned in Section 1.3, it is of better computational efficiency to write the Newton polynomial (3.2.1) in the nested multiplication form as

$$n_N(x) = ((\cdots(a_N(x - x_{N-1}) + a_{N-1})(x - x_{N-2}) + \cdots) + a_1)(x - x_0) + a_0 \quad (3.2.7)$$

and that the multiplication of two polynomials corresponds to the convolution of the coefficient vectors as mentioned in Section 1.1.6. We make the MATLAB function 'Newtonp()' to compose the divided difference table like Table 3.1 and construct the Newton polynomial for a set of data points.

For example, suppose we are to find a Newton polynomial matching the following data points

$$\{(-2, -6), (-1, 0), (1, 0), (2, 6), (4, 60)\}$$

From these data points, we construct the divided difference table as Table 3.2 and then use this table together with Eq. (3.2.1) to get the Newton polynomial as follows:

$$\begin{aligned}
n(x) &= y_0 + Df_0(x - x_0) + D^2f_0(x - x_0)(x - x_1) \\
&\quad + D^3f_0(x - x_0)(x - x_1)(x - x_2) + 0 \\
&= -6 + 6(x - (-2)) - 2(x - (-2))(x - (-1)) \\
&\quad + 1(x - (-2))(x - (-1))(x - 1) \\
&= -6 + 6(x + 2) - 2(x + 2)(x + 1) + (x + 2)(x^2 - 1) \\
&= x^3 + (-2 + 2)x^2 + (6 - 6 - 1)x - 6 + 12 - 4 - 2 = x^3 - x
\end{aligned}$$

INTERPOLATION BY NEWTON POLYNOMIAL

Table 3.2 Divided differences.

x_k	y_k	Df_k	D^2f_k	D^3f_k	D^4f_k
-2	-6	$\dfrac{0-(-6)}{-1-(-2)} = 6$	$\dfrac{0-6}{1-(-2)} = -2$	$\dfrac{2-(-2)}{2-(-2)} = 1$	$\dfrac{1-1}{4-(-2)} = 0$
-1	0	$\dfrac{0-0}{1-(-1)} = 0$	$\dfrac{6-0}{2-(-1)} = 2$	$\dfrac{7-2}{4-(-1)} = 1$	
1	0	$\dfrac{6-0}{2-1} = 6$	$\dfrac{27-6}{4-1} = 7$		
2	6	$\dfrac{60-6}{4-2} = 27$			
4	60				

We might begin with, not necessarily the first data point, but, say, the third one $(1,0)$ and proceed as follows to end up with the same result.

$$n(x) = y_2 + Df_2(x - x_2) + D^2f_2(x - x_2)(x - x_3)$$
$$+ D^3f_2(x - x_2)(x - x_3)(x - x_4) + 0$$
$$= 0 + 6(x - 1) + 7(x - 1)(x - 2) + 1(x - 1)(x - 2)(x - 4)$$
$$= 6(x - 1) + 7(x^2 - 3x + 2) + (x^2 - 3x + 2)(x - 4)$$
$$= x^3 + (7 - 7)x^2 + (6 - 21 + 14)x - 6 + 14 - 8 = x^3 - x$$

This process has been coded into the following MATLAB script 'do_Newtonp', which illustrates that the Newton polynomial (3.2.1) does not depend on the order of the data points, i.e. changing the order of the data points does not make any difference.

```
%do_Newtonp.m
x=[-2 -1 1 2  4]; y=[-6  0 0 6 60];
n=Newtonp(x,y) % l=Lagranp(x,y) for comparison
x=[-1 -2 1 2  4]; y=[0 -6 0 6 60];
n1=Newtonp(x,y) % with the order of data changed for comparison
xx=[-2:0.02: 2]; yy=polyval(n,xx);
clf, plot(xx,yy,'b-',x,y,'*')
```

Now, let us see the interpolation problem from the viewpoint of approximation. For this purpose, suppose we are to approximate some function, say,

$$f(x) = \frac{1}{1 + 8x^2} \tag{3.2.8}$$

by a polynomial. We first pick up some sample points on the graph of this function, such as listed later, and look for the polynomial functions $n_4(x)$, $n_8(x)$, and $n_{10}(x)$ to match each of the three sets of points, respectively.

We made the following MATLAB script "do_Newtonp1.m" to do this job and plot the graphs of the polynomial functions together with the graph of the true function $f(x)$ and their error functions separately for comparison as depicted in Figure 3.2, where the parts for $n_8(x)$ and $n_{10}(x)$ are omitted to provide the readers with some room for practice.

x_k	−1.0	−0.5	0	0.5	1.0
y_k	1/9	1/3	1	1/3	1/9

x_k	−1.0	−0.75	−0.5	−0.25	0	0.25	0.5	0.75	1.0
y_k	1/9	2/11	1/3	2/3	1	2/3	1/3	2/11	1/9

x_k	−1.0	−0.8	−0.6	−0.4	−0.2	0	0.2	0.4	0.6	0.8	1.0
y_k	1/9	25/153	25/97	25/57	25/33	1	25/33	25/57	25/97	25/153	1/9

```
%do_Newtonp1.m - plot Fig.3.2
x=[-1 -0.5  0  0.5  1.0]; y=f31(x);
n=Newtonp(x,y)
xx=[-1:0.02: 1]; % Interval to look over
yy=f31(xx); % Graph of the true function
yy1=polyval(n,xx); % Graph of the approximate polynomial function
subplot(221), plot(xx,yy,'k-', x,y,'o', xx,yy1,'b')
subplot(222), plot(xx,yy1-yy,'r') % Graph of the error function
```

```
function y=f31(x)
y=1./(1+8*x.^2); % Eq.(3.2.8)
```

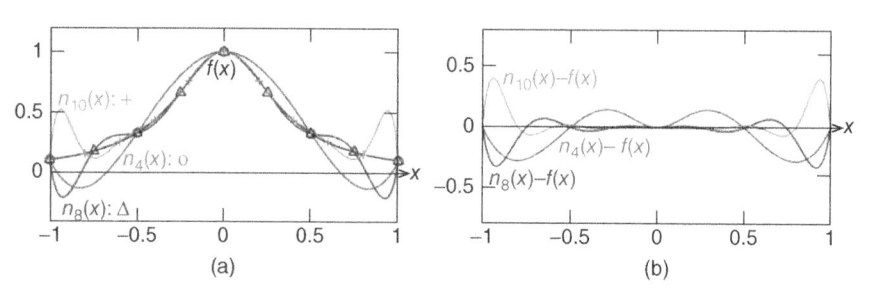

Figure 3.2 Interpolation from the viewpoint of approximation. (a) 4/8/10th-Degree polynomial approximation and (b) the error between the approximating polynomial and the true function.

Remark 3.1 Polynomial Wiggle and Runge Phenomenon.
Here is one thing to note. Strangely, increasing the degree of polynomial contributes little to reducing the approximation error. Rather contrary to our usual expectation, it tends to make the oscillation strikingly large, which is called the polynomial wiggle and the error gets bigger in the parts close to both ends as can be seen in Figure 3.2, which is called the Runge phenomenon. That is why polynomials of degree 5 or more are seldom used for the purpose of interpolation, unless they are sure to fit the data.

3.3 APPROXIMATION BY CHEBYSHEV POLYNOMIAL

At the end of the previous section, we considered a polynomial approximation problem of finding a polynomial close to a given (true) function $f(x)$ and have the freedom to pick up the target points $\{x_0, x_1, \ldots, x_N\}$ in our own way. Once the target points have been fixed, it is nothing but an interpolation problem that can be solved by the Lagrange or Newton polynomial.

In this section, we will think about how to choose the target points for better approximation, rather than taking equidistant points along the x-axis. Noting that the error tends to get bigger in the parts close to both ends of the interval when we chose the equidistant target points, it may be helpful to set the target points denser in the parts close to both ends than in the middle part. In this context, a possible choice is the projection (onto the x-axis) of the equidistant points on the circle centered at the middle point of the interval along the x-axis (see Figure 3.3). That is, we can choose in the normalized interval $[-1,+1]$

$$x'_k = \cos \frac{2N+1-2k}{2(N+1)}\pi \text{ for } k = 0, 1, \ldots, N \quad (3.3.1a)$$

and for an arbitrary interval $[a,b]$,

$$x_k = \frac{b-a}{2}x'_k + \frac{a+b}{2} = \frac{b-a}{2}\cos \frac{2N+1-2k}{2(N+1)}\pi + \frac{a+b}{2} \text{ for } k = 0, 1, \ldots, N \quad (3.3.1b)$$

which are referred to as the Chebyshev nodes. The approximating polynomial obtained on the basis of these Chebyshev nodes is called the Chebyshev polynomial.

Let us try the Chebyshev nodes on approximating the function

$$f(x) = \frac{1}{1+8x^2}$$

We can set the 5/9/11 Chebyshev nodes by Eq. (3.3.1) and get the Lagrange or Newton polynomials $c_4(x)$, $c_8(x)$, and $c_{10}(x)$ matching these target points,

138 INTERPOLATION AND CURVE FITTING

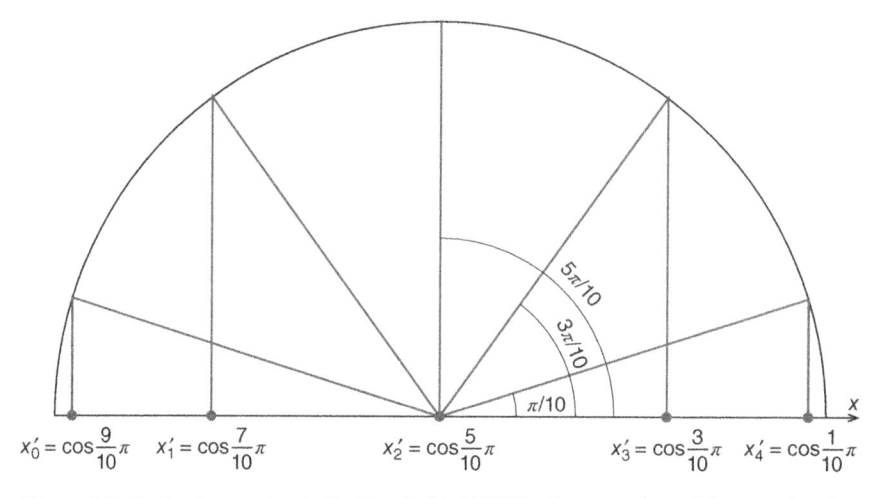

$x'_0 = \cos\frac{9}{10}\pi$ $x'_1 = \cos\frac{7}{10}\pi$ $x'_2 = \cos\frac{5}{10}\pi$ $x'_3 = \cos\frac{3}{10}\pi$ $x'_4 = \cos\frac{1}{10}\pi$

Figure 3.3 Chebyshev nodes (with $N = 4$). (a) 4/8/10th-degree polynomial approximation and (b) the error between the Chebyshev approximating polynomial and the true function.

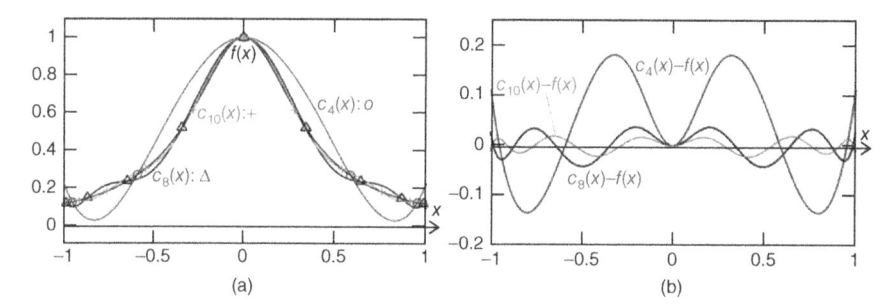

Figure 3.4 Approximation using the Chebyshev polynomial.

which are called the Chebyshev polynomial. We make the MATLAB script "do_lagnewch.m" to do this job and plot the graphs of the polynomial functions together with the graph of the true function $f(x)$ and their error functions separately for comparison as depicted in Figure 3.4. The parts for $c_8(x)$ and $c_{10}(x)$ are omitted to give the readers a chance to practice what they have learned in this section.

Comparing Figure 3.4 with Figure 3.2, we see that the maximum deviation of the Chebyshev polynomial from the true function is considerably less than that of Lagrange/Newton polynomial with equidistant nodes. It can also be seen that increasing the number of the Chebyshev nodes, or equivalently, increasing the degree of Chebyshev polynomial, makes a substantial contribution toward reducing the approximation error.

```
%do_lagnewch.m - plot Fig.3.4
N=4; k=[0:N];
x=cos((2*N+1-2*k)*pi/2/(N+1)); %Chebyshev nodes(Eq.(3.3.1))
y=f31(x);
c=Newtonp(x,y) %Chebyshev polynomial
xx=[-1:0.02: 1]; %the interval to look over
yy=f31(xx); %graph of the true function
yy1=polyval(c,xx); %graph of the approximate polynomial function
subplot(221), plot(xx,yy,'k-', x,y,'o', xx,yy1,'b')
subplot(222), plot(xx,yy1-yy,'r') %graph of the error function
```

There are several things to note about the Chebyshev polynomial.

Remark 3.2 Chebyshev Nodes and Chebyshev Coefficient Polynomials $T_m(x)$.

(1) The Chebyshev coefficient polynomial is defined as

$$T_{N+1}(x') = \cos\{(N+1)\cos^{-1}x'\} \text{ for } -1 \le x' \le +1 \qquad (3.3.2)$$

and the Chebyshev nodes defined by Eq. (3.3.1a) are actually zeros of this function:

$$T_{N+1}(x') = \cos\{(N+1)\cos^{-1}x'\} = 0; (N+1)\cos^{-1}x' = (2k+1)\pi/2$$

(2) Equation (3.3.2) can be written via the trigonometric formula in a recursive form as

$$T_{N+1}(x') = \cos(\cos^{-1}x' + N\cos^{-1}x')$$
$$= \cos(\cos^{-1}x')\cos(N\cos^{-1}x') - \sin(\cos^{-1}x')\sin(N\cos^{-1}x')$$
$$= x'T_N(x') + \frac{1}{2}\{\cos((N+1)\cos^{-1}x') - \cos((N-1)\cos^{-1}x')\}$$
$$= x'T_N(x') + \frac{1}{2}T_{N+1}(x') - \frac{1}{2}T_{N-1}(x');$$

$$T_{N+1}(x') = 2x'T_N(x') - T_{N-1}(x') \text{ for } N \ge 1 \qquad (3.3.3a)$$

$$T_0(x') = \cos 0 = 1, T_1(x') = \cos(\cos^{-1}x') = x' \qquad (3.3.3b)$$

(3) At the Chebyshev nodes x'_k defined by (3.3.1a), the set of Chebyshev coefficient polynomials

$$\{T_0(x'), T_1(x'), \ldots, T_N(x')\}$$

are orthogonal in the sense that

$$\sum_{k=0}^{N} T_m(x'_k)T_n(x'_k) = 0 \quad \text{for } m \neq n \quad (3.3.4a)$$

$$\sum_{k=0}^{N} T_m^2(x'_k) = \frac{N+1}{2} \quad \text{for } m \neq 0 \quad (3.3.4b)$$

$$\sum_{k=0}^{N} T_0^2(x'_k) = N+1 \quad \text{for } m = 0 \quad (3.3.4c)$$

(4) The Chebyshev coefficient polynomials $T_{N+1}(x')$ for up to $N = 6$ are collected in Table 3.3 and their graphs are depicted in Figure 3.5. As can be seen from the table or the graph, the Chebyshev coefficient polynomials of even/odd degree $(N+1)$ are even/odd functions, have an equiripple characteristic with the range of $[-1,+1]$ and the number of rising/falling (intervals) within the domain of $[-1,+1]$ is $N+1$.

We can make use of the orthogonality (Eq. (3.3.4)) of Chebyshev coefficient polynomials to derive the Chebyshev polynomial approximation formula.

$$f(x) \cong c_N(x) = \sum_{m=0}^{N} d_m T_m(x') \Big|_{x' = \frac{2}{b-a}\left(x - \frac{a+b}{2}\right)} \quad (3.3.5)$$

where

$$d_0 = \frac{1}{N+1} \sum_{k=0}^{N} f(x_k) T_0(x'_k) = \frac{1}{N+1} \sum_{k=0}^{N} f(x_k) \quad (3.3.6a)$$

$$d_m = \frac{2}{N+1} \sum_{k=0}^{N} f(x_k) T_m(x'_k)$$

$$= \frac{2}{N+1} \sum_{k=0}^{N} f(x_k) \cos \frac{m(2N+1-2k)}{2(N+1)} \pi \quad \text{for } m = 1, 2, \ldots, N \quad (3.3.6b)$$

Table 3.3 Chebyshev coefficient polynomials.

$T_0(x') = 1$
$T_1(x') = x'$
$T_2(x') = 2x'^2 - 1$
$T_3(x') = 4x'^3 - 3x'$
$T_4(x') = 8x'^4 - 8x'^2 + 1$
$T_5(x') = 16x'^5 - 20x'^3 + 5x'$
$T_6(x') = 32x'^6 - 48x'^4 + 18x'^2 - 1$
$T_7(x') = 64x'^7 - 112x'^5 + 56x'^3 - 7x'$

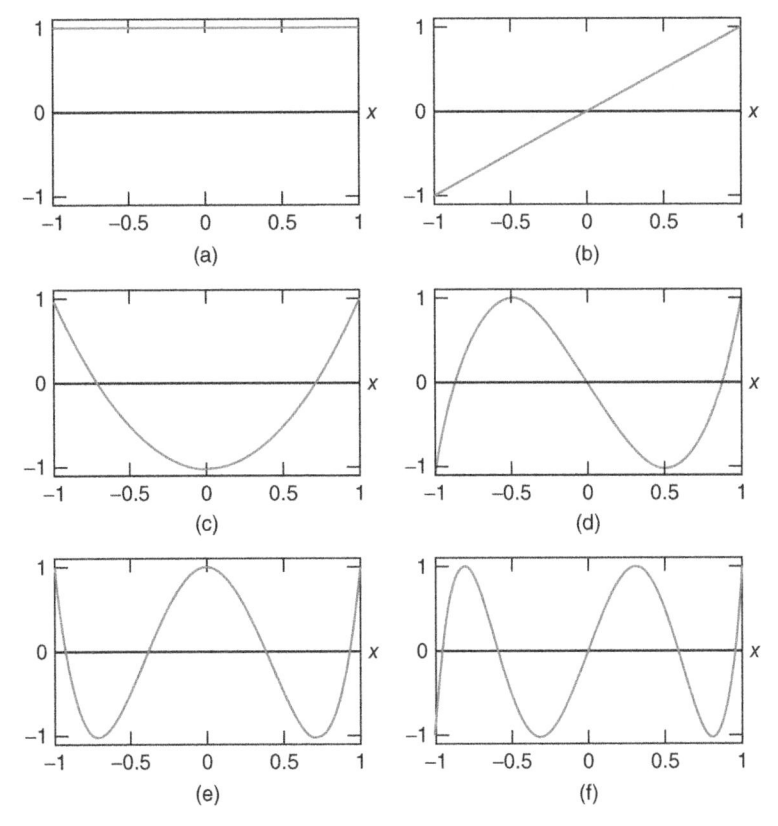

Figure 3.5 Chebyshev polynomial functions. (a) $T_0(x') = 1$, (b) $T_0(x') = x'$, (c) $T_2(x')$, (d) $T_2(x')$, (e) $T_4(x')$, and (f) $T_5(x')$.

We can apply this formula to get the polynomial approximation directly for a given function $f(x)$, without having to resort to the Lagrange or Newton polynomial. Given a function, the degree of the approximate polynomial and the left/right boundary points of the interval, the earlier MATLAB function 'cheby()' uses this formula to make the Chebyshev polynomial approximation.

The following example illustrates that this formula gives the same approximate polynomial function as could be obtained by applying the Newton polynomial with the Chebyshev nodes.

Example 3.1. Approximation by Chebyshev polynomial.
Consider the problem of finding the second-degree ($N = 2$) polynomial to approximate the function $f(x) = 1/(1 + 8x^2)$. We make the following script "do_Cheby.m", which uses the MATLAB function 'Cheby()' for this job and uses Lagrange/Newton polynomial with the Chebyshev nodes to do the same job. Readers can run this script to check if the results are the same.

```
function [c,x,y]=Cheby(f,N,a,b)
%Input : f=function name on [a,b]
%Output: n=Newton polynomial coefficients of degree N
%        (x,y)=Chebyshev nodes
if nargin==2, a=-1; b=1; end
k =[0: N];
theta =(2*N+1-2*k)*pi/(2*N+2);
xn =cos(theta);                    % Eq.(3.3.1a)
x =(b-a)/2*xn +(a+b)/2;            % Eq.(3.3.1b)
y =feval(f,x);
d(1) =y*ones(N+1,1)/(N+1);
for m=2: N+1
   cos_mth =cos((m-1)*theta);
   d(m) =y*cos_mth'*2/(N+1);       % Eq.(3.3.6b)
end
xn = [2 -(a+b)]/(b-a);             % The inverse of (3.3.1b)
T_0 =1;         T_1 =xn;           % Eq.(3.3.3b)
c =d(1)*[0 T_0] +d(2)*T_1;         % Eq.(3.3.5)
for  m=3: N+1
   tmp =T_1;
   T_1 =2*conv(xn,T_1) -[0  0  T_0]; % Eq.(3.3.3a)
   T_0 =tmp;
   c =[0  c] +d(m)*T_1;            % Eq.(3.3.5)
end
```

```
%do_Cheby.m
N=2; a=-2; b=2;
[c,x1,y1]=Cheby('f31',N,a,b) % Chebyshev polynomial ftn
% To compare with Lagrange/Newton polynomial ftn
k=[0:N]; xn=cos((2*N+1-2*k)*pi/2/(N+1));%Eq.(3.3.1a):Chebyshev nodes
x=((b-a)*xn +a+b)/2;   % Eq.(3.3.1b)
y=f31(x);
n=Newtonp(x,y), l=Lagranp(x,y)
```

```
»do_Cheby
   c = -0.3200    -0.0000    1.0000
```

3.4 PADE APPROXIMATION BY RATIONAL FUNCTION

Pade approximation tries to approximate a function $f(x)$ around a point x^o by a rational function

$$p_{M,N}(x - x^o) = \frac{Q_M(x - x^o)}{D_N(x - x^o)}$$

$$= \frac{q_0 + q_1(x - x^o) + q_2(x - x^o)^2 + \cdots + q_M(x - x^o)^M}{1 + d_1(x - x^o) + d_2(x - x^o)^2 + \cdots + d_N(x - x^o)^N} \quad (3.4.1)$$

$$(M = N \text{ or } M = N + 1)$$

where $f(x^o), f'(x^o), f^{(2)}(x^o), \ldots, f^{(M+N)}(x^o)$ are known.

How do we find such a rational function? We write the Taylor series expansion of $f(x)$ up to degree $M+N$ at $x = x^o$ as

$$f(x) \approx T_{M+N}(x - x^o) = f(x^o) + f'(x^o)(x - x^o)$$
$$+ \frac{f^{(2)}(x^o)}{2}(x - x^o)^2 + \cdots + \frac{f^{(M+N)}(x^o)}{(M+N)!}(x - x^o)^{M+N}$$
$$= a_0 + a_1(x - x^o) + a_2(x - x^o)^2 + \cdots + a_{M+N}(x - x^o)^{M+N} \quad (3.4.2)$$

Assuming $x^o = 0$ for simplicity, we get the coefficients of $D_N(x)$ and $Q_M(x)$ such that

$$T_{M+N}(x) - \frac{Q_M(x)}{D_N(x)} = 0;$$

$$\frac{(a_0 + a_1 x + \cdots + a_{M+N} x^{M+N})(1 + d_1 x + \cdots + d_N x^N)}{1 + d_1 x + d_2 x^2 + \cdots + d_N x^N} = 0;$$
$$-(q_0 + q_1 x + \cdots + q_M x^M)$$

$$(a_0 + a_1 x + \cdots + a_{M+N} x^{M+N})(1 + d_1 x + \cdots + d_N x^N) = q_0 + q_1 x + \cdots + q_M x^M \quad (3.4.3)$$

by solving the following equations:

$$\begin{array}{llllll}
a_0 & & & & & = q_0 \\
a_1 & + a_0 d_1 & & & & = q_1 \\
a_2 & + a_1 d_1 & + a_0 d_2 & & & = q_2 \quad (3.4.4a) \\
\vdots & \vdots & \vdots & \vdots & & \vdots \\
a_M & + a_{M-1} d_1 & + a_{M-2} d_2 & \cdots & + a_{M-N} d_N & = q_M \\
\\
a_{M+1} & + a_M d_1 & + a_{M-1} d_2 & \cdots & + a_{M-N+1} d_N & = 0 \\
a_{M+2} & + a_{M+1} d_1 & + a_M d_2 & \cdots & + a_{M-N+2} d_N & = 0 \quad (3.4.4b) \\
\vdots & \vdots & \vdots & \cdots & \vdots & \vdots \\
a_{M+N} & + a_{M+N-1} d_1 & + a_{M+N-2} d_2 & \cdots & + a_M d_N & = 0
\end{array}$$

Here, we must first solve Eq. (3.4.4b) for $d_1, d_2, \ldots, d_N$, and then, substitute d_is into Eq. (3.4.4a) to obtain $q_0, q_1, \ldots, q_M$.

144 INTERPOLATION AND CURVE FITTING

```
function [Q,D]=Padeap(f,xo,M,N,x0,xf)
%Input : f=function to be approximated around xo in [x0,xf]
%Output: Q=numerator coeffs of Pade approximation of degree M
%        D=denominator coeffs of Pade approximation of degree N
a(1)=feval(f,xo);
h=.005; tmp=1;
for i=1:M+N
   tmp=tmp*i*h; % i!h^i
   dix=difapx(i,[-i i])*feval(f,xo+[-i:i]*h)'; % Derivative (Sec.5.3)
   a(i+1)=dix/tmp; % Taylor series coefficient
end
for m=1:N
   n=1:N; A(m,n)=a(M+1+m-n);
   b(m)=-a(M+1+m);
end
d=A\b'; % Eq.(3.4.4b)
for m=1:M+1
   mm= min(m-1,N);
   q(m)=a(m:-1:m-mm)*[1; d(1:mm)]; % Eq.(3.4.4a)
end
Q=q(M+1:-1:1)/d(N); D=[d(N:-1:1)' 1]/d(N); % descending order
if nargout==0 % plot the true ftn, Pade ftn, and Taylor expansion
   if nargin<6, x0=xo-1; xf=xo+1; end
   x=x0+[xf-x0]/100*[0:100]; yt=feval(f,x);
   x1=x-xo; yp=polyval(Q,x1)./polyval(D,x1);
   yT=polyval(a(M+N+1:-1:1),x1);
   clf, plot(x,yt,'k', x,yp,'r', x,yT,'b')
end
```

The MATLAB function 'Padeap()' implements this scheme to find the coefficient vectors of the numerator/denominator polynomial $Q_M(x)/D_N(x)$ of the Pade approximation for a given function $f(x)$. Note the following things.

- The derivatives $f'(x^o), f^{(2)}(x^o), \ldots, f^{(M+N)}(x^o)$ up to order $(M+N)$ are computed numerically by using the function 'difapx()', that will be introduced in Section 5.3.
- In order to compute the values of the Pade approximate function, we substitute $(x-x^o)$ for x into $p_{M,N}(x)$ that has been obtained with the assumption that $x^o = 0$.

Example 3.2. Pade Approximation for $f(x) = e^x$.

Let us find the Pade approximation $p_{3,2}(x) = Q_3(x)/D_2(x)$ for $f(x) = e^x$ around $x^o = 0$. We make the MATLAB script "do_Pade.m", which uses the function 'Padeap()' for this job and uses it again with no output argument to see the graphic results as depicted in Figure 3.6.

```
»do_Pade %Pade approximation
     n = 0.3348    3.0088    12.0266    20.0354
     d = 1.0000   -8.0089    20.0354
```

PADE APPROXIMATION BY RATIONAL FUNCTION 145

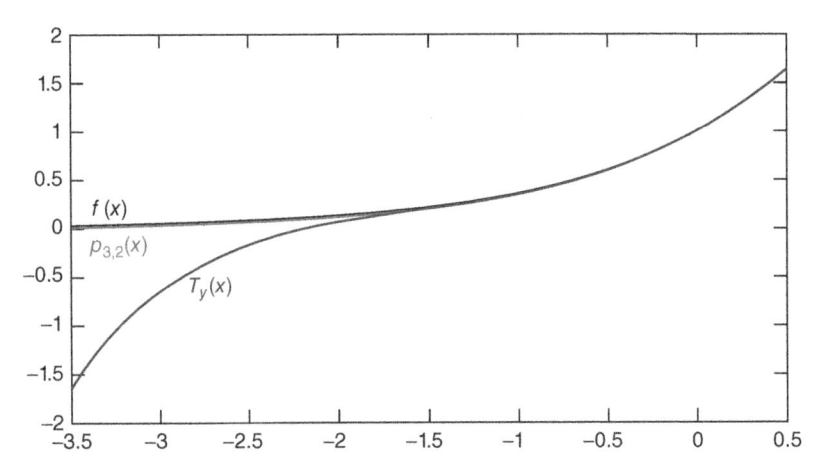

Figure 3.6 Pade approximation and Taylor series expansion for $f(x) = e^x$ (Example 3.2).

```
%do_Pade.m to get the Pade approximation for f(x)=e^x
f1=@(x)exp(x);
M=3; N=2; % Degrees of numerator Q(x) and denominator D(x)
xo=0; % Center of Taylor series expansion
[n,d]=Padeap(f1,xo,M,N) % to get the coefficients of Q(x)/P(x)
x0=-3.5; xf=0.5; % Left/Right boundary of the interval
Padeap(f1,xo,M,N,x0,xf) % to see the graphic results
```

To confirm and support this result from the analytical point of view and to help the readers understand the internal mechanism, we perform the hand-calculation procedure. First, we write the Taylor series expansion at $x = 0$ up to degree $M+N = 5$ for the given function $f(x) = e^x$ as

$$Ty(x) = \sum_{k=0}^{M+N} \frac{f^{(k)}(x)}{k!} x^k = 1 + x + \frac{1}{2}x^2 + \frac{1}{3!}x^3 + \frac{1}{4!}x^4 + \frac{1}{5!}x^5 + \cdots \quad \text{(E3.2.1)}$$

whose coefficients are

$$a_0 = 1, \quad a_1 = 1, \quad a_2 = \frac{1}{2}, \quad a_3 = \frac{1}{6}, \quad a_4 = \frac{1}{24}, \quad a_5 = \frac{1}{120}, \ldots \quad \text{(E3.2.2)}$$

We put this into Eq. (3.4.4b) with $M = 3$, $N = 2$ and solve it for d_is to get $D_2(x) = 1 + d_1 x + d_2 x^2$.

$$\begin{matrix} a_4 + a_3 d_1 + a_2 d_2 = 0 \\ a_3 + a_2 d_1 + a_1 d_2 = 0 \end{matrix}; \begin{bmatrix} 1/6 & 1/2 \\ 1/24 & 1/6 \end{bmatrix} \begin{bmatrix} d_1 \\ d_2 \end{bmatrix} = \begin{bmatrix} -1/24 \\ -1/120 \end{bmatrix}; \begin{bmatrix} d_1 \\ d_2 \end{bmatrix} = \begin{bmatrix} -2/5 \\ 1/20 \end{bmatrix} \quad \text{(E3.2.3)}$$

Substituting this to Eq. (3.4.4a) yields

$$q_0 = a_0 = 1$$
$$q_1 = a_1 + a_0 d_1 = 1 + 1 \times (-2/5) = 3/5 \qquad \text{(E3.2.4)}$$
$$q_2 = a_2 + a_1 d_1 + a_0 d_2 = 1/2 + 1 \times (-2/5) + 1 \times (1/20) = 3/20$$
$$q_3 = a_3 + a_2 d_1 + a_1 d_2 = 1/6 + (1/2) \times (-2/5) + 1 \times (1/20) = 1/60$$

With these coefficients, we write the Pade approximate function as

$$p_{3,2}(x) = \frac{Q_3(x)}{D_2(x)} = \frac{1 + (3/5)x + (3/20)x^2 + (1/60)x^3}{1 + (-2/5)x + (1/20)x^2}$$
$$= \frac{(1/3)x^3 + 3x^2 + 12x + 20}{x^2 - 8x + 20} \qquad \text{(E3.2.5)}$$

This agrees with the numerator and denominator coefficients obtained using 'Padeap()', though not exactly the same.

3.5 INTERPOLATION BY CUBIC SPLINE

If we use the Lagrange/Newton polynomial to interpolate a given set of $N + 1$ data points, the polynomial is usually of degree N and so has $N - 1$ local extrema (maxima/minima). Thus, it will show a wild swing/oscillation (called 'polynomial wiggle') particularly near the ends of the whole interval as the number of data points increases and so the degree of the polynomial gets higher, as illustrated in Figure 3.2. Then, how about a piecewise-linear approach, like assigning the individual approximate polynomial to every subinterval between data points? How about just a linear interpolation, i.e. connecting the data points by a straight-line? It is so simple, but too short of smoothness. Even with the second-degree polynomial, the piecewise-quadratic curve is not smooth enough to please our eyes, since the second-order derivatives of quadratic polynomials for adjacent subintervals cannot be made to conform with each other. In real life, there are many cases where the continuity of second-order derivatives is desirable. For example, it is very important to assure the smoothness up to order 2 for interpolation needed in computer-aided design (CAD)/computer-aided manufacturing (CAM), computer graphic and robot path/trajectory planning. That is why we often resort to the piecewise-cubic curve constructed by the individual third-degree polynomials assigned to each subinterval, which is called the cubic spline interpolation. (A spline is a kind of template that architects use to draw a smooth curve between two points.)

For a given set of data points $\{(x_k, y_k), k = 0:N\}$, the cubic spline $s(x)$ consists of N cubic polynomial $s_k(x)$'s assigned to each subinterval satisfying the following constraints (S0) to (S4).

INTERPOLATION BY CUBIC SPLINE   **147**

Table 3.4 Boundary conditions for a cubic spline.

First-order derivatives specified (clamped end condition)	$s'_0(x_0) = S_{0,1}, \quad s'_N(x_N) = S_{N,1}$
Second-order derivatives specified (end-curvature adjusted)	$s''_0(x_0) = 2S_{0,2}, \quad s''_N(x_N) = 2S_{N,2}$
Second-order derivatives extrapolated (not-a-knot)	$s''_0(x_0) \equiv s''_1(x_1) + \dfrac{h_0}{h_1}(s''_1(x_1) - s''_2(x_2))$ $s''_N(x_N) \equiv s''_{N-1}(x_{N-1}) + \dfrac{h_{N-1}}{h_{N-2}}(s''_{N-1}(x_{N-1}) - s''_{N-2}(x_{N-2}))$

(S0) $s(x) = s_k(x) = S_{k,3}(x - x_k)^3 + S_{k,2}(x - x_k)^2 + S_{k,1}(x - x_k) + S_{k,0}$
for $x \in [x_k, x_{k+1}], \; k = 0:N$

(S1) $s_k(x_k) = S_{k,0} = y_k$ for $k = 0:N$

(S2) $s_{k-1}(x_k) \equiv s_k(x_k) = S_{k,0} = y_k$ for $k = 1:N-1$

(S3) $s'_{k-1}(x_k) \equiv s'_k(x_k) = S_{k,1}$ for $k = 1:N-1$

(S4) $s''_{k-1}(x_k) \equiv s''_k(x_k) = 2S_{k,2}$ for $k = 1:N-1$

These constraints (S1) to (S4) amount to a set of $N + 1 + 3(N - 1) = 4N - 2$ linear equations having $4N$ coefficients of the N cubic polynomials

$$\{S_{k,0}, S_{k,1}, S_{k,2}, S_{k,3}, k = 0:N-1\}$$

as their unknowns. Two additional equations necessary for the equations to be solvable are supposed to come from the boundary conditions for the first/second-order derivatives at the end points (x_0, y_0) and (x_N, y_N) as listed in Table 3.4.

Now, noting from (S1) that $S_{k,0} = y_k$, we will arrange the constraints (S2)–(S4) and eliminate $S_{k,1}, S_{k,3}$,s to set up a set of equations with respect to (w.r.t.) the $N + 1$ unknowns $\{S_{k,2}, k = 0:N\}$. In order to do so, we denote each interval width by $h_k = x_{k+1} - x_k$ and substitute (S0) into (S4) to write

$$s''_k(x_{k+1}) = 6S_{k,3}h_k + 2S_{k,2} \equiv s''_{k+1}(x_{k+1}) = 2S_{k+1,2};$$

$$S_{k,3}h_k = \frac{1}{3}(S_{k+1,2} - S_{k,2}) \qquad (3.5.1a)$$

$$S_{k-1,3}h_{k-1} = \frac{1}{3}(S_{k,2} - S_{k-1,2}) \qquad (3.5.1b)$$

We substitute these equations into (S2) with $k + 1$ in place of k:

$$s_k(x_{k+1}) = S_{k,3}(x_{k+1} - x_k)^3 + S_{k,2}(x_{k+1} - x_k)^2 + S_{k,1}(x_{k+1} - x_k) + S_{k,0} \equiv y_{k+1};$$

$$S_{k,3}h_k^3 + S_{k,2}h_k^2 + S_{k,1}h_k + y_k \equiv y_{k+1}$$

148 INTERPOLATION AND CURVE FITTING

to eliminate $S_{k,3}$,s and rewrite it as

$$\frac{h_k}{3}(S_{k+1,2} - S_{k,2}) + S_{k,2}h_k + S_{k,1} = \frac{y_{k+1} - y_k}{h_k} = dy_k;$$

$$h_k(S_{k+1,2} + 2S_{k,2}) + 3S_{k,1} = 3dy_k \quad (3.5.2a)$$

$$h_{k-1}(S_{k,2} + 2S_{k-1,2}) + 3S_{k-1,1} = 3dy_{k-1} \quad (3.5.2b)$$

We also substitute Eq. (3.5.1b) into (S3)

$$s'_{k-1}(x_k) = 3S_{k-1,3}h_{k-1}^2 + 2S_{k-1,2}h_{k-1} + S_{k-1,1} \equiv s'_k(x_k) = S_{k,1}$$

to write

$$S_{k,1} - S_{k-1,1} = h_{k-1}(S_{k,2} - S_{k-1,2}) + 2h_{k-1}S_{k-1,2} = h_{k-1}(S_{k,2} + S_{k-1,2}) \quad (3.5.3)$$

In order to use this for eliminating $S_{k,1}$ from Eq. (3.5.2), we subtract (3.5.2b) from (3.5.2a) to write

$$h_k(S_{k+1,2} + 2S_{k,2}) - h_{k-1}(S_{k,2} + 2S_{k-1,2}) + 3(S_{k,1} - S_{k-1,1}) = 3(dy_k - dy_{k-1})$$

and then substitute Eq. (3.5.3) into this to write

$$h_k(S_{k+1,2} + 2S_{k,2}) - h_{k-1}(S_{k,2} + 2S_{k-1,2}) + 3h_{k-1}(S_{k,2} + S_{k-1,2}) = 3(dy_k - dy_{k-1});$$

$$h_{k-1}S_{k-1,2} + 2(h_{k-1} + h_k)S_{k,2} + h_kS_{k+1,2} = 3(dy_k - dy_{k-1}) \quad \text{for } k = 1 : N - 1 \quad (3.5.4)$$

Since these are $N-1$ equations w.r.t. $N+1$ unknowns $\{S_{k,2}, k = 0:N\}$, we need two more equations from the boundary conditions to be given as listed in Table 3.4.

How do we convert the boundary condition into equations? In case the first-order derivatives on the two boundary points are given as (i) in Table 3.4, we write Eq. (3.5.2a) for $k = 0$ as

$$h_0(S_{1,2} + 2S_{0,2}) + 3S_{0,1} = 3dy_0; \quad 2h_0S_{0,2} + h_0S_{1,2} = 3(dy_0 - S_{0,1}) \quad (3.5.5a)$$

We also write Eq. (3.5.2b) for $k = N$ as

$$h_{N-1}(S_{N,2} + 2S_{N-1,2}) + 3S_{N-1,1} = 3dy_{N-1}$$

and substitute (3.5.3) ($k = N$) into this to write

$$h_{N-1}(S_{N,2} + 2S_{N-1,2}) + 3S_{N,1} - 3h_{N-1}(S_{N,2} + S_{N-1,2}) = 3dy_{N-1};$$

$$h_{N-1}S_{N-1,2} + 2h_{N-1}S_{N,2} = 3(S_{N,1} - dy_{N-1}) \quad (3.5.5b)$$

Equations (3.5.5a) and (3.5.5b) are two additional equations that we need to solve Eq. (3.5.4) and that is it. In case the second-order derivatives on the two boundary points are given as (ii) in Table 3.4, $S_{0,2}$ and $S_{N,2}$ are directly known from the boundary conditions as

$$S_{0,2} = s_0''(x_0)/2, \quad S_{N,2} = s_N''(x_N)/2 \qquad (3.5.6)$$

and subsequently, we have just $N-1$ unknowns. In case the second-order derivatives on the two boundary points are given as (iii) in Table 3.4

$$s_0''(x_0) \equiv s_1''(x_1) + \frac{h_0}{h_1}(s_1''(x_1) - s_2''(x_2))$$

$$s_N''(x_N) \equiv s_{N-1}''(x_{N-1}) + \frac{h_{N-1}}{h_{N-2}}(s_{N-1}''(x_{N-1}) - s_{N-2}''(x_{N-2}))$$

we can instantly convert these into two equations w.r.t. $S_{0,2}$ and $S_{N,2}$ as

$$h_1 S_{0,2} - (h_0 + h_1) S_{1,2} + h_0 S_{2,2} = 0 \qquad (3.5.7a)$$

$$h_{N-2} S_{N,2} - (h_{N-1} + h_{N-2}) S_{N-1,2} + h_{N-1} S_{N-2,2} = 0 \qquad (3.5.7b)$$

Finally, we combine the two equations, say, ((3.5.5a) and (3.5.5b)) with Eq. (3.5.4) to write it in the matrix-vector form as

$$\begin{bmatrix} 2h_0 & h_0 & 0 & \vdots & & \vdots \\ h_0 & 2(h_0 + h_1) & h_1 & \vdots & & \vdots \\ 0 & \vdots & \vdots & \vdots & & 0 \\ \vdots & \vdots & h_{N-2} & 2(h_{N-2} + h_{N-1}) & h_{N-1} \\ \vdots & \vdots & & 0 & h_{N-1} & 2h_{N-1} \end{bmatrix} \begin{bmatrix} S_{0,2} \\ S_{1,2} \\ \vdots \\ S_{N-1,2} \\ S_{N,2} \end{bmatrix}$$

$$= \begin{bmatrix} 3(dy_0 - S_{0,1}) \\ 3(dy_1 - dy_0) \\ \vdots \\ 3(dy_{N-1} - dy_{N-2}) \\ 3(S_{N,1} - dy_{N-1}) \end{bmatrix} \qquad (3.5.8)$$

After solving this system of equation for $\{S_{k,2}, k = 0{:}N\}$, we substitute them into (S1), (3.5.2), and (3.5.1) to get the other coefficients of the cubic spline as

$$S_{k,0} \stackrel{(S1)}{=} y_k, \quad S_{k,1} \stackrel{(3.5.2)}{=} dy_k - \frac{h_k}{3}(S_{k+1,2} + 2S_{k,2}), \quad S_{k,3} \stackrel{(3.5.1)}{=} \frac{S_{k+1,2} - S_{k,2}}{3h_k} \qquad (3.5.9)$$

```
function [yi,S]=cspline(x,y,xi,KC,dy0,dyN)
%This function finds the cubic splines for the input data points (x.y)
%Input:  x=[x0 x1 ... xN], y=[y0 y1 ... yN], xi=interpolation points
%        KC=0 for zero derivatives at ends (Natural)
%        KC=1/2 for first/second derivatives on boundary specified
%        KC=3 for second derivative on boundary extrapolated (not-a-knot)
%        dy0 =S'(x0)=S01: initial derivative,
%        dyN =S'(xN)=SN1: final derivative
%Output: S(n,k); n=1:N, k=1,4 in descending order
if nargin<6, dyN=0; end, if nargin<5, dy0=0; end
if nargin<4, KC=0; end
N=length(x)-1;
% constructs a set of equations w.r.t. {S(n,2), n=1:N+1}
A=zeros(N+1,N+1); b=zeros(N+1,1);
S=zeros(N+1,4);  % Cubic spline coefficient matrix
k=1:N; h(k)=x(k+1)-x(k); dy(k)=(y(k+1)-y(k))/h(k);
% Boundary condition
if KC==0, dy0=0; dyN=0; end % Zero derivatives at ends (Natural)
if KC<=1 % first derivatives specified
   A(1,1:2)=[2*h(1) h(1)]; b(1)=3*(dy(1)-dy0); % Eq.(3.5.5a)
   A(N+1,N:N+1)=[h(N) 2*h(N)]; b(N+1)=3*(dyN-dy(N)); % Eq.(3.5.5b)
 elseif KC==2 % second derivatives specified
   A(1,1)=2; b(1)=dy0; A(N+1,N+1)=2; b(N+1)=dyN; % Eq.(3.5.6)
 else % second derivatives extrapolated
   A(1,1:3)=[h(2) -h(1)-h(2) h(1)]; %Eq.(3.5.7)
   A(N+1,N-1:N+1)=[h(N) -h(N)-h(N-1) h(N-1)];
end
for m=2:N % Eq.(3.5.8)
   A(m,m-1:m+1)=[h(m-1) 2*(h(m-1)+h(m)) h(m)];
   b(m)=3*(dy(m)-dy(m-1));
end
S(:,3)=A\b;
% Cubic spline coefficients
for m =1:N
   S(m,4) =(S(m+1,3)-S(m,3))/3/h(m);  %Eq.(3.5.9)
   S(m,2) =dy(m) -h(m)/3*(S(m+1,3)+2*S(m,3));
   S(m,1) =y(m);
end
S=S(1:N, 4:-1:1); %descending order
pp=mkpp(x,S); %make piecewise polynomial
yi=ppval(pp,xi); %values of piecewise polynomial ftn
```

The MATLAB function 'cspline()' constructs Eq. (3.5.8), solves it to get the cubic spline coefficients for given x,y coordinates of the data points and the boundary conditions, uses the 'mkpp()' function to get the piecewise polynomial expression and then uses the 'ppval()' function to obtain the value(s) of the piecewise polynomial function for xi, i.e. the interpolation over xi. The type of the boundary condition is supposed to be specified by the third input argument KC. In case the boundary condition is given as (i)/(ii) in Table 3.4, the input argument KC should be set to 1/2 and the fourth/fifth input arguments must be the first-/second-derivatives at the end points. In case the boundary condition is given as extrapolated like (iii) in Table 3.4, the input argument KC should be set to 3 and the fourth/fifth input arguments do not need to be fed.

(cf) See Problem 1.11 for the usages of the MATLAB functions 'mkpp()' and 'ppval()'.

Example 3.3. Cubic Spline.
Consider the problem of finding the cubic spline interpolation for the $N+1 = 4$ data points
$$\{(0,0),(1,1),(2,4),(3,5)\} \tag{E3.3.1}$$
subject to the boundary condition
$$s'_0(x_0) = s'_0(0) = S_{0,1} = 2, \quad s'_N(x_N) = s'_3(3) = S_{3,1} = 2 \tag{E3.3.2}$$
With the subinterval widths on the x-axis and the first divided differences as
$$h_0 = h_1 = h_2 = h_3 = 1$$
$$dy_0 = \frac{y_1 - y_0}{h_0} = 1, \quad dy_1 = \frac{y_2 - y_1}{h_1} = 3, \quad dy_2 = \frac{y_3 - y_2}{h_2} = 1 \tag{E3.3.3}$$
we write Eq. (3.5.8) as
$$\begin{bmatrix} 2 & 1 & 0 & 0 \\ 1 & 4 & 1 & 0 \\ 0 & 1 & 4 & 1 \\ 0 & 0 & 1 & 2 \end{bmatrix} \begin{bmatrix} S_{0,2} \\ S_{1,2} \\ S_{2,2} \\ S_{3,2} \end{bmatrix} = \begin{bmatrix} 3(dy_0 - S_{0,1}) \\ 3(dy_1 - dy_0) \\ 3(dy_2 - dy_1) \\ 3(S_{3,1} - dy_1) \end{bmatrix} = \begin{bmatrix} -3 \\ 6 \\ -6 \\ 3 \end{bmatrix} \tag{E3.3.4}$$

Then, we solve this equation to get
$$S_{0,2} = -3, \quad S_{1,2} = 3, \quad S_{2,2} = -3, \quad S_{3,2} = 3 \tag{E3.3.5}$$
and substitute this into Eq. (3.5.9) to obtain
$$S_{0,0} = 0, \quad S_{1,0} = 1, \quad S_{2,0} = 4 \tag{E3.3.6}$$
$$S_{0,1} = dy_0 - \frac{h_0}{3}(S_{1,2} + 2S_{0,2}) = 1 - \frac{1}{3}(3 + 2 \times (-3)) = 2 \tag{E3.3.7a}$$
$$S_{1,1} = dy_1 - \frac{h_1}{3}(S_{2,2} + 2S_{1,2}) = 3 - \frac{1}{3}(-3 + 2 \times 3) = 2 \tag{E3.3.7b}$$
$$S_{2,1} = dy_2 - \frac{h_2}{3}(S_{3,2} + 2S_{2,2}) = 1 - \frac{1}{3}(3 + 2 \times (-3)) = 2 \tag{E3.3.7c}$$
$$S_{0,3} = \frac{S_{1,2} - S_{0,2}}{3h_0} = \frac{3 - (-3)}{3} = 2 \tag{E3.3.8a}$$

152 INTERPOLATION AND CURVE FITTING

```
%do_csplines.m
KC=1; dy0=2; dyN=2; % with specified first derivatives on boundary
x=[0   1   2   3];   y=[0   1   4   5];
xi=x(1)+[0:200]*(x(end)-x(1))/200; % intermediate points
[yi_cs1,S_cs1]=cspline(x,y,xi,KC,dy0,dyN); S_cs1 % cubic spline
yi_spd=spline(x,[dy0 y dyN],xi); dis1=norm(yi_cs1-yi_spd)
[x_break,S_spd]=unmkpp(spline(x,[dy0 y dyN])); S_spd
yi_sp=spline(x,y,xi); % not-a-knot by default
yi_insp=interp1(x,y,xi,'spline'); dis2=norm(yi_sp-yi_insp)
KC=3; % with second derivatives extrapolated (not-a-knot)
[yi_cs3,S_cs3]=cspline(x,y,xi,KC);
plot(x,y,'ko', xi,yi_spd,'b', xi,yi_sp,'b', xi,yi_cs3,'r:')
```

$$S_{1,3} = \frac{S_{2,2} - S_{1,2}}{3h_1} = \frac{-3-3}{3} = -2 \qquad \text{(E3.3.8b)}$$

$$S_{2,3} = \frac{S_{3,2} - S_{2,2}}{3h_2} = \frac{3-(-3)}{3} = 2 \qquad \text{(E3.3.8c)}$$

Finally, we can write the cubic spline equations collectively from (S0) as

$$s_0(x) = S_{0,3}(x-x_0)^3 + S_{0,2}(x-x_0)^2 + S_{0,1}(x-x_0) + S_{0,0}$$
$$= 2x^3 - 3x^2 + 2x + 0$$

$$s_1(x) = S_{1,3}(x-x_1)^3 + S_{1,2}(x-x_1)^2 + S_{1,1}(x-x_1) + S_{1,0}$$
$$= -2(x-1)^3 + 3(x-1)^2 + 2(x-1) + 1$$

$$s_2(x) = S_{2,3}(x-x_2)^3 + S_{2,2}(x-x_2)^2 + S_{2,1}(x-x_2) + S_{2,0}$$
$$= 2(x-2)^3 - 3(x-2)^2 + 2(x-2) + 4$$

We make and run the script "do_csplines.m", which uses the function 'cspline()' to compute the cubic spline coefficients $\{S_{k,3}, S_{k,2}, S_{k,1}, S_{k,0}, k = 0:N-1\}$ and obtain the value(s) of the cubic spline function for xi, i.e. the interpolation over xi, and then plots the result as depicted in Figure 3.7. We also compare this result with that obtained by using the MATLAB built-in function 'spline(x,y,xi),' which works with the boundary condition of type (i) for the second input argument given as [dy0 y dyN], and with the boundary condition of type (iii) for the same lengths of x and y.

```
»do_csplines %cubic spline
  S =  2.0000   -3.0000    2.0000         0
      -2.0000    3.0000    2.0000    1.0000
       2.0000   -3.0000    2.0000    4.0000
```

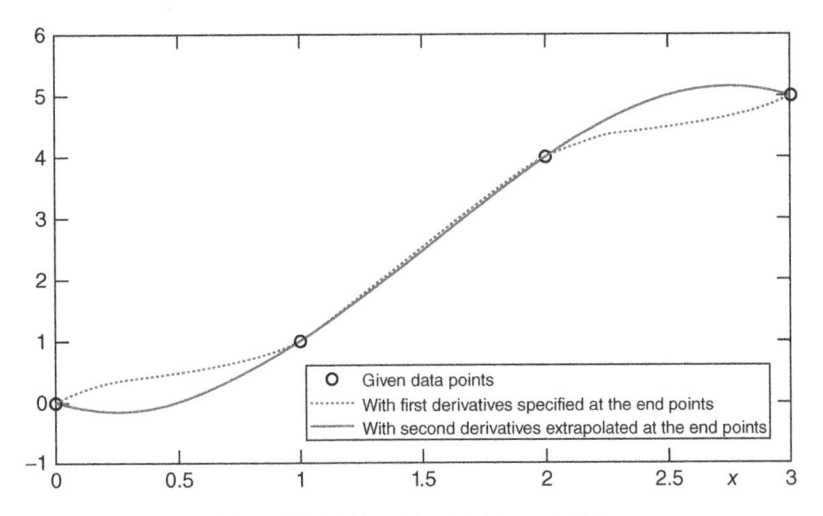

Figure 3.7 Cubic splines for Example 3.3.

3.6 HERMITE INTERPOLATING POLYNOMIAL

In some cases, we need to find the polynomial function that not only passes through the given points but also has the specified derivatives at every data point. We call such a polynomial the Hermite interpolating polynomial or the osculating polynomial.

For simplicity, we consider a third-degree polynomial

$$h(x) = H_3 x^3 + H_2 x^2 + H_1 x + H_0 \qquad (3.6.1)$$

matching just two points (x_0, y_0), (x_1, y_1) and having the specified first derivatives y_0', y_1' at the points. We can obtain the four coefficients H_3, H_2, H_1, and H_0 by solving

$$\begin{aligned} h(x_0) &= H_3 x_0^3 + H_2 x_0^2 + H_1 x_0 + H_0 = y_0 \\ h(x_1) &= H_3 x_1^3 + H_2 x_1^2 + H_1 x_1 + H_0 = y_1 \\ h'(x_0) &= 3H_3 x_0^2 + 2H_2 x_0 + H_1 = y_0' \\ h'(x_1) &= 3H_3 x_1^2 + 2H_2 x_1 + H_1 = y_1' \end{aligned} \qquad (3.6.2)$$

As an alternative, we approximate the specified derivatives at the data points by their differences

$$y_0' = \frac{h(x_0 + \varepsilon) - h(x_0)}{\varepsilon} = \frac{y_2 - y_0}{\varepsilon}, \quad y_1' = \frac{h(x_1) - h(x_1 - \varepsilon)}{\varepsilon} = \frac{y_1 - y_3}{\varepsilon} \qquad (3.6.3)$$

and find the Lagrange/Newton polynomial matching the four points

$$(x_0, y_0), \quad (x_2 = x_0 + \varepsilon, y_2 = y_0 + y_0'\varepsilon), \quad (x_3 = x_1 - \varepsilon, y_3 = y_1 - y_1'\varepsilon), \quad (x_1, y_1) \tag{3.6.4}$$

The MATLAB function 'Hermit()' constructs Eq. (3.6.2) and solves it to get the Hermite interpolating polynomial coefficients for a single interval given the two end points and the derivatives at them as the input arguments. The next function 'Hermits()' uses 'Hermit()' to get the Hermite coefficients for a set of multiple subintervals.

```
function H=Hermit(x0,y0,dy0,x1,y1,dy1)
A =[x0^3     x0^2    x0   1;   x1^3   x1^2   x1   1;
    3*x0^2   2*x0    1    0; 3*x1^2   2*x1   1    0];
b = [y0   y1   dy0 dy1]'; % Eq.(3.6.2)
H = (A\b)';
```

```
function H=Hermits(x,y,dy)
% find Hermite interpolating polynomials for multiple subintervals
%Input  : [x,y],dy - points and derivatives at the points
%Output: H=coefficients of cubic Hermite interpolating polynomials
for n=1:length(x)-1
   H(n,:)=Hermit(0,y(n),dy(n),x(n+1)-x(n),y(n+1),dy(n+1));
end
```

Note that the MATLAB built-in function 'pchip()' can be used to make a shape-preserving interpolation with piecewise Hermit interpolating polynomials or construct the structure of the polynomials for a given set of data points where the derivatives (slopes) at each data point are determined in such a way that the shape/monotonicity of the data can be preserved/respected.

Example 3.4. Hermite Interpolating Polynomial.
Consider the problem of finding the polynomial interpolation for the $N + 1 = 4$ data points

$$\{(0, 0), (1, 1), (2, 4), (3, 5)\} \tag{E3.4.1}$$

subject to the conditions

$$h_0'(x_0) = h_0'(0) = 2, \quad h_1'(1) = 0, \quad h_2'(2) = 0, \quad s_N'(x_N) = s_3'(3) = 2 \tag{E3.4.2}$$

For this problem, we only have to run the following statement:
```
»x=[0 1 2 3]; y=[0 1 4 5]; dy=[2 0 0 2];
H=Hermits(x,y,dy)
pp=pchip(x,y);
[xb,Hp]=unmkpp(pp) % with derivatives unspecified
```

3.7 TWO-DIMENSIONAL INTERPOLATION

In this section, we deal with only the simplest way of two-dimensional (2D) interpolation, that is, a generalization of piecewise linear interpolation called the bilinear interpolation. The bilinear interpolation for a point (x, y) on the rectangular subregion having (x_{m-1}, y_{n-1}) and (x_m, y_n) as its left-upper/right-lower corner points is described by the following formula

$$z(x, y_{n-1}) = \frac{x_m - x}{x_m - x_{m-1}} z_{m-1,n-1} + \frac{x - x_{m-1}}{x_m - x_{m-1}} z_{m,n-1} \qquad (3.7.1a)$$

$$z(x, y_n) = \frac{x_m - x}{x_m - x_{m-1}} z_{m-1,n} + \frac{x - x_{m-1}}{x_m - x_{m-1}} z_{m,n} \qquad (3.7.1b)$$

$$z(x, y) = \frac{y_n - y}{y_n - y_{n-1}} z(x, y_{n-1}) + \frac{y - y_{n-1}}{y_n - y_{n-1}} z(x, y_n)$$

$$= \frac{1}{(x_m - x_{m-1})(y_n - y_{n-1})} \{ (x_m - x)(y_n - y) z_{m-1,n-1}$$

$$+ (x - x_{m-1})(y_n - y) z_{m,n-1} + (x_m - x)(y - y_{n-1}) z_{m-1,n}$$

$$+ (x - x_{m-1})(y - y_{n-1}) z_{m,n} \} \quad \text{for } x_{m-1} \le x \le x_m, \ y_{n-1} \le y \le y_n$$
$$(3.7.2)$$

```
function Zi=interp2_my(x,y,Z,xi,yi)
%To interpolate Z(x,y) on (xi,yi)
M=length(x); N=length(y);
Mi=length(xi); Ni=length(yi);
for mi=1:Mi
 for ni=1:Ni
  for m=2:M
   for n=2:N
     break1=0;
     if xi(mi)<=x(m)&yi(ni)<=y(n)
       tmp=(x(m)-xi(mi))*(y(n)-yi(ni))*Z(n-1,m-1)...
          +(xi(mi)-x(m-1))*(y(n)-yi(ni))*Z(n-1,m)...
          +(x(m)-xi(mi))*(yi(ni)-y(n-1))*Z(n,m-1)...
          +(xi(mi)-x(m-1))*(yi(ni)-y(n-1))*Z(n,m);
       Zi(ni,mi)=tmp/(x(m)-x(m-1))/(y(n)-y(n-1)); % Eq.(3.7.2)
       break1=1;
     end
     if break1>0, break; end
   end
   if break1>0, break; end
  end
 end
end
```

This formula has been cast into the MATLAB function 'interp2_my()', which is so named in order to distinguish it from the MATLAB built-in function 'interp2()'. Note that in reference to Figure 3.8, the given values of data at

156 INTERPOLATION AND CURVE FITTING

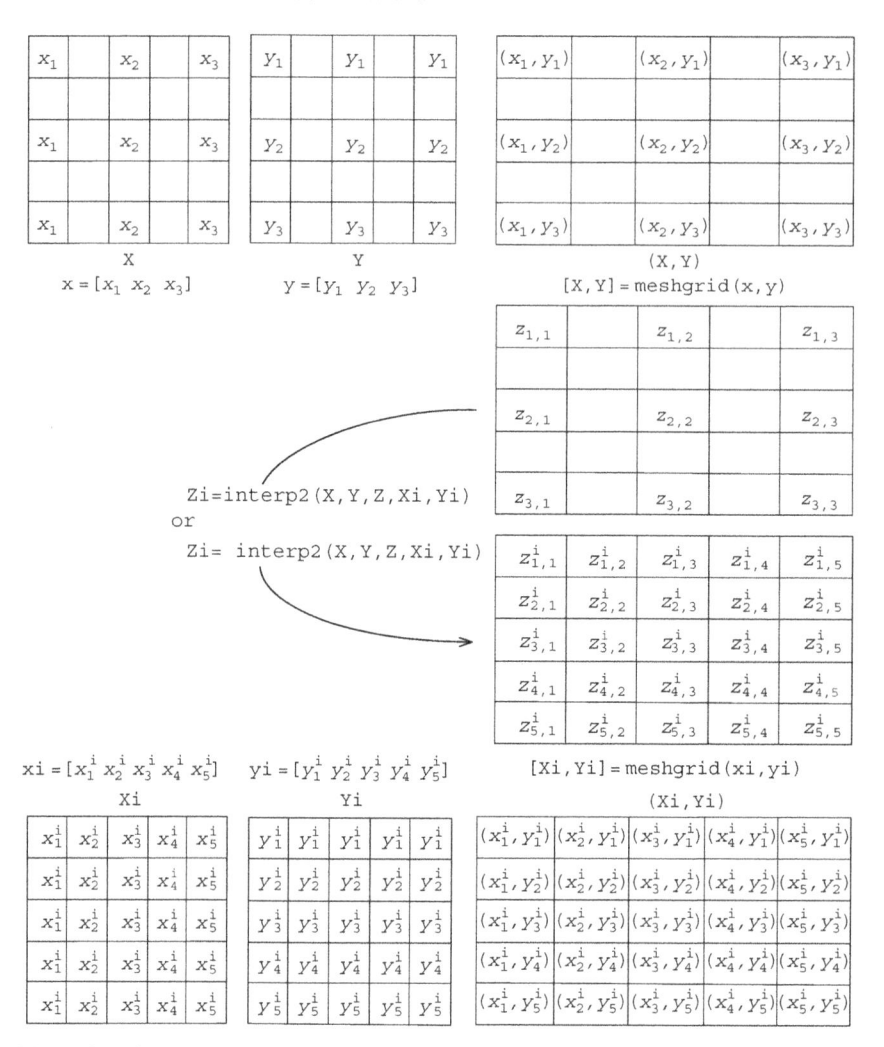

Figure 3.8 2D interpolation using `Zi=interp2()` on the grid array generated using `meshgrid()`.

grid points $(x(m),y(n))$ and the interpolated values for intermediate points $(xi(m),yi(n))$ are stored in $Z(n,m)$ and $Zi(n,m)$, respectively.

Example 3.5. 2D Bilinear Interpolation.
We consider interpolating the sample values of a function

$$f(x,y) = x^2 + y^2 \tag{E3.5.1}$$

for the 5×5 grid over the 21×21 grid on the domain $D = \{(x, y) \mid -2 \leq x \leq 2, -2 \leq y \leq 2\}$.

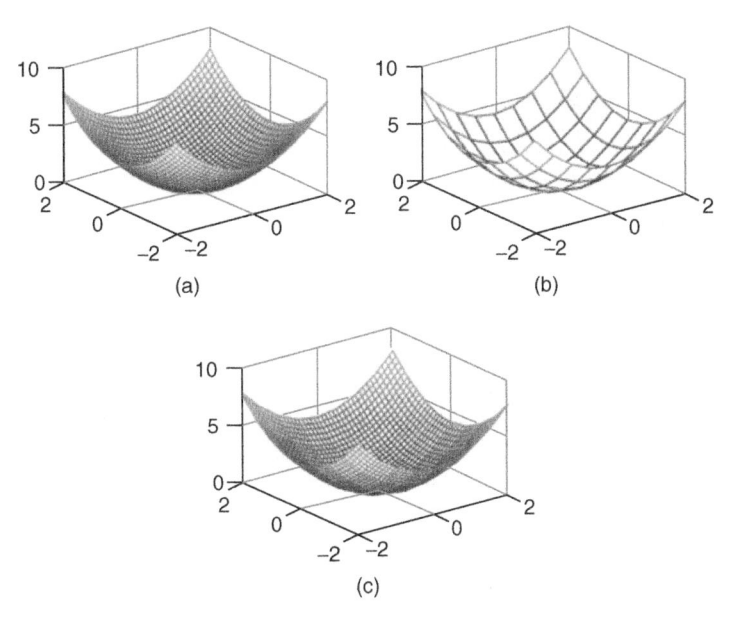

Figure 3.9 Two-dimensional approximation (Example 3.5). (a) True function, (b) the function over sample grid, and (c) bilinear interpolation.

We make the following MATLAB script "do_interp2.m", which uses the function 'interp2_my()' to do this job, compares its function with that of the MATLAB built-in function 'interp2()', and computes a kind of relative error to estimate how close the interpolated values are to the original values. The graphic results of running this script are depicted in Figure 3.9, which shows

```
%do_interp2.m
% 2-dimensional interpolation for Ex 3.5
f=@(x,y)x.^2+y.^2; % Function (E3.5.1)
x=-2:0.5:2; y=-2:0.5:2;
[X,Y]=meshgrid(x,y); % 5x5 grid
xi=-2:0.1:2; yi=-2:0.1:2;
[Xi,Yi]=meshgrid(xi,yi); % 21x21 grid
Z0=f(Xi,Yi);
subplot(131)
mesh(Xi,Yi,Z0)
Z=f(X,Y);
subplot(132)
mesh(X,Y,Z)
Zi1= interp2(x,y,Z,Xi,Yi); % built-in function
subplot(133), mesh(xi,yi,Zi1)
Zi2= interp2_my(x,y,Z,xi,yi); % our own function
pause, mesh(xi,yi,Zi2)
norm(Z0-Zi2)/norm(Z0)
```

that we obtained a reasonable approximation with the error of 2.6% from less than 1/16 of the original data. It is implied that the sampling may be a simple data compression method, as long as the interpolated data are little impaired.

3.8 CURVE FITTING

When many sample data pairs $\{(x_k, y_k), k = 0:M\}$ are available, we often need to grasp the relationship between the two variables or to describe the trend of the data, hopefully in a form of function $y = f(x)$. But, as mentioned in Remark 3.1, the polynomial approach meets with the polynomial wiggle and/or Runge phenomenon, which makes it not attractive for approximation purpose. Although the cubic spline approach may be a roundabout toward the smoothness as explained in Section 3.5, it has too many parameters and so does not seem to be an efficient way of describing the relationship or the trend, since every subinterval needs four coefficients. What other choices do we have? Noting that many data are susceptible to some error, we do not have to try to find a function passing exactly through every point. Instead of pursuing the exact matching at every data point, we look for an approximate function (not necessarily a polynomial) that describes the data points as a whole with the smallest error in some sense, which is called the curve fitting.

As a reasonable means, we consider the least-squares (LS) approach to minimizing the sum of squared errors, where the error is described by the vertical distance to the curve from the data points. We will look over various types of fitting functions in this section.

3.8.1 Straight-Line Fit – A Polynomial Function of Degree 1

If there is some theoretical basis on which we believe the relationship between the two variables to be

$$\theta_1 x + \theta_0 = y \qquad (3.8.1)$$

we should set up the following system of equations from the collection of many experimental data.

$$\theta_1 x_1 + \theta_0 = y_1$$
$$\theta_1 x_2 + \theta_0 = y_2$$
$$\vdots$$
$$\theta_1 x_M + \theta_0 = y_M;$$

CURVE FITTING   159

$$A\theta = y \text{ with } A = \begin{bmatrix} x_1 & 1 \\ x_2 & 1 \\ \vdots & \vdots \\ x_M & 1 \end{bmatrix}, \quad \theta = \begin{bmatrix} \theta_1 \\ \theta_0 \end{bmatrix}, \quad y = \begin{bmatrix} y_1 \\ y_2 \\ \vdots \\ y_M \end{bmatrix} \quad (3.8.2)$$

Noting that this apparently corresponds to the overdetermined case mentioned in Section 2.1.3, we resort to the LS solution (2.1.10)

$$\theta^o = \begin{bmatrix} \theta_1^o \\ \theta_0^o \end{bmatrix} = [A^T A]^{-1} A^T y, \quad (3.8.3)$$

which minimizes the objective function

$$J = \|e\|^2 = \|A\theta - y\|^2 = [A\theta - y]^T [A\theta - y] \quad (3.8.4)$$

Sometimes we have the information about the error bounds of the data, and it is reasonable to differentiate the data by weighing more/less each one according to its accuracy/reliability. This policy can be implemented by the WLS solution

$$\theta_W^o = \begin{bmatrix} \theta_{W1}^o \\ \theta_{W0}^o \end{bmatrix} = [A^T W A]^{-1} A^T W y, \quad (3.8.5)$$

which minimizes the weighted objective function

$$J_W = [A\theta - y]^T W [A\theta - y] \quad (3.8.6)$$

If the weighting matrix is $W = V^{-1} = R^{-T} R^{-1}$, then we can write the WLS solution (3.8.5) as

$$\theta_W^o = \begin{bmatrix} \theta_{W1}^o \\ \theta_{W0}^o \end{bmatrix} = [(R^{-1}A)^T (R^{-1}A)]^{-1} (R^{-1}A)^T R^{-1} y = [A_R^T A_R]^{-1} A_R^T y_R \quad (3.8.7)$$

where

$$A_R = R^{-1} A, \quad y_R = R^{-1} y, \quad W = V^{-1} = R^{-T} R^{-1} \quad (3.8.8)$$

One may use the MATLAB built-in function 'lscov(A,y,V)' to obtain this WLS solution.

160 INTERPOLATION AND CURVE FITTING

3.8.2 Polynomial Curve Fit – A Polynomial Function of Higher Degree

If there is no reason to limit the degree of fitting polynomial to one, then we may increase the degree of fitting polynomial to, say, *N* in expectation of decreasing the error. Still, we can use Eq. (3.8.4, 3.8.6), but with different definitions of *A* and *θ* as

$$A = \begin{bmatrix} x_1^N & \cdots & x_1 & 1 \\ x_2^N & \cdots & x_2 & 1 \\ \vdots & \cdots & \vdots & \vdots \\ x_M^N & \cdots & x_M & 1 \end{bmatrix}, \quad \boldsymbol{\theta} = \begin{bmatrix} \theta_N \\ \vdots \\ \theta_1 \\ \theta_0 \end{bmatrix} \quad (3.8.9)$$

The MATLAB function 'polyfits()' performs the WLS or LS scheme to find the coefficients of a polynomial fitting a given set of data points, depending on whether or not a vector (r) having the diagonal elements of the weighting matrix *W* is given as the fourth or fifth input argument. Note that in the case of a

```
function [th,err,yi]=polyfits(x,y,N,xi,r)
%x,y : Row vectors of data pairs
%N   : Order of polynomial(>=0)
%r   : Reverse weighting factor array of the same dimension as y
M=length(x); x=x(:); y=y(:); % Make all column vectors
if nargin==4
  if length(xi)==M, r=xi; xi=x; % With input argument (x,y,N,r)
   else r=1; % With input argument (x,y,N,xi)
  end
 elseif nargin==3, xi=x; r=1;  % With input argument (x,y,N)
end
A(:,N+1) =ones(M,1);
for n=N:-1:1, A(:,n) =A(:,n+1).*x; end % Eq.(3.8.9)
if length(r)==M
   for m=1:M, A(m,:)=A(m,:)/r(m); y(m)=y(m)/r(m); end % Eq.(3.8.8)
end
th=(A\y)' % Eq.(3.8.3) or (3.8.7)
ye= polyval(th,x); err=norm(y-ye)/norm(y); %estimated y values, error
yi= polyval(th,xi);
```

```
%do_polyfit.m
load xy1.dat
x=xy1(:,1); y=xy1(:,2);
[x,i]=sort(x); y=y(i); % sort the data for plotting
xi=min(x)+[0:100]/100*(max(x)-min(x)); % Intermediate points
for i=1:4
    [th,err,yi]=polyfits(x,y,2*i-1,xi); err %LS
    subplot(220+i)
    plot(x,y,'k*',xi,yi,'b:')
end
```

```
%xy1.dat
-3.0  -0.2774
-2.0   0.8958
-1.0  -1.5651
 0.0   3.4565
 1.0   3.0601
 2.0   4.8568
 3.0   3.8982
```

diagonal weighting matrix W, the WLS solution conforms to the LS solution with each row of the information matrix A and the data vector **y** multiplied by the corresponding element of the weighting matrix W. Let us see the following examples for its usage.

Example 3.6. Polynomial Curve Fit by LS (Least Squares).
Suppose we have an ASCII data file 'xy1.dat' containing a set of data pairs $\{(x_k, y_k), k = 0 : 6\}$ in two columns and we must fit these data into polynomials of degree 1, 3, 5, and 7.

x	−3	−2	−1	0	1	2	3
y	−0.2774	0.8958	−1.5651	3.4565	3.0601	4.8568	3.8982

We make the MATLAB script "do_polyfit.m", which uses the function 'polyfits()' to do this job and plot the results together with the given data points as depicted in Figure 3.10. We can observe the polynomial wiggle that the oscillation of the fitting curve between the data points becomes more pronounced with higher degree.

Example 3.7. Curve Fitting by WLS (Weighted Least Squares).
Most experimental data have some absolute and/or relative error bounds that are not uniform for all data. If we know the error bounds for each data, we may give each data a weight inversely proportional to the size of its error bound when extracting valuable information from the data. The WLS solution (3.8.7) enables us to reflect such a weighting strategy on estimating data trends. Consider the following two cases.

(a) Suppose there are two gauges A and B with the same function, but different absolute error bounds ± 0.2 and ± 1.0, respectively. We used them to get the input to output data pair (x_m, y_n) as

$\{(1, 0.0831), (3, 0.9290), (5, 2.4932), (7, 4.9292), (9, 7.9605)\}$ from gauge A
$\{(2, 0.9536), (4, 2.4836), (6, 3.4173), (8, 6.3903), (10, 10.2443)\}$ from gauge B

Let the fitting function be a second-degree polynomial function

$$y = a_2 x^2 + a_1 x + a_0 \tag{E3.7.1}$$

162 INTERPOLATION AND CURVE FITTING

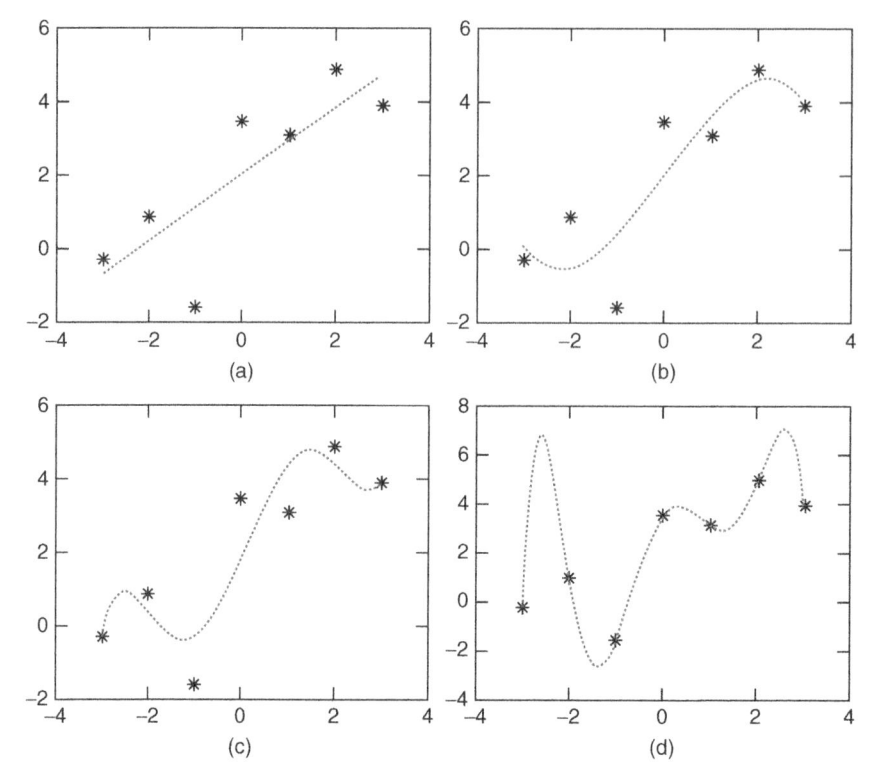

Figure 3.10 Polynomial curve fitting by the LS method. (a) Approximating polynomial of degree 1, (b) approximating polynomial of degree 3, (c) approximating polynomial of degree 5, and (d) Approximating polynomial of degree 7.

To find the parameters a_2, a_1, and a_0, we write the MATLAB script "do_wlse1.m", which uses the function 'polyfits()' twice, once without weighting coefficients and once with weighting coefficients. The results are depicted in Figure 3.11a, which shows that the WLS curve fitting tries to be closer to the data points with smaller error bound, while the LS curve fitting weights all data points equally, which may result in larger deviations from data points with small error bounds.

(b) Suppose we use one gauge that has relative error bound ±40% for measuring the output y for the input values $x = [1, 3, 5, \ldots, 19]$ and so the size of error bound of each output data is proportional to the magnitude of the output. We used it to get the input to output data pair (x_m, y_n) as

{(1,4.7334), (3,2.1873), (5,3.0067), (7,1.4273), (9,1.7787)
(11,1.2301), (13,1.6052), (15,1.5353), (17,1.3985), (19,2.0211)}

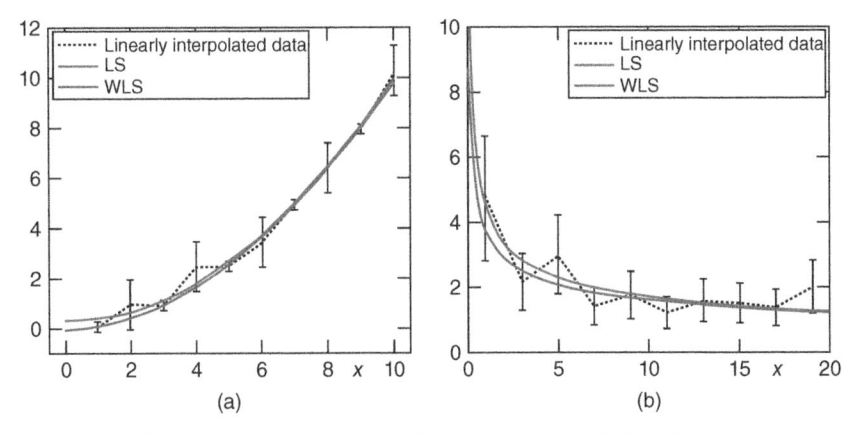

Figure 3.11 LS curve fitting and WLS curve fitting for Example 3.7. (a) Fitting to a polynomial $y = a_2 x^2 + a_1 x + a_0$ and (b) fitting to $y = ax^b$.

Let the fitting function be an exponential function

$$y = a x^b \qquad \text{(E3.7.2)}$$

To find the parameters a and b, we make the MATLAB script "do_wlse2.m", which uses the function 'curve()' without the weighting coefficients one time and with the weighting coefficients another time. The results depicted in Figure 3.11b show that the WLS curve fitting tries to get closer to the data points with smaller |y|, while the LS curve fitting pays equal respect to all data points, which may result in a larger deviation from data points with

```
%do_wlse1.m for Ex.3.7
clear, clf
x=[1 3 5 7 9 2 4 6 8 10]; %input data
y=[0.0831 0.9290 2.4932 4.9292 7.9605 ...
   0.9536 2.4836 3.4173 6.3903 10.2443]; % Output data
eb=[0.2*ones(5,1); ones(5,1)]; % Error bound for each y
[x,i]=sort(x); y=y(i); eb=eb(i); % sort the data for plotting
errorbar(x,y,eb,':'), hold on
N=2; % Degree of the approximate polynomial
xi=[0:100]/10; % Interpolation points
[th,err,yi]=polyfits(x,y,N,xi);
[thw,errw,yiw]=polyfits(x,y,N,xi,eb);
plot(xi,yi,'b', xi,yiw,'r')
%KC=0; thc=curve_fit(x,y,KC,N,xi); % for cross-check
%w=1./eb.^2; thwc=curve_fit(x,y,KC,N,xi,w);
```

164 INTERPOLATION AND CURVE FITTING

```
%do_wlse2.m
clear, clf
x=[1:2:20]; Nx=length(x); % Changing input
xi=[1:200]/10; % Interpolation points
y=[4.7334 2.1873 3.0067 1.4273 1.7787 1.2301 1.6052 1.5353 ...
   1.3985 2.0211];
eb=0.4*y; % Error bound for each y
[x,i]=sort(x); y=y(i); eb=eb(i); % sort the data for plotting
KC=6; [th,err,yi]=curve_fit(x,y,KC,0,xi);
w=1./eb.^2;
[thw,errw,yiw]=curve_fit(x,y,KC,0,xi,w);
errorbar(x,y,eb), hold on
plot(xi,yi,'b', xi,yiw,'r')
```

small error bound. Note that the MATLAB function 'curve_fit()' appears in Problem 3.11, which implements all of the schemes listed in Table 3.5 with the LS/WLS solution.

(cf) Note that the objective of the WLS scheme is to put greater emphasis on more reliable data.

Table 3.5 Linearization of nonlinear functions by parameter/data transformation.

Function to fit	Linearized function	Variable substitution/ parameter restoration
$y = \dfrac{a}{x} + b$	$y = a\dfrac{1}{x} + b \rightarrow y = ax' + b$	$x' = \dfrac{1}{x}$
$y = \dfrac{b}{x+a}$	$\dfrac{1}{y} = \dfrac{1}{b}x + \dfrac{a}{b} \rightarrow y' = a'x + b'$	$y' = \dfrac{1}{y},\ a = \dfrac{b'}{a'},\ b = \dfrac{1}{a'}$
$y = a\,b^x$	$\ln y = (\ln b)x + \ln a \rightarrow y' = a'x + b'$	$y' = \ln y,\ a = e^{b'},\ b = e^{a'}$
$y = b\,e^{ax}$	$\ln y = ax + \ln b \rightarrow y' = ax + b'$	$y' = \ln y,\ b = e^{b'}$
$y = C - b\,e^{-ax}$	$\ln(C-y) = -ax + \ln b \rightarrow y' = a'x + b'$	$y' = \ln(C-y)$ $a = -a',\ b = e^{b'}$
$y = a\,x^b$	$\ln y = b(\ln x) + \ln a \rightarrow y' = a'x' + b'$	$y' = \ln y,\ x' = \ln x$ $a = e^{b'},\ b = a'$
$y = ax\,e^{bx}$	$\ln y - \ln x = bx + \ln a \rightarrow y' = a'x + b'$	$y' = \ln(y/x)$ $a = e^{b'},\ b = a'$
$y = \dfrac{C}{1 + b\,e^{ax}}$ $(a<0, b>0,$ $C = y(\infty))$	$\ln\left(\dfrac{C}{y} - 1\right) = ax + \ln b$ $\rightarrow y' = ax + b'$	$y' = \ln\left(\dfrac{C}{y} - 1\right)$ $b = e^{b'}$
$y = a\ln x + b$	$\rightarrow y = ax' + b$	$x' = \ln x$

3.8.3 Exponential Curve Fit and Other Functions

Why do not we use functions other than the polynomial function as a candidate for fitting functions? There is no reason why we have to stick to the polynomial function, as illustrated in Example 3.7(b). In this section, we consider the case in which the data distribution or the theoretical background behind the data tells us that it is appropriate to fit the data into some nonpolynomial function.

Suppose it is desired to fit the data into the following exponential function.

$$c\, e^{ax} = y \tag{3.8.10}$$

Taking the natural logarithm of both sides, we linearize this as

$$a x + \ln c = \ln y \tag{3.8.11}$$

so that the LS algorithm (3.8.3) can be applied to estimate the parameters a and $\ln c$ based on the data pairs $\{(x_k, \ln y_k), k = 0{:}M\}$.

Like this, there are many other nonlinear relations that can be linearized to fit the LS algorithm, as listed in Table 3.5. This makes us believe in the extensive applicability of the LS algorithm. If you are interested in making a MATLAB function that implements what are listed in this table, see Problem 3.11, which lets you try the MATLAB built-in function 'lsqcurvefit(f,th0,x,y)' that enables one to use any type of function (f) for curve fitting.

Figure 3.12 shows the weighted curve fitting result for Example 3.7(b) obtained by using the interactive curve fitting tool (app) 'cftool' that can be used to fit curves/surfaces to given 1D/2D data and view the fitting results (together with the graph) instantly. You can perform the curve fitting by taking the following steps:

1. Enter

    »do_wlse

    at the MATLAB prompt to run the script "do_wlse2.m" so that the data x, y, and weight (vector) w can be saved in the workspace.

2. Enter

    »cftool

    at the MATLAB prompt to open the 'cftool' app.

3. Select the data x, y, and weight (vector) by clicking on the down arrow button in the corresponding fields.

4. Then the estimates of the parameters a and b and the curve fitting graph will appear in the corresponding panes.

166 INTERPOLATION AND CURVE FITTING

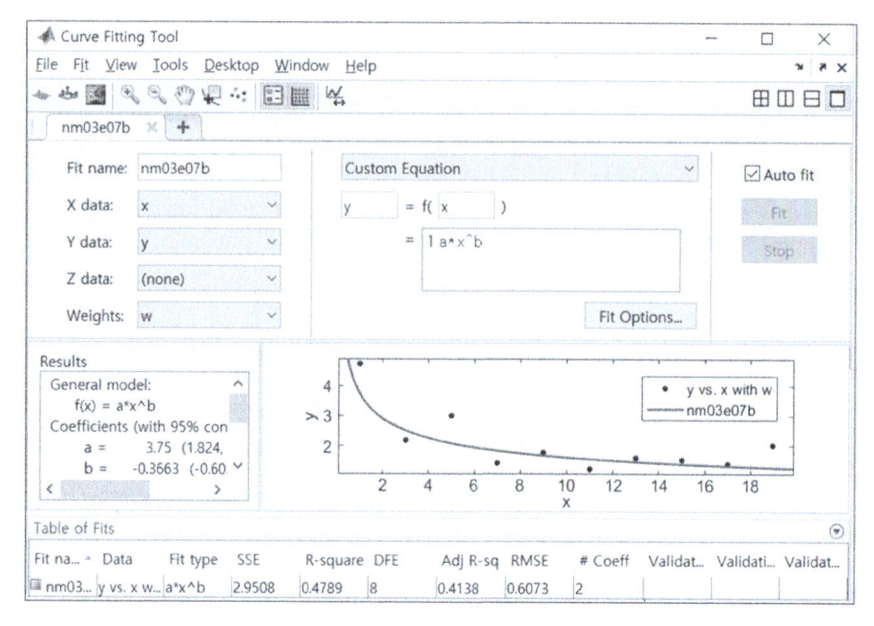

Figure 3.12 Using 'cftool' for curve fitting.

To make the MATLAB function 'curve_fit()' return the same parameter estimates as obtained from 'cftool', modify the part of the function corresponding to KC = 6 as follows where 'fminsearch()' is an optimizing function introduced in Chapter 7:

```
fx=@(x,th)th(1)*x.^th(2); % Eq.(E3.7.2)
fth=@(th)sum(((fx(x,th)-y).*w).^2); % Sum of weighted squared errors
th=fminsearch(fth,[1 1]); % Weighted LS solution
```

3.9 FOURIER TRANSFORM

Most signals existent in this world contain various frequency components, where rapidly/slowly changing one contains high/low frequency components. Fourier series/transform is a mathematical tool that can be used to analyze the frequency characteristic of periodic/aperiodic signals. There are four similar definitions of Fourier series/transform, i.e. continuous time Fourier series (CtFS), continuous time Fourier transform (CtFT), discrete time Fourier transform (DtFT), and discrete Fourier series (DFS)/discrete Fourier transform (DFT). Among these tools, DFT can easily and efficiently be programmed in computer languages and that is why we deal with just DFT in this section.

Suppose a sequence of data $\{x[n] = x(nT), n = 0:M-1\}$ (T: the sampling period) is obtained by sampling a continuous time/space signal once every T

seconds. The $N(\geq M)$-point DFT/inverse discrete Fourier transform (IDFT) pair is defined as

$$\text{DFT}: X(k) = \sum_{n=0}^{N-1} x[n]\, e^{-j2\pi nk/N}, \quad k = 0 : N-1 \qquad (3.9.1a)$$

$$\text{IDFT}: x[n] = \frac{1}{N} \sum_{k=0}^{N-1} X(k)\, e^{j2\pi nk/N}, \quad n = 0 : N-1 \qquad (3.9.1b)$$

Remark 3.3 DFS/DFT(Discrete Fourier Series/Transform).

(0) Note that the indices of the DFT/IDFT sequences appearing in MATLAB range from 1 to N.

(1) Generally, the DFT coefficient $X(k)$ is complex-valued and denotes the magnitude and phase of the signal component having digital frequency $\Omega_k = k\Omega_0 = 2\pi k/N$ (rad), which corresponds to analog frequency $\omega_k = k\omega_0 = 2\pi k/NT$ (rad/s). We call $\Omega_0 = 2\pi/N$ and $\omega_0 = 2\pi/NT$ (N: the DFT size) the digital/analog fundamental or resolution frequency, since it is the minimum digital/analog frequency difference that can be distinguished by the N-point DFT.

(2) The DFS and the DFT are essentially the same, but different in the range of time/frequency interval. More specifically, a signal $x[n]$ and its DFT $X(k)$ are of finite duration over the time/frequency range $\{0 \leq n \leq N-1\}$ and $\{0 \leq k \leq N-1\}$, respectively, while a signal $\tilde{x}[n]$ (to be analyzed by DFS) and its DFS $\tilde{X}(k)$ are periodic with period N over the whole set of integers.

(3) Fast Fourier transform (FFT) means the computationally-efficient algorithm developed by exploiting the periodicity and symmetry in the multiplying factor $e^{j2\pi nk/N}$ to reduce the number of complex number multiplications from N^2 to $(N/2)\log_2 N$ (N: the DFT size). The MATLAB built-in functions 'fft()'/'ifft()' implement the FFT/inverse fast Fourier transform (IFFT) algorithm for the data of length $N = 2^l$ (l: a nonnegative integer). If the length M of the original data sequence is not a power of 2, it can be extended by padding the tail part of the sequence with zeros, which is called zero-padding.

3.9.1 FFT vs. DFT

As mentioned in Remark 3.3(3), FFT/IFFT is the computationally efficient algorithm for computing the DFT/IDFT and is fabricated into the MATLAB functions 'fft()'/'ifft().' In order to practice the use of the MATLAB functions and realize the computational advantage of FFT/IFFT over DFT/IDFT,

168 INTERPOLATION AND CURVE FITTING

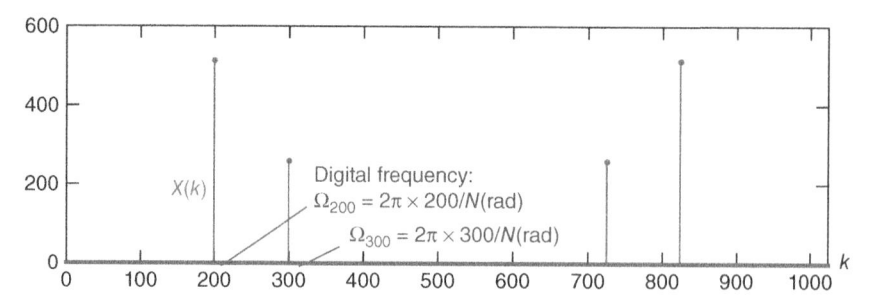

[Plot showing X(k) with peaks, annotated:
Digital frequency:
$\Omega_{200} = 2\pi \times 200/N$ (rad)
$\Omega_{300} = 2\pi \times 300/N$ (rad)]

Figure 3.13 The DFT(FFT) $\{X(k), k = 0: N - 1\}$ of $x[n] = \cos(2\pi \times 200n/N) + 0.5\sin(2\pi \times 300n/N)$ for $n = 0:N-1$ ($N = 2^{10} = 1024$).

```
%compare_DFT_FFT.m
clear, clf
N=2^10; n=[0:N-1];
x=cos(2*pi*200/N*n)+0.5*sin(2*pi*300/N*n);
tic
for k=0:N-1, X(k+1)=x*exp(-j*2*pi*k*n/N).'; end % DFT (Eq.(3.9.1a))
k=[0:N-1];
for n=0:N-1, xr(n+1)=X/N*exp(j*2*pi*k*n/N).'; end % IDFT (Eq.(3.9.1b))
time_dft=toc % Time to perform DFT computation
discrepancy1=norm(x-real(xr))
plot(k,abs(X)), hold on
tic
X1=fft(x); % FFT
xr1=ifft(X1); % IFFT
time_fft=toc % Time to perform FFT computation
discrepancy1=norm(x-real(xr))
plot(k,abs(X1),'r.:') % Magnitude spectrum in Fig.3.12
```

```
%do_fft.m
clear, clf
w1=1.5*pi; w2=3*pi; % Two tones
N=32; n=[0:N-1]; T=0.1; % Sampling period
t=n*T; xan=sin(w1*t)+0.5*sin(w2*t);
subplot(421), stem(t,xan,'.')
k=0:N-1; Xa=fft(xan);
dscrp=norm(xan-real(ifft(Xa))) %x[n] reconstructible from IFFT{X(k)}?
subplot(423), stem(k,abs(Xa),'.')
% Upsampling
N=64; n=[0:N-1]; T=0.05; % Sampling period
t=n*T; xbn=sin(w1*t)+0.5*sin(w2*t);
subplot(422), stem(t,xbn,'.')
k=0:N-1; Xb=fft(xbn);
subplot(424), stem(k,abs(Xb),'.')
% Zero-padding
N=64; n=[0:N-1]; T=0.1; % Sampling period
............................
```

FOURIER TRANSFORM   169

we make the MATLAB script "compare_dft_fft.m". Readers are recommended to run this script and compare the execution times consumed by the 1024-point DFT/IDFT computation and its FFT/IFFT scheme, seeing that the resulting spectra are exactly the same and so are overlapped onto each other as depicted in Figure 3.13.

3.9.2 Physical Meaning of DFT

In order to understand the physical meaning of FFT, we make the MATLAB script "do_fft.m" and run it to get Figure 3.14, which shows the magnitude spectra of the sampled data taken every T seconds from a two-tone analog signal

$$x(t) = \sin(1.5\pi t) + 0.5\cos(3\pi t) \qquad (3.9.2)$$

Readers are recommended to complete the part of this script to get Figure 3.14c,d and run the script to see the plotting results (see Problem 3.16).

What information do the four spectra for the same analog signal $x(t)$ carry? The magnitude of $X_a(k)$ (Figure 3.14a) is large at $k = 2$ and 5, each corresponding to $k\omega_0 = 2\pi k/NT = 2\pi k/3.2 = 1.25\pi \approx 1.5\pi$ and $3.125\pi \approx 3\pi$. The magnitude of $X_b(k)$ (Figure 3.14b) is also large at $k = 2$ and 5, each corresponding to $k\omega_0 = 1.25\pi \approx 1.5\pi$ and $3.125\pi \approx 3\pi$. The magnitude of $X_c(k)$ (Figure 3.14c) is large at $k = 4,5$ and 9,10 and they can be alleged to represent two tones of $k\omega_0 = 2\pi k/NT = 2\pi k/6.4 \approx 1.25\pi \sim 1.5625\pi$ and $2.8125\pi \sim 3.125\pi$. The magnitude of $X_d(k)$ (Figure 3.14d) is also large at $k = 5$ and 10, each corresponding to $k\omega_0 = 1.5625\pi \approx 1.5\pi$ and $3.125\pi \approx 3\pi$.

It is strange and interesting that we have many different DFT spectra for the same analog signal, depending on the DFT size, the sampling period, the whole interval, and zero-padding. Compared with spectrum (a), spectrum (b) obtained by decreasing the sampling period T from 0.1 to 0.05 seconds has wider analog frequency range $[0, 2\pi/T_b]$, but the same analog resolution frequency $\omega_0 = \Omega_0/T_b = 2\pi/N_bT_b = \pi/1.6 = 2\pi/N_aT_a$ and consequently, it does not present us with any new information over (a) for all increased number of data points. The shorter sampling period may be helpful in case the analog signal has some spectral contents of frequency higher than π/T_a. The spectrum (c) obtained by zero-padding has a better-looking, smoother shape, but the vividness is not much improved compared with (a) or (b), since the zeros essentially have no valuable information in the time-domain. In contrast with (b) and (c), spectrum (d) obtained by extending the whole time interval shows us the spectral information more distinctly.

170 INTERPOLATION AND CURVE FITTING

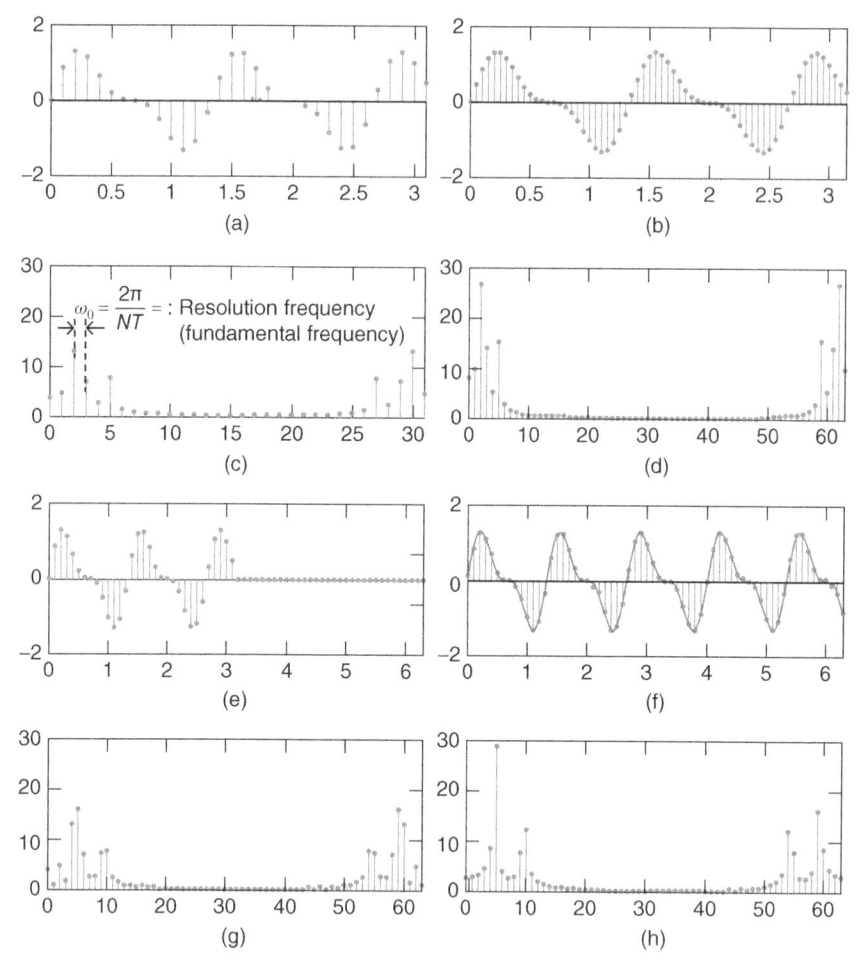

Figure 3.14 DFT spectra of a two-tone signal. (a) polar(th,r), (b) semilogx(x,y), (c) semilogy(x,y), (d) loglog(x,y), (e) stairs([1 3 2 0]), (f) stem([1 3 2 0]), (g) bar([2 3; 4 5]), and (h) barh([2 3; 4 5]).

Note the following things:

- Zero-padding in the time-domain yields the interpolation (smoothing) effect in the frequency domain and vice versa, which will be made use of for data smoothing in the next section (see Problem 3.15).
- If a signal is of finite duration and has the value of zeros outside its domain on the time axis, its spectrum is not discrete, but continuous along the frequency axis, while the spectrum of a periodic signal is discrete as can be seen in Figures 3.13 or 3.14.

- The DFT values $X(0)$ and $X(N/2)$ represent the spectra of the dc component ($\Omega_0 = 0$) and the virtually highest digital frequency components ($\Omega_{N/2} = N/2 \times 2\pi/N = \pi$ (rad)), respectively.

Here, we have something questionable. The DFT spectrum depicted in Figure 3.13 shows clearly the digital frequency components $\Omega_{200} = 200 \times 2\pi/N$ and $\Omega_{300} = 300 \times 2\pi/N$ (rad) ($N = 2^{10} = 1024$) contained in the discrete-time signal

$$x[n] = \cos(2\pi \times 200n/N) + 0.5\sin(2\pi \times 300n/N), \quad N = 2^{10} = 1024 \quad (3.9.3)$$

and so we can find the analog frequency components $\omega_k = \Omega_k/T$ as long as the sampling period T is known, while the DFT spectra depicted in Figure 3.14 are so unclear that we cannot discern even the prominent frequency contents. What is wrong with these spectra? It is never a 'right-or-wrong' problem. The only difference is that the digital frequencies contained in the discrete time signal described by Eq. (3.9.3) are multiples of the fundamental frequency $\Omega_0 = 2\pi/N$, but the analog frequencies contained in the continuous time signal described by Eq. (3.9.2) are not multiples of the fundamental frequency $\omega_0 = 2\pi/NT$; in other words, the whole time interval $[0, NT)$ is not a multiple of the period of each frequency to be detected. The phenomenon that the spectrum becomes blurred like this is said to be the 'leakage problem'. The leakage problem occurs in most cases because we cannot determine the length of the whole time interval in such a way that it is a multiple of the period of the signal as long as we do not know in advance the frequency contents of the signal. If we knew the frequency contents of a signal, why do we bother to find its spectrum that is already known? As a measure to alleviate the leakage problem, there is a windowing technique [O-1, Section 11.2]. Interested readers can see Problem 3.16.

Also note that the periodicity with period N (the DFT size) of the DFT sequence $X(k)$ as well as $x[n]$, as can be manifested by substituting $k+mN$ (m: any integer) for k in Eq. (3.9.1a) and also substituting $n+mN$ for n in Eq. (3.9.1a). A real-world example reminding us of the periodicity of DFT spectrum is the so-called stroboscopic effect that the wheel of a carriage driven by a horse in the scene of a western movie looks like spinning at lower speed than its real speed or even in the reverse direction. The periodicity of $x[n]$ is surprising, because we cannot imagine that every discrete time signal is periodic and that with the period of N, which is the DFT size to be determined by us. As a matter of fact, the 'weird' periodicity of $x[n]$ can be regarded as a kind of cost that we have to pay for computing the sampled DFT spectrum instead of the continuous spectrum $X(\omega)$ for a continuous time signal $x(t)$, which is originally defined as

$$X(\omega) = \int_{-\infty}^{\infty} x(t)e^{-j\omega t}\, dt \quad (3.9.4)$$

172 INTERPOLATION AND CURVE FITTING

Actually, this is to blame for the blurred spectra of the two-tone signal depicted in Figure 3.14.

3.9.3 Interpolation by Using DFS

We can use the DFS/DFT to interpolate a given sequence $x[n]$, which is supposed to have been obtained by sampling some signal at equidistant points (instants). The procedure consists of two steps; to take the N-point FFT $X(k)$ of $x[n]$ and to use the following formula:

```
function [xi,Xi]=interpolation_by_DFS(T,x,Ws,ti)
%T : sampling interval (sample period)
%x : discrete-time sequence
%Ws: normalized stop frequency (1.0=pi[rad])
%ti: interpolation time range or # of divisions for T
if nargin<4, ti=5; end
if nargin<3|Ws>1, Ws=1; end
N=length(x);
if length(ti)==1
   ti=0:T/ti:(N-1)*T; %subinterval divided by ti
end
ks=ceil(Ws*N/2);
Xi=fft(x);
Xi(ks+2:N-ks)=zeros(1,N-2*ks-1); %filtered spectrum
xi=zeros(1,length(ti));
for k=2:N/2
   xi=xi+Xi(k)*exp(j*2*pi*(k-1)*ti/N/T);
end
xi=real(2*xi+Xi(1)+Xi(N/2+1)*cos(pi*ti/T))/N; %Eq.(3.9.5)
```

```
%interpolate_by_DFS.m
clear, clf
w1=pi; w2=0.5*pi; % Two tones
N=32; n=[0:N-1]; T=0.1; t=n*T;
x=sin(w1*t)+0.5*sin(w2*t)+(rand(1,N)-0.5); % 0.2*sin(20*t);
ti=[0:T/5:(N-1)*T];
subplot(411), plot(t,x,'k.') % Original data sequence
title('original sequence and interpolated signal')
[xi,Xi]=interpolation_by_DFS(T,x,1,ti);
hold on, plot(ti,xi,'r') % Reconstructed signal
k=[0:N-1];
subplot(412), stem(k,abs(Xi),'k.') % Original spectrum
title('original spectrum')
[xi,Xi]=interpolation_by_DFS(T,x,1/2,ti);
subplot(413), stem(k,abs(Xi),'r.') % Filtered spectrum
title('filtered spectrum')
subplot(414), plot(t,x,'k.', ti,xi,'r') % Filtered signal
title('filtered/smoothed signal')
```

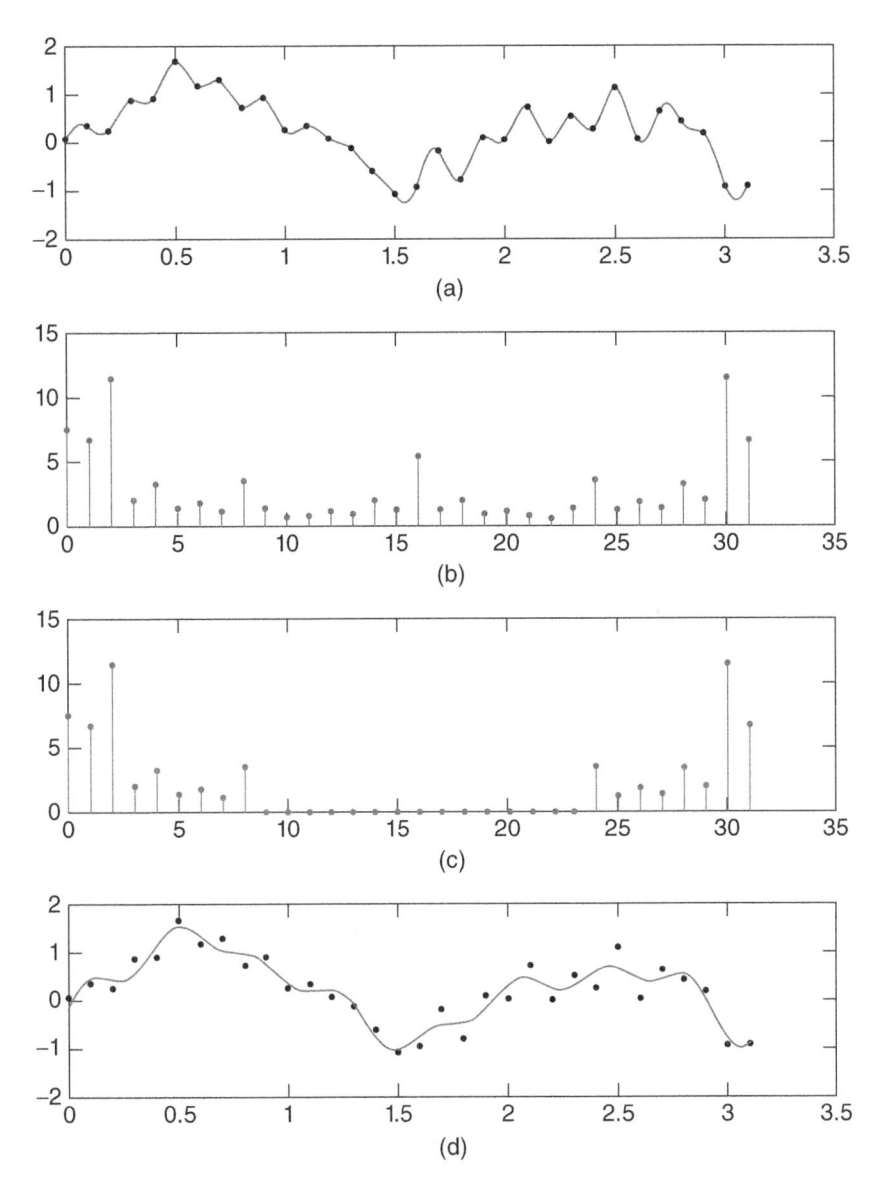

Figure 3.15 Interpolation/smoothing using DFS/DFT. (a) Original discrete-time sequence $x[n]$ and its interpolation using DFS. (b) The DFS/DFT spectrum of $x[n]$. (c) The zero-padded DFS/DFT spectrum of $x[n]$. (d) Original discrete-time sequence $x[n]$ and its interpolation using zero-padded DFS.

$$\hat{x}(t) = \frac{1}{N} \sum_{|k|<N/2} \widetilde{X}(k) \, e^{j2\pi \, kt/NT}$$

$$= \frac{1}{N} \left\{ X(0) + 2 \sum_{k=1}^{N/2-1} \text{Real}\{X(k) \, e^{j2\pi \, kt/NT}\} + X(N/2)\cos(\pi \, t/T) \right\}$$
(3.9.5)

This formula has been cast into the function 'interpolation_by_DFS()', which makes it possible to filter out the high-frequency portion over (wsπ,(2-ws)π) with ws given as the third input argument. The horizontal (time) range over which you want to interpolate the sequence can be given as the fourth input argument ti. We make the MATLAB script "interpolate_by_DFS.m", which applies the function to interpolate a set of data obtained by sampling at equidistant points along the spatial or temporal axis and run it to get Figure 3.15. Figure 3.15a shows a data sequence $x[n]$ of length $N = 32$ and its interpolation (reconstruction) $x(t)$ from the 32-point DFS/DFT $X(k)$ (Figure 3.15b), while Figure 3.15c,d show the (zero-padded) DFT spectrum $X'(k)$ with the digital frequency contents higher than $\pi/2$ (rad) ($N/4 < k < 3N/4$) removed and a smoothed interpolation (fitting curve) $x'(t)$ obtained from $X'(k)$, respectively. This can be viewed as the smoothing effect in the time domain by zero-padding in the frequency domain, in duality with the smoothing effect in the frequency domain by zero-padding in the time domain, which was observed in Figure 3.15c.

PROBLEMS

3.1 Quadratic Interpolation – Lagrange Polynomial and Newton Polynomial

(a) The second-degree Lagrange polynomial matching the three points (x_0,f_0), (x_1,f_1), and (x_2,f_2) can be written by substituting $N = 2$ into Eq. (3.1.3) as

$$l_2(x) = \sum_{m=0}^{2} f_m L_{2,m}(x) = \sum_{m=0}^{2} f_m \prod_{k \neq m}^{N} \frac{x - x_k}{x_m - x_k} \quad \text{(P3.1.1)}$$

Check if the zero of the derivative of this polynomial, i.e. the root of the equation $l_2'(x) = 0$ is found as

$$l_2'(x) = f_0 \frac{(x-x_1) + (x-x_2)}{(x_0-x_1)(x_0-x_2)} + f_1 \frac{(x-x_2) + (x-x_0)}{(x_1-x_2)(x_1-x_0)}$$
$$+ f_2 \frac{(x-x_0) + (x-x_1)}{(x_2-x_0)(x_2-x_1)};$$

$$f_0(2x - x_1 - x_2)(x_2 - x_1) + f_1(2x - x_2 - x_0)(x_0 - x_2)$$
$$+ f_2(2x - x_0 - x_1)(x_1 - x_0) = 0;$$

$$x = x_3 = \frac{f_0(x_1^2 - x_2^2) + f_1(x_2^2 - x_0^2) + f_2(x_0^2 - x_1^2)}{2\{f_0(x_1 - x_2) + f_1(x_2 - x_0) + f_2(x_0 - x_1)\}} \quad \text{(P3.1.2)}$$

You can use the symbolic computation capability of MATLAB by running the following statements:

```
»syms x x0 x1 x2 f0 f1 f2
»L2=f0*(x-x1)*(x-x2)/(x0-x1)/(x0-x2)+...
   f1*(x-x2)*(x-x0)/(x1-x2)/(x1-x0)+...
   f2*(x-x0)*(x-x1)/(x2-x0)/(x2-x1)
»pretty(solve(diff(L2)))
```

(b) The second-degree Newton polynomial matching the three points (x_0,f_0), (x_1,f_1), and (x_2,f_2) is Eq. (3.2.4).

$$n_2(x) = a_0 + a_1(x - x_0) + a_2(x - x_0)(x - x_1) \quad \text{(P3.1.3)}$$

where

$$a_0 = f_0, \quad a_1 = Df_0 = \frac{f_1 - f_0}{x_1 - x_0}$$

$$a_2 = D^2 f_0 = \frac{Df_1 - Df_0}{x_2 - x_0} = \frac{\frac{f_2 - f_1}{x_2 - x_1} - \frac{f_1 - f_0}{x_1 - x_0}}{x_2 - x_0} \quad \text{(P3.1.4)}$$

Find the zero of the derivative of this polynomial.

176 INTERPOLATION AND CURVE FITTING

(c) From Eq. (P3.1.1) with $x_0 = -1$, $x_1 = 0$, and $x_2 = 1$, find the coefficients of Lagrange coefficient polynomials $L_{2,0}(x)$, $L_{2,1}(x)$, and $L_{2,2}(x)$. You had better make use of the function 'Lagranp()' for this job.

(d) From the third-degree Lagrange polynomial matching the four points (x_0,f_0), (x_1,f_1), (x_2,f_2), and (x_3,f_3) with $x_0 = -3$, $x_1 = -2$, $x_2 = -1$, and $x_3 = 0$, find the coefficients of Lagrange coefficient polynomials $L_{3,0}(x)$, $L_{3,1}(x)$, $L_{3,2}(x)$, and $L_{3,3}(x)$. You had better make use of the function 'Lagranp()' for this job.

3.2 Error Analysis of Interpolation Polynomial

Consider the error between a true (unknown) function $f(x)$ and the interpolation polynomial $P_N(x)$ of degree N for some $(N+1)$ points of $y = f(x)$, i.e.

$$\{(x_0, y_0), (x_1, y_1), \ldots, (x_N, y_N)\}$$

where $f(x)$ is up to $(N+1)$th order differentiable. Noting that the error is also a function of x and becomes zero at the $(N+1)$ points, we can write it as

$$e(x) = f(x) - P_N(x) = (x-x_0)(x-x_1)\cdots(x-x_N)g(x) \quad \text{(P3.2.1)}$$

Technically, we define an auxiliary function $w(t)$ w.r.t. t as

$$w(t) = f(t) - P_N(t) - (t-x_0)(t-x_1)\cdots(t-x_N)g(x) \quad \text{(P3.2.2)}$$

Then, this function has the value of zero at the $(N+2)$ points $t = x_0, x_1, \ldots, x_N, x$ and the $1/2/\ldots/(N+1)$th order derivative has $(N+1)/N/\ldots/1$ zeros, respectively. For $t = t_0$ such that $w^{(N+1)}(t_0) = 0$, we have

$$w^{(N+1)}(t_0) = f^{(N+1)}(t_0) - 0 - (N+1)!\, g(x) = 0;\ g(x) = \frac{1}{(N+1)!}f^{(N+1)}(t_0) \quad \text{(P3.2.3)}$$

Based on this, show that the error function can be rewritten as

$$e(x) = f(x) - P_N(x) = (x-x_0)(x-x_1)\cdots(x-x_N)\frac{1}{(N+1)!}f^{(N+1)}(t_0) \quad \text{(P3.2.4)}$$

3.3 The Approximation of a Cosine Function

In the way suggested below, find an approximate polynomial of degree 4 for

$$y = f(x) = \cos x \quad \text{(P3.3.1)}$$

(a) Find the Lagrange/Newton polynomial of degree 4 matching the following five points and plot the resulting polynomial together with the true function $\cos x$ over $[-\pi, +\pi]$.

k	0	1	2	3	4
x_k	$-\pi$	$-\pi/2$	0	$+\pi/2$	$+\pi$
$f(x_k)$	-1	0	1	0	-1

(b) Find the Lagrange/Newton polynomial of degree 4 matching the following five points and plot the resulting polynomial on the same graph that has the result of (a).

k	0	1	2	3	4
x_k	$\pi \cos(9\pi/10)$	$\pi \cos(7\pi/10)$	0	$\pi \cos(3\pi/10)$	$\pi \cos(\pi/10)$
$f(x_k)$	−0.9882	−0.2723	1	−0.2723	−0.9882

(c) Find the Chebyshev polynomial of degree 4 for $\cos x$ over $[-\pi, +\pi]$ and plot the resulting polynomial on the same graph that has the result of (a) and (b).

3.4 Chebyshev Nodes

The current speed/pressure of the liquid flowing in the pipe, which has irregular radius, will be different from place to place. If you are to install seven speed/pressure gauges through the pipe of length 4 m as depicted in Figure P3.4, how would you determine the positions of the gauges so that the maximum error of estimating the speed/pressure over the interval [0,4] can be minimized?

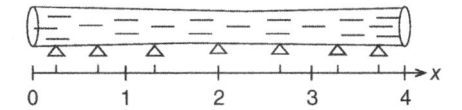

Figure P3.4 Chebyshev nodes.

3.5 Pade Approximation
For the Laplace transform

$$F(s) = e^{-sT} \tag{P3.5.1}$$

representing the delay of T (seconds), we can write its Maclaurin series expansion up to fifth order as

$$Mc(s) \cong 1 - sT + \frac{(sT)^2}{2!} - \frac{(sT)^3}{3!} + \frac{(sT)^4}{4!} - \frac{(sT)^5}{5!} \tag{P3.5.2}$$

(a) Show that we can solve Eq. (3.4.4) and use Eq. (3.4.1) to get the Pade approximation as

$$F(s) \cong p_{1,1}(s) = \frac{q_0 + q_1 s}{1 + d_1 s} = \frac{1 - (T/2)s}{1 + (T/2)s} \cong e^{-Ts} \tag{P3.5.3}$$

(b) Complete the following MATLAB script "nm03p05.m" so that it uses the function 'Padeap()' to generate the Pade approximation of (P3.5.1) with $T = 0.2$ and plots it together with the second-order Maclaurin series expansion and the true function (P3.5.1) for $s = [-5, 10]$. You also run it to see the result as

$$p_{1,1}(s) = \frac{1 - (T/2)s}{1 + (T/2)s} = \frac{-s + 10}{s + 10} \tag{P3.5.4}$$

178 INTERPOLATION AND CURVE FITTING

```
%nm03p05.m
% Pade approximation for f(x)=e^-sT (Eq.(P3.5.1))
T=0.2; f=@(x)exp(-T*x); % Eq.(P3.5.1)
xo=0; M=1; N=?;
[n,d]=Padeap(f,??,?,N)
s=[-5:0.05:10]; % Horizontal range of the graphs to be plotted
yy=exp(-s*T); % True function f(x): Eq.(P3.5.1)
yp=polyval(n,s)./polyval(d,s); % Pade approximate function
Mc=[1 -T T^2/2]; % Maclaurin series coefs up to 2nd order (P3.5.2)
yM=polyval(fliplr(Mc),s);
plot(s,yy,'k', s,yp,'r', s,yM,'b')
```

3.6 Rational Function Interpolation – Bulirsch-Stoer method [S-3]

Table P3.6 shows the Bulirsch-Stoer method, where its element in the mth row and the $(i+1)$th column is computed by the following formula:

$$R_m^{i+1} = R_{m+1}^i + \frac{(x - x_{m+i})(R_{m+1}^i - R_{m+1}^{i-1})(R_{m+1}^i - R_m^i)}{(x - x_m)(R_m^i - R_{m+1}^{i-1}) - (x - x_{m+i})(R_{m+1}^i - R_{m+1}^{i-1})} \quad (P3.6.1)$$

with $R_m^0 = 0$ and $R_m^1 = y_m$ for $i = 1{:}N$ and $i = 1{:}N - i$

Table P3.6 Bulirsch-Stoer method for rational function interpolation.

Data	$i=1$	$i=2$	$i=3$	$i=4$
(x_1, y_1)	$R_1^1 = y_1$	R_1^2	R_1^3	R_1^4
(x_2, y_2)	$R_2^1 = y_2$	R_2^2	R_2^3	⋮
(x_3, y_3)	$R_3^1 = y_3$	R_3^2	⋮	⋮
⋮	⋮	⋮	⋮	⋮
(x_m, y_m)				

(a) The following function 'rational_interpolation(x,y,xi)' uses the Bulirsch-Stoer method to interpolate the set of data pairs (x,y) given as its first/second input arguments over a set xi of intermediate points given as its third input argument. Complete the function, and use it to interpolate the four data points $\{(-1, f(-1)), (-0.2, f(-0.2)), (0.1, f(0.1)), (0.8, f(0.8))\}$ on the graph of

$$f(x) = \frac{1}{1 + 8x^2} \quad (P3.6.2)$$

for xi=[-100:100]/100 and plot the interpolated curve together with the graph of the true function (P3.6.2). Does it work well? How about doing the same job with another function 'rat_interp()' listed in Section 8.3 of [F-1]? What are the values of yi([95:97]) obtained from the two functions? If you come across anything odd in the graphic results and/or the output numbers, what is your explanation?

```
function yi=rational_interpolation(x,y,xi)
N=length(x); Ni=length(xi);
R(:,1)=y(:);
for n=1:Ni
   xn=xi(n);
   for i=1:N-1
      for m=1:N-i
         RR1=R(m+1,i); RR2=R(m,i);
         if i>1,
            RR1=RR1-R(m+1,???); RR2=RR2-R(???,i-1);
         end
         tmp1=(xn-x(???))*RR1;
         num=tmp1*(R(???,i)-R(m,?)); den=(xn-x(?))*RR2-tmp1;
         R(m,i+1)=R(m+1,i) ????/den;
      end
   end
   yi(n)=R(1,N);
end
```

(cf) MATLAB expresses the in-determinant 0/0 (zero-divided-by-zero) as not-a-number (NaN) and skips the value when plotting it on a graph. It may, therefore, be better off for the plotting purpose if we take no special consideration into the case of in-determinant.

(b) Use the Pade approximation function 'Padeap()' (with M=2 and N=2) to generate the rational function approximating (P3.6.2) and compare the result with the true function (P3.6.2).

(c) To compare the rational interpolation method with the Pade approximation scheme, apply the functions 'rational_interpolation()' and 'padeap()' (with M=3 and N=2) to interpolate the four data points $\{(-2, f(-2)), (-1, f(-1)), (1, f(1)), (2, f(2))\}$ on the graph of $f(x) = \sin x$ for xi=[-100:100]*pi/100 and plot the interpolated curve together with the graph of the true function. How do you compare the approximation/interpolation results?

3.7 Smoothness of a Cubic Spline Function

We claim that the cubic spline interpolation function $s(x)$ has the smoothness property of

$$\int_{x_k}^{x_{k+1}} (s''(x))^2 \, dx \leq \int_{x_k}^{x_{k+1}} (f''(x))^2 \, dx \tag{P3.7.1}$$

for any second-order differentiable function $f(x)$ matching the given grid points and having the same first-order derivatives as $s(x)$ at the grid points. This implies that the cubic spline functions are not so rugged. Prove it by doing the following:

(a) Check the validity of the following equality:

$$\int_{x_k}^{x_{k+1}} f''(x) s''(x) dx = \int_{x_k}^{x_{k+1}} (s''(x))^2 \, dx \tag{P3.7.2}$$

where the left-hand and right-hand sides of this equation are

$$\text{LHS}: \int_{x_k}^{x_{k+1}} f''(x)\, s''(x)\, dx = f'(x) s''(x)\big|_{x_k}^{x_{k+1}} - \int_{x_k}^{x_{k+1}} f'(x)\, s'''(x)\, dx$$

$$= f'(x_{k+1}) s''(x_{k+1}) - f'(x_k) s''(x_k) - C(f(x_{k+1}) - f(x_k)) \quad \text{(P3.7.3a)}$$

$$\text{RHS}: \int_{x_k}^{x_{k+1}} s''(x) s''(x)\, dx$$

$$= s'(x_{k+1}) s''(x_{k+1}) - s'(x_k) s''(x_k) - C(s(x_{k+1}) - s(x_k)) \quad \text{(P3.7.3b)}$$

(b) Check the validity of the following inequality:

$$0 \le \int_{x_k}^{x_{k+1}} (f''(x) - s''(x))^2\, dx$$

$$= \int_{x_k}^{x_{k+1}} (f''(x))^2\, dx - 2 \int_{x_k}^{x_{k+1}} f''(x)\, s''(x)\, dx + \int_{x_k}^{x_{k+1}} (s''(x))^2\, dx$$

$$\overset{(P3.7.2)}{=} \int_{x_k}^{x_{k+1}} (f''(x))^2\, dx - \int_{x_k}^{x_{k+1}} (s''(x))^2\, dx;$$

$$\int_{x_k}^{x_{k+1}} (s''(x))^2\, dx \le \int_{x_k}^{x_{k+1}} (f''(x))^2\, dx \quad \text{(P3.7.4)}$$

3.8 MATLAB Built-in Function for Cubic Spline

There are two MATLAB built-in functions

```
»yi=spline(x,y,xi);
»yi=interp1(x,y,xi,'spline');
```

Both receive a set of data points (x,y) and return the values of the cubic spline interploating function $s(x)$ for the (intermediate) points xi given as the third input argument. Write a script that uses these MATLAB functions to get the interpolation for the following set of data points:

$$\{(0,0), (0.5,2), (2,-2), (3.5,2), (4,0)\}$$

and plots the results for [0,4]. In this script, append the statements that do the same job by using the function 'cspline(x,y,KC)' (Section 3.5) with KC = 1, 2, and 3. Which one yields the same result as the MATLAB built-in function? What kind of boundary condition does the MATLAB built-in function assume?

3.9 Robot Path Planning using Cubic Spline

Every object having a mass is subject to the law of inertia and so its speed described by the first derivative of its displacement with respect to time must be continuous in any direction. In this context, the cubic spline having the continuous derivatives up to second-order presents a good basis for planning the robot path/trajectory. We will determine the path of a robot in such a way that the following conditions are satisfied:

- At time $t = 0$ s, the robot starts from its home position (0,0) with zero initial velocity, passing through the intermediate point (1,1) at $t = 1$ s and arriving at the final point (2,4) at $t = 2$ s.
- On arriving at (2,4), it starts the point at $t = 2$ s, stopping by the intermediate point (3,3) at $t = 3$ s and arriving at the point (4,2) at $t = 4$ s.
- On arriving at (4,2), it starts the point, passing through the intermediate point (2,1) at $t = 5$ s and then returning to the home position (0,0) at $t = 6$ s.

More specifically, what we need are the following:

- The spline interpolation matching the three points (0,0), (1,1), (2,2) and having zero velocity at both boundary points (0,0) and (2,2).
- The spline interpolation matching the three points (2,2), (3,3), (4,4) and having zero velocity at both boundary points (2,2) and (4,4).
- The spline interpolation matching the three points (4,4), (5,2), (6,0) and having zero velocity at both boundary points (4,4) and (6,0) on the tx-plane.

On the ty-plane, the following:

- The spline interpolation matching the three points (0,0), (1,1), (2,4) and having zero velocity at both boundary points (0,0) and (2,4).
- The spline interpolation matching the three points (2,4), (3,3), (4,2) and having zero velocity at both boundary points (2,4) and (4,2).
- The spline interpolation matching the three points (4,2), (5,1), (6,0) and having zero velocity at both boundary points (4,2) and (6,0).

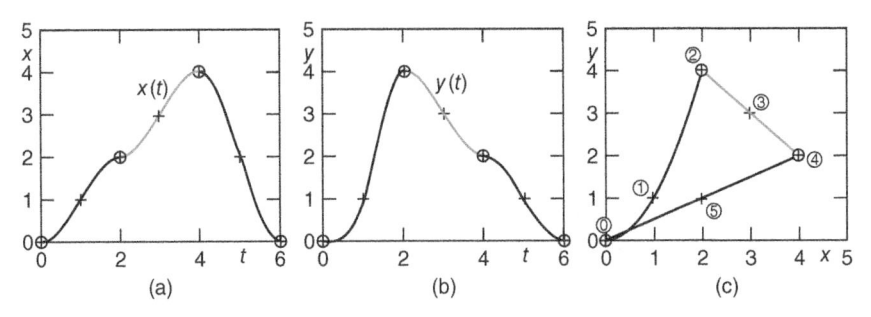

```
%robot_path.m
x1=[0 1 2]; y1=[0 1 4]; t1=[0 1 2]; ti1=[0:0.05: 2];
xi1=cspline(t1,x1,ti1);  yi1=cspline(t1,y1,ti1);
x2=[2 3 4]; y2=[4 3 2]; t2=[2 3 4]; ti2=[2:0.05: 4];
............................
plot(xi1,yi1,'k', xi2,yi2,'b', xi3,yi3,'k'), hold on
plot([x1(1) x2(1) x3(1) x3(end)],[y1(1) y2(1) y3(1) y3(end)],'o')
plot([x1 x2 x3],[y1 y2 y3],'k+'), axis([0 5 0 5])
```

Figure P3.9 Coordinates and path of a robot planned using the cubic spline. (a) x-coordinate varying with t, (b) y-coordinate varying with t, and (c) robot path on the xy-planes.

182 INTERPOLATION AND CURVE FITTING

Complete the above script 'robot_path' so that it makes the required spline interpolations and plots the whole robot path obtained through the interpolations on the xy plane. Run it to get the robot path as depicted in Figure P3.9c.

3.10 One-dimensional (1D) Interpolation

What do you have to give as the fourth input argument of the MATLAB built-in function 'interp1()' in order to get the same result as that would be obtained by using the following 1D interpolation function 'intrp1()'? What letter would you see if you apply this function to interpolate the data points $\{(0,3), (1,0), (2,3), (3,0), (4,3)\}$ for $[0,4]$?

```
function yi=intrp1(x,y,xi)
M=length(x); Mi=length(xi);
for mi=1: Mi
   if xi(mi)<x(1),  yi(mi)=y(1)-(y(2)-y(1))/(x(2)-x(1))*(x(1)-xi(mi));
   elseif xi(mi)>x(M)
       yi(mi)=y(M)+(y(M)-y(M-1))/(x(M)-x(M-1))*(xi(mi)-x(M));
   else
       for m=2:M
          if xi(mi)<=x(m)
             yi(mi)=y(m-1)+(y(m)-y(m-1))/(x(m)-x(m-1))*(xi(mi)-x(m-1));
             break;
          end
       end
   end
end
```

3.11 Least-Squares Curve Fitting

(a) There are several nonlinear relations listed in Table 3.5, which can be linearized to fit the LS algorithm. The MATLAB function 'curve_fit()' implements all the schemes which use the LS method to find the parameters for the template relations, but the parts for the relations (1), (2), (7), (8), (9) are missing. Supplement the missing parts to complete the function.

(b) The script 'nm03p11' generates the 12 sets of data pairs according to various types of relations (functions), applies the functions 'curve_fitting()'/'lsqcurvefit()' to find the parameters of the template relations and plots the data pairs on the fitting curves obtained from the template functions with the estimated parameters. Complete and run it to get the graphs like Figure P3.11. Answer the following questions.

(i) If any, find the case(s) where the results of using the two functions make a great difference. For the case(s), try with another initial guess th0 = [1 1] of parameters, instead of th0 = [0 0].

(ii) If the MATLAB built-in function 'lsqcurvefit()' yields a bad result, does it always give you a warning message? How do you compare the two functions?

```
function [th,err,yi]=curve_fit(x,y,KC,C,xi,w)
% Use LS method for curve-fitting
% KC = Approximating function number in Table 3.5
% C  = Optional constant (final value) for KC!=0 (nonlinear LS)
%      degree for KC=0 (standard LS)
% xi = Value(s) of x at which the interpolation is needed
% w  = Weighting factor proportional to (1/sgma^2)
Nx=length(x); x=x(:); y=y(:);
if nargin==6, w=w(:);
   elseif length(xi)==Nx, w=xi(:); xi=x;
   else w=ones(Nx,1);
end
w=sqrt(w);
if nargin<5, xi=x; end;   if nargin<4|C<1, C=1; end
switch KC
  case 1
       .........................
  case 2
       .........................
  case {3,4}
    A(1:Nx,:) =[x.*w  w];    RHS= log(y).*w;     th=A\RHS;
    yi =exp(th(1)*xi+th(2));  y2 =exp(th(1)*x+th(2));
    if KC==3, th=exp([th(2) th(1)]);
     else   th(2)=exp(th(2));
    end
  case 5
    if nargin<5, C=max(y)+1; end %final value
    A(1:Nx,:)=[x.*w  w];
    y1=y;  y1(find(y>C-0.01))=C-0.01;
    RHS=log(C-y1).*w;       th=A\RHS;
    yi=C-exp(th(1)*xi+th(2));  y2=C-exp(th(1)*x+th(2));
    th=[-th(1) exp(th(2))];
  case 6
    A(1:Nx,:) =[log(x).*w  w];   y1=y;  y1(find(y<0.01))=0.01;
    RHS=log(y1)./sig;      th=A\RHS;
    yi =exp(th(1)*log(xi)+th(2));  y2 =exp(th(1)*log(x)+th(2));
    th=[exp(th(2)) th(1)];
  case 7 .........................
  case 8 .........................
  case 9 .........................
  otherwise % Standard LS with degree C
    A(1:Nx,C+1)=w;
    for n=C:-1:1, A(1:Nx,n) =A(1:Nx,n+1).*x;  end
    RHS=y.*w;     th=A\RHS;
    yi=th(C+1);   tmp=ones(size(xi));
    y2=th(C+1);   tmp2=ones(size(x));
    for n=C:-1:1,
        tmp=tmp.*xi;  yi= yi+th(n)*tmp;
        tmp2=tmp2.*x; y2= y2+th(n)*tmp2;
    end
end
th=th(:)';  err=norm(y-y2);
if nargout==0, plot(x,y,'*', xi,yi,'k-'); end
```

184 INTERPOLATION AND CURVE FITTING

```
%nm03p11.m to plot Fig.P3.11 by curve fitting
x =[1:20]*2-0.1; Nx=length(x);
noise =rand(1,Nx)-0.5; % 1xNx random noise generator
xi=[1:40]-0.5; % interpolation points
figure(1), clf
a=0.1; b=-1; c=-50;            %Table 3.5(0)
y =a*x.^2+b*x+c +10*noise(1:Nx);
[th,err,yi]=curve_fit(x,y,0,2,xi); [a b c], th
[a b c],th %if you want parameters
f=@(th,x)th(1)*x.^2+th(2)*x+th(3);
[th,err]=lsqcurvefit(f,[0 0 0],x,y), yi1=f(th,xi);
subplot(621), plot(x,y,'*', xi,yi,'k', xi,yi1,'r')
a=2; b=1; y =a./x +b +0.1*noise(1:Nx);   %Table 3.5(1)
[th,err,yi]=curve_fit(x,y,1,0,xi); [a b], th
f=@(th,x)th(1)./x+th(2);
th0=[0 0]; [th,err]=lsqcurvefit(f,th0,x,y), yi1=f(th,xi);
subplot(622), plot(x,y,'*', xi,yi,'k', xi,yi1,'r')
a=-20; b=-9; y=b./(x+a) +0.4*noise(1:Nx); %Table 3.5(2)
[th,err,yi]=curve_fit(x,y,2,0,xi); [a b], th
f=@(th,x)th(2)./(x+th(1));
th0=[0 0]; [th,err]=lsqcurvefit(f,th0,x,y), yi1=f(th,xi);
subplot(623), plot(x,y,'*', xi,yi,'k', xi,yi1,'r')
a=2.; b=0.95; y=a*b.^x +0.5*noise(1:Nx); %Table 3.5(3)
[th,err,yi]=curve_fit(x,y,3,0,xi); [a b], th
f=@(th,x)th(1)*th(2).^x;
th0=[0 0]; [th,err]=lsqcurvefit(f,th0,x,y), yi1=f(th,xi);
subplot(624), plot(x,y,'*', xi,yi,'k', xi,yi1,'r')
a=0.1; b=1; y=b*exp(a*x) +2*noise(1:Nx); %Table 3.5(4)
[th,err,yi]=curve_fit(x,y,4,0,xi); [a b], th
f=@(th,x)th(2)*exp(th(1)*x);
th0=[0 0]; [th,err]=lsqcurvefit(f,th0,x,y), yi1=f(th,xi);
subplot(625), plot(x,y,'*', xi,yi,'k', xi,yi1,'r')
a=0.1; b=1;                %Table 3.5(5)
y =-b*exp(-a*x); C=-min(y)+1; y= C+y+0.1*noise(1:Nx);
[th,err,yi]=curve_fit(x,y,5,C,xi); [a b], th
f=@(th,x)1-th(2)*exp(-th(1)*x);
th0=[0 0]; [th,err]=lsqcurvefit(f,th0,x,y), yi1=f(th,xi);
subplot(626), plot(x,y,'*', xi,yi,'k', xi,yi1,'r')
figure(2), clf
a=0.5; b=0.5;   y=a*x.^b +0.2*noise(1:Nx); %Table 3.5(6a)
[th,err,yi]=curve_fit(x,y,0,2,xi); [a b], th
f=@(th,x)th(1)*x.^th(2);
th0=[0 0]; [th,err]=lsqcurvefit(f,th0,x,y), yi1=f(th,xi);
subplot(6,2,7), plot(x,y,'*', xi,yi,'k', xi,yi1,'r')
a=0.5; b=-0.5;             %Table 3.5(6b)
y=a*x.^b +0.05*noise(1:Nx);
[th,err,yi]=curve_fit(x,y,6,0,xi); [a b], th
f=@(th,x)th(1)*x.^th(2);
th0=[0 0]; [th,err]=lsqcurvefit(f,th0,x,y), yi1=f(th,xi);
subplot(6,2,8), plot(x,y,'*', xi,yi,'k', xi,yi1,'r')
```

PROBLEMS 185

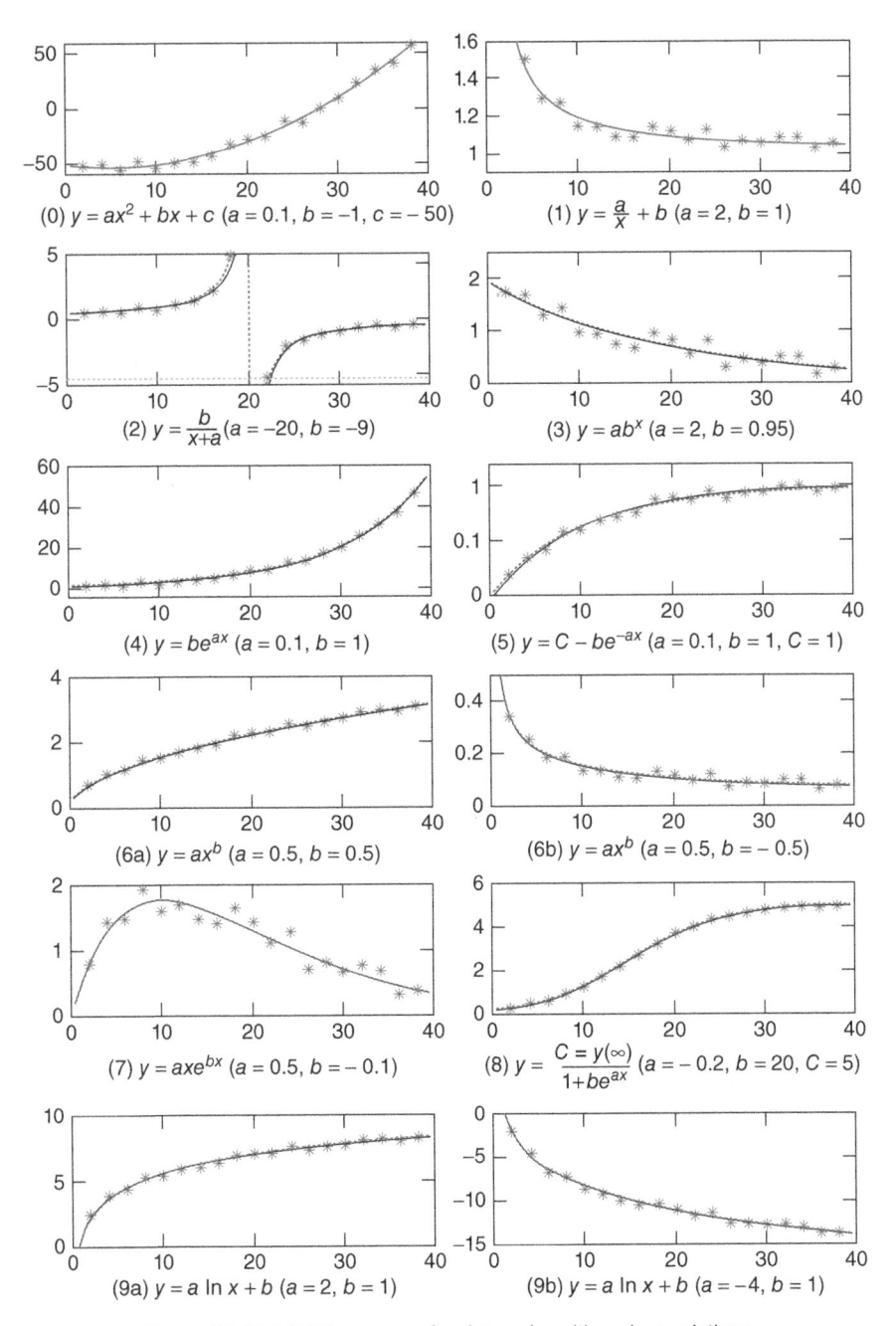

(0) $y = ax^2 + bx + c$ ($a = 0.1, b = -1, c = -50$)

(1) $y = \frac{a}{x} + b$ ($a = 2, b = 1$)

(2) $y = \frac{b}{x+a}$ ($a = -20, b = -9$)

(3) $y = ab^x$ ($a = 2, b = 0.95$)

(4) $y = be^{ax}$ ($a = 0.1, b = 1$)

(5) $y = C - be^{-ax}$ ($a = 0.1, b = 1, C = 1$)

(6a) $y = ax^b$ ($a = 0.5, b = 0.5$)

(6b) $y = ax^b$ ($a = 0.5, b = -0.5$)

(7) $y = axe^{bx}$ ($a = 0.5, b = -0.1$)

(8) $y = \frac{C = y(\infty)}{1 + be^{ax}}$ ($a = -0.2, b = 20, C = 5$)

(9a) $y = a \ln x + b$ ($a = 2, b = 1$)

(9b) $y = a \ln x + b$ ($a = -4, b = 1$)

Figure P3.11 LS fitting curves for data pairs with various relations.

186 INTERPOLATION AND CURVE FITTING

(cf) If there is no theoretical basis on which we can infer the physical relation between the variables, how do we determine the candidate function suitable for fitting the data pairs? We can plot the graph of data pairs, choose one of the graphs in Figure P3.11 that is closest to it, and choose the corresponding template function as the candidate fitting function.

3.12 Two-dimensional Interpolation

Compose a function 'z=find_depth(xi,yi)' that finds the depth z of a geological stratum at a point (xi,yi) given as the input arguments, based on the data in Problem 1.4.

(cf) If you have no idea, insert just one statement involving 'interp2()' into the script 'nm01p04' (Problem 1.4) and fit it into the format of a MATLAB function.

3.13 Polynomial Curve Fitting by Least Squares and Persistent Excitation

Suppose the theoretical (true) relationship between the input x and the output y is known as

$$y = x + 2 \tag{P3.13.1}$$

Charley measured the output data y 10 times for the same input value $x = 1$ by using a guage whose measurement errors has a uniform distribution $U[-0.5,+0.5]$. He made the following MATLAB script "nm03p13.m", which uses the function 'polyfits()' to find a straight-line fitting the data.

(a) Check the following script and modify it if needed. Then, run the script and see the result. Isn't it beyond your imagination? If you use the MATLAB built-in function 'polyfit()', does it get any better?

```
%nm03p13.m
tho=[1 2]; %true parameter
x=ones(1,10); %the unchanged input
y=tho(1)*x+tho(2)+(rand(size(x))-0.5);
th_ls=polyfits(x,y,1); %uses the MATLAB function in Sec.3.8.
polyfit(x,y,1) %uses MATLAB built-in function
```

(b) Note that substituting Eq. (3.8.2) into Eq. (3.8.3) yields

$$\theta^o = \begin{bmatrix} a^o \\ b^o \end{bmatrix} = [A^T A]^{-1} A^T y = \begin{bmatrix} \sum_{n=0}^{M} x_n^2 & \sum_{n=0}^{M} x_n \\ \sum_{n=0}^{M} x_n & \sum_{n=0}^{M} 1 \end{bmatrix}^{-1} \begin{bmatrix} \sum_{n=0}^{M} x_n y_n \\ \sum_{n=0}^{M} y_n \end{bmatrix} \tag{P3.13.2}$$

If $x_n = c$(constant) $\forall n = 0:M$, is the matrix $A^T A$ invertible?

(c) What conclusion can you derive based on (a) and (b), with reference to the identifiablity condition that the input must be rich in some sense or persistently exciting?

(cf) This problem implies that the performance of the identification/estimation scheme including the curve fitting depends on the characteristic of input as well as the choice of algorithm.

PROBLEMS

3.14 Scaled Curve Fitting for an Ill-conditioned Problem [M-2]

Consider Eq. (P3.13.2), which is a typical least squares (LS) solution. The matrix $A^T A$, which must be inverted for the solution to be obtained, may become ill-conditioned by the widely different orders of magnitude of its elements, if the magnitudes of all x_n's are too large or too small, being far from 1 (see Remark 2.3). You will realize something about this issue after solving this problem.

(a) Find a polynomial of degree 2 which fits four data points $(10^6, 1), (1.1 \times 10^6, 2),$ $(1.2 \times 10^6, 5)$, $(1.3 \times 10^6, 10)$ and plot the polynomial function (together with the data points) over the interval $[10^6, 1.3 \times 10^6]$ to check whether it fits the data points well. How big is the relative mismatch error? Does the polynomial do the fitting job well?

(b) Find a polynomial of degree 2 that fits four data points $(10^7, 1),$ $(1.1 \times 10^7, 2), (1.2 \times 10^7, 5)$, $(1.3 \times 10^7, 10)$ and plot the polynomial function (together with the data points) over the interval $[10^7, 1.3 \times 10^7]$ to check whether it fits the data points well. How big is the relative mismatch error? Does the polynomial do the fitting job well? Did you get any warning message on the MATLAB command window? What do you think about it?

(c) If you are not satisfied with the result obtained in (b), why don't you try the scaled curve fitting scheme described below?

1. Transform the x_ns of the data point (x_n, y_n)'s into the region $[-2, 2]$ by the following relation.

$$x'_n \leftarrow -2 + \frac{4}{x_{max} - x_{min}} (x_n - x_{min}) \quad \text{(P3.14.1)}$$

2. Find the LS polynomial $p(x')$ fitting the data point (x_n, y_n)s.

```
%nm03p14.m
clear, clf
format long e
x=1e6*[1 1.1 1.2 1.3]; y=[1 2 5 10];
xi=x(1)+[0:1000]/1000*(x(end)-x(1));
[p,err,yi]=curve_fit(x,y,0,2,xi);
p, err
subplot(220+iter)
plot(x,y,'o',xi,yi), hold on
xmin=min(x); xmax=max(x);
x1= -2+4*(x-xmin)/(xmax-xmin);
x1i=?????????????????????????;
[p1,err,yi]=?????????????????????????; p1, err
plot(x,y,'o',xi,yi)
%To get the coefficients of the original fitting polynomial
ps1=poly2sym(p1);
syms x;
ps0=subs(ps1,x,-2+4/(xmax-xmin)*(x-xmin));
p0=sym2poly(ps0)
format short
```

188 INTERPOLATION AND CURVE FITTING

3. Substitute

$$x' \leftarrow -2 + \frac{4}{x_{max} - x_{min}}(x - x_{min}) \qquad (P3.14.2)$$

for x' into $p(x')$.

(cf) You can complete the following script 'nm03p14' and run it to get the numeric answers.

3.15 Weighted Least-Squares Curve Fitting

As in Example 3.7, we want to compare the results of applying the LS approach and the WLS approach for finding a function that we can believe will describe the relation between the input x and the output y as

$$y = ax\,e^{bx} \qquad (P3.15.1)$$

where the data pair (x_m, y_m)'s are given as

$\{(1, 3.2908), (5, 3.3264), (9, 1.1640), (13, 0.3515), (17, 0.1140)\}$

from gauge A with error range ± 0.1

$\{(3, 4.7323), (7, 2.4149), (11, 0.3814), (15, -0.2396), (19, -0.2615)\}$

from gauge B with error range ± 0.5

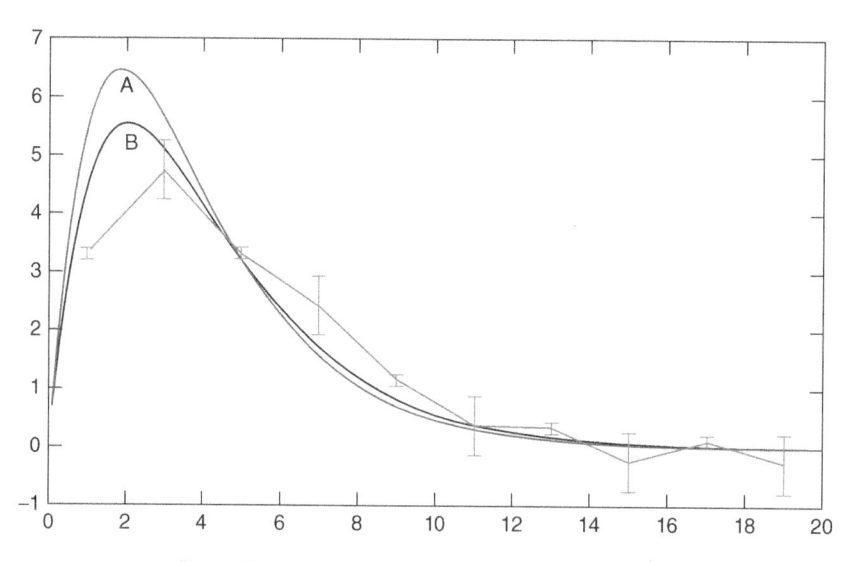

Figure P3.15 LS and WLS fitting curves to $y = axe^{bx}$.

PROBLEMS **189**

Noting that this corresponds to the case of Table 3.15(7), use the MATLAB function 'curve_fit()' for this job and get the result as depicted in Figure P3.15. Identify which one of the two lines A and B is the WLS fitting curve. How do you compare the results?

3.16 Discrete Fourier Transform (DFT) Spectrum
Complete the MATLAB script "do_fft.m" (Section 3.9.2) so that it computes the DFT spectra of the two-tone analog signal described by Eq. (3.9.2) for the cases of zero-padding and whole interval extension and plots them as in Figure 3.13c,d. Which is the clearest one among the four spectra depicted in Figure 3.13? If you can generalize this, which would you choose among up-sampling, zero-padding, and whole interval extension to get a clear spectrum?

3.17 Effect of Sampling Period, Zero-Padding, and Whole Time Interval on DFT Spectrum
In Section 3.9.2, we experienced the effect of zero-padding, sampling period reduction, and whole interval extension on the DFT spectrum of a two-tone signal that has two distinct frequency components. Here, we are going to investigate the effect of sampling period reduction, zero-padding, and repetition on the DFT spectrum of a triangular pulse depicted in Figure P3.17.1c. Additionally, we will compare the DFT with the Continuous-time Fourier Transform (CtFT) and the Discrete-time Fourier Transform (DtFT) [O-1].

(a) The definition of CtFT that is used for getting the spectrum of a continuous-time finite-duration signal $x(t)$ is

$$X(\omega) = \int_{-\infty}^{\infty} x(t) e^{-j\omega t}\, dt \qquad (P3.17.1)$$

The CtFT has several useful properties including the convolution property (Table E.1(5)) and the time-shifting property (Table E.1(3)) described as

$$x(t) * y(t) \underset{\text{Table E.1(5)}}{\overset{\text{CtFT}}{\rightarrow}} X(\omega) Y(\omega) \qquad (P3.17.2)$$

$$x(t - t_1) \underset{\text{Table E.1(3)}}{\overset{\text{CtFT}}{\rightarrow}} X(\omega) e^{-j\omega t_1} \qquad (P3.17.3)$$

Noting that the triangular pulse (Figure. P3.17.1b) is the convolution of two rectangular pulse $r(t)$s (Figure. P3.17.1a) whose CtFTs are

$$R(\omega) = \text{CtFT}\{r(t)\} = \int_{-1}^{1} e^{j\omega t}\, dt = 2\frac{\sin \omega}{\omega}$$

we can use the convolution property (P3.17.2) to get the CtFT of the triangular pulse as

190 INTERPOLATION AND CURVE FITTING

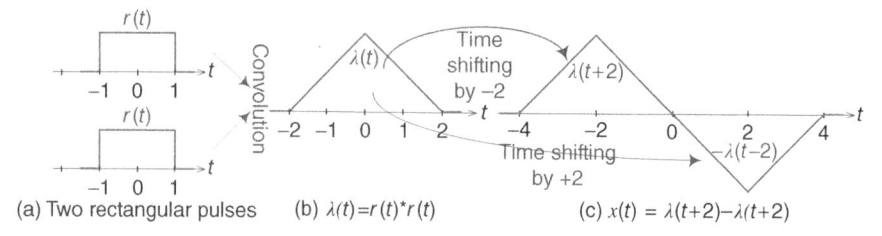

Figure P3.17.1 Two rectangular pulses, a triangular pulse, and a dual triangular base. (a) Two rectangular pulses, (b) $\lambda(t) = r(t)*r(t)$, and (c) $x(t) = \lambda(t+2) - \lambda(t-2)$.

$$\Lambda(\omega) = \text{CtFT}\{\lambda(t)\} = \text{CtFT}\{r(t) * r(t)\} \stackrel{(P3.17.2)}{=} R(\omega)R(\omega)$$

$$= 4\frac{\sin^2\omega}{\omega^2} = 4\sin c^2\left(\frac{\omega}{\pi}\right) \qquad (P3.17.4)$$

Successively, use the time shifting property (P3.17.3) to get the CtFT of

$$x(t) = \lambda(t+2) - \lambda(t-2) \qquad (P3.17.5)$$

as

$$X(\omega) \stackrel{(P3.17.3,4)}{=} \Lambda(\omega)e^{j2\omega} - \Lambda(\omega)e^{-j2\omega} = j\,8\sin(2\omega)\sin c^2\left(\frac{\omega}{\pi}\right) \qquad (P3.17.6)$$

Show that the CtFT $Y(\omega)$ of a triangular wave generated by repeating $x(t)$ two times:

$$y(t) = x(t+4) + x(t-4) \qquad (P3.17.7)$$

is

$$Y(\omega) \stackrel{(P3.17.3,6)}{=} j16\sin(2\omega)\sin c^2\left(\frac{\omega}{\pi}\right)\cos(4\omega) \qquad (P3.17.8)$$

(b) The definition of DtFT, which is used for getting the spectrum of a discrete-time signal $x[n]$, is

$$X(\Omega) = \sum_{n=-\infty}^{\infty} x[n]\,e^{-j\Omega n} \qquad (P3.17.9)$$

Noting that this formula has been cast into the MATLAB function 'DTFT()' (listed below), plot the magnitudes (absolute values) of the DtFTs of the following four discrete-time signals:

$$x[n] = x(t)|_{t=nT} \;(T = 1 \text{ s}) \text{ for } n = 0 : 7 \qquad (P3.17.10a)$$

$$x_b[n] = x(t)|_{t=nT} \;(T = 0.5 \text{ s}) \qquad (P3.17.10b)$$

for $n = 0 : 15$ (sampling period reduction)

```
%nm03p17.m
% DFT/DtFT/CtFT spectra of a triangular signal
clear, clf
W=[0:255]*pi/128; % Digital frequency range
tri=@(t)(mod(t,8)<4).*(2-abs(mod(t,8)-2)) ...
         +(4<=mod(t,8)).*(abs(mod(t,8)-6)-2); % Triangular wave
% (a) Original analog signal sampled with sampling period T=1
N=8; n=[0:N-1]; T=1; t=n*T;
xn=tri(t); Xk=fft(xn); % DFT
k=[0:N-1]; w0=2*pi/T; w=k*w0;
subplot(421), stem(t,xn,'o','Markersize',4)
axis([0 N*T -2.2 +2.2])
subplot(423), stem(k,abs(Xk),'o','Markersize',4)
axis([0 N 0 8])
XW=DTFT(xn,W); % DtFT
Wk=W*N/2/pi; w=W/T;
hold on, plot(Wk,abs(XW),'r:')
Xw=j*8*sin(2*w).*(sinc(w/pi)).^2; % CtFT Eq.(P3.17.6)
plot(Wk,abs(Xw),'k'), legend('DFT','DtFT','CtFT')
% (b) Upsampling with sampling period T=0.5
N=16; n=[0:N-1]; T=0.5; t=n*T;
xbn=tri(t); Xbk=fft(xbn); % DFT
k=[0:N-1]; w0=2*pi/T; w=k*w0;
subplot(422), stem(t,xbn,'o','Markersize',4), axis([0 N*T -2.2 +2.2])
subplot(424), stem(k,abs(Xbk),'o','Markersize',4), axis([0 N 0 14])
XbW=DTFT(xbn,W); % DtFT
Wk=W*N/2/pi; w=W/T;
hold on, plot(Wk,abs(XbW),'r:')
Xbw=j*16*sin(2*w).*(sinc(w/pi)).^2; % CtFT Eq.(P3.17.6)/T
plot(Wk,abs(Xbw),'k'), legend('DFT','DtFT','CtFT')
% (c) Zero-padding
N=16; T=1; t=n*T; xcn=[xn zeros(1,8)];
k=[0:N-1]; w0=2*pi/T;
w=k*w0; Xck=fft(xcn); % DFT
subplot(425), stem(t,xcn,'o','Markersize',4)
axis([0 N*T -2.2 +2.2])
subplot(427), stem(k,abs(Xck),'o','Markersize',4)
axis([0 N 0 8])
XcW=DTFT(xcn,W); % DtFT
Wk=W*N/2/pi; w=W/T;
hold on, plot(Wk,abs(XcW),'r:')
Xcw=j*8*???(2*w).*(sinc(w/pi)).^2; % CtFT Eq.(P3.17.6)/T
plot(Wk,abs(Xcw),'k'), legend('DFT','DtFT','CtFT')
% (d) Repetition (Sampling interval extension)
N=16; n=[0:N-1]; T=1; t=n*T;
yn=tri(t);
k=[0:N-1]; w0=2*pi/T; w=k*w0;
Yk=???(yn); % DFT
subplot(426), stem(t,yn,'o','Markersize',4)
axis([0 N*T -2.2 +2.2])
subplot(428), stem(k,abs(Yk),'o','Markersize',4)
axis([0 N 0 14])
YW=????(yn,W); % DtFT
Wk=W*N/2/pi; w=W/T;
hold on, plot(Wk,abs(YW),'r:')
Yw=2*Xw.*???(4*w); % CtFT Eq.(P3.17.8)
plot(Wk,abs(Yw),'k'), legend('DFT','DtFT','CtFT')
```

192 INTERPOLATION AND CURVE FITTING

```
function X=DTFT(x,W)
% W : A row vector of digital frequencies.
x=x(:).'; N=length(x); n=0:N-1;
X=x*exp(-j*n'*W);  % Eq.(P3.17.9)
```

$$x_c[n] = \begin{cases} x[n] & \text{for } n = 0 : 7 \\ 0 & \text{for } n = 8 : 15 \end{cases} \text{ (zero-padding)} \qquad \text{(P3.17.10c)}$$

$$y[n] = \begin{cases} x[n] & \text{for } n = 0 : 7 \\ x[n-8] & \text{for } n = 8 : 15 \end{cases} \text{ (repetition)} \qquad \text{(P3.17.10d)}$$

like the dotted lines in Figure P3.17.2a-d.

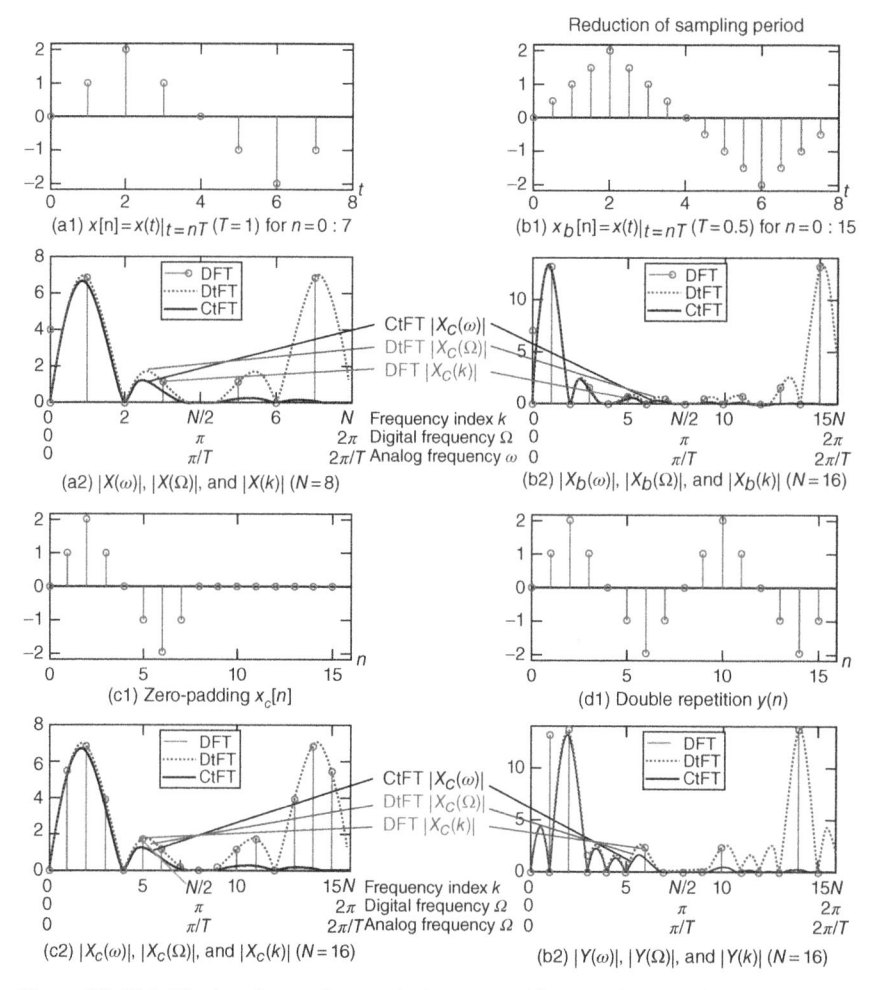

Figure P3.17.2 Effects of sampling period, zero-padding, and whole interval on DFT spectrum.

(c) Complete and run the above MATLAB script "nm03p17.m" to plot the DtFT and DFT spectra of the four discrete-time signals $x[n]$, $x_b[n]$, $x_c[n]$, and $y[n]$, together with the CtFT spectra $X(\omega)/Y(\omega)$ of $x(t)/y(t)$ for $0 \leq \omega \leq 2\pi/T$ as shown in Figure P3.17.2a-d. Make the following observations:

- Do the DFT spectra match the samples of the corresponding DtFTs at digital frequencies $\Omega_k = 2\pi k/N$ (rad/sample) or analog frequencies $\omega_k = \Omega_k/T = 2\pi k/NT$ (rad/s)?
- Are the DtFT spectra close to the CtFT spectra for the principal frequency range $0 \leq \Omega \leq 2\pi$ or $0 \leq \omega \leq 2\pi/T$ where T is the sampling period?

3.18 Windowing Techniques Against the Leakage of DFT Spectrum

There are several window functions ready to be used for alleviating the spectrum leakage problem or for other purposes. We have made a MATLAB function 'windowing()' for easy application of the various windows. Applying the Hamming window function to the discrete-time signal $x_a[n]$ in Figure 3.13a, get the DFT spectrum, plot its magnitude together with the windowed signal, check if they are the same as depicted in Figure P3.18b and compare it with the DFT spectra in Figure 3.13a or b. You can start with the incomplete MATLAB script "nm03p18.m" mentioned beneath. What is the effect of windowing

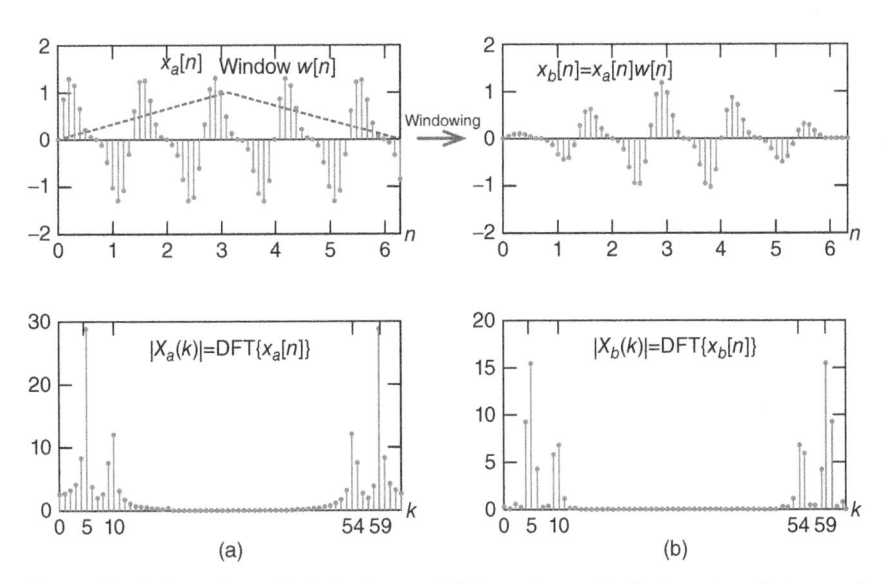

Figure P3.18 The effect of windowing on DFT spectrum. (a) Rectangular window and (b) Bartlett (triangular) window.

on the frequency resolution of the DFT spectrum, which seems to have two frequency components corresponding to around digital frequency indices $k = 5$ and 10?

```
function xw=windowing(x,w)
N= length(x);
if nargin<2|w=='rt'|isempty(w), xw=x;
    elseif w=='bt', xw=x.*bartlett(N)'; % Bartlett window
    elseif w=='bk', xw=x.*blackman(N)'; % Blackman window
    elseif w=='hm', xw=x.*hamming(N)'; % Hamming window
end
```

```
%nm03p18.m: windowing effect on DFT spectrum
w1=1.5*pi; w2=3*pi; % Two tones
N=64; n=1:N; T=0.1; t=(n-1)*T;
k=1:N; w0=2*pi/T; w=(k-1)*w0;
xan=sin(w1*t)+ 0.5*sin(w2*t);
xbn=windowing(xan,'bt');
Xa=fft(xan); Xb=fft(xbn);
subplot(421), stem(t,xan,'.')
subplot(423), stem(k,abs(Xa),'.')
.............
```

3.19 Interpolation Using DFS – Zero-padding on the frequency domain
The fitting curve in Figure 3.14d has been obtained by zeroing out all the digital frequency components higher than $\pi/2$ rad($N/4 < k < 3N/4$) of the sequence $x[n]$ in Figure 3.14a. Plot another fitting curve obtained by removing all the frequency components higher than $\pi/4$ rad ($N/8 < k < 7N/8$) and compare it with Figure 3.14d.

3.20 Interpolation Using Sinc Functions
Given a set of sample points $\{x(nT), n = n_0:n_f\}$, we can use the *Whittaker–Shannon interpolation formula*:

$$x(t) = \sum_{n=n0}^{n_f} x[n] \, sinc\left(\frac{t - nT}{T}\right) \qquad (P3.20.1)$$

to construct a continuous time function interpolating between the given points, which is bandlimited in the sense that its frequency components are below π/T (rad/s). This formula can be cast into the following MATLAB function 'sinc_interp()', which you are supposed to complete.

To try using this function for the interpolation between 30 points $\{ta^{-t}, t = [0, T, 2T, \ldots, 29T]$ with $T = 0.2\}$, complete and run the following script "nm03p20.m".

```
function [xt,xi]=sinc_interp(tn,xn,ti)
% Input:
%    tn=[t0 ... tf]: a set of time points
%    xn=[x(t0) ... x(tf)]: a set of sample points at the time points
%    ti=[ti0 ... tif]: a set of time points at which to interpolate
% Output:
%    xt= the interpolating function as a sum of sinc functions
%        (No vector operations can't be performed for this function.)
%    xi=[xi0 ... xif]: the set of interpolated points
T=tn(2)-tn(1); % Sampling period (interval)
xt=@(t)sum(xn.*sinc((t-tn)/T)); % Eq.(P3.20.1)
if nargin>2
  for n=1:numel(ti), xi(n)=xt(ti(n)); end
  else xi=NaN;
end
```

```
%nm03p20.m
N=30; n=0:N-1; T=0.2; tn=n*T;
a=2; xn=tn.*a.^-tn; ti=[tn(1):0.01:tn(end)];
[xt,xi]=sinc_interp(tn,xn,ti);
plot(tn,xn,'o', ti,xi,'Markersize',4)
```

3.21 Online Recursive Computation of DFT

For the case where you need to compute the DFT of a block of data every time a new sampled data replaces the oldest one in the block, we derive the following recursive algorithm for DFT computation.

Defining the first data block and the mth data block as

$$\{x_0[0], x_0[1], \ldots, x_0[N-1]\} = \{0, 0, \ldots, 0\} \qquad \text{(P3.21.1)}$$

$$\{x_m[0], x_m[1], \ldots, x_m[N-1]\} = \{x[m], x[m+1], \ldots, x[m+N-1]\} \qquad \text{(P3.21.2)}$$

the DFT for the $(m+1)$th data block

$$\{x_{m+1}[0], x_{m+1}[1], \ldots, x_{m+1}[N-1]\}$$
$$= \{x[m+1], x[m+2], \ldots, x[m+N]\} \qquad \text{(P3.21.3)}$$

can be expressed in terms of the DFT for the mth data block

$$X_m(k) = \sum_{n=0}^{N-1} x_m[n]\, e^{-j2\pi nk/N}, \quad k = 0: N-1 \qquad \text{(P3.21.4)}$$

as below:

$$X_{m+1}(k) = \sum_{n=0}^{N-1} x_{m+1}[n]\, e^{-j2\pi nk/N} = \sum_{n=0}^{N-1} x_m[n+1]\, e^{-j2\pi nk/N}$$

$$= \sum_{n=0}^{N-1} x_m[n+1] \, e^{-j2\pi(n+1)k/N} e^{j2\pi k/N} = \sum_{n=1}^{N} x_m[n] \, e^{-j2\pi nk/N} e^{j2\pi k/N}$$

$$= \left\{ \sum_{n=0}^{N-1} x_m[n] \, e^{-j2\pi nk/N} + x[N] - x[0] \right\} e^{j2\pi k/N}$$

$$= \{X_m(k) + x[N] - x[0]\} \, e^{j2\pi \, k/N}$$

You can compute the 128-point DFT for a block composed of 128 random numbers by using this RDFT algorithm and compare it with that obtained by using the MATLAB built-in function 'fft()'. You can start with the incomplete MATLAB script "do_RDFT.m" given below.

```
%do_RDFT.m
clear, clf
N=128; k=[0:N-1];
x=zeros(1,N); % initialize the data block
Xr=zeros(1,N); % and its DFT
for m=0:N
   xN=rand; % New data
   Xr=(Xr+xN-x(1)).*??????????????? % RDFT formula (P3.21.5)
   x=[x(2:N) xN];
end
dif=norm(Xr-fft(x)) % Difference between RDFT and FFT
```

4

NONLINEAR EQUATIONS

CHAPTER OUTLINE

4.1 Iterative Method toward Fixed Point	197
4.2 Bisection Method	201
4.3 False Position or Regula Falsi Method	203
4.4 Newton(-Raphson) Method	205
4.5 Secant Method	208
4.6 Newton Method for a System of Nonlinear Equations	209
4.7 Bairstow's Method for a Polynomial Equation	212
4.8 Symbolic Solution for Equations	215
4.9 Real-World Problems	216
Problems	223

4.1 ITERATIVE METHOD TOWARD FIXED POINT

Let us see the following theorem.

Applied Numerical Methods Using MATLAB®, Second Edition. Won Y. Yang, Jaekwon Kim, Kyung W. Park, Donghyun Baek, Sungjoon Lim, Jingon Joung, Suhyun Park, Han L. Lee, Woo June Choi, and Taeho Im.
© 2020 John Wiley & Sons, Inc. Published 2020 by John Wiley & Sons, Inc.
Companion website: www.wiley.com/go/yang/appliednumericalmethods

Fixed-Point Theorem – Contraction Theorem [K-3, Section 5.1]. *Suppose a function $g(x)$ is defined and its first derivative $g'(x)$ exists continuously on some interval $I = [x^o - r, x^o + r]$ around the fixed point x^o of $g(x)$ such that*

$$g(x^o) = x^o \qquad (4.1.1)$$

Then, if the absolute value of $g'(x)$ is less than or equal to a positive number α which is strictly less than 1, that is,

$$g'(x)| \le \alpha < 1 \qquad (4.1.2)$$

The fixed-point iteration starting from any point $x_0 \in I$

$$x_{k+1} = g(x_k) \quad \text{with } x_0 \in I \qquad (4.1.3)$$

converges to the (unique) fixed point x^o of $g(x)$.

Proof. The mean-value theorem (MVT) (Appendix A) says that for any two points x_0 and x^o, there exists a point x between the two points such that

$$g(x_0) - g(x^o) = g'(x)(x_0 - x^o); \quad x_1 - x^o \stackrel{(4.1.3),(4.1.1)}{=} g'(x)\,(x_0 - x^o) \quad (1)$$

Taking the absolute value of both sides of (1) and using the precondition (4.1.2) yields

$$|x_1 - x^o| \le \alpha \ |x_0 - x^o| < |x_0 - x^o| \qquad (2)$$

which implies that x_1 is closer to x^o than x_0 and so still stays inside the interval I. Applying this successively, we can get

$$|x_k - x^o| \le \alpha|x_{k-1} - x^o| \le \alpha^2|x_{k-2} - x^o| \le \cdots \le \alpha^k|x_0 - x^o| \to 0 \qquad (3)$$

$$\text{as } k \to \infty$$

which implies that the iterative sequence $\{x_k\}$ generated by (4.1.3) converges to x^o. ∎

(Q) Is there any possibility that the fixed point is not unique, i.e. more than one point satisfy Eq. (4.1.1) and so the iterative scheme may get confused among the several fixed points?

(A) It can never happen, because the points x^{o1} and x^{o2} satisfying Eq. (4.1.1) must be the same:

$$|x^{o1} - x^{o2}| = |g(x^{o1}) - g(x^{o2})| \le \alpha \ |x^{o1} - x^{o2}| \ (\alpha < 1);$$
$$|x^{o1} - x^{o2}| = 0; \quad x^{o1} \equiv x^{o2}$$

```
function [x,err,xx]=fixpt(g,x0,TolX,MaxIter)
% To solve x=g(x) starting from x0 by fixed-point iteration.
% Input :  g = Function to be given as a function handle or an M-file name
%          x0 = Initial guess
%          TolX = Upperbound of incremental difference |x(n+1)-x(n)|
%      MaxIter = Maximum # of iterations
% Output: x = Point which the algorithm has reached
%         err = Last value |x(k)-x(k-1)| achieved
%         xx = History of x
if nargin<4, MaxIter=100; end
if nargin<3, TolX=1e-6; end
xx(1)=x0;
for k=2:MaxIter
   xx(k)= feval(g,xx(k-1));  % Eq.(4.1.3)
   err= abs(xx(k)-xx(k-1));
   if err<TolX, break;  end  % Stopping criterion satisfied?
end
x= xx(k);
if k==MaxIter
   fprintf('Do not rely on me, though best in %d iterations\n',MaxIter)
end
```

In order to solve a nonlinear equation $f(x) = 0$ using the iterative method based on this fixed-point theorem, we must somehow arrange the equation into the form

$$x = g(x) \qquad (4.1.4)$$

and start the iteration (4.1.3) with an initial value x_0, then continue until some stopping criterion is satisfied; for example, the difference $|x_{k+1} - x_k|$ between the successive iteration values becomes smaller than some predefined number (TolX) or the iteration number exceeds some predetermined number (MaxIter). This scheme has been cast into the MATLAB function 'fixpt()'. Note that the second output argument (err) is never the real error, i.e. the distance to the true solution, but just the last value of $|x_{k+1} - x_k|$ as an error estimate. See the following remark and examples.

Remark 4.1 Fixed-Point Iteration.
Noting that Eq. (4.1.7) is not unique for a given $f(x) = 0$, it would be good to have $g(x)$ such that $|g'(x)| < 1$ inside the interval I containing its fixed point x^o which is the solution we are looking for. It may not be so easy, however, to determine whether $|g'(x)| < 1$ is satisfied around the solution point, if we don't have any rough estimate of the solution.

Example 4.1. Fixed-Point Iteration.
Consider the problem of solving the nonlinear equation

$$f_{41}(x) = x^2 - 2 = 0 \qquad (E4.1.1)$$

200 NONLINEAR EQUATIONS

To apply the fixed-point iteration for solving this equation, we need to convert it into a form like (4.1.7). Assuming that the solution is in the interval $I = [1,1.5]$, we try with the following three forms:

(a) How about

$$x^2 - 2 = 0 \rightarrow x^2 = 2 \rightarrow x = 2/x = g_a(x)? \qquad \text{(E4.1.2)}$$

Let us see if the absolute value of the first derivative of $g_a(x)$ is less than 1 for the solution interval, i.e. $|g'_a(x)| = \frac{2}{x^2} < 1 \;\; \forall x \in I$. This condition does not seem to be satisfied and so we must be pessimistic about the possibility of reaching the solution with (4.1.2). We don't need many iterations to confirm this.

$$x_0 = 1; \quad x_1 = \frac{2}{x_0} = 2; \quad x_2 = \frac{2}{x_1} = 1; \quad x_3 = \frac{2}{x_2} = 2;$$

$$x_4 = \frac{2}{x_3} = 1; \; \cdots \qquad \text{(E4.1.3)}$$

The iteration turned out to be swaying between 1 and 2, never approaching the solution.

(b) How about

$$x^2 - 2 = 0 \rightarrow (x-1)^2 + 2x - 3 = 0 \rightarrow x = -\frac{1}{2}\{(x-1)^2 - 3\} = g_b(x)?$$
$$\text{(E4.1.4)}$$

This form seems to satisfy the convergence condition

$$|g'_b(x)| = |x-1| \leq 0.5 < 1 \quad \forall x \in I \qquad \text{(E4.1.5)}$$

and so we may be optimistic about the possibility of reaching the solution with (4.1.4). To confirm this, we need just a few iterations, which can be performed by using the function 'fixpt()'.

```
»gb=@(x)-((x-1).^2-3)/2;
»[x,err,xx]=fixpt(gb,1,1e-4,50);
»xx
    1.0000    1.5000    1.3750    1.4297    1.4077    ...
```

The iteration is obviously converging to the true solution $\sqrt{2} = 1.414 \ldots$, which we already know in this case. This process is depicted in Figure 4.1a.

(c) How about $x^2 = 2 \rightarrow x = \frac{2}{x} \rightarrow x + x = \frac{2}{x} + x \rightarrow x = \frac{1}{2}\left(x + \frac{2}{x}\right) = g_c(x)?$
$$\text{(E4.1.6)}$$

This form seems to satisfy the convergence condition

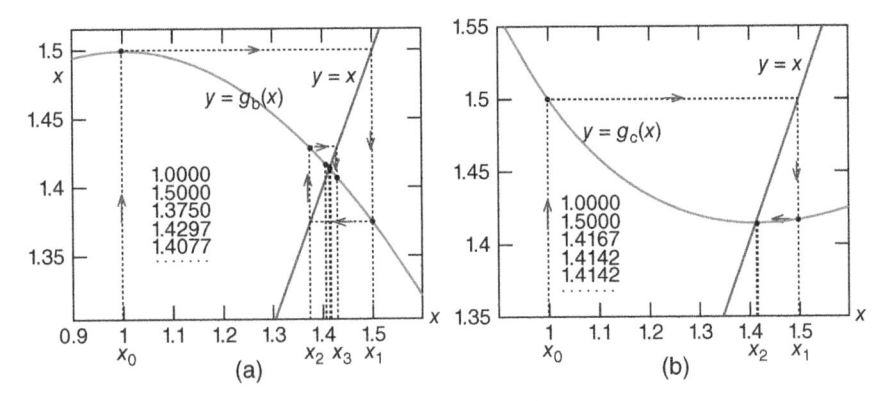

Figure 4.1 Chebyshev nodes. (a) $x_{k+1} = g_b(x_k) = -\frac{1}{2}\{(x_k - 1)^2 - 3\}$. (b) $x_{k+1} = g_c(x_k) = \frac{1}{2}\left(x_k + \frac{1}{x_k}\right)$.

$$|g_c'(x)| = \frac{1}{2}\left|1 - \frac{2}{x^2}\right| \le 0.5 < 1 \quad \forall x \in I \tag{E4.1.6}$$

which guarantees that the iteration will reach the solution. Moreover, since this derivative becomes zero at the solution of $x^2 = 2$, we may expect fast convergence, which is confirmed by using the function 'fixpt()'. The process is depicted in Figure 4.1b.

```
»gc=@(x) (x+2./x)/2;
» [x,err,xx]=fixpt(gc,1,1e-4,50);
»xx
      1.0000    1.5000    1.4167    1.4142    1.4142   ...
```

(cf) In fact, if the nonlinear equation that we must solve is a polynomial equation, then it is convenient to use the MATLAB built-in command 'roots()'.

(Q) How do we make the iteration converge to another solution $x = -\sqrt{2}$ of $x^2 - 2 = 0$?

4.2 BISECTION METHOD

The *bisection method* can be applied for solving nonlinear equations like $f(x) = 0$, only in case we know some interval $[a,b]$ on which $f(x)$ is continuous and the solution uniquely exists and most importantly, $f(a)$ and $f(b)$ have the opposite signs. The procedure toward the solution of $f(x) = 0$ is described as follows and has been cast into the MATLAB function 'bisct()'.

```
function [x,err,xx]= bisct(f,a,b,TolX,MaxIter)
% To solve f(x)=0 by using the bisection method.
% Input : f = Function to be given as a function handle or an M-file name
%         a/b = Initial left/right point of the solution interval
%         TolX = Upperbound of error |x(k)-xo|
%         MaxIter = Maximum # of iterations
% Output: x = Point which the algorithm has reached
%         err = (b-a)/2(half the last interval width)
%         xx = History of x
TolFun=eps; fa=feval(f,a); fb=feval(f,b);
if fa*fb>0, error('We must have f(a)f(b)<0!'); end
for k=1:MaxIter
   xx(k)=(a+b)/2;
   fx= feval(f,xx(k));   err=(b-a)/2;
   if abs(fx)<TolFun|abs(err)<TolX, break;
      elseif fx*fa>0, a=xx(k); fa=fx;
      else  b=xx(k);
   end
end
x=xx(k);
if k==MaxIter, fprintf('The best in %d iterations\n',MaxIter), end
```

(step 0) Initialize the iteration number $k = 0$.

(step 1) Let $m = \frac{1}{2}(a + b)$. If $f(m) \approx 0$ or $\frac{1}{2}(b - a) \approx 0$, then stop the iteration.

(step 2) If $f(a)f(m) > 0$, then let $a \leftarrow m$; otherwise, let $b \leftarrow m$. Go back to step 1.

Remark 4.2 Bisection Method vs. Fixed-Point Iteration.

(1) Only if the solution exists on some interval [a,b], the distance from the midpoint $(a + b)/2$ of the interval as an approximate solution to the true solution is at most one half of the interval width, i.e. $(b - a)/2$, which we take as a measure of error. Therefore, for every iteration of the bisection method, the upper-bound of the error in the approximate solution decreases by half.

(2) The bisection method and the false position method appearing in the next section will definitely give us the solution, only if the solution exists uniquely in some known interval. But the convergence of the fixed-point iteration depends on the derivative of $g(x)$ as well as the initial value x_0.

(3) The MATLAB built-in function 'fzero(f,x)' finds a zero of the function given as the first input argument, based on the interpolation and the bisection method with the initial solution interval vector x = [a b] given as the second input argument. The function is supposed to work even with an initial guess x = x_0 of the (scalar) solution, but it sometimes gives us a wrong result as illustrated in the following example. Therefore, it is safe to use the function 'fzero()' with the initial solution interval vector [a b] as the second input argument.

FALSE POSITION OR REGULA FALSI METHOD

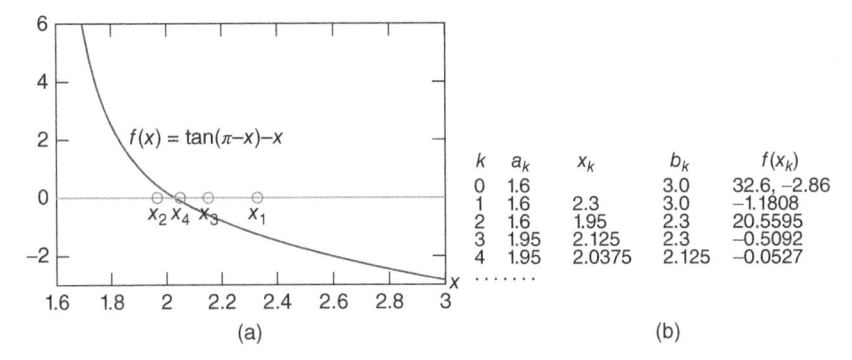

k	a_k	x_k	b_k	$f(x_k)$
0	1.6		3.0	32.6, −2.86
1	1.6	2.3	3.0	−1.1808
2	1.6	1.95	2.3	20.5595
3	1.95	2.125	2.3	−0.5092
4	1.95	2.0375	2.125	−0.0527

Figure 4.2 Bisection method for Example 4.2. (a) Graphic description of the process. (b) Numeric description of the process.

Example 4.2. Bisection Method.
Consider the problem of solving the nonlinear equation:

$$f_{42}(x) = \tan(\pi - x) - x = 0 \quad (E4.2.1)$$

Noting that $f_{42}(x)$ has the value of infinity at $x = \pi/2 = 1.57$, we set the initial solution interval to [1.6,3] excluding the singular point and use the MATLAB function 'bisct()' as follows. The iteration seems to be converging to the solution as we expect (see Figure 4.2):

```
»f42=@(x)tan(pi-x)-x;
»[x,err,xx]=bisct(f42,1.6,3,1e-4,50);
»xx
      2.3000   1.9500   2.1250   2.0375   1.9937   2.0156 ...   2.0287
```

But, if we start with the initial solution interval [a,b] such that f(a) and f(b) have the same sign, we will face the error message.

```
»[x,err,xx]=bisct(f42,1.5,3,1e-4,50);
     Error using bisct (line 15)
     We must have f(a)f(b)<0!
```

Now, let us see how the MATLAB built-in function 'fzero()' works.

```
»fzero(f42,[1.6 3])
     ans= 2.0287  % Good job!
»fzero(f42,[1.5 3])
     Error using fzero (line 274)
     The function values at interval endpoints must differ in sign.
»fzero(f42,1.8) % with an initial guess as the 2nd input argument
     ans= 1.5708  % Wrong result with no warning message
```

(cf) Not all the solutions given by computers are good, especially when we are careless.

4.3 FALSE POSITION OR REGULA FALSI METHOD

Similar to the bisection method, the *false position* or *regula falsi method* starts with the initial solution interval [a,b], that is believed to contain the solution of

$f(x) = 0$. Approximating the curve of $f(x)$ on $[a,b]$ by a straight-line connecting the two points $(a, f(a))$ and $(b, f(b))$, it guesses that the solution may be the point at which the straight-line crosses the x-axis:

$$x = a - \frac{f(a)}{f(a) - f(b)}(b-a) = b - \frac{f(b)}{f(b) - f(a)}(b-a) = \frac{af(b) - bf(a)}{f(a) - f(b)} \quad (4.3.1)$$

For this method, we take the larger of $|x - a|$ and $|b - x|$ as the measure of error. This procedure to search for the solution of $f(x) = 0$ has been cast into the MATLAB function 'falsp()'.

```
function [x,err,xx]=falsp(f,a,b,TolX,MaxIter)
% To solve f(x)=0 by using the false position method.
% Input : f = Function to be given as a function handle or an M-file name
%         a/b = Initial left/right point of the solution interval
%         TolX = Upperbound of error(max(|x(k)-a|,|b-x(k)|))
%     MaxIter = Maximum # of iterations
% Output: x = Point which the algorithm has reached
%         err = Max(x(last)-a|,|b-x(last)|)
%         xx = History of x
TolFun=eps; fa=feval(f,a); fb=feval(f,b);
if fa*fb>0, error('We must have f(a)f(b)<0!'); end
for k=1: MaxIter
   xx(k) = (a*fb-b*fa)/(fb-fa); % Eq.(4.3.1)
   fx= feval(f,xx(k));
   err= max(abs(xx(k)-a),abs(b-xx(k)));
   if abs(fx)<TolFun|err<TolX, break;
     elseif fx*fa>0, a=xx(k); fa=fx;
     else   b=xx(k); fb=fx;
   end
end
x=xx(k);
if k==MaxIter, fprintf('The best in %d iterations\n',MaxIter), end
```

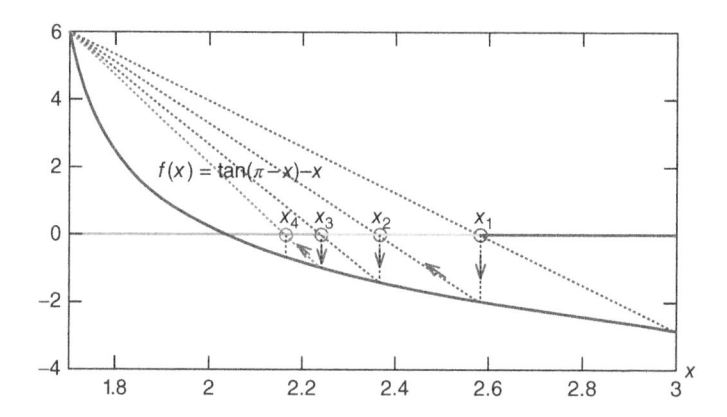

Figure 4.3 False position method for solving $f(x) = \tan(\pi - x) - x$.

Note that although the false position method aims to improve the convergence speed over the bisection method, it cannot always achieve the goal, especially when the curve of $f(x)$ on $[a,b]$ is not well approximated by a straight-line as depicted in Figure 4.3. Figure 4.3 shows how the false position method approaches the solution, performed by typing the following MATLAB statements at the MATLAB prompt:

```
>[x,err,xx]=falsp(f42,1.7,3,1e-4,50) % with initial interval [1.7,3]
```

4.4 NEWTON(-RAPHSON) METHOD

Consider the problem of finding numerically one of the solutions, x^o, for a nonlinear equation

$$f(x) = (x - x^o)^m g(x) = 0$$

where $f(x)$ has $(x - x^o)^m$ (m: an even number) as a factor and so its curve is tangential to the x-axis without crossing it at $x = x^o$. In this case, the signs of $f(x^o - \varepsilon)$ and $f(x^o + \varepsilon)$ are the same, and we cannot find any interval $[a,b]$ containing only x^o as a solution such that $f(a)f(b) < 0$. Consequently, bracketing methods such as the bisection or false position ones are not applicable to this problem. Neither can the MATLAB built-in function 'fzero()' be applied to solve as simple an equation as $x^2 = 0$, which you would not believe until you try it for yourself. Then, how do we solve it? The Newton(-Raphson) method can be used for this kind of problem as well as general nonlinear equation problems, only if the first derivative of $f(x)$ exists and is continuous around the solution.

The strategy behind the Newton(-Raphson) method is to approximate the curve of $f(x)$ by its tangential line at some estimate x_k:

$$y - f(x_k) = f'(x_k)(x - x_k) \qquad (4.4.1)$$

and set the zero (crossing the x-axis) of the tangent line to the next estimate x_{k+1}:

$$0 - f(x_k) = f'(x_k)(x_{k+1} - x_k);$$

$$x_{k+1} = x_k - \frac{f(x_k)}{f'(x_k)} \qquad (4.4.2)$$

This Newton iterative formula has been cast into the MATLAB function 'Newton()', which is designed to generate the numerical derivative (Chapter 5) in case the derivative function is not given as the second input argument.

```
function [x,fx,xx]=Newton(f,df,x0,TolX,MaxIter)
% To solve f(x)=0 by using the Newton method.
% Input:  f  = Function to be given as a function handle or an M-file name
%         df = df(x)/dx (If not given, numerical derivative is used.)
%         x0 = Initial guess of the solution
%         TolX = Upper limit of |x(k)-x(k-1)|
%         MaxIter = Maximum # of iteration
% Output: x  = Point which the algorithm has reached
%         fx = f(x(last)): Residual error, xx = History of x
h=1e-4; h2=2*h; TolFun=eps;
if nargin==4&isnumeric(df), MaxIter=TolX; TolX=x0; x0=df; end
xx(1)=x0; fx=feval(f,x0);
for k=1:MaxIter
    if ~isnumeric(df), dfdx=feval(df,xx(k)); % Derivative function
    else dfdx=(feval(f,xx(k)+h)-feval(f,xx(k)-h))/h2; %Numerical derivative
    end
    dx=-fx/dfdx;
    xx(k+1)=xx(k)+dx; % Eq.(4.4.2)
    fx= feval(f,xx(k+1));
    if abs(fx)<TolFun|abs(dx)<TolX, break; end
end
x=xx(k+1);
if k==MaxIter, fprintf('The best in %d iterations\n',MaxIter), end
```

Here, for the error analysis of the Newton method, we consider the second-degree Taylor polynomial (Appendix A) of $f(x)$ about $x = x_k$:

$$f(x) \approx f(x_k) + f'(x_k)(x - x_k) + \frac{f''(x_k)}{2}(x - x_k)^2$$

We substitute $x = x^o$ (the solution) into this and use $f(x^o) = 0$ to write

$$0 = f(x^o) \approx f(x_k) + f'(x_k)(x^o - x_k) + \frac{f''(x_k)}{2}(x^o - x_k)^2;$$

$$-f(x_k) \approx f'(x_k)(x^o - x_k) + \frac{f''(x_k)}{2}(x^o - x_k)^2$$

Substituting this into Eq. (4.4.2) and defining the error of the estimate x_k as $e_k = x_k - x^o$, we can get

$$x_{k+1} \approx x_k + (x^o - x_k) + \frac{f''(x_k)}{2f'(x_k)}(x^o - x_k)^2; \qquad (4.4.3)$$

$$|e_{k+1}| \approx \left|\frac{f''(x_k)}{2f'(x_k)}\right| e_k^2 = A_k e_k^2 = |A_k e_k||e_k| \qquad (4.4.4)$$

This implies that once the magnitude of initial estimation error $|e_0|$ is small enough to make $|Ae_0| < 1$, the magnitudes of successive estimation errors get smaller very quickly so long as A_k does not become large. The Newton method is said to be '*quadratically convergent*' on account of that the magnitude of the estimation error is proportional to the square of the previous estimation error.

Now, it is time to practice using the MATLAB function 'Newton()' for solving a nonlinear equation like that dealt with in Example 4.2. To this end, let us run

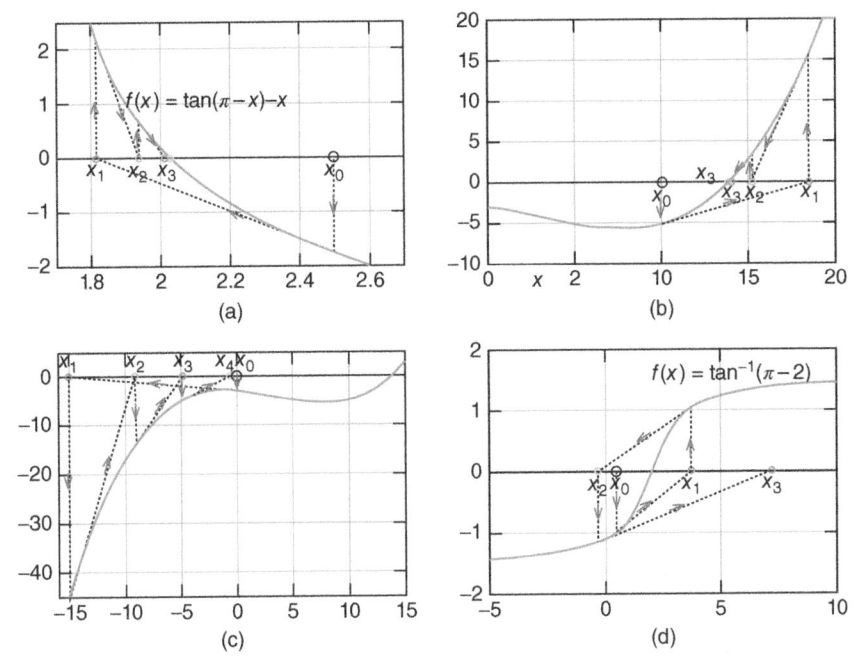

Figure 4.4 Solving nonlinear equations $f(x) = 0$ using the Newton method. (a) $f_{42}(x) = \tan(\pi - x) - x$. (b) $f_{44b}(x) = \frac{1}{125}(x^2 - 25)(x - 10) - 5$. (c) $f_{44b}(x) = \frac{1}{125}(x^2 - 25)(x - 10) - 5$. (d) $f_{44d}(x) = \tan^{-1}(x - 2)$.

the following statements and see the result:

```
»x0=1.8; TolX=1e-5; MaxIter=50; % with initial guess 1.8,...
» [x,err,xx]=Newton(f42,x0,1e-5,50) % with the numerical derivative
»df42=@(x)-(sec(pi-x)).^2-1; % the first derivative
» [x,err,xx1]=Newton(f42,df42,1.8,1e-5,50) % with the first derivative
```

Remark 4.3 Newton(-Raphson) Method.

(1) While bracketing methods such as the bisection method and the false position method converge in all cases, the Newton method is guaranteed to converge only in case where the initial value x_0 is sufficiently close to the solution x^o and $A(x) = |f''(x)/2f'(x)|$ is sufficiently small for $x \approx x^o$. Apparently, it is good for fast convergence if we have small $A(x)$, i.e. the relative magnitude of the second-order derivative $|f''(x)|$ over $|f'(x)|$ is small. In other words, the convergence of the Newton method is endangered if the slope of $f(x)$ is too flat or fluctuates too sharply.

(2) Note two drawbacks of the Newton(-Raphson) method. One is the effort and time required to compute the derivative $f''(x_k)$ at each iteration; the other is the possibility of going astray, especially when $f(x)$ has an abruptly changing slope around the solution (e.g. Figure 4.4c or d), whereas it converges to the solution quickly when $f(x)$ has a steady slope as illustrated in Figure 4.4a,b.

4.5 SECANT METHOD

The secant method can be regarded as a modification of the Newton method in the sense that the derivative is replaced by a difference approximation based on the successive estimates

$$f'(x_k) \approx \frac{f(x_k) - f(x_{k-1})}{x_k - x_{k-1}} \qquad (4.5.1)$$

which is expected to take less time than computing the analytical or numerical derivative. By this approximation, the iterative formula (4.5.2) becomes

$$x_{k+1} = x_k - \frac{f(x_k)}{df\,dx_k} \quad \text{with} \quad df\,dx_k = \frac{f(x_k) - f(x_{k-1})}{x_k - x_{k-1}} \qquad (4.5.2)$$

This secant iterative formula has been cast into the MATLAB function 'secant()', which never needs anything like the derivative as an input argument. We can use this function 'secant()' to solve a nonlinear equation like that dealt with in Example 4.2 by running the following statement. The process is depicted in Figure 4.5.

```
>>[x,err,xx]=secant(f42,2.5,1e-5,50) % with initial guess 2.5
```

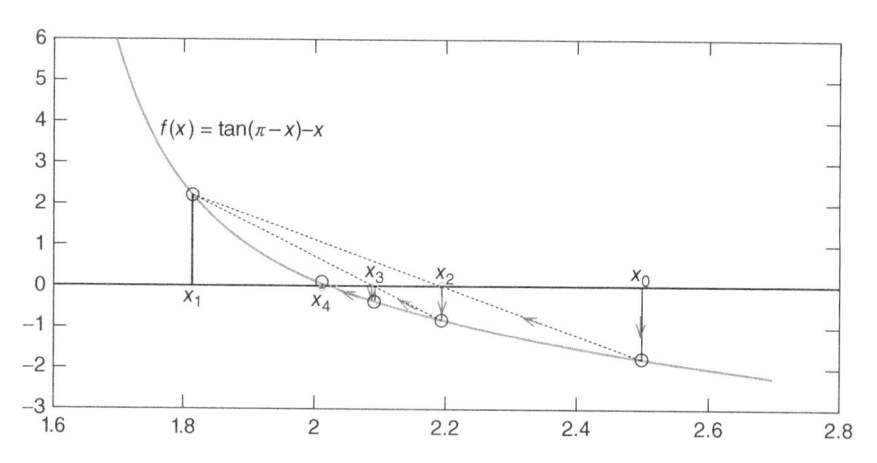

Figure 4.5 Secant method for solving $f(x) = \tan(\pi - x) - x$.

```
function [x,fx,xx]=secant(f,x0,TolX,MaxIter,varargin)
% To solve f(x)=0 by using the secant method.
% Input  : f = Function to be given as a function handle or an M-file name
%          x0 = Initial guess of the solution
%          TolX = Upper limit of |x(k)-x(k-1)|
%          MaxIter = Maximum # of iteration
% Output: x = Point which the algorithm has reached
%         fx = f(x(last)): Residual error, xx = History of x
h=1e-4; h2=2*h; TolFun=eps;
xx(1)=x0;  fx=feval(f,x0,varargin{:});
for k=1:MaxIter
   if k<=1, dfdx=(feval(f,xx(k)+h,varargin{:})-...
                  feval(f,xx(k)-h,varargin{:}))/h2;
     else   dfdx= (fx-fx0)/dx;
   end
   dx= -fx/dfdx;   xx(k+1)= xx(k)+dx; % Eq.(4.5.2)
   fx0= fx;
   fx= feval(f,xx(k+1));
   if abs(fx)<TolFun|abs(dx)<TolX,  break; end
end
x= xx(k+1);
if k==MaxIter, fprintf('The best in %d iterations\n',MaxIter), end
```

4.6 NEWTON METHOD FOR A SYSTEM OF NONLINEAR EQUATIONS

Note that the methods and the corresponding MATLAB functions mentioned so far can handle only one scalar equation with respect to one scalar variable. In order to see how a system of equations can be solved numerically, we rewrite the two equations:

$$f_1(x_1, x_2) = 0$$
$$f_2(x_1, x_2) = 0 \qquad (4.6.1)$$

by taking the first-order Taylor series expansion about some estimate point (x_{1k}, x_{2k}) as

$$f_1(x_1, x_2) \cong f_1(x_{1k}, x_{2k}) + \left.\frac{\partial f_1}{\partial x_1}\right|_{(x_{1k},x_{2k})}(x_1 - x_{1k}) + \left.\frac{\partial f_1}{\partial x_2}\right|_{(x_{1k},x_{2k})}(x_2 - x_{2k}) = 0$$

$$f_2(x_1, x_2) \cong f_2(x_{1k}, x_{2k}) + \left.\frac{\partial f_2}{\partial x_1}\right|_{(x_{1k},x_{2k})}(x_1 - x_{1k}) + \left.\frac{\partial f_2}{\partial x_2}\right|_{(x_{1k},x_{2k})}(x_2 - x_{2k}) = 0$$

$$(4.6.2)$$

This can be arranged into a matrix-vector form as

$$\begin{bmatrix} f_1(x_1,x_2) \\ f_2(x_1,x_2) \end{bmatrix} \approx \begin{bmatrix} f_1(x_{1,k},x_{2,k}) \\ f_2(x_{1,k},x_{2,k}) \end{bmatrix} + \begin{bmatrix} \partial f_1/\partial x_1 & \partial f_0/\partial x_2 \\ \partial f_2/\partial x_1 & \partial f_1/\partial x_2 \end{bmatrix}_{(x_{1,k},x_{2,k})} \begin{bmatrix} x_1 - x_{1,k+1} \\ x_2 - x_{2,k+1} \end{bmatrix} = \begin{bmatrix} 0 \\ 0 \end{bmatrix}$$

$$(4.6.3)$$

which we solve for (x_1, x_2) to get the updated vector estimate

$$\begin{bmatrix} x_{1,k+1} \\ x_{2,k+1} \end{bmatrix} = \begin{bmatrix} x_{1,k} \\ x_{2,k} \end{bmatrix} - \begin{bmatrix} \partial f_1/\partial x_1 & \partial f_0/\partial x_2 \\ \partial f_2/\partial x_1 & \partial f_1/\partial x_2 \end{bmatrix}^{-1}_{(x_{1,k}, x_{2,k})} \begin{bmatrix} f_1(x_{1,k}, x_{2,k}) \\ f_2(x_{1,k}, x_{2,k}) \end{bmatrix};$$

$$\mathbf{x}_{k+1} = \mathbf{x}_k - J_k^{-1} \mathbf{f}(\mathbf{x}_k) \text{ with Jacobian } J_k(m,n) = [\partial f_m/\partial x_n]|_{x_n = x_{n,k}} \quad (4.6.4)$$

This is not much different from the Newton iterative formula (4.4.2) and has been cast into the MATLAB function 'Newtons()'. See Eq. (C.9) for the definition of the Jacobian.

```
function [x,fx,xx]=Newtons(f,x0,TolX,MaxIter,varargin)
% to solve a set of nonlinear eqs f1(x)=0, f2(x)=0,..
% Input:  f = Vector function as a function handle or an M-file name
%         x0 = Initial guess of the solution
%         TolX = Upper limit of |x(k)-x(k-1)|
%         MaxIter = Maximum # of iteration
% Output: x = Point which the algorithm has reached
%         fx = f(x(last)): Residual errors
%         xx = History of x
h=1e-4; TolFun=eps; EPS=1e-6;
fx= feval(f,x0,varargin{:});
Nf=length(fx); Nx=length(x0);
if Nf~=Nx, error('Incompatible dimensions of f and x0!'); end
if nargin<4, MaxIter=100; end
if nargin<3, TolX=EPS; end
xx(1,:)=x0(:).'; % Initialize the solution as the initial row vector
%fx0= norm(fx);                                    %(1)
for k=1: MaxIter
    dx= -jacob(f,xx(k,:),h,varargin{:})\fx(:); %-[dfdx]^-1*fx
    %for l=1: 3 % Damping to avoid divergence   %(2)
    %dx= dx/2;                                    %(3)
    xx(k+1,:)= xx(k,:)+dx.';
    fx= feval(f,xx(k+1,:),varargin{:});
    fxn=norm(fx);
    % if fxn<fx0, break; end                      %(4)
    %end                                          %(5)
    if fxn<TolFun|norm(dx)<TolX, break; end
    %fx0= fxn;                                    %(6)
end
x= xx(k+1,:);
if k==MaxIter, fprintf('The best in %d iterations\n',MaxIter), end
```

```
function g=jacob(f,x,h,varargin) %Jacobian of f(x)
if nargin<3,  h=1e-4; end
h2= 2*h; N=length(x); x=x(:).'; I=eye(N);
for n=1:N
   g(:,n)= (feval(f,x+I(n,:)*h,varargin{:}) ...
          -feval(f,x-I(n,:)*h,varargin{:}))'/h2;
end
```

NEWTON METHOD FOR A SYSTEM OF NONLINEAR EQUATIONS

Now, let us use this function to solve the following system of nonlinear equations

$$x_1^2 + 4x_2^2 = 5$$
$$2x_1^2 - 2x_1 - 3x_2 = 2.5 \qquad (4.6.5)$$

In order to do so, we should first rewrite these equations into a form like Eq. (4.6.1) as

$$f_1(x_1, x_2) = x_1^2 + 4x_2^2 - 5 = 0$$
$$f_2(x_1, x_2) = 2x_1^2 - 2x_1 - 3x_2 - 2.5 = 0 \qquad (4.6.6)$$

and convert it into a MATLAB function defined in an M-file, say, "f46.m" as follows:

```
function y=f46(x)
y(1)= x(1)*x(1)+4*x(2)*x(2) -5;
y(2)= 2*x(1)*x(1)-2*x(1)-3*x(2) -2.5;
```

Then, we run the following statements

```
»x0=[0.8 0.2]; x=Newtons(@f46,x0) % with initial guess [.8 .2]
```

to get

```
x = 2.0000    0.5000
```

Figure 4.6 shows how the vector Newton iteration may proceed depending on the initial guess (x_{10}, x_{20}). With $(x_{10}, x_{20}) = (0.8, 0.2)$, it converges to $(2, 0.5)$, which is one of the two roots (Figure 4.6a) and with $(x_{10}, x_{20}) = (-1, 0.5)$, it converges to $(-1.2065, 0.9413)$, which is another root (Figure 4.6b). However, with $(x_{10}, x_{20}) = (0.5, 0.2)$, it wanders around as depicted in Figure 4.6c. From this figure, we can see that the iteration is jumping too far in the beginning and then going astray around the place where the curves of the two functions $f_1(x)$ and $f_2(x)$ are close, but not crossing. One idea for alleviating this problem is to modify the Newton algorithm in such a way that the step-size can be adjusted (decreased) to keep the norm of $\mathbf{f}(\mathbf{x}_k)$ from increasing at each iteration. The so-called damped Newton method based on this idea will be implemented in the MATLAB function 'Newtons()' if you activate the six statements numbered from (1) to (6) by deleting the comment mark(%) from the beginning of each line. With the same initial guess $(x_{10}, x_{20}) = (0.5, 0.2)$ as in Figure 4.6c, the damped Newton method successfully leads to the point $(2, 0.5)$, which is one of the two roots (Figure 4.6d).

MATLAB has the built-in function 'fsolve(f,x0)', which can give us a solution for a system of nonlinear equations. Let us try it for Eq. (4.6.5) or (4.6.6), which was already defined in the M-file named "f46.m".

```
»x=fsolve(@f46,x0,optimset('fsolve')) % with default parameters
    x = 2.0000    0.5000
```

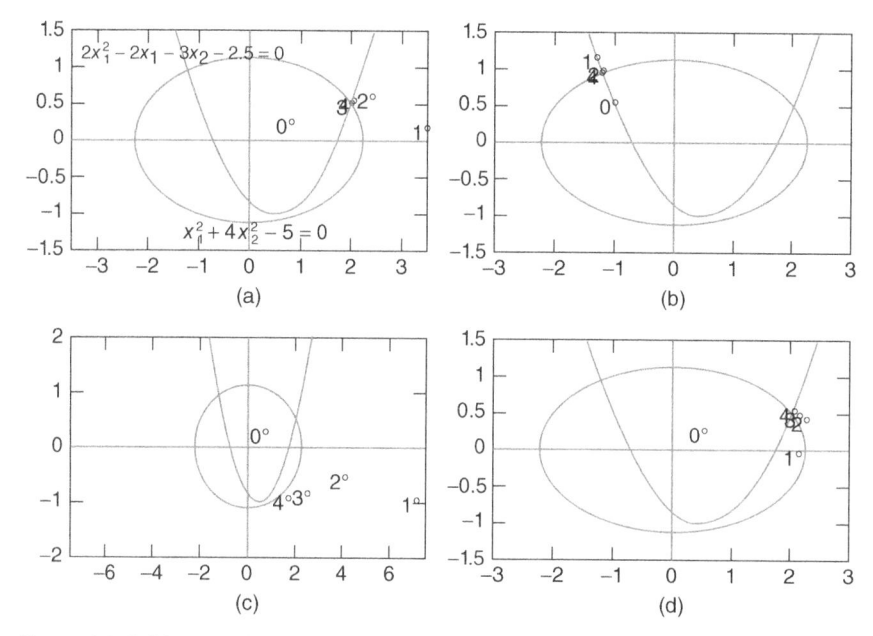

Figure 4.6 Solving nonlinear equations $f(x) = 0$ using the Newton method. (a) Newton method with $(x_{10}, x_{20}) = (0.8, 0.2)$. (b) Newton method with $(x_{10}, x_{20}) = (-1.0, 0.5)$. (c) Newton method with $(x_{10}, x_{20}) = (0.5, 0.2)$. (d) Damped Newton method with $(x_{10}, x_{20}) = (0.5, 0.2)$.

4.7 BAIRSTOW'S METHOD FOR A POLYNOMIAL EQUATION

Bairstow's method is an iterative procedure to find the roots of an Nth-degree real-coefficient polynomial equation $a(x)$

$$a(x) = a_0 x^N + a_1 x^{N-1} + a_2 x^{N-2} + \cdots + a_{N-1} x + a_N \quad \text{with } a_0 = 1 \quad (4.7.1)$$

by factorizing it into a product of a polynomial of degree 2 or less and a deflated polynomial $b(x)$:

$$\begin{aligned}
a(x) &= Q(x)b(x) + R(x) \\
&= (x^2 + q_1 x + q_2)(b_0 x^{N-2} + b_1 x^{N-3} + b_2 x^{N-4} + \cdots + b_{N-3} x + b_{N-2}) \\
&\quad + r_0 x + r_1 \\
&= b_0 x^N + (b_1 + b_0 q_1) x^{N-1} + (b_2 + b_1 q_1 + b_0 q_2) x^{N-2} \\
&\quad + \cdots + (b_{N-2} q_1 + b_{N-3} q_2 + r_0) x + (b_{N-2} q_2 + r_1) \quad (4.7.2)
\end{aligned}$$

In order for this factorization to be successful, we need to determine the coefficients $\{q_1, q_2\}$ of the polynomial factor so that the remainder polynomial becomes zero, i.e. $r_0 = r_1 = 0$, which can be considered as a nonlinear equation problem to solve for $\{q_1, q_2\}$, requiring some iterative procedure. Thus, we first use the equalities between the $(N-1)$ highest-degree terms of Eqs. (4.7.1) and (4.7.2) to set the coefficients of $b(x)$ as

$$
\begin{aligned}
a_0 &= b_0 \rightarrow b_0 = a_0 \\
a_1 &= b_1 + b_0 q_1 \rightarrow b_1 = a_1 - b_0 q_1 \\
a_2 &= b_2 + b_1 q_1 + b_0 q_2 \rightarrow b_2 = a_2 - b_1 q_1 - b_0 q_2 \\
&\vdots \\
a_m &= b_m + b_{m-1} q_1 + b_{m-2} q_2 \rightarrow b_m = a_m - b_{m-1} q_1 - b_{m-2} q_2
\end{aligned} \quad (4.7.3)
$$

for $m = 2 : N - 2$

and write the equalities between the two lowest-degree terms as

$$
\left.\begin{aligned}
a_{N-1} &= b_{N-2} q_1 + b_{N-3} q_2 + r_0 \\
a_N &= b_{N-2} q_2 + r_1
\end{aligned}\right\} \rightarrow
$$

$$
\begin{bmatrix} a_{N-1} - b_{N-2} q_1 - b_{N-3} q_2 = b_{N-1} \\ a_N - b_{N-2} q_2 = b_N \end{bmatrix} = \begin{bmatrix} r_0(q_1, q_2) \\ r_1(q_1, q_2) \end{bmatrix} \quad (4.7.4)
$$

Then, to find $q = [q_1 \ q_2]^T$ such that $r_0 = r_1 = 0$, the Newton method (Eq. (4.7.4)) can be applied to adjust the unknown vector $\mathbf{q}$ starting from some initial guess, say, $q_0 = [1 \ a_1]^T$ as

$$
\begin{bmatrix} q_{1,k+1} \\ q_{2,k+1} \end{bmatrix} = \begin{bmatrix} q_{1,k} \\ q_{2,k} \end{bmatrix} - \begin{bmatrix} \partial r_0/\partial q_1 & \partial r_0/\partial q_2 \\ \partial r_1/\partial q_1 & \partial r_1/\partial q_2 \end{bmatrix}^{-1} \begin{bmatrix} r_0(q_{1,k}, q_{2,k}) \\ r_1(q_{1,k}, q_{2,k}) \end{bmatrix};
$$

$$
\mathbf{q}_{k+1} = \mathbf{q}_k - J^{-1} \mathbf{r}(\mathbf{q}_k) \quad (4.7.5)
$$

where the partial derivatives constituting the Jacobian matrix J can be obtained iteratively as

$$
\begin{aligned}
\frac{\partial b_1}{\partial q_1} &= -b_0 - \frac{\partial b_0}{\partial q_1} q_1 = -b_0, & \frac{\partial b_1}{\partial q_2} &= -\frac{\partial b_0}{\partial q_2} q_1 = 0 \\
\frac{\partial b_2}{\partial q_1} &= -b_1 - \frac{\partial b_1}{\partial q_1} q_1 - \frac{\partial b_0}{\partial q_1} q_2, & \frac{\partial b_2}{\partial q_2} &= -\frac{\partial b_1}{\partial q_2} q_1 - b_0 - \frac{\partial b_0}{\partial q_2} q_2 \\
&\cdots\cdots\cdots\cdots & & \quad (4.7.6) \\
\frac{\partial b_m}{\partial q_1} &= -b_{m-1} - \frac{\partial b_{m-1}}{\partial q_1} q_1 - \frac{\partial b_{m-2}}{\partial q_1} q_2, & \frac{\partial b_m}{\partial q_2} &= -\frac{\partial b_{m-1}}{\partial q_2} q_1 - b_{m-2} - \frac{\partial b_{m-2}}{\partial q_2} q_2
\end{aligned}
$$

for $m = 3 : N$

214 NONLINEAR EQUATIONS

Equation (4.7.6) implies that the two series of partial derivatives $\{\partial b_1/\partial q_1, \partial b_2/\partial q_1, \partial b_3/\partial q_1, \ldots\}$ and $\{\partial b_2/\partial q_2, \partial b_3/\partial q_2, \partial b_4/\partial q_2, \ldots\}$, conforming to each other, can commonly be computed by using an identical recurrence relation:

$$c_0 = -b_0$$
$$c_1 = -b_1 - c_0 q_1$$
$$c_2 = -b_2 - c_1 q_1 - c_0 q_2 \qquad (4.7.7)$$
$$\vdots$$
$$c_m = -b_m - c_{m-1} q_1 - c_{m-2} q_2 \quad \text{for } m = 2 : N - 1$$

Substituting the last three elements of this series for the partial derivatives in the Jacobian matrix of Eq. (4.7.5) yields

$$\begin{bmatrix} q_{1,k+1} \\ q_{2,k+1} \end{bmatrix} = \begin{bmatrix} q_{1,k} \\ q_{2,k} \end{bmatrix} - \begin{bmatrix} c_{N-2} & c_{N-3} \\ c_{N-1} & c_{N-2} \end{bmatrix}^{-1} \begin{bmatrix} b_{N-1} \\ b_N \end{bmatrix} \qquad (4.7.8)$$

If this iteration converges to $\mathbf{q} = [q_1 \; q_2]^T$, we use the quadratic formula to find the roots of the quadratic factor polynomial equation $x^2 + q_1 x + q_2 = 0$ and repeat the same job for the deflated polynomial $b(x)$ till it ends up with a quadratic or first-degree polynomial. This algorithm has been cast into the following MATLAB function 'roots_Bairstow()' in a recursive way.

```
function x=roots_Bairstow(a,tol,MaxIter)
% Bairstow's method for finding the roots of an Nth-degree polynomial
%    a(1)*x^N +a(2)*x^(N-1) +...+a(N)*x +a(N+1) = 0
if nargin<3, MaxIter=100; end
if nargin<2, tol=1e-8; end
N=length(a)-1;
if a(1)~=1, a=a/a(1); end % Normalize so that a(1)=1
if N==1, x = -a(2);
 elseif N==2
   a2=a(2)/2;  sqD=sqrt(a2^2-a(3));
   x = a2+[sqD -sqD]; % Quadratic formula
  else
   q = reshape(a([1 2]),2,1); b(1) = a(1); c(1) = -b(1);
   for k=1:MaxIter
      b(2) = a(2) - b(1)*q(1); %b([1 2]) = 0;
      for m=3:N+1,  b(m) = a(m) - b(m-[1 2])*q; end  % Eq.(4.7.3)
      c(2) = -b(2) - c(1)*q(1);
      for m=3:N,   c(m) = -b(m) - c(m-[1 2])*q; end  % Eq.(4.7.7
      dq = [c(N-1) c(N-2); c(N) c(N-1)]\b([N N+1]).';
      q = q - dq; % Eq.(4.7.8)
      if norm(dq)/norm(q)<tol, break; end
   end
   a2=q(1)/2; sqD=sqrt(a2^2-q(2)); x=-a2+[sqD -sqD]; % Quadratic formula
   x0 = roots_Bairstow(b(1:N-1),tol);
   x = [x x0]; % Roots
end
```

4.8 SYMBOLIC SOLUTION FOR EQUATIONS

MATLAB has many commands and functions that can be very helpful in dealing with complex analytic (symbolic) expressions and equations as well as in getting numerical solutions. One of them is '`solve()`', which can be used for obtaining the symbolic or numeric roots of equations. According to what we could see by typing '`help solve`' into the MATLAB Command Window, its usages are as follows:

```
»solve('p*sin(x)=r')    %regarding x as an unknown variable and p as a parameter
   ans = asin(r/p)    %sin⁻¹(r/p)
»[x1,x2]=solve('x1^2+4*x2^2-5=0','2*x1^2-2*x1-3*x2-2.5=0')
    x1 = [                    2.]      x2 = [                  0.500000]
         [             -1.206459]           [                  0.941336]
         [0.603229  -0.392630*i]           [-1.095668  -0.540415e-1*i]
         [0.603229  +0.392630*i]           [-1.095668  +0.540415e-1*i]
»S=solve('x^3-y^3=2','x=-y')    % returns the solution in a structure.
    S = x: [3x1 sym]
        y: [3x1 sym]
»S.x
    ans = [                1]
          [ -1/2-1/2*i*3^(1/2)]
          [ -1/2+1/2*i*3^(1/2)]
»S.y
    ans = [               -1]
          [ 1/2+1/2*i*3^(1/2)]
          [ 1/2-1/2*i*3^(1/2)]
»[u,v]=solve('a*u^2+v^2=0','u-v=1')    %regarding u,v as unknowns and a
as a parameter
    u= [1/2/(a+1)*(-2*a+2*(-a)^(1/2))+1]  v= [1/2/(a+1)*(-2*a+2*(-a)^(1/2))]
       [1/2/(a+1)*(-2*a-2*(-a)^(1/2))+1]     [1/2/(a+1)*(-2*a-2*(-a)^(1/2))]
»[a,u]=solve('a*u^2+v^2','u-v=1','a,u')    % regards only v as a parameter
    a = -v^2/(v^2+2*v+1)              u = v+1
```

Note that, if the function '`solve()`' finds the symbols more than the equations in its input arguments, say, M symbols and N equations with $M > N$, it regards the N symbols closest alphabetically to 'x' as variables and the other $M - N$ symbols as constants, giving the priority of being a variable to the symbol after 'x' than to one before 'x' for two symbols that are at the same distance from 'x'. Consequently, the priority order of being treated as a symbolic variable is as follows:

$$x > y > w > z > v > u > t > s > r > q > \cdots$$

Actually, we can use the MATLAB built-in function '`findsym()`' to see the priority order.

```
»syms x y z q r s t u v w    % declare 10 symbols to consider
»findsym(x+y+z*q*r+s+t*u-v-w,10)    % symbolic variables?
   ans= x,y,w,z,v,u,t,s,r,q    % in lexicographical order
```

4.9 REAL-WORLD PROBLEMS

Let us see the following example.

Example 4.3. The Orbit of NASA's "Wind" Satellite.
One of the previous NASA plans is to launch a satellite, called Wind, which is to stay at a fixed position along a line from the Earth to the Sun as depicted in Figure 4.7 so that the solar wind passes around the satellite on its way to Earth. In order to find the distance of the satellite from Earth, we set up the following equation based on the related physical laws as

$$G\frac{M_s m}{r^2} = G\frac{M_e m}{(R-r)^2} + mr\omega^2 \rightarrow G\left(\frac{M_S}{r^2} - \frac{M_e}{(R-r)^2}\right) - r\omega^2 = 0 \qquad \text{(E4.3.1)}$$

Note that this function whose zero we want to find can be coded as a MATLAB function handle as follows:

```
»G=6.67e-11; Ms=1.98e30; Me=5.98e24; R=1.49e11; T=3.15576e7; w=2*pi/T;
»f=@(r)G*(Ms./(r.^2+eps)-Me./((R-r).^2+eps))-r*w^2; % Eq.(E4.3.1)
```

(a) This might be solved for *r* by using the (nonlinear) equation solvers like the function 'Newtons()' (Section 4.6) or the MATLAB built-in function 'fsolve()' without options (for default values of parameters) or with some options (for specifying the values of some parameters like the number of function evaluations, function/error tolerances, etc.). We can evaluate the residual error of a solution by substituting it for *r* into the nonlinear solution *f*, that is, the LHS of Eq. (E4.3.1):

```
r0=1e6; % Initial (starting) guess
rn=Newtons(f,r0,1e-4,100) % using Newtons()
rfs=fsolve(f,r0) % using fsolve()
options=optimoptions('fsolve','Display','off','MaxFunEvals',1000);
rfs1=fsolve(f,r0,options) % more iterations
r01=1e10 % with another initial guess closer to the solution
rfs2=fsolve(f,r01,options)
residual_errs=f([rn rfs rfs1 rfs2])
```

This yields

```
rn   = 1.4762e+011 <with residual error of -1.8908e-016>
rfs  = 5.6811e+007 <with residual error of  4.0919e+004>
rfs1 = 2.1610e+009 <with residual error of  2.8280e+001>
rfs2 = 1.0000e+010 <with residual error of  1.3203e+000>
```

It seems that, even with the increased number of function evaluations (as suggested in the warning message) and/or another initial guess, 'fsolve()' is not so successful as 'Newtons()' in this case.

REAL-WORLD PROBLEMS 217

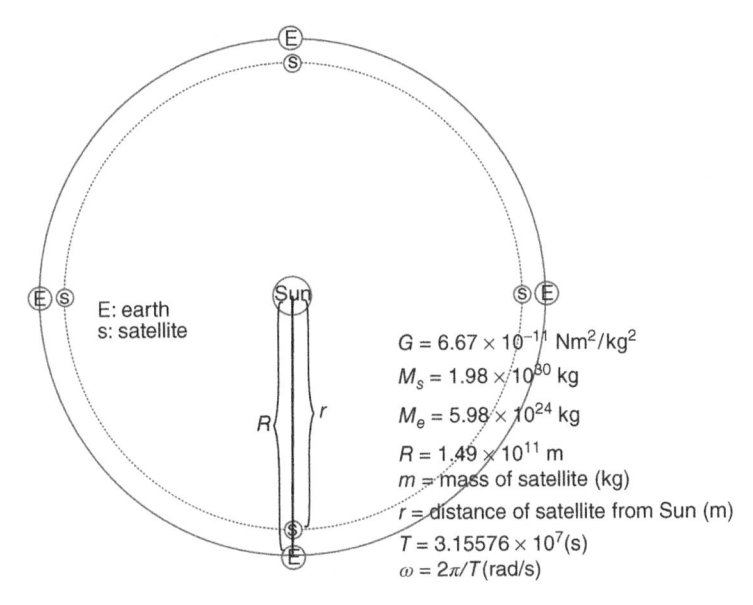

E: earth
s: satellite

$G = 6.67 \times 10^{-11}$ Nm^2/kg^2
$M_s = 1.98 \times 10^{30}$ kg
$M_e = 5.98 \times 10^{24}$ kg
$R = 1.49 \times 10^{11}$ m
m = mass of satellite (kg)
r = distance of satellite from Sun (m)
$T = 3.15576 \times 10^7$ (s)
$\omega = 2\pi/T$ (rad/s)

Figure 4.7 Solving nonlinear equations $f(x) = 0$ using the Newton method.

(b) Noting that Eq. (4.9.1) may cause 'division-by-zero' we multiply both sides of the equation by $r^2(R - r)^2$ to rewrite it as

$$r^3(R - r)^2\omega^2 - GM_S(R - r)^2 + GM_e r^2 = 0 \qquad \text{(E4.3.2)}$$

Note that this function whose zero we want to find can be coded as another MATLAB function handle:

```
»fb=@(r)r.^3.*(R-r).^2*w^2-G*Ms*(R-r).^2+G*Me*r.^2;
```

We can use 'Newtons()' or 'fsolve()' to get the solution of Eq. (E4.3.2) by running the following statements:

```
rnb=Newtons(fb,r0)
rfsb=fsolve(fb,r0)
residual_errs=f([rnb rfsb])
```

This yields

```
rnb  = 1.4762e+011 <with residual error of 4.3368e-018>
rfsb = 1.4762e+011 <with residual error of 4.3368e-018>
```

Both the functions 'Newtons()' and 'fsolve()' benefited from the function conversion and succeeded in finding the solution (with quite good accuracy).

218 NONLINEAR EQUATIONS

(c) The results obtained in (a) and (b) imply that the performance of the nonlinear equation solvers may depend on the shape of the (residual error) function whose zero they aim to find. Here, we try using them with scaling. On the assumption that the solution is known to be in the order of 10^{11}, we divide the unknown variable r by 10^{11} to scale it down into the order of 1. This can be done by substituting $r = r'/10^{11}$ into the equations and multiplying the resulting solution by 10^{11}. We can run the following statements:

```
fs=@(r,scale)f(r*scale);   scale=1e11;
rns=Newtons(fs,r0/scale,1e-6,100,scale)*scale
rfss=fsolve(fs,r0/scale,[],scale)*scale
residual_errs=f([rns rfss])
```

This yields

```
rns  = 1.4762e+011 <with residual error of -6.4185e-16>
rfss = 1.4763e+011 <with residual error of -3.3365e-06>
```

Compared with the results with no scaling obtained in (a), the function 'fsolve()' benefited from scaling and succeeded in finding the solution.

(d) Noting that Eq. (E4.3.2) can be formulated as a (fifth-degree) polynomial equation in r:

$$\omega^2 r^5 - 2R\omega^2 r^4 + \omega^2 R^2 r^3 + G(M_S - M_e)r^2 + 2GM_S Rr - GM_S R^2 = 0 \quad (E4.3.3)$$

so that it can be coded as

```
fp=[w^2 -2*R*w^2 w^2*R^2 G*(Me-Ms) 2*G*Ms*R -G*Ms*R^2];
```

and solved by using the MATLAB function 'roots()':

```
rrs=roots(fp)
residual_err=f(rrs(5))
```

This yields

```
rrs =
 -7.4676e+10 + 1.2934e+11i
 -7.4676e+10 - 1.2934e+11i
  1.4987e+11 + 1.2916e+09i
  1.4987e+11 - 1.2916e+09i
  1.4762e+11 + 0.0000e+00i    <with residual error of -2.6562e-14>
```

where the fifth root can be noticed to be the solution of Eq. (E4.3.3) only after the five roots have been obtained.

(cf) This example implies the following tips for solving nonlinear equations.
  – If you have some preliminary knowledge about the approximate value of the true solution, scale the unknown variable up/down to around one and then, scale the resulting solution back down/up to get the solution to the original equation.
  – It might be better for you to use at least two methods for solving a problem as a cross-check. It is suggested to use 'Newtons()' together with 'fsolve()' for finding a reasonably good solution of a system of nonlinear equations.

Example 4.4. DC Analysis of a BJT (Bipolar Junction Transistor) Circuit[Y-2].
Figure 4.8a,b show an NPN-BJT (bipolar junction transistor) biasing circuit and its PSpice schematic with DC (bias point) analysis result, respectively, where the BJT parameter values are

$$\beta_F = 100, \ \beta_R = 1, \ \alpha_F = \frac{\beta_F}{\beta_F + 1} = \frac{100}{101}, \ \alpha_R = \frac{\beta_R}{\beta_R + 1} = \frac{1}{2} \quad \text{(E4.4.1)}$$

$$I_S = 10^{-14}(\text{A}), \ I_{SC} = \frac{I_S}{\alpha_R} = 2 \times 10^{-14}(\text{A}) \quad \text{(E4.4.2)}$$

Note that we can apply KVL (Kirchhoff's voltage law) along the two paths V_{CC}-R_C-CBJ(collector-base junction)-BEJ(base-emitter junction)-R_E-V_{EE} and V_{CC}-R_C-CBJ-R_B-V_{BB} in the circuit of Figure 4.8a to write a system of equations in v_{BE} and v_{BC} as

$$V_{CC} - V_{EE} - R_C i_C(v_{BE}, v_{BC}) + v_{BC} - v_{BE} - R_E\{i_C(v_{BE}, v_{BC}) + i_B(v_{BE}, v_{BC})\} = 0 \quad \text{(E4.4.3a)}$$

$$V_{CC} - V_{BB} - R_C i_C(v_{BE}, v_{BC}) + v_{BC} - R_B i_B(v_{BE}, v_{BC}) = 0 \quad \text{(E4.4.3b)}$$

where the collector current i_C and base current i_B are

$$i_C(v_{BE}, v_{BC}) \stackrel{[Y-1],(3.1.7a)}{\underset{V_A=\infty}{=}} I_S e^{v_{BE}/V_T} - I_{SC} e^{v_{BC}/V_T} \left(V_T = \frac{273 + T \ (^\circ C)}{11\,605} \ (V) \right) \quad \text{(E4.4.4a)}$$

$$i_B(v_{BE}, v_{BC}) \stackrel{[Y-1],(3.1.7b)}{\underset{V_A=\infty}{=}} \frac{I_S}{\beta_F} e^{v_{BE}/V_T} + \frac{I_{SC}}{\beta_R + 1} e^{v_{BC}/V_T} \quad \text{(E4.4.4b)}$$

and the emitter current i_E is equal to the sum of i_C and i_B by Kirchhoff's current law (KCL) applied to the closed surface enclosing the BJT.

```
function [VC,VB,VE,IC,IB,IE]= ...
           BJT_DC_analysis_exp0(VCC,VBB,VEE,RC,RB,RE,bF,bR,Is,T)
% DC Analysis for an NPN-BJT biasing circuit.
% Copyleft: Won Y. Yang, wyyang53@hanmail.net, CAU for academic use only
if nargin<10, T=27; end % Temperature in degree Celsius (centigrade)
VT=(273+T)/11605; % Thermal voltage
alphaR=bR/(bR+1); Isc=Is/alphaR;
v0 = [0.7; 0.4]; % Initial guess for v=[vBE vBC]
iC = @(v) Is*exp(v(1)/VT)-Isc*exp(v(2)/VT); % v(1)=vBE, v(2)=vBC
iB = @(v) Is/bF*exp(v(1)/VT)+Isc/(bR+1)*exp(v(2)/VT);
eq = @(v) [VCC-VEE-v(1)+v(2)-RC*iC(v)-RE*(iC(v)+iB(v));
           VCC-VBB+v(2)-RC*iC(v)+RB*iB(v)]; % Eq. (E4.4.4a,b)
options=optimoptions('fsolve','Display','off','Diagnostics','off');
v = fsolve(eq,v0,options);
IC=iC(v); IB=iB(v); IE=IB+IC; VBE=v(1); VBC=v(2);
VC=VCC-RC*IC; VB=VC+VBC; VE=VB-VBE;
if nargout<1
   disp('  VCC    VEE    VBB    VBQ   VEQ    VCQ    IBQ       IEQ       ICQ')
   fprintf('%5.2f %5.2f %5.2f %5.2f %5.2f %5.2f %9.2e %9.2e %9.2e', ...
            VCC,VEE,VBB,VB,VE,VC,IB,IE,IC)
end
```

220 NONLINEAR EQUATIONS

(a)

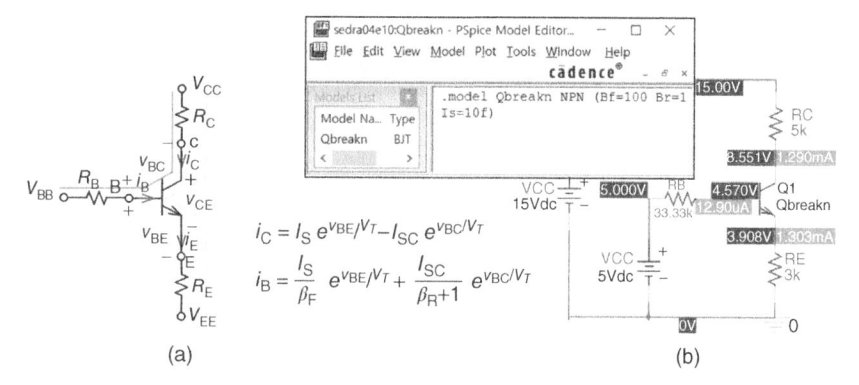

$$i_C = I_S\, e^{v_{BE}/V_T} - I_{SC}\, e^{v_{BC}/V_T}$$

$$i_B = \frac{I_S}{\beta_F}\, e^{v_{BE}/V_T} + \frac{I_{SC}}{\beta_R+1}\, e^{v_{BC}/V_T}$$

(b) PSpice Model Editor:
.model Qbreakn NPN (Bf=100 Br=1 Is=10f)

VCC = 15Vdc, VEE = 5Vdc
Node voltages: 15.00V, 8.551V (1.290mA), 4.570V (Q1 Qbreakn, 12.90uA), 3.908V (1.303mA), 0V
RC = 5k, RB = 33.33k, RE = 3k

Figure 4.8 A BJT circuit and its PSpice schematic with DC (bias point) analysis result. (a) A NPN-BJT biasing circuit. (b) PSpice schematic with bias point analysis results.

(a) We can compose the following function 'BJT_DC_analysis_exp0()' so that it can use the MATLAB built-in function 'fsolve()' to solve the system of nonlinear equations (E4.4.3a) and (E4.4.3b) with the necessary parameter values (like V_{CC}, V_{BB}, V_{EE}, R_C, R_B, R_C, ...) given as its input arguments and return the DC analysis results (such as V_C, V_B, V_E, I_C, I_B, ...) as its output arguments.

(b) We can use the function 'BJT_DC_analysis_exp0()' to analyze the circuit of Figure 4.8b by running the following statements:

```
»VCC=15; VBB=5; VEE=0;
  RC=5e3; RB=100e3/3; RE=3e3; % Circuit parameters
  bF=100; bR=1; Is=1e-14; % Device parameters
  BJT_DC_analysis_exp0(VCC,VBB,VEE,RC,RB,RE,bF,bR,Is);
```

This yields the following analysis result:

```
VCC    VEE    VBB    VBQ    VEQ    VCQ    IBQ       IEQ       ICQ
15.00  0.00   5.00   4.57   3.91   8.55   1.29e-05  1.30e-03  1.29e-03
```

which conforms with the PSpice simulation results shown in Figure 4.8b. Note that the values of the three BJT parameters, i.e. β_F, β_R, and I_S, have been set to 100, 1, and $10\,\text{fA} = 10^{-14}$ A, respectively, in the PSpice Model Editor window that was opened by selecting the menu Edit>PSpice Model after clicking on the BJT (so that it becomes pink-colored) in the PSpice schematic window.

Example 4.5. Analysis of a Complementary BJT Circuit[Y-2].

Figure 4.9a shows a circuit comprised of a complementary pair of NPN/PNP BJTs where the BJT parameter values are the same as those in Example 4.4 (see Eqs. (E4.4.1) and (E4.4.2)). Note that we can apply KCL at the two nodes 1 and 2 in the circuit to write a system of equations in v_1 and v_2 as

REAL-WORLD PROBLEMS   221

KCL at node 1 : $\dfrac{v_i - v_1}{R_i} = i_{B1}(v_{BE1}, v_{BC1}) - i_{B2}(v_{EB2}, v_{CB2})$ (E4.5.1a)

KCL at node 2 : $i_{E1}(v_{BE1}, v_{BC1}) - i_{E2}(v_{EB2}, v_{CB2}) - \dfrac{v_2}{R_L} = 0$ (E4.5.1b)

where the collector currents, base currents, and emitter currents of the two BJTs are

$$i_{C1}(v_{BE1}, v_{BC1}) = I_S e^{v_{BE1}/V_T} - I_{SC}\, e^{v_{BC1}/V_T} \left(V_T = \dfrac{273 + T(^\circ C)}{11\,605}(V)\right)$$

$$i_{B1}(v_{BE1}, v_{BC1}) = \dfrac{I_S}{\beta_F} e^{v_{BE1}/V_T} + \dfrac{I_{SC}}{\beta_R + 1}\, e^{v_{BC1}/V_T} \qquad \text{(E4.5.2)}$$

$$i_{E1} = i_{C1} + i_{B1}, \quad \text{with } v_{BE1} = v_1 - v_2 \text{ and } v_{BC1} = v_1 - V_{CC1}$$

$$i_{C2}(v_{EB2}, v_{CB2}) = I_S e^{v_{EB2}/V_T} - I_{SC} e^{v_{CB2}/V_T}$$

$$i_{B2}(v_{EB2}, v_{CB2}) = \dfrac{I_S}{\beta_F} e^{v_{EB2}/V_T} + \dfrac{I_{SC}}{\beta_R + 1}\, e^{v_{CB2}/V_T} \qquad \text{(E4.5.3)}$$

$$i_{E2} = i_{C2} + i_{B2}, \quad \text{with } v_{EB2} = v_2 - v_1 \text{ and } v_{CB2} = V_{CC2} - v_1$$

(a) We can compose the following function 'BJT2_complementary0()' so that it can use the MATLAB built-in function 'fsolve()' to solve the system of nonlinear equations (E4.5.1a) and (E4.5.1b) with the input voltage (v_i) and necessary parameter values (like R_i, R_L, V_{CC}, ...) given as its input arguments and return the analysis results (like v_o, i_i, and i_o) as its output arguments.

```
function [vs,iis,ios]= BJT2_complementary0(vis,Ri,RL,VCC,betaF,betaR,Is)
% analyzes a complementary BJT pair circuit like Fig.4.9(a)
% Copyleft: Won Y. Yang, wyyang53@hanmail.net, CAU for academic use only
VT=(273+27)/11605;   % Thermal voltage: VT=(273+T)/11605
alphaR=betaR/(betaR+1); Isc=Is/alphaR; % CB saturation current
VCC1=VCC(1); if numel(VCC)<2, VCC2=-VCC1; else VCC2=VCC(2); end
iC=@(v)Is*exp(v(1)/VT)-Isc*exp(v(2)/VT); % Eq.(E4.5.2a) with v=[vBE vBC]
iB=@(v)Is/betaF*exp(v(1)/VT)+Isc/(betaR+1)*exp(v(2)/VT); % Eq.(4.5.2b)
iE=@(v)Is*(1+1/betaF)*exp(v(1)/VT)-Isc*betaR/(betaR+1)*exp(v(2)/VT);
options=optimoptions('fsolve','Display','off','Diagnostics','off');
for n=1:length(vis)
   vi=vis(n);   v0=[vi/4 vi/8]; % Initial guess for v(1) and v(2)
   eq=@(v) [vi-Ri*(iB([v(1)-v(2),v(1)-VCC1])- ...
                   iB([v(2)-v(1),VCC2-v(1)]))  -v(1); % Eq.(E4.5.1a)
            v(2)-RL*(iE([v(1)-v(2),v(1)-VCC1])- ..
                   iE([v(2)-v(1),VCC2-v(1)]))]; % Eq.(E4.5.1b)
   [v,fe]=fsolve(eq,v0,options); vs(n,:)=v;
   iB1s(n)=iB([v(1)-v(2),v(1)-VCC1]); iC1s(n)=iC([v(1)-v(2),v(1)-VCC1]);
   iB2s(n)=iB([v(2)-v(1),VCC2-v(1)]); iC2s(n)=iC([v(2)-v(1),VCC2-v(1)]);
   iis(n)=(vi-v(1))/Ri; ios(n)=v(2)/RL;
end
if nargout<1, plot(vis,vs(:,2)); end
```

222 NONLINEAR EQUATIONS

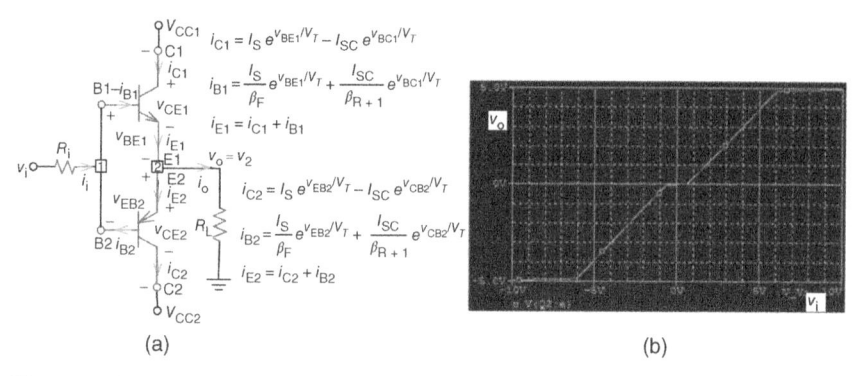

Figure 4.9 A complementary BJT circuit and its PSpice schematic. (a) A complementary BJT circuit. (b) v_o vs. v_i (from PSpice simulation with DC Sweep analysis).

Circuit equations shown in (a):
$$i_{C1} = I_S e^{v_{BE1}/V_T} - I_{SC} e^{v_{BC1}/V_T}$$
$$i_{B1} = \frac{I_S}{\beta_F} e^{v_{BE1}/V_T} + \frac{I_{SC}}{\beta_R + 1} e^{v_{BC1}/V_T}$$
$$i_{E1} = i_{C1} + i_{B1}$$
$$v_o = v_2$$
$$i_{C2} = I_S e^{v_{EB2}/V_T} - I_{SC} e^{v_{CB2}/V_T}$$
$$i_{B2} = \frac{I_S}{\beta_F} e^{v_{EB2}/V_T} + \frac{I_{SC}}{\beta_R + 1} e^{v_{CB2}/V_T}$$
$$i_{E2} = i_{C2} + i_{B2}$$

(b) We can use the function 'BJT2_complementary0()' to analyze the circuit of Figure 4.9a (with $R_i = 10\,\text{k}\Omega$, $R_L = 1\,\text{k}\Omega$, $V_{CC1} = 5\,\text{V}$, and $V_{CC2} = -5\,\text{V}$) for $v_i = -10 \sim 10\,\text{V}$ and plot v_o vs. v_i by running the following statements:

```
»Ri=1e4; RL=1e3; betaF=100; betaR=1; Is=1e-14;
VCC1=5; VCC2=-5; VCC=[VCC1 VCC2]; vis=[-10:0.1:10];
BJT2_complementary0(vis,Ri,RL,VCC,betaF,betaR,Is);
```

The plot of v_o vs. v_i is expected to conform with the PSpice simulation result (Figure 4.9b) obtained with the Analysis type of DC Sweep and Sweep variable v_i.

PROBLEMS

4.1 Fixed-Point Iterative Method
Consider the simple nonlinear equation

$$f(x) = x^2 - 3x + 1 = 0 \qquad (P4.1.1)$$

Knowing that this equation has two roots

$$x^o = 1.5 \pm \sqrt{1.25} \approx 2.6180 \text{ or } 0.382; \quad x^{o1} \approx 0.382, \quad x^{o2} \approx 2.6180 \qquad (P4.1.2)$$

investigate the practicability of the fixed-point iteration.

(a) First, consider the following iterative formula:

$$x_{k+1} = g_a(x_k) = \frac{1}{3}(x_k^2 + 1) \qquad (P4.1.3)$$

Noting that the first derivative of this iterative function $g_a(x)$ is

$$g_a'(x) = \frac{2}{3}x \qquad (P4.1.4)$$

determine which solution attracts this iteration and certify it in Figure P4.1a. In addition, use the MATLAB function 'fixpt()' to perform the iteration

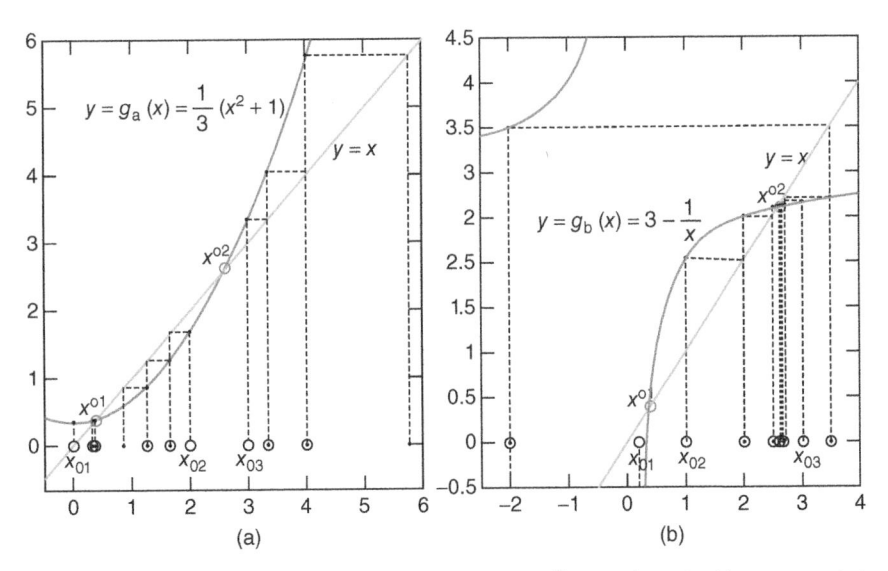

Figure P4.1 Fixed-point iteration for solving $f(x) = x^2 - 3x + 1 = 0$. (a) $x_{k+1} = g_a(x_k) = \frac{1}{3}(x_k^2 + 1)$. (b) $x_{k+1} = g_b(x_k) = 3 - \frac{1}{x_k}$.

(P4.1.3) with the initial points $x_{01} = 0$, $x_{02} = 2$, and $x_{03} = 3$. What does the function yield for each initial point? To figure out how the fixed iteration succeeds (converging to one of $\{x^{o1}, x^{o2}\}$) or fails (diverging to $+\infty$ or $-\infty$), plot some arrows denoting the iteration process from each initial point on Figure P4.1a.

(b) Now, consider the following iterative formula:

$$x_{k+1} = g_b(x_k) = 3 - \frac{1}{x_k} \qquad (P4.1.5)$$

Noting that the first derivative of this iterative function $g_b(x)$ is

$$g_b'(x) = -\frac{1}{x^2} \qquad (P4.1.6)$$

determine which solution attracts this iteration and certify it in Figure P4.1b. In addition, use the MATLAB function 'fixpt()' to carry out the iteration (P4.1.5) with the initial points $x_{01} = 0.2$, $x_{02} = 1$, and $x_{03} = 3$. What does the function yield for each initial point? Plot some arrows denoting the iteration process from each initial point in Figure P4.1b.

(c) If you want to get the other root x^{o1}, which could not be accessed by the iteration (P4.1.5) because $|g_b'(x)|_{x=x^{o1}} > 1$, how about using the inverse function of $g_b(x)$ as

$$x_{k+1} = g_c(x_k) = g_b^{-1}(x_k) = \frac{1}{3 - x_k}? \qquad (P4.1.7)$$

Use the MATLAB function 'fixpt()' to carry out the iteration (P4.1.7) with the initial points $x_0 = 0.2$, $x_0 = 1$, and $x_0 = 3$. What does the function yield for each initial point?

(cf) This illustrates that the outcome of the fixed-point iterative method depends on not only the starting point but also $g(x)$ for a given nonlinear equation $f(x) = 0$.

4.2 Fixed-Point Iterative Method

Consider the following system of two equations in two unknowns v_D and i_D:

$$\begin{cases} i_D = \dfrac{V_{DD} - v_D}{R} \\ i_D = I_s(e^{v_D/V_T} - 1) \end{cases} \qquad (P4.2.1a,b)$$

where $I_s = 10^{-14}$ A, $V_T = 25$ mV, $V_{DD} = 5$ V, and $R = 1\ k\Omega$

(a) First, we can rewrite Eq. (P4.2.1a) as

$$v_D = V_{DD} - Ri_D \qquad (P4.2.2)$$

and substitute this into Eq. (P4.2.1b) to write

$$i_{D,k+1} = I_s \{e^{(V_{DD}-R\,i_{D,k})/V_T} - 1\} \quad \text{(P4.2.3)}$$

To use the fixed-point iteration scheme with this equation, complete and run the following MATLAB script "nm04p02a.m". What result have you got?

```
%nm04p02a.m
Is=1e-14; VT=25e-3; VDD=5; R=1e3;
ga=@(iD)Is*(exp((VDD-R*??)/??)-1); % Eq.(P4.2.3)
tol=1e-9; kmax=3; iD(1)=4e-3;
for k=1:kmax
   iD(k+1)=??(iD(k)); if abs(iD(k+1)-iD(k))<tol, break; end
end
ID1=iD(end)
% Alternatively,
ID2=fixpt(??,iD(?),tol,kmax+1)
% To find the first derivative of ga w.r.t. iD
syms iD; double(subs(diff(ga,iD),iD,0.0043))
```

(b) Second, we can rewrite Eq. (P4.2.1b) as

$$v_D = V_T \ln\left(\frac{i_D}{I_s} + 1\right) \quad \text{(P4.2.4)}$$

and substitute Eq. (P4.2.1a) into this equation to write

$$i_D = V_T \ln\left(\frac{V_{DD} - v_D}{RI_s} + 1\right) \quad \text{(P4.2.5)}$$

To use the fixed-point iteration scheme with this equation, complete and run the following MATLAB script "nm04p02b.m". What result have you got?

```
%nm04p02b.m
Is=1e-14; VT=25e-3; VDD=5; R=1e3;
gb=@(vD)VT*log((VDD-??)/R/??+1); % Eq.(P4.2.5)
tol=1e-4; kmax=10; vD(1)=0.7; % iD(1)=(VDD-vD(1))/R;
for k=1:kmax
   vD(k+1) = ??(vD(k));
   if abs(vD(k+1)-vD(k))<tol, break; end
end
VD1=vD(end), ID1=(VDD-VD1)/R
% Alternatively,
VD2=fixpt(??,vD(?),tol,kmax+1), ID2=(VDD-VD2)/R
% To find the first derivative of gb w.r.t. vD
syms vD; double(subs(diff(gb,vD),vD,VD1))
```

(c) Referring to Remark 4.1, explain why you have got different results in (a) and (b).

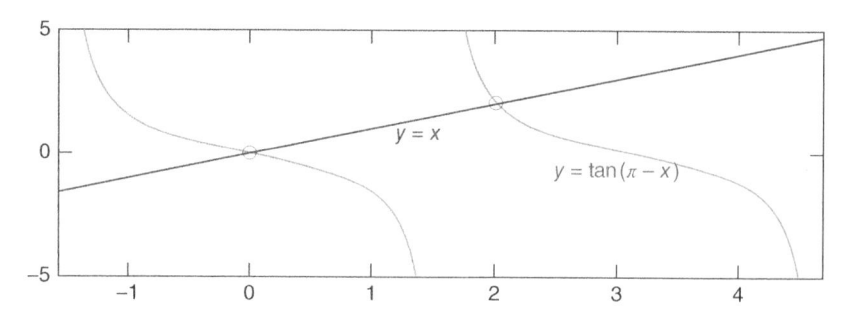

Figure P4.3 Bisection method for solving $f(x) = \tan(\pi - x) - x = 0$.

4.3 Bisection Method and Fixed-Point Iteration

Consider the nonlinear equation treated in Example 4.2.

$$f(x) = \tan(\pi - x) - x = 0 \qquad (P4.3.1)$$

Two graphical solutions of this equation are depicted in Figure P4.3, which can be obtained by running the following MATLAB statements:

```
»ezplot('tan(pi-x)',[-pi/2  3*pi/2])
  hold on, ezplot('x+0',[-pi/2  3*pi/2])
```

(a) In order to use the bisection method for finding the solution between 1.5 and 3, Charley typed the statements shown below. Could he get the right solution? If not, explain him why he failed and suggest him how to make it.

```
»fp43=@(x)tan(pi-x)-x;
  TolX=1e-4; MaxIter=50;
  x=bisct(fp43,1.5,3,TolX,MaxIter)
```

(b) In order to find some interval to which the bisection method is applicable, Jessica used the MATLAB command 'find()' as follows:

```
»x=[0:0.5:pi]; y=fp43(x);
»k=find(y(1:end-1).*y(2:end)<0);
»[x(k) x(k+1); y(k) y(k+1)]
  ans =   1.5000    2.0000    2.0000    2.5000
        -15.6014    0.1850    0.1850   -1.7530
```

This shows that the sign of $f(x)$ changes between $x = 1.5$ and 2.0 and also, between $x = 2.0$ and 2.5. Noting this, Jessica thought that she might use the bisection method to find a solution between 1.5 and 2.0 by typing the following command:

```
»x=bisct(fp43,1.5,2,TolX,MaxIter)
```

Check the validity of the solution, i.e. check if $f(x) = 0$ or not, by typing

```
»fp43(x)
```

If her solution is not good, explain the reason. If you are not sure about it, you can try plotting the graph in Figure P4.3 by running the following MATLAB statements:

```
»x=[-pi/2+0.05:0.05:3*pi/2-0.05];
»plot(x,tan(pi-x),x,x)
```

(cf) This helps us understand why 'fzero(fp42,1.8,tol)' leads to the wrong solution even without any warning message as mentioned in Example 4.2.

(c) To find the solution around $x = 2.0$ by using the fixed-point iteration with the initial point $x_0 = 2.0$, Vania defined the iterative function as

```
»gp431=@(x)tan(pi-x); %x = g₁(x) = tan(π - x)
```

and ran the following MATLAB statement.

```
»x=fixpt(gp431,2,TolX,MaxIter)
```

Could she reach the solution near 2? Will it be better if you start the routine with any different initial point? What is wrong?

(d) Itha, seeing what Vania did, decided to try with another iterative formula

$$\tan^{-1} x = \pi; \quad x = g_2(x) = \pi - \tan^{-1} x \tag{P4.3.2}$$

So she defined the iterative function as

```
»gp432=@(x)pi-atan(x); %x = g(x) = π - tan⁻¹(x)
```

and ran the following statement:

```
»x=fixpt(gp432,2,TolX,MaxIter)
```

What could she get? Is it the right solution? Does this command work with different initial value, like 0 or 6, that are far from the solution we want to find? Describe the difference between Vania's approach and Itha's.

4.4 Recursive (Self-Calling) Function for Bisection Method

As stated in Section 1.3, MATLAB allows us to make a nested (recursive) function, which calls itself. Modify the MATLAB function 'bisct()' (in Section 4.2) into a nested function 'bisct_r()' and run it to solve Eq. (P4.3.1).

```
function [x,err,xx]=bisct_r(f,a,b,TolX,k)
xm=(?+b)/2;    TolFun=eps;
fx=feval(f,xm);  err=(b-a)/2;
if (k<0|abs(fx)<TolFun)|(abs(err)<TolX),  x=xm; xx=x;
 else if fx*fa>0, [x,err,xx1]= bisct_r(f,xm,?,TolX,?-1);
       else  [x,err,xx1]= bisct_r(f,?,xm,TolX,k-?);
       end
       xx= [xm ???];
end
```

4.5 Newton Method and Secant Method

As can be seen in Figure 4.5, the secant method introduced in Section 4.5 was devised to remove the necessity of the derivative/gradient and improve the convergence. But it sometimes turns out to be worse than the Newton method. Apply the functions 'Newton()' and 'secant()' to solve

$$f_{p44}(x) = x^3 - x^2 - x + 1 = 0 \tag{P4.5.1}$$

starting with the initial point $x_0 = -0.2$ one time and $x_0 = -0.3$ for another shot.

4.6 Acceleration of Aitken – *Steffensen Method*

A sequence converging to a limit x^o can be described as

$$x^o - x_{k+1} = e_{k+1} \approx A e_k = A(x^o - x_k) \quad \text{with} \quad \lim_{k \to \infty} \frac{x^o - x_{k+1}}{x^o - x_k} = A(|A| < 1) \tag{P4.6.1}$$

In order to think about how to improve the convergence speed of this sequence, we define a new sequence p_k as

$$\frac{x^o - x_{k+1}}{x^o - x_k} \approx A \approx \frac{x^o - x_k}{x^o - x_{k-1}}; \quad (x^o - x_{k+1})(x^o - x_{k-1}) \approx A(x^o - x_k)^2;$$

$$(x^o)^2 - x_{k+1} x^o - x_{k-1} x^o + x_{k+1} x_{k-1} \approx (x^o)^2 - 2x^o x_k + x_k^2;$$

$$x^o \approx \frac{x_{k+1} x_{k-1} - x_k^2}{x_{k+1} - 2x_k + x_{k-1}} = p_k \tag{P4.6.2}$$

(a) Check that the error of this sequence p_k is as follows:

$$\begin{aligned}
x^o - p_k &= x^o - \frac{x_{k+1} x_{k-1} - x_k^2}{x_{k+1} - 2x_k + x_{k-1}} \\
&= x^o - \frac{x_{k+1}(x_{k+1} - 2x_k + x_{k-1}) - x_{k+1}^2 + 2x_{k+1} x_k - x_k^2}{x_{k+1} - 2x_k + x_{k-1}} \\
&= x^o - x_{k-1} + \frac{(x_k - x_{k-1})^2}{x_{k+1} - 2x_k + x_{k-1}} \\
&= x^o - x_{k-1} + \frac{(-(x^o - x_k) + (x^o - x_{k-1}))^2}{-(x^o - x_{k+1}) + 2(x^o - x_k) - (x^o - x_{k-1})} \\
&= x^o - x_{k-1} + \frac{(-A+1)^2(x^o - x_{k-1})^2}{(-A^2 + 2A - 1)(x^o - x_{k-1})} = 0 \tag{P4.6.3}
\end{aligned}$$

```
function [x,fx,xx]=Stfns(f,x0,TolX,MaxIter)
h=1e-5; h2=2*h; TolFun=eps; EPS=1e-6;
xx(1)= x0; xn(1)= x0;
for k=1: MaxIter
    dfdx=(feval(f,xn(k)+h)-feval(f,xn(k)-h))/h2;
    xn(k+1)= xn(k)-feval(f,xn(k))/????;
    if k>=2
        xx(k)=(xn(k+1)*xn(???)-xn(?)*xn(k))/(xn(k+?)-2*xn(?)+xn(k-?));
        fx= feval(f,xx(k));
        dx= abs(xx(k)-xx(k-1));
        if (abs(fx)<TolFun)|(dx<TolX), break; end
    end
end
x= xx(k);
```

```
%nm04p06.m
f=@(x)tan(??-x)-x;
x0=1.6; a=1; b=4; TolX=1e-5; MaxIter=100;
[x_n,e_n]=Newton(?,x0,TolX,MaxIter);
[x_sc,e_sc]=secant(f,??,TolX,MaxIter);
[x_st,e_st]=Stfns(f,x0,????,MaxIter);
```

(b) Modify the function 'Newton()' into a function 'Stfns()' which generates the sequence (P4.6.2) and use it to solve the following three nonlinear equations:

$$f_{42}(x) = \tan(\pi - x) - x = 0 \quad \text{(with } x_0 = 1.6\text{)} \tag{P4.6.4}$$

$$f_{p44}(x) = x^3 - x^2 - x + 1 = 0 \quad \text{(with } x_0 = 0\text{)} \tag{P4.6.5}$$

$$f_{p45}(x) = (x - 5)^4 = 0 \quad \text{(with } x_0 = 0\text{)} \tag{P4.6.6}$$

Fill in Table P4.6 with the results and those obtained by using the functions 'Newton()', 'secant()' (with the error tolerance TolX=10^{-5}), 'fzero()', and 'fsolve()'. To this end, you can complete/modify the above script "nm04p05.m" and run it.

Table P4.6 Comparison of various methods used for solving nonlinear equations.

		Newton	Secant	Steffensen	Schroder	fzero()	fsolve()
$x_0 = 1.6$	x^o	2.0288					
f_{42}	$f(x^o)$		1.1875×10^{-8}				1.7224×10^{-9}
	Flops	158	112	273	167	986	1454
$x_0 = 0$	x^o			1.0000			
f_{p44}	$f(x^o)$						
	Flops	53	30	63	31	391	364
$x_0 = 0$	x^o			5.0000		NaN	
f_{p45}	$f(x^o)$					NaN	
	Flops	536	434	42	19	3683	1978

4.7 Acceleration of Newton Method for Multiple Roots – Schroder Method
To improve the convergence speed, Schroder modifies the Newton iterative algorithm (4.4.2) as

$$x_{k+1} = x_k - M\frac{f(x_k)}{f'(x_k)} \qquad (P4.7.1)$$

with M : the order of multiplicity of the root we want to find

Based on this idea, modify the MATLAB function 'Newton()' into a function 'Schroder()' and use it to solve Eq. (P4.6.6). Fill in the corresponding blanks of Table P4.6 with the results.

4.8 Newton Method for Systems of Nonlinear Equations
Use the MATLAB function 'Newtons()' (Section 4.6) and the MATLAB built-in function 'fsolve()' (with [x0 y0] = [1 0.5]) to solve the following systems of equations. Fill in Table P4.8 with the results.

(a) $\begin{array}{l} x^2 + y^2 = 1 \\ x^2 - y = 0 \end{array}$ (P4.8.1)

(b) $\begin{array}{l} 5\cos\theta_1 + 6\cos(\theta_1 + \theta_2) = 10 \\ 5\sin\theta_1 + 6\sin(\theta_1 + \theta_2) = 4 \end{array}$ (P4.8.2)

(c) $\begin{array}{l} 3x^2 + 4y^2 = 3 \\ x^2 + y^2 = \sqrt{3/2} \end{array}$ (P4.8.3)

(d) $\begin{array}{l} x_1^3 + 10x_1 - x_2 = 5 \\ x_1 + x_2^3 - 10x_2 = -1 \end{array}$ (P4.8.4)

(e) $\begin{array}{l} x^2 - \sqrt{3}xy + 2y^2 = 10 \\ 4x^2 + 3\sqrt{3}xy + y = 22 \end{array}$ (P4.8.5)

(f) $\begin{array}{l} x^3y - y - 2x^3 = -16 \\ x - y^2 = -1 \end{array}$ (P4.8.6)

(g) $\begin{array}{l} x^2 + 4y^2 = 16 \\ xy^2 = 4 \end{array}$ (P4.8.7)

(h) $\begin{array}{l} xe^y - x^5 + y = 3 \\ x + y + \tan x - \sin y = 0 \end{array}$ (P4.8.8)

(i) $\begin{array}{l} 2\log y - x = 0 \\ xy - y = 1 \end{array}$ (P4.8.9)

(j) $\begin{array}{l} 12xy - 6x = -1 \\ 60x^2 - 180x^2y - 30xy = 1 \end{array}$ (P4.8.10)

Table P4.8 Using 'Newtons()'/'fsolve()' for systems of nonlinear equations.

		Newtons()	fsolve()
$x_0 = [1 \quad 0.5]$ (P4.8.1)	x^o		
	$\|f(x^o)\|$		
	Flops	1043	1393
$x_0 = [1 \quad 0.5]$ (P4.8.2)	x^o	[0.1560 0.4111]	
	$\|f(x^o)\|$	4.4409×10^{-15}	
	Flops	2489	3028
$x_0 = [1 \quad 0.5]$ (P4.8.3)	x^o		
	$\|f(x^o)\|$		
	Flops	1476	3821
$x_0 = [1 \quad 0.5]$ (P4.8.4)	x^o		[0.5024 0.1506]
	$\|f(x^o)\|$		8.8818×10^{-16}
	Flops	1127	1932
$x_0 = [1 \quad 0.5]$ (P4.8.5)	x^o		
	$\|f(x^o)\|$		
	Flops	2884	3153
$x_0 = [1 \quad 0.5]$ (P4.8.6)	x^o	[1.6922 −1.6408]	
	$\|f(x^o)\|$	6.4305×10^{-13}	
	Flops	9234	12 896
$x_0 = [1 \quad 0.5]$ (P4.8.7)	x^o		
	$\|f(x^o)\|$		
	Flops	2125	2378
$x_0 = [1 \quad 0.5]$ (P4.8.8)	x^o		[0.2321 1.5067]
	$\|f(x^o)\|$		1.0745
	Flops	6516	6492
$x_0 = [1 \quad 0.5]$ (P4.8.9)	x^o		
	$\|f(x^o)\|$		
	Flops	1521	1680
$x_0 = [1 \quad 0.5]$ (P4.8.10)	x^o	[0.2236 0.1273]	
	$\|f(x^o)\|$	1.1102×10^{-16}	
	Flops	1278	2566

4.9 Newton Method for Systems of Nonlinear Equations

Use the function 'Newtons()' (Section 4.6) and the MATLAB built-in function 'fsolve()' (with $[x0 \quad y0 \quad z0] = [1 \quad 1 \quad 1]$) to solve the following systems of equations. Fill in Table P4.9 with the results.

(a) $\begin{aligned} xyz &= -1 \\ x^2 + 2y^2 + 4z2 &= 7 \\ 2x^2 + y^3 + 6z &= 7 \end{aligned}$ (P4.9.1)

(b) $\begin{aligned} xyz &= 1 \\ x^2 + 2y^3 + z2 &= 4 \\ x + 2y^2 - z^3 &= 2 \end{aligned}$ (P4.9.2)

Table P4.9 Using 'Newtons()'/'fsolve()' for systems of nonlinear equations.

		Newtons()	fsolve()
$\mathbf{x}_0 = [1\ 1\ 1]$ (P4.9.1)	$\mathbf{x}^o$ $\|f(\mathbf{x}^o)\|$	[1.0000 −1.0000 1.0000] 1.1102×10^{-16}	
$\mathbf{x}_0 = [1\ 1\ 1]$ (P4.9.2)	$\mathbf{x}^o$ $\|f(\mathbf{x}^o)\|$		[1 1 1] 0
$\mathbf{x}_0 = [1\ 1\ 1]$ (P4.9.3)	$\mathbf{x}^o$ $\|f(\mathbf{x}^o)\|$		
$\mathbf{x}_0 = [1\ 1\ 1]$ (P4.9.4)	$\mathbf{x}^o$ $\|f(\mathbf{x}^o)\|$	[1.0000 −1.0000 1.0000] 4.5506×10^{-15}	
$\mathbf{x}_0 = [1\ 1\ 1]$ (P4.9.5)	$\mathbf{x}^o$ $\|f(\mathbf{x}^o)\|$		
$\mathbf{x}_0 = [1\ 1\ 1]$ (P4.9.6)	$\mathbf{x}^o$ $\|f(\mathbf{x}^o)\|$		[2.0000 1.0000 3.0000] 3.4656×10^{-8}

(c) $\quad \begin{aligned} x^2 + 4y^2 + 9z^2 &= 34 \\ x^2 + 9y^2 - 5z &= 40 \\ x^2 z - y &= 7 \end{aligned}$ \hfill (P4.9.3)

(d) $\quad \begin{aligned} x + 2\sin(y\pi/2) + z^2 &= 0 \\ -2xy + z &= 3 \\ e^{x+y} - z^2 &= 0 \end{aligned}$ \hfill (P4.9.4)

(e) $\quad \begin{aligned} x^2 + y^2 + z^2 &= 14 \\ x^2 + 2y^2 - z &= 6 \\ x - 3y^2 + z^2 &= -2 \end{aligned}$ \hfill (P4.9.5)

(f) $\quad \begin{aligned} x^3 - 12y + z^2 &= 5 \\ 3x^2 + y^3 - 2z &= 7 \\ x + 24y^2 - 2\sin(\pi z/18) &= 25 \end{aligned}$ \hfill (P4.9.6)

4.10 Newton Method to Solve a System of Nonlinear Equations for Electronic Circuit Analysis[Y-2]

Figure P4.10a1,a2 show a BJT amplifier circuit and its PSpice schematic, respectively, where the BJT parameter values are the same as those in Example 4.4 (see Eqs. (E4.4.1) and (E4.4.2)). Note that we can apply KVL (Kirchhoff's voltage law) along the two paths V_{BB}-R_B-v_{BE} and V_{CC}-R_C-v_{CB}-v_{BE} to write the following system of two equations in two unknowns v_{BE} and v_{BC} as

$$v_{BB} - R_B i_B(v_{BE}, v_{BC}) - v_{BE} = 0 \quad \text{(P4.10.1a)}$$

$$V_{CC} - R_C i_C(v_{BE}, v_{BC}) + v_{BC} - v_{BE} = 0 \quad \text{(P4.10.1b)}$$

where the collector current and base currents of the BJT are as expressed by Eqs. (E4.4.4a) and (E4.4.4b), respectively.

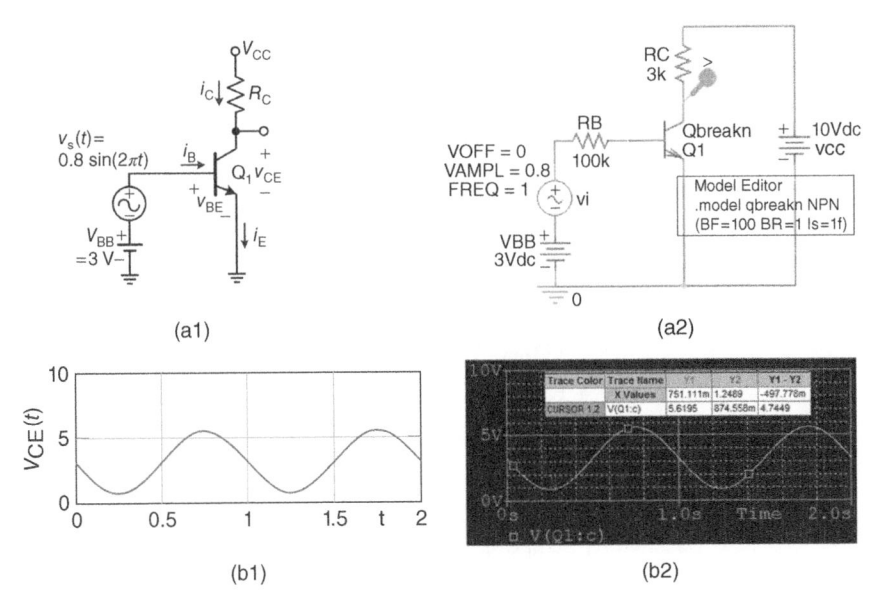

Figure P4.10 A BJT circuit and its PSpice simulation results with DC (bias point) analysis result. (a1) A BJT amplifier. (a2) PSpice schematic. (b1) $v_o(t) = v_{CE}(t)$ from MATLAB analysis. (b2) $v_o(t) = v_{CE}(t)$ from PSpice simulation.

Complete and run the above script "nm04p10.m" to solve the system of Eq. (P4.10.1) with $v_{BB}(t) = 3 + 0.8\sin(2\pi t)$ for $v_{BE}(t)$ and $v_{BC}(t)$, and then, plot $v_o(t) = v_{CE1}(t) - R_C i_C(t)$ for $t = 0 \sim 2$ s like Figure P4.10b1. Is it similar to the PSpice simulation result shown in Figure P4.10b2?

```
%nm04p10.m
% BJT parameters (Saturation current, Current gain, Early voltage)
bF=100; bR=1; aR=bR/(bR+1); bAC=bF; % BJT parameters
Is=1e-15; Isc=Is/aR; % CB(Collector-Base) saturation current
VT=(273+17)/11605; % Thermal voltage 25mV;
VBB=3; vbm=0.8; VCC=10; RB=100e3; RC=3e3; % Example 4.14 of Sedra
t=0:0.01:2; % Time vector
vBt = VBB + vbm*sin(2*pi*t); % DC+AC input voltage source
options=optimoptions('fsolve','Display','off');
iC=@(v)(Is*exp(v(1)/VT)-Isc*exp(v(2)/VT)); % v=[vBE vBC]
iB=@(v)(Is/bF*exp(v(1)/VT)+Isc/(bR+1)*exp(v(2)/VT)); % Eqs.(E4.4.4)
for n=1:length(t)
   %if n==1, v0 = [0.7 0.4]; else v0 = v; end % Initial guess
   if ~exist('v'), v0=[0.7 0.4]; else v0=v; end % Initial guess
   vBB=vBt(n);
   eq = @(v) [vBB-RB*iB(v)-v(?); % KVL for the CE path
              VCC-RC*iC(v)-v(?)+v(?)]; % KVL for the BE path
   v = fsolve(eq,v0,options); % v=Newtons(eq,v0) where v=[vBE vBC]
   iCt(n)=iC(v);
   vCEt(n)=VCC-RC*iCt(n);
end
plot(t,vBt, t,vCEt,'r'); xlabel('t'); ylabel('v_{CE}(t)')
```

4.11 Newton Method to Solve a System of Nonlinear Equations for Electronic Circuit Analysis[Y-2]

Figure P4.11a,b show a circuit (containing an NPN BJT and a PNP BJT) and its PSpice schematic with DC (bias point) analysis result, respectively, where the BJT parameter values are the same as those in Example 4.4 (see Eqs. (E4.4.1) and (E4.4.2)). Note that we can apply KVL (Kirchhoff's voltage law) along the four paths $V_{CC}\text{-}R_{C1}\text{-}v_{CB1}\text{-}R_{B1}\text{-}V_{BB1}$, $V_{BB1}\text{-}R_{B1}\text{-}v_{BE1}\text{-}R_{E1}$, $V_{CC}\text{-}R_{C1}\text{-}v_{BC2}\text{-}R_{C2}$, and $V_{CC}\text{-}R_{E2}\text{-}v_{EB2}\text{-}v_{BC2}\text{-}R_{C2}$ to write the following system of four equations in four unknowns v_{BE1}, v_{BC1}, v_{EB2}, and v_{CB2} as

$$V_{CC} - R_{C1}\{i_C(v_{BE1}, v_{BC1}) - i_B(v_{EB2}, v_{CB2})\}$$
$$+ v_{BC1} + R_{B1}i_B(v_{BE1}, v_{BC1}) - V_{BB1} = 0 \quad \text{(P4.11.1a)}$$

$$V_{BB1} - R_{B1}i_B(v_{BE1}, v_{BC1}) - v_{BE1}$$
$$- R_{E1}\{i_C(v_{BE1}, v_{BC1}) + i_B(v_{EB2}, v_{CB2})\} = 0 \quad \text{(P4.11.1b)}$$

$$V_{CC} - R_{C1}\{i_C(v_{BE1}, v_{BC1}) - i_B(v_{EB2}, v_{CB2})\}$$
$$+ v_{CB2} - R_{C2}i_C(v_{EB2}, v_{CB2}) = 0 \quad \text{(P4.11.1c)}$$

$$V_{CC} - R_{E2}\{i_C(v_{EB2}, v_{CB2}) + i_B(v_{EB2}, v_{CB2})\}$$
$$- v_{EB2} + v_{CB2} - R_{C2}i_C(v_{EB2}, v_{CB2}) = 0 \quad \text{(P4.11.1d)}$$

where the collector currents and base currents of the two BJTs are as expressed by Eqs. (E4.4.4a) and (E4.4.4b).

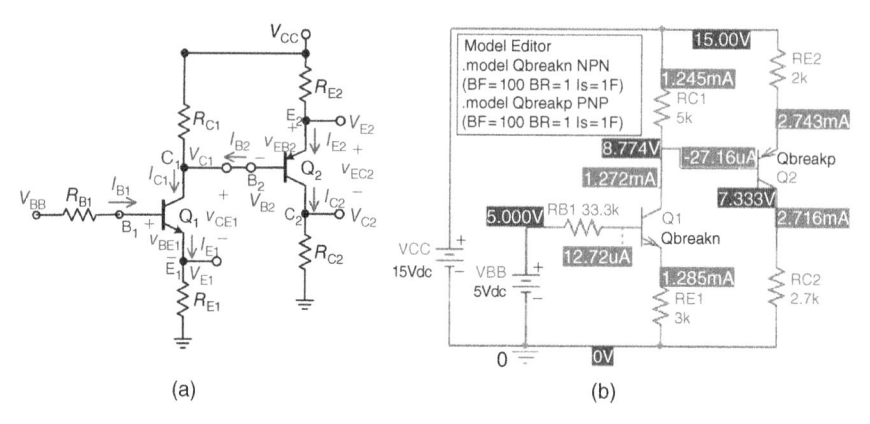

Figure P4.11 A BJT circuit (a) and its PSpice schematic with DC (bias point) analysis result (b).

Complete and run the following script "nm04p11.m" to solve the system of equations (P4.11.1) for v_{BE1}, v_{BC1}, v_{EB2}, and v_{CB2} and then, find I_{B1}, I_{C1}, I_{B2}, and I_{C2}. Are they close to those obtained from the PSpice simulation result shown in Figure P4.11b?

```
%nm04p11.m
clear
betaF=100; betaR=1; alphaR=betaR/(betaR+1);
Is=1e-14; Isc=Is/alphaR; % BJT parameters
VT=(273+27)/11605; % Thermal voltage: VT=(273+T)/11605
VCC=15; VBB1=5; RB1=33.3e3; RC1=5e3; RE1=3e3; RE2=2e3; RC2=2.7e3;
% Exponential model based approach
iC=@(v) Is*exp(v(1)/VT)-Isc*exp(v(2)/VT); % Eq.(E4.4.4a)
iB=@(v) Is/betaF*exp(v(1)/VT)+Isc/(betaR+1)*exp(v(2)/VT); % Eq.(E4.4.4b)
% Eq.(P4.10.1) with v=[vBE1 vBC1 vEB2 vCB2]
eq=@(v) [VCC-VBB1+v(2)-R??*(iC(v(1:?))-iB(v(?:4)))+RB1*iB(v(1:2));
         VBB1-RB1*iB(v(1:?))-v(?)-RE1*(iC(v(1:2))+iB(v(?:2)));
         VCC+v(4)-RC1*(iC(v(?:2))-iB(v(3:?)))-RC2*iC(v(3:?));
         VCC-v(?)+v(4)-RE2*(iC(v(?:4))+iB(v(3:4)))-RC2*iC(v(?:4))];
options=optimoptions('fsolve','Display','off','Diagnostics','off');
v0 = [0.7; 0.4; 0.7; 0.4]; % Initial guess for v=[vBE1 vBC1 vEB2 vCB2]
v = fsolve(eq,v0,options); % v = Newtons(eq,v0); % Alternatively,
VBE1=v(1); VBC1=v(2); VEB2=v(3); VCB2=v(4);
format short e
IB1=iB(v(1:2)), IC1=iC(v(1:2)), IB2=iB(v(3:4)), IC2=iC(v(3:4))
```

4.12 Newton Method to Solve a System of Nonlinear Equations for Electronic Circuit Analysis[Y-2]

Figure P4.12a,b show an NMOS (n-channel metal-oxide-semiconductor field-effect transistor) circuit and its PSpice schematic with DC (bias point) analysis result, respectively, where the device parameters of the NMOS are $K_p = 1$ mA/V^2, $V_t = 1$ V, and $\lambda = 0$ V^{-1}.

Noting that the drain current is determined as

$$i_D = \begin{cases} K_p \left\{ (v_{GS} - V_t)v_{DS} - \frac{1}{2}v_{DS}^2 \right\} & \text{(triode region)} & \text{for } 0 < v_{DS} \leq v_{GS} - V_t \\ \frac{1}{2}K_p(v_{GS} - V_t)^2(1 + \lambda v_{DS}) & \text{(saturation region)} & \text{for } 0 \leq v_{GS} - V_t \leq v_{DS} \end{cases}$$

(P4.12.1)

we can analyze this circuit as follows:

– First, V_G is determined by the voltage divider consisting of R_1 and R_2:

$$V_G = \frac{R_2}{R_1 + R_2} V_{DD} = \frac{10}{10 + 10} 10 = 5 \text{ V} \quad (P4.12.2)$$

– Then, to tell whether the NMOS operates in the saturation or triode mode, we assume that it operates in the *triode* mode where

$$i_D \overset{(P4.12.1a)}{=} K_p \left\{ (v_{GS} - V_t)v_{DS} - \frac{1}{2}v_{DS}^2 \right\} \quad \text{(A)} \quad (P4.12.3)$$

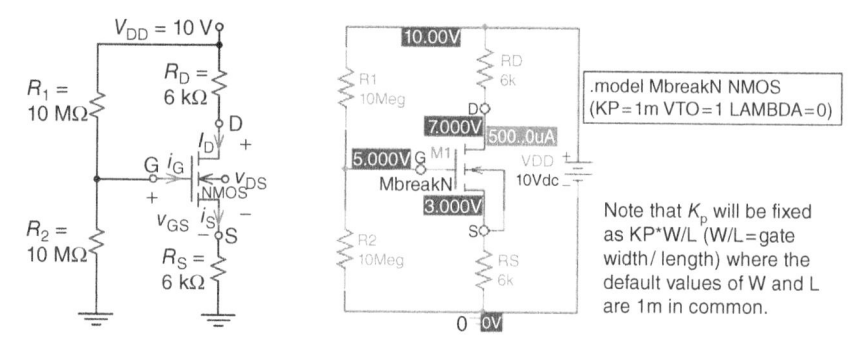

Figure P4.12 An NMOS circuit (a) and its PSpice schematic with DC (bias point) analysis result (b).

- We substitute $v_{GS} = V_G - R_S i_D = 5 - 6 i_D$ and $v_{DS} = V_{DD} - R_D i_D - R_S i_D = 10 - 12 i_D$ into this equation and solve it as

$$i_D = 1 \times \left\{ (5 - 6i_D - 1)(10 - 12i_D) - \frac{1}{2}(10 - 12i_D)^2 \right\};$$

$$11\, i_D - 10 = 0; \quad i_D = \frac{10}{11} \text{ mA} \qquad (P4.12.4)$$

- To check the validity, we should see if it satisfies the condition for the triode mode: $v_{GD} > V_t$ and $v_{GS} > 0$:

$$V_D = V_{DD} - R_D I_D = 10 - 6\frac{10}{11} = \frac{50}{11} \text{ V} \to$$

$$V_{GD} = V_G - V_D = 10 - \frac{50}{11} = \frac{5}{11} \text{ V} \overset{?}{>} V_t : \text{No} \qquad (P4.12.5)$$

This contradicts the assumption of triode mode.

- Then we assume the *saturation* mode where

$$i_D \overset{(P4.12.1b)}{=} \frac{1}{2} K_p (v_{GS} - V_t)^2 \qquad (P4.12.6)$$

substitute $v_{GS} = V_G - R_S i_D = 5 - 6 i_D$ into this equation, and solve it as

$$i_D = \frac{1}{2} K_p (V_G - V_S - V_t)^2 = \frac{1}{2}(5 - R_S i_D - 1)^2 = \frac{1}{2}(5 - 6i_D - 1)^2 \text{ mA};$$

$$18 I_D^2 - 25 I_D + 8 = 0 \to I_D = \frac{8}{9} \text{ or } \frac{1}{2} \text{ mA}; \quad I_D = \frac{1}{2} \text{ mA} \qquad (P4.12.7)$$

Here, the first root $I_D = (8/9)$ mA should be dumped because it does not satisfy even the turn-on condition: $v_{GS} = V_G - R_S I_D < 0$. How about the second one $I_D = (1/2)$ mA?

$$V_S = R_S I_D = 6 \times \frac{1}{2} = 3 \text{ V} \rightarrow v_{GS} = V_G - V_S = 5 - 3 = 2 \text{ V} \overset{?}{>} V_t:$$
$$\text{OK (turn-on)} \qquad (P4.12.8)$$

$$V_D = V_{DD} - R_D I_D = 10 - 6\frac{1}{2} = 7 \text{ V} \rightarrow v_{GD}$$

$$= 5 - V_D = 5 - 7 = -2 \text{ V} \overset{?}{\le} V_t : \text{ OK (saturation)} \qquad (P4.12.9)$$

This implies that the FET operates in the *saturation* mode with $v_{DS} = V_{DS,Q} = V_D - V_S = 7 - 3 = 4$ V.

(a) To avoid such a trial-and-error as this, we can write the following system of circuit equations:

$$V_G - v_{GS} - R_S i_D(v_{DS}, v_{GS}) = 0$$
$$V_{DD} - v_{DS} - (R_D + R_S) i_D(v_{DS}, v_{GS}) = 0 \qquad (P4.12.10)$$

and solve it by running the following MATLAB script "nm04p12a.m" where an M-file named "iD_NMOS_at_vDS_vGS.m" prescribing Eq. (P4.12.1) is supposed to have been saved as a user function on your PC.

```
%nm04p12a.m
VDD=10; R1=10e6; R2=10e6; RD=6e3; RS=6e3; Kp=1e-3; Vt=1;
VG=R2/(R1+R2)*VDD;   % Eq.(P4.12.2)
eq=@(v) [VG-v(2)-RS*iD_NMOS_at_vDS_vGS(v(1),v(2),Kp,Vt); %Eq.(P4.12.10)
         VDD-v(1)-(RD+RS)*iD_NMOS_at_vDS_vGS(v(1),v(2),Kp,Vt)];
v=fsolve(eq,[Vt Vt]); VDS=v(1), VGS=v(2), VS=VG-VGS, VD=VS+VDS
```
```
function iD=iD_NMOS_at_vDS_vGS(vDS,vGS,Kp,Vt)
vGD=vGS-vDS; ON=(vGS>=Vt); SAT=(vGD<Vt)&ON; TRI=~SAT&ON;
iD=Kp/2*(vGS-Vt).^2.*SAT+Kp*((vGS-Vt).*vDS-vDS.^2/2).*TRI; %Eq.(P4.12.1)
```

(b) If v_{GS} has somehow been determined as 2 V, the second equation of Eq. (P4.12.10) can be written as

$$\frac{V_{DD} - v_{DS}}{R_D + R_S} = i_D(v_{DS}, v_{GS}) \qquad (P4.12.11)$$

Plot the graph of the LHS (load line) and RHS (characteristic curve) for $v_{DS} = 0\text{–}10$ V and mark their intersection, called the operating or bias point $(V_{DS,Q}, I_{D,Q})$ and denoted by Q, by completing and running the following MATLAB script:

```
%nm04p12b.m
VDD=10; R1=10e6; R2=10e6; RD=6e3; RS=6e3; Kp=1e-3; Vt=1;
vDSs=[0:1e-3:VDD];
vGS=2; iDs=iD_NMOS_at_vDS_vGS(vDSs,vGS,Kp,Vt);
clf, plot(vDSs,???, vDSs,(VDD-vDSs)/(??+RS),'r')
[dmin,imin]=min(abs(iDs-(VDD-vDSs)/(RD+RS)));  %find the intersection
VDSQ=vDSs(imin); IDQ=iDs(imin); % Q-point
text(VDSQ,IDQ,sprintf('Q(%4.2f[V],%4.2f[mA])',VDSQ,IDQ*1e3))
```

4.13 Temperature Rising from Heat Flux in a Semiinfinite Slab

Consider a semiinfinite slab whose temperature rises as a function of position $x > 0$ and time $t > 0$ as

$$T(x,t) = \frac{Qx}{k}\left(\frac{e^{-s^2}}{\sqrt{\pi}s} - \mathrm{erfc}(s)\right) \quad \text{with } s^2 = \frac{x^2}{4at} \qquad \text{(P4.13.1)}$$

where

$$Q(\text{heat flux}) = 200\,\mathrm{J/m^2\,s}, k(\text{conductivity}) = 0.015\,\mathrm{J/m/s/°C}$$

$$a(\text{diffusivity}) = 2.5 \times 10^{-5}\,\mathrm{m^2/s}$$

In order to find the heat transfer speed, a heating system expert, Kyungwon, wants to solve the aforementioned equation to get the positions $x(t)$ with a temperature rise of $T = 30\,°C$ at $t = [10:10:200]$ seconds. Complete and run the following script "nm04p12.m" to do this job and plot $x(t)$ vs. t.

```
%nm04p13.m - Temperature Rise from Heat Flux in a Semi-infinite Slab
Q=200; k=0.015; a=2.5e-5; T=30;
eq=@(x,t) [Q*x(?)/k*(exp(-x(?)^2)/sqrt(pi)/x(2)-erfc(x(?)))-?;
           x(1)^?-x(2)^2*4*?*t]; % x=[x s], Eq.(P4.13.1)
options=optimoptions('fsolve','Display','off');
tt=[10:10:200];
for i=1:numel(tt)
   t=tt(i);   xs0=t/1000*[1; 100/sqrt(t)];
   [xs_fs,err_fs]=fsolve(eq,xs0,options,?);
   x_fs(i)=xs_fs(1);
end
plot(tt,x_fs,'r')
```

4.14 Single-Stub Impedance Matching of Load Impedance to Transmission Line

For impedance matching of a load impedance Z_L to a transmission line of characteristic impedance $Z_0 = R_0$ by inserting a single short-circuited stub of impedance $Z_s = jX_s$, the equation to be satisfied by the position d (the distance from the load) of a single stub ([C-2], Chapter 9) can be written as

$$Re\left\{y_B = \frac{Z_0 + Z_L \tanh \gamma d}{Z_L + Z_0 \tanh \gamma d}\right\} \stackrel{\text{lossless line}}{\underset{\gamma=j\beta, Z_0=R_0}{=}} Re\left\{\frac{R_0 + jZ_L \tan \beta d}{Z_L + jR_0 \tan \beta d}\right\} = 1 \quad \text{(P4.14.1)}$$

where $\gamma = \alpha + j\beta$, α, and β are the propagation constant, attenuation constant, and phase constant of the transmission line, respectively (see Figure P4.14). Once a value of d satisfying Eq. (P4.14.1) has been determined, say, as d_1, we can find another value of d, say, d_2 such that

$$Im\left\{y_B = \frac{R_0 + jZ_L \tan \beta d_2}{Z_L + jR_0 \tan \beta d_2}\right\} = -b_B \text{ with } b_B = Im\left\{\frac{R_0 + jZ_L \tan \beta d_1}{Z_L + jR_0 \tan \beta d_1}\right\}$$

(P4.14.2)

```
function [ds,ls]=imp_match_1stub0(Z0,ZL,r)
% Input:   Z0 = Characteristic impedance, ZL = Load impedance,
%          r = a+bi = Propagation constant
% Output: ds = Positions (Distances from the load) of single stubs
%         ls = Lengths of the stubs
YL=1/ZL; yL=YL*Z0; % Normalized load admittance
a=real(r); b=imag(r); lambda=2*pi/b; % Phase constant and Wavelength
options=optimoptions('fsolve','Display','off');
% To determine the positions (distances from ZL) of the single stubs
eq=@(d)real((Z?+i*Z?*tan(b*d))/(Z?+i*Z?*tan(b*d)))-?; % Eq.(P4.14.1)
d1=real(fsolve(eq,0.1,options)); % such that Yi(d1)=yB=1+j*bB
yB=(Z0+ZL*tanh(r*d1))/(ZL+Z0*tanh(r*d1)); bB=imag(yB);

eq=@(d)imag((Z0+i*ZL*tan(?*d))/(ZL+i*Z0*tan(b*?)))+??; % Eq.(P4.14.2)
d2=real(fsolve(eq,0.1,options));
ds=[d1 d2]; ds=ds+(ds<0)*pi/b;
yBs=(Z0+i*ZL*tan(b*ds))./(ZL+i*Z0*tan(b*ds)) % expected to be 1+/-j*bB
% To determine the lengths of the single stubs
eq=@(l)imag(1/(i*tan(?*1)))+bB; % Eq.(P4.14.3a): Yi(l1)=1-j*bB
l1=real(fsolve(eq,0.1,options));

eq=@(l)????(1/(i*tan(b*?))))?bB; % Eq.(P4.14.3b): Yi(l1)=1+j*bB
l2=real(fsolve(eq,0.1,options));
ls=[l1 l2];
```

If the value of d is negative, it should be increased by π/β to be positive since it is the distance of the single-stub from the load. Additionally, the length l of the short-circuited single stub can be determined by

$$Im\left\{y_s = \frac{R_0 + jZ_L \tan\beta l}{Z_L + jR_0 \tan\beta l}\right\} \stackrel{\text{short-circuited}}{\underset{Z_L=0}{=}} Im\left\{\frac{1}{j\tan\beta l}\right\} = \mp b_B \quad \text{(P4.14.3)}$$

so that

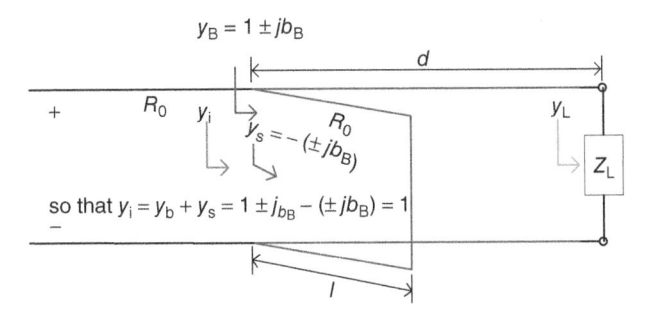

$$y_B + y_s = (1 + jb_B) - jb_B = 1 = y_i \quad \text{(P4.14.4)}$$

Figure P4.14 Single-stub impedance matching.

240 NONLINEAR EQUATIONS

(a) Complete the above MATLAB function 'imp_match_1stub0()' so that it can solve Eqs. (P4.14.1)-(P4.14.3) to find d_1 and d_2 and their corresponding l_1 and l_2. Then, use it to determine the positions/lengths of single stubs for impedance matching of $Z_L = 35 - j47.5\,\Omega$ to a transmission line of characteristic impedance $Z_0 = R_0 = 50\,\Omega$ where $\gamma = j\beta = j2\pi/\lambda$ with $\lambda = 1$.

(b) Cheng [C-2, Section 9.7.2] presented the formulas to determine the values of d's and l's for impedance matching as

$$d = \begin{cases} \dfrac{\lambda}{2\pi}\tan^{-1} t & \text{for } t \geq 0 \\ \dfrac{\lambda}{2} + \dfrac{\lambda}{2\pi}\tan^{-1} t & \text{for } t < 0 \end{cases} \quad (P4.14.5)$$

with

$$t = \begin{cases} \dfrac{1}{r_L - 1}\left\{ r_L \pm \sqrt{r_L\{(1 - r_L)^2 + x_L^2\}} \right\} & \text{if } r_L \neq 1 \\ -\dfrac{x_L}{2} & \text{if } r_L = 1 \end{cases}$$

$$l = \begin{cases} \dfrac{\lambda}{2\pi}\tan^{-1}\dfrac{1}{b_B} & \text{for } b_B \geq 0 \\ \dfrac{\lambda}{2} + \dfrac{\lambda}{2\pi}\tan^{-1}\dfrac{1}{b_B} & \text{for } b_B < 0 \end{cases} \quad (P4.14.6)$$

with

$$b_B = \frac{r_L^2 t - (1 - x_L\, t)(x_L + t)}{r_L^2 + (r_L + 1)^2}$$

Complete the following MATLAB function 'imp_match_1stub1()' so that it can use Eqs. (P4.14.5) and (P4.14.6) to determine d_1 and d_2 and their corresponding l_1 and l_2. Then use it to determine the positions/lengths of single stubs for impedance matching dealt with in (a).

```
function [ds,ls]=imp_match_1stub1(Z0,ZL,r)
% Input:   Z0 = Characteristic impedance, ZL = Load impedance,
%          r = a+bi = Propagation constant
% Output:  ds = Positions (Distances from the load) of single stubs
%          ls = Lengths of the stubs
b=imag(r); lambda=2*pi/b; % Phase constant and Wavelength
zL=ZL/Z0; rL=real(zL); xL=imag(zL); % Load impedance normalized
t=(rL~=?)*(xL+[1 -?]*sqrt(rL*((1-rL)^2+xL^2)))/(rL-1)-(rL??1)*xL/?;
bBs=(rL^2*t-(1-??*t).*(xL+t))./(rL^2+(xL+t).^2);
ds=lambda*(atan(t)/2/pi +(t<0)/2); % Eq.(P4.11.5)
ls=lambda*(atan(1./bBs)/2/pi +(bBs<0)/2); % Eq.(P4.11.6)
```

4.15 Impedance Matching for a Two-port Network ([Y-2], Problem 8.5)
Consider the circuit depicted in Figure P4.15, where $R_s = 75\,\Omega$, $R_L = 16\,\Omega$, and the a-parameter matrices of the three 2-port networks are

$$\begin{bmatrix} a_{11}^{(1)} & a_{12}^{(1)} \\ a_{21}^{(1)} & a_{22}^{(1)} \end{bmatrix} = \begin{bmatrix} -0.003 & -5 \\ 0 & -0.007 \end{bmatrix}, \quad \begin{bmatrix} a_{11}^{(R_{12})} & a_{12}^{(R_{12})} \\ a_{21}^{(R_{12})} & a_{22}^{(R_{12})} \end{bmatrix} = \begin{bmatrix} 1 & 0 \\ 1/R_{12} & 1 \end{bmatrix},$$

$$\text{and} \quad \begin{bmatrix} a_{11}^{(2)} & a_{12}^{(2)} \\ a_{21}^{(2)} & a_{22}^{(2)} \end{bmatrix} = \begin{bmatrix} 1 & 5.65 \\ 0 & 0.007 \end{bmatrix} \qquad \text{(P4.15.1)}$$

(a) Noting that the overall a-parameter matrix of a cascade connection of two-port networks is obtained by multiplying their a-parameter matrices, show that the overall a-parameter matrix of the cascade-connected two-port networks is

$$\begin{bmatrix} a_{11} & a_{12} \\ a_{21} & a_{22} \end{bmatrix} = \begin{bmatrix} -5/R_{12} - 0.003 & -113/4R_{12} - 0.5195 \\ -0.007/R_{12} & -0.03955/R_{12} - 0.000049 \end{bmatrix} \qquad \text{(P4.15.2)}$$

```
»syms R12; Rs=75; RL=16;
A_1=[-0.003 -5; 0 -0.007]; A_R12=[1 0; 1/R12 1];
A_2=[1 5.65; 0 0.007];
A=A_1*A_R12*A_2
```

(b) Noting that the output impedance of a two-port network with source impedance R_s can be obtained from its a-parameters as

$$Z_o = \frac{a_{22}R_s + a_{12}}{a_{21}R_s + a_{11}} \qquad \text{(P4.15.3)}$$

show that the output impedance of the circuit with $R_s = 75\,\Omega$ is

$$Z_o = \frac{a_{22}R_s + a_{12}}{a_{21}R_s + a_{11}}\bigg|_{R_s=75} = \frac{31.21625/R_{12} + 0.055625}{5.525/R_{12} + 0.003} \qquad \text{(P4.15.4)}$$

```
»Zo_=(A(2,2)*Rs+A(1,2))/(A(2,1)*Rs+A(1,1))
```

(c) Noting that Z_o should be equal to $R_L = 16\,\Omega$ for impedance matching, determine the value of R_{12} for impedance matching:

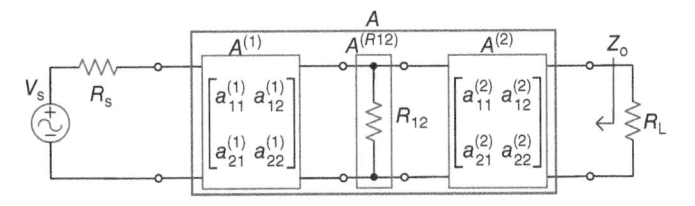

Figure P4.15 Two-port networks in cascade interconnection.

242 NONLINEAR EQUATIONS

$$Z_o = \frac{31.21625/R_{12} + 0.055625}{5.525/R_{12} + 0.003} = 16 \qquad \text{(P4.15.5)}$$

```
»R12=eval(solve(Zo_-16,R12)), Zo=eval(Zo_)
```

Alternatively, we can use 'fsolve()' instead of 'solve()' as follows:

```
»f=@(R12)(31.21625/R12+0.055625)/(5.525/R12+0.003)-16
  R12=fsolve(f,1)
```

4.16 Region of Two Variables Satisfying a Nonlinear Constraint
Plot the region of two variables (C, L) satisfying the following inequality on the C-L plane:

$$\{1 - (2\omega_0)^2 LC\}^2 + \left(\frac{2\omega_0 L}{R}\right)^2 \geq \left(\frac{2}{3r_{max}}\right)^2 \qquad \text{(P4.16.1)}$$

where $R = 10^4$, $\omega_0 = 2\pi 60$, and $r_{max} = 0.1$

It may be helpful to complete and run the following MATLAB script "nm04p16.m".

```
%nm04p16.m
R=1e4; w0=2*pi*60; rmax=0.1;
fL=@(L,C,rmax)(1-(2*w0)^2*L*C)^2+(2*w0*L/R)^2 -(2/3/rmax)^2; % Eq.(P4.15.1)
CC=logspace(-4,-1,90); % set the range of C to 90 points on [10^-4,10^-1]
opt=optimset('Display','off');
for m=1:length(CC)
   C=CC(m); L0=0.1;
   if m>1, L0=LL(m-1); end
   LL(m)= fsolve(fL,L0,opt,C,rmax);
end
loglog(CC,LL)
```

4.17 Plot of Van der Waals Isotherms
Consider the Van der Waals equation of state for one mole of an imperfect fluid/gas:

$$f(v,p) = \left(p + \frac{3}{v^2}\right)(v - 1/3) = \frac{8}{3}T \qquad \text{(P4.17.1)}$$

where p, v, and T are the pressure, the volume per particle, and the absolute temperature, respectively:

(a) Complete and run the following MATLAB script "nm04p17.m" to solve the equation to determine the values of v (l) for $p = 0.1 : 0.001 : 3$ (atm) and then plot p vs. v for $T = 0.9$ and 1.1 K. Do the two graphs look fine? If not, why?

```
%nm04p17a.m
Ts=[0.9 1.1]; ps=[0.1:0.001:3];
fv=@(v,p,T)(p+3./v.^2).*(v-1/3)-8/3*T;  % Eq. (P4.17.1)
v0_0=10; gss='brmk'; opt=optimset('Display','off');
for m=1:numel(Ts)
   T=Ts(m);
   for n=1:numel(ps)
      p=ps(n);
      if n<2, v0=v0_0; else v0=vs(n-1); end
      vs(n)=fsolve(fv,v0,opt,p,T);
   end
   plot(vs,ps,gss(m)); hold on
end
```

(b) Solve the equation to determine the values of p (atm) for $v = 0.5 : 0.01 : 10$ (l) and then plot p vs. v for $T = 0.9$ and 1.1 K. Do the two graphs look fine?

(c) Use the MATLAB function 'contour()' to plot the contours of the function $f(v,p)$ at the levels of $8/3 \times [0.5 : 0.1 : 1.5]$.

```
%nm04p17c.m
Ts1=[0.5:0.1:1.5]; Ls=8/3*Ts1;  % Contour levels
vs=[0:0.01:10]; ps=[0:0.01:3];
[Vs,Ps]=meshgrid(vs,ps);
F=(Ps+3./Vs.^2).*(Vs-1/3);
contour(Vs,Ps,F,Ls)
```

4.18 Intersection of Two Circles

Consider the following set of simultaneous equations to find the intersection point(s) of two circles, one with center $\mathbf{c}_1 = [c_{1x}\ c_{1y}]$ and radius r_1 and the other with center $\mathbf{c}_2 = [c_{2x}\ c_{2y}]$ and radius r_2:

$$(x - c_{1x})^2 + (y - c_{1y})^2 = r_1^2$$
$$(x - c_{2x})^2 + (y - c_{2y})^2 = r_2^2$$
(P4.18.1)

(a) Solve this set of equations to find the intersection point(s) of two circles, one with center $\mathbf{c}_1 = [2\ \ 1]$ and radius $r_1 = 4$ and the other with center $\mathbf{c}_2 = [-3\ -2]$ and radius $r_2 = \sqrt{10}$.

(b) Use the MATLAB function 'circcirc()' to find the two intersection points.

(c) To find the analytical solution of Eq. (P4.18.1), subtract Eq. (P4.18.1b) from Eq. (P4.18.1a) to get

$$y = -\frac{c_{2x} - c_{1x}}{c_{2y} - c_{1y}} x + \frac{r_1^2 - r_2^2 + c_{2x}^2 - c_{1x}^2 + c_{2y}^2 - c_{1y}^2}{2(c_{2y} - c_{1y})} = ax + b \quad \text{(P4.18.2)}$$

and substitute this into, say, Eq. (P4.18.1a) to write

$$(a^2 + 1)x^2 + 2(a(b - c_{1y}) - c_{1x})x + c_{1x}^2 + (b - c_{1y})^2 - r_1^2 = 0 \quad (P4.18.3)$$

Then we can use the MATLAB function 'roots()' to find the x-coordinates of the intersection points and substitute them into Eq. (P4.18.2) to obtain the y-coordinates of the intersection points. Make your own MATLAB function 'circcirc_my()' implementing this algorithm and use it to find the two intersection points obtained in (a) or (b).

4.19 Damped Newton Method for a Set of Nonlinear Equations
Consider the function 'Newtons()', which is made for solving a system of equations and introduced in Section 4.6.

(a) Use the function with the initial point $(x_{10}, x_{20}) = (0.5, 0.2)$ to solve Eq. (4.6.5) and certify that it does not yield the right solution as depicted in Figure 4.6c.

(b) In order to keep the step-size adjusted in case the norm of the vector function $\mathbf{f}(\mathbf{x}_{k+1})$ at iteration $k+1$ is larger than that of $\mathbf{f}(\mathbf{x}_k)$ at iteration k, insert(activate) the statements numbered from (1) to (6) of the function 'Newtons()' (Section 4.VI.) by deleting the comment mark (%) at the beginning of each line to make a modified function 'Newtonds()', which implements the damped Newton method. Run it with the initial point $(x_{10}, x_{20}) = (0.5, 0.2)$ to solve Eq. (4.6.5) and certify that it yields the right solution as depicted in Figure 4.6d.

(c) Use the MATLAB built-in function 'fsolve()' with the initial point $(x_{10}, x_{20}) = (0.5, 0.2)$ to solve Eq. (4.6.5). Does it present you a right solution?

5

NUMERICAL DIFFERENTIATION/ INTEGRATION

CHAPTER OUTLINE

5.1	Difference Approximation for the First Derivative	246
5.2	Approximation Error of the First Derivative	248
5.3	Difference Approximation for Second and Higher Derivative	253
5.4	Interpolating Polynomial and Numerical Differential	258
5.5	Numerical Integration and Quadrature	259
5.6	Trapezoidal Method and Simpson Method	263
5.7	Recursive Rule and Romberg Integration	265
5.8	Adaptive Quadrature	268
5.9	Gauss Quadrature	272
	5.9.1 Gauss-Legendre Integration	272
	5.9.2 Gauss-Hermite Integration	275
	5.9.3 Gauss-Laguerre Integration	277
	5.9.4 Gauss-Chebyshev Integration	277
5.10	Double Integral	278
5.11	Integration Involving PWL Function	281
	Problems	285

Applied Numerical Methods Using MATLAB®, Second Edition. Won Y. Yang, Jaekwon Kim, Kyung W. Park, Donghyun Baek, Sungjoon Lim, Jingon Joung, Suhyun Park, Han L. Lee, Woo June Choi, and Taeho Im.
© 2020 John Wiley & Sons, Inc. Published 2020 by John Wiley & Sons, Inc.
Companion website: www.wiley.com/go/yang/appliednumericalmethods

5.1 DIFFERENCE APPROXIMATION FOR THE FIRST DERIVATIVE

For a function $f(x)$ of a variable x, its first derivative is defined as

$$f'(x) = \lim_{h \to 0} \frac{f(x+h) - f(x)}{h} \tag{5.1.1}$$

However, this gives our computers a headache, since they do not know how to take a limit. Any input number given to computers must be a definite number and can be neither too small nor too large to be understood by the computer. The 'theoretically' infinitesimal number h involved in this equation is a problem.

A simple approximation that computers might be happy with is the forward difference approximation.

$$D_{f1}(x, h) = \frac{f(x+h) - f(x)}{h} \quad (h : \text{step-size}) \tag{5.1.2}$$

How far away is this approximation from the true value of Eq. (5.1.1)? In order to do the error analysis, we take the Taylor series expansion of $f(x+h)$ about x as follows:

$$f(x+h) = f(x) + hf'(x) + \frac{h^2}{2}f^{(2)}(x) + \frac{h^3}{3!}f^{(3)}(x) + \cdots \tag{5.1.3}$$

Subtracting $f(x)$ from both sides and dividing both sides by the step-size h yields

$$D_{f1}(x, h) = \frac{f(x+h) - f(x)}{h} = f'(x) + \frac{h}{2}f^{(2)}(x) + \frac{h^2}{3!}f^{(3)}(x) + \cdots = f'(x) + O(h) \tag{5.1.4}$$

where $O(g(h))$, called 'big Oh of $g(h)$,' denotes a truncation error term proportional to $g(h)$ for $|h| \ll 1$. This means that the error of the forward difference approximation (5.1.2) of the first derivative is proportional to the step-size h, or equivalently, in the order of h.

Now, in order to derive another approximation formula for the first derivative having a smaller error, let us remove the first-order term with respect to h from Eq. (5.1.4) by substituting $2h$ for h in the equation

$$D_{f1}(x, 2h) = \frac{f(x+2h) - f(x)}{2h} = f'(x) + \frac{2h}{2}f^{(2)}(x) + \frac{4h^2}{3!}f^{(3)}(x) + \cdots$$

and subtracting this result from two times the equation. Then, we get

$$2D_{f1}(x, h) - D_{f1}(x, 2h) = 2\frac{f(x+h) - f(x)}{h} - \frac{f(x+2h) - f(x)}{2h}$$

$$= f'(x) - \frac{2h^2}{3!}f^{(3)}(x) + \cdots ;$$

DIFFERENCE APPROXIMATION FOR THE FIRST DERIVATIVE

$$D_{f2}(x,h) = \frac{2D_{f1}(x,h) - D_{f1}(x,2h)}{2-1} = \frac{-f(x+2h) + 4f(x+h) - 3f(x)}{2h} \quad (5.1.5)$$

$$= f'(x) + O(h^2)$$

which can be regarded as an improvement over Eq. (5.1.4), since it has the truncation error of $O(h^2)$ for $|h| \ll 1$.

How about the backward difference approximation?

$$D_{b1}(x,h) = \frac{f(x) - f(x-h)}{h} \equiv D_{f1}(x,-h) \; (h : \text{step-size}) \quad (5.1.6)$$

This also has an error of $O(h)$ and can be processed to yield an improved version having a truncation error of $O(h^2)$.

$$D_{b2}(x,h) = \frac{2D_{b1}(x,h) - D_{b1}(x,2h)}{2-1} = \frac{3f(x) - 4f(x-h) + f(x-2h)}{2h} \quad (5.1.7)$$

$$= f'(x) + O(h^2)$$

In order to derive another approximation formula for the first derivative, we take the Taylor series expansion of $f(x+h)$ and $f(x-h)$ up to the fifth-order to write

$$f(x+h) = f(x) + hf'(x) + \frac{h^2}{2}f^{(2)}(x) + \frac{h^3}{3!}f^{(3)}(x) + \frac{h^4}{4!}f^{(4)}(x) + \frac{h^5}{5!}f^{(5)}(x) + \cdots$$

$$f(x-h) = f(x) - hf'(x) + \frac{h^2}{2}f^{(2)}(x) - \frac{h^3}{3!}f^{(3)}(x) + \frac{h^4}{4!}f^{(4)}(x) - \frac{h^5}{5!}f^{(5)}(x) + \cdots$$

and divide the difference between these two equations by $2h$ to get the central difference approximation for the first derivative as

$$D_{c2}(x,h) = \frac{f(x+h) - f(x-h)}{2h} = f'(x) + \frac{h^2}{3!}f^{(3)}(x) + \frac{h^4}{5!}f^{(5)}(x) + \cdots$$

$$= f'(x) + O(h^2) \quad (5.1.8)$$

which has an error of $O(h^2)$ similarly to Eqs. (5.1.5) and (5.1.7). This can also be processed to yield an improved version having a truncation error of $O(h^4)$.

$$2^2 D_{c2}(x,h) - D_{c2}(x,2h) = 4\frac{f(x+h) - f(x-h)}{2h} - \frac{f(x+2h) - f(x-2h)}{2 \cdot 2h}$$

$$= 3f'(x) - \frac{12h^4}{5!}f^{(5)}(x) - \cdots ;$$

$$D_{c4}(x,h) = \frac{2^2 D_{c1}(x,h) - D_{c1}(x,2h)}{2^2 - 1}$$

$$= \frac{8f(x+h) - 8f(x-h) - f(x+2h) + f(x-2h)}{12h} \quad (5.1.9)$$

$$= f'(x) + O(h^4);$$

Furthermore, this procedure can be formularized into a general formula, called 'Richardson's extrapolation,' for improving the difference approximation of the derivatives as follows:

RICHARDSON'S EXTRAPOLATION

$$D_{f,n+1}(x,h) = \frac{2^n D_{f,n}(x,h) - D_{f,n}(x,2h)}{2^n - 1} \quad (n: \text{the order of error}) \quad (5.1.10a)$$

$$D_{b,n+1}(x,h) = \frac{2^n D_{b,n}(x,h) - D_{b,n}(x,2h)}{2^n - 1} \quad (5.1.10b)$$

$$D_{c,2(n+1)}(x,h) = \frac{2^{2n} D_{c,2n}(x,h) - D_{c,2n}(x,2h)}{2^{2n} - 1} \quad (5.1.10c)$$

5.2 APPROXIMATION ERROR OF THE FIRST DERIVATIVE

In the previous section, we derived some difference approximation formulas for the first derivative. Since their errors are proportional to some power of the step-size h, it seems that the errors continue to decrease as h gets smaller. However, this is only half of the story since we considered only the truncation error caused by truncating the high-order terms in the Taylor series expansion and did not take account of the round-off error caused by quantization.

In this section, we will discuss the round-off error as well as the truncation error so as to gain a better understanding of how the computer really works. For this purpose, suppose that the function values

$$f(x+2h), f(x+h), f(x), f(x-h), f(x-2h)$$

are quantized (rounded-off) to

$$y_2 = f(x+2h) + e_2, \quad y_1 = f(x+h) + e_1$$
$$y_0 = f(x) + e_0 \tag{5.2.1}$$
$$y_{-1} = f(x-h) + e_{-1}, \quad y_{-2} = f(x-2h) + e_{-2}$$

where the magnitudes of the round-off (quantization) errors e_2, e_1, e_0, e_{-1}, and e_{-2} are all smaller than some positive number ε, i.e. $|e_i| \leq \varepsilon$. Then, the total error of the forward difference approximation (5.1.4) can be derived as follows:

$$D_{f1}(x,h) = \frac{y_0 - y_0}{h} = \frac{f(x+h) + e_1 - f(x) - e_0}{h} \stackrel{(5.1.4)}{=} f'(x) + \frac{e_1 - e_0}{h} + \frac{K_1}{2}h;$$

$$|D_{f1}(x,h) - f'(x)| \leq \left|\frac{e_1 - e_0}{h}\right| + \frac{|K_1|}{2}h \leq \frac{2\varepsilon}{h} + \frac{|K_1|}{2}h \text{ with } K_1 = f^{(2)}(x)$$

Look at the RHS of this inequality, i.e. the upper bound of error. It consists of two parts: the first one is due to the round-off error and in inverse proportion to the step-size h, while the second one is due to the truncation error and in direct proportion to h. Therefore, the upper bound of the total error can be minimized with respect to the step-size h to give the optimum step-size h_o as

$$\frac{d}{dh}\left(\frac{2\varepsilon}{h} + \frac{|K_1|}{2}h\right) = -\frac{2\varepsilon}{h^2} + \frac{|K_1|}{2} = 0; \quad h_o = 2\sqrt{\frac{\varepsilon}{|K_1|}} \tag{5.2.2}$$

The total error of the central difference approximation (5.1.8) can also be derived as follows:

$$D_{c2}(x,h) = \frac{y_1 - y_{-1}}{2h} = \frac{f(x+h) + e_1 - f(x-h) - e_{-1}}{2h}$$

$$\stackrel{(5.1.8)}{=} f'(x) + \frac{e_1 - e_{-1}}{2h} + \frac{K_2}{6}h^2;$$

$$|D_{c2}(x,h) - f'(x)| \leq \left|\frac{e_1 - e_{-1}}{2h}\right| + \frac{|K_1|}{6}h^2 \leq \frac{2\varepsilon}{2h} + \frac{|K_2|}{6}h^2 \text{ with } K_2 = f^{(3)}(x)$$

The RHS of this inequality is minimized to yield the optimum step-size h_o as

$$\frac{d}{dh}\left(\frac{\varepsilon}{h} + \frac{|K_2|}{6}h^2\right) = -\frac{\varepsilon}{h^2} + \frac{|K_2|}{3}h = 0; \quad h_o = \sqrt[3]{\frac{3\varepsilon}{|K_2|}} \tag{5.2.3}$$

Similarly, we can derive the total error of the central difference approximation (5.1.9) as

$$|D_{c4}(x,h) - f'(x)| \leq \left|\frac{8e_1 - 8e_{-1} - e_2 + e_{-2}}{12h}\right| + \frac{|K_4|}{30}h^4$$

$$\leq \frac{18\varepsilon}{12h} + \frac{|K_4|}{30}h^4 \quad \text{with } K_4 = f^{(5)}(x)$$

and find out the optimum step-size h_o as follows:

$$\frac{d}{dh}\left(\frac{3\varepsilon}{2h} + \frac{|K_4|}{30}h^4\right) = -\frac{3\varepsilon}{2h^2} + \frac{2|K_4|}{15}h^3 = 0; \quad h_o = \sqrt[5]{\frac{45\varepsilon}{4|K_4|}} \quad (5.2.4)$$

From what we have seen so far, we can tell that, as we make the step-size h smaller, the round-off error may increase, while the truncation error decreases. This is called 'step-size dilemma.' Therefore, there must be some optimal step-size h_o for the difference approximation formulas, as derived analytically in Eqs. (5.2.2) to (5.2.4). However, these equations are only of theoretical value and cannot be used practically to determine h_o because we usually do not have any information about the high-order derivatives and consequently, we cannot estimate $K_1, K_2, \ldots$. Besides, noting that h_o minimizes not the real error, but its upper bound, we can never expect the true optimal step-size to be uniform for all x even with the same approximation formula.

Now, we can verify the step-size dilemma and the existence of some optimal step-size h_o by computing the numerical derivative of a function, say, $f(x) = \sin x$, whose analytical derivatives are well known. To see how the errors of the difference approximation formulas (5.1.4) and (5.1.8) depend on the step-size h, we computed their values for $x = \pi/4$ together with their errors as summarized in Tables 5.1.1 and 5.1.2. From these results, it appears that the errors of (5.1.4) and (5.1.8) are minimized with $h \approx 10^{-8}$ and 10^{-5}, respectively. This may be justified by the following facts:

- Noting that the number of significant bits is 52, which is the number of mantissa bits (Section 1.2.1) or equivalently, the number of significant digits is about $52 \times 3/10 \approx 16$ (since $2^{10} \approx 10^3$), and the value of $f(x) = \sin x$ is less than or equal to one, the round-off error is roughly

$$\varepsilon \approx 10^{-16}/2$$

- Accordingly, Eqs. (5.2.2), and (5.2.3) give the theoretical optimal values of step-size h as

$$h_o = 2\sqrt{\frac{\varepsilon}{|K_1|}} = 2\sqrt{\frac{\varepsilon}{|f'(\pi/4)|}} = 2\sqrt{\frac{10^{-16}/2}{|-\sin(\pi/4)|}} = 1.68 \times 10^{-8}$$

$$h_o = \sqrt[3]{\frac{3\varepsilon}{|K_2|}} = \sqrt[3]{\frac{3\varepsilon}{|f^{(3)}(\pi/4)|}} = \sqrt[3]{\frac{3 \times 10^{-16}/2}{|-\cos(\pi/4)|}} = 0.5964 \times 10^{-5}$$

APPROXIMATION ERROR OF THE FIRST DERIVATIVE

Table 5.1.1 The forward difference approximation (5.1.4) for the first derivative of $f(x) = \sin x$ and its error from the true value ($\cos \pi/4 = 0.7071067812$) depending on the step size h.

$h_k = 10^{-k}$	$D_{f1}(x, h_k)\|_{x=\pi/4}$	$D_{f1}(x, h_k) - D_{f1}(x, h_{k-1})\|_{x=\pi/4}$	$D_{f1}(x, h_k)\|_{x=\pi/4} - \cos \frac{\pi}{4}$
$h_1 = 0.1000000000$	0.6706029729		-0.03650380828
$h_2 = 0.0100000000$	0.7035594917	0.0329565188	-0.00354728950
$h_3 = 0.0010000000$	0.7067531100	0.0031936183	-0.00035367121
$h_4 = 0.0001000000$	0.7070714247	0.0003183147	-0.00003535652
$h_5 = 0.0000100000$	0.7071032456	0.0000318210	-0.00000353554
$h_6 = 0.0000010000$	0.7071064277	0.0000031821	-0.00000035344
$h_7 = 0.0000001000$	0.7071067465	0.0000003187	-0.00000003470
$h_8 = 0.0000000100$*	0.7071067842	0.0000000377	0.000000000305*
$h_9 = 0.0000000010$	0.7071068175	0.0000000333*	0.000000003636
$h_{10} = 0.0000000001$	0.7071077057	0.0000008882	0.000000092454
$h_o = 0.0000000168$ (the optimal value of h obtained from Eq. (5.2.2))			

Table 5.1.2 The central difference approximation (5.1.8) for the first derivative of $f(x) = \sin x$ and its error from the true value ($\cos \pi/4 = 0.7071067812$) depending on the step size h.

$h_k = 10^{-k}$	$D_{c2}(x, h_k)\|_{x=\pi/4}$	$D_{c2}(x, h_k) - D_{c2}(x, h_{k-1})\|_{x=\pi/4}$	$D_{c2}(x, h_k)\|_{x=\pi/4} - \cos \frac{\pi}{4}$
$h_1 = 0.1000000000$	0.7059288590		-0.00117792219
$h_2 = 0.0100000000$	0.7070949961	0.0011661371	-0.00001178505
$h_3 = 0.0010000000$	0.7071066633	0.0000116672	-0.00000011785
$h_4 = 0.0001000000$	0.7071067800	0.0000001167	-0.00000000118
$h_5 = 0.0000100000$*	0.7071067812	0.0000000012	-0.00000000001*
$h_6 = 0.0000010000$	0.7071067812	0.0000000001*	0.00000000005
$h_7 = 0.0000001000$	0.7071067809	-0.0000000003	-0.00000000028
$h_8 = 0.0000000100$	0.7071067842	0.0000000033	0.00000000305
$h_9 = 0.0000000010$	0.7071067620	-0.0000000222	-0.00000001915
$h_{10} = 0.0000000001$	0.7071071506	0.0000003886	0.00000036942
$h_o = 0.0000059640$ (the optimal value of h obtained from Eq. (5.2.3))			

Figure 5.1a,b show how the error bounds of the difference approximations (5.1.4)/(5.1.8) for the first derivative vary with the step-size h, implying that there is some optimal value of step-size h with which the error bound of the numerical derivative is minimized. It seems that we might be able to get the optimal step-size h_o by using this kind of graph or directly using Eq. (5.2.2), (5.2.3) or (5.2.4). But, as mentioned before, it is not possible, as long as the high-order derivatives are unknown (as is usually the case). Very fortunately,

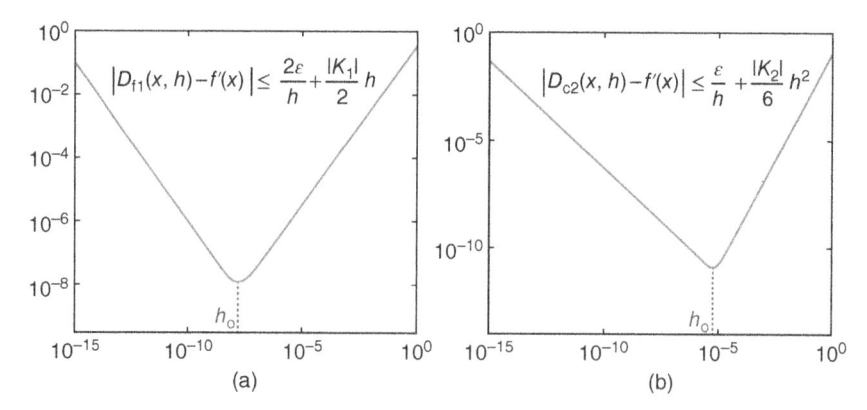

Figure 5.1 Forward/central difference approximation error of the first derivative of $f(x) = \sin x$ vs. stepsize h. (a) Error bound of Eq. (5.1.4) vs. stepsize h. (b) Error bound of Eq. (5.1.8) vs. stepsize h.

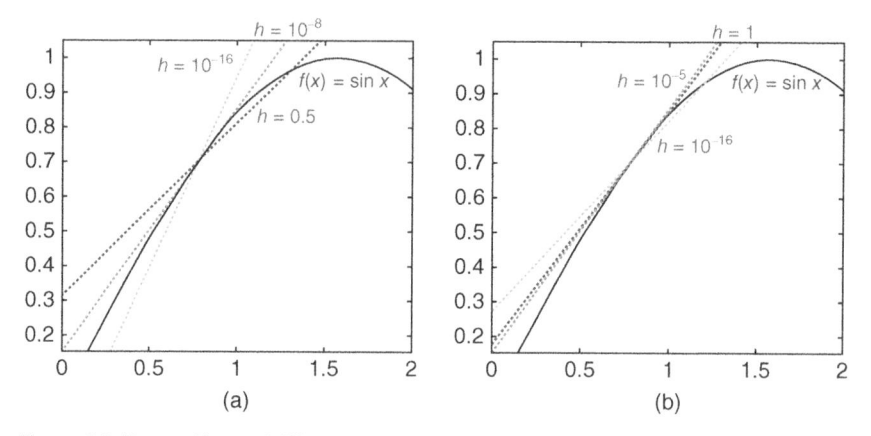

In figure 5.1(a) the bound shown is $|D_{f1}(x,h) - f'(x)| \leq \dfrac{2\varepsilon}{h} + \dfrac{|K_1|}{2} h$, and in figure 5.1(b) the bound shown is $|D_{c2}(x,h) - f'(x)| \leq \dfrac{\varepsilon}{h} + \dfrac{|K_2|}{6} h^2$.

Figure 5.2 Forward/central difference approximation error of the first derivative of $f(x) = \sin x$. (a) Forward difference approximated by Eq. (5.1.4). (b) Forward difference approximated by Eq. (5.1.8).

Tables 5.1.1 and 5.1.2 suggest that we might be able to guess the good value of h by watching how small $|D_{ik} - D_{i(k-1)}|$ is for a given problem. On the other hand, Figure 5.2a,b shows the tangential lines based on the forward/central difference approximations (5.1.4)/(5.1.8) of the first derivative at $x = \pi/4$ with the three values of step-size h. They imply that there is some optimal step-size h_o, and the numerical approximation error becomes larger if we make the step-size h larger or smaller than the value.

5.3 DIFFERENCE APPROXIMATION FOR SECOND AND HIGHER DERIVATIVE

In order to obtain an approximation formula for the second derivative, we take the Taylor series expansion of $f(x+h)$ and $f(x-h)$ up to the fifth-order to write

$$f(x+h) = f(x) + hf'(x) + \frac{h^2}{2}f^{(2)}(x) + \frac{h^3}{3!}f^{(3)}(x) + \frac{h^4}{4!}f^{(4)}(x) + \frac{h^5}{5!}f^{(5)}(x) + \cdots$$

$$f(x-h) = f(x) - hf'(x) + \frac{h^2}{2}f^{(2)}(x) - \frac{h^3}{3!}f^{(3)}(x) + \frac{h^4}{4!}f^{(4)}(x) - \frac{h^5}{5!}f^{(5)}(x) + \cdots$$

Adding these two equations (to remove the $f'(x)$ terms), then subtracting $2f(x)$ from both sides and dividing both sides by h^2 yields the central difference approximation for the second derivative as

$$D_{c2}^{(2)}(x,h) = \frac{f(x+h) - 2f(x) + f(x-h)}{h^2} = f^{(2)}(x) + \frac{h^2}{12}f^{(4)}(x) + \frac{2h^4}{6!}f^{(6)}(x) + \cdots \quad (5.3.1)$$

which has a truncation error of $O(h^2)$.

Richardson's extrapolation can be used for manipulating this equation to remove the h^2 term, which yields an improved version

$$\frac{2^2 D_{c2}^{(2)}(x,h) - D_{c2}^{(2)}(x,2h)}{2^2 - 1} = \frac{-f(x+2h) + 16f(x+h) - 30f(x) + 16f(x-h) - f(x-2h)}{12h^2}$$

$$= f^{(2)}(x) - \frac{h^4}{90}f^{(5)}(x) + \cdots;$$

$$D_{c4}^{(2)}(x,h) = \frac{-f(x+2h) + 16f(x+h) - 30f(x) + 16f(x-h) - f(x-2h)}{12h^2} = f^{(2)}(x) + O(h^4) \quad (5.3.2)$$

which has a truncation error of $O(h^4)$.

The difference approximation formulas for the first and second derivatives derived so far are summarized in Table 5.2, where the following notations are used.

$D_{fi}^{(N)}/D_{bi}^{(N)}/D_{ci}^{(N)}$: the forward/backward/central difference approximation for the Nth derivative having an error of $O(h^i)$ (h: the step-size)

$f_k = f(x + kh)$

Now, we turn our attention to the high-order derivatives. But, instead of deriving the specific formulas, let us make an algorithm to generate whatever difference approximation formula we want. For instance, if we want to get the approximation formula of the second derivative based on the function values f_2, $f_1, f_0, f_{-1},$ and f_{-2}, we write

Table 5.2 The difference approximation formulas for the first and second derivatives.

$O(h)$ forward difference approximation for the first derivative: $$D_{f1}(x,h) = \frac{f_1 - f_0}{h} \quad (5.1.4)$$	$O(h)$ backward difference approximation for the first derivative: $$D_{b1}(x,h) = \frac{f_0 - f_{-1}}{h} \quad (5.1.6)$$
$O(h^2)$ forward difference approximation for the first derivative: $$D_{f2}(x,h) = \frac{2D_{f1}(x,h) - D_{f1}(x,2h)}{2-1}$$ $$= \frac{-f_2 + 4f_1 - 3f_0}{2h} \quad (5.1.5)$$	$O(h^2)$ backward difference approximation for the first derivative: $$D_{b2}(x,h) = \frac{2D_{b1}(x,h) - D_{b1}(x,2h)}{2-1}$$ $$= \frac{3f_0 - 4f_{-1} + f_{-2}}{2h} \quad (5.1.7)$$
$O(h^2)$ central difference approximation for the first derivative: $$D_{c2}(x,h) = \frac{f_1 - f_{-1}}{2h} \quad (5.1.8)$$	
$O(h^4)$ forward difference approximation for the first derivative: $$D_{c4}(x,h) = \frac{2^2 D_{c2}(x,h) - D_{c2}(x,2h)}{2^2 - 1} = \frac{-f_2 + 8f_1 - 8f_{-1} + f_{-2}}{12h} \quad (5.1.9)$$	
$O(h^2)$ central difference approximation for the second derivative: $$D_{c2}^{(2)}(x,h) = \frac{f_1 - 2f_0 + f_{-1}}{h^2} \quad (5.3.1)$$	
$O(h^4)$ forward difference approximation for the second derivative: $$D_{c4}^{(2)}(x,h) = \frac{2^2 D_{c2}^{(2)}(x,h) - D_{c2}^{(2)}(x,2h)}{2^2 - 1} = \frac{-f_2 + 16f_1 - 30f_0 + 16f_{-1} - f_{-2}}{12h^2} \quad (5.3.2)$$	
$O(h^2)$ central difference approximation for the fourth derivative: $$D_{c2}^{(4)}(x,h) = \frac{f_{-2} - 4f_{-1} + 6f_0 - 4f_1 + f_2}{h^4} \quad \text{(from difapx(4,[-2 2 1]) (5.3.6)}$$	

254

DIFFERENCE APPROXIMATION FOR SECOND AND HIGHER DERIVATIVE 255

$$D_{c4}^{(2)}(x, h) = \frac{c_2 f_2 + c_1 f_1 + c_0 f_0 + c_{-1} f_{-1} + c_{-2} f_{-2}}{h^2} \quad (5.3.3)$$

and take the Taylor series expansion of f_2, f_1, f_{-1}, and f_{-2}, excluding f_0 on the RHS of this equation to rewrite it as

$$D_{c4}^{(2)}(x,h) = \frac{1}{h^2} \left\{ \begin{array}{l} c_2 \left(f_0 + 2hf_0' + \frac{(2h)^2}{2} f_0^{(2)} + \frac{(2h)^3}{3!} f_0^{(3)} + \frac{(2h)^4}{4!} f_0^{(4)} + \cdots \right) \\ + c_1 \left(f_0 + hf_0' + \frac{h^2}{2} f_0^{(2)} + \frac{h^3}{3!} f_0^{(3)} + \frac{h^4}{4!} f_0^{(4)} + \cdots \right) + c_0 f_0 \\ + c_{-1} \left(f_0 - hf_0' + \frac{h^2}{2} f_0^{(2)} - \frac{h^3}{3!} f_0^{(3)} + \frac{h^4}{4!} f_0^{(4)} - \cdots \right) \\ + c_{-2} \left(f_0 - 2hf_0' + \frac{(2h)^2}{2} f_0^{(2)} - \frac{(2h)^3}{3!} f_0^{(3)} + \frac{(2h)^4}{4!} f_0^{(4)} - \cdots \right) \end{array} \right\}$$

$$= \frac{1}{h^2} \left\{ \begin{array}{l} (c_2 + c_1 + c_0 + c_{-1} + c_{-2}) f_0 + h(2c_2 + c_1 - c_{-1} - 2c_{-2}) f_0' \\ + h^2 \left(\frac{2^2}{2} c_2 + \frac{1}{2} c_1 + \frac{1}{2} c_{-1} + \frac{2^2}{2} c_{-2} \right) f_0^{(2)} \\ + h^3 \left(\frac{2^3}{3!} c_2 + \frac{1}{3!} c_1 - \frac{1}{3!} c_{-1} - \frac{2^3}{3!} c_{-2} \right) f_0^{(3)} \\ + h^4 \left(\frac{2^4}{4!} c_2 + \frac{1}{4!} c_1 + \frac{1}{4!} c_{-1} + \frac{2^4}{4!} c_{-2} \right) f_0^{(4)} + \cdots \end{array} \right\}$$

(5.3.4)

We should solve the following set of equations to determine the coefficients c_2, c_1, c_0, c_{-1}, and c_{-2}, so as to make the expression conform to the second derivative $f_0^{(2)}$ at $x + 0h = x$.

$$\begin{bmatrix} 1 & 1 & 1 & 1 & 1 \\ 2 & 1 & 0 & -1 & -2 \\ 2^2/2! & 1/2! & 0 & 1/2! & 2^2/2! \\ 2^3/3! & 1/3! & 0 & -1/3! & -2^3/3! \\ 2^4/4! & 1/4! & 0 & 1/4! & 2^4/4! \end{bmatrix} \begin{bmatrix} c_2 \\ c_1 \\ c_0 \\ c_{-1} \\ c_{-2} \end{bmatrix} = \begin{bmatrix} 0 \\ 0 \\ 1 \\ 0 \\ 0 \end{bmatrix} \quad (5.3.5)$$

The procedure of setting up this equation and solving it has been cast into the following MATLAB function 'difapx()', which can be used to generate the coefficients of, say, the approximation formulas (5.1.7), (5.1.9), and (5.3.2) just for practice/verification/fun, whatever your purpose is.

```
»format rat % to make all numbers represented in rational form
»difapx(1,[0 -2]) % first derivative based on {f_0,f_-1,f_-2}
   ans =   3/2      -2       1/2 % Eq.(5.1.7)
»difapx(1,[-2 2]) % first derivative based on {f_-2,f_-1,f_0,f_1,f_2}
   ans =   1/12    -2/3      0       2/3    -1/12 % Eq.(5.1.9)
»difapx(2,[2 -2]) % second derivative based on {f_2,f_1,f_0,f_-1,f_-2}
   ans =  -1/12    4/3    -5/2    4/3    -1/12 % Eq.(5.3.2)
```

```
function [c,err,eoh,A,b]=difapx(N,points)
% to get the difference approximation for the Nth derivative
l= max(points);
L= abs(points(1)-points(2))+1;
if L<N+1, error('More points are needed!'); end
for n=1: L
   A(1,n)= 1;
   for m=2:L+2
      A(m,n)= A(m-1,n)*l/(m-1); % Eq.(5.3.5)
   end
   l= l-1;
end
b= zeros(L,1); b(N+1)= 1;
c=(A(1:L,:)\b)'; % Coefficients of difference approximation formula
err=A(L+1,:)*c'; eoh=L-N; % Coefficient & order of error term
if abs(err)<eps, err=A(L+2,:)*c'; eoh=L-N+1; end
if points(1)<points(2), c=fliplr(c); end
```

Example 5.1. Numerical/Symbolic Differentiation for Taylor Series Expansion.

Consider how to use MATLAB to get the Taylor series expansion of a function, say, e^{-x}, about $x = 0$, which we already know is

$$e^{-x} = 1 - x + \frac{1}{2}x^2 - \frac{1}{3!}x^3 + \frac{1}{4!}x^4 - \frac{1}{5!}x^5 + \cdots \qquad (E5.1.1)$$

As a numerical method, we can use the MATLAB function 'difapx()'. On the other hand, we can also use the MATLAB command 'taylor()', which is a symbolic approach. Readers may put 'help taylor' at the MATLAB prompt to see its usage, which is restated in the box below.

- `taylor(f)` gives the fifth-order Maclaurin series expansion of `f` about $x = 0$.
- `taylor(f,x,'expansionpoint',a)`: the fifth-order Taylor series expansion of `f` w.r.t. x about $x = a$.
- `taylor(sin(x),'order',N)`: the Nth-order Taylor series expansion of `f` about a.
  (cf) The target function `f` given directly as the first input argument must be a legitimate function defined as a symbolic expression like `exp(x)` (with x declared as a symbolic variable).
  (cf) Before using the command '`taylor()`' with the target function defined as a symbolic expression, one should declare the arguments of the function as symbols by running the statement like '`syms x t`.'
  (cf) If the independent variable is not given as the second input argument, the default independent variable will be one closest (alphabetically) to 'x' among the (symbolic or literal) variables contained as arguments of the function `f`.
- You should use the MATLAB command '`sym2poly()`' if you want to extract the coefficients from the Taylor series expansion obtained as a symbolic expression.

The following MATLAB script "nm05e01.m" finds us the coefficients of fifth-order Taylor series expansion of e^{-x} about $x = 0$ by using the two methods.

```
%nm05e01.m
% Nth-order Taylor series expansion for e^-x about xo in Example 5.1
f=@(x)exp(-x);
N=5; xo=0; % Order and Center of Taylor series
% Numerical computation method
T(1)=feval(f,xo);
h=0.005 %.01 or 0.001 make it worse
tmp=1;
for i=1:N
   tmp=tmp*i*h; % i!(factorial i)*h^i
   c=difapx(i,[-i i]); % Coeffs of numerical derivative of ith order
   dix=c*feval(f,xo+[-i:i]*h)'; % Numerical derivative of ith order
   T(i+1)=dix/tmp; % Taylor series coefficient in h
end
format rat
Tn=fliplr(T) % polynomial coefficients in descending order
% Symbolic computation method
syms x;
Ts=sym2poly(taylor(exp(-x),x,'expansionpoint',xo,'order',N+1))
% Discrepancy between the two Taylor series expansions
format short, discrepancy=norm(Tn-Ts)
```

5.4 INTERPOLATING POLYNOMIAL AND NUMERICAL DIFFERENTIAL

The difference approximation formulas derived in the previous sections are applicable only when the target function $f(x)$ to differentiate is somehow given. In this section, we think about how to get the numerical derivatives when we are given only the data file containing several data points. A possible measure is to make the interpolating function by using one of the methods explained in Chapter 3 and get the derivative of the interpolating function.

For simplicity, let us reconsider the problem of finding the derivative of $f(x) = \sin x$ at $x = \pi/4$, where the function is given as one of the following data point sets.

$$\left\{ \left(\frac{\pi}{8}, \sin\frac{\pi}{8}\right), \left(\frac{\pi}{4}, \sin\frac{\pi}{4}\right), \left(\frac{3\pi}{8}, \sin\frac{3\pi}{8}\right) \right\}$$

$$\left\{ (0, \sin 0), \left(\frac{\pi}{8}, \sin\frac{\pi}{8}\right), \left(\frac{\pi}{4}, \sin\frac{\pi}{4}\right), \left(\frac{3\pi}{8}, \sin\frac{3\pi}{8}\right), \left(\frac{4\pi}{8}, \sin\frac{4\pi}{8}\right) \right\}$$

$$\left\{ \left(\frac{2\pi}{16}, \sin\frac{2\pi}{16}\right), \left(\frac{3\pi}{16}, \sin\frac{3\pi}{16}\right), \left(\frac{4\pi}{16}, \sin\frac{4\pi}{16}\right), \left(\frac{5\pi}{16}, \sin\frac{5\pi}{16}\right), \left(\frac{6\pi}{16}, \sin\frac{6\pi}{16}\right) \right\}$$

We make the MATLAB script "nm0504.m", which uses the function 'lagranp()' to find the interpolating polynomial, uses the function 'polyder()' to differentiate the polynomial, and computes the error of the resulting derivative from the true value. Let us run it with x defined appropriately according to the given set of data points and see the results.

```
»nm0504
   dfx( 0.78540)= 0.689072 (error: -0.018035) %with x=[1:3]*pi/8
   dfx( 0.78540)= 0.706556 (error: -0.000550) %with x=[0:4]*pi/8
   dfx( 0.78540)= 0.707072 (error: -0.000035) %with x=[2:6]*pi/16
```

```
%nm0504.m
% to interpolate by Lagrange polynomial and get the derivative
x0=pi/4;
df0=cos(x0); % True value of derivative of sin(x) at x0=pi/4
for m=1:3
   if m==1, x= [1:3]*pi/8;
   elseif  m==2, x= [0:4]*pi/8;
   else     x= [2:6]*pi/16;
   end
   y=sin(x);
   px= lagranp(x,y); % Lagrange polynomial interpolating (x,y)
   dpx= polyder(px); % derivative of polynomial px
   dfx= polyval(dpx, x0);
   fprintf(' dfx(%6.4f)=%10.6f (error:%10.6f)\n', x0,dfx,dfx-df0);
end
```

```
»load xy.dat   %input the contents of 'xy.dat' as a matrix named xy
»dydx = diff(xy(:,2))./diff(xy(:,1));  dydx'  %divided difference
   dydx =   2.0000    0.50000    2.0000
```

k	x_k xy(:,1)	$f(x_k)$ xy(:,2)	$x_{k+1}-x_k$ diff((xy(:,1))	$f(x_{k+1})-f(x_k)$ diff((xy(:,2))	$D_k = \dfrac{f(x_{k+1})-f(x_k)}{x_{k+1}-x_k}$
1	−1	2	1	2	2
2	0	4	2	1	1/2
3	2	5	−1	−2	2
4	1	3			

This illustrates that if we have more points which are distributed closer to the target point, we may get better result.

One more thing to mention before closing this section is that we have the MATLAB built-in function '`diff()`', which finds us the difference vector for a given vector. When the data points $\{(x_k, f(x_k)), k = 1,2, \ldots\}$ are given as an ASCII data file named 'xy.dat', we can use the function '`diff()`' to get the divided difference, which is similar to the derivative of a continuous function.

5.5 NUMERICAL INTEGRATION AND QUADRATURE

The general form of numerical integration of a function $f(x)$ over some interval $[a, b]$ is a weighted sum of the function values at a finite number $(N + 1)$ of sample points (nodes), so-called 'quadrature', as

$$\int_a^b f(x)dx \cong \sum_{k=0}^{N} w_k f(x_k) \text{ with } a = x_0 < x_1 < \cdots < x_N = b \qquad (5.5.1)$$

Here, the sample points are equally spaced for the midpoint rule, the trapezoidal rule, and Simpson's rule, while they are chosen to be zeros of certain polynomials for Gaussian quadrature.

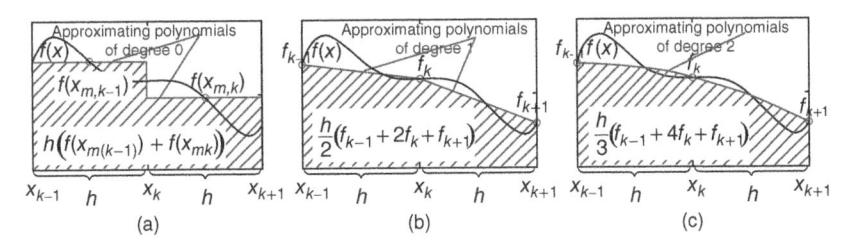

Figure 5.3 Various methods of numerical integration using Newton–Cotes rules. (a) Midpoint rule. (b) Trapezoidal rule. (c) Simpson's rule.

Figure 5.3 shows the integrations over two segments by the midpoint rule, the trapezoidal rule, and Simpson's rule, that are referred to as Newton-Cotes formulas for being based on the approximate polynomial and are implemented by the following formulas.

<midpoint rule> $\int_{x_k}^{x_{k+1}} f(x)dx \cong hf_{mk}$ (5.5.2)

with $h = x_{k+1} - x_k$, $f_{mk} = f(x_{mk})$, $x_{mk} = \dfrac{x_k + x_{k+1}}{2}$

<trapezoidal rule> $\int_{x_k}^{x_{k+1}} f(x)dx \cong \dfrac{h}{2}(f_k + f_{k+1})$ (5.5.3)

with $h = x_{k+1} - x_k$, $f_k = f(x_k)$

<Simpson's rule> $\int_{x_{k-1}}^{x_{k+1}} f(x)dx \cong \dfrac{h}{3}(f_{k-1} + 4f_k + f_{k+1})$ (5.5.4)

with $h = \dfrac{x_{k+1} - x_{k-1}}{2}$

These three integration rules are based on approximating the target function (integrand) to the 0th, first, and second-degree polynomial, respectively. Since the first two integrations are obvious, we are going to derive just Simpson's rule (5.5.4). For simplicity, we shift the graph of $f(x)$ by $-x_k$ along the x-axis, or equivalently, make the variable substitution $t = x - x_k$ so that the abscissas of the three points on the curve of $f(x)$ change from $x = \{x_k - h, x_k, x_k + h\}$ to $t = \{-h, 0, +h\}$. Then, in order to find the coefficients of the second-degree polynomial

$$p_2(t) = c_1 t^2 + c_2 t + c_3 \quad (5.5.5)$$

matching the points $(-h, f_{k-1})$, $(0, f_k)$, $(+h, f_{k+1})$, we should solve the following set of equations:

$$p_2(-h) = c_1(-h)^2 + c_2(-h) + c_3 = f_{k-1}$$
$$p_2(0) = c_1 0^2 + c_2 0 + c_3 = f_k$$
$$p_2(+h) = c_1(+h)^2 + c_2(+h) + c_3 = f_{k+1}$$

to determine the coefficients c_1, c_2, and c_3 as follows:

$$c_3 = f_k, \quad c_2 = \dfrac{f_{k+1} - f_{k-1}}{2h}, \quad c_1 = \dfrac{1}{h^2}\left(\dfrac{f_{k+1} + f_{k-1}}{2} - f_k\right)$$

Integrating the second-degree polynomial (5.5.5) with these coefficients from $t = -h$ to $t = h$ yields

$$\int_{-h}^{h} p_2(t)\, dt = \frac{1}{3}c_1 t^3 + \frac{1}{2}c_2 t^2 + c_3 t \Big|_{-h}^{h} = \frac{2}{3}c_1 h^3 + 2c_3 h$$

$$= \frac{2h}{3}\left(\frac{f_{k+1} + f_{k-1}}{2} - f_k + 3f_k\right) = \frac{h}{3}(f_{k-1} + 4f_k + f_{k+1})$$

This is the Simpson integration formula (5.5.4).

Now, as a preliminary work toward diagnosing the errors of the aforementioned integration formulas, we take the Taylor series expansion of the integral function

$$g(x) = \int_{x_k}^{x} f(t)\, dt \quad \text{with} \quad g'(x) = f(x), \quad g^{(2)}(x) = f'(x), \quad g^{(3)}(x) = f^{(2)}(x)$$

(5.5.6)

about the lower bound x_k of the integration interval to write

$$g(x) = g(x_k) + g'(x_k)(x - x_k) + \frac{1}{2}g^{(2)}(x_k)(x - x_k)^2 + \frac{1}{3!}g^{(3)}(x_k)(x - x_k)^3 + \cdots$$

Substituting Eq. (5.5.6) together with $x = x_{k+1}$ and $x_{k+1} - x_k = h$ into this yields

$$\int_{x_k}^{x_{k+1}} f(x)\, dx = 0 + hf(x_k) + \frac{h^2}{2}f'(x_k) + \frac{h^3}{3!}f^{(2)}(x_k) + \frac{h^4}{4!}f^{(3)}(x_k) + \frac{h^5}{5!}f^{(4)}(x_k) + \cdots$$

(5.5.7)

First, for the error analysis of the midpoint rule, we substitute x_{k-1} and $-h = x_{k-1} - x_k$ in place of x_{k+1} and h in this equation to write

$$\int_{x_k}^{x_{k-1}} f(x)\, dx = 0 - hf(x_k) + \frac{h^2}{2}f'(x_k) - \frac{h^3}{3!}f^{(2)}(x_k) + \frac{h^4}{4!}f^{(3)}(x_k) - \frac{h^5}{5!}f^{(4)}(x_k) + \cdots$$

and subtract this equation from Eq. (5.5.7) to write

$$\int_{x_k}^{x_{k+1}} f(x)\, dx - \int_{x_k}^{x_{k-1}} f(x)\, dx = \int_{x_k}^{x_{k+1}} f(x)\, dx + \int_{x_{k-1}}^{x_k} f(x)\, dx = \int_{x_{k-1}}^{x_{k+1}} f(x)\, dx$$

$$= 2hf(x_k) + \frac{2h^3}{3!}f^{(2)}(x_k) + \frac{2h^5}{5!}f^{(4)}(x_k) + \cdots \quad (5.5.8)$$

Substituting x_k and $x_{mk} = (x_k + x_{k+1})/2$ in place of x_{k-1} and x_k in this equation and noting that $x_{k+1} - x_{mk} = x_{mk} - x_k = h/2$, we obtain

$$\int_{x_k}^{x_{k+1}} f(x)\, dx = hf(x_{mk}) + \frac{h^3}{3 \times 2^3}f^{(2)}(x_{mk}) + \frac{h^5}{5 \times 4 \times 3 \times 2^5}f^{(4)}(x_{mk}) + \cdots ;$$

$$\int_{x_k}^{x_{k+1}} f(x)\, dx - hf(x_{mk}) = \frac{h^3}{24}f^{(2)}(x_{mk}) + \frac{h^5}{1920}f^{(4)}(x_{mk}) + \cdots = O(h^3) \quad (5.5.9)$$

This, together with Eq. (5.5.2), implies that the error of integration over one segment by the midpoint rule is proportional to h^3.

Second, for the error analysis of the trapezoidal rule, we subtract Eq. (5.5.3) from Eq. (5.5.7) to write

$$\int_{x_k}^{x_{k+1}} f(x)dx - \frac{h}{2}(f(x_k) + f(x_{k+1}))$$

$$= -\frac{h^3}{12}f^{(2)}(x_k) - \frac{h^4}{24}f^{(3)}(x_k) - \frac{h^5}{80}f^{(4)}(x_k) + O(h^6) = O(h^3)$$

$$= hf(x_k) + \frac{h^2}{2}f'(x_k) + \frac{h^3}{3!}f^{(2)}(x_k) + \frac{h^4}{4!}f^{(3)}(x_k) + \frac{h^5}{5!}f^{(4)}(x_k) + \cdots$$

$$- \frac{h}{2}\left\{ f(x_k) + f(x_k) + hf'(x_k) + \frac{h^2}{2}f^{(2)}(x_k) + \frac{h^3}{3!}f^{(3)}(x_k) + \frac{h^4}{4!}f^{(4)}(x_k) + \cdots \right\}$$

$$= -\frac{h^3}{12}f^{(2)}(x_k) - \frac{h^4}{24}f^{(3)}(x_k) - \frac{h^5}{80}f^{(4)}(x_k) + O(h^6) = O(h^3) \quad (5.5.10)$$

This implies that the error of integration over one segment by the trapezoidal rule is proportional to h^3.

Third, for the error analysis of Simpson's rule, we subtract the Taylor series expansion of Eq. (5.5.4)

$$\frac{h}{3}\{f(x_{k-1}) + 4f(x_k) + f(x_{k+1})\}$$

$$= \frac{h}{3}\left\{ f(x_k) + 4f(x_k) + f(x_k) + \frac{2h^2}{2}f^{(2)}(x_k) + \frac{2h^4}{4!}f^{(4)}(x_k) + \cdots \right\}$$

$$= 2hf(x_k) + \frac{h^3}{3}f^{(2)}(x_k) + \frac{h^5}{36}f^{(4)}(x_k) + \cdots$$

from Eq. (5.5.8) to write

$$\int_{x_{k-1}}^{x_{k+1}} f(x)dx - \frac{h}{3}\{f(x_{k-1}) + 4f(x_k) + f(x_{k+1})\} = -\frac{h^5}{90}f^{(4)}(x_k) + O(h^7) = O(h^5)$$

$$(5.5.11)$$

This implies that the error of integration over two segments by Simpson's rule is proportional to h^5.

Before closing this section, let us make use of these error equations to find a way of estimating the error of the numerical integral from the true integral without knowing the derivatives of the target (integrand) function $f(x)$. For this purpose, we investigate how the error of numerical integration by Simpson's rule

$$I_S(x_{k-1}, x_{k+1}, h) = \frac{h}{3}\{f(x_{k-1}) + 4f(x_k) + f(x_{k+1})\}$$

will change if the segment width h is halved to $h/2$. Noting that, from Eq. (5.5.11),

$$E_S(h) = \int_{x_{k-1}}^{x_{k+1}} f(x)dx - I_S(x_{k-1}, x_{k+1}, h) \approx -\frac{h^5}{90} f^{(4)}(c) \quad (c \in [x_{k-1}, x_{k+1}])$$

$$E_S\left(\frac{h}{2}\right) = \int_{x_{k-1}}^{x_{k+1}} f(x)dx - I_S\left(x_{k-1}, x_{k+1}, \frac{h}{2}\right)$$

$$= \int_{x_{k-1}}^{x_k} f(x)dx - I_S\left(x_{k-1}, x_k, \frac{h}{2}\right) + \int_{x_k}^{x_{k+1}} f(x)dx - I_S\left(x_k, x_{k+1}, \frac{h}{2}\right)$$

$$\approx -2\frac{(h/2)^5}{90} f^{(4)}(c) = \frac{1}{16} E_S(h) \quad (c \in [x_{k-1}, x_{k+1}])$$

we can express the change of the error caused by halving the segment width as

$$\left|E_S(h) - E_S\left(\frac{h}{2}\right)\right| = \left|I_S(x_{k-1}, x_{k+1}, h) - I_S\left(x_{k-1}, x_{k+1}, \frac{h}{2}\right)\right| \approx \frac{15}{16}|E_S(h)|$$

$$\approx 15\left|E_S\left(\frac{h}{2}\right)\right| \quad (5.5.12)$$

This suggests the error estimate of numerical integration by Simpson's rule as

$$\left|E_S\left(\frac{h}{2}\right)\right| \approx \frac{1}{2^4 - 1}\left|I_S(x_{k-1}, x_{k+1}, h) - I_S\left(x_{k-1}, x_{k+1}, \frac{h}{2}\right)\right| \quad (5.5.13)$$

Also for the trapezoidal rule, similar result can be derived:

$$\left|E_T\left(\frac{h}{2}\right)\right| \approx \frac{1}{2^2 - 1}\left|I_T(x_{k-1}, x_{k+1}, h) - I_T\left(x_{k-1}, x_{k+1}, \frac{h}{2}\right)\right| \quad (5.5.14)$$

5.6 TRAPEZOIDAL METHOD AND SIMPSON METHOD

In order to get the formulas for numerical integration of a function $f(x)$ over some interval $[a, b]$, we divide the interval into N segments of equal length $h = (b - a)/N$ so that each nodes (sample points) can be expressed as $\{x = a + kh, k = 0, 1, 2, \ldots, N\}$. Then, we have the numerical integration of $f(x)$ over $[a, b]$ by the trapezoidal rule (5.5.3) as follows:

$$\int_a^b f(x)dx = \sum_{k=0}^{N-1} \int_{x_k}^{x_{k+1}} f(x)dx \cong \frac{h}{2}\{(f_0+f_1)+(f_1+f_2)+\cdots+(f_{N-2}+f_{N-1})+(f_{N-1}+f_N)\};$$

$$I_{T2}(a,b,h) = h\left\{\frac{f(a)+f(b)}{2} + \sum_{k=1}^{N-1} f(x_k)\right\} \quad (5.6.1)$$

whose error is proportional to h^2 as N times the error for one segment (Eq. (5.5.10)), i.e.

$$NO(h^3) = (b-a)/h \times O(h^3) = O(h^2)$$

On the other hand, we have the numerical integration of $f(x)$ over $[a,b]$ by Simpson's rule (5.5.4) with an even number of segments N as follows:

$$\int_a^b f(x)dx = \sum_{m=0}^{N/2-1} \int_{x_{2m}}^{x_{2m+2}} f(x)dx$$

$$\cong \frac{h}{3}\{(f_0+4f_1+f_2)+(f_2+4f_3+f_4)+\cdots+(f_{N-2}+4f_{N-1}+f_N)\}$$

$$I_{S4}(a,b,h) = \frac{h}{3}\left\{f(a)+f(b)+4\sum_{m=0}^{N/2-1} f(x_{2m+1}) + 2\sum_{m=1}^{N/2-1} f(x_{2m})\right\}$$

$$= \frac{h}{3}\left\{f(a)+f(b)+2\left(\sum_{m=0}^{N/2-1} f(x_{2m+1}) + \sum_{k=1}^{N-1} f(x_k)\right)\right\} \quad (5.6.2)$$

whose error is proportional to h^4 as N times the error for one segment (Eq. (5.5.11)), i.e.

$$\left(\frac{N}{2}\right) O(h^5) = (b-a)/2h \times O(h^5) = O(h^4)$$

These two integration formulas by the trapezoidal rule and Simpson's rule are cast into the MATLAB functions 'trpzds()' and 'smpsns()', respectively.

```
function INTf=Smpsns(f,a,b,N,varargin)
%integral of f(x) over [a,b] by Simpson's rule with N segments
if nargin<4, N=100; end
if abs(b-a)<1e-12|N<=0, INTf=0; return; end
if mod(N,2)~=0, N=N+1; end  % make N even
h=(b-a)/N; x=a+[0:N]*h; % the boundary nodes for N segments
fx=fevel(f,x,varargin{:}); % values of f for all nodes
fx(find(fx==inf))=realmax; fx(find(fx==-inf))=-realmax;
kodd=2:2:N; keven=3:2:N-1; % the set of odd/even indices
INTf=h/3*(fx(1)+fx(N+1)+4*sum(fx(kodd))+2*sum(fx(keven))); %Eq.(5.6.2)
```

5.7 RECURSIVE RULE AND ROMBERG INTEGRATION

```
function INTf=trpzds(f,a,b,N)
%integral of f(x) over [a,b] by trapezoidal rule with N segments
if abs(b-a)<eps|N<=0, INTf=0; return; end
h=(b-a)/N; x=a+[0:N]*h; fx=feval(f,x); % values of f for all nodes
INTf=h*((fx(1)+fx(N+1))/2+sum(fx(2:N))); % trapz(x,fx) Eq.(5.6.1)
```

In this section, we are going to look for a recursive formula that enables us to use some numerical integration with the segment width h to produce another (hopefully better) numerical integration with half the segment width ($h/2$). Additionally, we use Richardson extrapolation (Section 5.I.) together with the two successive numerical integrations to make a Romberg table that can be used to improve the accuracy of the numerical integral step by step.

Let us start with halving the segment width h to $h/2$ for the trapezoidal method. Then, the numerical integration formula (5.6.1) can be written in recursive form as

$$I_{T2}\left(a,b,\frac{h}{2}\right) = \frac{h}{2}\left\{\frac{f(a)+f(b)}{2} + \sum_{k=1}^{2N-1} f(x_{k/2})\right\}$$

$$= \frac{h}{2}\left\{\frac{f(a)+f(b)}{2} + \sum_{m=1}^{N-1} f(x_{2m/2}) + \sum_{m=0}^{N-1} f(x_{(2m+1)/2})\right\}$$

$$= \frac{1}{2}\left\{I_{T2}(a,b,h) + h\sum_{m=0}^{N-1} f(x_{(2m+1)/2}) \text{ (terms for inserted nodes)}\right\}$$

(5.7.1)

Noting that the error of this formula is proportional to h^2 ($O(h^2)$), we apply a Richardson extrapolation (Eq. (5.1.10)) to write a higher level integration formula having an error of $O(h^4)$ as

$$I_{T4}(a,b,h) = \frac{2^2 I_{T2}(a,b,h) - I_{T2}(a,b,2h)}{2^2 - 1}$$

$$\stackrel{(5.6.1)}{=} \frac{1}{3}\left\{4\frac{h}{2}\left(f(a)+f(b) + 2\sum_{k=1}^{N-1} f(x_k)\right)\right.$$

$$\left. - \frac{2h}{2}\left(f(a)+f(b) + 2\sum_{m=1}^{N/2-1} f(x_{2m})\right)\right\}$$

266 NUMERICAL DIFFERENTIATION/INTEGRATION

$$= \frac{h}{3}\left\{f(a) + f(b) + 4\sum_{m=1}^{N/2} f(x_{2m-1}) + 2\sum_{m=1}^{N/2-1} f(x_{2m})\right\} \quad (5.7.2)$$

$$\overset{(5.6.2)}{\equiv} I_{S4}(a,b,h)$$

which coincides with the Simpson's integration formula. This implies that we do not have to distinguish the trapezoidal rule from Simpson's rule. Anyway, replacing h by $h/2$ in this equation yields

$$I_{T4}\left(a,b,\frac{h}{2}\right) = \frac{2^2 I_{T2}(a,b,h/2) - I_{T2}(a,b,h)}{2^2 - 1}$$

which can be generalized to the following formula.

$$I_{T,2(n+1)}(a,b,2^{-(k+1)}h) = \frac{2^{2n} I_{T,2n}(a,b,2^{-(k+1)}h) - I_{T,2n}(a,b,2^{-k}h)}{2^{2n} - 1} \quad (5.7.3)$$

$$\text{for } n \geq 1, k \geq 0$$

Now, it is time to introduce a systematic way, called Romberg integration, of improving the accuracy of the integral step by step and estimating the (truncation) error at each step to determine when to stop. It is implemented by a Romberg table (Table 5.3), i.e. a lower-triangular matrix that we construct one row per iteration by applying Eq. (5.7.1) in halving the segment width h to get the next-row element (downward in the first column), and applying Eq. (5.7.3) in upgrading the order of error to get the next-column elements (rightward in the row) based on the up-left (north-west) one and the left (west) one. At each iteration k, we use Eq. (5.5.14) to estimate the truncation error as

$$\left| E_{T,2(k+1)}\left(2^{-k}h\right) \right| \approx \frac{1}{2^{2k}-1} \left| I_{T,2k}(2^{-k}h) - I_{T,2k}(2^{-(k-1)}h) \right| \quad (5.7.4)$$

Table 5.3 Romberg table.

Iteration k	Segment width h	$n=1$	$n=2$	$n=3$	...
0	h_0	$I_{T,2}(h_0)$			
0	$2^{-1} h_0$	(5.7.1)↓ (5.7.3)↘ $I_{T,2}(2^{-1}h_0)$ → $I_{T,4}(2^{-1}h_0)$			
0	$2^{-2} h_0$	(5.7.1)↓ (5.7.3)↘ $I_{T,2}(2^{-2}h_0)$ → $I_{T,4}(2^{-2}h_0)$ (5.7.3)↘ → $I_{T,6}(2^{-2}h_0)$			
.	.	.	.	.	.

RECURSIVE RULE AND ROMBERG INTEGRATION

and stop the iteration when the estimated error becomes less than some prescribed tolerance. Then, the last diagonal element is taken to be 'supposedly' the best estimate of the integral. This sequential procedure of Romberg integration is cast into the MATLAB function 'Rmbrg()'.

```
function [x,R,err,N]=Rmbrg(f,a,b,tol,K)
%construct Romberg table to find definite integral of f over [a,b]
h=b-a;  N=1;
if nargin<5, K=10; end
R(1,1)=h/2*(feval(f,a)+feval(f,b));
for k=2:K
    h=h/2;  N=N*2;
    R(k,1)=R(k-1,1)/2 +h*sum(feval(f,a+[1:2:N-1]*h));   %Eq.(5.7.1)
    tmp=1;
    for n=2:k
        tmp= tmp*4;
        R(k,n)= (tmp*R(k,n-1)-R(k-1,n-1))/(tmp-1);      %Eq.(5.7.3)
    end
    err= abs(R(k,k-1)-R(k-1,k-1))/(tmp-1);              %Eq.(5.7.4)
    if err<tol, break; end
end
x=R(k,k);
```

Before closing this section, we test and compare the trapezoidal method ('trpzds()'), Simpson method ('Smpsns()') and Romberg integration ('Rmbrg()') by trying them on the following integral

$$\int_0^4 400x(1-x)e^{-2x}\,dx = 100\left\{-2e^{-2x}x(1-x)|_0^4 + \int_0^4 2e^{-2x}(1-2x)dx\right\}$$

$$= 100\left\{-2e^{-2x}x(1-x)|_0^4 - e^{-2x}(1-2x)|_0^4 - 2\int_0^4 e^{-2x}\,dx\right\}$$

$$= 200x^2 e^{-2x}|_0^4 = 1.073\,480\,409\,29 \qquad (5.7.5)$$

Here are the MATLAB statements for this job listed together with the running results.

```
»f=@(x)400*x.*(1-x).*exp(-2*x);
»a=0;  b=4;  N=80;
»format short e
»true_I=3200*exp(-8);
»It= trpzds(f,a,b,N), errt=It-true_I  % trapezoidal
   It= 9.9071e-001,  errt=-8.2775e-002
»Is= Smpsns(f,a,b,N), errs= Is-true_I  %Simpson
   INTfs= 1.0731e+000, error=-3.3223e-004
»[IR,R,err,N1]= Rmbrg(f,a,b,.0005), errR=IR-true_I  %Romberg
   INTfr= 1.0734e+000,  N1= 32
   error=-3.4943e-005
```

As expected from the fact that the errors of numerical integration by the trapezoidal method and Simpson method are $O(h^2)$ and $O(h^4)$, respectively, Simpson method presents better result (with smaller error) than the trapezoidal one with the same number of segments $N = 80$. Moreover, Romberg integration with $N = 32$ shows better result than both of them.

5.8 ADAPTIVE QUADRATURE

The numerical integration methods in the previous sections divide the integration interval uniformly into the segments of equal width, making the error nonuniform over the interval, i.e. small/large for smooth/swaying portion of the curve of integrand $f(x)$. In contrast, the strategy of the *adaptive quadrature* is to divide the integration interval nonuniformly into segments of (generally) unequal lengths, that is, short/long segments for swaying/smooth portion of the curve of integrand $f(x)$, aiming at having smaller error with fewer segments for a reasonable balance between numerical error and computation time.

The algorithm of adaptive quadrature scheme starts with a numerical integral (INTf) for the whole interval and the sum of numerical integrals (INTf12=INTf1+INTf2) for the two segments of equal width. Based on the difference between the two successive estimates INTf and INTf12, it estimates the error of INTf12 by using Eq. (5.5.13)/(5.5.14) depending on the basic integration rule. Then, if the error estimate is within a given tolerance (tol), it terminates with INTf12. Otherwise, it digs into each segment by repeating the same procedure with half of the tolerance (tol/2) assigned to both segments, until the deepest level satisfies the error condition. This is how the adaptive scheme forms sections of nonuniform width, as illustrated in Figure 5.4. In fact, this algorithm really fits the recursive (self-calling) structure introduced

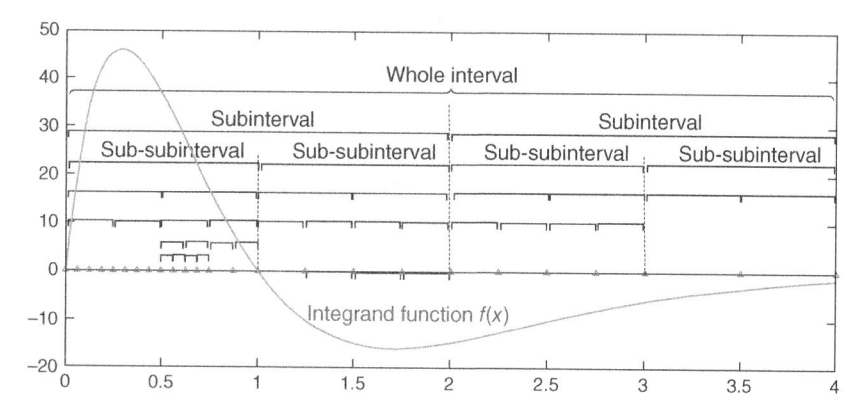

Figure 5.4 Subintervals (segments) and their boundary points (nodes) determined by the adaptive Simpson method.

ADAPTIVE QUADRATURE

in Section 1.3 and has been cast into the following MATLAB function 'adap_smpsn()', which needs the calling function 'adapt_smpsn()' for start-up.

We can apply these functions to get the approximate value of integration (5.7.5) by running the following MATLAB statements:

```
»f=@(x)400*x.*(1-x).*exp(-2*x);
»a=0; b=4; tol=0.001;
»format short e
»true_I=3200*exp(-8);
»Ias=adapt_smpsn(f,a,b,tol), erras=Ias-true_I
   Ias= 1.0735e+00, erras= -8.9983e-06
```

```
function [INTf,nodes,err]=adap_smpsn(f,a,b,INTf,tol,varargin)
% adaptive recursive Simpson method
c=(a+b)/2;
INTf1= smpsns(f,a,c,1,varargin{:});
INTf2= smpsns(f,c,b,1,varargin{:});
INTf12= INTf1+INTf2;
err= abs(INTf12-INTf)/15; % Error estimate by Eq.(5.5.13)
if err>tol
   [INTf1,nodes1,err1]=adap_smpsn(f,a,c,INTf1,tol/2,varargin{:});
   [INTf2,nodes2,err2]=adap_smpsn(f,c,b,INTf2,tol/2,varargin{:});
   INTf=INTf1+INTf2;
   nodes=[nodes1 nodes2(2:length(nodes2))];
   err=err1+err2;
else
   INTf=INTf12; points=[a c b];
end
```

```
function [INTf,nodes,err]=adapt_smpsn(f,a,b,tol,varargin)
%apply adaptive recursive Simpson method
INTf= smpsns(f,a,b,1,varargin{:});
[INTf,nodes,err]=adap_smpsn(f,a,b,INTf,tol,varargin{:});
```

Figure 5.4 shows the curve of the integrand $f(x) = 400x(1-x)e^{-2x}$ together with the 25 nodes determined by the function 'adapt_smpsn()', which yields better result (having smaller error) with fewer segments than other methods discussed so far. From this figure, we see that the nodes are dense/sparse in the swaying/smooth portion of the curve of the integrand.

Here, we introduce the MATLAB built-in functions adopting the adaptive recursive integration scheme together with the illustrative example of their usage.

```
'quad(f,a,b,tol,trace,p1,p2,..)'/ 'quadl(f,a,b,tol,trace,p1,p2,..)'

»Iq=quad(f,a,b,tol), errq=Iq-true_I
   Iq= 1. 0735e+00, errq=  4.0107e-05
»Iql=quadl(f,a,b,tol), errql=Iql-true_I
   Iql= 1.0735e+00, errql= -1.2168e-08
```

(cf) These functions are capable of passing the parameter values (p1,p2,..) to the integrand (target) function f and can be asked to show the list of intermediate subintervals by setting the fifth input argument as trace=1. If you want to pass some parameter values (p1,p2) to the integrand function f without specifying the error tolerance (to use its default value) and/or without seeing the list of intermediate subintervals, you should set tol and/or trace to an empty set [] as a place holder.

Additionally, note that MATLAB has a symbolic integration function 'int(f,a,b)'. Readers may type 'help int' into the MATLAB Command window to see its usage, which is restated below.

- int(f) gives the indefinite integral of f w.r.t. its independent variable (closest to 'x')
- int(f,v) gives the indefinite integral of f(v) with respect to v given as the second input argument
- int(f,a,b) gives the definite integral of f over [a,b] w.r.t. its independent variable
- int(f,v,a,b) gives the definite integral of f(v) w.r.t. v over [a,b]

(cf) The target function f must be a legitimate expression given directly as the first input argument and the upper/lower bound a,b of the integration interval can be a symbolic scalar or a numeric.

Example 5.2. Numerical/Symbolic Integration for CtFS by using quad()/quadl() or int().

Consider how to use MATLAB for finding the continuous-time Fourier series (CtFS) coefficients:

$$X_k = \int_{-P/2}^{P/2} x(t) e^{-jk\omega_0 t/P} dt = \int_{-P/2}^{P/2} x(t) e^{-j2\pi kt/P} dt \qquad (E5.2.1)$$

For simplicity, let us try to get just the 16th CtFS coefficient of a rectangular wave

$$x(t) = \begin{cases} 1 & \text{for } -1 \leq t < 1 \\ 0 & \text{for } -2 \leq t < 1 \text{ or } 1 \leq t < 2 \end{cases} \qquad (E5.2.2)$$

which is periodic in t with period $P = 4$. We can compute it analytically as follows:

ADAPTIVE QUADRATURE

$$X_{16} = \int_{-2}^{2} x(t)e^{-j2\pi 16t/4}\,dt \stackrel{(E5.2.2)}{=} \int_{-1}^{1} e^{-j8\pi t}\,dt$$

$$= \frac{1}{-j8\pi}e^{-j8\pi t}\Big|_{-1}^{1} = \frac{1}{8\pi}\sin(8\pi t)\Big|_{-1}^{1} = 0 \qquad (E5.2.3)$$

As a numerical approach, we can use the MATLAB function 'quad()'/'quadl()'. On the other hand, we can also use the MATLAB function 'int()', which is a symbolic approach. We put all the statements together to make the MATLAB script "nm05e02.m", in which the fifth input argument (trace) of 'quad()'/'quadl()' is set to 1 so that we can see their nodes and tell how different they are. Let us run it and see the results:

```
»nm05e02
    X16_quad = 0.8150 + 0.0000i % Betrayal of MATLAB?
    X16_quadl = 7.4771e-008 + 6.9389e-18i  % Almost zero, OK!
    X16_as = 2 % Shame on me, See Problem 5.5 to make up
    Iexp = (exp(-pi*t*8i)*1i)/(8*pi)% (E5.2.3) by symbolic computation
    Icos = sin(8*pi*t)/(8*pi) % (E5.2.3) by symbolic computation
    X16_sym = 0 % Exact answer by symbolic computation
```

What a surprise! It is totally unexpected that the MATLAB function 'quad()' gives us a quite eccentric value (0.8150), even without any warning message. The function 'quad()' must be branded as a betrayer for a piecewise-linear function multiplied by a periodic function. This seems to imply that 'quadl()' is better than 'quad()' and that 'int()' is the best of the three commands. It should, however, be noted that 'int()' can directly accept and handle only the functions composed of basic mathematical functions, rejecting functions defined in the form of string and ones defined by the 'inline()' command or through an m-file and besides, it takes more time to execute.

(cf) What about our lovely function 'adapt_smpsn()'? Regrettably, you had better not count on it, since it will give the wrong answer for this problem. Actually, 'quadl()' is much more reliable than 'quad()' and 'adapt_smpsn()'.

```
%nm05e02.m
% use quad()/quadl() and int() to get CtFS coefficient X16 in Ex 5.2
fcos=@(t,k,w0)exp(-j*k*w0*t);
P=4; k=16; w0=2*pi/P; a=-1; b=1; tol=0.001; trace=0;
X16_quad=quad(fcos,a,b,tol,trace,k,w0) % with k and w0 as parameters
X16_quadl=quadl(fcos,a,b,tol,trace,k,w0) %with k and w0 as parameters
X16_as=adapt_smpsn(fcos,a,b,tol,k,w0)
syms t; % declare symbolic variable
Iexp=int(exp(-j*k*w0*t),t) % symbolic indefinite integral
Icos=int(cos(k*w0*t),t) % symbolic indefinite integral
X16_sym=int(cos(k*w0*t),t,-1,1) % symbolic definite integral
fcos0=@(t)exp(-j*k*w0*t);
X16_integral = integral(fcos0,-1,1) % using integral()
```

5.9 GAUSS QUADRATURE

In this section, we cover several kinds of Gauss quadrature methods, i.e. Gauss-Legendre, Gauss-Hermite, Gauss-Laguerreon, and Gauss-Chebyshev I,II integrations. Each tries to approximate one of the following integrations, respectively.

$$\int_a^b f(t)\,dt, \quad \int_{-\infty}^{+\infty} e^{-t^2} f(t)\,dt, \quad \int_0^{+\infty} e^{-t} f(t)\,dt, \quad \int_{-1}^{1} \frac{1}{\sqrt{1-t^2}} f(t)\,dt,$$

$$\int_{-1}^{1} \sqrt{1-t^2}\, f(t)\,dt \approx \sum_{i=1}^{N} w_i f(t_i)$$

The problem is how to fix the weight w_i's and the (Gauss) grid points t_i's.

5.9.1 Gauss-Legendre Integration

If the integrand $f(t)$ is a polynomial of degree $\leq 3(=2N-1)$, then its integration

$$I(-1,1) = \int_{-1}^{+1} f(t)\,dt \tag{5.9.1}$$

can exactly be obtained from just $2(N)$ points by using the following formula

$$I[t_1, t_2] = w_1 f(t_1) + w_2 f(t_2) \tag{5.9.2}$$

How marvelous it is! It is almost a magic. Do you doubt it? Then, let us find the weights w_1, w_2, and the grid points t_1, t_2 such that the approximating formula (5.9.2) equals the integration (5.9.1) for $f(t) = 1$ (of degree 0), t (of degree 1), t^2 (of degree 2), and t^3 (of degree 3). In order to do so, we should solve the following system of equations:

$$f(t) = 1: \quad w_1 f(t_1) + w_2 f(t_2) = w_1 + w_2 \equiv \int_{-1}^{1} 1\,dt = 2 \tag{5.9.3a}$$

$$f(t) = t: \quad w_1 f(t_1) + w_2 f(t_2) = w_1 t_1 + w_2 t_2 \equiv \int_{-1}^{1} t\,dt = 0 \tag{5.9.3b}$$

$$f(t) = t^2: \quad w_1 f(t_1) + w_2 f(t_2) = w_1 t_1^2 + w_2 t_2^2 \equiv \int_{-1}^{1} t^2\,dt = \frac{2}{3} \tag{5.9.3c}$$

$$f(t) = t^3: \quad w_1 f(t_1) + w_2 f(t_2) = w_1 t_1^3 + w_2 t_2^3 \equiv \int_{-1}^{1} t^3\,dt = 0 \tag{5.9.3d}$$

Multiplying Eq. (5.9.3b) by t_1^2 and subtracting the result from Eq. (5.9.3d) yields

$$w_2(t_2^3 - t_1^2 t_2) = w_2 t_2(t_2 + t_1)(t_2 - t_1) = 0 \rightarrow t_2 = -t_1, \quad t_2 = t_1 \text{(meaningless)}$$

$$t_2 = -t_1 \rightarrow (5.9.3b); \quad (w_1 - w_2) t_1 = 0; \quad w_1 = w_2 \rightarrow (5.9.3a); \quad w_1 + w_1 = 2$$

$$w_1 = w_2 = 1 \rightarrow (5.9.3c); \quad t_1^2 + (-t_1)^2 = \frac{2}{3}; \quad t_1 = -t_2 = -\frac{1}{\sqrt{3}}$$

so that Eq. (5.9.2) becomes

$$I[t_1, t_2] = f\left(-\frac{1}{\sqrt{3}}\right) + f\left(\frac{1}{\sqrt{3}}\right) \tag{5.9.4}$$

We can expect this approximating formula to give us the exact value of the integral (5.9.1) when the integrand $f(t)$ is a polynomial of degree ≤ 3.

Now, you are concerned about how to generalize this 2-point Gauss-Legendre integration formula to an N-point case, since a system of nonlinear equation like Eq. (5.9.3) can be very difficult to solve as the dimension increases. But do not worry about it. The N grid points (t_is) of Gauss-Legendre integration formula

$$I_{GL}[t_1, t_2, \ldots, t_N] = \sum_{i=1}^{N} w_{N,i} f(t_i) \tag{5.9.5}$$

giving us the exact integral of an integrand polynomial of degree $\leq (2N-1)$ can be obtained as the zeros of the Nth-degree Legendre polynomial [K-2, Section 4.3]

$$L_N(t) = \sum_{i=0}^{\lfloor N/2 \rfloor} (-1)^i \frac{(2N - 2i)!}{2^N i!(N - i)!(N - 2i)!} t^{N-2i} \tag{5.9.6a}$$

$$L_N(t) = \frac{1}{N}\{(2N - 1)tL_{N-1}(t) - (N - 1)L_{N-2}(t)\} \tag{5.9.6b}$$

Given the N grid point t_i's, we can get the corresponding weight $w_{N,i}$'s of the N-point Gauss-Legendre integration formula by solving the system of linear equations

$$\begin{bmatrix} 1 & 1 & 1 & - & 1 \\ t_1 & t_2 & t_n & - & t_N \\ t_1^{n-1} & t_2^{n-1} & t_n^{n-1} & - & t_N^{n-1} \\ - & - & - & - & - \\ t_1^{N-1} & t_2^{N-1} & t_n^{N-1} & - & t_N^{N-1} \end{bmatrix} \begin{bmatrix} w_{N,1} \\ w_{N,2} \\ w_{N,n} \\ - \\ w_{N,N} \end{bmatrix} = \begin{bmatrix} 2 \\ 0 \\ (1 - (-1)^n)/n \\ - \\ (1 - (-1)^N)/N \end{bmatrix} \tag{5.9.7}$$

where the nth element of the right-hand side (RHS) vector is

274   NUMERICAL DIFFERENTIATION/INTEGRATION

$$\text{RHS}(n) = \int_{-1}^{1} t^{n-1}\, dt = \frac{1}{n} t^n \Big|_{-1}^{1} = \frac{1 - (-1)^n}{n} \qquad (5.9.8)$$

This procedure of finding the Ngrid point t_i's and the weight $w_{N,i}$'s of the N-point Gauss-Legendre integration formula is cast into the MATLAB function 'Gausslp()'. We can get the two grid point t_i's and the weight $w_{N,i}$'s of the two-point Gauss-Legendre integration formula by just running the following statement:

```
»[t,w]=Gausslp(2)
   t=    0.5774   -0.5774
   w=    1         1
```

Even though we are happy with the N-point Gauss-Legendre integration formula (5.9.1) giving the exact integral of polynomials of degree $\leq(2N-1)$, we do not feel comfortable with the fixed integration interval $[-1, +1]$. But we can be relieved from the stress because any arbitrary finite interval $[a, b]$ can be transformed into $[-1, +1]$ by the variable substitution known as the Gauss-Legendre translation

$$x = \frac{(b-a)t + a + b}{2}, \quad dx = \frac{b-a}{2} dt \qquad (5.9.9)$$

Then, we can write the N-point Gauss-Legendre integration formula for the integration interval $[a, b]$ as

```
function [t,w]=Gausslp(N)
if N<0, fprintf('\nGauss-Legendre polynomial of negative order??\n');
 else
   t=roots(Lgndrp(N))'; % make it a row vector
   A(1,:)= ones(1,N);  b(1)=2;
   for n=2:N % Eq.(5.9.7)
      A(n,:)=A(n-1,:).*t;
      if mod(n,2)==0, b(n)=0;
       else  b(n)=2/n; % Eq.(5.9.8)
      end
   end
   w= b/A';
end
```

```
function p=Lgndrp(N) %Legendre polynomial
if N<=0, p=1; %n*Ln(t)=(2n-1)t Ln-1(t)-(n-1)Ln-2(t) Eq.(5.9.6b)
 elseif N==1, p=[1 0];
 else p=((2*N-1)*[Lgndrp(N-1) 0]-(N-1)*[0 0 Lgndrp(N-2)])/N;
end
```

```
function I=Gauss_Legendre(f,a,b,N,varargin)
%Gauss_Legendre integration of f over [a,b] with N grid points
% Never try N larger than 25
[t,w]=Gausslp(N); x=((b-a)*t+a+b)/2; % Eq.(5.9.9)
fx=feval(f,x,varargin{:});
I=w*fx'*(b-a)/2; % Eq.(5.9.10)
```

$$I[a,b] = \int_a^b f(x)dx = \frac{b-a}{2}\int_{-1}^1 f(x(t))dt;$$

$$I[x_1, x_2, \ldots, x_N] = \frac{b-a}{2}\sum_{i=1}^N w_{N,i}f(x_i) \text{ with } x_i = \frac{(b-a)t_i + a + b}{2} \quad (5.9.10)$$

The scheme of integrating $f(x)$ over $[a, b]$ by the N-point Gauss-Legendre formula has been cast into the above MATLAB function 'Gauss_Legendre()'. We can get the integral (5.7.5) by just running the following statements. The result shows that the 10-point Gauss-Legendre formula yields better accuracy (smaller error), even with fewer nodes/segments than other methods discussed so far.

```
»f=@(x)400*x.*(1-x).*exp(-2*x); % Eq.(5.7.5)
»format short e
»true_I=3200*exp(-8);
»a=0; b=4; N=10; % Integration interval, Nnumber of nodes(grid points)
»IGL=Gauss_Legendre(f,a,b,N), errGL=IGL-true_I
   IGL= 1.0735e+00, errGL= 1.6289e-09
```

5.9.2 Gauss-Hermite Integration

The Gauss-Hermite integration formula is expressed by Eq. (5.9.5) as

$$I_{GH}[t_1, t_2, \ldots, t_N] = \sum_{i=1}^N w_{N,i}f(t_i) \quad (5.9.11)$$

and is supposed to give us the exact integral of the exponential e^{-t^2} multiplied by a polynomial $f(t)$ of degree $\leq(2N-1)$ over $(-\infty, +\infty)$:

$$I = \int_{-\infty}^{+\infty} e^{-t^2}f(t)dt \quad (5.9.12)$$

The N grid point t_i's can be obtained as the zeros of the N-point Hermite polynomial [K-2, Section 4.8]

$$H_N(t) = \sum_{i=0}^{\lfloor N/2 \rfloor} \frac{(-1)^i}{i!}N(N-1)\cdots(N-2i+1)(2t)^{N-2i}; \quad (5.9.13a)$$

$$H_N(t) = 2tH_{N-1}(t) - \frac{d}{dt}H_{N-1}(t) \quad (5.9.13b)$$

Given the N grid point t_i's, we can get the weight $w_{N,i}$'s of the N-point Gauss-Hermite integration formula by solving the system of linear equations like Eq. (5.9.7), but with the right-hand side (RHS) vector as

$$\text{RHS}(1) = \int_{-\infty}^{\infty} e^{-t^2}\, dt = \sqrt{\int_{-\infty}^{\infty} e^{-x^2} dx \int_{-\infty}^{\infty} e^{-y^2} dy} = \sqrt{\int_{-\infty}^{\infty}\int_{-\infty}^{\infty} e^{-(x^2+y^2)}\, dx\, dy}$$

$$= \sqrt{\int_{-\infty}^{\infty} e^{-r^2}\, 2\pi r\, dr} = \sqrt{-\pi\, e^{-r^2}\big|_0^{\infty}} = \sqrt{\pi} \qquad (5.9.14\text{a})$$

$$\text{RHS}(n) = \int_{-\infty}^{\infty} e^{-t^2} t^{n-1}\, dt = \int_{-\infty}^{\infty} (-2t)e^{-t^2} \frac{1}{-2} t^{n-2}\, dt \,(= 0 \quad \text{if } n \text{ is even})$$

$$= -\frac{1}{2} e^{-t^2} t^{n-2}\bigg|_{-\infty}^{\infty} + \frac{1}{2}(n-2) \int_{-\infty}^{\infty} e^{-t^2} t^{n-3}\, dt = \frac{1}{2}(n-2)\text{RHS}(n-2)$$

$$(5.9.14\text{b})$$

The procedure for finding the N grid point t_i's and the corresponding weight $w_{N,i}$'s of the N-point Gauss–Hermite integration formula has been cast into the above MATLAB function 'Gausshp()'. Note that, even though the integrand function ($g(t)$) does not have e^{-t^2} as a multiplying factor, we can multiply it by $e^{-t^2} e^{t^2} = 1$ to fabricate it as if it were like in Eq. (5.9.12):

$$I = \int_{-\infty}^{\infty} g(t)\, dt = \int_{-\infty}^{\infty} e^{-t^2} (e^{t^2} g(t))\, dt = \int_{-\infty}^{\infty} e^{-t^2} f(t)\, dt \qquad (5.9.15)$$

```
function [t,w]=Gausshp(N)
if N<0
 error('Gauss-Hermite polynomial of negative degree??');
end
t=roots(Hermitp(N))';
A(1,:)= ones(1,N);  b(1)=sqrt(pi);
for n=2:N
   A(n,:)=A(n-1,:).*t; % Eq.(5.9.7)
   if mod(n,2)==1, b(n)=(n-2)/2*b(n-2); % Eq.(5.9.14)
     else  b(n)= 0;
   end
end
w= b/A';
```

```
function p=Hermitp(N)
% Hn+1(x)=2xHn(x)-Hn'(x) from 'Advanced Engineering Math' by Kreyszig
if N<=0,  p=1;
 else p=[2 0];
      for n=2:N, p= 2*[p 0]-[0 0 drvtp(p)]; end %Eq.(5.9.13b)
end
```

5.9.3 Gauss-Laguerre Integration

The Gauss-Laguerre integration formula is also expressed by Eq. (5.9.5) as

$$I_{\text{GLa}}[t_1, t_2, \ldots, t_N] = \sum_{i=1}^{N} w_{N,i} f(t_i) \qquad (5.9.16)$$

and is supposed to give us the exact integral of the exponential e^{-t} multiplied by a polynomial $f(t)$ of degree $\leq (2N-1)$ over $[0, \infty)$:

$$I = \int_0^{\infty} e^{-t} f(t) dt \qquad (5.9.17)$$

The N grid point t_is can be obtained as the zeros of the Nth-degree Laguerre polynomial [K-2, Section 4.7]

$$L_N(t) = \sum_{i=0}^{N} \frac{(-1)^i}{i!} \frac{N^i}{(N-i)! i!} t^i \qquad (5.9.18)$$

Given the N grid point t_i's, we can get the corresponding weight $w_{N,i}$'s of the N-point Gauss-Laguerre integration formula by solving the system of linear equations like Eq. (5.9.7), but with the RHS vector as

$$\text{RHS}(1) = \int_0^{\infty} e^{-t} dt = -e^{-t} \big|_0^{\infty} = 1 \qquad (5.9.19a)$$

$$\text{RHS}(n) = \int_0^{\infty} e^{-t} t^{n-1} dt = -e^{-t} t^{n-1} \big|_0^{\infty} + (n-1) \int_0^{\infty} e^{-t} t^{n-2} dt$$

$$= (n-1) \text{RHS}(n-1) \qquad (5.9.19b)$$

5.9.4 Gauss-Chebyshev Integration

The Gauss-Chebyshev I integration formula is also expressed by Eq. (5.9.5) as

$$I_{\text{GC1}}[t_1, t_2, \ldots, t_N] = \sum_{i=1}^{N} w_{N,i} f(t_i) \qquad (5.9.20)$$

and is supposed to give us the exact integral of $1/\sqrt{1-t^2}$ multiplied by a polynomial $f(t)$ of degree $\leq (2N-1)$ over $[-1, 1]$:

$$I = \int_{-1}^{+1} \frac{1}{\sqrt{1-t^2}} f(t) dt \qquad (5.9.21)$$

The N grid point t_i's are the zeros of the Nth-degree Chebyshev polynomial (Section 3.3).

$$t_i = \cos\frac{(2i-1)\pi}{2N} \quad \text{for } i = 1, 2, \ldots, N \tag{5.9.22}$$

and the corresponding weight $w_{N,i}$'s are uniformly selected as

$$w_{N,i} = \pi/N, \forall i = 1, \ldots, N \tag{5.9.23}$$

The Gauss-Chebyshev II integration formula is also expressed by Eq. (5.9.5) as

$$I_{GC2}[t_1, t_2, \ldots, t_N] = \sum_{i=1}^{N} w_{N,i} f(t_i) \tag{5.9.24}$$

and is supposed to give us the exact integral of $\sqrt{1-t^2}$ multiplied by a polynomial $f(t)$ of degree $\leq(2N-1)$ over $[-1, 1]$:

$$I = \int_{-1}^{+1} \sqrt{1-t^2} f(t) dt \tag{5.9.25}$$

The N grid point t_i's and the corresponding weight $w_{N,i}$'s are

$$t_i = \cos\left(\frac{i\pi}{N+1}\right), w_{N,i} = \frac{\pi}{N+1}\sin^2\left(\frac{i\pi}{N+1}\right) \quad \text{for } i = 1, 2, \ldots, N \tag{5.9.26}$$

5.10 DOUBLE INTEGRAL

In this section, we consider the numerical integration of a function $f(x,y)$ with respect to two variables x and y over the integration region $R = \{(x,y) | a \leq x \leq b, c(x) \leq y \leq d(x)\}$ as depicted in Figure 5.5.

$$I = \iint_R f(x,y) dx\, dy = \int_a^b \left\{ \int_{c(x)}^{d(x)} f(x,y) dy \right\} dx \tag{5.10.1}$$

The numerical formula for this double integration over a 2D (two-dimensional) region takes the form

$$I(a, b, c(x), d(x)) = \sum_{m=1}^{M} w_m \sum_{n=1}^{N} v_n f(x_m, y_{m,n}) \tag{5.10.2}$$

where the weights w_m and v_n depend on the method of 1D integration we choose.

DOUBLE INTEGRAL   279

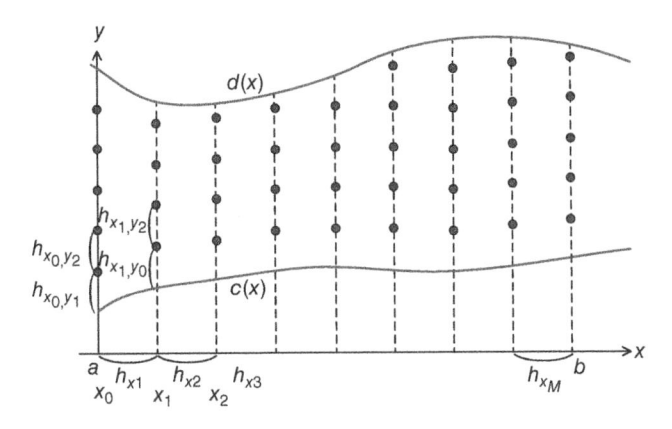

Figure 5.5 Region for a double integral.

(cf) The MATLAB built-in function 'dblquad()' can accept the boundaries a, b, c, and d of integration region along only given as numbers. Therefore, if we want to use the function in computing a double integral for a nonrectangular region D, we should define the integrand function $f(x,y)$ for a rectangular region $R \supseteq D$ (containing the actual integration region D) in such a way that $f(x,y) = 0$ for $(x, y) \notin D$, i.e. the value of the function becomes zero outside the integration region D, which may result in more computations. In contrast, another MATLAB built-in function 'integral2()' can accept the boundaries c and d of integration region (along the y-axis) given as functions of x.

Although the integration rules along the x-axis and along the y-axis do not need to be the same, we make a double integration function 'int2s(f,a,b,c,d,M,N)', which uses the Simpson method in common for both integrations and calls another function 'smpsns_fxy()' for 1D (one-dimensional) integration along the y-axis. The left/right boundary a/b of integration region given as the second/third input argument must be a number, while the lower/upper boundary c/d of integration region given as the fourth/fifth input argument may be either a number or a function of x. If the sixth/seventh input argument M/N is given as a positive integer, it will be accepted as the number of segments; otherwise, it will be interpreted as the segment width h_x/h_y. We also compose the following MATLAB script "nm05f06.m" to use the function 'int2s()' for finding one-fourth of the volume of a sphere with the radius $r = 1$ depicted in Figure 5.6:

$$I = \int_{-1}^{1} \int_{0}^{\sqrt{1-x^2}} \sqrt{1 - x^2 - y^2} \, dy \, dx = \frac{\pi}{3} = 1.047\,197\,55 \ldots \qquad (5.10.3)$$

```
function INTfxy=int2s(f,a,b,c,d,M,N)
% Double integral of f(x,y) over R={(x,y)|a<=x<=b,c(x)<=y<=d(x)}
% using Simpson's rule
if ceil(M)~=floor(M) % fixed width of segments on x
   hx=M; M=ceil((b-a)/hx);
end
if mod(M,2)~=0, M=M+1; end
hx=(b-a)/M; m=1:M+1; x=a+(m-1)*hx;
if isnumeric(c), cx(m)=c;   % if c is given as a constant number
  else  cx(m)=feval(c,x(m)); % in case c is given as a function of x
end
if isnumeric(d), dx(m)=d;   % if c is given as a constant number
  else  dx(m)=feval(d,x(m)); % in case d is given as a function of x
end
if ceil(N)~=floor(N) % fixed width of segments on y
  hy=N; Nx(m)=ceil((dx(m)-cx(m))/hy);
  ind=find(mod(Nx(m),2)~=0); Nx(ind)=Nx(ind)+1;
 else     % fixed number of subintervals
  if mod(N,2)~=0,  N=N+1; end
  Nx(m)=N;
end
for m=1:M+1 sx(m)= Smpsns_fxy(f,x(m),cx(m),dx(m),Nx(m)); end
kodd=2:2:M; keven=3:2:M-1; % Set of odd/even indices
INTfxy= hx/3*(sx(1)+sx(M+1)+4*sum(sx(kodd))+2*sum(sx(keven)));
```

```
function INTf=Smpsns_fxy(f, x, c, d, N)
% 1D integration of f(x,y) for Ry={c<=y<=d}
if nargin<5, N=100; end
if abs(d-c)<eps|N<=0, INTf=0; return; end
if mod(N,2)~=0,  N=N+1; end
h=(d-c)/N; y=c+[0:N]*h; fxy=feval(f,x,y);
fxy(find(fxy==inf))=realmax; fxy(find(fxy==-inf))=-realmax;
kodd=2:2:N; keven=3:2:N-1; % Set of odd/even indices
INTf= h/3*(fxy(1)+fxy(N+1)+4*sum(fxy(kodd))+2*sum(fxy(keven)));
```

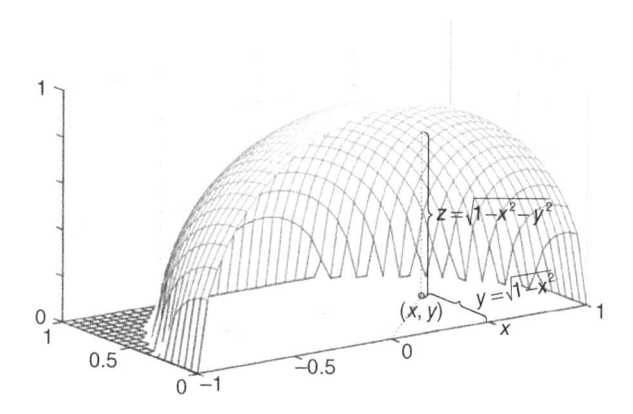

Figure 5.6 One-fourth (1/4) of a unit sphere (with the radius of 1).

INTEGRATION INVOLVING PWL FUNCTION   281

Interested readers are recommended to work on these functions and run the script "nm05f06.m" to see the result where the MATLAB built-in functions 'dblquad()' and 'integral2()' as well as 'in2s()' have been tried.

```
»nm05f06
   Vs1 = 1.0470
   Vs2 = 1.0470
   Vsi = 1.0472
   Vsd = 1.0472
```

```
%nm05f06.m: the volume of a sphere
x=[-1:0.05:1]; y=[0:0.05:1]; [X,Y]=meshgrid(x,y);
f510=@(x,y)sqrt(max(1-x.^2-y.^2,0)); Z=f510(X,Y); mesh(x,y,Z);
a=-1; b=1; c=0; d=@(x)sqrt(max(1-x.^2,0));
Vs1=int2s(f510,a,b,c,d,100,100) % with the numbers of segments given
Vs2=int2s(f510,a,b,c,d,0.01,0.01) % with the segment width given
Vsi=integral2(f510,a,b,c,d) % Using integral2()
Vsd=dblquad(f510,-1,1,0,1) % Using dblquad()
```

5.11 INTEGRATION INVOLVING PWL FUNCTION

In this section, we will see how an integrand involving a piecewise linear (PWL) waveform consisting of many data points, that is difficult to express as a mathematical function, can be dealt with.

For example, consider computing the (exponential) continuous-time Fourier series (CtFS) coefficients

$$X_k = \int_P x(t)e^{-jk\omega_0 t}\, dt \text{ with } \omega_0 = \frac{2\pi}{P} \text{ (integration over one period } P) \quad (5.11.1)$$

of the periodic PWL function $x(t)$ with period $P = 6$ as shown in Figure 5.7. Since they have been obtained analytically in Chapter 9 of [Y-1] as

$$a_0 = 0,\ a_k = 0\ \forall k,\ b_k \stackrel{[Y-1]}{\underset{(E9.6.3)}{=}} 12(-1)^{m+1}\frac{\sin(k\pi/2)+\sin(k\pi/6)}{k\pi} \text{ for } k = 2m-1$$

$$\to X_k = \frac{P}{2}(a_k - jb_k) \stackrel{a_k=0}{\underset{P=6}{=}} -j36(-1)^{m+1}\frac{\sin(k\pi/2)+\sin(k\pi/6)}{k\pi} \text{ for } k = 2m-1$$
(5.11.2)

let us use the MATLAB function 'CTFS_exp()' (listed below) to find the Fourier coefficients. To this end, we cast the function $x(t)$ into the following MATLAB function 'nm05f07_f()' and compose the following script "find_CTFS_PWL.m":

282 NUMERICAL DIFFERENTIATION/INTEGRATION

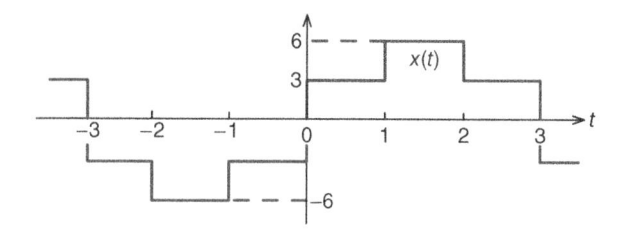

Figure 5.7 A piecewise linear (PWL) waveform.

```
function [X,kk]=CTFS_exp(x,P,N)
% Find the complex exponential Fourier coefficients X(k) for k=-N:N
% x: A periodic function with period P
% P: Period
% N: Highest frequency index of Fourier coefficients to be computed
w0=2*pi/P; % Fundamental frequency [rad/s]
xexp_jkw0t_=[x '(t).*exp(-j*k*w0*t)'];
xexp_jkw0t=inline(xexp_jkw0t_,'t','k','w0'); % Integrand of Eq.(5.11.1)
tol=1e-6; % toleration on numerical error
kk=-N:N; % Frequency indices in range
for k=kk
    X(k+N+1)=quadl(xexp_jkw0t,-P/2,P/2,tol,[],k,w0); % Eq.(5.11.1)
end
```

We run the script "find_CTFS_PWL.m" to get

```
ans =
  0.0000 + 0.0000i  -0.0000 -17.1887i  -0.0000 + 0.0000i  -0.0000 + 0.0000i
 -0.0000 - 0.0000i  -0.0000  -3.4377i   0.0000 - 0.0000i  -0.0000 - 2.4555i
ans =
  0.0000 + 0.0000i   0.0000 -17.1887i   0.0000 + 0.0000i   0.0000 + 0.0000i
  0.0000 + 0.0000i   0.0000 - 3.4377i   0.0000 + 0.0000i   0.0000 - 2.4555i
```

These results imply the following:

- The values of the Fourier coefficients computed using 'CTFS_exp()' agree with those obtained using the analytical approach (Eq. (5.11.2)).
- Even a waveform difficult to express as a mathematical function can be made into a MATLAB function using the PWL interpolation so that we can use 'CTFS_exp()' to find the Fourier coefficients of the waveform with sufficient accuracy.

```
function x=nm05f07_f(t)
% A PWL (piecewise linear) function of multi-step type
EPS=1e-8; EPS2=[-EPS EPS];
tt=[0 EPS 1+EPS2 2+EPS2 3+EPS2 4+EPS2 5+EPS2 6-EPS 6];
xx=[0  3   3  6   6  3   3 -3  -3 -6  -6 -3  -3   0];
P=6; % Period
x=interp1(tt,xx,mod(t,P)); % PWL interpolation
```

INTEGRATION INVOLVING PWL FUNCTION **283**

```
%find_CTFS_PWL.m
P=6; w0=2*pi/P; % Period and Fundamental angular frequency
t=[-P:0.01:P]; % Time interval of two periods
x='nm05f07_f';
% Plot x(t) to see if the function defines the waveform well,
xt=feval(x,t);
subplot(221), plot(t,xt), hold on
% Fourier analysis using the numerical approach
N=7; % The highest frequency index to be computed
[X,kk]=CTFS_exp(x,P,N); % kk=[-N:N]
Xk=X(find(kk>=0)) % complex exponential Fourier coefficients for k=0:N
% Fourier analysis using the analytical approach
m=1:ceil(N/2); k=2*m-1;
X_(k)=-j*P/2*12*(-1).^(m+1).*(sin(k*pi/2)+sin(k*pi/6))./k/pi; % Eq.(5.11.2)
a0=0;
Xk_=[P*a0 X_] % complex exponential Fourier coefficients for k=0:N
```

Example 5.3. CtFS for a Triangular Wave.

Compute the (exponential) CtFS coefficients of the periodic PWL function $x(t)$ with period $P = 3$ as shown in Figure 5.8 where they have been obtained analytically in Chapter 9 of [Y-1] as

$$a_0 = 0, \quad a_k = 0\,\forall k, \quad b_k \stackrel{[Y-1]}{\underset{(P9.1.4)}{=}} \frac{8A}{d(P-d)} \frac{\sin(k\omega_0 d/2)}{(k\omega_0)^2}$$

$$\to X_k = \frac{P}{2}(a_k - jb_k) \stackrel{a_k=0}{\underset{A=1,P=3,d=2}{=}} -j6\frac{\sin(2k\pi/3)}{(2k\pi/3)^2} \quad (E5.3.1)$$

To this end, we cast the function $x(t)$ into the following MATLAB function 'nm05e03_f()', compose the following script "nm05e03.m", and run it to get

```
ans =
  -0.0000 + 0.0000i  -0.0000 - 1.1846i   0.0000 + 0.2961i  -0.0000 - 0.0000i
  -0.0000 - 0.0740i  -0.0000 + 0.0474i  -0.0000 - 0.0000i  -0.0000 - 0.0242i
ans =
   0.0000 + 0.0000i   0.0000 - 1.1846i   0.0000 + 0.0000i   0.0000 + 0.0000i
   0.0000 + 0.0000i   0.0000 + 0.0474i   0.0000 + 0.0000i   0.0000 - 0.0242i
```

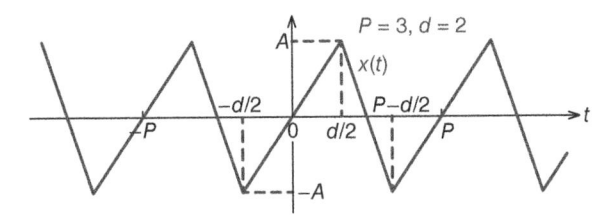

Figure 5.8 A piecewise linear (PWL) waveform.

NUMERICAL DIFFERENTIATION/INTEGRATION

```
function x=nm05e03_f(t)
% A PWL (piecewise linear) function
tt=[0 1  2 3];
xx=[0 1 -1 0];
P=3; % Period
x=interp1(tt,xx,mod(t,P)); % PWL interpolation
```

```
%nm05e03.m
clear, clf
P=3; w0=2*pi/P; % Period and Fundamental angular frequency
t=[-P:0.01:P]; % Time interval of two periods
x='nm05e03_f';
% Plot x(t) to see if the function defines the waveform well,
xt=feval(x,t);
subplot(221), plot(t,xt), hold on
% Fourier analysis using the numerical approach
N=7; % The highest frequency index to be computed
[X,kk]=CTFS_exp(x,P,N); % kk=[-N:N]
X(find(kk>=0)) % complex exponential Fourier coefficients for k=0:N
% Fourier analysis using the analytical approach
m=1:ceil(N/2); k=2*m-1;
X_(k)=-j*P/2*4*sin(2*k*pi/3)./(2*k*pi/3).^2; % Eq.(E5.3.1)
a0=0; [P*a0 X_] % complex exponential Fourier coefficients for k=0:N
```

PROBLEMS

5.1 Numerical Differentiation of Basic Functions
If we want to find the derivative of a polynomial/trigonometric/exponential function, it would be more convenient and accurate to use an analytical computation (by hand) than to use a numerical computation (by computer). But, in order to test the accuracy of the numerical derivative formulas, consider the three basic functions as

$$f_1(x) = x^3 - 2x, \quad f_2(x) = \sin x, \quad f_3(x) = e^{-x} \quad \text{(P5.1.1)}$$

(a) To find the first derivatives of these functions by using the formulas (5.1.8) and (5.1.9) listed in Table 5.2 (Section 5.3), modify the script "nm05p01.m", which uses the MATLAB routine 'difapx()' (Section 5.3) for generating the coefficients of the numerical derivative formulas. Fill in the following table with the error results obtained from running the script:

First derivatives	h	$\dfrac{f_1 - f_{-1}}{2h}$	$\dfrac{-f_2 + 8f_1 - 8f_{-1} + f_{-2}}{12h}$
$(x^3 - 2x)'\|_{x=1}$ = 1.0000000000	0.1	1.0000e − 02	
	0.01		8.8818e − 15
$(\sin x)'\|_{x=\pi/3}$ = 0.5000000000	0.1	−8.3292e − 04	
	0.01	−8.3333e − 06	
$(e^{-x})'\|_{x=0}$ = −1.0000000000	0.1		3.3373e − 06
	0.01	−1.6667e − 05	

```
%nm05p01.m
f= @(x)x.*(x.*x-2);
n=[1 -1];   x0=1;    h=0.1;    DT=1;
c= difapx(1,n);   i=1:length(c);
num= c*feval(f,x0+(n(1)+1-i)*h)';  drv=num/h;
fprintf('with h=%6.4f, %12.6f %12.4e\n', h,drv,drv-DT);
```

(b) Likewise in (a), modify the script "nm05p01.m" in such a way that the formulas (5.3.1) and (5.3.2) in Table 5.2 are generated and used to find the second numerical derivatives. Fill in the following table with the error results obtained from running the script:

Second derivatives	h	$\dfrac{f_1 - 2f_0 + f_{-1}}{h^2}$	$\dfrac{-f_2 + 16f_1 - 30f_0 + 16f_{-1} - f_{-2}}{12h^2}$
$(x^3 - 2x)^{(2)}\|_{x=1}$ = 6.0000000000	0.1	2.6654e − 14	
	0.01		2.9470e − 12
$(\sin x)''\|_{x=\pi/3}$ = −0.8660254037	0.1		9.6139e − 07
	0.01	7.2169e − 06	
$(e^{-x})''\|_{x=0}$ = 1.0000000000	0.1	8.3361e − 04	
	0.01		−1.1419e − 10

Table P5.2 Three functions each given as a set of five data pairs.

x	$f_1(x)$	x	$f_2(x)$	x	$f_3(x)$
0.8000	−1.0880	0.8472	0.7494	−0.2000	1.2214
0.9000	−1.0710	0.9472	0.8118	−0.1000	1.1052
1.0000	−1.0000	1.0472	0.8660	0	1.0000
1.1000	−0.8690	1.1472	0.9116	0.1000	0.9048
1.2000	−0.6720	1.2472	0.9481	0.2000	0.8187

5.2 Numerical Differentiation of a Function given as a Set of Data Pairs

Consider the three (numerical) functions each given as a set of five data pairs in Table P5.2.

(a) Use the formulas (5.1.8) and (5.1.9) to find the first derivatives of the three numerical functions (at $x = 1$, 1.0472 and 0, respectively) and fill in the following table with the results. Also use the formulas (5.3.1) and (5.3.2) to find the second derivatives of the three functions (at $x = 1$, 1.0472, and 0, respectively) and fill in the following table with the results.

	$f'_1(x)|_{x=1}$	$f'_2(x)|_{x=1.0472}$	$f'_3(x)|_{x=0}$
First derivative by Eq. (5.1.8)	1.010000		−1.002000
First derivative by Eq. (5.1.9)		0.499750	
	$f^{(2)}_1(x)|_{x=1}$	$f^{(2)}_2(x)|_{x=1.0472}$	$f^{(2)}_3(x)|_{x=0}$
Second derivative by Eq. (5.3.1)		−0.860000	
Second derivative by Eq. (5.3.2)	6.000000		0.999167

(b) Based on the Lagrange/Newton polynomial matching the three/five points around the target point, find the first/second derivatives of the three functions (at $x = 1$, 1.0472, and 0, respectively) and fill in the following table with the results.

	$f'_1(x)|_{x=1}$	$f'_2(x)|_{x=1.0472}$	$f'_3(x)|_{x=0}$
First derivative on $l_2(x)$		0.499000	
First derivative on $l_4(x)$	1.000000		−1.000417
	$f^{(2)}_1(x)|_{x=1}$	$f^{(2)}_2(x)|_{x=1.0472}$	$f^{(2)}_3(x)|_{x=0}$
Second derivative on $l_2(x)$	6.000000		1.000000
Second derivative on $l_4(x)$		−0.859167	

5.3 First Derivative and Step-size

Consider the routine 'jacob()' in Section 4.6, which is used for computing the Jacobian, i.e. the first derivative of a vector function with respect to a vector variable.

(a) Which one is used for computing the Jacobian in the routine 'jacob()' among the first derivative formulas in Section 5.1?

(b) Expecting that smaller step-size h would yield a better solution to the problem given in Example 4.3, Bush changed h=1e-4 to h=1e-5 in the routine 'Newtons()' and then ran the following statement. What solution could he get?

```
»G=667e-13; Ms=198e28; Me=598e22; R=149e9; T=315576e2; w=2*pi/T;
  f=@(r)G*(Ms./(r.^2+eps)-Me./((R-r).^2+eps))-r*w^2;
  rn1=Newtons(f,1e6,1e-4,100)
```

(c) What gave rise to such a disappointing result? Jessica diagnosed the trouble as caused by a singular Jacobian and modified the statement 'dx=-jacob()\fx(:)' in the routine 'Newtons()' as follows:

```
J=jacob(f,xx(k,:),h,varargin{:});
if rank(J)<Nx
   k=k-1;
   fprintf('Jacobian singular! det(J)=%10.3e\n', det(J)); break;
else
   dx= -J\fx(:); %-[dfdx]^-1*fx;
end
```

With 'Newtons()' (with h=1e-5) modified like this, what solution (to the problem in Example 4.3) would she get by running the same statement as in (b)? Do you agree to her diagnosis? How about her complementary measure against the case of a singular Jacobian?

```
»rn2=Newtons(f,1e6,1e-4,100), res_err=f(rn2)
```

(d) To investigate how the accident of Jacobian singularity happened, add/subtract h=1e-5 to/from the (tentative) solution (rn2) obtained in (c). Does the result differ from rn2? If not, why? (see Section 1.2.2 and Problem 1.19). Explain how this happening is related with the Jacobian singularity.

```
»rn2+1e-5~=rn2, rn2-1e-5~=rn2
```

(e) Charley thought that Jessica just circumvented the Jacobian singularity problem. To remove the source of singularity due to a small step-size h, he modified the formula (5.1.8) into

$$D'_{c2}(x,h) = \frac{f(x+jh)-f(x-jh)}{j2h} \text{ or } \text{Im}\left\{\frac{f(x+jh)-f(x-jh)}{2h}\right\} \quad (P5.3.1)$$

$$\text{with } j = \sqrt{-1}$$

and implemented it in another routine 'jacob1()' as follows:

```
function g=jacob1(f,x,h,varargin)
% Jacobian of f(x)
if nargin<3, h=.0001; end
N=length(x); h2=2*h;
x=x(:); Ih=i*h*eye(N);
for n=1:N
   f1=feval(f,x+Ih(:,n),varargin{:});
   f2=feval(f,x-Ih(:,n),varargin{:});
   g(:,n)=imag(f1-f2)/h2;
end
```

With `h=1e-5` or `h=1e-6` and 'jacob()' replaced with 'jacob1()' in the function 'Newtons()', type the same statement as in (c) to get a solution to the problem in Example 4.3 together with its residual error and check if his scheme works fine.

```
»rn3=Newtons(f,1e6,1e-4,100), f(rn3)
```

5.4 Numerical Integration of Basic Functions

Compute the following integrals by using the trapezoidal rule, the Simpson's rule, and Romberg method, and fill in Table P5.4 with the resulting errors.

(i) $\int_0^2 (x^3 - 2x)dx$

(ii) $\int_0^{\pi/2} \sin x \, dx$

(iii) $\int_0^1 e^{-x} dx$

Table P5.4 Results of using various numerical integration methods.

	N	Trapezoidal rule	Simpson rule	Romberg(tol = 0.0005)
$\int_0^2 (x^3 - 2x)dx = 0$	4		0.0000e + 0	
	8	6.2500e − 1		
$\int_0^{\pi/2} \sin x \, dx = 1$	4	1.2884e − 2		8.4345e − 6
	8		8.2955e − 6	
$\int_0^1 e^{-x} dx = 0.63212055883$	4		1.3616e − 5	
	8	8.2286e − 4		

5.5 Improvement of Adaptive Integrator 'adapt_smpsns()'

Consider the adaptive integrator 'adapt_smpsns()' that was introduced in Section 5.8 and used to find the integral (E5.2.3) in Example 5.2.

(a) What are the values of the integral returned by 'adapt_smpsns()' and 'quad()'? To see how the integration interval has been divided by the integrator 'adapt_smpsns()', run the following MATLAB statements:

```
»f=@(t)cos(8*pi*t); [I_as,points]=adapt_smpsn(f,-1,1,1e-4)
```

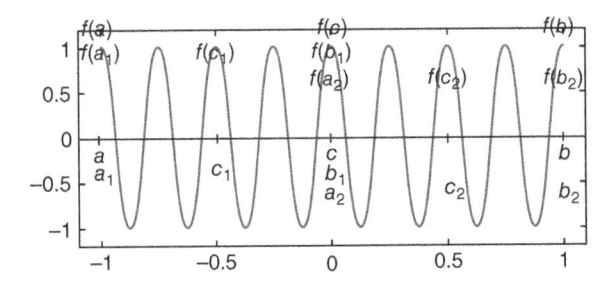

Figure P5.5 The graph of the integrand function of the integral (P5.5.1)

(b) To see why 'adapt_smpsns()' stopped its iteration so prematurely to return 2, which is far from the true integral value 0, perform the adaptive integration iteration by hand (referring to the graph of the integrand function shown in Figure P5.5) and/or by using the breakpoint and F10 (step) key in the MATLAB Editor. Find the following:

  (i) The values of the integral obtained using the Simpson's rule based on the three sets of three function values $\{f(-1) = 1, f(0) = 1, f(1) = 1\}$, $\{f(-1) = 1, f(-0.5) = 1, f(0) = 1\}$, and $\{f(0) = 1, f(0.5) = 1, f(1) = 1\}$.

  (ii) The first value of err computed inside 'adap_smpsns()', which is the difference between the successive numerical integration values.

(c) There would be no problem if the graph of the integrand function is a straight-line. The problem is that the result is the same with the case of straight-line just because all the sample points make a straight-line even though if the graph of the integrand function is not really a straight-line. How could you complement 'adapt_smpsns()' so that it can get out of such a pitfall? Will it be of a help to insert the following statements somewhere? If so, explain how they work.

```
d2=a+rand(1,2)*(b-a); fd2=feval(f,d2,varargin{:});
straight_line=(norm(fd2-interp1([a b],[fa fb],d2))<1e-10);
|(err<eps&~straight_line)
```

5.6 Adaptive Quadrature and Gaussian Quadrature for Improper Integral
Consider the following two integrals.

(i) $\int_0^1 \frac{1}{\sqrt{x}} dx = 2x^{1/2}\big|_0^1 = 2$

(ii) $\int_{-1}^1 \frac{1}{\sqrt{x}} dx = \int_{-1}^0 \frac{1}{\sqrt{x}} dx + \int_0^1 \frac{1}{\sqrt{x}} dx = 2 - 2i$

(a) Run the following MATLAB statements to use the integrating routines for the above integral. What did you get? If something is wrong, what do you think caused it?

```
f=@(x)1./sqrt(x); % define the integrand function
Smpsns(f,0,1,100) % integral over [0,1] with 100 segments
Rmbrg(f,0,1,1e-4) % with error tolerance=0.0001
adapt_smpsn(f,0,1,1e-4) % with error tolerance=0.0001
Gauss_Legendre(f,0,1,20) % Gauss-Legendre with N=20 grid points
quad(f,0,1) % MATLAB built-in routine
quadl(f,0,1) % Another MATLAB built-in routine
adapt_smpsn(f,-1,1,1e-4) % Integral over [-1,1]
quad(f,-1,1) % MATLAB built-in routine
quadl(f,-1,1) % MATLAB built-in routine
```

(b) Itha decided to retry the routine 'Smpsns()', but with the singular point excluded from the integration interval. In order to do that, she replaced the singular point (0) which is the lower bound of the integration interval [0,1] by 10^{-4} or 10^{-5}, running the following MATLAB statements:

```
Smpsns(f,1e-4,1,100)
Smpsns(f,1e-5,1,100)
Smpsns(f,1e-5,1,1e4)
Smpsns(f,1e-4,1,1e3)
Smpsns(f,1e-4,1,1e4)
```

What are the results? Will it be better if you make the lower-bound of the integration interval closer to zero (0), without increasing the number of segments or (equivalently) decreasing the segment width? How about increasing the number of segments without making the lower-bound of the integration interval closer to the original lower-bound which is zero (0)?

(c) For the purpose of improving the performance of 'adap_smpsn()', Vania would put the following statements into both of the routines 'Smpsns()' and 'adap_smpsn()'. Supplement the routines and check whether her idea works or not.

```
EPS=1e-12;
fa=feval(f,a,varargin{:});
if isnan(fa)|abs(fa)==inf,  a=a+max(abs(a)*EPS,EPS); end
fb=feval(f,b,varargin{:});
if isnan(fb)|abs(??)==inf,  b=b-max(abs(?)*EPS,EPS); end
```

5.7 Various Numerical Integration Methods and Improper Integral Consider the following integrals:

$$\int_0^\infty \frac{\sin x}{x} dx = \frac{\pi}{2} \cong \int_0^{100} \frac{\sin x}{x} dx \quad \text{(P5.7.1)}$$

$$\int_0^\infty e^{-x^2} dx = \frac{1}{2}\sqrt{\pi} \quad \text{(P5.7.2)}$$

Note that the true values of these integrals can be obtained by using the symbolic computation command 'int()' as follows:

```
»syms x, int(sin(x)/x,0,inf)
»int(exp(-x^2),0,inf)
```

(cf) Don't you believe it without seeing it? Blessed are those who have not seen and yet believe.

(a) To use the routines like 'Smpsns()', 'adapt_smpsn()', 'Gauss_Legendre()', and 'quadl()' for evaluating the integral (P5.7.1), do the following:

(i) Note that the integration interval $[0,\infty)$ can be changed into a finite interval as follows:

$$\int_0^\infty \frac{\sin x}{x} dx = \int_0^1 \frac{\sin x}{x} dx + \int_1^\infty \frac{\sin x}{x} dx$$

$$= \int_0^1 \frac{\sin x}{x} dx + \int_1^0 \frac{\sin(1/y)}{1/y}\left(-\frac{1}{y^2}\right) dy$$

$$= \int_0^1 \frac{\sin x}{x} dx + \int_0^1 \frac{\sin(1/y)}{y} dy \qquad (P5.7.3)$$

(ii) Add the block of statements in P5.7(c) into the routines 'Smpsns()' and 'adap_smpsn()' to make them cope with the cases of NaN (Not-a-Number) and Inf (Infinity).

(iii) Supplement the following script "nm05p07a.m" so that the various routines are applied for computing the integrals (P5.7.1) and (P5.7.3), where the parameters like the number of segments ($N = 200$), the error tolerance (tol = 1e − 4), and the number of grid points (MGL = 20) are supposed to be used as they are in the script. Noting that the second integrand function in Eq. (P5.7.3) oscillates like crazy with higher frequency and larger amplitude as y gets closer to zero (0), set the lower-bound of the integration interval to a2=0.001.

```
%nm05p07a.m
fp56a=@(x)sin(x)./x; fp56a2=@(y)sin(1./y)./y;
IT=pi/2; % True value of the integral
a=0; b=100; N=200; tol=1e-4; MGL=20; a1=0; b1=1; a2=0.001; b2=1;
format short e
e_s=Smpsns(fp56a,a,b,N)-IT
e_as=adapt_smpsn(fp56a,a,b,tol)-IT
e_ql=quadl(fp56a,a,b,tol)-IT
e_GL=Gauss_Legendre(fp56a,a,b,MGL)-IT
e_ss=Smpsns(fp56a,a1,b1,N)+Smpsns(fp56a2,a2,b2,N)-IT
e_Iasas=adapt_smpsn(fp???,a?,b?,tol)+ ...
        adapt_smpsn(fp56a2,a2,b2,tol)-IT
e_Iqq=quadl(fp56a,a1,b1,tol)+quad(fp????,a?,b?,tol)-IT
```

292 NUMERICAL DIFFERENTIATION/INTEGRATION

```
%nm05p07b.m
fp57b=@(x)exp(-x.*x); fp57b1=@(x)ones(size(x));
fp57b2=@(y)exp(-1./y./y)./y./y;
a=0; b=200; N=200; tol=1e-4; IT=sqrt(pi)/2;
a1=0; b1=1; a2=0; b2=1; MGH=2;
e_s= Smpsns(fp57b,a,b,N)-IT
e_as= adapt_smpsn(fp57b,a,b,tol)-IT
e_q= quad(fp57b,a,b,tol)-IT
e_GH= Gauss_Hermite(fp57b1,MGH)/2-IT
e_ss= Smpsns(fp57b,a1,b1,N)+Smpsns(fp56b2,a2,b2,N)-IT
Iasas= adapt_smpsn(fp57b,??,b1,tol)+ ...
       + adapt_smpsn(??????,a2,??,tol) -IT
e_qq= quad(?????,a1,??,tol)+quad(fp57b2,??,b2,tol) -IT
```

(iv) Run the supplemented script and fill in Table P5.7 with the absolute errors of the results.

(b) To use the routines like 'Smpsns()', 'adapt_smpsn()', 'quad()', and 'Gauss_Hermite()' for evaluating the integral (P5.7.2), do the following:

(i) Note that the integration interval $[0,\infty)$ can be changed into a finite interval as follows:

$$\int_0^\infty e^{-x^2}\,dx = \int_0^1 e^{-x^2}\,dx + \int_1^\infty e^{-x^2}\,dx$$
$$= \int_0^1 e^{-x^2}\,dx + \int_1^0 e^{-1/y^2}\left(-\frac{1}{y^2}\right)dy = \int_0^1 e^{-x^2}\,dx + \int_0^1 \frac{e^{-1/y^2}}{y^2}\,dy$$
(P5.7.4)

(ii) Referring to 'Gauss_Legendre()' (Section 5.9.1), complete the MATLAB function 'Gauss_Hermite()' below so that it can perform the Gauss-Hermite integration introduced in Section 5.9.2.

(iii) Supplement the script "nm05p06b.m" to use the various routines for computing the integrals (P5.7.2) and (P5.7.4), where the parameters like the number of segments ($N = 200$), the error tolerance (tol = 1e−4), and the number of grid points (MGH = 2) are supposed to be used as they are in the script. Note that the integration interval is not $(-\infty, \infty)$ like that of Eq. (5.9.12), but $[0,\infty)$ and so you should cut the result of 'Gauss_Hermite()' by half to get the right answer for the integral (P5.7.2).

Table P5.7 Results of using various numerical integration methods for improper integrals.

	Simpson	Adaptive	Quad	Gauss	S&S	a&a	q&q
(P5.7.1)	8.5740e − 3		1.9135e − 1		1.1969e + 0		2.4830e − 1
(P5.7.2)		6.6730e − 6		0.0000e + 0		3.3546e − 5	

(iv) Run the supplemented script and fill in Table P5.7 with the absolute errors of the results.

(c) Based on the results listed in Table P5.7, answer the following questions:

(i) Among the routines 'Smpsns()', 'adapt_smpsn()', 'quad()', and 'Gauss()', choose the best two ones for (P5.7.1) and (P5.7.2), respectively.

(ii) The routine 'Gauss_Legendre()' works (badly, perfectly) even with as many as 20 grid points for (P5.7.1), while the routine 'Gauss_Hermite()' works (perfectly, badly) just with two grid points for (P5.7.2). It is because the integrand function of (P5.7.1) is (far from, just like) a polynomial, while (P5.7.2) matches Eq. (5.9.11) and the part of it excluding e^{-x^2} is (just like, far from) a polynomial.

```
function I=Gauss_Hermite(f,N,varargin)
[t,w]=???????(N); ft=feval(f,t,varargin{:}); I= w*ft';
```

(iii) Run the following script "nm05p07c.m" to see the shapes of the integrand functions of (P5.7.1) and (P5.7.2) and the second integrals of (P5.7.3) and (P5.7.4). You can zoom in/out the graphs by selecting the Tools>Zoom-In/Out menu (from the top menubar) and then clicking any point on the graphs with the left/right mouse button in the MATLAB graphic window. Which one is oscillating furiously? Which one is oscillating moderately? Which one is just changing abruptly?

```
%nm05p07c.m
fp571=@(x)sin(x)./x;     fp5712=@(y)sin(1./y)./y;
fp572=@(x)exp(-x.^2);    fp5722=@(y)exp(-1./y.^2)./y.^2;
subplot(221), x=[eps:1000]/10; plot(x,fp571(x));
subplot(222), y=logspace(-5,0,500); loglog(y,abs(fp5712(y)))
subplot(223), x=[0:1000]/100; plot(x,fp572(x)),
subplot(224), y=logspace(-5,0,500); loglog(y,abs(fp5722(y)))
```

(iv) The adaptive integration routines like 'adapt_smpsn()' and 'quad()' work (badly, fine) for (P5.7.1), but (fine, badly) for (P5.7.2). From this fact, we might conjecture that the adaptive integration routines may be (ineffective, effective) for the integrand functions having many oscillations, while they may be (effective, ineffective) for the integrand functions with abruptly changing slope. To support this conjecture, run the following script "nm05p07d.m", which uses the 'quad()' routine for the following integrals:

$$\int_1^b \frac{\sin x}{x} dx \text{ with } b = 5, 10, 100, 1000 \quad \text{(P5.7.5a)}$$

$$\int_a^1 \frac{\sin(1/y)}{y} dy \text{ with } a = 0.1, 0.01, 0.001, 0.0001 \quad \text{(P5.7.5b)}$$

Does the MATLAB built-in routine 'quad()' work stably for (P5.7.5a) with the changing value of the upper-bound of the integration interval? Does it work stably for (P5.7.5b) with the changing value of the lower-bound of the integration interval? Do the results support or defy the conjecture?

```
%nm05p07d.m
fp571=@(x)sin(x)./x;
fp5712=@(y)sin(1./y)./y;
syms x
IT2=pi/2-double(int(sin(x)/x,0,1)) % True value of integral
disp('Change of upper limit of the integration interval')
a=1; b=[5 10 100 1e3]; tol=1e-4;
for i=1:length(b)
   Iq2=quad(fp571,a,b(i),tol);
   fprintf('With b=%8.0f, err_Iq=%12.4e\n', b(i),Iq2-IT2);
end
disp('Change of lower limit of the integration interval')
a2=[1e-1 1e-2 1e-3 1e-4 0]; b2=1;
tol=1e-4;
for i=1:numel(a2)
   Iq2=quad(fp5712,a2(i),b2,tol);
   fprintf('With a2=%6.6f, err_Iq=%12.4e\n', a2(i),Iq2-IT2);
end
```

(cf) This problem warns us that it may be not good to use only one routine for a computational work and suggests us to use more than one method for cross check.

5.8 Gauss–Hermite Integration Method
Consider the following integral

$$\int_0^\infty e^{-x^2} \cos x \, dx = \frac{\sqrt{\pi}}{2} e^{-1/4} \quad (P5.8.1)$$

Select a Gauss quadrature suitable for this integral and apply it with the number $N = 4$ of grid points as well as the routines 'Smpsns()', 'adapt_smpsn()', 'quad()', and 'quadl()' to evaluate the integral. In order to compare the number of floating-point operations required to achieve almost the same level of accuracy, set the number of segments for Simpson method to $N = 700$ and the error tolerance for all other routines to tol $= 10^{-5}$. Fill in Table P5.8 with the error results.

Table P5.8 Results of using various numerical integration methods.

		Simpson (N = 700)	Adaptive (tol = 10^{-5})	Gauss	Quad (tol = 10^{-5})	Quadl (tol = 10^{-5})
(P5.7)	|error|		1.0001×10^{-3}		1.0000×10^{-3}	
	flops	4 930	5 457	1 484	11 837	52 590 (with quadl)
(P5.8)	|error|	1.3771×10^{-2}		0		4.9971×10^{-7}
	flops	5 024	7 757	131	28 369	75 822

5.9 Gaus-Laguerre Integration Method

(a) As in Sections 5.9.1 and 5.9.2 and Problem 5.6(b), compose the MATLAB routines 'Laguerp()', which generates the Laguerre polynomial (5.9.17), 'Gausslp()', which finds the grid point t_i's and the coefficient $w_{N,i}$'s for Gauss-Laguerre integration formula (5.9.15) and 'Gauss_Laguerre(f,N)', which uses these two routines to carry out the Gauss-Laguerre integration method.

(b) Consider the following integral

$$\int_0^\infty e^{-t} t \, dt = -e^{-t} t \Big|_0^\infty + \int_0^\infty e^{-t} \, dt = -e^{-t} \Big|_0^\infty = 1 \qquad (P5.9.1)$$

Noting that, since this integral matches Eq. (5.9.16) with $f(t) = t$, Gauss-Laguerre method is the right choice, apply the routine 'Gauss_Laguerre(f,N)' (manufactured in (a)) with N = 2 as well as the routines 'smpsns()', 'adapt_smpsn()', 'quad()', and 'quadl()' for evaluating the integral and fill in Table P5.7 with the error results. Which turns out to be the best? Is the performance of 'quad()' improved by lowering the error tolerance?

(cf) This illustrates that the routine 'adapt_smpsn()' sometimes outperforms the MATLAB built-in routine 'quad()' with fewer computations. On the other hand, Table P5.7 shows that it is most desirable to apply the Gauss quadrature schemes only if one of them is applicable to the integration problem.

5.10 Numerical Integrals

Consider the following integrals:

(1) $\int_0^{\pi/2} x \sin x \, dx = 1$

(2) $\int_0^1 x \ln(\sin x) \, dx = -\frac{1}{2} \pi^2 \ln 2$

(3) $\int_0^1 \frac{1}{x(1 - \ln x)^2} dx = 1$

(4) $\int_1^\infty \frac{1}{x(1 + \ln x)^2} dx = 1$

(5) $\int_0^1 \frac{1}{\sqrt{x}(1 + x)} dx = \frac{\pi}{2}$

(6) $\int_1^\infty \frac{1}{\sqrt{x}(1+x)} dx = \frac{\pi}{2}$

(7) $\int_0^1 \sqrt{\ln \frac{1}{x}} \, dx = \frac{\sqrt{\pi}}{2}$

(8) $\int_0^\infty \sqrt{x} \, e^{-x} \, dx = \frac{\sqrt{\pi}}{2}$

(9) $\int_0^\infty x^2 e^{-x} \cos x \, dx = -\frac{1}{2}$

(a) Apply the integration routines 'smpsns()' (with N = 10^4), 'adapt_smpsn()', 'quad()', 'quadl()' (tol = 10^{-6}), and 'Gauss_legendre()' (Sec. 5.9.1) or 'Gauss_Laguerre()' (Problem 5.9) (with N = 15) to compute the above integrals and fill in Table P5.10 with the relative errors. Use the upper-bound/lower-bounds of the integration interval in Table P5.10 if they are specified in the table.

(b) Based on the results listed in Table P5.10, answer the following questions or circle the right answer.

  (i) Since the Gauss-Legendre integration scheme worked best only for (1), it is implied that the scheme is (recommendable, not recommendable) for the case where the integrand function is far from being approximated by a polynomial.

  (ii) Since that the Gauss-Laguerre integration scheme worked best only for (9), it is implied that the scheme is (recommendable, not recommendable) for the case where the integrand function excluding the multiplying term e^{-x} is far from being approximated by a polynomial.

  (iii) Note that

  - The integrals (3) and (4) can be converted into each other by a variable substitution of $x = u^{-1}$, $dx = -u^{-2} du$. The integrals (5) and (6) have the same relationship.
  - The integrals (7) and (8) can be converted into each other by a variable substitution of $u = e^{-x}$, $dx = -u^{-1} du$.

  From the results for (3) to (8), it can be conjectured that the numerical integration may work (better, worse) if the integration interval is changed from $[1, \infty)$ into $(0, 1]$ through the substitution of variable like

  $$x = u^{-n}, \quad dx = -nu^{-(n+1)} du \quad \text{or} \quad u = e^{-nx}, \quad dx = -(nu)^{-1} du$$
  (P5.10.1)

Table P5.10 Relative error results of suing various numerical integration methods.

	Simpson (N=10^4)	Adaptive (tol=10^{-6})	Gauss (N=10)	Quad (tol=10^{-6})	Quadl (tol=10^{-6})
(1)	1.9984×10^{-15}		1.1102e−16		7.5719×10^{-11}
(2)		2.8955×10^{-8}		1.5343×10^{-6}	
(3)	9.7850×10^{-2} ($a = 10^{-4}$)		1.2713×10^{-1}		2.2352×10^{-2}
(4), $b = 10^4$		9.7940×10^{-2}		9.7939×10^{-2}	
(5)	1.2702×10^{-2} ($a = 10^{-4}$)		3.5782×10^{-2}		2.6443×10^{-7}
(6), $b = 10^3$		4.0250×10^{-2}		4.0250×10^{-2}	
(7)	6.8678×10^{-5}		5.1077×10^{-4}		3.1781×10^{-7}
(8), $b = 10$		1.6951×10^{-4}		1.7392×10^{-4}	
(9), $b = 10$	7.8276×10^{-4}		2.3916×10^{-3}		7.8276×10^{-4}

5.11 The Bit Error Rate (BER) Curve of Communication with Multidimensional Signaling

For a communication system with multidimensional (orthogonal) signaling, the BER, i.e. the probability of bit error is derived as

$$P_e = \frac{2^{b-1}}{2^b - 1}\left\{1 - \frac{1}{\sqrt{\pi}}\int_{-\infty}^{\infty}(Q^{M-1}(-\sqrt{2}x - \sqrt{b\text{SNR}}))e^{-x^2}\,dy\right\} \quad (P5.11.1)$$

where b is the number of bits, $M = 2^b$ is the number of orthogonal waveforms, SNR is the signal-to-noise-ratio, and $Q(\cdot)$ is the complementary error function defined by

$$Q(x) = \frac{1}{\sqrt{2\pi}}\int_x^{\infty} e^{-y^2/2}\,dy \quad (P5.11.2)$$

We want to plot the BER curves for SNR = 0 : 10 dB and $b = 1 : 4$.

(a) Complete the following script "nm05p11.m", whose objective is to compute the values of $P_e(\text{SNR},b)$ for SNR = 0 : 10 dB and $b = 1 : 4$ by using the function 'Gauss_Hermite()' (Problem 5.6) and also by using the MATLAB

```
%nm05p11.m: plots the probability of bit error versus SNRdB
clear, clf
Q=@(x)erfc(x/sqrt(2))/2;
f=@(x,SNR,b)Q(-sqrt(2)*x-sqrt(b*SNR)).^(2^b-1);
fex2=@(x,SNR,b)f(x,SNR,b).*exp(-x.^2);
SNRdB=0:10; tol=1e-4; % SNR[dB] and tolerance used for 'quad'
for b=1:4
   tmp=2^(b-1)/(2^b-1); spi=sqrt(pi);
   for i=1:length(SNRdB),
      SNR=10^(SNRdB(i)/10);
      Pe(i)=tmp*(1-Gauss_Hermite(?,10,???,b)/spi);
      Pe1(i)=tmp*(1-quad(f???,-??,10,tol,[],SNR,?)/spi);
      Pe2(i)=tmp*(1-quad(f???,-1e3,???,tol,[],SNR,b)/spi);
   end
   semilogy(SNRdB,Pe,'ko',SNRdB,Pe1,'b+:',SNRdB,Pe2,'r.-'), hold on
end
```

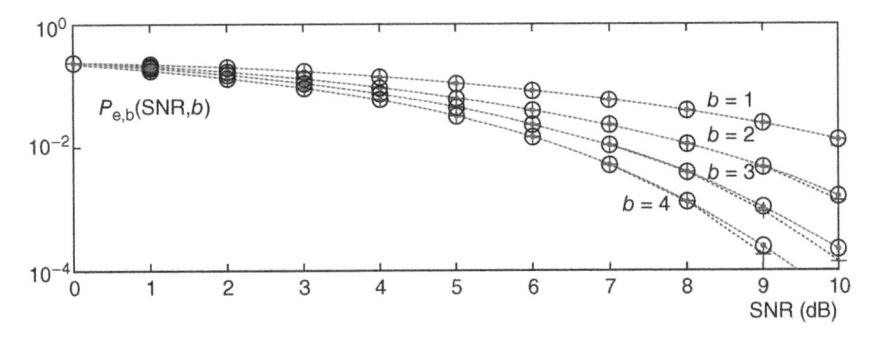

Figure P5.11 Bit error rate (BER) vs. SNR curves for multidimensional (orthogonal) signaling.

built-in function 'quad()' two times, once with the integration interval [−10, 10] and once with [−1000, 1000] instead of [−∞, ∞], and plot them vs. SNR (dB) =10log$_{10}$SNR. Then, run the script to get the BER curves like Figure P5.11.

(b) Of the two functions, which one is faster and which one presents us with more reliable values of the integral in Eq. (P5.11.1)?

5.12 Length of Curve/Arc – Superb Harmony of Numerical Derivative and Numerical Integral

The graph of a function $y = f(x)$ of a variable x is generally a curve and its length over the interval $[a, b]$ on the x-axis can be described by a line integral as

$$I = \int_a^b dl = \int_a^b \sqrt{dx^2 + dy^2} = \int_a^b \sqrt{1 + (dy/dx)^2}\, dx$$

$$= \int_a^b \sqrt{1 + (f'(x))^2}\, dx \qquad (P5.12.1)$$

For example, the length of the half circumference of a circle with the radius of unit length (Figure P5.14) can be obtained from this line integral with

$$y = f(x) = \sqrt{1 - x^2}, \quad a = -1, \quad b = 1 \qquad (P5.12.2)$$

```
%nm05p12.m
% To find the curve length
fp512=@(x)sqrt(max(1-x.^2,0));
a=-1; b=1; tol=1e-6; N=1000; M=20;
IT1=integral(@(x)sqrt(1+x.^2./(1-x.^2)),a,b) % True integral value =pi
hs=[1e-3 1e-4 1e-5]; % Step-size for the numerical derivative
format short e
df1dx=@(f,x,h)(f(x+h)-f(x-h))/h/2;
for n=1:numel(hs)+1
   if n<=numel(hs), h=hs(n), flength=@(x)sqrt(1+df1dx(fp512,x,h).^2);
      else flength=@(x)sqrt(1+x.^2./(1-x.^2)); % Using true derivative
   end
   h=hs(n)
   flength=@(x)sqrt(1+df1dx(fp512,x,?).^2); % Integrand (P5.12.1)
   err_Is= Smpsns(@(x)???????(fp512,x),a,b,N)-IT
   err_Ias= adapt_smpsn(flength,?,b,tol)-IT
   err_Iq= quad(flength,a,?,tol)-IT
   err_Iql= quadl(flength,a,b,???)-IT
   err_IGL =Gauss_Legendre(flength,a,b,?)-IT
   err_Iq2=f_length(fp512,a,b,h,tol)-IT
end
```

```
function length=f_length(f,a,b,h,tol,varargin)
if nargin<5, tol=1e-4; end
if nargin<4, h=1e-4; end
df1dx=@(f,x,h)(feval(f,x+h,varargin{:})-feval(f,x-h,varargin{:}))/h/2;
length=quadl(@(x)sqrt(1+df1dx(f,x,h).^2),a,b,tol);
```

PROBLEMS   **299**

Complete the following script "nm05p12.m" to use the numerical integration routines 'Smpsns()', 'adapt_smpsn()', 'quad()', 'quadl()', and 'Gauss_Legendre()' for evaluating the integral (P5.12.1 and P5.12.12) with the first derivative approximated by Eq. (5.1.8), where the parameters like the number of segments (N), the error tolerance (tol), and the number of grid points (M) are supposed to be as they are in the script. Run the script with the step-size $h = 0.001$, 0.0001, and 0.000 01 in the numerical derivative and fill in Table P5.12 with the errors of the results, noting that the true value of the half circumference of a unit circle is π.

Table P5.12 Results of using various numerical integration methods for (P5.12.1 and P5.12.12)/(P5.13.1 and P5.13.12).

	Step-size h	Simpson	Adaptive	Quad	Quadl	Gauss
(P5.12.1) and (P5.12.2)	0.001	4.6212e−2		2.9822e−2		8.4103e−2
	0.0001		9.4278e−3		9.4277e−3	
	0.00001	2.1853e−1		2.9858e−3		8.4937e−2
(P5.13.1) and (P5.13.2)	0.001		1.2393e−5		1.3545e−05	
	0.0001	8.3626e−3		5.0315e−6		6.4849e−6
	0.00001		1.4489e−09		8.8255e−7	
(P5.14.1)	N/A	1.7764e−15		0		8.8818e−16

5.13 Surface Area of Revolutionary 3D (Cubic) Object

The upper/lower surface area of a three-dimensional (3D) structure formed by one revolution of the graph (curve) of a function $y = f(x)$ around the x-axis over the interval $[a,b]$ can be described by the following integral.

$$I = 2\pi y \int_a^b dl = 2\pi \int_a^b f(x)\sqrt{1 + (f'(x))^2}\, dx \qquad (P5.13.1)$$

For example, the surface area of a sphere with the radius of unit length can be obtained from this equation with

$$y = f(x) = \sqrt{1 - x^2}, \quad a = -1, \quad b = 1 \qquad (P5.13.2)$$

Starting from the script "nm05p12.m", make a script "nm05p13.m" to use the numerical integration routines 'Smpsns()' (with the number of segments N = 1000), 'adapt_smpsn()', 'quad()', 'quadl()' (with the error tolerance tol = 10^{-6}), and 'Gauss_Legendre()' (with the number of grid points M=20) for evaluating the integral (P5.13.1,2) with the first derivative approximated by Eq. (5.1.8), where the parameters like the number of segments(N), the error tolerance(tol), and the number of grid points(M) are supposed to be as they are in the script. Run the script with the step-size $h = 0.001$, 0.0001, and 0.000 01 in the numerical derivative and fill in Table P5.12 with the errors of the results, noting that the true value of the surface area of a unit sphere is 4π.

```
%nm05p13.m: To find the surface area
% .. .. .. .. .. .. .. .. .. .. ..
IT1=2*pi*integral(@(x)sqrt((1-x.^2).*(1+x.^2./(1-x.^2))),a,b)
  fsarea=@(x)2*pi*?????(x).*sqrt(1+?????(fp512,?,h).^2);
% .. .. .. .. .. .. .. .. .. .. ..
  err_Iq2=f_sarea(fp512,a,b,h,tol)-IT
% .. .. .. .. .. .. .. .. .. .. ..
```

```
function sarea=f_sarea(f,a,b,h,tol,varargin)
% To find the surface area of the revolutionary 3D object
% formed by one revolution of the graph of a function f(x)
if nargin<5, tol=1e-6; end
if nargin<4, h=1e-4; end
df1dx=@(f,x)(feval(f,x+h,varargin{:})-feval(f,x-h,varargin{:}))/h/2;
sarea=quad(@(x)feval(f,x,varargin{:}).*sqrt(1+df1dx(f,x).^2),a,b,tol);
sarea=sarea*2*pi;
```

5.14 Volume of Revolutionary 3D (Cubic) Object

The volume of a 3D structure formed by one revolution of the graph (curve) of a function $y = f(x)$ around the x-axis over the interval $[a, b]$ can be described by the following integral.

$$I = 2\pi \int_a^b f^2(x)\,dx \qquad (P5.14.1)$$

For example, the volume of a sphere with the radius of unit length can be obtained from this equation with Eq. (P5.12.2).

Starting from the script "nm05p12.m", make a script "nm05p14.m", which uses the numerical integration routines 'Smpsns()' (with the number of segments $N = 100$), 'adapt_smpsn()', 'quad()', 'quadl()' (with the error tolerance $tol = 10^{-6}$), and 'Gauss_Legendre()' (with the number of grid points $M = 2$) to evaluate the integral (P5.14.1). Run the script and fill in Table P5.12 with the errors of the results, noting that the volume of a unit sphere is $4\pi/3$ (Figure P5.14).

```
%nm05p14.m: To find the surface area
% .. .. .. .. .. .. .. .. .. .. ..
IT1 = pi*integral(@(x)1-x.^2,a,b) % True value of integral
fvolume=@(x)pi*fp512(x).^?;
% .. .. .. .. .. .. .. .. .. .. ..
err_Iq2=f_volume(fp512,a,b,tol)-IT
% .. .. .. .. .. .. .. .. .. .. ..
```

```
function volume=f_volume(f,a,b,tol,varargin)
% To find the volume of the revolutionary 3D object
% formed by one revolution of the graph of a function f(x)
if nargin<4, tol=1e-6; end
volume=pi*quad(@(x)feval(f,x,varargin{:}).^2,a,b,tol);
```

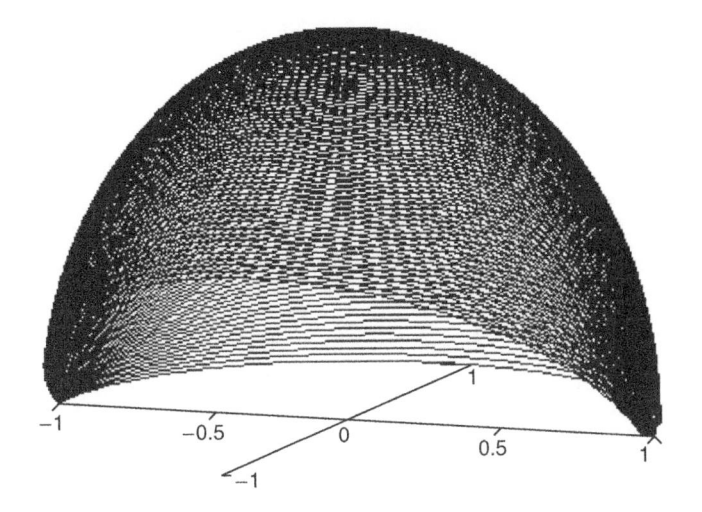

Figure P5.14 A unit sphere.

5.15 Double Integral

(a) Consider the following double integral

$$I = \int_0^2 \int_0^\pi y \sin x \, dx \, dy = \int_0^2 -y \cos x \big|_0^\pi \, dy = \int_0^2 2y \, dy = y^2 \big|_0^2 = 4 \quad \text{(P5.15.1)}$$

Use the routine 'int2s()' (Section 5.10) (with M=N=20, M=N=100, and M=N=200) and the MATLAB built-in routine 'dblquad()' to compute this double integral. Fill in Table P5.15.1 with the results and the times measured by using the commands tic/toc to be taken for carrying out each computation. Based on the results listed in Table P5.15.1, can we say that the numerical error becomes smaller as we increase the numbers (M,N) of segments along the x-axis and y-axis for the routine 'int2s()'?

(b) Consider the following double integral

$$I = \int_0^1 \int_0^1 \frac{1}{1-xy} dx \, dy = \frac{\pi^2}{6} \quad \text{(P5.15.2)}$$

Table P5.15.1 Results of using 'int2s()' and "dblquad()' for the integral (P5.15.1).

	int2s(),M=N=20	int2s(),M=N=100	int2s(),M=N=200	dblquad()
|error|		2.1649×10^{-8}		1.3250×10^{-8}
time				

NUMERICAL DIFFERENTIATION/INTEGRATION

```
%nm05p15.m
% Double integral
fp515a=@(x,y)y.*???(x);
x=[0:0.1:pi]; y=[0:0.1:2]; [X,Y]= meshgrid(x,y); Z=fp515a(X,Y);
subplot(221), mesh(x,y,Z)
N=[20 100 200];
for i=1:numel(N)
   tic, DI(i)=int2s(fp515a,0,pi,0,2,N(i),N(i)); times1(i)=toc;
end
tic, DI(i+1)= dblquad(fp515a,0,??,?,2); times1(i+1)=toc;
format short e, err1=DI-4, format short, times1
fp515b=@(x,y)1./(1-?.*?);
x=[0:0.001:0.999]; y=[0:0.001:0.999];
[X,Y]= meshgrid(x,y); Z= fp515b(X,Y);
subplot(222), mesh(x,y,Z)
d=[0.999 0.9999 0.99999 0.999999]; N=2000;
for i=1:numel(d)
   tic, I(1)=int2s(fp515b,0,?,0,d(i),N,N); times2(1,i)=toc;
   tic, I(2)=dblquad(fp515b,?,1,0,d(i),1e-4); times2(2,i)=toc;
   err2(:,i)=I'-pi^2/6;
end
err2, times2
N=[1000 2000 5000];
for i=1:3
   tic, I= int2s(fp515b,0,1,?,.9999,N(i),N(i)); times3(i)=toc;
   err3(i)=I-pi^2/6;
end
err3, times3
```

Noting that the integrand function is singular at $(x, y) = (1, 1)$, use the routine 'int2s()' and the MATLAB built-in routine 'dblquad()' with the upper limit (d) of the integration interval along the y-axis d = 0.999, 0.9999, 0.99999, and 0.999999 to compute this double integral. Fill in Tables P5.15.2 and P5.15.3 with the results and the times measured by using the commands tic/toc to be taken for carrying out each computation. Based on the results listed in Tables P5.15.2 and P5.15.3, answer the following questions.

Table P5.15.2 Results of using 'int2s()' and 'dblquad()' for the integral (P5.15.2).

		a=0, b=1 c=0, d=1-10-3	a=0, b=1 c=0, d=1-10-4	a=0, b=1 c=0, d=1-10-5	a=0, b=1 c=0, d=1-10-6
int2s() M=2000 N=2000	|error|	0.0079		0.0024	
	time				
dblquad	|error|		0.0004		0.0006
	time				

Table P5.15.3 Results of using the double integration routine 'int2s()' for (P5.15.2).

		M=1000,N=1000	M=2000,N=2000	M=5000,N=5000
int2s()	lerrorl	0.0003		
a=0,b=1,c=0, d=1-10^{-4}	time			

(i) Can we say that the numerical error becomes smaller as we set the upper limit (d) of the integration interval along the y-axis closer to the true limit 1?

(ii) Can we say that the numerical error becomes smaller as we increase the numbers (M,N) of segments along the x-axis and y-axis for the routine 'int2s()'? If this is contrary to the case of (a), can you blame the weird shape of the integrand function in Eq. (P5.15.2) for such a mess-up?

(cf) Note that the computation times listed in l.l Tables P5.15.1 to P5.15.3 may vary with the speed of CPU as well as the computational jobs that are concurrently processed by the CPU. To get reliable computation times, take the average over, say, 10 trials.

5.16 Area of a Triangle

Consider how to find the area between the graph (curve) of a function $f(x)$ and the x-axis. For example, let $f(x)=x$ for $0 \le x \le 1$ in order to find the area of a right-angled triangle with two equal sides of unit length. We might use either the one-dimensional (1D) integration or the two-dimensional (2D) integration, i.e. the double integral for this job.

(a) Use any integration method that you like best to evaluate the integral

$$I_1 = \int_0^1 x\, dx = \frac{1}{2} \quad \text{(P5.16.1)}$$

(b) Use any double integration routine that you like best to evaluate the integral

$$I_2 = \int_0^1 \int_0^{f(x)} 1\, dy\, dx = \int_0^1 \int_0^x 1\, dy\, dx \quad \text{(P5.16.2)}$$

You may get puzzled with some problem when applying the routine 'int2s()' if you define the integrand function as

```
»fp516b=@(x,y)1;
```

It is because this function, being called inside the routine 'smpsns_fxy()', yields just a scalar output even for the vector-valued input argument. There are two remedies for this problem. One is to define the integrand function in such a way that it can generate the output of the same dimension as the input.

```
»fp516b=@(x,y)1+0*(x+y);
```

But this will cause a waste of computation time due to the dead multiplication for each element of the input arguments x and y. The other is to modify the routine 'Smpsns_fxy()' in such a way that it can avoid the vector operation. More specifically, you can replace some part of the routine with the following. But this remedy also increases the computation time due to the abandonment of vector operation, which takes less time than scalar operation (see Section 1.3).

```
function INTf=Smpsns_fxy(f,x,c,d,N)
% .. .. .. .. .. .. .. .. .. ..
sum_odd=f(x,y(2));  sum_even=0;
for n=4:2:N
   sum_odd=sum_odd+f(x,y(n)); sum_even=sum_even+f(x,y(n-1));
end
INTf=(f(x,y(1))+f(x,y(N+1))+4*sum_odd+2*sum_even)*h/3;
% .. .. .. .. .. .. .. .. .. ..
```

(cf) This problem illustrates that we must be provident to use the vector operation, especially in defining a MATLAB function.

5.17 Volume of a Cone

Likewise in Section 5.10, modify the script "nm0510.m" so that it uses the routines 'int2s()' and 'dblquad()' to compute the volume of a cone that has a unit circle as its base side and a unit height, and run it to obtain the values of the volume up to four digits below the decimal point.

6

ORDINARY DIFFERENTIAL EQUATIONS

CHAPTER OUTLINE

6.1 Euler's Method	306
6.2 Heun's Method – Trapezoidal Method	309
6.3 Runge-Kutta Method	310
6.4 Predictor-Corrector Method	312
6.4.1 Adams-Bashforth-Moulton Method	312
6.4.2 Hamming Method	316
6.4.3 Comparison of Methods	317
6.5 Vector Differential Equations	320
6.5.1 State Equation	320
6.5.2 Discretization of LTI State Equation	324
6.5.3 High-order Differential Equation to State Equation	327
6.5.4 Stiff Equation	328
6.6 Boundary Value Problem (BVP)	333
6.6.1 Shooting Method	333
6.6.2 Finite Difference Method	336
Problems	341

Applied Numerical Methods Using MATLAB®, Second Edition. Won Y. Yang, Jaekwon Kim, Kyung W. Park, Donghyun Baek, Sungjoon Lim, Jingon Joung, Suhyun Park, Han L. Lee, Woo June Choi, and Taeho Im.
© 2020 John Wiley & Sons, Inc. Published 2020 by John Wiley & Sons, Inc.
Companion website: www.wiley.com/go/yang/appliednumericalmethods

306 ORDINARY DIFFERENTIAL EQUATIONS

Differential equations (DEs) are mathematical descriptions of how the variables and their derivatives (rates of change) with respect to one or more independent variable affect each other in a dynamical way. Their solutions show us how the dependent variable(s) will change with the independent variable(s). Many problems in natural sciences and engineering fields are formulated into a scalar differential equation or a vector differential equation, i.e. a system of differential equations.

In this chapter, we look into several methods of obtaining the numerical solutions to ordinary differential equations (ODEs) in which all dependent variables (x) depend on a single independent variable (t). First, the initial value problems (IVPs) will be handled with several methods including Runge-Kutta method and predictor-corrector methods in Sections 6.1 to 6.5. The last section (Section 6.6) will introduce the shooting method and the finite difference method for solving the two-point boundary value problem (BVP). ODEs are called an IVP if the values $x(t_0)$ of dependent variables are given at the initial point t_0 of the independent variable, while they are called a BVP if the values $x(t_0)/x(t_f)$ are given at the initial/final points t_0 and t_f.

6.1 EULER'S METHOD

When talking about the numerical solutions to ODEs, everyone starts with the Euler's method, since it is easy to understand and simple to program. Even though its low accuracy keeps it from being widely used for solving ODEs, it gives us a clue to the basic concept of numerical solution for a DE simply and clearly. Let us consider a first-order DE

$$y'(t) + a\, y(t) = r \quad \text{with } y(0) = y_0 \tag{6.1.1}$$

It has the following form of analytical solution

$$y(t) = \left(y_0 - \frac{r}{a}\right) e^{-at} + \frac{r}{a} \tag{6.1.2}$$

which can be obtained by using a conventional method or the Laplace transform technique. However, such a nice analytical solution does not exist for every DE and even if it exists, it is not easy to find even by using a computer equipped with the capability of symbolic computation. That is why we should study the numerical solutions to DEs.

Then, how do we translate the DE into a form that can easily be handled by computer? First of all, we have to replace the derivative $y'(t) = dy/dt$ in the differential equation by a numerical derivative (introduced in Chapter 5), where the step-size h is determined based on the accuracy requirements and the computation time constraints. Euler's method approximates the derivative in Eq. (6.1.1) with Eq. (5.1.2) as

$$\frac{y(t+h)-y(t)}{h} + ay(t) = r; \quad y(t+h) = (1-ah)y(t) + hr \quad \text{with} \quad y(0) = y_0$$
(6.1.3)

and solves this difference equation step-by-step with increasing t by h each time from $t = 0$.

$$y(h) = (1-ah)y(0) + hr = (1-ah)y_0 + hr$$

$$y(2h) = (1-ah)y(h) + hr = (1-ah)^2 y_0 + (1-ah)hr + hr$$

$$y(3h) = (1-ah)y(2h) + hr = (1-ah)^3 y_0 + \sum_{m=0}^{2}(1-ah)^m hr \quad (6.1.4)$$

This is a numeric sequence $\{y(kh)\}$, which we call a numerical solution of Eq. (6.1.1).

To be specific, let us define the parameters and the initial value of Eq. (6.1.1) be $a = 1$, $r = 1$, and $y_0 = 0$. Then, the analytical solution (6.1.2) becomes

$$y(t) = 1 - e^{-at} \quad (6.1.5)$$

and the numerical solution (6.1.4) with the step-size $h = 0.5$ and 0.25 are as listed in Table 6.1 and depicted in Figure 6.1. We make a MATLAB script "nm0601.m", which uses Euler's method for the DE (6.1.1), actually solving the DE (6.1.3) and plots the graphs of the numerical solutions in Figure 6.1. The graphs seem to tell us that a small step-size helps reduce the error so as to make the numerical solution closer to the (true) analytical solution. But, as will be investigated thoroughly in Section 6.2, it is only partially true. In fact, a too

```
%nm0601.m: Euler method to solve a first-order DE
clear, clf
a=1; r=1; y0=0;   tf=2;
t=[0:0.01:tf]; yt=1-exp(-a*t); % Eq.(6.1.5): true analytical solution
plot(t,yt,'k'),  hold on
klast=[8 4 2];   h(1:3)=tf./klast(1:3);
y(1)=y0;
for itr=1:length(klast) % with various step-size h=1/8,1/4,1/2
    for k=1:klast(itr)
        y(k+1)=(1-a*h(itr))*y(k) +h(itr)*r; % Eq.(6.1.3):
        plot([k-1 k]*h(itr),[y(k) y(k+1)],'b', k*h(itr),y(k+1),'ro')
        if k<4, pause; end
    end
end
```

Table 6.1 A numerical solution of the DE (6.1.1) obtained by the Euler's method.

t	$h = 0.5$	$h = 0.25$
0.25		$y(0.25) = (1-ah)y_0 + hr = 1/4 = 0.25$
0.50	$y(0.50) = (1-ah)y_0 + hr = 1/2 = 0.5$	$y(0.50) = (3/4)y(0.25) + 1/4 = 0.4375$
0.75		$y(0.75) = (3/4)y(0.50) + 1/4 = 0.5781$
1.00	$y(1.00) = (1/2)y(0.5) + 1/2 = 3/4 = 0.75$	$y(1.00) = (3/4)y(0.75) + 1/4 = 0.6836$
1.25		$y(1.25) = (3/4)y(1.00) + 1/4 = 0.7627$
1.50	$y(1.50) = (1/2)y(1.0) + 1/2 = 7/8 = 0.875$	$y(1.50) = (3/4)y(1.25) + 1/4 = 0.8220$
⋮	⋮	⋮

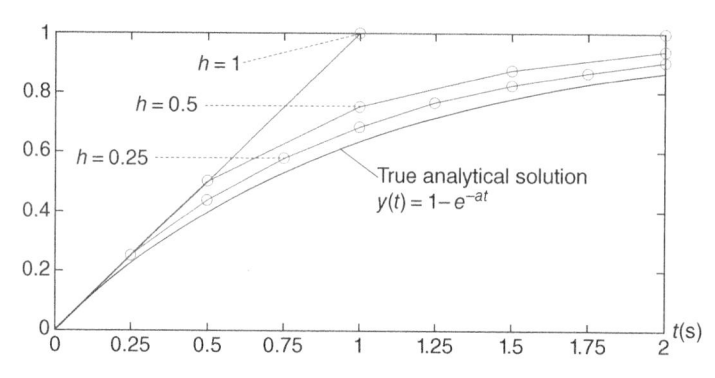

Figure 6.1 Numerical solutions of the DE $y'(t) + y(t) = 1$ with $y(0) = 0$ obtained by the Euler's method.

small step-size not only makes the computation time longer (proportional as $1/h$) but also results in rather larger errors due to the accumulated round-off effect. This is why we should look for other methods to decrease the errors rather than simply reduce the step-size.

Euler's method can also be applied for solving a first-order vector DE

```
function [t,y]=ode_Euler(f,tspan,y0,N)
%Euler's method to solve vector DE y'(t)=f(t,y(t))
% for tspan=[t0,tf] and with the initial value y0 and N time steps
if nargin<4|N<=0, N=100; end
if nargin<3, y0=0; end
h=(tspan(2)-tspan(1))/N; % Step-size
t=tspan(1)+[0:N]'*h; % Time vector
y(1,:)=y0(:)'; % always make the initial value a row vector
for k=1:N
    y(k+1,:)= y(k,:) +h*feval(f,t(k),y(k,:)); % Eq. (6.1.7)
end
```

$$\mathbf{y}'(t) = \mathbf{f}(t,\mathbf{y}) \quad \text{with} \quad \mathbf{y}(t_0) = \mathbf{y}_0 \qquad (6.1.6)$$

which is equivalent to a high-order scalar DE. The algorithm can be described by

$$\mathbf{y}_{k+1} = \mathbf{y}_k + h\mathbf{f}(t_k,\mathbf{y}_k) \quad \text{with} \quad \mathbf{y}(t_0) = \mathbf{y}_0 \qquad (6.1.7)$$

and has been cast into the MATLAB function 'ode_Euler()'.

6.2 HEUN'S METHOD – TRAPEZOIDAL METHOD

Another method of solving a first-order vector DE like Eq. (6.1.6) comes from integrating both sides (numerically).

$$\mathbf{y}'(t) = \mathbf{f}(t,\mathbf{y}); \quad \mathbf{y}(t)\big|_{t_k}^{t_{k+1}} = \mathbf{y}(t_{k+1}) - \mathbf{y}(t_k) = \int_{t_k}^{t_{k+1}} \mathbf{f}(t,\mathbf{y})\,dt$$

$$; \quad \mathbf{y}(t_{k+1}) = \mathbf{y}(t_k) + \int_{t_k}^{t_{k+1}} \mathbf{f}(t,\mathbf{y})\,dt \quad \text{with} \quad \mathbf{y}(t_0) = \mathbf{y}_0 \qquad (6.2.1)$$

If we assume that the value of the (derivative) function $\mathbf{f}(t,\mathbf{y})$ is constant as $\mathbf{f}(t_k,\mathbf{y}(t_k))$ within one time step $[t_k,t_{k+1}]$, this becomes Eq. (6.1.7) (with $h = t_{k+1} - t_k$), amounting to Euler's method. If we use the trapezoidal rule (5.5.3), it becomes

$$\mathbf{y}_{k+1} = \mathbf{y}_k + \frac{h}{2}\{\mathbf{f}(t_k,\mathbf{y}_k) + \mathbf{f}(t_{k+1},\mathbf{y}_{k+1})\} \qquad (6.2.2)$$

But the RHS of this equation has $\mathbf{y}_{k+1}$ which is unknown at t_k. To resolve this problem, we replace the $\mathbf{y}_{k+1}$ on the RHS by the following approximation.

$$\mathbf{y}_{k+1} \cong \mathbf{y}_k + h\,\mathbf{f}(t_k,\mathbf{y}_k) \qquad (6.2.3)$$

so that it becomes

$$\mathbf{y}_{k+1} = \mathbf{y}_k + \frac{h}{2}\{\mathbf{f}(t_k,\mathbf{y}_k) + \mathbf{f}(t_{k+1},\mathbf{y}_k + h\,\mathbf{f}(t_k,\mathbf{y}_k))\} \qquad (6.2.4)$$

This is the Heun's method, which is implemented in the MATLAB function 'ode_Heun()'. It is a kind of predictor-and-corrector method in that it predicts the value of $\mathbf{y}_{k+1}$ by Eq. (6.2.3) at t_k and then corrects the predicted value by Eq. (6.2.4) at t_{k+1}. The truncation error of Heun's method is $O(h^2)$ (proportional to h^2) as shown in Eq. (5.6.1), while the error of Euler's method is $O(h)$.

```
function [t,y]=ode_Heun(f,tspan,y0,N)
%Heun method to solve vector DE y'(t)=f(t,y(t))
% for tspan=[t0,tf] and with the initial value y0 and N time steps
if nargin<4|N<=0, N=100; end
if nargin<3, y0=0; end
h=(tspan(2)-tspan(1))/N; % Step-size
t=tspan(1)+[0:N]'*h; % Time vector
y(1,:)=y0(:)'; % always make the initial value a row vector
for k=1:N
   fk= feval(f,t(k),y(k,:)); y(k+1,:)= y(k,:)+h*fk; % Eq.(6.2.3)
   y(k+1,:)= y(k,:) +h/2*(fk +feval(f,t(k+1),y(k+1,:))); % Eq.(6.2.4)
end
```

6.3 RUNGE-KUTTA METHOD

Although the Heun's method is a little better than the Euler's method, it is still not accurate enough for most real-world problems. The fourth-order Runge-Kutta (RK4) method having a truncation error of $O(h^4)$ is one of the most widely used methods for solving differential equations. Its algorithm is described as follows:

$$\mathbf{y}_{k+1} = \mathbf{y}_k + \frac{h}{6}(\mathbf{f}_{k1} + 2\mathbf{f}_{k2} + 2\mathbf{f}_{k3} + \mathbf{f}_{k4}) \quad (6.3.1)$$

where

$$\mathbf{f}_{k1} = \mathbf{f}(t_k, \mathbf{y}_k) \quad (6.3.2a)$$

$$\mathbf{f}_{k2} = \mathbf{f}\left(t_k + \frac{h}{2}, \mathbf{y}_k + \mathbf{f}_{k1}\frac{h}{2}\right) \quad (6.3.2b)$$

$$\mathbf{f}_{k3} = \mathbf{f}\left(t_k + \frac{h}{2}, \mathbf{y}_k + \mathbf{f}_{k2}\frac{h}{2}\right) \quad (6.3.2c)$$

$$\mathbf{f}_{k4} = \mathbf{f}(t_k + h, \mathbf{y}_k + \mathbf{f}_{k3}h) \quad (6.3.2d)$$

Equation (6.3.1) is the core of RK4 method, which may be obtained by substituting Simpson's rule (5.5.4)

$$\int_{t_k}^{t_{k+1}} f(x)\,dx \cong \frac{h'}{3}(f_k + 4f_{k+1/2} + f_{k+1}) \text{ with } h' = \frac{x_{k+1} - x_k}{2} = \frac{h}{2} \quad (6.3.3)$$

into the integral form (6.3.1) of differential equation and replacing $f_{k+1/2}$ with the average of the successive function values $(f_{k2}+f_{k3})/2$. Accordingly, the RK4 method has a truncation error of $O(h^4)$ as Eq. (5.6.2) and so is expected to work better than the previous two methods.

```
function [t,y]=ode_RK4(f,tspan,y0,N,varargin)
%Runge-Kutta method to solve vector DE y'(t)=f(t,y(t))
% for tspan=[t0,tf] and with the initial value y0 and N time steps
if nargin<4|N<=0, N=100; end
if nargin<3, y0=0; end
y(1,:)=y0(:)'; %make it a row vector
h=(tspan(2)-tspan(1))/N; t=tspan(1)+[0:N]'*h;
for k=1:N
   f1=h*feval(f,t(k),y(k,:),varargin{:}); f1=f1(:)'; % Eq.(6.3.2a)
   f2=h*feval(f,t(k)+h/2,y(k,:)+f1/2,varargin{:}); f2=f2(:)';%Eq.(6.3.2b)
   f3=h*feval(f,t(k)+h/2,y(k,:)+f2/2,varargin{:}); f3=f3(:)';%Eq.(6.3.2c)
   f4=h*feval(f,t(k)+h,y(k,:)+f3,varargin{:}); f4=f4(:)'; % Eq.(6.3.2d)
   y(k+1,:)= y(k,:) +(f1+2*(f2+f3)+f4)/6; % Eq.(6.3.1)
end
```

```
%nm0603.m: Heun/Euer/RK4 method to solve a DE
clear, clf
tspan=[0 2];
t=tspan(1)+[0:100]*(tspan(2)-tspan(1))/100;
a=1; yt=1-exp(-a*t); % Eq.(6.1.5): true analytical solution
plot(t,yt,'k'), hold on
df61=@(t,y)-y+1; % Eq.(6.1.1): DE to be solved
y0=0; N=4;
[t1,ye]=oed_Euler(df61,tspan,y0,N);
[t1,yh]=ode_Heun(df61,tspan,y0,N);
[t1,yr]=ode_RK4(df61,tspan,y0,N);
plot(t,yt,'k', t1,ye,'r:', t1,yh,'g:', t1,yr,'b:')
plot(t1,ye,'ro', t1,yh,'g+', t1,yr,'b*')
N=1e3; %to estimate the time for N iterations
tic, [t1,ye]=ode_Euler(df61,tspan,y0,N); time_Euler=toc
tic, [t1,yh]=ode_Heun(df61,tspan,y0,N); time_Heun=toc
tic, [t1,yr]=ode_RK4(df61,tspan,y0,N); time_RK4=toc
```

The RK4 method has been cast into the MATLAB function 'ode_RK4()'. The script "nm0603.m" uses this function to solve Eq. (6.3.1) with the step-size $h = (t_f - t_0)/N = 2/4 = 0.5$ and plots the numerical result together with the (true) analytical solution. Comparison of this result with those of the Euler's method ('ode_Euler()') and the Heun's method ('ode_Heun()') is given in Figure 6.2, which shows that the RK4 method is better than the Heun's method, while the Euler's method is the worst in terms of accuracy with the same step-size. But, in terms of computational load, the order is reversed because the Euler's method, the Heun's method, and the RK4 method need 1, 2, and 4 function evaluations (calls) per iteration, respectively.

(cf) Note that a function call takes much more time than a multiplication and so the number of function calls should be a criterion in estimating and comparing computational time.

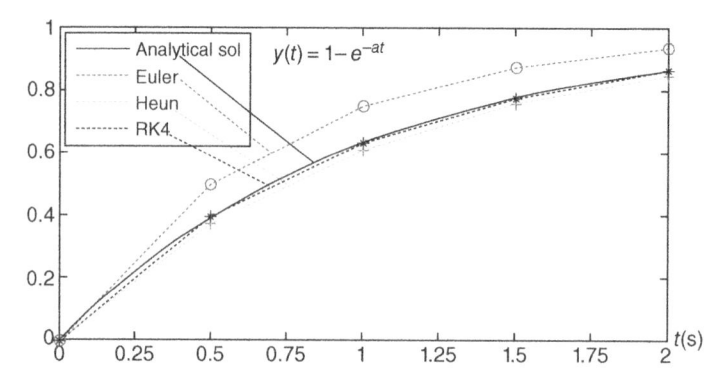

Figure 6.2 Numerical solutions of the DE $y'(t) + y(t) = 1$ with $y(0) = 0$.

The MATLAB built-in functions 'ode23()' and 'ode45()' implement the Runge-Kutta method with an adaptive step-size adjustment, which uses a large/small step-size depending on whether $f(t)$ is smooth or rough. In Section 6.4.3, we will try using these functions together with our functions to solve a differential equation for practice rather than for comparison.

6.4 PREDICTOR-CORRECTOR METHOD

6.4.1 Adams-Bashforth-Moulton Method

The Adams-Bashforth-Moulton (ABM) method consists of two steps. The first step is to approximate $\mathbf{f}(t,\mathbf{y})$ by the (Lagrange) polynomial of degree 3 matching the four points

$$\{(t_{k-3}, \mathbf{f}_{k-3}), (t_{k-2}, \mathbf{f}_{k-2}), (t_{k-1}, \mathbf{f}_{k-1}), (t_k, \mathbf{f}_k)\}$$

and substitute the polynomial into the integral form (6.2.1) of differential equation to get a predicted estimate of $\mathbf{y}_{k+1}$.

$$\mathbf{p}_{k+1} = \mathbf{y}_k + \int_0^h l_3(t)\,dt = \mathbf{y}_k + \frac{h}{24}(-9\mathbf{f}_{k-3} + 37\mathbf{f}_{k-2} - 59\mathbf{f}_{k-1} + 55\mathbf{f}_k) \quad (6.4.1a)$$

The second step is to repeat the same work with the updated four points

$$\{(t_{k-2}, \mathbf{f}_{k-2}), (t_{k-1}, \mathbf{f}_{k-1}), (t_k, \mathbf{f}_k), (t_{k+1}, \mathbf{f}_{k+1})\}\, (\mathbf{f}_{k+1} = \mathbf{f}(t_{k+1}, \mathbf{p}_{k+1}))$$

to get a corrected estimate of $\mathbf{y}_{k+1}$.

$$\mathbf{c}_{k+1} = \mathbf{y}_k + \int_0^h l_3'(t)\,dt = \mathbf{y}_k + \frac{h}{24}(\mathbf{f}_{k-2} - 5\mathbf{f}_{k-1} + 19\mathbf{f}_k + 9\mathbf{f}_{k+1}) \quad (6.4.1b)$$

The coefficients of Eqs. (6.4.1a) and (6.4.1b) can be obtained by using the MATLAB functions 'lagranp()' and 'polyint()', each of which generates Lagrange (coefficient) polynomials and integrates a polynomial, respectively. Let us try running the script "ABMc.m".

```
»abmc
   cAP =    -3/8     37/24    -59/24    55/24
   cAC =    1/24     -5/24    19/24     3/8
```

```
%ABMc.m
% Predictor/Corrector coefficients in Adams-Bashforth-Moulton method
clear
format rat
[l,L]=lagranp([-3 -2 -1 0],[0 0 0 0]); %only coefficient polynomial L
for m=1:4
    iL=polyint(L(m,:)); %indefinite integral of polynomial
    cAP(m)=polyval(iL,1)-polyval(iL,0); %definite integral over [0,1]
end
cAP %Predictor coefficients
[l,L]=lagranp([-2 -1 0 1],[0 0 0 0]); %only coefficient polynomial L
for m=1:4
    iL=polyint(L(m,:)); %indefinite integral of polynomial
    cAC(m)=polyval(iL,1)-polyval(iL,0); %definite integral over [0,1]
end
cAC %Corrector coefficients
format short
```

Alternatively, we write the Taylor series expansion of $\mathbf{y}_{k+1}$ about t_k and that of $\mathbf{y}_k$ about t_{k+1} as

$$\mathbf{y}_{k+1} = \mathbf{y}_k + h\mathbf{f}_k + \frac{h^2}{2}\mathbf{f}_k' + \frac{h^3}{3!}\mathbf{f}_k^{(2)} + \frac{h^4}{4!}\mathbf{f}_k^{(3)} + \frac{h^5}{5!}\mathbf{f}_k^{(4)} + \cdots \quad (6.4.2a)$$

$$\mathbf{y}_k = \mathbf{y}_{k+1} - h\mathbf{f}_{k+1} + \frac{h^2}{2}\mathbf{f}_{k+1}' - \frac{h^3}{3!}\mathbf{f}_{k+1}^{(2)} + \frac{h^4}{4!}\mathbf{f}_{k+1}^{(3)} - \frac{h^5}{5!}\mathbf{f}_{k+1}^{(4)} + \cdots;$$

$$\mathbf{y}_{k+1} = \mathbf{y}_k + h\mathbf{f}_{k+1} - \frac{h^2}{2}\mathbf{f}_{k+1}' + \frac{h^3}{3!}\mathbf{f}_{k+1}^{(2)} - \frac{h^4}{4!}\mathbf{f}_{k+1}^{(3)} + \frac{h^5}{5!}\mathbf{f}_{k+1}^{(4)} - \cdots \quad (6.4.2b)$$

and replace the first, second, and third-derivatives by their difference approximations.

$$\mathbf{y}_{k+1} = \mathbf{y}_k + h\mathbf{f}_k + \frac{h^2}{2}\left(\frac{-\frac{1}{3}\mathbf{f}_{k-3} + \frac{3}{2}\mathbf{f}_{k-2} - 3\mathbf{f}_{k-1} + \frac{11}{6}\mathbf{f}_k}{h} + \frac{1}{4}h^3\mathbf{f}_k^{(4)} + \cdots\right)$$

$$+ \frac{h^3}{3!}\left(\frac{-\mathbf{f}_{k-3} + 4\mathbf{f}_{k-2} - 5\mathbf{f}_{k-1} + 2\mathbf{f}_k}{h^2} + \frac{11}{12}h^2\mathbf{f}_k^{(4)} + \cdots\right)$$

$$+ \frac{h^4}{4!}\left(\frac{-\mathbf{f}_{k-3} + 3\mathbf{f}_{k-2} - 3\mathbf{f}_{k-1} + \mathbf{f}_k}{h^3} + \frac{3}{2}h\mathbf{f}_k^{(4)} + \cdots\right) + \frac{h^5}{120}\mathbf{f}_k^{(4)} + \cdots$$

$$= \mathbf{y}_k + \frac{h}{24}(-9\mathbf{f}_{k-3} + 37\mathbf{f}_{k-2} - 59\mathbf{f}_{k-1} + 55\mathbf{f}_k) + \frac{251}{720}h^5\mathbf{f}_k^{(4)} + \cdots$$

$$\stackrel{(6.4.1a)}{\approx} \mathbf{p}_{k+1} + \frac{251}{720}h^5\mathbf{f}_k^{(4)} \qquad (6.4.3a)$$

$$\mathbf{y}_{k+1} = \mathbf{y}_k + h\mathbf{f}_{k+1} - \frac{h^2}{2}\left(\frac{-\frac{1}{3}\mathbf{f}_{k-2} + \frac{3}{2}\mathbf{f}_{k-1} - 3\mathbf{f}_k + \frac{11}{6}\mathbf{f}_{k+1}}{h} + \frac{1}{4}h^3\mathbf{f}_{k+1}^{(4)} + \cdots\right)$$

$$+ \frac{h^3}{3!}\left(\frac{-\mathbf{f}_{k-2} + 4\mathbf{f}_{k-1} - 5\mathbf{f}_k + 2\mathbf{f}_{k+1}}{h^2} + \frac{11}{12}h^2\mathbf{f}_{k+1}^{(4)} + \cdots\right)$$

$$- \frac{h^4}{4!}\left(\frac{-\mathbf{f}_{k-2} + 3\mathbf{f}_{k-1} - 3\mathbf{f}_k + \mathbf{f}_{k+1}}{h^3} + \frac{3}{2}h\mathbf{f}_{k+1}^{(4)} + \cdots\right) + \frac{h^5}{120}\mathbf{f}_{k+1}^{(4)} + \cdots$$

$$= \mathbf{y}_k + \frac{h}{24}(\mathbf{f}_{k-2} - 5\mathbf{f}_{k-1} + 19\mathbf{f}_k + 9\mathbf{f}_{k+1}) - \frac{19}{720}h^5\mathbf{f}_{k+1}^{(4)} + \cdots$$

$$\stackrel{(6.4.1b)}{\approx} \mathbf{c}_{k+1} - \frac{19}{720}h^5\mathbf{f}_{k+1}^{(4)} \qquad (6.4.3b)$$

These derivations are supported by running the MATLAB script "ABMc1.m".

```
% ABMc1.m
% another way to get the ABM coefficients together with the error term
clear, format rat
for i=1:3, [ci,erri]=difapx(i,[-3 0]); c(i,:)=ci; err(i)=erri; end
cAP=[0 0 0 1]+[1/2 1/6 1/24]*c, errp=-[1/2 1/6 1/24]*err'+1/120
cAC=[0 0 0 1]+[-1/2 1/6 -1/24]*c, errc=-[-1/2 1/6 -1/24]*err'+1/120
```

From these equations and under the assumption that $\mathbf{f}_{k+1}^{(4)} \cong \mathbf{f}_k^{(4)} \cong K$, we can write the predictor/corrector errors as

$$E_{P,k+1} = \mathbf{y}_{k+1} - \mathbf{p}_{k+1} \approx \frac{251}{720}h^5\mathbf{f}_k^{(4)} \cong \frac{251}{720}Kh^5 \qquad (6.4.4a)$$

$$E_{C,k+1} = \mathbf{y}_{k+1} - \mathbf{c}_{k+1} \approx -\frac{19}{720}h^5\mathbf{f}_{k+1}^{(4)} \cong -\frac{19}{720}Kh^5 \qquad (6.4.4b)$$

We still cannot use these formulas to estimate the predictor/corrector errors, since K is unknown. But from the difference between these two formulas

$$E_{P,k+1} - E_{C,k+1} = \mathbf{c}_{k+1} - \mathbf{p}_{k+1} \cong \frac{270}{720} K h^5 \equiv \frac{270}{251} E_{P,k+1} \equiv -\frac{270}{19} E_{C,k+1}$$
(6.4.5)

we can get the practical formulas for estimating the errors as

$$E_{P,k+1} = \mathbf{y}_{k+1} - \mathbf{p}_{k+1} \cong \frac{251}{270}(\mathbf{c}_{k+1} - \mathbf{p}_{k+1})$$
(6.4.6a)

$$E_{C,k+1} = \mathbf{y}_{k+1} - \mathbf{c}_{k+1} \cong -\frac{19}{270}(\mathbf{c}_{k+1} - \mathbf{p}_{k+1})$$
(6.4.6b)

These formulas give us rough estimates of how close the predicted/corrected values are to the true value and so can be used to improve them as well as to adjust the step-size.

$$\mathbf{p}_{k+1} \rightarrow \mathbf{p}_{k+1} + \frac{251}{270}(\mathbf{c}_k - \mathbf{p}_k) \Rightarrow \mathbf{m}_{k+1}$$
(6.4.7a)

$$\mathbf{c}_{k+1} \rightarrow \mathbf{c}_{k+1} - \frac{19}{270}(\mathbf{c}_{k+1} - \mathbf{p}_{k+1}) \Rightarrow \mathbf{y}_{k+1}$$
(6.4.7b)

These modification formulas are expected to reward the efforts which we put into deriving them.

The ABM method with the modification formulas can be described by Eqs. (6.4.1a), (6.4.1b) and (6.4.7a), (6.4.7b) summarized below and has been cast into the MATLAB function 'ode_ABM()'. This scheme, having a truncation error of $O(h^5)$ (Eq. (6.4.3)), needs only two function evaluations (calls) per iteration and so is expected to work better than the methods discussed so far. It is implemented by the MATLAB built-in function 'ode113()' with many additional sophisticated techniques.

ADAMS-BASHFORTH-MOULTON METHOD WITH MODIFICATION FORMULAS

Predictor : $\mathbf{p}_{k+1} = \mathbf{y}_k + \dfrac{h}{24}(-9\mathbf{f}_{k-3} + 37\mathbf{f}_{k-2} - 59\mathbf{f}_{k-1} + 55\,\mathbf{f}_k)$ (6.4.8a)

Modifier : $\mathbf{m}_{k+1} = \mathbf{p}_{k+1} + \dfrac{251}{270}(\mathbf{c}_k - \mathbf{p}_k)$ (6.4.8b)

Corrector : $\mathbf{c}_{k+1} = \mathbf{y}_k + \dfrac{h}{24}\{\mathbf{f}_{k-2} - 5\mathbf{f}_{k-1} + 19\mathbf{f}_k + 9\mathbf{f}(t_{k+1}, \mathbf{m}_{k+1})\}$ (6.4.8c)

$\mathbf{y}_{k+1} = \mathbf{c}_{k+1} - \dfrac{19}{270}(\mathbf{c}_{k+1} - \mathbf{p}_{k+1})$ (6.4.8d)

```
function [t,y]=ode_ABM(f,tspan,y0,N,KC,varargin)
%Adams-Bashforth-Moulton method to solve vector DE y'(t)=f(t,y(t))
% for tspan=[t0,tf] and with the initial value y0 and N time steps
% using the modifier based on the error estimate depending on KC=1/0
if nargin<5, KC=1; end % with modifier by default
if nargin<4|N<=0, N=100; end % Default maximum number of iterations
y0=y0(:)'; % make it a row vector
h=(tspan(2)-tspan(1))/N;   tspan0=tspan(1)+[0 3]*h;
[t,y]=ode_RK4(f,tspan0,y0,3,varargin{:}); % Initialize by Runge-Kutta
t=[t(1:3)'  t(4):h:tspan(2)]';
for k=1:4, F(k,:)= feval(f,t(k),y(k,:),varargin{:}); end
p= y(4,:); c= y(4,:); KC22= KC*251/270; KC12= KC*19/270;
h24=h/24; h241=h24*[1 -5 19 9]; h249=h24*[-9 37 -59 55];
for k=4:N
  p1 = y(k,:) + h249*F; % Eq.(6.4.8a)
  m1 = pk1 + KC22*(c-p); % Eq.(6.4.8b)
  fk1 = feval(f,t(k+1),m1,varargin{:})];
  c1 = y(k,:) + h241*[F(2:4,:); fk1(:)']; % Eq.(6.4.8c)
  y(k+1,:) = c1 - KC12*(c1-p1); % Eq.(6.4.8d)
  p=p1;  c=c1; % update the predicted/corrected values
  fk11 = feval(f,t(k+1),y(k+1,:),varargin{:});
  F = [F(2:4,:); fk11(:)'];
end
```

6.4.2 Hamming Method

```
function [t,y]=ode_Ham(f,tspan,y0,N,KC,varargin)
% Hamming method to solve vector DE y'(t)=f(t,y(t))
% for tspan=[t0,tf] and with the initial value y0 and N time steps
% using the modifier based on the error estimate depending on KC=1/0
if nargin<5, KC=1; end % with modifier by default
if nargin<4|N<=0, N=100; end % Default maximum number of iterations
if nargin<3, y0=0; end % Default initial value
y0=y0(:)'; end % make it a row vector
h=(tspan(2)-tspan(1))/N;   tspan0=tspan(1)+[0 3]*h;
[t,y]= ode_RK4(f,tspan0,y0,3,varargin{:}); % Initialize by Runge-Kutta
t=[t(1:3)'  t(4):h:tspan(2)]';
for k=2:4, F(k-1,:)= feval(f,t(k),y(k,:),varargin{:}); end
p= y(4,:); c= y(4,:); h34=h/3*4; KC11=KC*112/121; KC91=KC*9/121;
h312= 3*h*[-1 2 1];
for k=4:N
  p1 = y(k-3,:)+h34*(2*(F(1,:)+F(3,:))-F(2,:)); % Eq.(6.4.9a)
  m1 = p1+KC11*(c-p);            % Eq.(6.4.9b)
  fk1 = feval(f,t(k+1),m1,varargin{:})];
  c1 = (-y(k-2,:)+9*y(k,:)+h312*[F(2:3,:); fk1(:)'])/8; % Eq.(6.4.9c)
  y(k+1,:) = c1 -KC91*(c1-p1);        % Eq.(6.4.9d)
  p=p1;  c=c1; %update the predicted/corrected values
  fk11 = feval(f,t(k+1),y(k+1,:),varargin{:});
  F = [F(2:3,:); fk11(:)'];
end
```

In this section, we introduce just the algorithm of the Hamming method[H-1] summarized in the box beneath and the corresponding function 'ode_Ham()', which is another multistep predictor-corrector method like the ABM method.

This scheme also needs only two function evaluations (calls) per iteration, while having the error of $O(h^5)$ and so is comparable with the ABM method discussed in the previous Section 6.4.1.

HAMMING METHOD WITH MODIFICATION FORMULAS

Predictor : $\mathbf{p}_{k+1} = \mathbf{y}_{k-3} + \dfrac{4h}{3}(2\mathbf{f}_{k-2} - \mathbf{f}_{k-1} + 2\mathbf{f}_k)$ (6.4.9a)

Modifier : $\mathbf{m}_{k+1} = \mathbf{p}_{k+1} + \dfrac{112}{121}(\mathbf{c}_k - \mathbf{p}_k)$ (6.4.9b)

Corrector : $\mathbf{c}_{k+1} = \dfrac{1}{8}\{9\mathbf{y}_k - \mathbf{y}_{k-2} + 3h(-\mathbf{f}_{k-1} + 2\mathbf{f}_k + \mathbf{f}(t_{k+1}, \mathbf{m}_{k+1}))\}$ (6.4.9c)

$\mathbf{y}_{k+1} = \mathbf{c}_{k+1} - \dfrac{9}{121}(\mathbf{c}_{k+1} - \mathbf{p}_{k+1})$ (6.4.9d)

6.4.3 Comparison of Methods

The major factors to be considered in evaluating/comparing different numerical methods are the accuracy of the numerical solution and its computation time. In this section, we will compare the functions 'ode_RK4()', 'ode_ABM()', 'ode_Ham()', 'ode23()', 'ode45()', and 'ode113()' by trying them out on the same differential equations, hopefully to make some conjectures about their performances. It is important to note that the evaluation/comparison of numerical methods is not so simple because their performances may depend on the characteristic of the problem at hand. It should also be noted that there are other factors to be considered, such as stability, versatility, proof against run-time error, etc. These points are being considered in most of the MATLAB built-in functions.

The first thing we are going to do is to validate the effectiveness of the modifiers (Eqs. ((6.4.8b) and (6.4.8d)) and ((6.4.9b) and (6.4.9d))) in the ABM method and the Hamming method. For this job, we write and run the script "nm0643_1.m" to get the results depicted in Figure 6.3 for the differential equation

$$y'(t) = -y(t) + 1 \quad \text{with} \quad y(0) = 0 \quad (6.4.10)$$

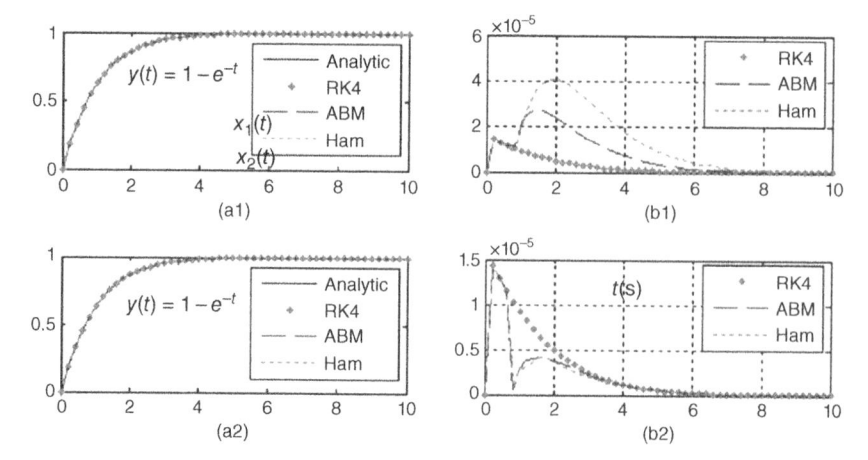

Figure 6.3 Numerical solutions and their errors for the DE $y'(t) = -y(t) + 1$. (a1) Numerical solutions without modifier, (a2) numerical solutions with modifier, (b1) relative errors without modifier, and (b2) relative errors with modifier.

which was given at the beginning of this chapter. Figure 6.3 shows us an interesting fact that although the ABM method and the Hamming method, even without modifiers, are theoretically expected to have better accuracy than the RK4 method, they turn out to work better than RK4 only with modifiers. Of course, it is not always the case, as illustrated in Figure 6.4, which we obtained by applying the same functions to solve another differential equation

$$y'(t) = y(t) + 1 \text{ with } y(0) = 0 \quad (6.4.11)$$

where the true analytical solution is

$$y(t) = e^t - 1 \quad (6.4.12)$$

Readers are invited to supplement the script "nm0643_2.m" in such a way that 'ode_Ham()' is also used to solve Eq. (6.4.11). Running the script yields the results depicted in Figure 6.4 and listed in Table 6.2. From Figure 6.4, it is noteworthy that, without the modifiers, the ABM method seems to be better than the Hamming method; however, with the modifiers, it is the other way around or at least they run a neck-and-neck race. Anyone will see that the predictor-corrector methods such as the ABM method ('ode_ABM()') and the Hamming method ('ode_Ham()') give us a better numerical solution with less error and shorter computation time than the MATLAB built-in functions 'ode23()', 'ode45()', and 'ode113()' as well as the RK4 method ('ode_RK4()'), as listed in Table 6.2. But a general conclusion should not be deduced just from one example.

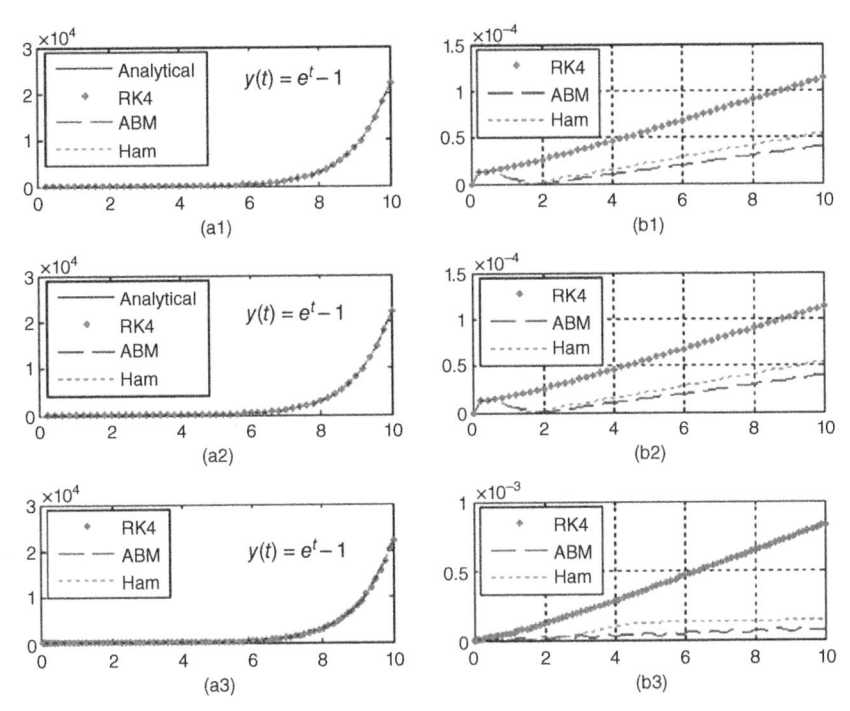

Figure 6.4 Numerical solutions and their errors for the DE $y'(t) = y(t) + 1$. (a1) Numerical solutions without modifier, (a2) numerical solutions with modifier, (a3) numerical solutions by ode23, ode45, ode113, (b1) relative errors without modifier, (b2) relative errors with modifier, and (b3) relative errors.

Plot labels: $y(t) = e^t - 1$

```
%nm0643_1.m: RK4/Adams/Hamming method to solve a DE
clear, clf
t0=0; tf=10; y0=0; % Starting/Final time, Initial value
N=50; % Number of segments
df643=@(t,y)-y+1; % DE to solve
f643=@(t)1-exp(-t); % True analytical solution
for KC=0:1
   tic, [t1,yR]= ode_RK4(df643,[t0 tf],y0,N); tR=toc
   tic, [t1,yA]= ode_ABM(df643,[t0 tf],y0,N,KC); tA=toc
   tic, [t1,yH]= ode_Ham(df643,[t0 tf],y0,N,KC); tH=toc
   yt1= f643(t1); % True analytical solution to plot
   subplot(221+KC*2) % plot analytical/numerical solutions
   plot(t1,yt1,'k', t1,yR,'r.', t1,yA,'b-', t1,yH,'m:')
   tmp= abs(yt1)+eps; l_t1= length(t1);
   eR=abs(yR-yt1)./tmp; e_R=norm(eR)/lt1
   eA=abs(yA-yt1)./tmp; e_A=norm(eA)/lt1
   eH=abs(yH-yt1)./tmp; e_H=norm(eH)/lt1
   subplot(222+KC*2) %plot relative errors
   plot(t1,eR,'r.', t1,eA,'b--', t1, eH,'m:')
end
```

Table 6.2 Results of applying several functions for solving a simple DE.

	ode_RK4()	ode_ABM()	ode_Ham()	ode23()	ode45()	ode113()
Relative error	0.0925×10^{-4}	0.0203×10^{-4}	0.0179×10^{-4}	0.4770×10^{-4}	0.0422×10^{-4}	0.1249×10^{-4}
Computing time (s)	0.05	0.03	0.03	0.07	0.05	0.05

```
%nm0643_2.m: ode23()/ode45()/ode113() to solve a DE
clear, clf
t0=0; tf=10; y0=0; N=50; % Starting/Final time, Initial value
df643=@(t,y)y+1; % DE to solve
f643=@(t)exp(t)-1; % True analytical solution
tic, [t1,yR]= ode_RK4(df643,[t0 tf],y0,N); time(1)=toc;
tic, [t1,yA]= ode_ABM(df643,[t0 tf],y0,N); time(2)=toc;
yt1= f643(t1);
tmp= abs(yt1)+eps; l_t1= length(t1);
eR=abs(yR-yt1)./tmp; err(1)=norm(eR)/l_t1;
eA=abs(yA-yt1)./tmp; err(2)=norm(eA)/l_t1;
options=odeset('RelTol',1e-4); %set the tolerance of relative error
tic, [t23,yode23]= ode23(df643,[t0 tf],y0,options); time(3)=toc;
tic, [t45,yode45]= ode45(df643,[t0 tf],y0,options); time(4)=toc;
tic, [t113,yode113]= ode113(df643,[t0 tf],y0,options); time(5)=toc;
yt23= f643(t23); tmp= abs(yt23)+eps;
eode23=abs(yode23-yt23)./tmp; err(3)=norm(eode23)/length(t23);
yt45= f643(t45); tmp= abs(yt45)+eps;
eode45=abs(yode45-yt45)./tmp; err(4)=norm(eode45)/length(t45);
yt113= f643(t113); tmp= abs(yt113)+eps;
eode113=abs(yode113-yt113)./tmp; err(5)=norm(eode113)/length(t113);
subplot(221), plot(t23,yode23,'r.', t45,yode45,'b--', t113,yode113,'m:')
subplot(222), plot(t23,eode23,'r.',t45,eode45,'b--', t113,eode113,'m:')
err, time
```

6.5 VECTOR DIFFERENTIAL EQUATIONS

6.5.1 State Equation

Although we have tried using the MATLAB functions only for scalar differential equations, all the functions made by us or built inside MATLAB are ready to entertain first-order vector differential equations, called state equations, as

$$x_1'(t) = f_1(t, x_1(t), x_2(t), \ldots) \quad \text{with} \quad x_1(t_0) = x_{10}$$
$$x_2'(t) = f_2(t, x_1(t), x_2(t), \ldots) \quad \text{with} \quad x_2(t_0) = x_{20}$$
$$\ldots\ldots\ldots\ldots\ldots\ldots\ldots\ldots\ldots\ldots\ldots\ldots\ldots\ldots\ldots\ldots\ldots\ldots$$
$$; \quad \mathbf{x}'(t) = \mathbf{f}(t, \mathbf{x}(t)) \quad \text{with} \quad \mathbf{x}(t_0) = \mathbf{x}_0 \quad (6.5.1)$$

For example, we can define the following system of first-order DE

VECTOR DIFFERENTIAL EQUATIONS

$$x_1'(t) = x_2(t) \text{ with } x_1(0) = 1 \qquad (6.5.2)$$
$$x_2'(t) = -x_2(t) + 1 \text{ with } x_2(0) = -1$$

in a file named "df651.m" and solve it by running the MATLAB script "nm0651_1.m", which uses the functions 'ode_Ham()'/'ode45()' to get the numerical solutions and plots the results as shown in Figure 6.5. Note that the function given as the first input argument of 'ode45()' must be fabricated to generate its value in a column vector or at least, in the same form of vector as the input argument 'x' so long as it is a vector-valued function.

```
%nm0651_1.m to solve a system of DEs, i.e., state equation
df='df651';
%df=@(t,x)[x(2); -x(2)+1]; % Alternative definition of the DE
t0=0; tf=2; tspan=[t0 tf]; % Start/Final time and Time span
x0=[1 -1]; N=45; % Initial value of x and Number of segments
[tH,xH]=ode_Ham(df,[t0 tf],x0,N); % with N=number of segments
[t45,x45]=ode45(df,[t0 tf],x0);
plot(tH,xH), hold on, pause, plot(t45,x45)
```

```
function dx=df651(t,x)
dx=zeros(size(x)); %row/column vector depending on the shape of x
dx(1)=x(2);   dx(2)=-x(2)+1;
```

Especially for the state equations having only constant coefficients like Eq. (6.5.2), we can change it into a matrix-vector form as

$$\begin{bmatrix} x_1'(t) \\ x_2'(t) \end{bmatrix} = \begin{bmatrix} 0 & 1 \\ 0 & -1 \end{bmatrix} \begin{bmatrix} x_1(t) \\ x_2(t) \end{bmatrix} + \begin{bmatrix} 0 \\ 1 \end{bmatrix} u_s(t) \qquad (6.5.3)$$

$$\text{with } \begin{bmatrix} x_1(0) \\ x_2(0) \end{bmatrix} = \begin{bmatrix} 1 \\ -1 \end{bmatrix} \text{ and } u_s(t) = 1 \ \forall \ t \geq 0;$$

$$\mathbf{x}'(t) = A\,\mathbf{x}(t) + B\,u(t) \text{ with the initial state } \mathbf{x}(0) \text{ and the input } u(t) \qquad (6.5.4)$$

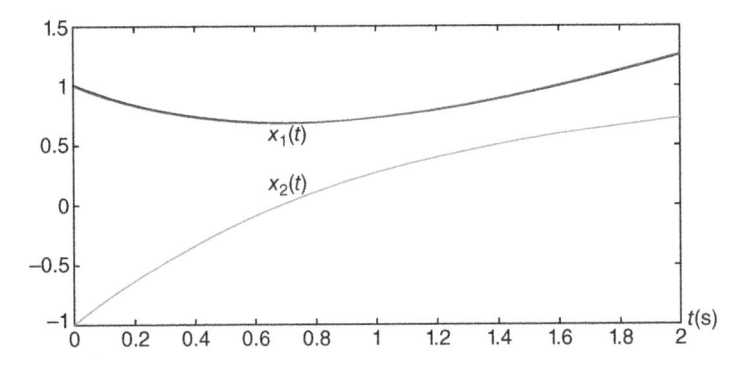

Figure 6.5 Numerical/analytical solutions of the state equation (6.5.2)/(6.5.3).

which is called a linear time-invariant (LTI) state equation, and then try to find the analytical solution. For this purpose, we take the Laplace transform of both sides to write

$$sX(s) - \mathbf{x}(0) = A\,X(s) + \mathbf{b}U(s) \text{ with } X(s) = \mathrm{L}\{\mathbf{x}(t)\},\ U(s) = \mathrm{L}\{u(t)\};$$

$$[sI - A]X(s) = \mathbf{x}(0) + \mathbf{b}U(s);\ X(s) = [sI - A]^{-1}\mathbf{x}(0) + [sI - A]^{-1}\mathbf{b}U(s) \tag{6.5.5}$$

where $\mathscr{L}\{\mathbf{x}(t)\}$ and $\mathscr{L}^{-1}\{X(s)\}$ denote the Laplace transform of $\mathbf{x}(t)$ and the inverse Laplace transform of $X(s)$, respectively. Note that

$$[sI - A]^{-1} = s^{-1}[I - As^{-1}]^{-1} = s^{-1}[I + As^{-1} + A^2 s^{-2} + \cdots];$$

$$\phi(t) = \mathscr{L}^{-1}\{[sI - A]^{-1}\} = I + At + \frac{A^2}{2}t^2 + \frac{A^3}{3!}t^3 + \cdots = e^{At} \text{ with } \phi(0) = I \tag{6.5.6}$$

By applying the convolution property of Laplace transform (Table D.2(4))

$$\mathscr{L}^{-1}\{[sI - A]^{-1}BU(s)\} \overset{\text{Convolution property}}{=} \mathscr{L}^{-1}\{[sI - A]^{-1}\} * \mathscr{L}^{-1}\{BU(s)\}$$

$$\overset{(6.5.6)}{=} \phi(t) * Bu(t) = \int_{-\infty}^{\infty} \varphi(t - \tau)Bu(\tau)d\tau$$

$$\overset{u(\tau)=0 \text{ for } \tau<0 \text{ or } \tau>t}{=} \int_0^t \varphi(t - \tau)Bu(\tau)d\tau, \tag{6.5.7}$$

we can take the inverse Laplace transform of Eq. (6.5.5) to write

$$\mathbf{x}(t) = \varphi(t)\mathbf{x}(0) + \varphi(t) * Bu(t) = \varphi(t)\mathbf{x}(0) + \int_0^t \varphi(t - \tau)B\,u(\tau)\,d\tau \tag{6.5.8}$$

For Eq. (6.5.3), we use Eq. (6.5.6) to find

$$\phi(t) = \mathscr{L}^{-1}\{[sI - A]^{-1}\} = \mathscr{L}^{-1}\left\{\left[\begin{bmatrix} s & 0 \\ 0 & s \end{bmatrix} - \begin{bmatrix} 0 & 1 \\ 0 & -1 \end{bmatrix}\right]^{-1}\right\}$$

$$= \mathscr{L}^{-1}\left\{\begin{bmatrix} s & -1 \\ 0 & s+1 \end{bmatrix}^{-1}\right\} = \mathscr{L}^{-1}\left\{\frac{1}{s(s+1)}\begin{bmatrix} s+1 & 1 \\ 0 & s \end{bmatrix}\right\}$$

$$= \mathscr{L}^{-1}\left\{\begin{bmatrix} 1/s & 1/s - 1/(s+1) \\ 0 & 1/(s+1) \end{bmatrix}\right\}$$

$$= \begin{bmatrix} 1 & 1 - e^{-t} \\ 0 & e^{-t} \end{bmatrix} \tag{6.5.9}$$

and use Eqs. (6.5.8), (6.5.9), and $u(t) = u_s(t) = 1 \ \forall t \geq 0$ to obtain

$$\mathbf{x}(t) = \begin{bmatrix} 1 & 1-e^{-t} \\ 0 & e^{-t} \end{bmatrix} \begin{bmatrix} 1 \\ -1 \end{bmatrix} + \int_0^t \begin{bmatrix} 1 & 1-e^{-(t-\tau)} \\ 0 & e^{-(t-\tau)} \end{bmatrix} \begin{bmatrix} 0 \\ 1 \end{bmatrix} 1 \, d\tau$$

$$= \begin{bmatrix} e^{-t} \\ -e^{-t} \end{bmatrix} + \begin{bmatrix} \tau - e^{-(t-\tau)} \\ e^{-(t-\tau)} \end{bmatrix} \Big|_0^t = \begin{bmatrix} t - 1 + 2e^{-t} \\ 1 - 2e^{-t} \end{bmatrix} \quad (6.5.10)$$

Alternatively, we can directly take the inverse transform of Eq. (6.5.5) to get

$$X(s) = [sI - A]^{-1}\{\mathbf{x}(0) + BU(s)\} = \frac{1}{s(s+1)} \begin{bmatrix} s+1 & 1 \\ 0 & s \end{bmatrix} \left\{ \begin{bmatrix} 1 \\ -1 \end{bmatrix} + \begin{bmatrix} 0 \\ 1 \end{bmatrix} \frac{1}{s} \right\}$$

$$= \frac{1}{s^2(s+1)} \begin{bmatrix} s+1 & 1 \\ 0 & s \end{bmatrix} \begin{bmatrix} s \\ -s+1 \end{bmatrix} = \frac{1}{s^2(s+1)} \begin{bmatrix} s^2+1 \\ s(1-s) \end{bmatrix} \quad (6.5.11)$$

$$; \quad X_1(s) = \frac{s^2+1}{s^2(s+1)} = \frac{1}{s^2} - \frac{1}{s} + \frac{2}{s+1}; \quad x_1(t) = t - 1 + 2e^{-t} \quad (6.5.12a)$$

$$X_2(s) = \frac{1-s}{s(s+1)} = \frac{1}{s} - \frac{2}{s+1}; \quad x_2(t) = 1 - 2e^{-t} \quad (6.5.12b)$$

which conforms with Eq. (6.5.10).

The MATLAB script "nm0651_2.m" uses a symbolic computation routine 'ilaplace()' to get the inverse Laplace transforms $x_1(t)$ and $x_2(t)$, which supports the above computation. Then, it uses 'eval()' to evaluate them for some time range t and plots the results as shown in Figure 6.5. Additionally, it uses the symbolic DE solver 'dsolve()' to get the analytical solution directly.

```
%nm0651_2.m
% Analytical solution for state eq. x'(t)=Ax(t)+Bu(t)  (6.5.3)
syms s t % declare s,t as symbolic variables
A=[0 1;0 -1]; B=[0 1]'; % Eq.(6.5.3)
x0=[1 -1]'; % Initial value
Xs=(s*eye(size(A))-A)^-1*(x0+B/s)  % Eq.(6.5.5)
for n=1:size(A,1)
   xt(n)=ilaplace(Xs(n)); % Inverse Laplace transform
end
disp('Solution of DE based on Laplace transform')
xt=transpose(xt) % Eq.(6.5.12)
t0=0; tf=2; N=45; % Initial/final time
t=t0+[0:N]'*(tf-t0)/N; % Time vector
xtt=eval([xt(1) xt(2)]); % evaluate the inverse Laplace transform
plot(t,xtt)
disp('Analytical solution')
xt= dsolve('Dx1=x2, Dx2=-x2+1', 'x1(0)=1, x2(0)=-1');
xt1=xt.x1, xt2=xt.x2 % Eq.(6.5.10)
```

```
»nm0651_2
  Solution of DE based on Laplace transform
  Xs = [ 1/s+1/s/(s+1)*(-1+1/s) ]
       [         1/(s+1)*(-1+1/s) ]
  xt = [ -1+t+2*exp(-t) ]
       [   -2*exp(-t)+1 ]
  Analytical solution
  xt1 =   -1+t+2*exp(-t)
  xt2 =   -exp(-t)+exp(-t)*(exp(t)-1)
```

6.5.2 Discretization of LTI State Equation

In this section, we consider a discretization method of converting a continuous time LTI state equation

$$\mathbf{x}'(t) = A\,\mathbf{x}(t) + B\,u(t) \text{ with the initial state } \mathbf{x}(0) \text{ and the input } u(t) \quad (6.5.13)$$

into an equivalent discrete time LTI state equation with the sampling period T:

$$\mathbf{x}[n+1] = A_d\,\mathbf{x}[n] + B_d\,u[n] \quad (6.5.14)$$

with the initial state $\mathbf{x}[0]$ and the input $u[n] = u(nT)$ for $nT \le t < (n+1)T$

which can be solved easily by an iterative scheme mobilizing just simple multiplications and additions.

For this purpose, we rewrite the solution (6.5.8) of the continuous time LTI state equation with the initial time t_0 as

$$\mathbf{x}(t) = \varphi(t - t_0)\mathbf{x}(t_0) + \int_{t_0}^{t} \varphi(t - \tau)\,Bu(\tau)\,d\tau \quad (6.5.15)$$

Under the assumption that the input is constant as the initial value within each sampling interval, i.e. $u[n] = u(nT)$ for $nT \le t < (n+1)T$, we substitute $t_0 = nT$ and $t = (n+1)T$ into this equation to write the discrete time LTI state equation as

$$\mathbf{x}((n+1)T) = \varphi(T)\mathbf{x}(nT) + \int_{nT}^{(n+1)T} \varphi((n+1)T - \tau)B\,u(nT)\,d\tau$$

$$;\quad \mathbf{x}[n+1] = \varphi(T)\mathbf{x}[n] + \int_{nT}^{(n+1)T} \varphi(nT + T - \tau)\,d\tau\,B\,u[n]$$

$$;\quad \mathbf{x}[n+1] = A_d\,\mathbf{x}[n] + B_d\,u[n] \quad (6.5.16)$$

where the discretized system matrices are

$$A_d = \varphi(T) = e^{AT} \quad (6.5.17a)$$

$$B_d = \int_{nT}^{(n+1)T} \varphi(nT + T - \tau)d\tau B \stackrel{\sigma=nT+T-\tau}{=} -\int_{T}^{0} \varphi(\sigma)\,d\sigma\, B = \int_{0}^{T} \varphi(\tau)\,d\tau\, B \quad (6.5.17b)$$

Here, let us consider another way of computing these system matrices, which is to the taste of digital computers. It comes from making use of the definition of a matrix exponential function in Eq. (6.5.6) to rewrite Eq. (6.5.17) as

$$A_d = e^{AT} = \sum_{m=0}^{\infty} \frac{A^m T^m}{m!} = I + AT \sum_{m=0}^{\infty} \frac{A^m T^m}{(m+1)!} = I + AT\,\Psi \quad (6.5.18a)$$

$$B_d = \int_{0}^{T} \varphi(\tau)\,d\tau\, B = \int_{0}^{T} \sum_{m=0}^{\infty} \frac{A^m \tau^m}{m!}\,d\tau\, B = \sum_{m=0}^{\infty} \frac{A^m T^{m+1}}{(m+1)!} B = \Psi T B \quad (6.5.18b)$$

where

$$\Psi = \sum_{m=0}^{\infty} \frac{A^m T^m}{(m+1)!}$$

$$\cong I + \frac{AT}{2}\left\{I + \frac{AT}{3}\left\{I + \cdots + \frac{AT}{N-1}\left(I + \frac{AT}{N}\right)\right\}\cdots\right\} \text{ for } N \gg 1 \quad (6.5.19)$$

Now, we apply these discretization formulas for the continuous time state equation (6.5.3)

$$\begin{bmatrix} x_1'(t) \\ x_2'(t) \end{bmatrix} = \begin{bmatrix} 0 & 1 \\ 0 & -1 \end{bmatrix}\begin{bmatrix} x_1(t) \\ x_2(t) \end{bmatrix} + \begin{bmatrix} 0 \\ 1 \end{bmatrix} u_s(t)$$

$$\text{with } \begin{bmatrix} x_1(0) \\ x_2(0) \end{bmatrix} = \begin{bmatrix} 1 \\ -1 \end{bmatrix} \text{ and } u_s(t) = 1 \; \forall t \geq 0$$

to get the discretized system matrices and the discretized state equation as

$$\varphi(t) = \mathcal{L}^{-1}\{[sI - A]^{-1}\} = \mathcal{L}^{-1}\left\{\begin{bmatrix} s & -1 \\ 0 & s+1 \end{bmatrix}^{-1}\right\} \stackrel{(6.5.9)}{=} \begin{bmatrix} 1 & 1-e^{-t} \\ 0 & e^{-t} \end{bmatrix} \quad (6.5.20a)$$

$$A_d \stackrel{(6.5.17a)}{=} \varphi(T) \stackrel{(6.5.20a)}{=} \begin{bmatrix} 1 & 1-e^{-T} \\ 0 & e^{-T} \end{bmatrix} \quad (6.5.20b)$$

$$B_d \stackrel{(6.5.17b)}{=} \int_{0}^{T} \varphi(\tau)\,d\tau\, B \stackrel{(6.5.20a)}{=} \int_{0}^{T} \begin{bmatrix} 1 & 1-e^{-\tau} \\ 0 & e^{-\tau} \end{bmatrix} d\tau \begin{bmatrix} 0 \\ 1 \end{bmatrix} = \begin{bmatrix} T-1+e^{-T} \\ 1-e^{-T} \end{bmatrix} \quad (6.5.20c)$$

ORDINARY DIFFERENTIAL EQUATIONS

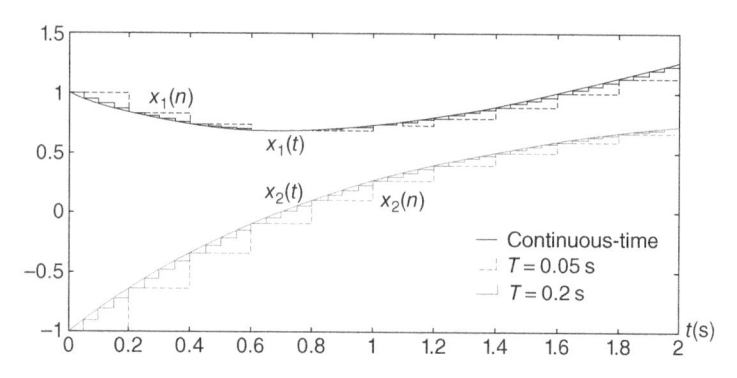

Figure 6.6 Solutions of the discretized state equation (6.5.21).

$$\mathbf{x}[n+1] \stackrel{(6.5.16)}{=} A_d\,\mathbf{x}[n] + B_d u[n]$$

$$;\begin{bmatrix} x_1[n+1] \\ x_2[n+1] \end{bmatrix} = \begin{bmatrix} 1 & 1-e^{-T} \\ 0 & e^{-T} \end{bmatrix} \begin{bmatrix} x_1[n] \\ x_2[n] \end{bmatrix} + \begin{bmatrix} T-1+e^{-T} \\ 1-e^{-T} \end{bmatrix} u[n] \quad (6.5.21)$$

```
%nm0652.m
% discretize a state eqn x'(t)=Ax(t)+Bu(t) to x[n+1]=Ad*x[n]+Bd*u[n]
clear, clf
A=[0 1;0 -1]; B=[0;1]; % Eq.(6.5.3)
x0=[1 -1]; t0=0; tf=2; % Initial value and Time span
T= 0.2; % Sampling interval(period)
eT= exp(-T);
AD=[1 1-eT; 0 eT] % discretized system matrices obtained analytically
BD=[T+eT-1; 1-eT] % Eq.(6.5.20)
[Ad,Bd]= c2d_steq(A,B,T,100) % Continuous-to-Discerte conversion
[Ad1,Bd1]= c2d(A,B,T) % Alternatively, using the built-in function
% To solve the discretized state equation
t(1)=0; xd(1,:)=x0; % Initial time and initial value
for k=1:(tf-t0)/T % solve the discretized state equation
   t(k+1)=k*T; xd(k+1,:)=xd(k,:)*Ad'+Bd'; % Eq.(6.5.21)
end
stairs([0; t'],[x0; xd]), hold on % Stairstep graph
N=100; t=t0+[0:N]'*(tf-t0)/N; % Time (column) vector
x(:,1)=t-1+2*exp(-t); % Analytical solution
x(:,2)=1-2*exp(-t); % Eq.(6.5.12)
plot(t,x)
```

```
function [Ad,Bd]=c2d_steq(A,B,T,N)
if nargin<4, N=100; end
I= eye(size(A,2)); PSI=I;
for m=N:-1:1
   PSI = I+A*PSI*T/(m+1); % Eq.(6.5.19)
end
Ad= I+A*PSI*T; Bd= PSI*T*B; % Eq.(6.5.18)
```

We do not need any special algorithm other than an iterative scheme to solve this discrete time state equation. The formulas (6.5.18a) and (6.5.18b) for computing the discretized system matrices are cast into the function 'c2d_steq()'. The script "nm0652.m" discretizes the continuous time state equation (6.5.3) by using the function and alternatively, the MATLAB built-in function 'c2d()'. It solves the discretized state equation and plots the results as in Figure 6.6. As long as the assumption that $u[n] = u(nT)$ for $nT \leq t < (n+1)T$ is valid, the solution $(x[n])$ of the discretized state equation is expected to match that $(x(t))$ of the continuous time state equation at every sampling instant $t = nT$ and also becomes closer to $x(t)$ $\forall t$ as the sampling interval T gets shorter (see Figure 6.6).

6.5.3 High-order Differential Equation to State Equation

Suppose we are given an Nth-order scalar differential equation together with the initial values of the variable and its derivatives of up to order $N-1$, which is called an IVP:

$$[\text{IVP}]_N : \quad x^{(N)}(t) = f(t, x(t), x'(t), x^{(2)}(t), \ldots, x^{(N-1)}(t)) \quad (6.5.22)$$

with the initial values $x(t_0) = x_{10}$, $x'(t_0) = x_{20}, \ldots, x^{(N-1)}(t_0) = x_{N0}$

Defining the state vector and the initial state as

$$\mathbf{x}(t) = \begin{bmatrix} x_1 = x \\ x_2 = x' \\ x_3 = x^{(2)} \\ \vdots \\ x_N = x^{(N-1)} \end{bmatrix}, \quad \mathbf{x}(t_0) = \begin{bmatrix} x_{10} \\ x_{20} \\ x_{30} \\ \vdots \\ x_{N0} \end{bmatrix} \quad (6.5.23)$$

we can rewrite Eq. (6.5.22) in the form of a first-order vector differential equation, i.e. a state equation as

$$\begin{bmatrix} x_1'(t) \\ x_2'(t) \\ x_3'(t) \\ \vdots \\ x_N'(t) \end{bmatrix} = \begin{bmatrix} x_2(t) \\ x_3(t) \\ x_4(t) \\ \vdots \\ f(t, x(t), x'(t), x^{(2)}(t), \ldots, x^{(N-1)}(t)) \end{bmatrix}$$

$$; \mathbf{x}'(t) = \mathbf{f}(t, \mathbf{x}(t)) \quad \text{with} \quad \mathbf{x}(t_0) = \mathbf{x}_0 \quad (6.5.24)$$

For example, we can write a third-order scalar differential equation

$$x^{(3)}(t) + a_2 x^{(2)}(t) + a_1 x'(t) + a_0 x(t) = u(t)$$

in state-space form, i.e. as a state equation of the form

$$\begin{bmatrix} x_1'(t) \\ x_2'(t) \\ x_3'(t) \end{bmatrix} = \begin{bmatrix} 0 & 1 & 0 \\ 0 & 0 & 1 \\ -a_0 & -a_1 & -a_2 \end{bmatrix} \begin{bmatrix} x_1(t) \\ x_2(t) \\ x_3(t) \end{bmatrix} + \begin{bmatrix} 0 \\ 0 \\ 1 \end{bmatrix} u(t) \quad (6.5.25a)$$

$$x(t) = \begin{bmatrix} 1 & 0 & 0 \end{bmatrix} \begin{bmatrix} x_1(t) \\ x_2(t) \\ x_3(t) \end{bmatrix} \quad (6.5.25b)$$

6.5.4 Stiff Equation

Suppose that we are given a vector differential equation involving more than one dependent variable with respect to the independent variable t. If the magnitudes of the derivatives of the dependent variables with respect to t (corresponding to their changing rates) are significantly different, such a differential equation is said to be stiff on account of that it is difficult to be solved numerically. For such a stiff differential equation, we should be very careful in choosing the step-size to avoid numerical instability problem and get a reasonably accurate solution within a reasonable computation time. Why? Because we should use a small step-size to grasp rapidly changing variables, and it requires a lot of computation to cover slowly changing variables for such a long time as it lasts.

Actually, there is no clear distinction between stiff and nonstiff differential equations, since stiffness of a differential equation is a matter of degree. Then, is there any way to estimate the degree of stiffness for a given differential equation? The answer is yes, if the differential equation can be arranged into an LTI state equation like Eq. (6.5.4), the solution of which consists of components having the time constants (modes) equal to the eigenvalues of the system matrix A. For example, the system matrix of Eq. (6.5.3) has the eigenvalues

$$|sI - A| = 0; \quad \det\left\{\begin{bmatrix} s & -1 \\ 0 & s+1 \end{bmatrix}\right\} = s(s+1) = 0; \quad s = 0 \text{ and } s = -1$$

which can be observed as the time constants of two terms $1 = e^{0t}$ and e^{-t} in the solution (6.5.12b). In this context, a measure of stiffness is the ratio of the maximum over the minimum among the absolute values of (negative) real parts of the eigenvalues of the system matrix A:

$$\eta(A) = \frac{\text{Max}\{|Re(\lambda_i)|\}}{\text{Min}\{|Re(\lambda_i)| \neq 0\}} \quad (6.5.26)$$

This can be thought of as the degree of unbalance between the fast mode and the slow mode.

VECTOR DIFFERENTIAL EQUATIONS

Now, what we must know is how to handle stiff differential equations. Fortunately, MATLAB has several built-in functions like 'ode15s()', 'ode23s()', 'ode23t()', and 'ode23tb()', which are fabricated to deal with stiff differential equations efficiently. One may use the help command to see their detailed usages. Let us try them through the following examples:

Example 6.1. A Stiff Differential Equation.

Consider the following DE:

$$\begin{cases} x'(t) = 998x(t) - 1998y(t) \\ y'(t) = 999x(t) - 1999y(t) \\ \text{with } x(0) = 1 \text{ and } y(0) = 0 \end{cases} \rightarrow$$

$$\begin{bmatrix} x_1'(t) \\ x_2'(t) \end{bmatrix} = A \begin{bmatrix} x_1(t) \\ x_2(t) \end{bmatrix} = \begin{bmatrix} 998 & -1998 \\ 999 & -1999 \end{bmatrix} \begin{bmatrix} x_1(t) \\ x_2(t) \end{bmatrix} \text{ with } \begin{bmatrix} x_1(0) \\ x_2(0) \end{bmatrix} = \begin{bmatrix} 1 \\ 0 \end{bmatrix} \quad (E6.1.1)$$

where the eigenvalues of the coefficient matrix A are $\lambda_1 = -1$ and $\lambda_2 = -1000$, yielding the measure of stiffness (defined by Eq. (6.5.26)) as 1000.

(a) Find the analytical solution two times, once using Eq. (6.5.5) and once using the MATLAB built-in function 'dsolve()'.

To this end, we run the following MATLAB script "em06e01a.m" to get

```
x  = [ 2*exp(-t) - exp(-1000*t), exp(-t) - exp(-1000*t)]
x1 =   2*exp(-t) - exp(-1000*t),   x2 = exp(-t) - exp(-1000*t)
```

both of which imply the same analytical solution

$$\begin{bmatrix} x(t) \\ y(t) \end{bmatrix} = \begin{bmatrix} x_1(t) \\ x_2(t) \end{bmatrix} = \begin{bmatrix} 2e^{-t} - e^{-1000t} \\ e^{-t} - e^{-1000t} \end{bmatrix} \quad (E6.1.2)$$

```
%nm06e01a.m
% Analytical solutions for the stiff DE (E6.1.1)
syms s % declare s as a symbolic variable
A=[998 -1998; 999 -1999]; B=[0; 0]; % Eq.(E6.1.1)
x0=[1 0]'; % Initial value
eigs_A=real(eig(A)); abs_eigs=abs(eigs_A(find(eigs_A)~=0));
Stiffness=max(abs_eigs)/min(abs_eigs) % Eq.(6.5.26)
disp('Analytical solution based on Laplace transform')
Xs=(s*eye(size(A))-A)^-1*(x0+B/s) % Eq.(6.5.5)
x=ilaplace(Xs); % Inverse Laplace transform
x
disp('Analytical solution Using dsolve()')
x_ds=dsolve('Dx1=998*x1-1998*x2,Dx2=999*x1-1999*x2','x1(0)=1,x2(0)=0');
x1=x_ds.x1, x2=x_ds.x2
```

330 ORDINARY DIFFERENTIAL EQUATIONS

(b) Find and plot the numerical solutions by using 'ode_Ham()' two times, each time with the number of segments $N = 9206$ and $10\,000$, respectively, 'ode45()', 'ode23()', 'ode15s()', 'ode23s()', 'ode23t()', and 'ode23tb()'.

To this end, we compose the following MATLAB script "em06e01b.m" and run it to get Figure 6.7 (implying that in order for 'ode_Ham()' to work properly for the DE, the number of segments should be roughly at least 9210) and the runtimes of each function such as

Hamming	ode45	ode23	ode15s	ode23s	ode23t	ode23tb
0.5696	0.4453	0.2375	0.0143	0.0154	0.0143	0.0113

This implies that the stiff DE solvers 'ode15s()', 'ode23s()', 'ode23t()', and 'ode23tb()' take less time than the regular DE solvers like 'ode_Ham()', 'ode45()', and 'ode23()'.

```
%nm06e01b.m
% Numerical Solutions for the stiff DE (E6.1.1)

df=@(t,x)[998 -1998; 999 -1999]*x(:); % DE (E6.1.1)
t0=0; tf=8; tspan=[t0 tf]; x0=[1; 0];

subplot(421)
N=9206; [tH1,xH1]=ode_Ham(df,tspan,x0,N);
plot(tH1,xH1), axis([t0 tf -0.1 2.1])
title(['ode-Ham() with N=' num2str(N)])

subplot(422)
N=10000; tic, [tH2,xH2]=ode_Ham(df,tspan,x0,N); time_H2=toc;
% To evaluate the inverse Laplace transform solution x obtained in (a)
t=tH2(:); xt=eval(x(:).'); % Only when x is available from (a)
plot(tH2,xt, tH2,xH2), axis([t0 tf -0.1 2.1])
title(['Analytical solution x(t) and ode-Ham() with N=' num2str(N)])

subplot(423)
tic, [t45,x45]=ode45(df,tspan,x0); time_45=toc;
tic, [t23,x23]=ode23(df,tspan,x0); time_23=toc;
plot(t45,x45), hold on
plot(t23,x23(:,1),'r:', t23,x23(:,2),'m:')
axis([t0 tf -0.1 2.1]), title('ode45() and ode23()')

subplot(424)
tic,[t15s,x15s]=ode15s(df,tspan,x0); time_15s=toc;
tic,[t23s,x23s]=ode23s(df,tspan,x0); time_23s=toc;
tic,[t23t,x23t]=ode23t(df,tspan,x0); time_23t=toc;
tic,[t23tb,x23tb]=ode23tb(df,tspan,x0); time_23tb=toc;
plot(t15s,x15s, t23s,x23s, t23t,x23t, t23tb,x23tb)
axis([t0 tf -0.1 2.1])
title('ode15s(), ode23s(), ode23t(), and ode23tb()')

disp(' Hamming   ode45   ode23   ode15s   ode23s   ode23t   ode23tb')
[time_H2 time_45 time_23 time_15s time_23s time_23t time_23tb]
```

VECTOR DIFFERENTIAL EQUATIONS   331

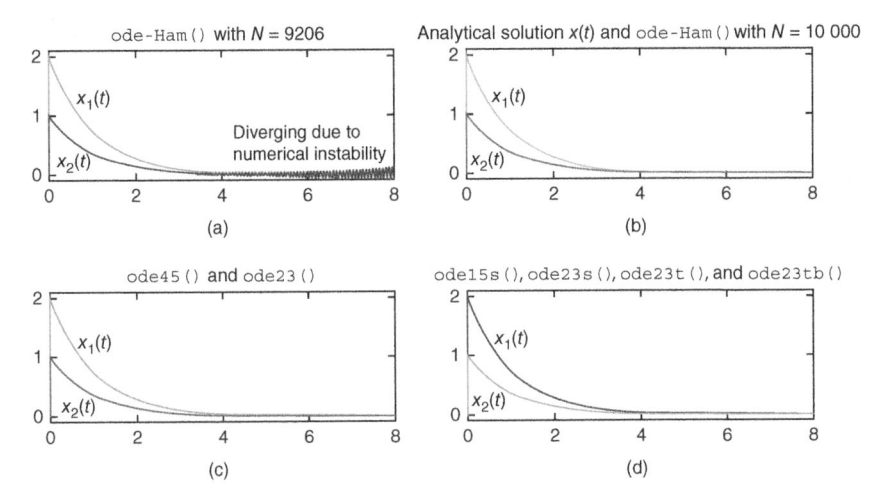

Figure 6.7 Solution graphs for Example 6.1. (a) Numerical solution using `ode_Ham()` with $N = 9206$. (b) Numerical solution using `ode_Ham()` with $N = 104$. (c) Numerical solutions using `ode45()` and `ode23()`. (d) Numerical solutions using `ode15s()`, `ode23s()`, ...

Example 6.2. Another Stiff Differential Equation.
Consider the Van der Pol equation:

$$\frac{d^2y(t)}{dt^2} - \mu(1-y^2(t))\frac{dy(t)}{dt} + y(t) = 0 \text{ with } y(0)=2 \text{ and } \frac{dy(t)}{dt}=0, \quad \text{(E6.2.1)}$$

which can be written in state equation form as

$$\begin{bmatrix} x_1'(t) \\ x_2'(t) \end{bmatrix} = \begin{bmatrix} x_2(t) \\ \mu(1-x_1^2(t))x_2(t) - x_1(t) \end{bmatrix} \text{ with } \begin{bmatrix} x_1(0) \\ x_2(0) \end{bmatrix} = \begin{bmatrix} 2 \\ 0 \end{bmatrix} \quad \text{(E6.2.2)}$$

To solve this DE, we define it as below in an M-file named "df_van.m" and run the following MATLAB script "nm06e02.m", where the parameter μ (mu) has been declared as a global variable so that it could be passed on to any related routines/functions as well as "df_van.m". In the beginning of the script, we set the global parameter μ to 25 and use 'ode_Ham()' with the number of segments $N = 8700$ and 9000. The results of running the script are shown in Figure 6.7a,b where 'ode_Ham()' seems to have failed/succeeded with $N = 8700/9000$, respectively, showing how crucial the choice of step-size or number of segments may be for a stiff DE. Next, we used 'ode45()' and 'ode23()' to obtain the solution shown in Figure 6.7c, which is almost the same as Figure 6.7b. This reveals the merit of the MATLAB built-in functions that may save the computation time as well as spare our trouble to choose the step-size because the step-size is adaptively determined inside the functions. Then, setting $\mu = 200$, we used the MATLAB built-in ODE solvers 'ode45()'/'ode23()'/'ode15s()'/'ode23s()'/ 'ode23t()'/'ode23tb()' to get the results that are little different as shown in Figure 6.7d, each having taken the computation time as

```
         Times =  5.0976    2.9859    0.0856    0.1220    0.1363    0.1201
```

The computation time-efficiency of the stiff ODE solvers 'ode15s()'/'ode23()'/'ode23t()'/'ode23tb()' (designed deliberately for handling stiff DEs) over the regular ODE solvers 'ode45()'/'ode23()' becomes prominent as the value of parameter μ (mu) gets large, reflecting high stiffness. See [B-3] for the comparison of the ODE solvers (Figure 6.8a,b).

```
%nm06e02.m
% to solve a stiff DE called Van der Pol equation (E6.2.1 or 2)
clear, clf
global mu
df='df_van';
mu=25; t0=0; tf=100; tspan=[t0 tf]; x0=[2 0];

N=8700; [tH1,xH1]=ode_Ham(df,tspan,x0,N);
subplot(221)
plot(tH1,xH1); xlim([t0 tf]); N=9000;
tic, [tH2,xH2]=ode_Ham(df,tspan,x0,N); t_H2=toc
tic, [t45,x45]=ode45(df,tspan,x0); t_45=toc
[t15s,x15s]=ode15s(df,tspan,x0);

subplot(222)
plot(tH2,xH2, t45,x45, t15s,x15s)
mu=200; tf=200; tspan=[t0 tf];
tic, [t45,x45]=ode45(df,tspan,x0); times(1)=toc;
tic, [t23,x23]=ode23(df,tspan,x0); times(2)=toc;

subplot(223)
plot(t45,x45, t23,x23(:,1),'m:', t23,x23(:,2),'g:')
axis([t0 tf -3 2])

tic, [t15s,x15s]=ode15s(df,tspan,x0); times(3)=toc;
tic, [t23s,x23s]=ode23s(df,tspan,x0); times(4)=toc;
tic, [t23t,x23t]=ode23t(df,tspan,x0); times(5)=toc;
tic, [t23tb,x23tb]=ode23tb(df,tspan,x0); times(6)=toc;

subplot(224)
plot(t15s,x15s, t23s,x23s, t23t,x23t, t23tb,x23tb)
axis([t0 tf -3 2])
disp('  ode45   ode23   ode15s   ode23s   ode23t   ode23tb')
times % Runtimes taken by each ODE solver
```

```
function dx=df_van(t,x)
%Van der Pol DE (E6.2.2)
global mu
dx=zeros(size(x)); % To make the size (dimension) of dx with x
dx(1)=x(2);
dx(2)=mu*(1-x(1).^2).*x(2)-x(1);
```

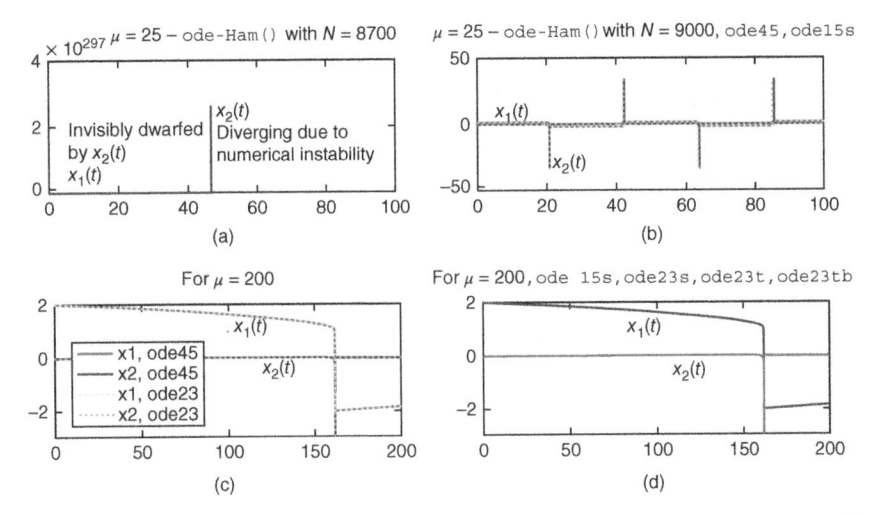

Figure 6.8 Solution graphs for Example 6.2. (a) Numerical solution using `ode_Ham()` with $N = 8700$. (b) Numerical solution using `ode_Ham()` with $N = 9000$. (c) Numerical solutions using `ode45()` and `ode23()`. (d) Numerical solutions using `ode15s()`, `ode23s()`,....

6.6 BOUNDARY VALUE PROBLEM (BVP)

A BVP is an Nth-order differential equation with some of the values of dependent variable $x(t)$ and its derivative specified at the initial time t_0 and others specified at the final time t_f.

$$[\text{BVP}]_N : \quad x^{(N)}(t) = f(t, x(t), x'(t), x^{(2)}(t), \ldots, x^{(N-1)}(t)) \quad (6.6.1)$$

with the boundary values $x(t_1) = x_{10}, x'(t_2) = x_{21}, \ldots, x^{(N-1)}(t_N) = x_{N,N-1}$

In some cases, some relations between the initial values and the final values may be given as a mixed-boundary condition instead of the initial/final values specified. This section covers the shooting method and the finite difference method that can be used to solve a second-order BVP as

$$[\text{BVP}]_2 : \quad x''(t) = f(t, x(t), x'(t)) \text{ with } x(t_0) = x_0, \ x(t_f) = x_f \quad (6.6.2)$$

6.6.1 Shooting Method

The idea of this method is to assume the value of $x'(t_0)$, then solve the differential equation (IVP) with the initial condition $[x(t_0) \ x'(t_0)]$, and keep adjusting the value of $x'(t_0)$ and solving the IVP repetitively until the final value $x(t_f)$ of the solution matches the given boundary value x_f with enough accuracy. It is similar to adjusting the angle of firing a cannon so that the shell will eventually hit the target and that is why this method is named the *shooting method*. This can be viewed as a nonlinear equation problem, if we regard $x'(t_0)$ as an independent variable and the difference between the resulting final value $x(t_f)$ and the desired one x_f as a (mismatching residual error) function of $x'(t_0)$. So the solution scheme can be systemized by using the secant method (Section 4.5) and has been cast into the MATLAB function 'bvp2_shoot()'.

(cf) We might have to adjust the shooting position with the angle fixed, instead of adjusting the shooting angle with the position fixed or deal with the mixed-boundary conditions (see Problems 6.8, 6.9, and 6.10).

```
function [t,x]=bvp2_shoot(f,t0,tf,x0,xf,N,tol,kmax)
%To solve BVP2: [x1,x2]'=f(t,x1,x2) with x1(t0)=x0, x1(tf)=xf
if nargin<8, kmax= 10; end
if nargin<7, tol= 1e-8; end
if nargin<6, N=100; end

dx0(1)=(xf-x0)/(tf-t0); % Initial guess of x'(t0)
[t,x]=ode_RK4(f,[t0 tf],[x0 dx0(1)],N); % start up with RK4
plot(t,x(:,1)), hold on

e(1)=x(end,1)-xf; % x(tf)-xf : the first mismatching (deviation)
dx0(2)=dx0(1)-sign(e(1)); % 2nd guess of x'(t0)

for k=2:kmax-1
  [t,x]=ode_RK4(f,[t0 tf],[x0 dx0(k)],N);
  plot(t,x(:,1))

  % Difference between the resulting final value and the target one
  e(k)=x(end,1)-xf; % x(tf)-xf
  ddx= dx0(k)-dx0(k-1); % Difference between successive derivatives

  if abs(e(k))<tol|abs(ddx)<tol, break; end

  deddx= (e(k)-e(k-1))/ddx; % Gradient of mismatching error
  dx0(k+1)= dx0(k)-e(k)/deddx; % move by secant method
end
```

Example 6.3. Using the Shooting Method for a BVP.
Consider a BVP consisting of the second-order differential equation

$$x''(t) = 2x^2(t) + 4t\,x(t)x'(t) \quad \text{with } x(0) = \frac{1}{4},\ x(1) = \frac{1}{3} \qquad \text{(E6.3.1)}$$

The solution $x(t)$ and its derivative $x'(t)$ are known as

$$x(t) = \frac{1}{4-t^2} \quad \text{and } x'(t) = \frac{2t}{(4-t^2)^2} = 2t\,x^2(t) \qquad \text{(E6.3.2)}$$

Note that this second-order differential equation can be written in the form of state equation as

```
%nm06e03.m
% to solve BVP2 by the shooting method

df=@(t,x)[x(2);
          (2*x(1)+4*t*x(2))*x(1)]; % Eq.(E6.3.1)
t0=0; tf=1;      % initial/final times
x0=1/4; xf=1/3;  % initial/final positions

N=100; tol=1e-8; kmax=10;
[t,x]=bvp2_shoot(df,t0,tf,x0,xf,N,tol,kmax);

xo=1./(4-t.*t); % True solution (E6.3.2)
err=norm(x(:,1)-xo)/(N+1)    % Size of the error
plot(t,x(:,1), t,xo,'r')  % To compare with true solution (6.6.4)
```

$$\begin{bmatrix} x_1'(t) \\ x_2'(t) \end{bmatrix} = \begin{bmatrix} x_2(t) \\ 2x_1^2(t) + 4t\,x_1(t)x_2(t) \end{bmatrix} \text{ with } \begin{bmatrix} x_1(t_0) \\ x_1(t_f) \end{bmatrix} = \begin{bmatrix} x_0 = 1/4 \\ x_f = 1/3 \end{bmatrix} \quad (E6.3.3)$$

where $t_0 = 0$ and $t_f = 1$.

To apply the shooting method, we set the initial guess of $x_2(0) = x'(0)$ to

$$dx0[1] = x_2(0) = \frac{x_f - x_0}{t_f - t_0} \quad (E6.3.4)$$

and solve the state equation with the initial condition $[x_1(0)\ x_2(0) = dx0[1]]$. Then, depending on the sign of the difference e(1) between the final value $x_1(1)$ of the solution and the target final value x_f, we make the next guess of $x_2(0) = x'(0)$:

$$dx0[2] = x_2(0) = dx0[1] - sign(e(1)) = dx0[1] - sign(x(t_f) - x_f) \quad (E6.3.5)$$

larger/smaller than the initial guess of $dx0[1]$ and solve the state equation again with the initial condition $[x_1(0)\ dx0[2]]$. We can start up the secant method with the two initial values $dx0[1]$ and $dx0[2]$ to make the next guess of $x_2(0) = x'(0)$:

$$dx0[3] = dx0[2] - \frac{e(2) = x(t_f) - x_f}{(e(2) - e(1))/(dx0[2] - dx0[1])} \quad (E6.3.6)$$

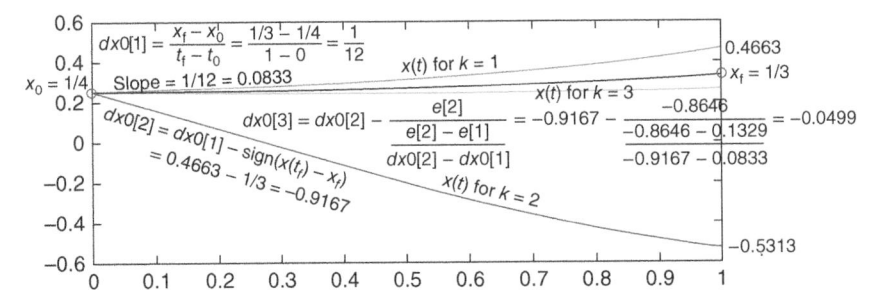

Figure 6.9 The solution of the BVP (6.6.1) obtained by using the shooting method.

and repeat the iteration until the difference (error) `e(k)` becomes sufficiently small. To this end, we run the above MATLAB script "nm06e03.m", which uses the function '`bvp2_shoot()`' to get the numerical solution and compares it with the true analytical solution. Figure 6.9 shows that the numerical solution gets closer to the true analytical solution after each round of adjustment.

(Q) Why don't we use the Newton method (Section 4.4)?

(A) Because, in order to use the Newton method in the shooting method, we need IVP solutions instead of function evaluations to find the numerical Jacobian at every iteration, which will require much longer computation time.

6.6.2 Finite Difference Method

The idea of this method is to divide the whole interval $[t_0, t_f]$ into N segments of width $h = (t_f - t_0)/N$ and approximate the first/second derivatives in the differential equations for each grid point by the central difference formulas. This leads to a tri-diagonal system of equations with respect to $(N-1)$ variables $\{x_i = x(t_0 + ih), i = 1, \ldots, N-1\}$. However, in order for this system of equations to be solved easily, it should be linear, implying that its coefficients may not contain any term of x.

For example, let us consider a BVP consisting of the second-order linear differential equation

$$x''(t) + a_1(t)x'(t) + a_0(t)x(t) = u(t) \text{ with } x(t_0) = x_0, \; x(t_f) = x_f \quad (6.6.3)$$

```
function [t,x]=bvp2_fdf(a1,a0,u,t0,tf,x0,xf,N)
% solve BVP2: x"+a1*x'+a0*x=u with x(t0)=x0, x(tf)=xf
%   by the finite difference method
if ~isnumeric(a1), a1=a1(t(2:N)); end % if a1 was a t-function name
if numel(a1)==1, a1=a1*ones(N-1,1); end
if ~isnumeric(a0), a0=a0(t(2:N)); end % if a0 was a t-function name
if numel(a0)==1, a0=a0*ones(N-1,1); end
if ~isnumeric(u), u=u(t(2:N)); end
if numel(u)==1, u=u*ones(N-1,1); end % if u was a t-function name
A=zeros(N-1,N-1);   b=h2*u(:);
ha=h*a1(1);
A(1,1:2)=[-4+h2*a0(1)   2+ha];   b(1)=b(1)+(ha-2)*x0;
for m=2:N-2 % Eq.(6.6.5)
    ha=h*a1(m);
    A(m,m-1:m+1)= [2-ha   -4+h2*a0(m)   2+ha];
end
ha=h*a1(N-1);
A(N-1,N-2:N-1) = [2-ha   -4+h2*a0(N-1)];
b(N-1) = b(N-1)-(ha+2)*xf;
x=[x0 trid(A,b)' xf]';
```

According to the finite difference method, we divide the solution interval $[t_0, t_f]$ into N segments and convert the differential equation for each grid point $t_i = t_0 + ih$ into a difference equation as

$$\frac{x_{i+1} - 2x_i + x_{i-1}}{h^2} + a_{1i}\frac{x_{i+1} - x_{i-1}}{2h} + a_{0i}x_i = u_i$$

$$; (2 - ha_{1i})x_{i-1} + (-4 + 2h^2 a_{0i})x_i + (2 + ha_{1i})x_{i+1} = 2h^2 u_i \quad (6.6.4)$$

Then, taking account of the boundary condition (BC) that $x_0 = x(t_0)$ and $x_N = x(t_f)$, we collect all the $(N-1)$ equations to construct a tri-diagonal system of equations as

$$\begin{bmatrix} -4+2h^2a_{01} & 2+ha_{11} & 0 & \bullet & 0 & 0 & 0 \\ 2-ha_{12} & -4+2h^2a_{02} & 2+ha_{12} & \bullet & 0 & 0 & 0 \\ 0 & 2-ha_{13} & -4+2h^2a_{03} & \bullet & 0 & 0 & 0 \\ \bullet & \bullet & \bullet & \bullet & \bullet & \bullet & \bullet \\ 0 & 0 & 0 & \bullet & -4+2h^2a_{0,N-3} & 2+ha_{1,N-3} & 0 \\ 0 & 0 & 0 & \bullet & 2-ha_{1,N-2} & -4+2h^2a_{0,N-2} & 2+ha_{1,N-2} \\ 0 & 0 & 0 & \bullet & 0 & 2-ha_{1,N-1} & -4+2h^2a_{0,N-1} \end{bmatrix}$$

$$\times \begin{bmatrix} x_1 \\ x_2 \\ x_2 \\ \bullet \\ x_{N-3} \\ x_{N-2} \\ x_{N-1} \end{bmatrix} = \begin{bmatrix} 2h^2 u_1 - (2 - ha_{11})x_0 \\ 2h^2 u_2 \\ 2h^2 u_3 \\ \bullet \\ 2h^2 u_{N-3} \\ 2h^2 u_{N-2} \\ 2h^2 u_{N-1} - (2 + ha_{1,N-1})x_N \end{bmatrix} \quad (6.6.5)$$

This can be solved efficiently by using the MATLAB function 'trid()', which is dedicated to solving a tri-diagonal system of linear equations.

```
function x=trid(A,b)
% solve a tridiagonal system of equations
N= size(A,2);
for m =2:N % Upper Triangularization
   tmp = A(m,m-1)/A(m-1,m-1);
   A(m,m)  = A(m,m) - A(m-1,m)*tmp;
   A(m,m-1) =0;
   b(m,:)  = b(m,:) - b(m-1,:)*tmp;
end
x(N,:)=b(N,:)/A(N,N);
for m =N-1:-1: 1 % Back Substitution
   x(m,:) =(b(m,:) -A(m,m+1)*x(m+1))/A(m,m);
end
```

The whole procedure of the finite difference method for solving a second-order linear differential equation with BCs has been cast into the MATLAB function 'bvp2_fdf()'. This function is designed to accept the two coefficients a_1 and a_0 and the RHS input u of Eq. (6.6.3) as its first three input arguments, where any of those three input arguments can be given as the function name in case the corresponding term is not a numeric value, but a function of time t.

Example 6.4. Using the Finite Difference Method for a BVP.
Consider a BVP consisting of the second-order differential equation

$$x''(t) + \frac{2}{t}x'(t) - \frac{2}{t^2}x(t) = 0 \text{ with } x(1) = 5, \ x(2) = 3 \qquad \text{(E6.4.1)}$$

The solution $x(t)$ and its derivative $x'(t)$ are known as

$$x(t) = t + \frac{4}{t^2} \text{ and } x'(t) = 1 - \frac{8}{t^3} \qquad \text{(E6.4.2)}$$

We compose the following MATLAB script "nm06e04.m", which uses the above MATLAB function 'bvp2_fdf()' for solving the second-order BVP and additionally does the following:

- It uses the MATLAB built-in function 'bvp4c()' with a BC function fbc (specifying the BC) and an initial guess solinit (created using 'bvpinit()') as the second and third input arguments, respectively,) to get a numerical solution structure sol. Then, it uses another built-in function 'deval(sol,t)' to obtain the solution x_bvp for a given time vector t.
- If the BC of the BVP (6.6.1) were given as $x(1) = 5$ and $x'(2) = 0$, the eighth line of the script "nm06e04.m" should be changed as
    ```
    dxf=0; fbc=@(xdx0,xdxf) [xdx0(1)-x0; xdxf(2)-dxf];
    ```
    where the two input arguments, xdx0 and xdxf, of the BC function fbc can be regarded as $[x_1(t_0) = x(t_0), x_2(t_0) = x'(t_0)]$ and $[x_1(t_f) = x(t_f), x_2(t_f) = x'(t_f)]$, respectively.
- The MATLAB built-in function 'bvpinit()', given some initial mesh (grid) and a rough (constant) guess of the solution, generates an initial guess of the solution for the mesh. It seems to be reasonable to set its second input argument to $[(x_0 + x_f)/2(x_f - x_0)/(t_f - t_0)]$.
- It uses the symbolic computation commands 'dsolve()' and 'subs()' to get the analytical solution and substitute the time vector t into the analytical solution to obtain its numeric values for check.

We run it to get the result shown in Figure 6.10 and the analytical solution:
```
xo =    t + 4/t^2
```

BOUNDARY VALUE PROBLEM (BVP)

```
%nm06e04.m
% to solve BVP2 by the finite difference method
t0=1; x0=5; tf=2; xf=3; N=100;
a1=@(t)2./t; a0=@(t)-2./t./t; u=0; % Eq.(E6.4.1)
[t,x]=bvp2_fdf(a1,a0,u,t0,tf,x0,xf,N);
% To use the MATLAB built-in function 'bvp4c()'
df=@(t,x)[x(2); 2./t.*(x(1)./t-x(2))];
fbc=@(xdx0,xdxf)[xdx0(1)-x0; xdxf(1)-xf]; % BCs (boundary conditions)
x_guess=[mean([x0 xf]) (xf-x0)/(tf-t0)]; % Guess of [x1(t) x2(t)]
solinit=bvpinit(linspace(t0,tf,5),x_guess); % Guess of solution (t,x)
sol=bvp4c(df,fbc,solinit,bvpset('RelTol',1e-4));
x_bvp=deval(sol,t); xbv=x_bvp(1,:)'; % To evaluate the sol for t
% To use the symbolic DE solver 'dsolve()'
xo=dsolve('D2x+2*(t*Dx-x)/t^2=0','x(1)=5, x(2)=3')
xot=subs(xo,'t',t); % xot=4./tt./tt +tt; % True analytical solution
err_fd=double(norm(x-xot))/(N+1) % Error from analytical solution
plot(t,x, t,xbv,'r', t,xot,'k') % compare with analytical solution
```

Note the following:

- While the shooting method is applicable to linear/nonlinear BVPs, the finite difference method is suitable for linear BVPs. However, we can also apply the finite difference method in an iterative manner to solve nonlinear BVPs (see Problems 6.9 and 6.10). Both methods can be modified to solve BVPs with mixed-boundary conditions (see Problems 6.8 and 6.9).
- The MATLAB built-in function 'bvp4c(df,fbc,solinit,options)' can be used to solve linear/nonlinear BVPs with mixed-boundary conditions (see Problems 6.9 to 6.11). Its first input argument (df) is the (RHS of) the DE arranged in the form of state equation (first-order vector DE). Its second input argument (fbc) is the residual error function (required to be zero by the BCs) in the initial/final values of the unknown function $\mathbf{x}(t)$, i.e. $\mathbf{x}(t_0)$ and $\mathbf{x}(t_f)$. For example, if mixed BCs are given as

$$c_{01}x(t_0) + c_{02}x'(t_0) = c_{03} \quad (6.6.6a)$$

$$c_{f1}x(t_f) + c_{f2}x'(t_f) = c_{f3} \quad (6.6.6b)$$

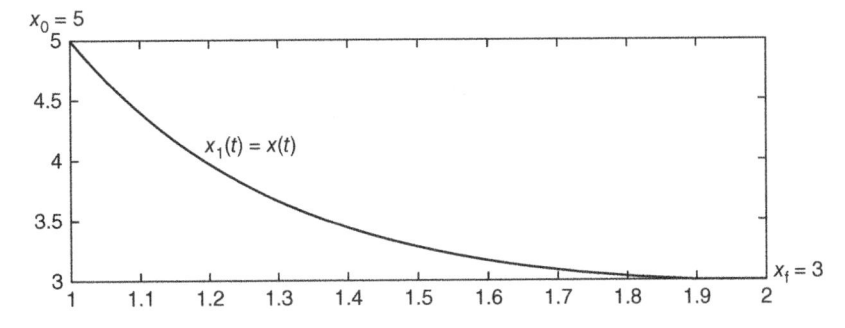

Figure 6.10 The solution of the BVP (6.6.1) obtained by using the finite difference method.

it can be coded as

```
fbc=@(xa,xb) [c01*xa(1)+c02*xa(2)-c03; cf1*xb(1)+cf2*xb(2)-cf3];
```

The third input argument (solinit) is an initial guess of the solution in the form of structure made by using 'bvpinit(t,xinit)' with t (a vector of ordered several values of t in $[t_0, t_f]$) and xinit usually given as a guess for (the mean of) $\mathbf{x}(t)$. The fourth input argument (options) is a structure of options for specifying the performance-related parameters like the error tolerances 'AbsTol' and 'RelTol' that can be made by using 'bvpset()'. Its output argument (sol) is a structure consisting of sol.x (mesh for t as a row vector selected by 'bvp4c'), sol.y ($\mathbf{x}(t)$ for the mesh sol.x), sol.yp ($\mathbf{x}'(t)$ for the mesh), etc. In Example 6.4, the value of sol returned by the 'bvp4c()' is

```
solver: 'bvp4c'
     x: [1 1.0625 1.1250 1.1875 1.2500 … 2]
     y: [2x18 double]
    yp: [2x18 double]
 stats: [1x1 struct]
```

where you can run

```
[sol.y; sol.yp]
```

to see

```
ans =  5.0000    4.6058    4.2855    4.0241  …   3.0000
      -7.0000   -5.6697   -4.6187   -3.7774  …  -0.0000
      -7.0000   -5.6697   -4.6187   -3.7774  …  -0.0000
      24.0000   18.8320   14.9831   12.0692  …   1.5000
```

- The symbolic DE solver 'dsolve()' introduced in Section 6.5.1 can be used to solve a BVP so long as the DE is linear, i.e. its coefficients may depend on time t but not on the (unknown) dependent variable $x(t)$.
- The usages of 'bvp4c()' and 'dsolve()' are illustrated in the script "nm06e04.m".

PROBLEMS

6.1 MATLAB Commands 'quiver()' and 'quiver3()' and Differential Equation

(a) Usage of 'quiver()'

Type 'help quiver' in the MATLAB command window and you will see the following script showing you how to use the 'quiver()' command for plotting gradient vectors. You can also get Figure P6.1.1 by running the block of statements in the box below. Try it and note that the heights (function values) plotted by the contours is expressed by colors (as illustrated by the color bar) and the size of the gradient vector at each point is proportional to the slope at the point.

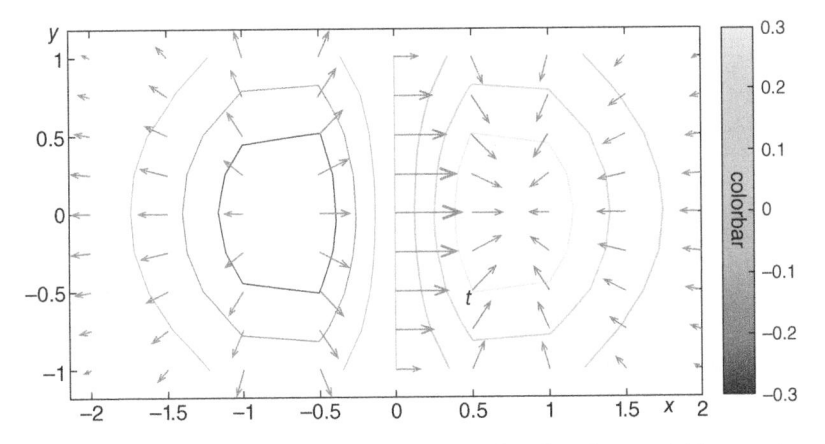

```
%do_quiver.m
[x,y]=meshgrid(-2:.5:2,-1:.25:1);
z=x.*exp(-x.^2 - y.^2);
[px,py]=gradient(z,.5,.25);
contour(x,y,z), hold on, quiver(x,y,px,py)
axis image % the same as AXIS EQUAL except that
           % the plot box fits tightly around the data
colorbar
```

Figure P6.1.1 Contours and gradient vectors for a function of two variables x and y.

(b) Usage of 'quiver3()'

You can obtain Figure P6.1.2 by running the block of statements that you see after typing 'help quiver3' in the MATLAB Command window. Note that the 'surfnorm()' command generates normal vectors at points specified by (x,y) on the surface drawn by 'surf()' and the 'quiver3()' command plots the normal vectors.

342 ORDINARY DIFFERENTIAL EQUATIONS

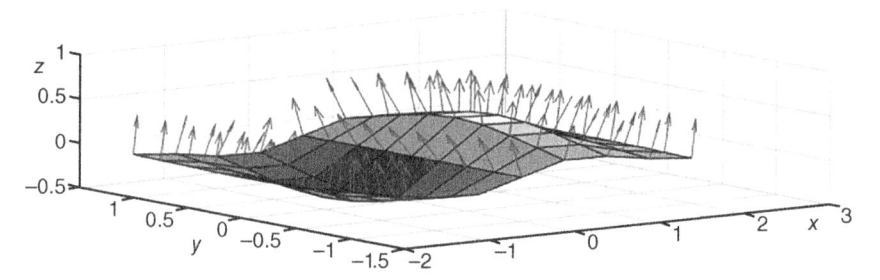

Figure P6.1.2 A surface and its normal vectors each plotted using 'surf()' and 'quiver3()'.

```
%do_quiver3.m
[x,y]=meshgrid(-2:.5:2,-1:.25:1);
z=x.*exp(-x.^2-y.^2);
surf(x,y,z), hold on
[u,v,w]=surfnorm(x,y,z);
quiver3(x,y,z,u,v,w);
```

(c) Gradient vectors and one-variable DE (differential equation)
We might get the meaning of the solution of a DE by using the 'quiver()' command, which is used in the following script "do_ode.m" for drawing the time derivatives at grid points as defined by the DE

$$\frac{dy(t)}{dt} = -y(t) + 1 \text{ with the initial condition } y(0) = 0 \qquad (P6.1.1)$$

The slope/direction field together with the numerical solution in Figure P6.1.3a has been obtained by running the script, and it can be regarded as a set of possible solution curve segments. Starting from the initial point and moving along the slope vectors, you can get the solution curve. Modify the script and run it to plot the slope/direction field ($dx_2(t)$ vs. $dx_1(t)$) and the numerical solution for the following DE as depicted in Figure P6.1.3b.

```
%do_ode.m
t0=0; tf=2; tspan=[t0 tf]; x0=0;
[t,y]=meshgrid(t0:(tf-t0)/10:tf,0:.1:1);
pt=ones(size(t));
py=(1-y).*pt; % dy=(1-y)dt from Eq.(P6.1.1)
quiver(t,y,pt,py) % y(displacement) vs. t(time)
axis([t0 tf+0.2 0 1.05]), hold on
dy=@(t,y) -y+1; % Eq.(P6.0.1)
[tR,yR]=ode_RK4(dy,tspan,x0,40);
for k=1:length(tR)
    plot(tR(k),yR(k),'rx'), pause(0.001);
end
```

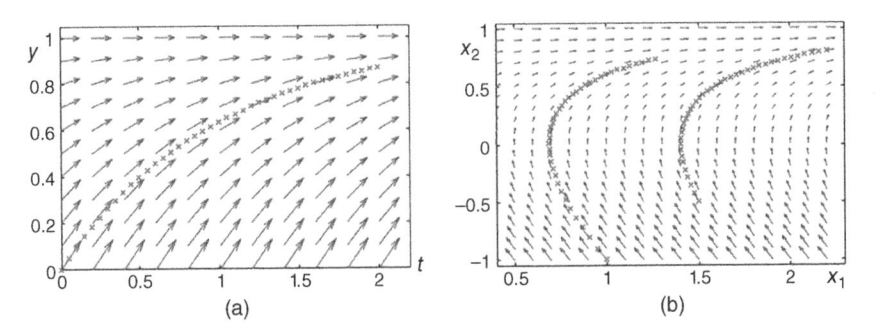

Figure P6.1.3 Slope/direction fields and possible solutions for differential equations. (a) Tangent lines (dy/dt) for the solution curves to $y'(t) = -y(t) + 1$ and a solution with $y(0) = 0$.
(b) Tangent lines (dx2/dx1) for the solution curves to $\begin{cases} x_1'(t) = x_2(t) \\ x_2'(t) = -x_2(t) + 1 \end{cases}$ and a solution with $[x_1(0)\, x_2(0)] = [1\,{-}1]$ or $[1.5\,{-}0.5]$.

$$\begin{aligned} x_1'(t) &= x_2(t) \\ x_2'(t) &= -x_2(t) + 1 \end{aligned} \text{ with } \begin{bmatrix} x_1(0) \\ x_2(0) \end{bmatrix} = \begin{bmatrix} 1 \\ -1 \end{bmatrix} \text{ or } \quad \text{(P6.1.2)}$$

6.2 A System of Linear Time-Invariant DEs – an LTI State Equation
Consider the following state equation:

$$\begin{bmatrix} x_1'(t) \\ x_2'(t) \end{bmatrix} = \begin{bmatrix} 0 & 1 \\ -2 & -3 \end{bmatrix} \begin{bmatrix} x_1(t) \\ x_2(t) \end{bmatrix} + \begin{bmatrix} 0 \\ 1 \end{bmatrix} u_s(t) \text{ with } \begin{bmatrix} x_1(0) \\ x_2(0) \end{bmatrix} = \begin{bmatrix} 1 \\ 0 \end{bmatrix} \quad \text{(P6.2.1)}$$

(a) Check the process and result of obtaining the analytical solution by using the Laplace transform technique.

$$X(s) = [sI - A]^{-1}\{\mathbf{x}(0) + B\, U(s)\}$$

$$= \frac{1}{s(s+3)+2} \begin{bmatrix} s+3 & 1 \\ -2 & s \end{bmatrix} \left\{ \begin{bmatrix} 1 \\ 0 \end{bmatrix} + \begin{bmatrix} 1 \\ 0 \end{bmatrix} \frac{1}{s} \right\}$$

$$= \frac{1}{(s+1)(s+2)} \begin{bmatrix} s+3+1/s \\ -2+1 \end{bmatrix} = \begin{bmatrix} (s^2+3s+1)/s(s+1)(s+2) \\ -1/(s+1)(s+2) \end{bmatrix}$$

$$; X_1(s) = \frac{1/2}{s} + \frac{1}{s+1} - \frac{1/2}{s+2}; \quad x_1(t) = \frac{1}{2} + e^{-t} - \frac{1}{2}e^{-2t} \quad \text{(P6.2.2a)}$$

$$X_2(s) = \frac{-1}{s+1} + \frac{1}{s+2}; \quad x_2(t) = -e^{-t} + e^{-2t}; \quad \text{(P6.2.2b)}$$

(b) Find the numerical solution of the aforementioned state equation (P6.2.1) by using the function 'ode_RK4()' (with the number of segments $N = 50$) and the MATLAB built-in function 'ode45()'. Compare the two numerical solutions in terms of their execution time (measured by using tic and toc) and closeness to the analytical solution.

```
%nm06p01.m
syms s t
A=[0 1;-2 -3]; B=[0 1]'; x0=[1; 0];
Xs=(s*eye(size(A))-?)^-1*(x?+?/s)
x=ilaplace(Xs);
dx=@(t,x)[x(2); -2*x(?)-3*x(?)+1]; % Eq.(P6.2.1)
t0=0; tf=2; tspan=[t0 tf]; N=50;
tic, [tR,xR]=ode_RK4(dx,tspan,x0,N); t_RK4=toc
x=@(t)[0.5+exp(-t)-0.5*exp(-2*t) -exp(-?)+exp(-2*t)]; %Eq.(P6.2.2a,b)
xaR=x(tR(:)); e_RK4=norm(xR-xaR)/length(tR)
tic, [t45,x45]=ode45(dx,tspan,x0); t_45=toc
xa45=x(t45); e_45=norm(x45-xa45)/length(t45)
plot(tR,xR, t45,x45,':')
```

6.3 A Second-order Linear Time-Invariant Differential Equation
Consider the following second-order differential equation (DE)

$$x''(t) + 3x'(t) + 2x(t) = 1 \text{ with } x(0) = 1, \quad x'(0) = 0 \quad \text{(P6.3.1)}$$

(a) Check the procedure and the result of obtaining the analytical solution by using the Laplace transform technique.

$$s^2 X(s) - x'(0) - sx(0) + 3(sX(s) - x(0)) + 2X(s) = \frac{1}{s};$$

$$X(s) = \frac{s^2 + 3s + 1}{s(s+1)(s+2)}; \quad x(t) = \frac{1}{2} + e^{-t} - \frac{1}{2}e^{-2t} \quad \text{(P6.3.2)}$$

(b) Do the same job as asked in Problem 6.3(b) (see Section 6.5.1).

6.4 Ordinary Differential Equation and State Equation

(a) Van der Pol equation
Consider a nonlinear DE

$$\frac{d^2}{dt^2}y(t) - \mu(1 - y^2(t))\frac{d}{dt}y(t) + y(t) = 0 \text{ with } \mu = 2 \quad \text{(P6.4.1)}$$

```
%nm06p04a.m
f=@(t,x,u)[x(2); -x(1)+?*(1-x(?).^2).*x(?)];
t0=0; tf=20; tspan=[t0 tf]; N=100;
x0s=[0.5 0; -1 ?]; % A matrix consisting of initial points
u=2; % The parameter of the Van de Pol equation (P6.4.1)
for iter=1:size(x0s,1)
   x0=x0s(iter,:);
   tic, [tR,xR]=ode_RK4(f,tspan,x0,N,u); t_RK4=toc
   tic, [t,x]= ode45(f,tspan,x0,[],?); t_45=toc
   subplot(220+iter)
   plot(tR,xR,'b', t,x,':')
   legend('ode_ RK4','','ode45')
   subplot(222+iter), plot(x(:,1),x(:,2), x0(1),x0(2),'mo')
end
```

Compose and run a script, named "nm06p03a", to solve this equation with the initial condition [y(0) y'(0)] = [0.5 0] and [−1 2] for the time interval [0, 20] and plot y'(t) vs. y(t) as well as y(t) and y'(t) along the t-axis.

(b) Lorenz equation – turbulent flow and chaos
Consider a nonlinear state equation.

$$x_1'(t) = \sigma(x_2(t) - x_1(t)) \qquad \sigma = 10$$
$$x_2'(t) = (1 + \lambda - x_3(t))x_1(t) - x_2(t) \text{ with } \lambda = 20 \sim 100 \qquad (P6.4.2)$$
$$x_3'(t) = x_1(t)x_2(t) - \gamma\, x_3(t) \qquad \gamma = 2$$

Compose and run a script, named "nm06p03b", to solve this equation with $\lambda = 20$ and 100 for the time interval [0, 10] and plot $x_3(t)$ vs. $x_1(t)$. Let the initial condition be $[x_1(0)\ \ x_2(0)\ \ x_3(0)] = [-8\ -16\ \ 80]$.

```
%nm06p04b.m
% To solve DEs of Lorenz Attractor - (Caotic) Turbulent Flow & Caos
f=@(t,x,s,r,lambda) [s*(x(2)-x(?));  (1+lambda-x(3)).*x(3)-x(?);
                    x(1).*x(2)-r*x(?)];
t0=0; tf=10; tspan=[t0 tf];
x0=[-8 -16 80]; % Initial point
s=10; r=2; lambdas=[20 100];
for i=1:length(lambdas)
   lambda=lambdas(i);
   [t,x]= ode45(f,tspan,x0,[],s,?,lambda);
   subplot(219+i*2), plot(t,x)
   subplot(220+i*2), plot(x(:,1),x(:,3))
end
```

(c) Chemical Reactor
Consider a nonlinear state equation describing the concentrations of two reactants and one product in the chemical process:

$$x_1'(t) = a(u_1 - x_1(t)) - b\,x_1(t)x_2(t) \qquad a = 5$$
$$x_2'(t) = a(u_2 - x_2(t)) - b\,x_1(t)x_2(t) \text{ with } b = 2 \qquad (P6.4.3)$$
$$x_3'(t) = -a\,x_3(t) + b x_1(t)x_2(t) \qquad u_1 = 3,\ \ u_2 = 5$$

Compose and run a script, named "nm06p03c", to solve this equation for the time interval [0, 1] and plot $x_1(t)$, $x_2(t)$, and $x_3(t)$. Let the initial condition be $[x_1(0)\ \ x_2(0)\ \ x_3(0)] = [1\ \ 2\ \ 3]$.

```
%nm06p04c.m: to solve DE of Chemical Reactor
a=5; b=2; u1=3; u2=5;
f=@(t,x) [a*(u1-x(?))-b*x(1).*x(2); a*(u2-x(2))-b*x(1).*x(?);
         -a*x(?)+b*x(1).*x(2)];
t0=0; tf=1; tspan=[t0 tf]; x0=[1 2 3];
[t,x]=ode45(f,tspan,x0);
plot(t,x)
```

(d) *Cantilever beam*: a differential equation with respect to a spatial variable
Consider a nonlinear state equation describing the vertical deflection of a beam due to its own weight

$$JE\frac{d^2y}{dx^2} = \rho g \left(1 + \left(\frac{dy}{dx}\right)^2\right)^{3/2} \left\{x\left(x - \frac{L}{2}\right) + \frac{L^2}{2}\right\}$$ (P6.4.4)

where JE = 2000 kg m^3/s^2, ρ = 10 kg/m, g = 9.8 m/s^2, and L = 2 m. Compose and run a script, named "nm06p03d", to solve this equation for the interval [0 L] and plot $y(t)$. Let the initial condition be $[y(0)\ y'(0)]$ =[0 0]. Note that the physical meaning of the independent variable for which we usually use the symbol 't' in writing the differential function is not a time, but the x-coordinate of the cantilever beam along the horizontal axis in this problem.

```
%nm06p04d.m: to solve DE of Cantilever
L=2; rho=10; g=9.8; JE=2000;
f=@(t,x)[x(2);      % Eq.(P6.4.4)
         rho*g/JE*(?+x(2).^?).^(?/2).*(t.*(t-L/2)+L^?/2)];
x0=0; xf=L; y0=[0 0]; N=100;
[x,y]=ode45(f,[x0 xf],y0);
plot(x,y)
```

(e) Phase-locked loop (PLL)
Consider a nonlinear state equation describing the behavior of a PLL circuit shown in Figure P6.4.1.

$$\begin{aligned} x_1'(t) &= \frac{au(t)\cos(x_2(t)) - x_1(t)}{\tau} \\ x_2'(t) &= x_1(t) + \omega_c \end{aligned} \quad \text{with } \begin{aligned} a &= 1500 \\ \tau &= 0.002 \\ u(t) &= \sin(\omega_o t) \end{aligned}$$ (P6.4.5a)

$$y(t) = x_1(t) + \omega_c$$ (P6.4.5b)

where $\omega_0 = 2100\pi$ rad/s and $\omega_c = 2100\pi$ rad/s. Compose and run a script, named "nm06p03e", to solve this equation with $[x_1(0)\ x_2(0)]$ = [0 0] for the time interval [0, 0.03] and plot $y(t)$ together with $r(t) = \omega_0$ to see $y(t)$ tracking the frequency ω_0 of the input $u(t)$?

```
%nm06p04e.m: PLL(Phase-Locked Loop)
wo=2*pi*1050; wc=2*pi*1000;
a=1500; tau=0.002;
f=@(t,x)[(a*sin(??*t)*cos(x(?))-x(?))/tau; x(?)+wc];  % Eq.(P6.4.5)
t0=0; tf=0.03; tspan=[t0 tf]; x0=[0 0];
[t,x]=ode45(f,tspan,x0);
plot(t,x(:,1)+wc, t,wo*ones(size(t)),'m')
```

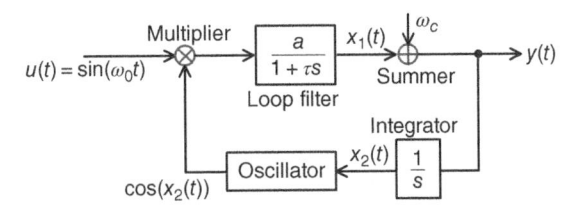

Figure P6.4.1 Block diagram of a PLL circuit.

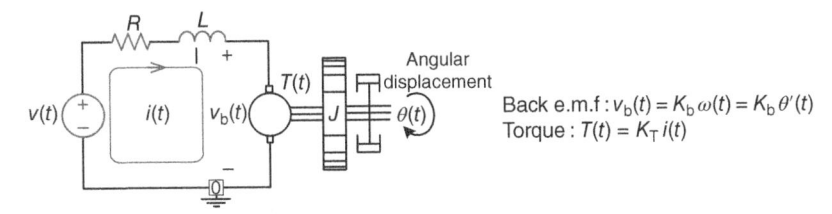

Figure P6.4.2 A DC motor system.

Back e.m.f: $v_b(t) = K_b \omega(t) = K_b \theta'(t)$
Torque: $T(t) = K_T i(t)$

(f) *DC motor*

Consider a linear DE describing the behavior of a DC motor system (Figure P6.4.2)

$$J\frac{d^2\theta(t)}{dt^2} + B\frac{d\theta(t)}{dt} = T(t) = K_T i(t)$$

$$L\frac{di(t)}{dt} + R\,i(t) + K_b\frac{d\theta(t)}{dt} = v(t)$$

(P6.4.6)

Convert this system of equations into a first-order vector DE, i.e. a state equation with respect to (w.r.t.) the state vector $[\theta(t)\ \ \theta'(t)\ \ i(t)]$.

(g) *RC circuit*: a Stiff system

Consider a two-mesh RC circuit depicted in Figure P6.4.3. We can write the mesh equation with respect to the two mesh currents $i_1(t)$ and $i_2(t)$ as

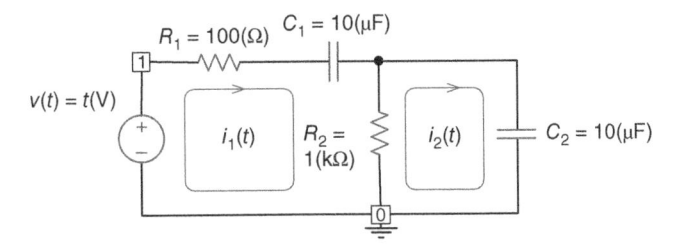

Figure P6.4.3 A two-mesh *RC* circuit.

348 ORDINARY DIFFERENTIAL EQUATIONS

$$R_1 i_1(t) + \frac{1}{C_1} \int_{-\infty}^{t} i_1(\tau) d\tau + R_2(i_1(t) - i_2(t)) = v(t) = t$$

$$R_2(i_2(t) - i_1(t)) + \frac{1}{C_2} \int_{-\infty}^{t} i_2(\tau) d\tau = 0$$

(P6.4.7.1)

To convert this system of DEs into a state equation, we differentiate both sides and rearrange them to get

$$(R_1 + R_2) \frac{di_1(t)}{dt} - R_2 \frac{di_2(t)}{dt} + \frac{1}{C_1} i_1(t) = \frac{dv(t)}{dt} = 1$$

$$-R_2 \frac{di_1(t)}{dt} + R_2 \frac{di_2(t)}{dt} + \frac{1}{C_2} i_2(t) = 0$$

$$; \begin{bmatrix} R_1 + R_2 & -R_2 \\ -R_2 & R_2 \end{bmatrix} \begin{bmatrix} i_1'(t) \\ i_2'(t) \end{bmatrix} = \begin{bmatrix} 1 - i_1(t)/C_1 \\ -i_2(t)/C_2 \end{bmatrix}$$

(P6.4.7.2)

$$; \begin{bmatrix} i_1'(t) \\ i_2'(t) \end{bmatrix} = \begin{bmatrix} R_1 + R_2 & -R_2 \\ -R_2 & R_2 \end{bmatrix}^{-1} \begin{bmatrix} 1 - i_1(t)/C_1 \\ -i_2(t)/C_2 \end{bmatrix}$$

$$= \begin{bmatrix} -G_1/C_1 & -G_1/C_2 \\ -G_1/C_1 & -(G_1 + G_2)/C_2 \end{bmatrix} \begin{bmatrix} i_1(t) \\ i_2(t) \end{bmatrix} + \begin{bmatrix} G_1 \\ G_1 \end{bmatrix} u_s(t) \quad (P6.4.7.3)$$

where $u_s(t)$ denotes the unit step function whose value is 1 $\forall$ $t \geq 0$.

(i) After constructing an M-file function "df6p03g", which defines Eq. (P6.4.7.4) with $R_1 = 100\,\Omega$, $C_1 = 10\,\mu F$, $R_2 = 1\,k\Omega$, $C_2 = 10\,\mu F$, use the MATLAB built-in functions 'ode45()' and 'ode23s()' to solve the state equation with the zero initial condition $i_1(0) = i_2(0) = 0$ and plot the numerical solution $i_2(t)$ for $0 \leq t \leq 0.05$ seconds. For possible change of parameters, you may declare R_1, C_1, R_2, C_2 as global variables both in the function and in the script named, say, "nm06p03g". Do you see any symptom of stiffness from the results?

(ii) If we apply the Laplace transform technique to solve this equation with zero initial condition $\mathbf{i}(0) = 0$, we can get

$$\begin{bmatrix} I_1(s) \\ I_2(s) \end{bmatrix} \overset{(6.5.5)}{=} [sI - A]^{-1} BU(s)$$

$$= \begin{bmatrix} s + G_1/C_1 & G_1/C_2 \\ G_1/C_2 & s + (G_1 + G_2)/C_2 \end{bmatrix}^{-1} \begin{bmatrix} G_1 \\ G_1 \end{bmatrix} \frac{1}{s}$$

$$;I_2(s) = \frac{G_1}{s^2 + (G_1/C_1 + (G_1+G_2)/C_2)s + G_1G_2/C_1C_2}$$

$$= \frac{1/100}{s^2 + 2100s + 100\,000} \cong \frac{1/100}{(s+2051.25)(s+48.75)}$$

$$\cong \frac{1}{200\,250}\left(\frac{1}{s+48.75} - \frac{1}{s+2051.25}\right)$$

$$;i_2(t) \cong \frac{1}{200\,250}(e^{-48.75t} - e^{-2051.25t}) \quad \text{(P6.4.7.4)}$$

where $\lambda_1 = -2051.25$ and $\lambda_2 = -48.75$ are actually the eigenvalues of the system matrix A in Eq. (P6.4.7.4). Find the measure of stiffness defined by Eq. (6.6.26).

(iii) Using the MATLAB symbolic DE solver 'dsolve()', find the analytical solution of the differential equation (P6.4.7.2) and plot $i_2(t)$ for $0 \le t \le 0.05$ seconds. For possible change of parameters, you may declare R_1, C_1, R_2, C_2 as global together with (P6.4.7.5) for $0 \le t \le 0.05$ seconds. Which of the two numerical solutions obtained in (i) is better? You may refer to the following code:

```
syms R1 R2 C1 C2
i=dsolve('(R1+R2)*Di1-R2*Di2+i1/C1=1',...
  '-R2*Di1+R2*Di2+i2/C2','i1(0)=0','i2(0)=0'); % Eq.(P6.3.7.2)
R1=100; R2=1000; C1=1e-5; C2=1e-5;
t0=0; tf=0.05; t=t0+(tf-t0)/100*[0:100];
i2t=eval(i.i2); plot(t,i2t,'m')
```

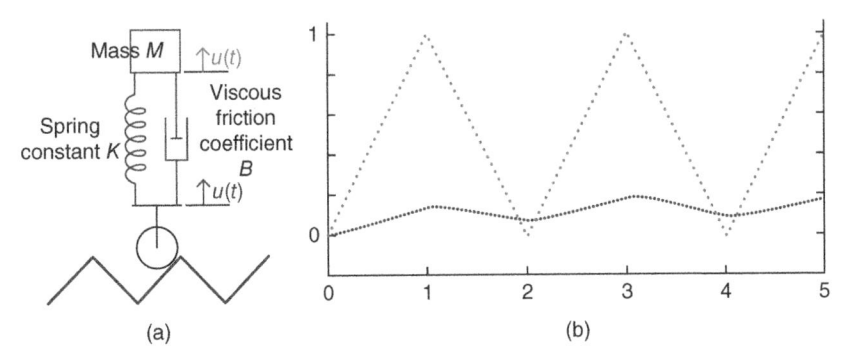

Figure P6.5 A model for vehicle suspension system. (a) The block diagram and (b) the graphs of the input $u(t)$ and the output $y(t)$.

6.5 Physical Meaning of a Solution for Differential Equation and its Animation
Suppose we are going to simulate how a vehicle vibrates when it moves with a constant speed on a rugged way, as depicted in Figure P6.5a. Based on Newton's second law, the situation is modeled by the following DE:

```
%do_MBK.m
global M B K g
t0=0; tf=5; x0=[0 0];
[t1,x]=ode_Ham('f_MBK',[t0 tf],x0);
%[t1,x]=ode45('f_MBK',[t0 tf],x0);
dt=t1(2)-t1(1);
u= udu_MBK(t1);
animation=1;
if animation
  subplot(211), cla
  draw_MBK(5,1,x(1,2),u(1))
  axis([-2 2 -1 14]), axis('equal')
  pause
  for n=1:length(t1)
    cla, draw_MBK(5,1,x(n,2),u(n),'b')
    axis([-2 2 -1 14]), axis('equal')
    F(n) = getframe; pause(dt)
    subplot(212)
    plot(t1(n),u(n),'r.', t1(n),x(n,2),'b.')
    axis([0 tf -0.2 1.2]), hold on
    subplot(211)
  end
  draw_MBK(5,1,x(n,2),u(n)), axis([-2 2 -1 14]), axis('equal')
end
movie(F)
```

```
function [u,du]=udu_MBK(t)
t1 = mod(t,2);
u = (t1<=1).*t1 + (t1>1).*(2-t1);
du = (t1<=1) - (t1>1);
```

```
function draw_MBK(n,w,y,u,color)
% n: Number of spring windings
% w: Width of each object
% y: Displacement of the top of MBK
% u: Displacement of the bottom of MBK
if nargin<5, color='k'; end
p1=[-w  u+4];  p2=[-w  9+y];
xm=0; ym=(p1(2)+p2(2))/2;
xM= xm+w*1.2*[-1 -1  1  1 -1];
yM= p2(2)+w*[1 3 3 1 1];
plot(xM,yM,color), hold on % Mass
spring(n,p1,p2,w,color) % Spring
damper(xm+w,p1(2),p2(2),w,color) % Damper
wheel_my(xm,p1(2)-3*w,w,color) % Wheel
```

```
function dx=f_MBK(t,x)
global M B K g
[u,du]= udu_MBK(t);
if size(x,1)>1, dx=A*x+[0;(K*u+B*du)/M]; % B=[0; -g+(K*u+B*du)/M];
   else    dx = x*A' + [0 (K*u+B*du)/M];
end
```

```
function spring(n,p1,p2,w,color)
% draws a spring of n windings, width w from p1 to p2
if nargin<5, color='k'; end
c= (p2(1)-p1(1))/2;  d= (p2(2)-p1(2))/2;
f= (p2(1)+p1(1))/2;  g= (p2(2)+p1(2))/2;
y= -1:0.01:1;  t= (y+1)*pi*(n+0.5);
x=-0.5*w*sin(t);  y= y+0.15*(1-cos(t));
a=y(1); b=y(length(x));
y= 2*(y-a)/(b-a)-1;
yyS= d*y - c*x +g;  xxS= x+f;  xxS1= [f f];
yyS1= yyS(length(yyS))+[0 w]; yyS2= yyS(1)-[0 w];
plot(xxS,yyS,color, xxS1,yyS1,color, xxS1,yyS2,color)
```

```
function damper(xm,y1,y2,w,color)
% draws a damper in (xm-0.5 xm+0.5 y1 y2)
if nargin<5, color='k'; end
ym=(y1+y2)/2;
xD1= xm+w*[0.3*[0  0  -1  1]];   yD1= [y2+w  ym  ym  ym];
xD2= xm+w*[0.5*[-1 -1  1  1]];   yD2= ym+w*[1  -1  -1  1];
xD3= xm+[0  0];     yD3= [y1 ym]-w;
plot(xD1,yD1,color, xD2,yD2,color, xD3,yD3,color)
```

```
function wheel_my(xm,ym,w,color)
%draws a wheel of size w at center (xm,ym)
if nargin<5, color='k'; end
xW1= xm+w*1.2*[-1  1];   yW1= ym+w*[2  2];
xW2= xm*[1  1];          yW2= ym+w*[2  0];
plot(xW1,yW1,color, xW2,yW2,color)
th=[0:100]/50*pi;
plot(xm+j*ym+w*exp(j*th),color)
```

$$M\frac{d^2}{dt^2}y(t) + B\frac{d}{dt}(y(t) - u(t)) + K(y(t) - u(t)) = 0 \qquad (P6.5.1)$$

with $y(0) = 0$, $y'(0) = 0$

where the values of the mass, the viscous friction coefficient, and the spring constant are given as $M = 1$ kg, $B = 0.1$ N s/m, and $K = 0.1$ N/m, respectively. The input to this system is the movement $u(t)$ of the wheel part causing the movement $y(t)$ of the body as the output of the system and is approximated to a triangular wave of height 1 m, duration 1 second and period 2 seconds as depicted in Figure P6.5b. After converting this equation into a state equation as

$$\begin{bmatrix} x_1'(t) \\ x_2'(t) \end{bmatrix} = \begin{bmatrix} 0 & 1 \\ -K/M & -B/M \end{bmatrix} \begin{bmatrix} x_1(t) \\ x_2(t) \end{bmatrix} + \begin{bmatrix} 0 \\ (B/M)u'(t) + (K/M)u(t) \end{bmatrix} \quad \text{(P6.5.2)}$$

$$\text{with } \begin{bmatrix} x_1(0) \\ x_2(0) \end{bmatrix} = \begin{bmatrix} 0 \\ 0 \end{bmatrix}$$

we can use such MATLAB functions as 'ode_Ham()', 'ode45()', ... to solve this state equation and use some graphic functions to draw not only the graphs of $y(t)$ and $u(t)$ but also the animated simulation diagram. You can run the above MATLAB script "do_MBK.m" to see the results. Does the suspension system made of a spring and a damper as depicted in Figure P6.5a absorbs effectively the shock caused by the rolling wheel so that the amplitude of vehicle body oscillation is less than 1/5 times that of wheel oscillation?

(cf) If one is interested in graphic visualization with MATLAB, he/she can refer to [N-1].

6.6 A Nonlinear Differential Equation for an Orbit of a Satellite

Consider the problem of an orbit of a satellite, whose position and velocity are obtained as the solution of the following state equation:

$$\begin{aligned} x_1'(t) &= x_3(t) \\ x_2'(t) &= x_4(t) \\ x_3'(t) &= -GM_E x_1(t)/(x_1^2(t) + x_2^2(t))^{3/2} \\ x_4'(t) &= -GM_E x_2(t)/(x_1^2(t) + x_2^2(t))^{3/2} \end{aligned} \quad \text{(P6.6.1)}$$

where $G = 6.672 \times 10^{-11} \, \text{N m}^2/\text{kg}^2$ is the gravitational constant, and $G_M = 5.97 \times 10^{24}$ kg is the mass of the Earth. Note that (x_1, x_2) and (x_3, x_4) denote the position and velocity, respectively, of the satellite on the plane having the Earth at its origin. This state equation is defined in the M-file "df_sat.m" below.

(a) Supplement the MATLAB script "nm06p05.m", which uses the three functions 'ode_RK4()', 'ode45()', and 'ode23()' to find the paths of the satellite with the following initial positions/velocities for one day.

  (i) $(x_{10}, x_{20}) = (4.223 \times 10^7, 0)$ (m)
  and $(x_{30}, x_{40}) = (v_{10}, v_{20}) = (0, 3071)$ (m/s).
  (ii) $(x_{10}, x_{20}) = (4.223 \times 10^7, 0)$ (m)
  and $(x_{30}, x_{40}) = (v_{10}, v_{20}) = (0, 3500)$ (m/s).
  (iii) $(x_{10}, x_{20}) = (4.223 \times 10^7, 0)$ (m)
  and $(x_{30}, x_{40}) = (v_{10}, v_{20}) = (0, 2000)$ (m/s).

Run the script and check if the plotting results are as shown in Figure P6.6.

```
%nm06p05.m to solve a nonlinear DE on the orbit of a satellite
global G Me Re
G=6.67e-11; Me=5.97e24; Re=64e5;
dx='df_sat';
t0=0; T=24*60*60; tf=T; tspan=[t0 tf]; N=2000;
v20s=[3071 3500 2000];
for iter=1:length(v20s)
   x10=4.223e7; x20=0; v10=0; v20=v20s(iter);
   x0= [x10 x20 v10 v20]; tol=1e-6;
   [tR,xR]=ode_RK4(dx,????,x0,?);
   [t45,x45]= ode45(dx,tspan,??);
   [t23s,x23s]= ode23s(??,tspan,x0);
   plot(xR(:,1),xR(:,2), x45(:,1),x45(:,2),'k.', ...
        x23s(:,1),x23s(:,2),'rx'), hold on
   [t451,x451]= ode45(dx,tspan,x0,odeset('RelTol',???));
   [t23s1,x23s1]= ode23s(dx,tspan,x0,odeset('??????',tol));
   plot(x451(:,1),x451(:,2),'m:', x23s1(:,1),x23s1(:,2),'g.-')
   ers=[error_DE_sol(dx,xR,tR) error_DE_sol(dx,x45,t45) ...
        error_DE_sol(dx,x23s,t23s) ...
          error_DE_sol(dx,x451,t451) error_DE_sol(dx,x23s1,t23s1)]
end
```

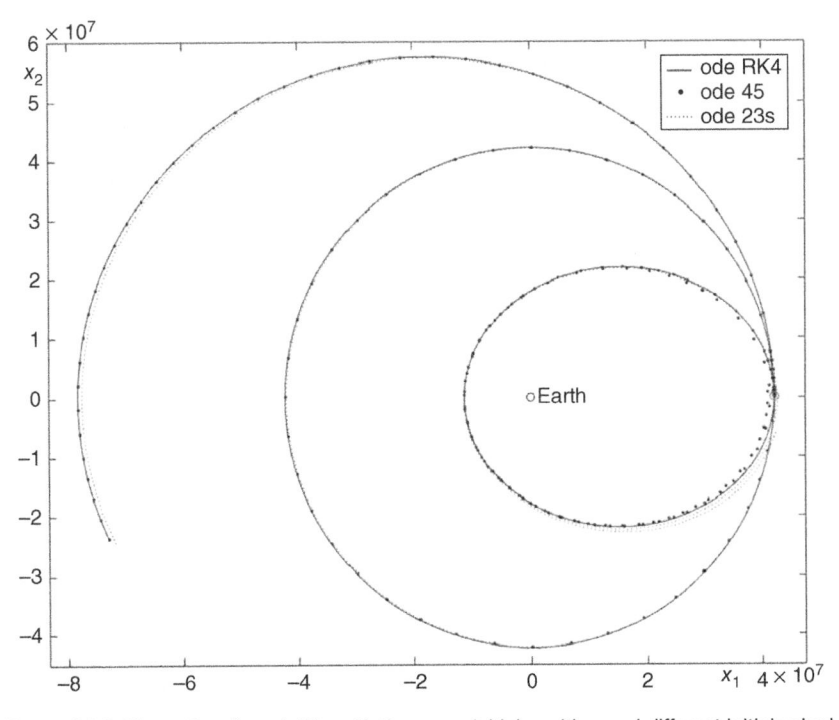

Figure P6.6 The paths of a satellite with the same initial position and different initial velocities. (a) An RC diode circuit for half-wave rectification. (b) $v_o(t)$ obtained from PSpice simulation. (c) $v_o(t)$ obtained from MATLAB simulation.

```
function dx=df_sat(t,x)
global G Me Re
dx=zeros(size(x));
r=sqrt(sum(x(1:2).^2)); GMr3=G*Me/r^3;
if r<=Re, return; end % when colliding against the Earth surface
dx(1)=x(3); dx(2)=x(4); dx(3)=-GMr3*x(1); dx(4)=-GMr3*x(2);
```

```
function err=error_DE_sol(dx,xn,tt,varargin)
%Input:    dx = DE defined as an inline function or in an M-file
%          xn = Numerical solution,
%          tt = Time vector (as a column vector)
%Output:  err = Relative error between dxdt and dxdt
%               computed from xn
M=size(xn,2); tt=tt(:); dt=repmat(tt(3:end)-tt(1:end-2),1,M);
dxdt_approximate=(xn(3:end,:)-xn(1:end-2,:))./dt;
for n=1:length(tt)-2
   tmp=feval(dx,tt(n+1),xn(n+1,:),varargin{:});
   dxdt(n,:)=reshape(tmp(1:M),1,M);
end
err=norm(dxdt-dxdt_approximate)/norm(dxdt); % Relative error
```

(b) In Figure P6.6, we see that the solution path obtained from 'ode23s()' differs from the others for case (ii), and the paths from 'ode45()' and 'ode23s()' differ from the one from 'ode_RK4()' for case (iii). But we do not know which one is more accurate. To find which one is closer to the true solution, use the two MATLAB functions 'ode45()' and 'ode23s()' with smaller relative error tolerance of tol=1e-6 to find the paths for the three cases. Which one do you think is the closest to the true solution among the paths obtained in (a)? If helpful, use the MATLAB function 'error_DE_sol()' listed below.

(cf) The purpose of this problem is not to compare the several MATLAB functions but to warn the users of the danger of abusing them where it is sometimes very important for obtaining a reasonably accurate solution to set the parameters such as the relative error tolerance (RelTol).

6.7 Solving a Half-Wave Rectifier Circuit

Figure P6.7a shows an *RC* diode circuit for half-wave rectification for which the circuit equation can be written as

$$C\frac{dv_o(t)}{dt} + \frac{v_o(t)}{R} - i_D(t) = 0 \qquad (P6.7.1)$$

with $i_D = I_s(e^{v_D/V_T} - 1) = I_s(e^{(v_s-v_o)/V_T} - 1)$

where $R = 10\,k\Omega$, $C = 5\,\mu F$, $v_s(t) = 5\sin(2\pi 60 t)$ V, $I_s = 10^{-15}$ A, and $V_T = 25$ mV.

(a) Noting that replacing the first derivative in the DE (P6.7.1) with its forward difference approximation (5.1.2) yields the following difference equation:

```
%nm06p07a.m
% To solve the RC diode circuit in Fig P6.7(a)
R=1e4; C=5e-6;
VT=25e-3; Is=1e-15; % Thermal voltage, Scale current of D
iD=@(vD)Is*(exp(vD/VT)-1);
vs=@(t)5*sin(2*pi*60*t); % Input source voltage
t0=0; tf=0.05; N=1000; tt=linspace(t0,tf,N+1); % Time vector
vst=vs(tt);  % Input voltage waveform,
vo(1)=0; % initial output voltage
for n=2:length(tt)
    t=tt(n); dt=t-tt(n-1);
    vo(n)=vo(n-1)+dt/?*(iD(vs(t)-vo(?-1))-vo(n-?)/?);
end
ttm=1e3*tt; plot(ttm,vst, ttm,vo,'k'), hold on
xlabel('Time[ms]'); axis([t0 tf*1000 -5.1 5.1])
```

$$C\frac{v_o(t+\Delta t)-v_o(t)}{\Delta t}+\frac{v_o(t)}{R}-I_s(e^{(v_s-v_o)/V_T}-1)=0$$

$$\xrightarrow{\text{Discretization}} v_o[n+1]=v_o[n]+\frac{\Delta t}{C}\left\{I_s(e^{(v_s[n+1]-v_o[n])/V_T}-1)-\frac{v_o[n]}{R}\right\}$$

(P6.7.2)

complete and run the above MATLAB script "nm06p07a.m" to plot the output voltage $v_o(t)$.

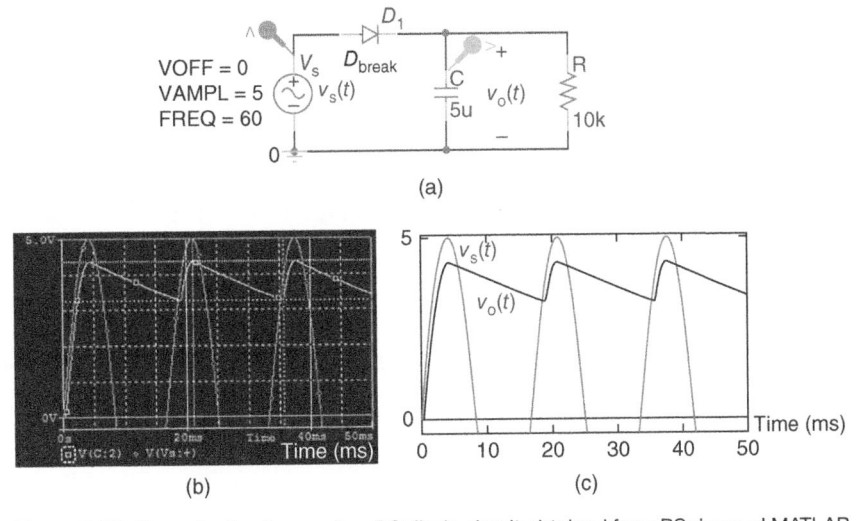

Figure P6.7 The output voltages of an *RC* diode circuit obtained from PSpice and MATLAB simulations.

(b) To solve the nonlinear DE (P6.7.1) by using the numerical DE solvers 'ode_RK4()', 'ode45()', and 'ode15s()' (with the relative error tolerance smaller than its default value 10^{-3}), complete and run the following script "nm06p06b.m" to plot the output voltage $v_o(t)$. Do you see any of the five statements in the 4th, 6th, 8th, 11th, and 13th lines fails to yield a proper solution?

```
%nm06p07b.m
dv=@(t,v)(iD(vs(t)-v)-v/R)/C;  % DE to be solved
tspan=[t0 tf]; N=1000;
[t,vo_RK4]=ode_RK4(dv,tspan,vo(?),N);
tm_RK4=t*1000; plot(tm_RK4,vo_RK4,'m:'), pause
[t,vo_45]=ode45(??,tspan,vo(1));  % yielding a weird solution
tm_45=t*1000; plot(tm_45,vo_45,'rx'), pause
[t,vo_15s]=ode15s(dv,tspan,vo(1));  % yielding a wrong sol
tm_15s=t*1000; plot(tm_15s,vo_15s,'g:'), pause
options=odeset('RelTol',1e-6);
[t,vo_451]=ode45(dv,?????,vo(1),options);  % Another weird sol
tm_45=t*1000; plot(tm_45,vo_451,'ro'), pause
[t,vo_15s1]=ode15s(dv,tspan,vo(1),????????);
tm_15s=t*1000; plot(tm_15s,vo_15s1,'gs')
```

6.8 Shooting Method for BVP with adjustable position and fixed angle (derivative) Suppose the boundary condition for a second-order BVP is given as

$$x'(t_0) = x_{20}, \quad x(t_f) = x_{1f} \quad \text{(P6.8.1)}$$

Consider how to modify the MATLAB functions 'bvp2_shoot()' and 'bvp2_fdf()' so that they can accommodate this kind of problem.

(a) As for 'bvp2_shootp()' that you should make, the variable quantity to adjust for improving the approximate solution is not the derivative $x'(t_0)$, but the position $x(t_0)$ and what should be made close to zero is still $f(x(t_0)) = x(t_f) - x_f$. Modify the MATLAB function in such a way that $x(t_0)$ is adjusted to make this quantity close to zero and make its declaration part have the initial derivative (dx0) instead of the initial position (x0) as the fourth input argument as follows:

```
function [t,x]= bvp2_shootp(f,t0,tf,dx0,xf,N,tol,kmax)
```

Noting that the initial derivative of the true solution for Eq. (E6.3.11) is zero, apply this MATLAB function to solve the BVP by inserting the following statement into the script "do_shoot".

```
[t,x1]= bvp2_shootp(df,t0,tf,0,xf,N,tol,kmax);
```

and plot the result to check if it conforms with that (Figure 6.8) obtained by 'bvp2_shoot()'.

(b) As for 'bvp2_fdfp()' implementing the finite difference method, you have to approximate the boundary condition as

$$x'(t_0) = x_{20} \rightarrow \frac{x_1 - x_{-1}}{2h} = x_{20}; \quad x_{-1} = x_1 - 2hx_{20} \quad (P6.8.2)$$

substitute this into the finite difference equation corresponding to the initial time as

$$\frac{x_1 - 2x_0 + x_{-1}}{h^2} + a_{10}\frac{x_1 - x_{-1}}{2h} + a_{00}x_0 = u_0 \quad (P6.8.3)$$

$$; \frac{x_1 - 2x_0 + x_1 - 2hx_{20}}{h^2} + a_{10}x_{20} + a_{00}x_0 = u_0$$

$$; (a_{00}h^2 - 2)x_0 + 2x_1 = h^2 u_0 + h(2 - ha_{10})x_{20} \quad (P6.8.4)$$

and augment the matrix-vector equation with this equation. Also, make its declaration part to have the initial derivative (dx0) instead of the initial position (x0) as the sixth input argument as follows:

```
function [t,x]=bvp2_fdfp(a1,a0,u,t0,tf,dx0,xf,N)
```

Noting that the initial derivative of the true solution for Eq. (E6.4.1) is -7, apply this MATLAB function to solve the BVP by inserting the following statement into the script "do_fdf".

```
[t,x1]= bvp2_fdfp(a1,a0,u,t0,tf,-7,xf,N);
```

and plot the result to check if it conforms with that obtained by using 'bvp2_fdf()' and depicted in Figure 6.9.

6.9 BVP with Mixed-Boundary Conditions I
Suppose the BC for a second-order BVP is given as

$$x(t_0) = x_{10}, \quad c_1 x(t_f) + c_2 x'(t_f) = c_3 \quad (P6.9.1)$$

Consider how to modify the MATLAB functions 'bvp2_shoot()' and 'bvp2_fdf()' so that they can accommodate this kind of problem.

(a) As for 'bvp2_shoot()' that you should modify, the variable quantity to adjust for improving the approximate solution is still the derivative $x'(t_0)$, but what should be made close to zero is

$$f(x'(t_0)) = c_1 x(t_f) + c_2 x'(t_f) - c_3 \quad (P6.9.2)$$

If you do not know where to begin, modify the MATLAB function 'bvp2_shoot()' in such a way that $x'(t_0)$ is adjusted to make this quantity close to zero. Regarding the quantity (P6.9.2) as a function of $x'(t_0)$, you may feel as if you were going to solve a nonlinear equation $f(x'(t_0)) = 0$. Here are a few hints for this job:

- Make the declaration part have the boundary coefficient vector cf=[c1 c2 c3] instead of the final position (xf) as the fifth input argument as follows:

    function [t,x]=bvp2m_shoot(f,t0,tf,x0,cf,N,tol,kmax)

- Pick up the first two guesses of $x'(t_0)$ arbitrarily.
- You may need to replace a couple of statements in 'bvp2_shoot()' by

    e(1)=cf*[x(end,:)';-1];
    e(k)=cf*[x(end,:)';-1];

Now that you have the MATLAB function 'bvp2m_shoot()' of your own making, do not hesitate to try using the weapon to attack the following problem:

$$x''(t) - 4t\,x(t)x'(t) + 2x^2(t) = 0 \quad \text{with } x(0) = \frac{1}{4}, \quad 2x(1) - 3x'(1) = 0$$
(P6.9.3)

For this job, you only have to modify one statement of the script "nm06e03.m" (Section 6.6.1) into

    [t,x]=bvp2m_shoot(df,t0,tf,x0,[2 -3 0],N,tol,kmax);

If you run it to obtain the same solution as depicted in Figure 6.8, you deserve to be proud of yourself having this book as well as MATLAB; otherwise, just keep trying until you succeed.

(b) As for 'bvp2_fdf()' that you should modify, you have only to augment the matrix-vector equation with one row corresponding to the approximate version of the boundary condition $c_1 x(t_f) + c_2 x'(t_f) = c_3$, i.e.

$$c_1 x_N + c_2 \frac{x_N - x_{N-1}}{h} = c_3; \quad -c_2 x_{N-1} + (c_1 h + c_2) x_N = c_3 h \quad \text{(P6.9.4)}$$

Needless to say, you should increase the dimension of the matrix A to N and move the x_N term on the RHS of the $(N-1)$th row back to the LHS by incorporating the corresponding statement into the for loop. What you have to do with 'bvp2m_fdf()' for this job is as follows:

- Make the declaration part have the boundary coefficient vector cf=[c1 c2 c3] instead of the final position (xf) as the seventh input argument.

    function [t,x]=bvp2m_fdf(a1,a0,u,t0,tf,x0,cf,N)

- Replace some statement by A=zeros(N,N);
- Increase the last index of the for loop to N-1.
- Replace the statements corresponding to the $(N-1)$th row equation by

    A(N,N-1:N) = [-cf(2) cf(1)*h+cf(2)]; b(N)=cf(3)*h;

which implements Eq. (P6.9.4).

- Modify the last statement arranging the solution as

    ```
    x=[x0 trid(A,b)']';
    ```

Now that you have the MATLAB function 'bvp2m_fdf()' of your own making, do not hesitate to try it on the following problem:

$$x''(t) + \frac{2}{t}x'(t) - \frac{2}{t^2}x(t) = 0 \quad \text{with} \quad x(1) = 5, \quad x(2) + x'(2) = 3 \quad \text{(P6.9.5)}$$

For this job, you only have to modify one statement of the script "do_fdf" (Section 6.6.2) into

```
[t,x]=bvp2m_fdf(a1,a0,u,t0,tf,x0,[1 1 3],N);
```

You might need to increase the number of segments N to improve the accuracy of the numerical solution. If you run it to obtain the same solution as depicted in Figure 6.9, be happy with it.

6.10 BVP with Mixed-Boundary Conditions II
Suppose the boundary condition for a second-order BVP is given as

$$c_{01}x(t_0) + c_{02}x'(t_0) = c_{03} \quad \text{(P6.10.1a)}$$

$$c_{f1}x(t_f) + c_{f2}x'(t_f) = c_{f3} \quad \text{(P6.10.1b)}$$

Consider how to modify the MATLAB functions 'bvp2m_shoot()' and 'bvp2m_fdf()' so that they can accommodate this kind of problems.

(a) As for 'bvp2mm_shoot()' that you should make, the variable quantity to be adjusted for improving the approximate solution is $x'(t_0)$ or $x(t_0)$ depending on whether or not $c_{01} \neq 0$, while the quantity to be made close to zero is still

$$f(x(t_0), x'(t_0)) = c_{f1}x(t_f) + c_{f2}x'(t_f) - c_{f3} \quad \text{(P6.10.2)}$$

If you do not have your own idea, modify the MATLAB function 'bvp2m_shoot()' in such a way that $x'(t_0)$ or $x(t_0)$ is adjusted to make this quantity close to zero and $x(t_0)$ or $x'(t_0)$ is set by (P6.10.1a), making its declaration as

```
function [t,x]=bvp2mm_shoot(f,t0,tf,c0,cf,N,tol,kmax)
```

where the boundary coefficient vectors c0=[c01 c02 c03] and cf=[cf1 cf2 cf3] are supposed to be given as the fourth and fifth input arguments, respectively.

Now that you get the MATLAB function 'bvp2mm_shoot()' of your own making, try it for the following problem:

360 ORDINARY DIFFERENTIAL EQUATIONS

$$x''(t) - \frac{2t}{t^2+1}x'(t) + \frac{2}{t^2+1}x(t) = t^2 + 1 \quad \text{(P6.10.3)}$$

with $x(0) + 6x'(0) = 0$, $x(1) + x'(1) = 0$

(b) As for 'bvp2_fdf()' implementing the finite difference method, you only have to augment the matrix-vector equation with two rows corresponding to the approximate versions of the boundary conditions $c_{01}x(t_0) + c_{02}x'(t_0) = c_{03}$ and $c_{f1}x(t_f) + c_{f2}x'(t_f) = c_{f3}$, i.e.

$$c_{01}x_0 + c_{02}\frac{x_1 - x_0}{h} = c_{03}; \quad (c_{01}h - c_{02})x_0 + c_{02}x_1 = c_{03}h \quad \text{(P6.10.4a)}$$

$$c_{f1}x_N + c_{f2}\frac{x_N - x_{N-1}}{h} = c_{f3}; \quad -c_{f2}x_{N-1} + (c_{f1}h + c_{f2})x_N = c_{f3}h$$
(P6.10.4b)

Now that you have the MATLAB function 'bvp2mm_fdf()' of your own making, try it for the problem described by Eq. (P6.10.3).

(c) Overall, you will need to make the scripts like "nm06p09a" and "nm06p09b" that use the MATLAB functions 'bvp2mm_shoot()' and 'bvp2mm_fdf()' to get the numerical solutions of Eq. (P6.10.3) and plot them. Additionally, use the MATLAB function 'bvp4c()' to get another solution and plot it together for crosscheck.

```
%nm06p10a.m: to solve BVP2 with mixed boundary conditions
f=@(t,x)[x(2); 2*(t*x(2)-x(1))./(t.^2+1)+(t.^2+1)];
    % Eq.(P6.8.3)
t0=0; tf=1; N=100; tol=1e-8; kmax=10;
c0=[1 6 0]; cf=[1 1 0]; % Coefficient vectors of BC
[tt,x_sh]=bvp2mm_shoot(f,t0,tf,c0,cf,N,tol,kmax);
plot(tt,x_sh(:,1))
```

```
%nm06p10b.m
a1=@(t)-2*t./(t.^2+1); a0=@(t)2./(t.^2+1); u=@(t)t.^2+1;
t0=0; tf=1; N=500;
c0=[1 6 0]; cf=[1 1 0]; % Coefficient vectors of BC
[tt,x_fd]=bvp2mm_fdf(a1,a0,u,t0,tf,c0,cf,N);
```

```
%nm06p10c.m
fbc=@(x0,xf)[c0(1)*x0(1)+c0(2)*x0(2); cf(1)*xf(1)+cf(2)*xf(2)];
x_guess=[x0 0]; solinit=bvpinit(linspace(t0,tf,5),x_guess);
sol = bvp4c(df,fbc,solinit);
```

6.11 Shooting Method and Finite Difference Method for Linear BVPs

Use the MATLAB functions 'bvp2_shoot()', 'bvp2_fdf()', and 'bvp4c()' to solve the following BVPs. Plot the solutions and fill in Table P6.11 with the mismatching errors (of the numerical solutions) that are defined for

$$y''(x) = f(y'(x), y(x), u(x)) \qquad \text{(P6.11.1a)}$$

with the boundary condition $y(x_0) = y_0$, $y(x_f) = y_f$

as

$$\text{err} = \frac{1}{N-1} \sum_{i=1}^{N-1} \{D^{(2)}y(x_i) - f(Dy(x_i), y(x_i), u(x_i))\}^2 \qquad \text{(P6.11.1b)}$$

with

$$D^{(2)}y(x_i) = \frac{y(x_{i+1}) - 2y(x_i) + y(x_{i-1})}{h^2}, \quad Dy(x_i) = \frac{y(x_{i+1}) - y(x_{i-1})}{2h} \qquad \text{(P6.11.1c)}$$

$$x_i = x_0 + ih, \qquad h = \frac{x_f - x_0}{N} \qquad \text{(P6.11.1d)}$$

and can be computed by using the following function 'error_DE2_sols()'.

Overall, which one works the best for linear BVPs among the three MATLAB functions?

(a) $y''(x) = y'(x) - y(x) + 3e^{2x} - 2\sin x$ with $y(0) = 5$, $y(2) = -10$ (P6.11.2)

(b) $y''(x) = -4y(x)$ with $y(0) = 5$, $y(1) = -5$ (P6.11.3)

(c) $y''(t) = 10^{-6}y(t) + 10^{-7}(t^2 - 50t)$ with $y(0) = 0$, $y(50) = 0$ (P6.11.4)

(d) $y''(t) = -2y(t) + \sin t$ with $y(0) = 0$, $y(1) = 0$ (P6.11.5)

(e) $y''(x) = y'(x) + y(x) + e^x(1 - 2x)$ with $y(0) = 1$, $y(1) = 3e$ (P6.11.6)

(f) $\dfrac{d^2y(r)}{dr^2} + \dfrac{1}{r}\dfrac{dy(r)}{dr} = 0$ with $y(1) = \ln 1$, $y(2) = \ln 2$ (P6.11.7)

```
function err=error_DE2_sols(f,t,x,varargin)
% estimates the errors of solutions to,
% possibly multiple 2nd-order DEs:
%     x"=f(t,[x x'])
[Nt,Nx]=size(x);
if Nt<Nx, x=x.'; [Nt,Nx]=size(x); end
n1=2:Nt-1;   t=t(:); h2s=t(n1+1)-t(n1-1);
dx= (x(n1+1,:)-x(n1-1,:))./(h2s*ones(1,Nx));
num= x(n1+1,:)-2*x(n1,:)+x(n1-1,:);
den= (h2s/2).^2*ones(1,Nx);
d2x=num./den;
for m=1:Nx
  for n=n1(1):n1(end)
    fx=feval(f,t(n),[x(n,m) dx(n-1,m)],varargin{:});
    errm(n-1,m)=d2x(n-1,m)-fx(end);
  end
end
err=sum(errm.^2)/(Nt-2);
```

362 ORDINARY DIFFERENTIAL EQUATIONS

```
%nm06p11a.m
%y"-y'+y=3*e^2x-2sin(x) with y(0)=5 & y(2)=-10
x0=0; xf=2; y0=5; yf=-10; N=100; tol=1e-6; kmax=10;
df=@(x,y)[y(2); y(2)-y(1)+3*exp(2*x)-2*sin(x)];
a1=??; a0=1; u=@(x)3*exp(2*x)-2*sin(x);

y_guess=[mean([y0 yf]) (yf-y0)/(xf-x0)];
solinit = bvpinit(linspace(x0,xf,5),y_guess); %[1 9]
fbc=@(ydy0,ydyf)[ydy0(?)-y0; ydyf(1)-y?];

% Shooting method
tic, [xx,y_sh]=bvp2_shoot(df,x0,x?,y?,yf,N,tol,kmax);
times(1)=toc;

% Finite difference method
tic, [xx,y_fd]=bvp2_fdf(a?,a0,u,x?,xf,y0,y?,N); times(2)=toc;

% MATLAB built-in function bvp4c
tic, sol=bvp4c(df,fbc,solinit,bvpset('RelTol',1e-6));
y_bvp=deval(sol,xx); times(3)=toc

% Error evaluation
ys=[y_sh(:,1) y_fd y_bvp(1,:)']; plot(xx,ys)
err=error_DE2_sols(df,xx,yy)
```

Table P6.11 Comparison of the BVP solvers 'bvp2_shoot()' and 'bvp2_fdf()'.

BVP	MATLAB functions	Mismatching error (P6.10.1b)	Times
(P6.11.1)	bvp2_shoot()	1.5×10^{-6}	
N=100, tol=1e-6,	bvp2_fdf()		
kmax=10	bvp4c()	2.9×10^{-6}	
(P6.11.2)	bvp2_shoot()		
N=100, tol=1e-6,	bvp2_fdf()	1.6×10^{-23}	
kmax=10	bvp4c()		
(P6.11.3)	bvp2_shoot()	1.7×10^{-17}	
N=100, tol=1e-6,	bvp2_fdf()		
kmax=10	bvp4c()	7.8×10^{-14}	
(P6.11.4)	bvp2_shoot()		
N=100, tol=1e-6,	bvp2_fdf()	4.4×10^{-27}	
kmax=10	bvp4c()		
(P6.11.5)	bvp2_shoot()	8.9×10^{-9}	
N=100, tol=1e-6,	bvp2_fdf()		
kmax=10	bvp4c()	8.9×10^{-7}	
(P6.11.6)	bvp2_shoot()		
N=100, tol=1e-6,	bvp2_fdf()	4.4×10^{-25}	
kmax=10	bvp4c()		

6.12 Shooting Method and Finite Difference Method for Nonlinear BVPs

(a) Consider a nonlinear boundary value problem of solving

$$\frac{d^2T}{dx^2} = 1.9 \times 10^{-9}(T^4 - T_a^4) \quad \text{(P6.12.1)}$$

with the boundary condition $T(x_0) = T_0$, $T(x_f) = T_f$

to find the temperature distribution $T(x)$(K) in a rod of 4 m long, where $[x_0, x_f] = [0, 4]$. Use the MATLAB functions 'bvp2_shoot()', 'bvp2_fdf()', and 'bvp4c()' to solve this differential equation for the two sets of boundary conditions $\{T(0) = 500, T(4) = 300\}$ and $\{T(0) = 550, T(4) = 300\}$ as listed in Table P6.12. Fill in the table with the mismatching errors defined by Eq. (P6.11.1b) for the three numerical solutions

$$\{T(x_i), i = 0:N\} \ (x_i = x_0 + ih = x_0 + i(x_f - x_0)/N) \text{ with } N = 500$$

```
%nm06p12a.m
K=1.9e-9; Ta=400; Ta4=Ta^4;
df=@(t,T)[T(?); K*(T(?).^4-Ta4)];
x0=0; xf=4; T0=500; Tf=300; N=500; tol=1e-5; kmax=10;

% Shooting method
tic, [xx,T_sh]=bvp2_shoot(df,x?,xf,T0,T?,N,tol,kmax);
times(1)=toc;

% Iterative finite difference method
% We should have the initial guess of u, dependent on T.
a1=0; a0=?; N=500; TT=T0+[1:N-1]*(Tf-T0)/N; u=K*(TT.^4-Ta4);
tic
for k=1:100
   [xx,T_fd]=bvp2_fdf(a1,a?,u,x0,xf,T0,Tf,N);
   u=K*(T_fd(2:N).^4-Ta4); % RHS of Eq.(P6.10.1)
   if k>1&norm(T_fd-T_fd0)/norm(T_fd0)<tol
      k, break;
   end
   T_fd0=T_fd;
end
times(2)=toc;

% MATLAB built-in function bvp4c
T_guess=[mean([T0 Tf]) (Tf-T0)/(xf-x0)]; % Guess of solution
solinit = bvpinit(linspace(x0,xf,5),T_?????);
fbc=@(TdT0,TdTf)[TdT0(1)-T?; TdTf(?)-Tf];
tic, sol=bvp4c(df,fbc,solinit,bvpset('RelTol',1e-6));
T_bvp = deval(sol,xx); times(3)=toc;
% The set of three solutions
Ts=[T_sh(:,1) T_fd T_bvp(1,:)'];

% Evaluates the errors and plot the graphs of the solutions
err=error_DE2_sols(df,xx,Ts)
subplot(321), plot(xx,Ts)
```

```
function [t,x]=bvp2_shoot(f,t0,tf,x0,xf,N,tol,kmax,Kg)
% Shooting method to solve BVP2: [x1,x2]'=f(t,x1,x2)
%   with x1(t0)=x0, x1(tf)=xf
% Note that you can give x0 as [x0 dx0] or [x0 dx0 ddx0L]
%if you want to specify the initial guess of dx0 and the maximum limit
%   ddx0L of the change of dx0.
if nargin<9, Kg=0; end
if nargin<8, kmax=10; end
if nargin<7, tol=1e-8; end
if nargin<6, N=100; end
Nx0=length(x0); %dx0L=-1e4; dx0H=1e4;
if Nx0<2, dx0(1)=(xf-x0)/(tf-t0); end
if Nx0>2, ddx0L=abs(x0(3)); else ddx0L=1e6; end
if Nx0>1, dx0(1)=x0(2); x0=x0(1); end
tt=t0+[0:N]*(tf-t0)/N;
[t,x]=ode_RK4(f,[t0 tf],[x0 dx0(1)],N);
if Kg>0, plot(t,x(:,1)), hold on; end
e(1)=x(end,1)-xf;
dx0(2)= dx0(1)-sign(e(1));
for k=2:kmax-1
   % To limit the abrupt change of dx0
   if dx0(k)-dx0(k-1)>ddx0L
      dx0(k)=dx0(k-1)+ddx0L; ddx0L=ddx0L*2;
    elseif dx0(k)-dx0(k-1)<-ddx0L
      dx0(k)=dx0(k-1)-ddx0L; ddx0L=ddx0L*2;
   end
   [t,x]=ode_RK4(f,[t0 tf],[x0 dx0(k)],N);
   %[t,x]=ode45(f,tt,[x0 dx0(k)]);
   if k<Kg, shg, pause, plot(t,x(:,1));
      text(tf,x(end,1),['k=' num2str(k)]);
   end
   e(k)=x(end,1)-xf; % Difference between resulting xf and desired xf
   ddx= dx0(k)-dx0(k-1);
   if abs(e(k))<tol|abs(ddx)<tol, break; end
   deddx= (e(k)-e(k-1))/ddx;
   dx0(k+1)= dx0(k)-e(k)/deddx; %move by secant method
end
if Kg>0, plot(t,x, [t0 tf],x([1 end],1),'ro'); end
```

Note that the MATLAB function 'bvp2_fdf()' should be applied in an iterative way to solve a nonlinear BVP because it has been fabricated to accommodate only linear BVPs. You may start with the above script "nm06p12a.m". Which MATLAB function works the best for the first case and the second case, respectively?

(b) Use the MATLAB functions 'bvp2_shoot()', 'bvp2_fdf()', and 'bvp4c()' to solve the following BVPs. Fill in Table P6.12 with the mismatching errors defined by Eq. (P6.11.1b) (implemented by the MATLAB function 'error_DE2_sols()') for the three numerical solutions and plot the solution graphs if they are reasonable solutions.

(i) $y'' - e^y = 0$ with the BC $y(0) = 0$, $y(1) = 0$ \hfill (P6.12.2)

(ii) $y'' - \frac{1}{t}y' - \frac{2}{y}(y')^2 = 0$ with the BC $y(1) = 4$, $y(2) = 8$ \hfill (P6.12.3)

(iii) $y'' - \dfrac{2}{y'+1} = 0$ with the BC $y(1) = \dfrac{1}{3}$, $y(4) = \dfrac{20}{3}$ (P6.12.4)

(iv) $y'' = t\,(y')^2$ with the BC $y(0) = \pi/2$, $y(2) = \pi/4$ (P6.12.5)

(v) $y'' + \dfrac{1}{y^2}y' = 0$ with the BC $y(2) = 2$, $y'(8) = 1/4$ (P6.12.6)

Note that for the BVP (P6.12.6), the MATLAB function 'bvp2m_shoot()' or 'bvp2mm_hoot()' (in Problems 6.9 to 6.10) should be used instead of 'bvp2_shoot()', since it has a mixed-boundary condition I.

(cf) Originally, the shooting method was developed for solving nonlinear BVPs, while the finite difference method is designed as a one-shot method for solving linear BVPs. But the finite difference method can also be applied in an iterative way to handle nonlinear BVPs, producing more accurate solutions in less computation time.

Table P6.12 Comparison of the BVP solvers 'bvp2_shoot()' and 'bvp2_fdf()'.

Boundary conditions	MATLAB function	Mismatching error (P6.10.1b)	Times (s)
(P6.12.1) with $T_a = 400$, $T(0) = 500$, $T(4) = 300$	bvp2_shoot() bvp2_fdf() bvp4c()	3.6×10^{-6}	
(P6.12.1) with $T_a = 400$ $T(0) = 550$, $T(4) = 300$	bvp2_shoot() modified bvp2_fdf() bvp4c()	NaN (divergent) 3.0×10^{-5}	N/A
(P6.12.2) with $y(0) = 0$, $y(1) = 0$	bvp2_shoot() bvp2_fdf() bvp4c()	3.2×10^{-13}	
(P6.12.3) with $y(1) = 4$, $y(2) = 8$	bvp2_shoot() modified bvp2_fdf() bvp4c()	NaN (divergent) 3.5×10^{-6}	N/A
(P6.12.4) with $y(1) = 1/3$, $y(4) = 20/3$	bvp2_shoot() bvp2_fdf() bvp4c()	3.4×10^{-10}	
(P6.12.5) with $y(0) = \pi/2$, $y(2) = \pi/4$	bvp2_shoot() bvp2_fdf() bvp4c()	3.7×10^{-14} 2.2×10^{-9}	
(P6.12.6) with $y(2) = 2$, $y'(8) = 1/4$	bvp2_shoot() bvp2_fdf() bvp4c()	5.0×10^{-14}	

(c) If you have been disappointed at the poor performance of 'bvp_shoot()' for the BVPs (P6.12.1) and (P6.12.3), modify the MATLAB function 'bvp2_shoot()' as follows so that it can use a good guess of the initial derivative and its maximum change limit to avoid the divergence problem where its fourth input argument x0 is supposed to be augmented with a guess of $x'(0)$ and its maximum change limit. Then do the following:

(c1) From the first element of the second row of the solution of BVP (P6.12.1) (with the BCs $\{T(0) = 550, T(4) = 300\}$) obtained using 'bvp4c()' and 'deval()', have a guess of $x'(0)$ as, say, -100 and use the modified MATLAB function 'bvp_shoot()' together with the guess to solve the BVP again. Fill in the corresponding space of Table P6.12 with the error of the solution.

(c2) From the first element of the second row of the solution of BVP (P6.12.3) obtained using 'bvp4c()' and 'deval()', have a guess of $x'(0)$ as, say, 1 and use the modified MATLAB function 'bvp_shoot()' together with the guess to solve the BVP again. Fill in the corresponding space of Table P6.12 with the error of the solution.

6.13 Eigenvalue BVPs

(a) A homogeneous second-order BVP to an eigenvalue problem
Consider an eigenvalue boundary value problem (BVP) of solving

$$y''(x) + \omega^2 y = 0 \tag{P6.13.1}$$

with $c_{01} y(x_0) + c_{02} y'(x_0) = 0$, $c_{f1} y(x_f) + c_{f2} y'(x_f) = 0$

to find $y(x)$ for $x \in [x_0, x_f]$ with the (possible) angular frequency ω.

To use the finite-difference method, we divide the solution interval $[x_0, x_f]$ into N subintervals to have the grid points $x_i = x_0 + ih = x_0 + i(x_f - x_0)/N$ and then replace the derivatives in the differential equation and the BCs by their finite-difference approximations (5.3.1) and (5.1.8) to write

$$\frac{y_{i-1} - 2y_i + y_{i+1}}{h^2} + \omega^2 y_i = 0;$$

$$y_{i-1} - (2 - \lambda) y_i + y_{i+1} = 0 \text{ with } \lambda = h^2 \omega^2 \tag{P6.13.2}$$

with

$$c_{01} y_0 + c_{02} \frac{y_1 - y_{-1}}{2h} = 0 \rightarrow y_{-1} = 2h \frac{c_{01}}{c_{02}} y_0 + y_1 \tag{P6.13.3a}$$

$$c_{f1} y_N + c_{f2} \frac{y_{N+1} - y_{N-1}}{2h} = 0 \rightarrow y_{N+1} = y_{N-1} - 2h \frac{c_{f1}}{c_{f2}} y_N \tag{P6.13.3b}$$

Substituting the discretized BC (P6.13.3) into (P6.13.2) yields

$$y_{-1} - 2y_0 + y_1 = -\lambda y_0 \overset{(P6.12.3a)}{\rightarrow} \left(2 - 2h\frac{c_{01}}{c_{02}}\right) y_0 - 2y_1 = \lambda y_0 \quad (P6.13.4a)$$

$$y_{i-1} - 2y_i + y_{i+1} = -\lambda y_i \rightarrow -y_{i-1} + 2y_i - y_{i+1} = \lambda y_i \quad (P6.13.4b)$$

for $i = 1, \ldots, N-1$

$$y_{N-1} - 2y_N + y_{N+1} = -\lambda y_N \overset{(P6.12.3b)}{\rightarrow}$$

$$-2y_{N-1} + \left(2 + 2h\frac{c_{f1}}{c_{f2}}\right) y_N = \lambda y_N \quad (P6.13.4c)$$

which can be formulated in a compact form as

$$\begin{bmatrix} 2 - 2hc_{01}/c_{02} & -2 & 0 & 0 & 0 \\ -1 & 2 & -1 & 0 & 0 \\ 0 & -1 & 2 & -1 & 0 \\ 0 & 0 & -1 & 2 & -1 \\ 0 & 0 & 0 & -2 & 2 + 2hc_{f1}/c_{f2} \end{bmatrix} \begin{bmatrix} y_0 \\ y_1 \\ \vdots \\ y_{N-1} \\ y_N \end{bmatrix} = \lambda \begin{bmatrix} y_0 \\ y_1 \\ \vdots \\ y_{N-1} \\ y_N \end{bmatrix};$$

$$A\mathbf{y} = \lambda \mathbf{y}; \quad [A - \lambda I]\mathbf{y} = \mathbf{0} \quad (P6.13.5)$$

For this equation to have a nontrivial solution $\mathbf{y} \neq \mathbf{0}$, λ must be one of the eigenvalues of the matrix A, and the corresponding eigenvectors are possible solutions.

Note the following:

- The angular frequency corresponding to the eigenvalue λ can be obtained as

$$\omega = \frac{\sqrt{\lambda/a_0}}{h} \quad (P6.13.6)$$

- The eigenvalues and the eigenvectors of a matrix A can be obtained by using the MATLAB command '`[V,D]=eig(A)`'.
- The following MATLAB function '`bvp2_eig()`' implements the aforementioned scheme to solve the second-order eigenvalue problem (P6.13.1).
- Especially, a second-order eigenvalue BVP

$$y''(x) + \omega^2 y = 0 \quad \text{with } y(x_0) = 0, \quad y(x_f) = 0 \quad (P6.13.7)$$

corresponds to the problem (6.6.1) with $\mathbf{c}_0 = [c_{01} \ c_{02}] = [1 \ 0]$ and $\mathbf{c}_f = [c_{f1} \ c_{f2}] = [1 \ 0]$ and has the following analytical solutions:

$$y(x) = a \sin \omega x \quad \text{with } \omega = \frac{k\pi}{x_f - x_0}, k = 1, 2, \ldots \quad (P6.13.8)$$

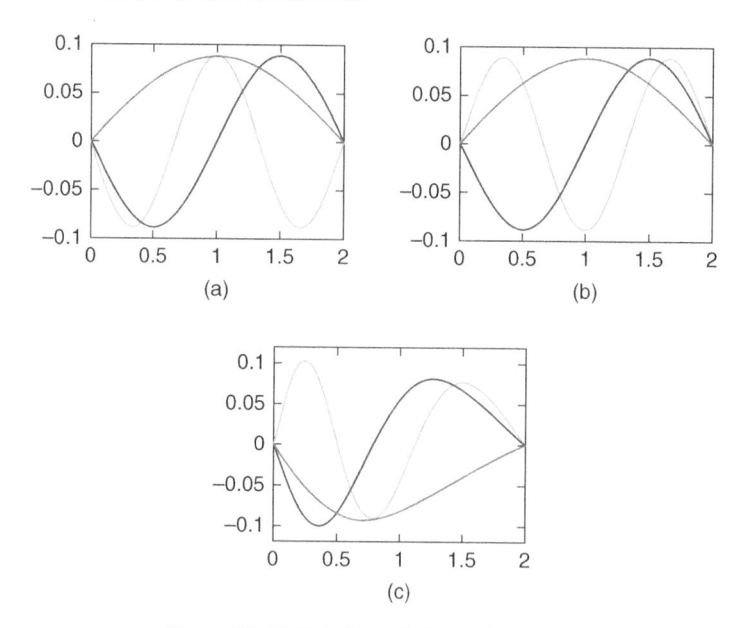

Figure P6.13 Solutions of eigenvalue BVPs.

Now, use the MATLAB function 'bvp2_eig()' with the number of grid points $N = 256$ to solve the BVP2 (P6.13.8) with $x_0 = 0$ and $x_f = 2$, find the lowest three angular frequencies (ω_i,s) and plot the corresponding eigenvector solutions as depicted in Figure P6.13a.

(b) A homogeneous forth-order BVP to an eigenvalue problem
Consider an eigenvalue boundary value problem of solving

$$\frac{d^4 y}{dx^4} - \omega^4 y = 0 \quad \text{with} \quad y(x_0) = 0, \frac{d^2 y}{dx^2}(x_0) = 0, y(x_f) = 0, \frac{d^2 y}{dx^2}(x_f) = 0$$
(P6.13.9)

to find $y(x)$ for $x \in [x_0, x_f]$ with the (possible) angular frequency ω.

To use the finite-difference method, we divide the solution interval $[x_0, x_f]$ into N subintervals to have the grid points $x_i = x_0 + ih = x_0 + i(x_f - x_0)/N$ and then, replace the derivatives in the differential equation and the boundary conditions by their finite-difference approximations to write

$$\frac{y_{i-2} - 4y_{i-1} + 6y_i - 4y_{i+1} + y_{i+2}}{h^4} - \omega^4 y_i = 0;$$

$$y_{i-2} - 4y_{i-1} + 6y_i - 4y_{i+1} + y_{i+2} = \lambda y_i \quad \text{with} \quad \lambda = h^4 \omega^4 \quad \text{(P6.13.10)}$$

with

```
function [x,Y,ws,eigvals]=bvp2_eig(x0,xf,c0,cf,N)
% use the finite difference method to solve an eigenvalue BVP4:
%    y"+w^2*y=0 with c01y(x0)+c02y'(x0)=0, cf1y(xf)+cf2y'(xf)=0
%input: x0/xf= the initial/final boundaries
%       c0/cf=the initial/final boundary condition coefficients
%       N-1 = the number of internal grid points.
%output: x   = the vector of grid points
%        Y   = the matrix composed of the eigenvector solutions
%        ws  = angular frequencies corresponding to eigenvalues
%        eigvals = the eigenvalues
if nargin<5|N<3, N=3; end
h=(xf-x0)/N; h2=h*h; x=x0+[0:N]*h;
N1=N+1;
if abs(c0(2))<eps, N1=N1-1; A(1,1:2)=[2 -1];
  else A(1,1:2)=[2*(1-c0(?)/c0(2)*h) -2];     % Eq.(P6.13.4a)
end
if abs(cf(2))<eps, N1=N1-1; A(N1,N1-1:N1)=[-1 2];
  else A(N1,N1-1:N1)=[-2 2*(1+cf(1)/cf(?)*h)];  % Eq.(P6.13.4c)
end
if N1>2
  for m=2:ceil(N1/2), A(m,m-1:m+1)=[-1 ? -1]; end % Eq.(P6.13.4b)
end
for m=ceil(N1/2)+1:N1-1, A(m,:)=fliplr(A(N1+1-m,:)); end
[V,LAMBDA]=eig(A); eigvals=diag(LAMBDA)';
[eigvals,I]=sort(eigvals); % sorting in the ascending order
V=V(:,I);
ws=sqrt(eigvals)/h;
if abs(c0(2))<eps, Y=zeros(1,N1); else Y=[]; end
Y=[Y; V];
if abs(cf(2))<eps, Y=[Y; zeros(1,N1)]; end
```

$$y_0 = 0, \quad \frac{y_{-1} - 2y_0 + y_1}{h^2} = 0 \rightarrow y_{-1} = -y_1 \quad \text{(P6.13.11a)}$$

$$y_N = 0, \quad \frac{y_{N-1} - 2y_N + y_{N+1}}{h^2} = 0 \rightarrow y_{N+1} = -y_{N-1} \quad \text{(P6.13.11b)}$$

Substituting the discretized boundary condition (P6.13.3) into (P6.13.2) yields

$$y_{-1} - 4y_0 + 6y_1 - 4y_2 + y_3 = \lambda y_1 \quad \overset{\text{(P6.12.12a)}}{\rightarrow}$$

$$5y_1 - 4y_2 + y_3 = \lambda y_1$$

$$y_0 - 4y_1 + 6y_2 - 4y_3 + y_4 = \lambda y_2 \quad \overset{\text{(P6.12.12a)}}{\rightarrow}$$

$$-4y_1 + 6y_2 - 4y_3 + y_4 = \lambda y_2$$

$$y_i - 4y_{i+1} + 6y_{i+2} - 4y_{i+3} + y_{i+4} = \lambda y_{i+2} \quad \text{for } i = 1, \ldots, N-5 \quad \text{(P6.13.12)}$$

$$y_{N-4} - 4y_{N-3} + 6y_{N-2} - 4y_{N-1} + y_N = \lambda y_{N-2} \quad \overset{\text{(P6.12.12b)}}{\rightarrow}$$

$$y_{N-4} - 4y_{N-3} + 6y_{N-2} - 4y_{N-1} = \lambda y_{N-2}$$

$$y_{N-3} - 4y_{N-2} + 6y_{N-1} - 4y_N + y_{N+1} = \lambda y_{N-1} \quad \overset{\text{(P6.12.12b)}}{\rightarrow}$$

$$y_{N-3} - 4y_{N-2} + 5y_{N-1} = \lambda y_{N-1}$$

which can be formulated in a compact form as

$$\begin{bmatrix} 5 & -4 & 1 & 0 & 0 & 0 & 0 \\ -4 & 6 & -4 & 1 & 0 & 0 & 0 \\ 1 & -4 & 6 & -4 & 1 & 0 & 0 \\ 0 & \cdot & \cdot & \cdot & \cdot & \cdot & 0 \\ 0 & 0 & 1 & -4 & 6 & -4 & 1 \\ 0 & 0 & 0 & 1 & -4 & 6 & -4 \\ 0 & 0 & 0 & 0 & 1 & -4 & 5 \end{bmatrix} \begin{bmatrix} y_1 \\ y_2 \\ y_3 \\ \cdot \\ y_{N-3} \\ y_{N-2} \\ y_{N-1} \end{bmatrix} = \lambda \begin{bmatrix} y_1 \\ y_2 \\ y_3 \\ \cdot \\ y_{N-3} \\ y_{N-2} \\ y_{N-1} \end{bmatrix};$$

$$A\mathbf{y} = \lambda \mathbf{y}; \quad [A - \lambda I]\mathbf{y} = \mathbf{0} \quad (P6.13.13)$$

For this equation to have a nontrivial solution $\mathbf{y} \neq \mathbf{0}$, λ must be one of the eigenvalues of the matrix A and the corresponding eigenvectors are possible solutions. Note that the angular frequency corresponding to the eigenvalue λ can be obtained as

$$\omega = \sqrt[4]{\lambda}/h \quad (P6.13.14)$$

(i) Complete the following MATLAB function 'bvp4_eig()' so that it can implement the aforementioned scheme to solve the fourth-order eigenvalue problem (P6.12.10).

(ii) Use the MATLAB function 'bvp4_eig()' with the number of grid points $N = 256$ to solve the BVP4 (P6.12.19) with $x_0 = 0$ and $x_f = 2$, find the lowest three angular frequencies (ω_is) and plot the corresponding eigenvector solutions as depicted in Figure P6.13b.

(c) The Sturm-Liouville Equation.
Consider an eigenvalue boundary value problem of solving

$$\frac{d}{dx}(f(x)y') + r(x)y = \lambda q(x)y \quad \text{with } y(x_0) = 0, \quad y(x_f) = 0 \quad (P6.13.15)$$

to find $y(x)$ for $x \in [x_0, x_f]$ with the (possible) angular frequency ω.

In order to use the finite-difference method, we divide the solution interval $[x_0, x_f]$ into N subintervals to have the grid points $x_i = x_0 + ih = x_0 + i(x_f - x_0)/N$ and then, replace the derivatives in the differential equation and the boundary conditions by their finite-difference approximations (with the stepsize $h/2$) to write

$$\frac{f(x_i + h/2)y'(x_i + h/2) - f(x_i - h/2)y'(x_i - h/2)}{2(h/2)} + r(x_i)y_i = \lambda q(x_i)y(x_i);$$

$$\frac{1}{h}\left\{f\left(x_i + \frac{h}{2}\right)\frac{y_{i+1} - y_i}{h} - f\left(x_i - \frac{h}{2}\right)\frac{y_i - y_{i-1}}{h}\right\} + r(x_i)y_i = \lambda q(x_i)y(x_i);$$

$$a_i y_{i-1} + b_i y_i + c_i y_{i+1} = \lambda y_i \quad \text{for } i = 1, 2, \ldots, N-1 \quad (P6.13.16)$$

```
function [x,Y,ws,eigvals]=bvp4_eig(x0,xf,N)
% Finite difference method to solve an eigenvalue BVP4:
%    y""-w^4*y=0 with y(x0)=0,y"(x0)=0, y(xf)=0,y"(xf)=0
%input: x0/xf= the initial/final boundary
%         N-1 = the number of internal grid points.
%output:  x   = the vector of grid points
%         Y   = the matrix composed of the eigenvector solutions
%         ws  = angular frequencies corresponding to eigenvalues
%     eigvals = the eigenvalues
%Copyleft: Won Y. Yang, wyyang53@hanmail.net, CAU for
%          academic use only
if nargin<3|N<4, N=4; end
h=(xf-x0)/N; h2=h*h;
x=x0+[0:N]*h;
A=zeros(N-1,N-1);
A(1,1:3)=[5 -? 1]; % Eq.(P6.13.3.1)
A(2,1:3)=[-4 6 -?]; % Eq.(P6.13.3.2,...)
if N>4, A(2,4)=1; end % Eq.(P6.13.3.2,...)
if N>5
  for m=3:ceil((N-1)/2)
    A(m,m-2:m+2)=[1 -? 6 -4 1]; % Eq.(P6.13.3)
  end
end
for m=ceil(N/2):N-1, A(m,:)=fliplr(A(N-m,:)); end
[V,LAMBDA]=eig(A); eigvals=diag(LAMBDA)';
[eigvals,I]=sort(eigvals); % sorting in the ascending order
V=V(:,I);
ws=eigvals.^0.25/h;
Y=[zeros(1,N-1); V; zeros(1,N-1)];
```

with

$$a_i = \frac{f(x_i - h/2)}{h^2 q(x_i)}, \quad c_i = \frac{f(x_i + h/2)}{h^2 q(x_i)}, \quad \text{and} \quad b_i = \frac{r(x_i)}{q(x_i)} - a_i - c_i$$
(P6.13.17)

```
function [x,Y,ws,eigvals]=sturm(f,r,q,x0,xf,N)
% use the finite difference method to solve an eigenvalue BVP2:
%    (d/dx)(f(x)y') + r(x)y = lambda q(x)y with y(x0)=y(xf)=0
if nargin<6|N<3, N=3; end
h=(xf-x0)/N; h2=h^2;
N=N-1; x=x0+[1:N]*h;
qx=feval(q,x); h2q=h^2*qx;
a=feval(f,x-?/2)./h2q;
c=feval(f,x+h/?)./h2q;
b=feval(r,x)./qx-?-c;
A(1,1:2)= [b(1) c(1)];  A(N,N-1:N)= [a(N) b(N)];
for m=2:N-1, A(m,m-1:m+1)=[a(m) b(m) c(m)]; end
[V,LAMBDA]=eig(A); eigvals=diag(LAMBDA)';
[eigvals,I]=sort(eigvals); % sorting in the ascending order
V=V(:,I);
ws=sqrt(eigvals)/h;
x=[x0 x xf];
Y=[zeros(1,N); V; zeros(1,N)];
```

(i) Complete the above MATLAB function 'sturm()' so that it can implement the aforementioned scheme to solve the Sturm-Liouville BVP (P6.13.16).

(ii) Use the MATLAB function 'sturm()' with the number of grid points $N = 256$ to solve the following BVP2.

$$\frac{d}{dx}((1+x^2)y') = -2\lambda y \text{ with } y(x_0) = 0, \quad y(x_f) = 0 \quad \text{(P6.13.18)}$$

Plot the eigenvector solutions corresponding to the lowest three angular frequencies (ω_i's).

```
%nm06p13c.m to solve an eigenvalue BVP4
% (d/dx)((1+x^2)y') + 2 *lambda*y =0 with y(x0)=0, y(xf)=0,
x0=0; xf=2; N=256;
f=@(x)1+x.^2;
r=@(x)zeros(size(x));
q=@(x)-?*ones(size(x));
[x,Y,ws,eigvals]=sturm(f,r,q,x0,xf,N);
w3=ws(1:3), freqs=w3/2/pi
subplot(233)
plot(x,Y(:,1:3)), axis([0 2 -0.12 0.12])
```

6.14 Danger of Gauss-Seidel Iteration for the Finite Difference Method
Consider a nonlinear BVP of solving

$$\frac{d^2T}{dx^2} = h(T - T_a) + \sigma(T^4 - T_a^4) \text{ with the BCs } T(0) = T_0, \; T(10) = T_f$$
(P6.14.1)

to find the temperature distribution $T(x)$ K in a rod of 10 m long where $h = 0.05 \text{ m}^{-2}$, $\sigma = 2.7 \times 10^{-9}$ K^{-3} m^{-2}], $[x_0, x_f] = [0, 10]$ m, $T_a = 200$ K, $T_0 = 300$ K, and $T_f = 400$ K.

(a) Note that the DE (P6.14.1) can be discretized into the following difference equation:

$$\frac{T_{i-1} - 2T_i + T_{i+1}}{\Delta x^2} = h(T_i - T_a) + \sigma(T_i^4 - T_a^4) \rightarrow$$

$$T_i = \frac{h\Delta x^2 T_a + \sigma \Delta x^2(T_a^4 - T_i^4) + T_{i-1} + T_{i+1}}{2 + h\Delta x^2} \quad \text{(P6.14.2)}$$

With the length of the rod into $N = 5$ segments each of length $\Delta x = (x_f - x_0)/N = 10/5 = 2$ m, apply the Gauss-Seidel iteration to find the solution to Eq. (P6.14.2) for $\{x_i = x_0 + i\Delta x, i = 0:N\}$.

(b) Use the MATLAB function 'bvp4c()' to find the numerical solution of the BVP for $\{x_i = x_0 + i\Delta x = x_0 + i(x_f - x_0)/N, i = 0:N\}$ with $N = 500$. Compare the solution with the numerical solution obtained in (a) in terms of the numerical error computed using 'error_DE_sols()'.

```
%nm06p14a.m
N=5; x0=0; xf=10; dx=(xf-x0)/N; xx=x0+[0:N]*dx;
h=0.05; s=2.7e-9; K1=h*dx^2; K2=s*dx^2; Ta=200; T0=300; Tf=400;
df=@(x,T) [T(?); -h*(Ta-T(?))-s*(Ta^4-T(?)^4)];
tol=1e-6;
disp('Gauss-Seidel method')
Ts=[T0 zeros(1,N-1) Tf].'; % Ts=[T0 ones(1,N-1)*(T0+Tf)/2 Tf].';
for k=1:100
   Ts0=Ts;
   for i=2:N
      Ts(i)=(K1*Ta+K2*(Ta^4-Ts(?).^4)+Ts(i-?)+Ts(i+?))/(2+K1);
   end
   if norm((Ts-Ts0)./Ts)<tol, break; end
end
er_GS=error_DE2_sols(df,xx,Ts)
plot(xx,Ts,'r');
if N<11, Ts.', end % See Example 24.7 of Chapra
```

```
%nm06p14b.m
disp('Using bvp4c()')
fbc=@(TdT0,TdTf) [TdT0(?)-T?; TdTf(?)-T?];
% Note that TdT0(2)=dT(x0), TdTf(2)=dT(xf)
T_guess=[mean([T0 Tf]) (Tf-T0)/(xf-x0)];
solinit=bvpinit(linspace(x0,xf,10),T_guess); % Guess of solution
sol=bvp4c(df,fbc,solinit,bvpset('RelTol',tol));
T_bv=deval(sol,xx); Tbv=T_bv(1,:)';
dTdx0_bv=T_bv(2,1) % Initial derivative dTdx(x0)
er_bv=error_DE2_sols(df,xx,Tbv.')
plot(xx,Tbv,'g')
```

(c) With the rod divided into $N = 100$ and 1000 segments, repeat the same jobs as done in (a). How about beginning with a different initial guess, say, `Ts=[T0 ones(1,N-1)*(T0+Tf)/2 Tf].'`? Discuss the reliability of the Gauss-Seidel iteration.

6.15 'ode45()' (with adaptive step-size) vs. 'ode_RK4()' (with fixed step-size)
Consider the following DE modeling the Pliny's intermittent fountain[C-1]:

$$\frac{dy}{dt} = \frac{Q_{in} - Q_{out}}{\pi R_t^2} \quad \text{with } Q_{out} = s_i \times C\sqrt{2gy}\pi r^2 \quad \text{(P6.15.1)}$$

where

$$s_i = \begin{cases} 1 & \text{if } y \geq y_h \\ \text{keep the previous value} & \text{if } y_l < y < y_h \\ 0 & \text{if } y \leq y_l \end{cases} \quad \text{(P6.15.2)}$$

Noting that the value of a variable declared as 'persistent' in a MATLAB function is retained in the global memory between calls to the function even if it is accessible only inside the function, complete and run the MATLAB script "solve_fountain.m" to solve the DE (P6.15.1) five times, each time using 'ode45()', 'ode15s()', 'ode23s()', 'ode_RK45()', and without using any DE solver where the last two methods use a fixed step-size, while the other ones use adaptively adjusted step-size. Which of the three methods, i.e. Euler's method, Heun's method, and Runge-Kutta method is used in the for loop? Is it good to adjust the step-size adaptively depending on the error estimates for this DE with an abruptly changing parameter?

```
%solve_fountain.m
global Rt r yh yl C g Qin;
Rt=0.05; r=0.007; yh=0.1; yl=0.025;
C=0.6; g=9.81; Qin=5e-5; Cpr2=C*pi*r^2; pR2=pi*Rt^2;
f=@Plinyode; t0=0; tf=100; tspan=[t0 tf]; y0=0;
[t45,y45]=ode45(f,tspan,y0);
[t15s,y15s]=ode15s(f,tspan,y0);
[t23s,y23s]=ode23s(f,tspan,y0);
N=200; [tR,yR]=ode_RK4(f,tspan,y0,N);
subplot(211), hold on
plot(t45,y45,'k', t15s,y15s,'r:', t23s,y23s,'g', tR,yR,'b')
legend('ode45','ode15s','ode23s','ode-RK4')
% A recursive way to solve the DE
si=0; y(1)=0;
dt=tf/N; tt=t0+[0:N]*dt; % Time vector
for n=1:N
    f1=(Qin-si*Cpr2*sqrt(2*?*y(n)))/pR2; % Eq.(P6.15.1)
    y(n+1)=y(n)+f1*??;
    if y(n+1)<=yl, si=0; elseif y(n+1)>=yh, si=1; end % Eq.(P6.15.2)
    f2=(Qin-??*Cpr2*sqrt(2*g*y(n+?)))/pR2;
    y(n+1)=y(n)+(f?+f?)/2*dt;
end
[tR,yR]=ode_RK4(f,tspan,y0,N);
subplot(212), plot(tR,yR,'b', tt,y,'r'), hold on
plot([t0 tf],[yh yh],'k:', [t0 tf],[yl yl],'k:')
legend('ode-RK4 with dt=0.5','recursive with dt=0.5')
```

```
function dy=Plinyode(t,y)
global Rt r yh yl C g Qin;
persistent si
if isempty(si), si=(y>yh); end
if y<=yl, si=0; elseif y>=yh, si=1; end   % Eq.(P6.15.2)
Qout=si*C*sqrt(2*g*y)*pi*r^2;
dy=(Qin-Qout)/(pi*Rt^2); % Eq.(P6.15.1)
```

7

OPTIMIZATION

CHAPTER OUTLINE

7.1 Unconstrained Optimization	376
7.1.1 Golden Search Method	376
7.1.2 Quadratic Approximation Method	378
7.1.3 Nelder-Mead Method	380
7.1.4 Steepest Descent Method	383
7.1.5 Newton Method	385
7.1.6 Conjugate Gradient Method	387
7.1.7 Simulated Annealing	389
7.1.8 Genetic Algorithm	393
7.2 Constrained Optimization	399
7.2.1 Lagrange Multiplier Method	399
7.2.2 Penalty Function Method	406
7.3 MATLAB Built-In Functions for Optimization	409
7.3.1 Unconstrained Optimization	409
7.3.2 Constrained Optimization	413
7.3.3 Linear Programming (LP)	416
7.3.4 Mixed Integer Linear Programming (MILP)	423
7.4 Neural Network[K-1]	433

Applied Numerical Methods Using MATLAB®, Second Edition. Won Y. Yang, Jaekwon Kim, Kyung W. Park, Donghyun Baek, Sungjoon Lim, Jingon Joung, Suhyun Park, Han L. Lee, Woo June Choi, and Taeho Im.
© 2020 John Wiley & Sons, Inc. Published 2020 by John Wiley & Sons, Inc.
Companion website: www.wiley.com/go/yang/appliednumericalmethods

7.5 Adaptive Filter[Y-3] 439
7.6 Recursive Least Square Estimation (RLSE)[Y-3] 443
Problems 448

Optimization involves finding the minimum/maximum of an objective (or cost) function $f(x)$ subject to some constraint $x \in S$. If there is no constraint for x to satisfy, or equivalently, S is the universe, it is called an unconstrained optimization and otherwise, it is a constrained optimization. In this chapter, we will cover several unconstrained optimization techniques such as the golden search method, the quadratic approximation method, Nelder-Mead method, the steepest descent method, Newton method, simulated-annealing (SA) method, and genetic algorithm (GA). As for constrained optimization, we will only introduce the MATLAB built-in functions together with the functions for unconstrained optimization. Note that we do not have to distinguish maximization and minimization because maximizing $f(x)$ is equivalent to minimizing $-f(x)$ and so, without loss of generality, we deal only with the minimization problems.

7.1 UNCONSTRAINED OPTIMIZATION

7.1.1 Golden Search Method

This method is applicable to an unconstrained minimization problem such that the solution interval $[a,b]$ is known and the objective function $f(x)$ is unimodal within the interval, i.e. the sign of its derivative $f'(x)$ changes at most once in $[a,b]$ so that $f(x)$ decreases/increases monotonically for $[a, x^o]/[x^o, b]$, where x^o is the solution that we are looking for. The so-called golden search procedure is summarized below and has been cast into the MATLAB function 'opt_gs()'. We composed the following MATLAB script "nm0711.m", which uses this function

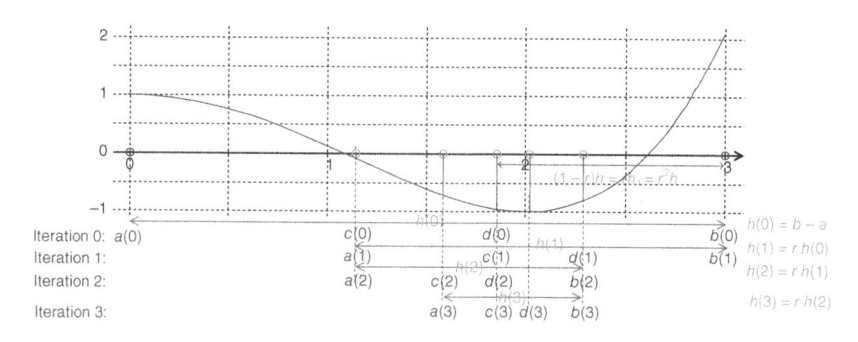

Figure 7.1 Illustration of the golden search method.

to find the minimum point of the objective function:

$$f(x) = (x^2 - 4)^2/8 - 1 \tag{7.1.1}$$

Figure 7.1 shows how the function 'opt_gs()' proceeds toward the minimum point step by step.

GOLDEN SEARCH PROCEDURE

(Step 1) Pick up the two points $c = a + (1-r)h$ and $d = a + rh$ inside the interval $[a,b]$, where $r = (\sqrt{5} - 1)/2$ and $h = b - a$.

(Step 2) If the values of $f(x)$ at the two points are almost equal, i.e. $f(a) \approx f(b)$ and the width of the interval is sufficiently small, i.e. $h \approx 0$, then stop the iteration to exit the loop and declare $x^o = c$ or $x^o = d$ depending on whether $f(c) < f(d)$ or not. Otherwise, go to Step 3.

(Step 3) If $f(c) < f(d)$, let the new upper-bound of the interval: $b \leftarrow d$; otherwise, let the new lower-bound of the interval: $a \leftarrow c$. Then, go to Step 1.

```
function [xo,fo,k0]=opt_gs(f,a,b,r,TolX,TolFun,k)
if k<=0|(abs(h)<TolX&abs(fa-fb)<TolFun)
   if k==0, fprintf('Not reliable, though the best!\n'), end
   if fa<=fb, xo=a; fo=fa;   else   xo=b; fo=fb;   end
   k0=k;
 else
   if nargin<8, Kg=0; end
   if nargout==0, Kg=2; end
   rh=r*h; c=b-rh; d=a+rh;
   fc=feval(f,c); fd=feval(f,d);
   if fc<fd, [xo,fo,k0]=opt_gs(f,a,d,r,TolX,TolFun,k-1,Kg-1);
     else    [xo,fo,k0]=opt_gs(f,c,b,r,TolX,TolFun,k-1,Kg-1);
   end
end
```

```
%nm0711.m   to perform the golden search method
f711=@(x)(x.^2-4).^2/8-1; % Eq.(7.1.1): The objective function
a=0; b=3; r=(sqrt(5)-1)/2;
TolX=1e-4; TolFun=1e-4; MaxIter=100;
[xo,fo]=opt_gs(f711,a,b,r,TolX,TolFun,MaxIter)
```

Note the following points about the golden search procedure:

- At every iteration, the new interval width is

$$b - c = b - \{a + (1-r)(b-a)\} = rh \quad \text{or} \quad d - a = a + rh - a = rh \tag{7.1.2}$$

so that it becomes r times the old interval width ($b - a = h$).

- The golden ratio r is fixed so that a point $c_1 = b_1 - rh_1 = b - r^2h$ in the new interval $[c,b]$ conforms with $d = a + rh = b - (1-r)h$; i.e.

$$r^2 = 1 - r; \quad r^2 + r - 1 = 0; \quad r = \frac{-1 + \sqrt{1+4}}{2} = \frac{-1 + \sqrt{5}}{2} \quad (7.1.3)$$

7.1.2 Quadratic Approximation Method

The idea of this method is to approximate the objective function $f(x)$ by a quadratic function $p_2(x)$ matching the previous three (estimated solution) points, and to keep updating the three points by replacing one of them with the minimum point of $p_2(x)$. More specifically, for the three points

$$\{(x_0,f_0),(x_1,f_1),(x_2,f_2)\} \quad \text{with} \quad x_0 < x_1 < x_2$$

we find the interpolation polynomial $p_2(x)$ of degree 2 to fit them and replace one of them with the zero of the derivative, i.e. the root of $p_2'(x) = 0$ (see Eq. (P3.1.2))

$$x = x_3 = \frac{f_0(x_1^2 - x_2^2) + f_1(x_2^2 - x_0^2) + f_2(x_0^2 - x_1^2)}{2\{f_0(x_1 - x_2) + f_1(x_2 - x_0) + f_2(x_0 - x_1)\}} \quad (7.1.4)$$

Especially if the previous estimated solution points are equidistant with an equal distance h, i.e. $x_2 - x_1 = x_1 - x_0 = h$, then this formula becomes

$$x_3 = \frac{f_0(x_1^2 - x_2^2) + f_1(x_2^2 - x_0^2) + f_2(x_0^2 - x_1^2)}{2\{f_0(x_1 - x_2) + f_1(x_2 - x_0) + f_2(x_0 - x_1)\}} \bigg|_{\substack{x_1 = x+h \\ x_2 = x_1+h}}$$

$$= x_0 + h\frac{3f_0 - 4f_1 + f_2}{2(-f_0 + 2f_1 - f_2)} \quad (7.1.5)$$

We keep updating the three points this way until $|x_2 - x_0| \approx 0$ and/or $|f(x_2) - f(x_0)| \approx 0$, when we stop the iteration and declare x_3 as the minimum point. The rule for updating the three points is as follows:

1. In case $x_0 < x_3 < x_1$, we take $\{x_0,x_3,x_1\}$ or $\{x_3,x_1,x_2\}$ as the new set of three points depending on whether $f(x_3) < f(x_1)$ or not.
2. In case $x_1 < x_3 < x_2$, we take $\{x_1,x_3,x_2\}$ or $\{x_0,x_1,x_3\}$ as the new set of three points depending on whether $f(x_3) < f(x_1)$ or not.

UNCONSTRAINED OPTIMIZATION

```
function [xo,fo]=opt_quad(f,x0,TolX,TolFun,MaxIter)
% search for the minimum of f(x) by quadratic approximation method
if length(x0)>2 x012=x0(1:3);
 else
    if length(x0)==2 a=x0(1); b=x0(2);
     else a=x0-10; b=x0+10;
    end
    x012= [a (a+b)/2 b];
end
f012= f(x012);
[xo,fo]=opt_quad0(f,x012,f012,TolX,TolFun,MaxIter);
```

```
function [xo,fo]=opt_quad0(f,x012,f012,TolX,TolFun,k)
x0= x012(1); x1= x012(2); x2= x012(3);
f0= f012(1); f1= f012(2); f2= f012(3);
nd= [f0-f2 f1-f0 f2-f1]*[x1*x1 x2*x2 x0*x0; x1 x2 x0]';
x3= nd(1)/2/nd(2); f3=feval(f,x3); %Eq.(7.1.4)

if k<=0|abs(x3-x1)<TolX|abs(f3-f1)<TolFun
   xo=x3; fo=f3;
   if k==0, fprintf('Just the best in given # of iterations'), end
 else
   if x3<x1
     if f3<f1, x012=[x0 x3 x1]; f012= [f0 f3 f1];
      else  x012=[x3 x1 x2]; f012= [f3 f1 f2];
     end
    else
     if f3<=f1, x012=[x1 x3 x2]; f012= [f1 f3 f2];
      else  x012=[x0 x1 x3]; f012= [f0 f1 f3];
     end
   end
   [xo,fo]=opt_quad0(f,x012,f012,TolX,TolFun,k-1);
end
```

This procedure, called the quadratic approximation method, has been cast into the MATLAB function 'opt_quad()', which has a recursive (self-calling) structure. We composed the following MATLAB script "nm0712.m", which uses this function to find the minimum point of the objective function (7.1.1) and also uses the MATLAB built-in function 'fminbnd()' to find it for crosscheck. Figure 7.2 shows how the function 'opt_quad()' proceeds toward the minimum point step by step.

```
%nm0712.m
% performs the quadratic approximation method
f711=@(x)(x.*x-4).^2/8-1; % Eq.(7.1.1): The objective function
a=0; b=3; TolX=1e-5; TolFun=1e-8; MaxIter=100;
[xoq,foq]=opt_quad(f711,[a b],TolX,TolFun,MaxIter)

% Minimum point and its function value
[xob,fob]=fminbnd(f711,a,b) % MATLAB built-in function
```

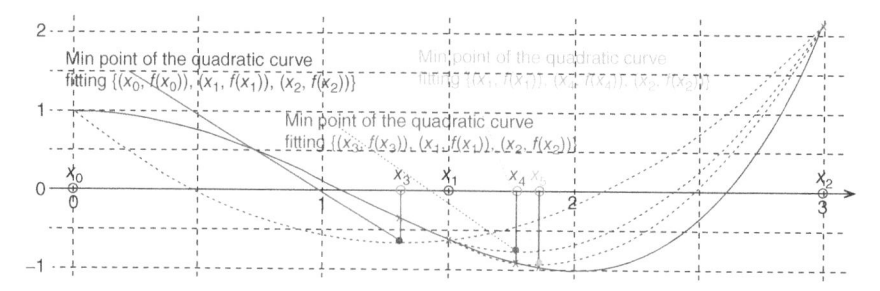

Figure 7.2 Illustration of the quadratic approximation method.

7.1.3 Nelder-Mead Method

The Nelder-Mead method is applicable to the minimization of a multivariable objective function, for which neither the golden search method nor the quadratic approximation method can be applied. The algorithm of the Nelder-Mead method summarized in the box beneath has been cast into the following MATLAB function 'Nelder0()'. Note that in case of N-dimensional case ($N > 2$), this algorithm should be repeated for each sub-plane as implemented in the outer function 'opt_Nelder()'.

NELDER-MEAD ALGORITHM

(Step 1) Let the initial three estimated solution points be a, b, and c, where $f(a) < f(b) < f(c)$.

(Step 2) If the three points or their function values are sufficiently close to each other, then declare a to be the minimum and terminate the procedure.

(Step 3) Otherwise, expecting that the minimum we are looking for may be at the opposite side of the worst point c over the line $\overline{ab}$ (see Figure 7.3a), take

$$e = m + 2(m - c) \text{ where } m = (a + b)/2$$

and if $f(e) < f(b)$, take e as the new c; otherwise, take

$$r = (m + e)/2 = 2m - c$$

and if $f(r) < f(c)$, take r as the new c; if $f(r) \geq f(b)$, take

$$s = (c + m)/2$$

and if $f(s) < f(c)$, take s as the new c; otherwise, give up the two points b, c and take m and $c_1 = (a + c)/2$ as the new b and c, reflecting our expectation that the minimum would be around a.

(Step 4) Go back to Step 1.

UNCONSTRAINED OPTIMIZATION   **381**

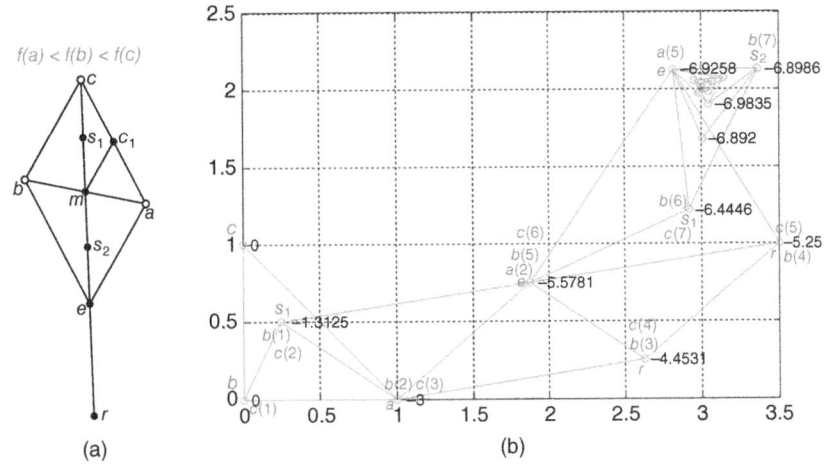

Figure 7.3 Illustration of the Nelder–Mead method. (a) Notations used in the Nelder–Mead method. (b) Process of the Nelder–Mead method.

```
function [xo,fo]=Nelder0(f,abc,fabc,TolX,TolFun,k)
%Copyleft: Won Y. Yang, wyyang53@hanmail.net, CAU for academic use only
[fabc,I]=sort(fabc);
a=abc(I(1),:); b=abc(I(2),:); c=abc(I(3),:);
fa=fabc(1); fb=fabc(2); fc=fabc(3);
fba=fb-fa; fcb=fc-fb;

if k<=0|abs(fba)+abs(fcb)<TolFun|abs(b-a)+abs(c-b)<TolX
   xo=a; fo=fa;
   if k==0, fprintf('Just best in given # of iterations'), end

else
   m=(a+b)/2;   e=3*m-2*c;   fe=feval(f,e);  % Step 3(i)

   if fe<fb, c=e;  fc=fe;
    else
       r=(m+e)/2;   fr=feval(f,r);
       if fr<fc,  c=r;   fc=fr;   end
       if fr>=fb
         s=(c+m)/2;   fs=feval(f,s);
         if fs<fc,   c=s;   fc=fs;           % Step 3(iii)
           else b=m;  c=(a+c)/2;  fb=feval(f,b); fc=feval(f,c); % Step 3(ii)
         end
       end
    end

   [xo,fo]= Nelder0(f,[a;b;c],[fa fb fc],TolX,TolFun,k-1);
end
```

```
function [xo,fo]=opt_Nelder(f,x0,TolX,TolFun,MaxIter)
if numel(x0)==1 % for 1-dimensional case
  [xo,fo]=opt_quad(f,x0,TolX,TolFun); return
end
N=size(x0,2); S=eye(N);
for i=1:N % repeat the procedure for each directional subplane
    Nr=size(x0,1); i1=mod(i,N)+1;
    if Nr==3, abc=x0;
     else abc=[x0; x0+S(i,:); x0+S(i1,:)];
    end
    abc=abc(:,rotate_l([1:N],i-1));
    fabc=[feval(f,abc(1,:)); feval(f,abc(2,:)); feval(f,abc(3,:))];
    [x0,fo]=Nelder0(f,abc,fabc,TolX,TolFun,MaxIter);
    x0=x0(:,rotate_r([1:N],i-1));
    if N<3, break; end % No repetition needed for a 2D case
end
xo=x0;
```

We made the following MATLAB script "nm0713.m" to minimize a two-variable objective function

$$f(x_1,x_2) = x_1^2 - x_1 x_2 - 4x_1 + x_2^2 - x_2 \qquad (7.1.6)$$

whose minimum can be found in an analytical way, i.e. by setting the partial derivatives of $f(x_1, x_2)$ with respect to x_1 and x_2 to zero as

$$\frac{\partial}{\partial x_1} f(x_1,x_2) = 2x_1 - x_2 - 4 = 0$$
$$\frac{\partial}{\partial x_2} f(x_1,x_2) = 2x_2 - x_1 - 1 = 0$$; $\mathbf{x}_o = (x_{1o}, x_{2o}) = (3, 2)$

```
%nm0713.m: do_Nelder
f713=@(x)x(1)*(x(1)-4-x(2)) +x(2)*(x(2)-1); % Eq.(7.1.6)
x0=[0 0], TolX=1e-4; TolFun=1e-9; MaxIter=100;
[xon,fon]=opt_Nelder(f713,x0,TolX,TolFun,MaxIter)
%minimum point and its function value
[xos,fos]=fminsearch(f713,x0) % use the MATLAB built-in function
```

This script also uses the MATLAB built-in function 'fminsearch()' to minimize the same objective function for practice and confirmation. The minimization process is illustrated in Figure 7.3b.

(cf) The MATLAB built-in function 'fminsearch()' uses the Nelder-Mead algorithm to minimize a multi-variable objective function.

7.1.4 Steepest Descent Method

This method searches for the minimum of an N-dimensional objective function in the direction of a negative gradient

$$-\mathbf{g}(\mathbf{x}) = -\nabla f(\mathbf{x}) = -\left[\frac{\partial f(\mathbf{x})}{\partial x_1} \ \frac{\partial f(\mathbf{x})}{\partial x_2} \cdots \frac{\partial f(\mathbf{x})}{\partial x_N}\right]^{\mathrm{T}} \qquad (7.1.7)$$

with the step size α_k (at iteration k) adjusted so that the function value is minimized along the direction by a (one-dimensional) line search technique like the quadratic approximation method. The algorithm of the steepest descent method is summarized in the box below and has been cast into the following MATLAB function 'opt_steep()'.

We composed the following MATLAB script "nm0714.m" to minimize the objective function (7.1.6) by using the steepest descent method. The minimization process is illustrated in Figure 7.4.

```
function [xo,fo]=opt_steep(f,x0,TolX,TolFun,alpha0,MaxIter)
% minimize the ftn f by the steepest descent method.
%input:  f = ftn to be given as a string 'f'
%        x0= the initial guess of the solution
%output: xo= the minimum point reached,  fo= f(x(o))
if nargin<6, MaxIter=100; end %maximum # of iteration
if nargin<5, alpha0=10; end %initial step size
if nargin<4, TolFun=1e-8; end %|f(x)|<TolFun wanted
if nargin<3, TolX=1e-6; end %|x(k)-x(k-1)|<TolX wanted
x=x0;  fx0= feval(f,x0); fx=fx0;
alpha= alpha0;  kmax1=25;
warning= 0; %the # of vain wanderings to find the optimum step size
for k=1: MaxIter
    g= grad(f,x); g= g/norm(g); %gradient as a row vector
    alpha= alpha*2; %for trial move in negative gradient direction
    fx1 =feval(f,x-alpha*2*g);
    for k1=1:kmax1 %find the optimum step size(alpha) by line search
        fx2= fx1; fx1= feval(f,x-alpha*g);
        if fx0>fx1+TolFun&fx1<fx2-TolFun %fx0>fx1<fx2
            den=4*fx1-2*fx0-2*fx2; num=den-fx0+fx2; %Eq.(7.1.5)
            alpha= alpha*num/den;
            x= x-alpha*g;  fx= feval(f,x); %Eq.(7.1.9)
            break;
          else  alpha= alpha/2;
        end
    end
    if k1>=kmax1, warning=warning+1; %failed to find optimum step-size
      else  warning= 0;
    end
    if warning>=2|(norm(x-x0)<TolX&abs(fx-fx0)<TolFun), break;  end
    x0= x;  fx0= fx;
end
xo= x;  fo= fx;
if k==MaxIter, fprintf('Just best in %d iterations',MaxIter), end
```

STEEPEST DESCENT ALGORITHM

(Step 0) With the iteration number $k = 0$, find the function value $f_0 = f(\mathbf{x}_0)$ at the initial point $\mathbf{x}_0$.

(Step 1) Increment the iteration number k by one, find the step size α_{k-1} along the direction of the negative gradient $-\mathbf{g}_{k-1}$ by a (one-dimensional) line search like the quadratic approximation method.

$$\alpha_{k-1} = \text{ArgMin}_\alpha f(\mathbf{x}_{k-1} - \alpha \mathbf{g}_{k-1}/\|\mathbf{g}_{k-1}\|) \qquad (7.1.8)$$

(Step 2) Move the approximate minimum by the step-size α_{k-1} along the direction of the negative gradient $-\mathbf{g}_{k-1}$ to get the next point

$$\mathbf{x}_k = \mathbf{x}_{k-1} - \alpha_{k-1} \mathbf{g}_{k-1}/\|\mathbf{g}_{k-1}\| \qquad (7.1.9)$$

(Step 3) If $\mathbf{x}_k \approx \mathbf{x}_{k-1}$ and $f(\mathbf{x}_k) \approx f(\mathbf{x}_{k-1})$, then declare $\mathbf{x}_k$ to be the minimum and terminate the procedure. Otherwise, go back to Step 1.

```
%nm0714.m
f713=@(x)x(1)*(x(1)-4-x(2)) +x(2)*(x(2)-1); % Eq.(7.1.6)
x0=[0 0], TolX=1e-4; TolFun=1e-9; alpha0=1; MaxIter=100;
[xo,fo]= opt_steep(f713,x0,TolX,TolFun,alpha0,MaxIter)
```

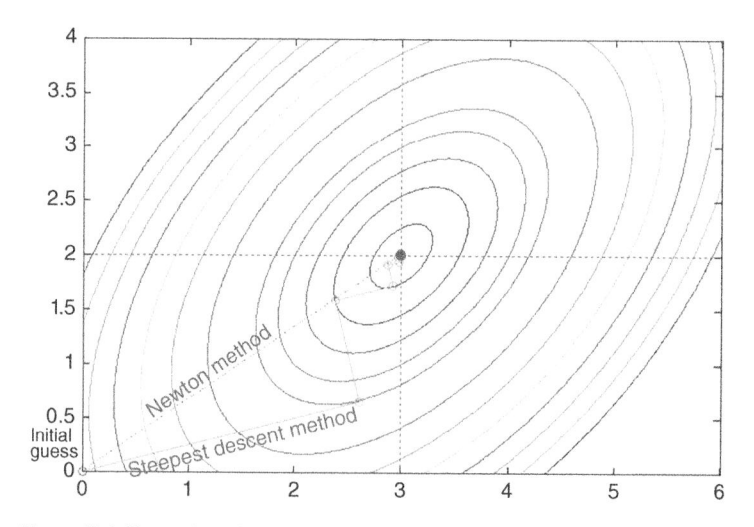

Figure 7.4 Illustration of the Newton's method and steepest descent method.

7.1.5 Newton Method

Like the steepest descent method, this method also uses the gradient to search for the minimum point of an objective function. Such gradient-based optimization methods are supposed to reach a point at which the gradient is (close to) zero. In this context, the optimization of an objective function $f(\mathbf{x})$ is equivalent to finding a zero of its gradient $\mathbf{g}(\mathbf{x})$, which in general is a vector-valued function of a vector-valued independent variable $\mathbf{x}$. Therefore, if we have the gradient function $\mathbf{g}(\mathbf{x})$ of the objective function $f(\mathbf{x})$, we can solve the system of nonlinear equations $\mathbf{g}(\mathbf{x}) = 0$ to get the minimum of $f(\mathbf{x})$ by using the Newton method explained in Section 4.4.

The backgrounds of this method as well as the steepest descent method can be shown by taking the Taylor series of, say, a two-variable objective function $f(x_1, x_2)$:

$$f(x_1, x_2) \cong f(x_{1k}, x_{2k}) + \begin{bmatrix} \dfrac{\partial f}{\partial x_1} & \dfrac{\partial f}{\partial x_2} \end{bmatrix}_{(x_{1k}, x_{2k})} \begin{bmatrix} x_1 - x_{1k} \\ x_2 - x_{2k} \end{bmatrix}$$

$$+ \frac{1}{2}\begin{bmatrix} x_1 - x_{1k} & x_2 - x_{2k} \end{bmatrix} \begin{bmatrix} \partial^2 f/\partial x_1^2 & \partial^2 f/\partial x_1 \partial x_2 \\ \partial^2 f/\partial x_2 \partial x_1 & \partial^2 f/\partial x_2^2 \end{bmatrix}_{(x_{1k}, x_{2k})} \begin{bmatrix} x_1 - x_{1k} \\ x_2 - x_{2k} \end{bmatrix};$$

$$f(\mathbf{x}) \cong f(\mathbf{x}_k) + \nabla f(\mathbf{x})^T|_{\mathbf{x}_k}[\mathbf{x} - \mathbf{x}_k] + \frac{1}{2}[\mathbf{x} - \mathbf{x}_k]^T \nabla^2 f(\mathbf{x})|_{\mathbf{x}_k}[\mathbf{x} - \mathbf{x}_k];$$

$$f(\mathbf{x}) \cong f(\mathbf{x}_k) + \mathbf{g}_k^T[\mathbf{x} - \mathbf{x}_k] + \frac{1}{2}[\mathbf{x} - \mathbf{x}_k]^T H_k[\mathbf{x} - \mathbf{x}_k] \qquad (7.1.10)$$

with the gradient vector $\mathbf{g}_k = \nabla f(\mathbf{x})|_{\mathbf{x}_k}$ and the Hessian matrix $H_k = \nabla^2 f(\mathbf{x})|_{\mathbf{x}_k}$. In the light of this equation, we can see that the value of the objective function at point $\mathbf{x}_{k+1}$ updated by the steepest descent algorithm described by Eq. (7.1.9)

$$\mathbf{x}_{k+1} \stackrel{(7.1.9)}{=} \mathbf{x}_k - \alpha_k \mathbf{g}_k / \|\mathbf{g}_k\|$$

is most likely smaller than that at the old point $\mathbf{x}_k$, with the third term in Eq. (7.1.10) neglected.

$$f(\mathbf{x}_{k+1}) \cong f(\mathbf{x}_k) + \mathbf{g}_k^T[\mathbf{x}_{k+1} - \mathbf{x}_k] \stackrel{(7.1.9)}{=} f(\mathbf{x}_k) - \alpha_k \mathbf{g}_k^T \mathbf{g}_k / \|\mathbf{g}_k\|;$$

$$f(\mathbf{x}_{k+1}) - f(\mathbf{x}_k) \cong -\alpha_k \mathbf{g}_k^T \mathbf{g}_k / \|\mathbf{g}_k\| \leq 0; \quad f(\mathbf{x}_{k+1}) \leq f(\mathbf{x}_k) \qquad (7.1.11)$$

Slightly different from this strategy of the steepest descent algorithm, the Newton method tries to go straight to the zero of the gradient of the approximate objective function (7.1.10)

$$\frac{df(\mathbf{x})}{d\mathbf{x}} \stackrel{(7.1.10)}{\underset{(C.1),(C.2),(C.6)}{=}} \mathbf{g}_k + H_k[\mathbf{x} - \mathbf{x}_k] = 0; \quad \mathbf{x} = \mathbf{x}_k - H_k^{-1}\mathbf{g}_k \qquad (7.1.12)$$

by the updating rule

$$\mathbf{x}_{k+1} = \mathbf{x}_k - H_k^{-1}\mathbf{g}_k \qquad (7.1.13)$$

with the gradient vector $\mathbf{g}_k = \nabla f(\mathbf{x})|_{\mathbf{x}_k}$ and the Hessian matrix $H_k = \nabla^2 f(\mathbf{x})|_{\mathbf{x}_k}$ (Appendix C).

This algorithm is essentially to find the zero of the gradient function $\mathbf{g}(\mathbf{x})$ of the objective function and consequently, it can be implemented by using any vector nonlinear equation solver. What we have to do is just to define the gradient function $\mathbf{g}(\mathbf{x})$ and put the function name as an input argument of any function like 'Newtons()' or 'fsolve()' for solving a system of nonlinear equations (see Section 4.6).

Now, we make a MATLAB script "nm0715.m", which actually solves $\mathbf{g}(\mathbf{x}) = 0$ for the gradient function

$$\mathbf{g}(\mathbf{x}) = \nabla f(\mathbf{x}) = \left[\frac{\partial f}{\partial x_1} \; \frac{\partial f}{\partial x_2}\right]^T = \left[2x_1 - x_2 - 4 \; \; 2x_2 - x_1 - 1\right]^T \qquad (7.1.14)$$

of the objective function (7.1.6)

$$f(\mathbf{x}) = f(x_1, x_2) = x_1^2 - x_1 x_2 - 4x_1 + x_2^2 - x_2$$

Figure 7.4 illustrates the process of searching for the minimum point by the Newton algorithm (7.1.13) as well as the steepest descent algorithm (7.1.9), where the steepest descent algorithm proceeds in the negative gradient direction, mostly moving zigzag toward a (local) minimum point, while the Newton algorithm approaches the minimum point almost straightly and reaches it in a few iterations.

```
»nm0715
    xo= [3.0000   2.0000], ans= -7
```

```
%nm0715.m to minimize an objective ftn f(x) by the Newton method.
f713=@(x)x(1).^2-4*x(1)-x(1).*x(2)+x(2).^2-x(2); % Eq.(7.1.6)
g713=@(x) [2*x(1)-x(2)-4  2*x(2)-x(1)-1];
x0=[0 0],   TolX=1e-4; TolFun=1e-6;  MaxIter= 50;
[xo,go,xx]=Newtons(g713,x0,TolX,MaxIter);
xo, f713(xo) %an extremum point reached and its function value
```

Remark 7.1 Weak Point of Newton Method.
Newton method is usually more efficient than the steepest descent method if only it works as illustrated above. However, regrettably, it is not guaranteed to reach the minimum point. The decisive weak point of Newton method is that it may approach one of the extrema having zero gradient, which is not necessarily a (local) minimum, but possibly a maximum or a saddle point.

7.1.6 Conjugate Gradient Method

Like the steepest descent method or Newton method, this method also uses the gradient to search for the minimum point of an objective function, but in a different way. It has two versions, the Polak-Ribiere (PR) method and the Fletcher-Reeves (FR) method, that are slightly different only in the search direction vector. This algorithm, summarized in the following box, has been cast into the MATLAB function 'opt_conjg()', which implements PR or FR depending on the last input argument KC=1 or 2. The quasi-Newton algorithm used in the MATLAB built-in function 'fminunc()' is similar to the conjugate gradient method.

CONJUGATE GRADIENT ALGORITHM

(Step 0) With the iteration number $k = 0$, find the objective function value $f_0 = f(\mathbf{x}_0)$ at the initial point $\mathbf{x}_0$.

(Step 1) Initialize the inside loop index, the temporary solution, and the search direction vector to $n = 0$, $\mathbf{x}(n) = \mathbf{x}_k$, and $\mathbf{s}(n) = -\mathbf{g}_k = -\mathbf{g}(\mathbf{x}_k)$, respectively, where $\mathbf{g}(\mathbf{x})$ is the gradient of the objective function $f(\mathbf{x})$.

(Step 2) For $n = 0$ to $N-1$, repeat the following:
Find the (optimal) step-size

$$\alpha_n = \text{ArgMin}_\alpha f(\mathbf{x}(n) + \alpha \mathbf{s}(n)) \qquad (7.1.15)$$

and update the temporary solution point to

$$\mathbf{x}(n+1) = \mathbf{x}(n) + \alpha_n \mathbf{s}(n) \qquad (7.1.16)$$

and the search direction vector to

$$\mathbf{s}(n+1) = -\mathbf{g}_{n+1} + \beta_n \mathbf{s}(n) \qquad (7.1.17)$$

with

$$\beta_n = \frac{[\mathbf{g}_{n+1} - \mathbf{g}_n]^T \mathbf{g}_{n+1}}{\mathbf{g}_n^T \mathbf{g}_n} \text{ (PR)} \quad \text{or} \quad \frac{\mathbf{g}_{n+1}^T \mathbf{g}_{n+1}}{\mathbf{g}_n^T \mathbf{g}_n} \text{ (FR)} \qquad (7.1.18)$$

(Step 3) Update the approximate solution point to $\mathbf{x}_{k+1} = \mathbf{x}(N)$, which is the last temporary one.

(Step 4) If $\mathbf{x}_k \approx \mathbf{x}_{k-1}$ and $f(\mathbf{x}_k) \approx f(\mathbf{x}_{k-1})$, then declare $\mathbf{x}_k$ to be the minimum and terminate the procedure. Otherwise, increment k by one and go back to Step 1.

```
function [xo,fo]=opt_conjg(f,x0,TolX,TolFun,alpha0,MaxIter,KC)
%KC=1: Polak-Ribiere Conjugate Gradient method
%KC=2: Fletcher-Reeves Conjugate Gradient method
if nargin<7, KC=0; end
if nargin<6, MaxIter=100; end
if nargin<5, alpha0=10; end
if nargin<4, TolFun=1e-8; end
if nargin<3, TolX=1e-6; end
N=length(x0); nmax1=20; warning=0; h=1e-4; % Dimension of variable
x=x0; fx=feval(f,x0); fx0=fx;
for k=1: MaxIter
   xk0= x; fk0= fx; alpha= alpha0;
   g= grad(f,x,h); s= -g;
   for n=1:N
      alpha= alpha0;
      fx1= feval(f,x+alpha*2*s); % Trial move in search direction
      for n1=1:nmax1 % To find the optimum step-size by line search
         fx2= fx1; fx1= feval(f,x+alpha*s);
         if fx0>fx1+TolFun & fx1<fx2-TolFun %fx0>fx1<fx2
            den=4*fx1-2*fx0-2*fx2; num=den-fx0+fx2; % Eq.(7.1.5)
            alpha= alpha*num/den;
            x= x+alpha*s; fx= feval(f,x);
            break;
           elseif n1==nmax1/2
            alpha= -alpha0; fx1= feval(f,x+alpha*2*s);
           else
            alpha= alpha/2;
         end
      end
      x0= x; fx0= fx;
      if n<N
        g1= grad(f,x,h);
        if KC<=1, s= -g1 +(g1-g)*g1'/(g*g'+1e-5)*s; % Eq.(7.1.18a)
          else  s= -g1 +g1*g1'/(g*g'+1e-5)*s; % Eq.(7.1.18b)
        end
        g= g1;
      end
      if n1>=nmax1, warning=warning+1; % can't find optimum step-size
       else  warning=0;
      end
   end
   if warning>=2|(norm(x-xk0)<TolX&abs(fx-fk0)<TolFun), break; end
end
xo= x; fo= fx;
if k==MaxIter, fprintf('Just best in %d iterations',MaxIter), end
```

```
%nm0716.m to minimize f(x) by the conjugate gradient method.
f713=@(x)x(1).^2-4*x(1)-x(1).*x(2)+x(2).^2-x(2); % Eq.(7.1.6)
x0=[0 0], TolX= 1e-4; TolFun= 1e-4; alpha0=10; MaxIter=100;
[xo,fo]= opt_conjg(f713,x0,TolX,TolFun,alpha0,MaxIter,1)
[xo,fo]= opt_conjg(f713,x0,TolX,TolFun,alpha0,MaxIter,2)
```

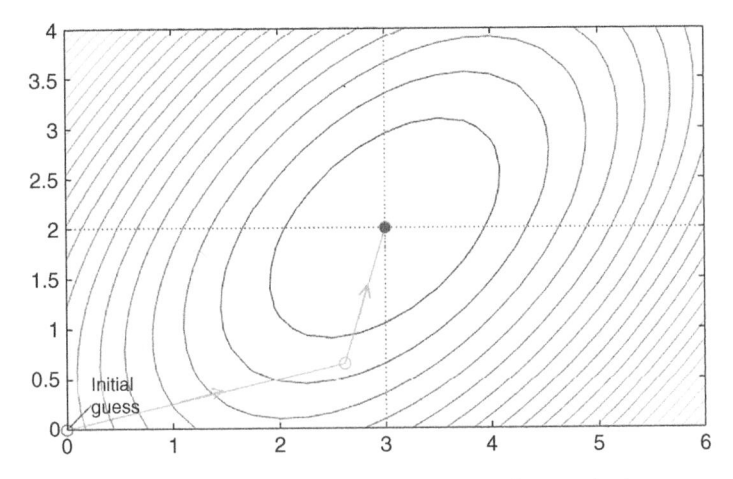

Figure 7.5 Illustration of the conjugate gradient method.

This method borrows the framework of the steepest descent method and needs a bit more effort for computing the search direction vector $s(n)$. It takes at most N iterations to reach the minimum point in case the objective function is quadratic with a positive-definite Hessian matrix H as

$$f(\mathbf{x}) = \frac{1}{2}\mathbf{x}^T H \mathbf{x} + \mathbf{b}^T \mathbf{x} + c \text{ where } \mathbf{x} \text{ is an } N\text{-dimensional vector} \quad (7.1.19)$$

Based on the fact that minimizing this quadratic objective function is equivalent to solving the linear equation

$$\mathbf{g}(\mathbf{x}) = \nabla f(\mathbf{x}) = H\mathbf{x} + \mathbf{b} = 0 \quad (7.1.20)$$

MATLAB has several built-in functions such as 'cgs()', 'pcg()', and 'bicg()', which use the conjugate gradient method to solve a set of linear equations.

We composed the above MATLAB script "nm0716.m" to minimize the objective function (7.1.6) by the conjugate gradient method, and the minimization process is illustrated in Figure 7.5.

7.1.7 Simulated Annealing

All of the optimization methods discussed so far may be more or less efficient in finding the minimum point if only they start from the initial point sufficiently close to it. But the point they reach may be one of several local minima, and we often cannot be sure that it is the global minimum. How about repeating the procedure to search for all local minima starting from many different initial guesses and taking the best one as the global minimum? This would be a computationally formidable task, since there is no systematic way to determine a suitable sequence of initial guesses, each of which leads to its own (local) minimum so that all the local minima can be exhaustively found to compete with each other for the global minimum.

An interesting alternative is based on the analogy between annealing and minimization. Annealing is the physical process of heating up a solid metal above its melting point and then cooling it down so slowly that the highly excited atoms can settle into a (global) minimum energy state, yielding a single crystal with a regular structure. Fast cooling by rapid quenching may result in widespread irregularities and defects in the crystal structure – analogous to being too hasty to find the global minimum. The simulated annealing process can be implemented using the Boltzman probability distribution of an energy level $E(\geq 0)$ at temperature T described by

$$p(E) = \alpha \exp(-E/KT) \text{ with the Boltzmann constant } K \text{ and } \alpha = 1/KT \quad (7.1.21)$$

Note that at high temperature, the probability distribution curve is almost flat over a wide range of E, implying that the system can be in a high energy state as equally well as in a low energy state, while at low temperature, the probability distribution curve gets higher/lower for lower/higher E, implying that the system will most probably be in a low energy state, but still have a slim chance to be in a high energy state so that it can escape from a local minimum energy state.

The idea of simulated annealing is summarized in the box beneath and cast into the MATLAB function 'sim_anl()'. This function has two parts that vary with the iteration number as the temperature falls down. One is the size of step $\Delta \mathbf{x}$ from the previous guess to the next guess, which is made by generating a random vector $\mathbf{y}$ (having uniform distribution $U(-1,+1)$ and the same dimension as the variable $\mathbf{x}$) and multiplying $\mu^{-1}(\mathbf{y})$ (term wisely) by the difference vector $(\mathbf{u} - \mathbf{l})$ between the upper-bound $\mathbf{u}$ and the lower-bound $\mathbf{l}$ of the domain of $\mathbf{x}$. The μ^{-1}-law

$$g_\mu^{-1}(y) \stackrel{(P1.9b)}{=} \frac{(1+\mu)^{|y|} - 1}{\mu} sign(y) \quad \text{for } |y| \leq 1 \quad (7.1.22)$$

implemented in the function 'mu_inv()' has the parameter μ that is increased according to a rule

$$\mu = 10^{100(k/k_{max})^q} \quad \text{with } q > 0 : \text{ the } \textit{quenching factor} \quad (7.1.23)$$

as the iteration number k increases, reaching $\mu = 10^{100}$ at the last iteration $k = k_{max}$. Note that

- the quenching factor $q > 0$ is made large/small for slow/fast quenching; and
- the value of μ^{-1}-law function becomes small for $|y| < 1$ as μ increases (see Figure 7.6a).

UNCONSTRAINED OPTIMIZATION

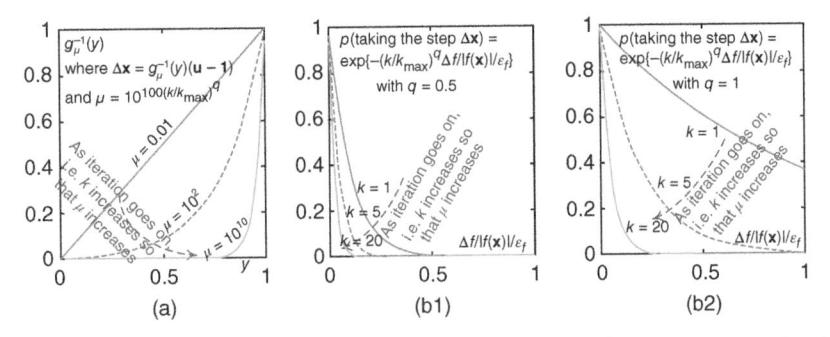

Figure 7.6 Some illustrative functions used for controlling the randomness – temperature in SA. (a) μ-inverse law $g_\mu^{-1}(y)$. (b1) p(taking $\Delta \mathbf{x}$) with $q = 0.5$. (b2) p(taking $\Delta \mathbf{x}$) with $q = 1$.

Simulated Annealing (SA)

(Step 0) Pick the initial guess $\mathbf{x}_0$, the lower-bound $\mathbf{l}$, the upper-bound $\mathbf{u}$, the maximum number of iterations $k_{max} > 0$, the quenching factor $q > 0$ (to be made large/small for slow/fast quenching) and the relative tolerance ε_f of function value fluctuation.

(Step 1) Let $\mathbf{x} = \mathbf{x}_0$, $\mathbf{x}^o = \mathbf{x}$, and $f^o = f(\mathbf{x})$.

(Step 2) For $k = 1$ to k_{max}, do

{Generate an $N \times 1$ uniform random vector of $U(-1, +1)$ and transform it by the inverse μ-law (with $\mu = 10^{100(k/k_{max})^q}$) to make $\Delta \mathbf{x}$ and then take $\mathbf{x}_1 \leftarrow \mathbf{x} + \Delta \mathbf{x}$, confining the next guess inside the admissible region $\{\mathbf{x} \mid \mathbf{l} \leq \mathbf{x} \leq \mathbf{u}\}$ as needed.

If $\Delta f = f(\mathbf{x}_1) - f(\mathbf{x}) < 0$,

{set $\mathbf{x} \leftarrow \mathbf{x}_1$ and if $f(\mathbf{x}) < f^o$, set $\mathbf{x}^o \leftarrow \mathbf{x}$ and $f^o \leftarrow f(\mathbf{x}^o)$}

Otherwise,

{generate a uniform random number z of $U(0,1)$ and set $\mathbf{x}_1 \leftarrow \mathbf{x}$ only in case

$$z < p(\text{taking the step } \Delta \mathbf{x}) \stackrel{(7.1.24)}{=} \exp\{-(k/k_{max})^q \Delta f / |f(\mathbf{x})| / \varepsilon_f\}$$

}
}

(Step 3) Regarding $\mathbf{x}^o$ as close to the minimum point that we are looking for, we may set $\mathbf{x}^o$ as the initial value and apply any (local) optimization algorithm to search for the minimum point of $f(\mathbf{x})$.

```
function [xo,fo]=sim_anl(f,x0,l,u,kmax,q,TolFun,Kg)
% simulated annealing method to minimize f(x) s.t. l<=x<=u
N=length(x0);
x=x0; fx=feval(f,x);
xo=x; fo=fx;
if nargin<8, Kg=0; end %# of iterations to plot the graph
if nargin<7, TolFun=1e-8; end
if nargin<6, q=1; end %quenching factor
if nargin<5, kmax=100; end %maximum iteration number
for k=0:kmax
   Ti=(k/kmax)^q; %inverse of temperature from 0 to 1
   mu=10^(Ti*100); % Eq.(7.1.23)
   dx= mu_inv(2*rand(size(x))-1,mu).*(u-l);
   x1=x+dx; %next guess
   x1=(x1<l).*l +(l<=x1).*(x1<=u).*x1 +(u<x1).*u;
   % To confine it inside the admissible region bounded by l and u.
   fx1=feval(f,x1); df=fx1-fx;
   if df<0|rand<exp(-Ti*df/(abs(fx)+eps)/TolFun) Eq.(7.1.24)
      x=x1; fx=fx1;
   end
   if fx<fo, xo=x; fo=fx1; end
end

function x=mu_inv(y,mu) % inverse of mu-law Eq.(7.1.22)
x=(((1+mu).^abs(y)-1)/mu).*sign(y); % Eq.(P1.9b)
```

The other is the probability of taking a step $\Delta \mathbf{x}$ that would result in change $\Delta f > 0$ of the objective function value $f(\mathbf{x})$. Similarly to Eq. (7.1.21), this is determined by

$$p(\text{taking the step } \Delta \mathbf{x}) = \exp\{-(k/k_{\max})^q \Delta f / |f(\mathbf{x})|/\varepsilon_f\} \quad \text{for } \Delta f > 0 \quad (7.1.24)$$

This remains as big as e^{-1} for $|\Delta f/f(\mathbf{x})| = \varepsilon_f$ at the last iteration $k = k_{\max}$, meaning that the probability of taking a step hopefully to escape from a local minimum and find the global minimum at the risk of increasing the value of objective function by the amount $\Delta f = |f(\mathbf{x})|\varepsilon_f$ is still that high. The shapes of the two functions related with the temperature are depicted in Figure 7.6.

We make the MATLAB script "nm0717.m", which uses the function 'sim_anl()' to minimize a function

$$f(\mathbf{x}) = x_1^4 - 16x_1^2 - 5x_1 + x_2^4 - 16x_2^2 - 5x_2 \quad (7.1.25)$$

and tries other functions such as 'opt_Nelder()', 'fminsearch()', 'fminunc()', and 'simulannealbnd()' for cross-check. The results of running the script are summarized in Table 7.1, which shows that the function 'sim_anl()' may give us the global minimum more possibly than the other functions. But even this function based on the idea of simulated annealing (SA) cannot always succeed

Table 7.1 Results of using the several optimization functions with various initial values.

x_0	opt_Nelder()	fminsearch()	fminunc()	sim_anl()
[0, 0]	[2.9035, 2.9035] ($f^o = -156.66$)	[2.9035, 2.9036] ($f^o = -156.66$)	[2.9036, 2.9036] ($f^o = -156.66$)	[2.8966, 2.9036] ($f^o = -156.66$)
[−0.5, −1.0]	[2.9035, −2.7468] ($f^o = -128.39$)	[−2.7468, −2.7468] ($f^o = -100.12$)	[−2.7468, −2.7468] ($f^o = -100.12$)	[2.9029, 2.9028] ($f^o = -156.66$)

and its success/failure depends partially on the initial guess and partially on luck, while the success/failure of the other functions depends solely on the initial guess.

```
%nm0717.m to minimize an objective function f(x) by various methods.
f=@(x)x(1)^4-16*x(1)^2-5*x(1)+x(2)^4-16*x(2)^2-5*x(2); % Eq.(7.1.25)
l=[-5 -5]; u=[5 5]; % Lower/Upper bounds of the solution domain
x0=[0 0]; % Initial guess
[xo_nd,fo]= opt_Nelder(f,x0)
[xos,fos]=fminsearch(f,x0) % to compare with MATLAB built-in functions
[xou,fou]=fminunc(f,x0)
kmax=500; q=1; TolFun=1e-9;
[xo_sa,fo_sa]=sim_anl(f,x0,l,u,kmax,q,TolFun)
opt = optimoptions('simulannealbnd', 'PlotFcn',@saplotbestf, ...
       'FunctionTolerance',TolFun);
[xo_sa1,fo_sa1]=simulannealbnd(f,x0,l,u,opt)
```

7.1.8 Genetic Algorithm

HYBRID GENETIC ALGORITHM

(Step 0) Pick the initial guess $\mathbf{x}_0 = [x_{01} \ldots x_{0N}]$ (N: the dimension of the variable), the lower-bound $\mathbf{l} = [l_1 \ldots l_N]$, the upper-bound $\mathbf{u} = [u_1 \ldots u_N]$, the population size N_p, the vector $\mathbf{N}_b = [N_{b1} \ldots N_{bN}]$ consisting of the numbers of bits assigned for the representation of each variable x_i, the probability of crossover P_c, the probability of mutation P_m, the learning rate η ($0 < \eta \leq 1$, to be made small/large for slow/fast learning) and the maximum number of iterations $k_{max} > 0$. Note that the dimensions of $\mathbf{x}_0$, $\mathbf{u}$, and $\mathbf{l}$ are all the same as N, which is the dimension of the variable $\mathbf{x}$ to be found and the population size N_p cannot be greater than 2^{N_b} in order to avoid duplicated chromosomes and should be an even integer for constituting the mating pool in the crossover stage.

(Step 1) Random Generation of Initial Population
Set $\mathbf{x}^o = \mathbf{x}_0, f^o = f(\mathbf{x}^o)$ and construct in a random way the initial population array X_1 that consists of N_p states (in the admissible region bounded by $\mathbf{u}$ and $\mathbf{l}$) including the initial state $\mathbf{x}_0$, by setting

$$X_1(1) = \mathbf{x}_0 \quad \text{and} \quad X_1(2) = \mathbf{l} + \mathbf{rand}.*(\mathbf{u}-\mathbf{l}), \ldots, \quad (7.1.26)$$
$$X_1(N_p) = \mathbf{l} + \mathbf{rand}.*(\mathbf{u}-\mathbf{l})$$

where **rand** is a random vector of the same dimension N as $\mathbf{x}_0$, $\mathbf{u}$, and $\mathbf{l}$. Then, encode each number of this population array into a binary string by

$$P_1\left(n, 1 + \sum_{i=1}^{m-1} N_{bi} : \sum_{i=1}^{m} N_{bi}\right)$$
$$= \text{binary representation of } X_1(n,m) \text{ with } N_{bm} \text{ bits}$$
$$= (2^{N_{bm}} - 1)\frac{X_1(n,m) - l(m)}{u(m) - l(m)} \quad \text{for } n = 1 : N_p \text{ and } m = 1 : N \quad (7.1.27)$$

so that the whole population array becomes a pool array each row of which is a chromosome represented by a binary string of $\sum_{i=1}^{N} N_{bi}$ bits.

(Step 2) For $k = 1$ to $k_{\max}$, do
{ 1. Decode each number in the pool into a (decimal) number by

$$X_k(n,m) = \text{decimal representation of}$$
$$P_k\left(n, 1 + \sum_{i=1}^{m-1} N_{bi} : \sum_{i=1}^{m} N_{bi}\right) \text{ with } N_{bm} \text{ bits}$$
$$= P_k(n,\cdot)\frac{u(m) - l(m)}{2^{N_{bm}} - 1} + l(m) \quad \text{for } n = 1 : N_p \text{ and } m = 1 : N \quad (7.1.28)$$

and evaluate the value $f(n)$ of function for every row $X_k(n,:) = \mathbf{x}(n)$ corresponding to each chromosome and find the minimum $f_{\min} = f(n_b)$ corresponding to $X_k(n_b,:) = \mathbf{x}(n_b)$.

2. If $f_{\min} = f(n_b) < f^o$, then set $f^o = f(n_b)$ and $\mathbf{x}^o = \mathbf{x}(n_b)$.
3. Convert the function values into the values of fitness by

$$f_1(n) = \text{Max}_{n=1}^{N_p}\{f(n)\} - f(n) \quad (7.1.29)$$

which is nonnegative $\forall n = 1 : N_p$ and is large for a good chromosome.

UNCONSTRAINED OPTIMIZATION

4. If $\text{Max}_{n=1}^{N_p} \{f_1(n)\} \approx 0$, then terminate this procedure, declaring $\mathbf{x}^o$ as the best.

   Otherwise, in order to make more chromosomes around the best point $\mathbf{x}(n_b)$ in the next generation, use the reproduction rule

   $$\mathbf{x}(n) \leftarrow \mathbf{x}(n) + \eta \frac{f_1(n_b) - f_1(n)}{f_1(n_b)} \{\mathbf{x}(n_b) - \mathbf{x}(n)\} \quad (7.1.30)$$

   to get a new population X_{k+1} with $X_{k+1}(n,:) = \mathbf{x}(n)$ and encode it to reconstruct a new pool array P_{k+1} by Eq. (7.1.27).

5. Shuffle the row indices of the pool array for random mating of the chromosomes.

6. With the crossover probability P_c, exchange the tail part starting from some random bit of the numbers in two randomly paired chromosomes (rows of P_{k+1}) with each other's to get a new pool array P'_{k+1}.

7. With the mutation probability P_m, reverse a random bit of each number represented by chromosomes (rows of P'_{k+1}) to make a new pool array P_{k+1}.

}

```
function [xo,fo]=genetic(f,x0,l,u,Np,Nb,Pc,Pm,eta,kmax)
% Genetic Algorithm to minimize f(x) s.t. l<=x<=u
N=length(x0);
if nargin<10, kmax=100; end % Number of iterations (generations)
if nargin<9|eta>1|eta<=0, eta=1; end % Learning rate(0<eta<1)
if nargin<8, Pm=0.01; end % Probability of mutation
if nargin<7, Pc=0.5; end % Probability of crossover
if nargin<6, Nb=8*ones(1,N); end % # of genes(bits) for each variable
if nargin<5, Np=10; end % Population size(number of chromosomes)
% Initialize the population pool
NNb=sum(Nb);
xo=x0(:)'; l=l(:)'; u=u(:)';
fo=feval(f,xo);
X(1,:)=xo;
for n=2:Np
   X(n,:)=l+rand(size(x0)).*(u-l); % Eq.(7.1.26)
end
P=gen_encode(X,Nb,l,u); % Eq.(7.1.27)
for k=1:kmax
   X=gen_decode(P,Nb,l,u); % Eq.(7.1.28)
   for n=1:Np, fX(n)=feval(f,X(n,:)); end
   [fxb,nb]=min(fX); % Selection of the fittest
   if fxb<fo, fo=fxb; xo=X(nb,:); end
```

```
   fX1=max(fxs)-fX; %make the nonnegative fitness vector by Eq.(7.1.29)
   fXm=fX1(nb);
   if fXm<eps, return; end %terminate if all the chromosomes are equal
   % Reproduction of next generation
   for n=1:Np
      X(n,:)=X(n,:)+eta*(fXm-fX1(n))/fXm*(X(nb,:)-X(n,:)); %Eq.(7.1.30)
   end
   P=gen_encode(X,Nb,l,u);
   % Mating/Crossover
   is=shuffle([1:Np]);
   for n=1:2:Np-1
     if rand<Pc, P(is(n:n+1),:)=crossover(P(is(n:n+1),:),Nb); end
   end
   % Mutation
   P=mutation(P,Nb,Pm);
end
```

```
function P=gen_encode(X,Nb,l,u)
% Encode a population(X) of state into an array(P) of binary strings
Np=size(X,1); % Population size
N=length(Nb); % Dimension of the variable(state)
for n=1:Np
   b2=0;
   for m=1:N
      b1=b2+1; b2=b2+Nb(m);
      Xnm=(2^Nb(m)-1)*(X(n,m)-l(m))/(u(m)-l(m)); %Eq.(7.1.27)
      P(n,b1:b2)=dec2bin(Xnm,Nb(m)); %encoding to binary strings
   end
end
```

```
function X=gen_decode(P,Nb,l,u)
% Decode an array of binary strings(P) into a population(X) of state
Np=size(P,1); % Population size
N=length(Nb); % Dimension of the variable(state)
for n=1:Np
   b2=0;
   for m=1:N
      b1=b2+1; b2=b1+Nb(m)-1; % Eq.(7.1.28)
      X(n,m)=bin2dec(P(n,b1:b2))*(u(m)-l(m))/(2^Nb(m)-1)+l(m);
   end
end
```

```
function P=mutation(P,Nb,Pm) % Mutation
Nbb=length(Nb);
for n=1:size(P,1)
   b2=0;
   for m=1:Nbb
      if rand<Pm
         b1=b2+1; bi=b1+mod(floor(rand*Nb(m)),Nb(m)); b2=b2+Nb(m);
         P(n,bi)=num2str(1-eval(P(n,bi)));
      end
   end
end
```

```
function chrms2=crossover(chrms2,Nb)
% Crossover between two chromosomes
Nbb=length(Nb);
b2=0;
for m=1:Nbb
   b1=b2+1; bi=b1+mod(floor(rand*Nb(m)),Nb(m)); b2=b2+Nb(m);
   tmp=chrms2(1,bi:b2);
   chrms2(1,bi:b2)=chrms2(2,bi:b2);
   chrms2(2,bi:b2)=tmp;
end
```

```
function is=shuffle(is) % Shuffle
N=length(is);
for n=N:-1:2
   in=ceil(rand*(n-1)); tmp=is(in);
   is(in)=is(n); is(n)=tmp; % swap the n-th element with the in-th one
end
```

GA is a directed random search technique that is modeled on the natural evolution/selection process toward the survival of the fittest. The genetic operators deal with the individuals in a population over several generations to improve their fitness gradually. Individuals standing for possible solutions are often compared to chromosomes and represented by strings of binary numbers. Like the SA method, GA is also expected to find the global minimum solution even in the case where the objective function has several extrema, including local maxima, saddle points as well as local minima.

A so-called hybrid GA[P-2] consists of initialization, evaluation, reproduction (selection), crossover, and mutation as depicted in Figure 7.7 and is summarized in the box below. The reproduction/crossover process is illustrated in Figure 7.8. This algorithm is cast into the function 'genetic()'. To use this function and the MATLAB built-in GA function 'ga()' for minimizing the function defined by Eq. (7.1.25), we append the following statements to the MATLAB script "nm0717.m", run the script, and compare the results obtained using various functions. The following is some introductory usage of the function 'ga()'. Note that like the simulated annealing, the function based on the idea of GA cannot always succeed, and its success/failure depends partially on the initial guess and partially on luck.

```
%do_genetic.m
  Np=30; % Population size
  Nb=[10 10]; % Numbers of bits for representing each variable
  Pc=1; Pm=0.01; % Probability of crossover/mutation
  eta=0.1; kmax=100; % Learning rate and the maximum # of iterations
  [xo_gen,fo_gen]=genetic(f,x0,l,u,Np,Nb,Pc,Pm,eta,kmax)
  Nv=2; % Number of variables
  [xo_ga,fo_ga]=ga(f,Nv) % Using the MATLAB built-in function 'ga()'
```

OPTIMIZATION

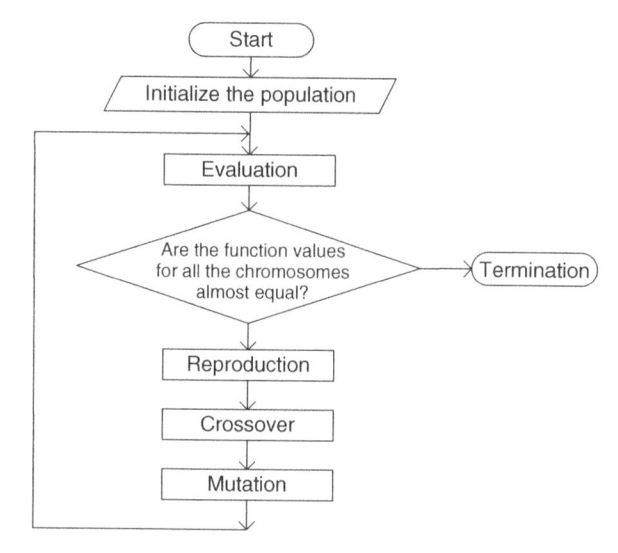

Figure 7.7 Flowchart of genetic algorithm (GA).

$N_p = 30$, $N = 2$, $N_b = [10\ 10]$

n	Pool $\mathbf{P}_{30\times 20}$			Population $\mathbf{X}_{30\times 2}$		Function value $f\mathbf{X}$
1	0111111111	0111111111		−0.0049	−0.0049	0.05
2	1000001100	0111010111		0.1222	−0.3959	−1.35
3	0101100110	0001100001		−1.5005	−4.0518	3.66
4	0110111011	1011010101	Decode	−0.6696	2.0870	Evaluate −64.78
5	0001110110	0001001111		−3.8465	−4.2278	56.05
6	0101111001	0000100010		−1.3148	−4.6676	131.32
7	0011000100	0111100010		−3.0841	−0.2884	−46.18
ℓ	0010010100	1011011110		−3.5533	2.1750	−89.02
30	.. ..	.. ..		.. ..	.. ..	..

Random pairing: 2-6, 3-7, ... ↓Reproduction

1 a_1	1000001010	0111110100 b_1		0.1026	−0.1124	−0.32
2 a_2	1000010110	0111001101 b_2		0.2239	−0.4861	−3.21
3 a_3	0101110111	0001100101 b_3		−1.3327	−4.0058	2.18
4 a_4	0111000001	1011001011 b_4	Encode	−0.6109	1.9892	−60.38
5 a_5	0010011010	0001010110 b_5		−3.4946	−4.1593	14.51
6 a_6	0110011001	0000110000 b_6		−0.9983	−4.5308	105.64
7 a_7	0011010010	0111011011 b_7		−2.9421	−0.3492	−49.05
ℓ	0010011011	1011011000		−3.4803	2.1180	−91.93
30	.. ..	.. ..		.. ..	.. ..	..

↓Crossover/mutation

1 a_1	1000001010	0111110100 b_1		0.1026	−0.1124	−0.32
2 a_2	1000010001	0000110000 b_2		0.1711	−4.5308	114.28
3 a_3	0101010010	0001101011 b_3		−1.6960	−3.9541	−15.21
4 a_4	0111000001	1010001011 b_4	Decode	−0.6109	1.3636	−35.89
5 a_5	0010011010	0001010110 b_5		−3.4946	−4.1593	14.51
6 a_6	0110011110	0111001101 b_6		−0.9531	−0.4936	−10.31
7 a_7	0011110111	0111010101 b_7		−2.5855	−0.4154	−50.00
ℓ	0010011011	1011011000		−3.4803	2.1180	−91.93
30	.. ..	.. ..		.. ..	.. ..	..

Figure 7.8 Reproduction/crossover/mutation in one iteration of genetic algorithm (GA).

Usage of the MATLAB built-in function 'ga()' for genetic algorithm

```
[xo,fo,.]=ga('ftn',Nx,A,b,Aeq,beq,l,u,'nlcon',options,p1,p2,...)
```

- **Input arguments** (At least, two input arguments 'ftn', Nx required)

    'ftn' : objective (fitness) function $f(\mathbf{x})$ to be minimized, usually defined in an M-file, but can be defined as an inline function or a function handle, which will remove the necessity of quotes(").

    Nx : number of variables

    A,b : linear inequality constraints $A\mathbf{x} \leq \mathbf{b}$ (to be given as [] if not applied).

    Aeq,beq : linear equality constraints $A_{eq}\mathbf{x} = \mathbf{b}_{eq}$ (to be given as [] if not applied).

    l,u : lower/upper bound vectors such that $\mathbf{l} \leq \mathbf{x} \leq \mathbf{u}$; to be given as [] if not applied, set l(i)=-inf/u(i)=inf if x(i) is not bounded below/above.

    'nlcon' : nonlinear constraint function defined in an M-file, supposed to return the two output arguments for a given $\mathbf{x}$; the first one being the LHS (vector) of inequality constraints $\mathbf{c}(\mathbf{x}) \leq 0$ and the second one being the LHS (vector) of equality constraints $\mathbf{c}_{eq}(\mathbf{x}) = 0$; to be given as [] if not applied.

    options : can be created with 'optimoptions' to be used for setting the display parameter, the tolerances for $\mathbf{x}^o$, $f(\mathbf{x}^o)$, etc.; to be given as [] if not applied.

- **Output arguments**

    xo : minimum point ($\mathbf{x}^o$) reached in the permissible region satisfying the constraints

    fo : minimized function value $f(\mathbf{x}^o)$

7.2 CONSTRAINED OPTIMIZATION

In this section, only the concept of constrained optimization is introduced. The functions and usages of the corresponding MATLAB functions like 'fmincon()' will be explained in the next section.

7.2.1 Lagrange Multiplier Method

A class of common optimization problems subject to equality constraints may be nicely handled by the Lagrange multiplier method. Consider an optimization problem with M equality constraints.

$$\text{Min} f(\mathbf{x}) \tag{7.2.1a}$$

$$\text{subject to (s.t.) } \mathbf{h}(\mathbf{x}) = \begin{bmatrix} h_1(\mathbf{x}) \\ h_2(\mathbf{x}) \\ \vdots \\ h_M(\mathbf{x}) \end{bmatrix} = \mathbf{0} \tag{7.2.1b}$$

According to the Lagrange multiplier method, this problem can be converted to the following unconstrained optimization problem.

$$\text{Min} \, l(\mathbf{x}, \boldsymbol{\lambda}) = f(\mathbf{x}) + \boldsymbol{\lambda}^T \mathbf{h}(\mathbf{x}) = f(\mathbf{x}) + \sum_{m=1}^{M} \lambda_m h_m(\mathbf{x}) \tag{7.2.2}$$

The solution of this problem, if it exists, can be obtained by setting the derivatives of this new (augmented) objective function $l(\mathbf{x},\boldsymbol{\lambda})$ with respect to (w.r.t.) $\mathbf{x}$ and $\boldsymbol{\lambda}$ to zero:

$$\frac{\partial}{\partial \mathbf{x}} l(\mathbf{x}, \boldsymbol{\lambda}) = \frac{\partial}{\partial \mathbf{x}} f(\mathbf{x}) + \boldsymbol{\lambda}^T \frac{\partial}{\partial \mathbf{x}} \mathbf{h}(\mathbf{x}) = \nabla f(\mathbf{x}) + \sum_{m=1}^{M} \lambda_m \nabla h_m(\mathbf{x}) = \mathbf{0} \tag{7.2.3a}$$

$$\frac{\partial}{\partial \boldsymbol{\lambda}} l(\mathbf{x}, \boldsymbol{\lambda}) = \mathbf{h}(\mathbf{x}) = \mathbf{0} \tag{7.2.3b}$$

Note that the solutions for this system of equations are the extrema of the objective function. We may know if they are minima/maxima, from the positive/negative-definiteness of the second derivative (Hessian matrix) of $l(\mathbf{x},\boldsymbol{\lambda})$ w.r.t. $\mathbf{x}$ and $\boldsymbol{\lambda}$. Let us see the following examples.

Remark 7.2 Inequality Constraints with the Lagrange Multiplier Method. Even though the optimization problem involves inequality constraints like $g_n(\mathbf{x}) \leq 0$, we can convert them to equality constraints by introducing the (nonnegative) slack variables y_j^2 as

$$g_j(\mathbf{x}) + y_j^2 \leq 0 \tag{7.2.4}$$

Then, we can use the Lagrange multiplier method to handle it like an equality constrained problem. An alternative is not to consider any *inactive inequality* for which the value of the Lagrange multiplier (obtained as the solution to Eq. (7.2.3)) is negative or positive depending on whether the problem is a minimization one or maximization one.

Example 7.1. Minimization (with Equality Constraint) by the Lagrange Multiplier Method.
Consider the following minimization problem subject to a single equality constraint.

CONSTRAINED OPTIMIZATION **401**

$$\text{Min } f(\mathbf{x}) = x_1^2 + x_2^2 \quad \text{(E7.1.1a)}$$

$$\text{s.t. } h(\mathbf{x}) = x_1 + x_2 - 2 = 0 \quad \text{(E7.1.1b)}$$

We can substitute the equality constraint $x_2 = 2 - x_1$ into the objective function (E7.1.1a) so that this problem becomes an unconstrained optimization problem as

$$\text{Min } f(x_1) = x_1^2 + (2 - x_1)^2 = 2x_1^2 - 4x_1 + 4 \quad \text{(E7.1.2)}$$

which can be easily solved by setting the derivative of this new objective function w.r.t. x_1 to zero.

$$\frac{\partial}{\partial x_1} f(x_1) = 4x_1 - 4 = 0; \quad x_1 = 1, \quad x_2 \stackrel{\text{(E7.1.1b)}}{=} 2 - x_1 = 1 \quad \text{(E7.1.3)}$$

Alternatively, we can apply the Lagrange multiplier method as follows:

$$\text{Min } l(\mathbf{x}, \lambda) \stackrel{(7.2.2)}{=} x_1^2 + x_2^2 + \lambda(x_1 + x_2 - 2) \quad \text{(E7.1.4)}$$

$$\frac{\partial}{\partial x_1} l(\mathbf{x}, \lambda) \stackrel{(7.2.3a)}{=} 2x_1 + \lambda = 0; \quad x_1 = -\lambda/2 \quad \text{(E7.1.5a)}$$

$$\frac{\partial}{\partial x_2} l(\mathbf{x}, \lambda) \stackrel{(7.2.3a)}{=} 2x_2 + \lambda = 0; \quad x_2 = -\lambda/2 \quad \text{(E7.1.5b)}$$

$$\frac{\partial}{\partial \lambda} l(\mathbf{x}, \lambda) \stackrel{(7.2.3b)}{=} x_1 + x_2 - 2 = 0 \quad \text{(E7.1.5c)}$$

$$x_1 + x_2 \stackrel{\text{(E7.1.5c)}}{=} 2 \stackrel{\text{(E7.1.5a,b)}}{\rightarrow} -\lambda/2 - \lambda/2 = -\lambda = 2; \quad \lambda = -2 \quad \text{(E7.1.6)}$$

$$x_1 \stackrel{\text{(E7.1.5a)}}{=} -\lambda/2 = 1, \quad x_2 \stackrel{\text{(E7.1.5b)}}{=} -\lambda/2 = 1 \text{ (Figure 7.9)} \quad \text{(E7.1.7)}$$

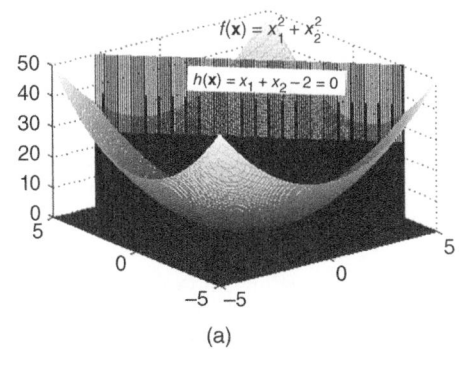

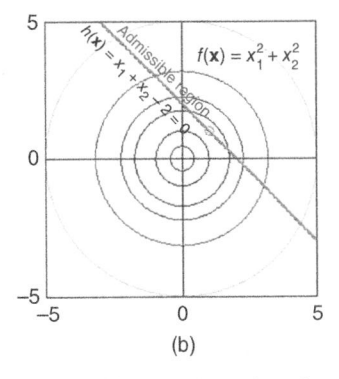

Figure 7.9 The objective and constraint functions for Example 7.1. (a) Mesh-shaped graphs. (b) Contour-shaped graphs.

In this example, the substitution of (linear) equality constraints is more convenient than the Lagrange multiplier method. However, it is not always the case, as illustrated by the next example.

Example 7.2. Optimization (with Inequality Constraint) by the Lagrange Multiplier Method.

(a) Consider the following minimization problem subject to a single inequality ($\leq$) constraint:

$$\text{Min } f(\mathbf{x}) = x_1^2 + x_2^2 \tag{E7.2.1a}$$

$$\text{s.t. } g(\mathbf{x}) = x_1 + x_2 - 2 \leq 0 \tag{E7.2.1b}$$

As in the previous example (to minimize the same objective function subject to an equality constraint), we can apply the Lagrange multiplier method to get

$$x_1 = 1, x_2 = 1, \text{ and } \lambda = -2 \tag{E7.2.2}$$

This negative value of $\lambda = -2$ implies that the corresponding constraint is inactive (even though barely satisfied with equality) in the sense that there may be a room for improving (decreasing) the value of the objective function without having to consider the inequality ($\leq$). It can be seen that the global minimum point (0,0) is in the admissible region satisfying the inequality (E7.2.1b) so that it is the solution to this problem as if the constraint were not given.

(b) Consider the following minimization problem subject to a single inequality ($\geq$) constraint:

$$\text{Min } f(\mathbf{x}) = x_1^2 + x_2^2 \tag{E7.2.3a}$$

$$\text{s.t. } g(\mathbf{x}) = x_1 + x_2 - 2 \geq 0 \tag{E7.2.3b}$$

To fit this problem into the frame of standard minimization problem, let us put the negative sign to both sides of the inequality constraint:

$$\text{Min } f(\mathbf{x}) = x_1^2 + x_2^2 \tag{E7.2.4a}$$

$$\text{s.t. } g_-(\mathbf{x}) = -x_1 - x_2 + 2 \leq 0 \tag{E7.2.4b}$$

Then, we can apply the Lagrange multiplier method to get

$$x_1 = 1, x_2 = 1, \text{ and } \lambda = 2 \tag{E7.2.5}$$

This positive value of $\lambda = 2$ implies that the corresponding constraint is active in the sense that there exists no better solution (with smaller value of the objective function) in the admissible region satisfying the inequality (E7.2.4b).

Example 7.3. Minimization (with Two Inequality Constraints) by the Lagrange Multiplier Method.

Consider the following minimization problem subject to two inequality ($\leq$) constraints:

$$\text{Min } f(\mathbf{x}) = x_1^2 + x_2^2 \qquad \text{(E7.3.1a)}$$

$$\text{s.t.} \quad \begin{aligned} g_1(\mathbf{x}) &= -x_1 - x_2 + 2 \leq 0 \\ g_2(\mathbf{x}) &= x_1 - x_2 - 1 \leq 0 \end{aligned} \qquad \text{(E7.3.1b)}$$

We make the augmented objective function incorporating the two inequalities like equalities as

$$\text{Min } l(\mathbf{x}) = x_1^2 + x_2^2 + \lambda_1(-x_1 - x_2 + 2) + \lambda_2(x_1 - x_2 - 1) \qquad \text{(E7.3.2)}$$

Then, we set the derivatives of this objective function w.r.t. x_1, x_2, λ_1, and λ_2 to zero:

$$\frac{\partial l}{\partial x_1} = 2x_1 - \lambda_1 + \lambda_2 = 0 \qquad \text{(E7.3.3a)}$$

$$\frac{\partial l}{\partial x_2} = 2x_2 - \lambda_1 - \lambda_2 = 0 \qquad \text{(E7.3.3b)}$$

$$\frac{\partial l}{\partial \lambda_1} = x_1 + x_2 - 2 = 0 \qquad \text{(E7.3.3c)}$$

$$\frac{\partial l}{\partial \lambda_2} = x_1 - x_2 - 1 = 0 \qquad \text{(E7.3.3d)}$$

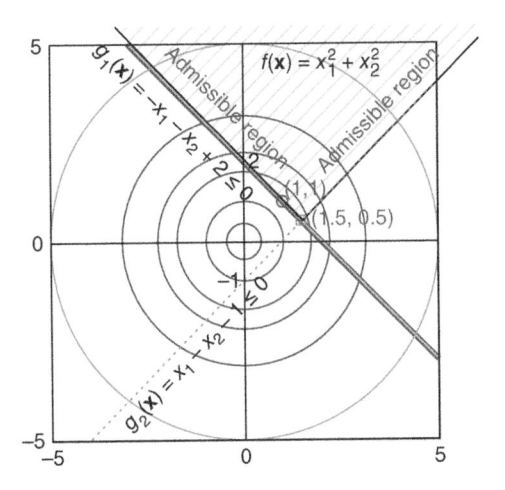

Figure 7.10 The objective and constraint functions for Example 7.3.

Solving this set of equations yields

$$\begin{bmatrix} 2 & 0 & -1 & 1 \\ 0 & 2 & -1 & -1 \\ 1 & 1 & 0 & 0 \\ 1 & -1 & 0 & 0 \end{bmatrix} \begin{bmatrix} x_1 \\ x_2 \\ \lambda_1 \\ \lambda_2 \end{bmatrix} = \begin{bmatrix} 0 \\ 0 \\ 2 \\ 1 \end{bmatrix}; \quad \begin{bmatrix} x_1 \\ x_2 \\ \lambda_1 \\ \lambda_2 \end{bmatrix} = \begin{bmatrix} 1.5 \\ 0.5 \\ 2 \\ -1 \end{bmatrix} \quad \text{(E7.3.4)}$$

This seems to denote that $(x_1, x_2) = (1.5, 0.5)$ is the minimum point. However, the sign of the Lagrange multiplier value $\lambda_2 = -1$ is negative, and therefore, the second constraint (E7.3.1b-2) corresponding to λ_2, having turned out to be inactive, should be discarded (see Remark 7.2). Thus the problem becomes virtually the same as Example 7.2(b) (see Figure 7.10).

Example 7.4. Minimization (with Equality Constraint) by the Lagrange Multiplier Method.

Consider the following minimization problem subject to a single nonlinear equality constraint.

$$\text{Min } f(\mathbf{x}) = x_1 + x_2 \quad \text{(E7.4.1a)}$$

$$\text{s.t. } h(\mathbf{x}) = x_1^2 + x_2^2 - 2 = 0 \quad \text{(E7.4.1b)}$$

Noting that it is absurd to substitute the equality constraint (E7.4.1b) into the objective function (E7.4.1a), we apply the Lagrange multiplier method as follows:

$$\text{Min } l(\mathbf{x}, \lambda) \overset{(7.2.2)}{=} x_1 + x_2 + \lambda(x_1^2 + x_2^2 - 2) \quad \text{(E7.4.2)}$$

$$\frac{\partial}{\partial x_1} l(\mathbf{x}, \lambda) \overset{(7.2.3a)}{=} 1 + 2\lambda x_1 = 0; \quad x_1 = -1/2\lambda \quad \text{(E7.4.3a)}$$

$$\frac{\partial}{\partial x_2} l(\mathbf{x}, \lambda) \overset{(7.2.3a)}{=} 1 + 2\lambda x_2 = 0; \quad x_2 = -1/2\lambda \quad \text{(E7.4.3b)}$$

$$\frac{\partial}{\partial \lambda} l(\mathbf{x}, \lambda) \overset{(7.2.3b)}{=} x_1^2 + x_2^2 - 2 = 0 \quad \text{(E7.4.3c)}$$

$$x_1^2 + x_2^2 \overset{(E7.4.4c)}{=} 2 \overset{(E7.4.4a,b)}{\rightarrow} \left(-\frac{1}{2\lambda}\right)^2 + \left(-\frac{1}{2\lambda}\right)^2 = 2; \quad \lambda = \pm\frac{1}{2} \quad \text{(E7.4.4)}$$

$$x_1 \overset{(E7.4.3a)}{=} -\frac{1}{2\lambda} = \mp 1, \quad x_2 \overset{(E7.4.3b)}{=} -\frac{1}{2\lambda} = \mp 1 \quad \text{(E7.4.5)}$$

Now, to tell whether each of these is a minimum or a maximum, we should determine the positive/negative-definiteness of the second derivative (Hessian matrix) of $l(\mathbf{x}, \lambda)$ w.r.t. $\mathbf{x}$.

$$H = \frac{\partial^2}{\partial \mathbf{x}^2} l(\mathbf{x}, \lambda) = \begin{bmatrix} \partial^2 l/\partial x_1^2 & \partial^2 l/\partial x_1 \partial x_2 \\ \partial^2 l/\partial x_2 \partial x_1 & \partial^2 l/\partial x_2^2 \end{bmatrix} = \begin{bmatrix} 2\lambda & 0 \\ 0 & 2\lambda \end{bmatrix} \quad \text{(E7.4.6)}$$

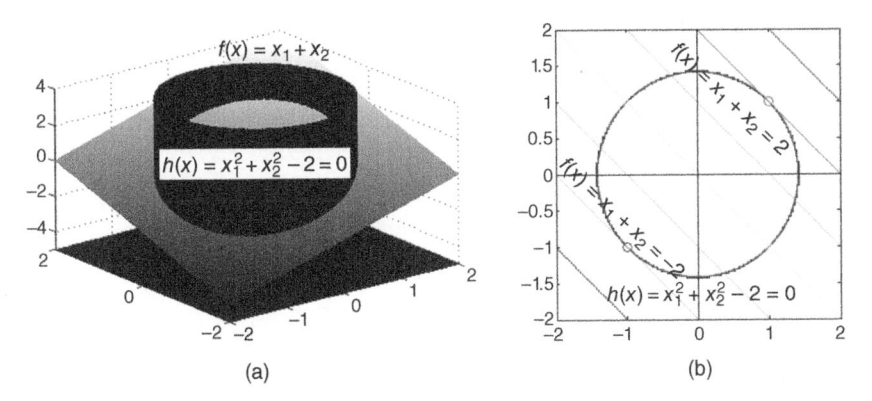

Figure 7.11 The objective and constraint functions for Example 7.4. (a) Mesh-shaped graphs. (b) Contour-shaped graphs.

This matrix is positive/negative-definite if the sign of λ is positive/negative. Therefore, the solution $(x_1,x_2) = (-1,-1)$ corresponding to $\lambda = 1/2$ is a (local) minimum that we want to get, while the solution $(x_1,x_2) = (1,1)$ corresponding to $\lambda = -1/2$ is a (local) maximum (see Figure 7.11). Note that as mentioned in Remark 7.2, the positive/negative sign of the Lagrange multiplier value implies that the corresponding inequality constraint is active for minimization/maximization in the sense that it shows its *raison d'être*, playing its role (of increasing/decreasing the value of the objective function compared with the case where it does not exist).

Example 7.5. Minimization (with Inequality Constraint) by the Lagrange Multiplier Method.

Consider the following minimization problem subject to a single inequality ($\leq$) constraint:

$$\text{Min} f(\mathbf{x}) = x_1 + x_2 \quad \text{(E7.5.1a)}$$

$$\text{s.t.} \; g(\mathbf{x}) = x_1^2 + x_2^2 - 2 \leq 0 \quad \text{(E7.5.1b)}$$

As in the previous example (to minimize the same objective function subject to an equality constraint), we can apply the Lagrange multiplier method to get

$$x_1 = \mp 1, \quad x_2 = \mp 1, \quad \text{and} \quad \lambda = \pm 1/2 \quad \text{(E7.5.2)}$$

Here, since this is a minimization problem, $(x_1,x_2) = (-1,-1)$ corresponding to the positive Lagrange multiplier value $\lambda = 1/2$ is meaningful while $(x_1,x_2) = (+1,+1)$ corresponding to the negative Lagrange multiplier value $\lambda = -1/2$ is not. In order for $(x_1,x_2) = (+1,+1)$ to be a meaningful solution, the problem should be not to minimize but to maximize the same objective function.

7.2.2 Penalty Function Method

This method is practically very useful for dealing with the general constrained optimization problems involving equality/inequality constraints. It is really attractive for optimization problems with fuzzy or loose constraints that are not so strict with zero tolerance.

Consider the following problem:

$$\text{Min } f(\mathbf{x}) \tag{7.2.5a}$$

$$\text{s.t. } \mathbf{h}(\mathbf{x}) = \begin{bmatrix} h_1(\mathbf{x}) \\ \vdots \\ h_M(\mathbf{x}) \end{bmatrix} = 0, \quad \mathbf{g}(\mathbf{x}) = \begin{bmatrix} g_1(\mathbf{x}) \\ \vdots \\ g_L(\mathbf{x}) \end{bmatrix} \leq 0 \tag{7.2.5b}$$

The penalty function method consists of two steps. The first step is to construct a new objective function

$$\text{Min } l(\mathbf{x}) = f(\mathbf{x}) + \sum_{m=1}^{M} w_m h_m^2(\mathbf{x}) + \sum_{l=1}^{L} v_l \psi_l(g_l(\mathbf{x})) \tag{7.2.6}$$

by including the constraint terms in such a way that violating the constraints would be penalized through the large value of the constraint terms in the objective function, while satisfying the constraints would not affect the objective function. The second step is to minimize the new objective function with no constraints by using the method that is applicable to unconstrained optimization problems, but not a gradient-based approach like the Nelder method. Why do not we use a gradient-based optimization method? Because the inequality constraint terms $v_l \psi(g_l(\mathbf{x}))$ attached to the objective function are often determined to be zero as long as $\mathbf{x}$ stays inside the (permissible) region satisfying the corresponding constraint ($g_l(\mathbf{x}) \leq 0$) and to increase very steeply (like $\psi(g_l(\mathbf{x})) = \exp(e_l g_l(\mathbf{x}))$) as $\mathbf{x}$ goes out of the region; consequently, the gradient of the new objective function may not carry useful information about the direction along which the value of the objective function decreases.

From application point of view, it might be a good feature of this method that we can make the weighting coefficient (w_m, v_m, and e_m) on each penalizing constraint term large/small depending on how strictly it should be satisfied.

Let us see the following example:

Example 7.6. Minimization by the Penalty Function Method.

Consider the following minimization problem subject to several nonlinear inequality constraints:

$$\text{Min } f(\mathbf{x}) = \{(x_1 + 1.5)^2 + 5(x_2 - 1.7)^2\}\{(x_1 - 1.4)^2 + 0.6(x_2 - 0.5)^2\} \quad \text{(E7.6.1a)}$$

$$\text{s.t. } \mathbf{g}(\mathbf{x}) = \begin{bmatrix} -x_1 \\ -x_2 \\ 3x_1 - x_1x_2 + 4x_2 - 7 \\ 2x_1 + x_2 - 3 \\ 3x_1 - 4x_2^2 - 4x_2 \end{bmatrix} \leq \begin{bmatrix} 0 \\ 0 \\ 0 \\ 0 \\ 0 \end{bmatrix} \quad \text{(E7.6.1b)}$$

According to the penalty function method, we construct a new objective function (7.2.6) as

$$\text{Min } l(\mathbf{x}) = \{(x_1 + 1.5)^2 + 5(x_2 - 1.7)^2\}\{(x_1 - 1.4)^2 + 0.6(x_2 - 0.5)^2\}$$

$$+ \sum_{m=1}^{5} v_m \psi_m(g_m(\mathbf{x})) \quad \text{(E7.6.2a)}$$

where

$$v_m = 1, \psi_m(g_m(\mathbf{x})) = \begin{cases} 0 & \text{if } g_m(\mathbf{x}) \leq 0 \text{ (constraint satisfied)} \\ \exp(e_m g_m(\mathbf{x})) & \text{if } g_m(\mathbf{x}) > 0 \text{ (constraint vilolated)} \end{cases},$$

$$e_m = 1 \; \forall \; m = 1, \ldots, 5 \quad \text{(E7.6.2b)}$$

```
%nm0722.m for Example 7.6
% to solve a constrained optimization problem by penalty ftn method.
f='f722p';
x0=[0.4 0.5] % Initial guess
TolX=1e-4; TolFun=1e-9; alpha0=1;
[xo_Nelder,fo_Nelder]= opt_Nelder(f,x0) %Nelder method
[fc_Nelder,fo_Nelder,co_Nelder]= f722p(xo_Nelder) %its results
[xo_s,fo_s]=fminsearch(f,x0) %MATLAB built-in fminsearch()
[fc_s,fo_s,co_s]= f722p(xo_s) %its results

% including how the constraints are satisfied or violated
xo_steep= opt_steep(f,x0,TolX,TolFun,alpha0) %steepest descent method
[fc_steep,fo_steep,co_steep]= f722p(xo_steep) %its results
[xo_u,fo_u]= fminunc(f,x0); % MATLAB built-in fminunc()
[fc_u,fo_u,co_u]= f722p(xo_u) %its results
```

```
function [fc,f,c]=f722p(x)
f=((x(1)+1.5)^2+5*(x(2)-1.7)^2)*((x(1)-1.4)^2+.6*(x(2)-.5)^2); (E7.6.1a)
c=[-x(1);
   -x(2);
   3*x(1)-x(1)*x(2)+4*x(2)-7;
   2*x(1)+x(2)-3;
   3*x(1)-4*x(2)^2-4*x(2)]; % Constraint vector (E7.6.1b)
v=[1 1 1 1 1]; e=[1 1 1 1 1]'; % Weighting coefficient vector (E7.6.2b)
fc= f +v*((c>0).*exp(e.*c)); % New objective function (E7.6.2a)
```

```
»nm0722
xo_Nelder =   1.2118      0.5765
fo_Nelder =   0.5322 %min value
co_Nelder =  -1.2118
             -0.5765
             -1.7573 %high margin
             -0.0000 %no margin
             -0.0000 %no margin
xo_s =        1.2118      0.5765
fo_s =        0.5322 %min value
```

```
xo_steep =   1.2768      0.5989
fo_steep =   0.2899 %not a minimum
co_steep =  -1.2768
            -0.5989
            -1.5386
             0.1525 %violating
            -0.0001

Solver stopped prematurely.
fminunc stopped because it exceeded
the function evaluation limit,
options.MaxFunctionEvaluations = 200
(the default value).

xo_u =       1.1024      0.7951
fo_u =       1.5300 % not a minimum
```

Note that the shape of the penalty function as well as the values of the weighting coefficients is set by the users to cope with their own problems.

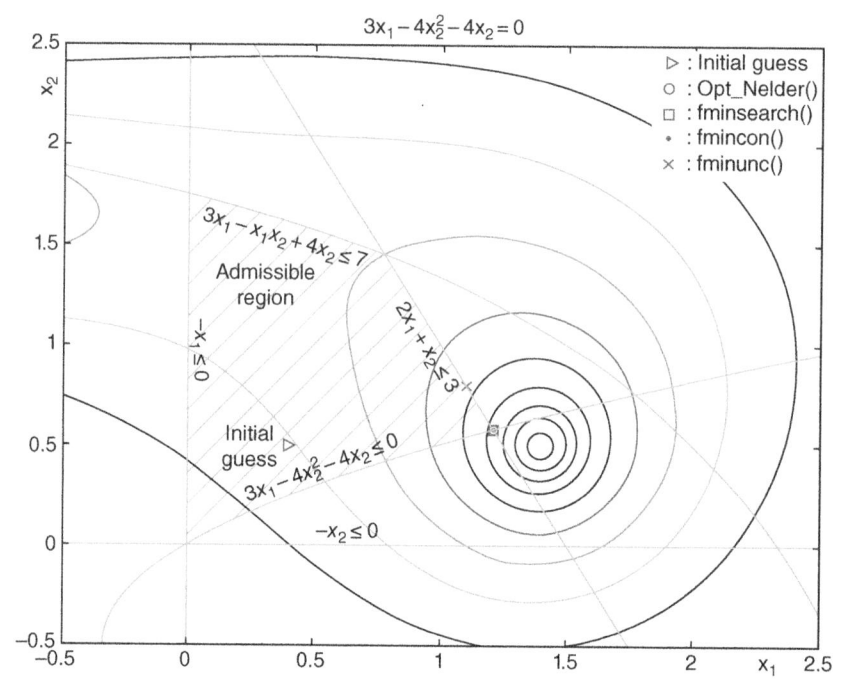

Figure 7.12 Minimum points in the admissible region (satisfying the constraints) for Example 7.6.

Then, we apply an unconstrained optimization technique like the Nelder-Mead method, which is not a gradient-based approach. Here, we make the script "nm0722.m", which applies not only the function 'opt_Nelder()' and the MATLAB built-in function 'fminsearch()' for crosscheck but also the function 'opt_steep()' and the MATLAB built-in function 'fminunc()' in order to show that the gradient-based methods do not work well. To our expectation, the running results listed above and depicted in Figure 7.12 show that, for the objective function (E7.6.2a) augmented with the penalized constraint terms, the gradient-based functions 'opt_steep()' and 'fminunc()' are not so effective as the nongradient-based functions 'opt_Nelder()' and 'fminsearch()' in finding the constrained minimum, which is on the intersection of the two boundary curves corresponding to the fourth and fifth constraints of (E7.6.1b).

7.3 MATLAB BUILT-IN FUNCTIONS FOR OPTIMIZATION

In this section, we apply several MATLAB built-in unconstrained optimization functions including 'fminsearch()' and 'fminunc()' to the same problem, expecting that their nuances will be clarified. Our intention is not to compare or evaluate the performances of these sophisticated functions, but rather to give the readers some feelings for their functional differences. We also introduce the function 'linprog()' implementing Linear Programming (LP) scheme and 'fmincon()' designed for attacking the (most challenging) constrained optimization problems. Interested readers are encouraged to run the tutorial functions 'optdemo' or 'tutdemo', which demonstrate the usages and performances of the representative built-in optimization functions such as 'fminunc()' and 'fmincon()'.

7.3.1 Unconstrained Optimization

To try applying the unconstrained optimization functions introduced in Section 7.1 and see how they work, we composed the following MATLAB script "nm0731_1.m", which uses those functions for solving the problem

$$\text{Min } f(\mathbf{x}) = (x_1 - 0.5)^2(x_1 + 1)^2 + (x_2 + 1)^2(x_2 - 1)^2 \quad (7.3.1)$$

where the contours and the (local) maximum/minimum/saddle points of this objective function are depicted in Figure 7.13. Note that a point is said to be a *saddle*, a *minmax*, or *stationary point* if the partial derivatives w.r.t. every variable are zero, but it is a (local) maximum w.r.t. some variable(s), while it is a (local) minimum w.r.t. other variables so that it cannot be called a local maximum or minimum w.r.t. all the variables.

410 OPTIMIZATION

$$f(x) = (x_1 - 0.5)^2 (x_1 + 1)^2 + (x_2 + 1)^2 (x_2 - 1)^2$$

(a)

(b)

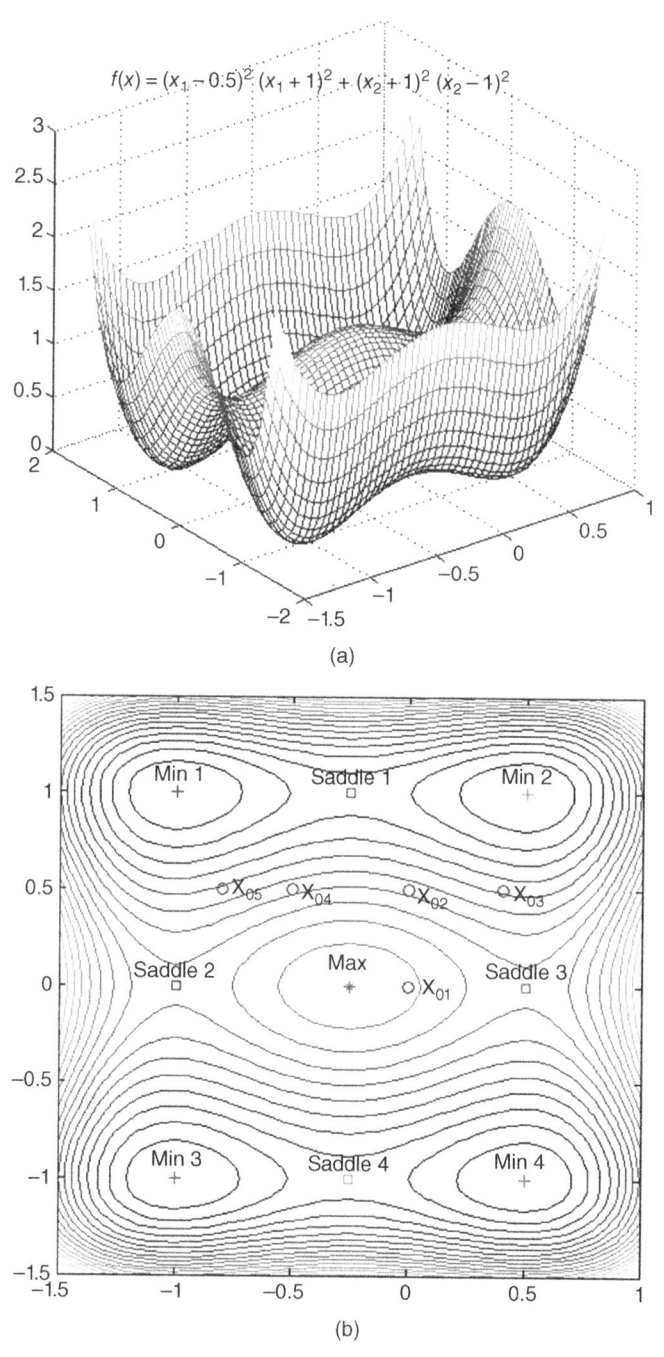

Figure 7.13 The contours, local minima/maxima, and saddle points of the objective function (7.3.1). (a) The mesh-type graph of objective function (7.3.1). (b) The contour-type graph of objective function (7.3.1).

MATLAB BUILT-IN FUNCTIONS FOR OPTIMIZATION

Table 7.2 Results of using several unconstrained optimization functions with various initial values.

x_0	opt_Nelder	fminsearch	opt_steep	Newtons	opt_conjg	fminunc
[0, 0]	[−1, 1] (minimum)	[0.5, 1] (minimum)	[0.5, 0] (saddle)	[−0.25, 0] (maximum)	[0.5, 0] (saddle)	[0.5, 0] (saddle)
[0, 0.5]	[0.5, 1] (minimum)	[0.02, 1] (lost)	[0.5, 1] (minimum)	[−0.25, −1] (saddle)	[0.5, 1] (minimum)	[0.5, 1] (minimum)
[0.4, 0.5]	[0.5, 1] (minimum)	[0.5, 1] (minimum)	[0.5, 1] (minimum)	[0.5, −1] (minimum)	[0.5, 1] (minimum)	[0.5, 1] (minimum)
[−0.5, 0.5]	[0.5, 1] (minimum)	[−1, 1] (minimum)	[−1, 1] (minimum)	[−0.25, −1] (saddle)	[−1, 1] (minimum)	[−1, 1] (minimum)
[−0.8, 0.5]	[−1, 1] (minimum)	[−1, 1] (minimum)	[−1, 1] (minimum)	[−1, −1] (minimum)	[−1, 1] (minimum)	[−1, 1] (minimum)

```
%nm0731_1.m
% An objective function and its gradient function
f=@(x)(x(1)-0.5).^2.*(x(1)+1).^2+(x(2)+1).^2.*(x(2)-1).^2;  (E7.6.1a)
g=@(x)[2*(x(1)-0.5)*(x(1)+1)*(2*x(1)+0.5) 4*(x(2)^2-1).*x(2)];
x0=[0 0.5] % Initial guess
[xon,fon]=opt_Nelder(f,x0) % Min point, its ftn value by opt_Nelder
[xos,fos]=fminsearch(f,x0) % Min point, its ftn value by fminsearch()
[xost,fost]=opt_steep(f,x0) % Min point, its ftn value by opt_steep()
TolX=1e-4; MaxIter=100;
xont=Newtons(g,x0,TolX,MaxIter);
xont,f(xont) % Minimum point and its function value by Newtons()
[xocg,focg]=opt_conjg(f,x0) % Min point, its ftn value by opt_conjg()
[xou,fou]=fminunc(f,x0) % Minimum point, its ftn value by fminunc()
```

To show that it depends mainly on the initial value x_0 whether each function succeeds in finding a minimum point, the results of running those MATLAB functions with various initial values are summarized in Table 7.2. It can be seen from this table that the gradient-based optimization functions like 'opt_steep()', 'fminunc()', and 'Newtons()' sometimes get to a saddle point or even a maximum point (Remark 7.1) and that the functions do not always approach the extremum that is closest to the initial point. It is interesting to note that even the nongradient-based MATLAB built-in function 'fminsearch()' may get lost, while our function 'opt_Nelder()' works well for this case. We cannot, however, conclude that this function is better than that one based on only one trial because there may be many problems for which the MATLAB built-in function works well, but our function does not. What can be stated over this happening is that no human work is free from defect.

Now, we try using a MATLAB built-in function 'lsqnonlin(f,a0,..)', which presents a nonlinear least squares (NLLS) solution to the following minimization problem:

$$\text{Min}_{\mathbf{a}} \sum_{n=1}^{N} f_n^2(\mathbf{a}) \qquad (7.3.2)$$

The function needs at least a vector or a matrix function **f(a)** and an initial guess **a**$_0$ (of the minimum point) as its first and second input arguments, respectively, where the components of $\mathbf{f(a)} = [f_1(\mathbf{a}) \, f_2(\mathbf{a}) \cdots f_N(\mathbf{a})]^T$ are squared, summed, and then minimized over **x**. To learn the function and usage of 'lsqnonlin()', we compose the following MATLAB script "test_opt_lsq.m", which uses it to find a second-degree polynomial approximating (or fitting) the following function at $x = \{-2, -1, 0, 1, 1\}$ as shown in Figure 7.14:

$$y = g(x) = \frac{1}{1+8x^2} \qquad (7.3.3)$$

Two other MATLAB functions 'polyfit(x,y,M)' and 'lsqcurvefit(gax,a0,x,y)' are used to find the solutions to the following curve fitting problems, respectively:

```
%test_opt_lsq.m
% try using lsqnonlin() for a vector-valued objective ftn F(x)
g=@(x)1./(1+8*x.^2); % True function to approximate by a polynomial ftn
x=-2+[0:4]; % The vector of points where curve fitting is evaluated.
y=g(x); % True function values at x
f=@(a)polyval(a,x)-y; % Error vector ftn whose squared sum to minimize
M=2; % Degree of the curve-fitting polynomial
% To make the parameter estimate using lsqnonlin()
a0=zeros(1,M+1); % Initial guess of polynomial coefficient vector
a_lsq=lsqnonlin(f,a0)
% To make the parameter estimate using polyfit() (polynomial fitting)
a_poly=polyfit(x,y,M)
% To make the parameter estimate using lsqcurvefit()
gax=@(a,x)polyval(a,x); % Polynomial as a trial curve-fitting function
a_lsqcf=lsqcurvefit(gax,a0,x,y) % LS curve-fitting by gax(a,x)
% To plot the curve fitting result together with the target function
xx=-2+[0:400]/100; gxx=g(xx);
plot(xx,polyval(a_lsq,xx),'r:', xx,gxx,'b', x,y,'x')
```

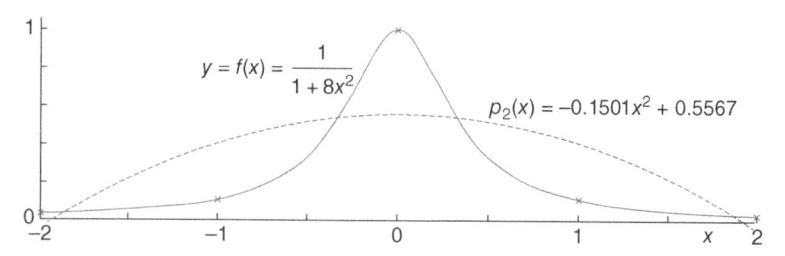

Figure 7.14 The result of using the MATLAB built-in function 'lsqnonlin()' for polynomial approximation (curve fitting)

$$\operatorname*{Min}_{\mathbf{a}} \sum_{n=1}^{N} \left\{ \sum_{m=1}^{M} a_m x_n^{M-m} - y_n \right\}^2 \quad \text{with } y_n = g(x_n) \qquad (7.3.4)$$

$$\operatorname*{Min}_{\mathbf{a}} \sum_{n=1}^{N} \{\hat{g}(\mathbf{a}, x_n) - y_n\}^2 \quad \text{with } y_n = g(x_n) \qquad (7.3.5)$$

where the curve-fitting functions with 'polyfit()' are restricted to only polynomials. Running the script "test_opt_lsq.m" yields the following:

```
»test_opt_lsq
  a_lsq   =      -0.1501    0.0000    0.5567
  a_poly  =      -0.1501    0.0000    0.5567
  a_lsqcf =      -0.1501    0.0000    0.5567
```

7.3.2 Constrained Optimization

Generally, constrained optimization is very complicated and difficult to deal with. So we will not cover the topic in details here and instead, will just introduce the powerful MATLAB built-in function 'fmincon()', which makes us relieved from a big headache.

This function is well-designed for attacking the optimization problems subject to some constraints:

$$\operatorname{Min} f(\mathbf{x}) \qquad (7.3.6)$$

s.t. $A\mathbf{x} \le \mathbf{b}$, $A_{eq}\mathbf{x} = \mathbf{b}_{eq}$, $\mathbf{c}(\mathbf{x}) \le 0$, $\mathbf{c}_{eq}(\mathbf{x}) = 0$, and $\mathbf{l} \le \mathbf{x} \le \mathbf{u}$ (7.3.7)

A part of its usage can be seen by typing 'help fmincon' into the MATLAB Command window as summarized in the box below. We make the MATLAB script "nm0732_1.m", which uses the function 'fmincon()' to solve the problem presented in Example 7.6. Interested readers are welcomed to run it and observe the result to check if it agrees with that of Example 7.6.

```
%nm0732_1.m to solve a constrained optimization problem by fmincon()
f722o=@(x)((x(1)+1.5)^2+5*(x(2)-1.7)^2)* ...
           ((x(1)-1.4)^2+.6*(x(2)-.5)^2); %Objective function (E7.6.1a)
x0=[0 0.5] % Initial guess
A=[]; B=[]; Aeq=[]; Beq=[]; % No linear constraints
l=-inf*ones(size(x0)); u=inf*ones(size(x0)); % No lower/upperbound
options=optimoptions('fmincon','Display','final'); % just [] is OK.
[xo_con,fo_con]=fmincon(f722o,x0,A,B,Aeq,Beq,l,u,'f722c',options)
[co,ceqo]=f722c(xo_con) % to see how constraints are.
```

```
function [c,ceq]=f722c(x)
c=[-x(1);
   -x(2);
   3*x(1)-x(1)*x(2)+4*x(2)-7; % Inequality constraints (E7.6.1b)
   2*x(1)+x(2)-3;
   3*x(1)-4*x(2)^2-4*x(2)]; % desired to be nonnegative
ceq=[]; % Equality constraints (desired to be zero) if any
```

Usage of the MATLAB built-in function '`fmincon()`'

`[xo,fo,.]=fmincon('ftn',x0,A,b,Aeq,beq,l,u,'nlcon',options,p1,p2,.)`

- *Input arguments* (At least, four input arguments '`ftn`', `x0`, `A`, and `B` required)

  '`ftn`' : objective function $f(\mathbf{x})$ to be minimized, usually defined in an M-file, but can be defined as an inline function or a function handle, which will remove the necessity of quotes(").

  `x0` : initial guess $\mathbf{x}_0$ of the solution.

  `A,b` : linear inequality constraints $A\mathbf{x} \le \mathbf{b}$ (to be given as `[]` if not applied).

  `Aeq,beq`: linear equality constraints $A_{eq}\mathbf{x} = \mathbf{b}_{eq}$ (to be given as `[]` if not applied).

  `l,u` : lower/upper bound vectors such that $\mathbf{l} \le \mathbf{x} \le \mathbf{u}$ (to be given as `[]` if not applied and set `l(i)=-inf`/`u(i)=inf` if `x(i)` is not bounded below/above.

  '`nlcon`': nonlinear constraint function defined in an M-file, supposed to return the two output arguments for a given $\mathbf{x}$; the first one being the LHS (vector) of inequality constraints $\mathbf{c}(\mathbf{x}) \le \mathbf{0}$ and the second one being the LHS (vector) of equality constraints $\mathbf{c}_{eq}(\mathbf{x}) = \mathbf{0}$ (to be given as `[]` if not applied).

  `options`: used for setting the display parameter, the tolerances for $\mathbf{x}^o$ and $f(\mathbf{x}^o)$, etc; to be given as `[]` if not applied. For details, see the function of '`optimset`' or '`optimoptions`' in the MATLAB Help browser.

  `p1,p2,…`: problem-dependent parameters to be passed to the objective function $f(\mathbf{x})$ and the nonlinear constraint functions $\{\mathbf{c}(\mathbf{x}), \mathbf{c}_{eq}(\mathbf{x})\}$.

- Output *arguments*

  `xo` : minimum point $\mathbf{x}^o$ reached in the permissible region satisfying the constraints

  `fo` : minimized function value $f(\mathbf{x}^o)$

There are two more MATLAB built-in functions to be introduced in this section. One is

'fminimax('ftn',x0,A,b,Aeq,beq,l,u,'nlcon',options,p1,..)',

which is focused on minimizing the maximum among several components of the vector/matrix-valued objective function $\mathbf{f}(\mathbf{x}) = [f_1(\mathbf{x}) \cdots f_N(\mathbf{x})]^T$ subject to some constraints as described beneath. Its usage is almost the same as that of 'fmincon()'.

$$\operatorname*{Min}_{\mathbf{x}}\{\operatorname*{Max}_{n}\{f_n(\mathbf{x})\}\} \quad (7.3.8)$$

s.t. $A\mathbf{x} \leq \mathbf{b}$, $A_{eq}\mathbf{x} = \mathbf{b}_{eq}$, $\mathbf{c}(\mathbf{x}) \leq \mathbf{0}$, $\mathbf{c}_{eq}(\mathbf{x}) = \mathbf{0}$, and $\mathbf{l} \leq \mathbf{x} \leq \mathbf{u}$ (7.3.9)

The other is the constrained linear least squares (LLS) function

'lsqlin(C,d,A,b,Aeq,beq,l,u,x0,options,p1,..)',

whose job is to solve the problem

$$\operatorname*{Min}_{\mathbf{x}} \|C\mathbf{x} - \mathbf{d}\|^2 \quad (7.3.10)$$

s.t. $A\mathbf{x} \leq \mathbf{b}$, $A_{eq}\mathbf{x} = \mathbf{b}_{eq}$, and $\mathbf{l} \leq \mathbf{x} \leq \mathbf{u}$ (7.3.11)

To learn the usage and function of this function, we make and run the MATLAB script "nm0732_2.m", which uses both 'fminimax()' and 'lsqlin()' to find a second-degree polynomial approximating the function (7.3.3) and compares the results with that of applying the function 'lsqnonlin()' introduced in the previous section for verification. From the plotting result depicted in Figure 7.15, note the following:

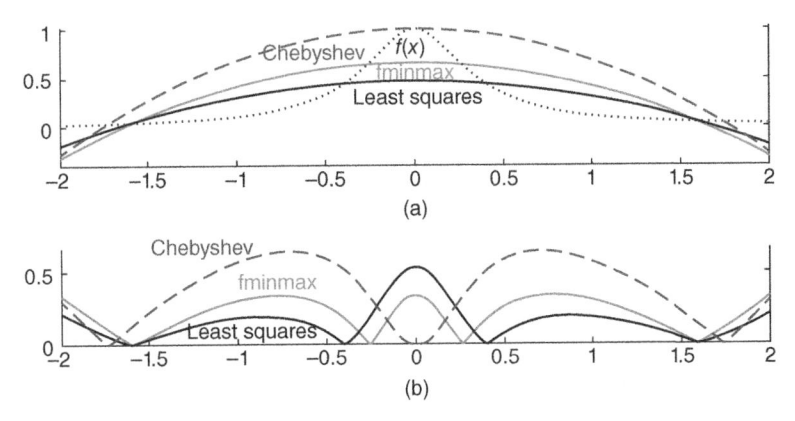

Figure 7.15 Illustration of the Newton's method and steepest descent method. (a) Approximations of f(x). (b) Approximation errors.

```
%nm0732_2.m
% uses fminimax() for a vector-valued objective ftn f(x)
f=@(x)1./(1+8*x.*x); % Target function to approximate
f73221=@(a,x,fx)abs(polyval(a,x)-fx);
f73222=@(a,x,fx)polyval(a,x)-fx;

N=2; % Degree of approximating polynomial
a0=zeros(1,N+1); % Initial guess of polynomial coefficients
xx=-2+[0:200]'/50; % Intermediate points
fx=feval(f,xx);        % and their function values f(xx)
ao_m=fminimax(f73221,a0,[],[],[],[],[],[],[],[],xx,fx) %fminimax sol
for n=1:N+1, C(:,n)=xx.^(N+1-n); end
ao_ll=lsqlin(C,fx) % Linear LS to minimize (Ca-fx)^2 with no constraint
ao_ln=lsqnonlin(f73222,a0,[],[],[],xx,fx) % Nonlinear LS
c2=Cheby(f,N,-2,2) % Chebyshev polynomial over [-2,2]
plot(xx,fx,':', xx,polyval(ao_m,xx),'m', xx,polyval(ao_ll,xx),'r')
hold on, plot(xx,polyval(ao_ln,xx),'b', xx,polyval(c2,xx),'-')
```

- Since no constraints have been put to the function 'fminimax()', it yielded the approximate polynomial coefficient vector **a** minimizing the maximum deviation from $f(x)$.
- We attached no constraints to the constrained linear least squares function 'lsqlin()' either, so it yielded the approximate polynomial curve minimizing the sum (integral) of squared deviation from $f(x)$, which is the same as the (unconstrained) least squares solution obtained by using the function 'lsqnonlin()'.
- Another MATLAB built-in function 'lsqnonneg()' gives us a nonnegative least squares (NLS) solution to the problem (7.3.10).

7.3.3 Linear Programming (LP)

An LP problem is a constrained minimization problem with linear objective and constraint functions such as

$$\text{Min } f(\mathbf{x}) = \mathbf{f}^T \mathbf{x} \qquad (7.3.12a)$$

$$\text{s.t. } A\mathbf{x} \leq \mathbf{b}, \ A_{eq}\mathbf{x} = \mathbf{b}_{eq}, \text{ and } \mathbf{l} \leq \mathbf{x} \leq \mathbf{u} \qquad (7.3.12b)$$

LP problems can be solved by using the MATLAB built-in function 'linprog()' as

'[xo,fo]=linprog(f,A,b,Aeq,Beq,l,u,x0,options)'.

This MATLAB function produces the solution $\mathbf{x}_o$ (in column vector form if it is a vector) and the minimized value $f(\mathbf{x}_o)$ of the objective function as its first and second output arguments xo and fo, where the objective function and constraints excluding the constant term are linear w.r.t. the independent (decision) variables.

It works for such linear optimization problems as Eq. (7.3.12) more efficiently than the general constrained optimization function 'fmincon()'.

The usage of 'linprog()' is exemplified by the following MATLAB script "test_LP.m", which uses the function for solving an LP problem described as follows:

$$\text{Min } f(\mathbf{x}) = \mathbf{f}^T \mathbf{x} = [-3 \ -2] \begin{bmatrix} x_1 \\ x_2 \end{bmatrix} = -3x_1 - 2x_2 \quad (7.3.13a)$$

$$\text{s.t. } A\mathbf{x} = \begin{bmatrix} 3 & 4 \\ 2 & 1 \\ -3 & 2 \end{bmatrix} \begin{bmatrix} x_1 \\ x_2 \end{bmatrix} \begin{matrix} \leq \\ \leq \\ = \end{matrix} \begin{bmatrix} 7 \\ 3 \\ 2 \end{bmatrix} = \mathbf{b} \text{ and } \mathbf{l} = \begin{bmatrix} 0 \\ 0 \end{bmatrix} \leq \mathbf{x} = \begin{bmatrix} x_1 \\ x_2 \end{bmatrix} \leq \begin{bmatrix} 10 \\ 10 \end{bmatrix} = \mathbf{u}$$

(7.3.13b)

The script also applies the general constrained minimization function 'fmincon()' to solve the same problem for crosscheck. Running the script yields

```
»test_LP
    xo_lp= 0.3333      % Solution using 'linprog()'
           1.5000
    fo_lp=-4.0000      % Minimum value of the objective function f(x)
    cons_satisfied=  -0.0000 %<=0(inequality)
                     -0.8333 %<=0(inequality)
                     -0.0000 %=0(equality)
    xo_con=[0.3333  1.5000] % Solution using 'fmincon()'
```

In this result, the solutions obtained by using the two functions 'linprog()' and 'fmincon()' agree with each other, satisfying the inequality/equality constraints, as can be seen in Figure 7.16.

Here, the *simplex method*, which is the standard algorithm (developed by George Dantzig in 1946) for solving LP problems, is introduced. It begins with constructing the *simplex tableau* from the coefficients of the linear inequality/equality constraints and objective function as

```
%test_LP.m
% to solve the Linear Programming problem (7.3.13).
% Min f*x=-3*x(1)-2*x(2) s.t. A*x<=b, Aeq*x=beq, and l<=x<=u
f=[-3 -2]; % Coefficient vector of the objective function
A=[3 4; 2 1]; b=[7; 3]; % Inequality constraint A*x<=b
Aeq=[-3 2]; beq=2; % Equality constraint Aeq*x=beq
l=[0 0]; u=[10 10]; % Lower/upper bound l<=x<=u
[xo_lp,fo_lp]=linprog(f,A,b,Aeq,beq,l,u)
cons_satisfied=[A; Aeq]*xo_lp-[b; beq] % How constraints are satisfied
% To apply fmincon() for the LP problem,
f_090309a=@(x)-3*x(1)-2*x(2); % The objective function
x0=[0 0]; % Initial point
[xo_con, fo_con]=fmincon(f_090309a,x0,A,b,Aeq,beq,l,u)
```

418 OPTIMIZATION

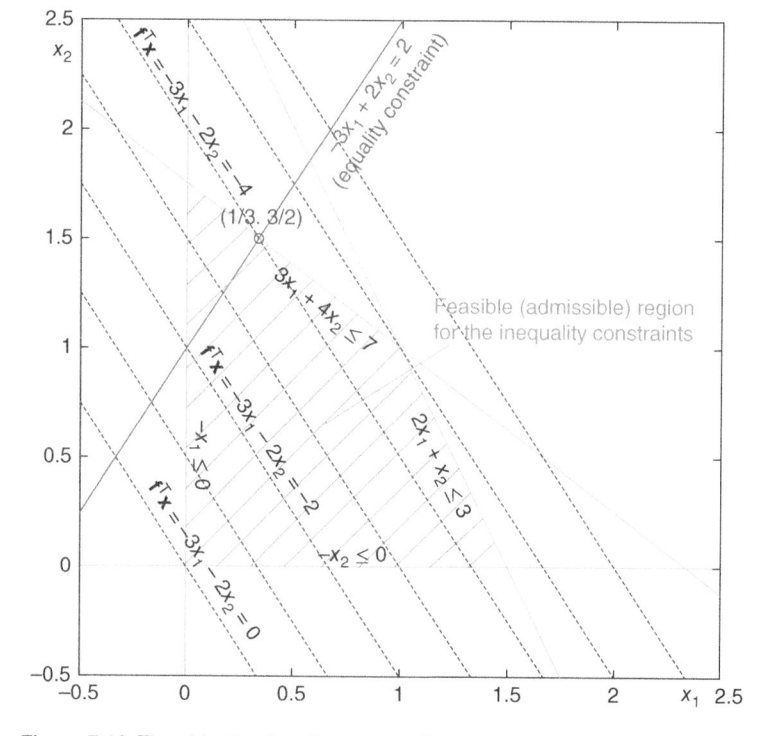

Figure 7.16 The objective function, constraints, and solution of an LP problem.

$$
\begin{array}{c}
\phantom{A_{eq}} \\
A \\
A_{eq} \\
-A_{eq} \\
\mathbf{f}^T(\text{price})
\end{array}
\begin{array}{c}
x_1 \quad x_2 \quad s_1 \quad s_2 \quad s_3 \quad s_4 \quad \mathbf{b} \\
\begin{bmatrix}
3 & 4 & 1 & 0 & 0 & 0 & 7 \\
2 & 1 & 0 & 1 & 0 & 0 & 3 \\
-3 & 2 & 0 & 0 & 1 & 0 & 2(\mathbf{b}_{eq}) \\
3 & -2 & 0 & 0 & 0 & 1 & -2(-\mathbf{b}_{eq}) \\
\hdashline
-3 & -2 & 0 & 0 & 0 & 0 & \boxed{0}
\end{bmatrix}
\end{array}
\begin{array}{c}
\text{Basic variables} \\
s_1 \\
s_2 \\
s_3 \\
s_4 \\
\text{Minus current cost} \\
\text{(objective function value)}
\end{array}
$$

(7.3.14)

where the bound constraints are not included. Note that the equality constraint

$$A_{eq}\mathbf{x} = \mathbf{b}_{eq}$$

has been split into two opposite inequality constraints:

$$A_{eq}\mathbf{x} \le \mathbf{b}_{eq} \tag{7.3.15a}$$

$$A_{eq}\mathbf{x} \ge \mathbf{b}_{eq}, \text{ or equivalently, } -A_{eq}\mathbf{x} \le -\mathbf{b}_{eq} \tag{7.3.15b}$$

This simplex tableau amounts to writing the constraints as

$$\begin{aligned} 3x_1 + 4x_2 & +s_1 & & & =7 \\ 2x_1 + x_2 & & +s_2 & & =3 \\ -3x_1 + 2x_2 & & & +s_3 & =2 \\ 3x_1 - 2x_2 & & & & +s_4 =-2 \end{aligned} \tag{7.3.16}$$

where $\{s_1, \ldots, s_4\}$ are called the *slack variables*. Currently, this set of linear equations has the basic solution

$$[\underbrace{x_1 x_2}_{\text{Nonbasic variables}} \underbrace{s_1 s_2 s_3 s_4}_{\text{Basic variables}}] = [0 \; 0 \; 7 \; 3 \; 2 \; -2] \tag{7.3.17}$$

where the variables $\{s_1, \ldots, s_4\}$ (corresponding single-one columns) and others $\{x_1, x_2\}$ are said to be *basic variables* and *nonbasic variables*, respectively, and the corresponding value of objective function (with opposite sign) can be found as zero at the lower-right corner of the simplex tableau.

The simplex method continues as follows where the simplex tableau (7.3.10) is denoted by A_{bf}:

(Step1) Find the minimum negative entry $A_{bf,\min} = \text{Min}\{A_{bf}(\text{end},1:N) < 0\}$ among the first N (the number of variables $\{x_j\text{'s}\}$) of the last (objective function) row. The column having that entry is chosen as the *pivot column* since the (nonbasic) variable corresponding to that column j_p (having the negative highest price) is expected to decrease the objective function value the most by being a new basic variable (what we call the *entering (incoming) variable*) – the *steepest descent rule*. If all $A_{bf}(\text{end},1:N)$'s are nonnegative, it implies that there is no candidate for entering variable that can contribute decreasing the objective value and consequently, go to Step 4 to find the optimal solution in the current tableau.

(Step2) Find the pivot element a_{i_p,j_p} with minimum ratio $\text{Min}\{b_i/a_{i,j_p}$ such that $a_{i_p,j_p} > 0\}$ in the pivot column – the *minimum ratio rule*. If there is no positive a_{i,j_p} in the pivot column, the process should be stopped and declared as an *unbounded case*, where the objective function value can be decreased no more, without violating the constraints.

420 OPTIMIZATION

(Step 3) Do the pivot operation on the pivot element a_{i_p,j_p}, i.e. make elementary row operations to make $a_{i_p,j_p} = 1$ so that it can be only one nonzero entry in the pivot column as done with Gauss elimination (Section 1.3.4). This operation amounts to selecting the basic variable corresponding to the pivot row i_p as a *departing (leaving) variable* and replacing it by the entering variable (corresponding to the pivot column j_p determined in Step 1) in the new set of basic variables. Then go to Step 1.

(Step 4) Set the values of basic variables (corresponding to single-one columns) to the value (corresponding to the single one) in the last (RHS) column. Also set the optimal value of the objective function to the opposite-signed value of the entry at the lower-right corner of the tableau. Then, the process is terminated.

Now, let us use the simplex method for the simplex tableau (7.3.10):

$$\begin{array}{c|cccccc|c|c}
\text{Basic variables} & x_1 & x_2 & s_1 & s_2 & s_3 & s_4 & b_i & b_i/a_{i1} \\
\hline
s_1 & 3 & 4 & 1 & 0 & 0 & 0 & 7 & 7/3 \\
s_2 & 2 & 1 & 0 & 1 & 0 & 0 & 3 & 3/2 \\
s_3 & -3 & 2 & 0 & 0 & 1 & 0 & 2 & \times \\
s_4 & 3 & -2 & 0 & 0 & 0 & 1 & -2 & -2/3 \\
& -3 & -2 & 0 & 0 & 0 & 0 & 0 &
\end{array} \quad (7.3.18\text{-}1)$$

Choosing the first column (having the minimum negative entry among the first $N = 2$ entries in the last row) as the pivot column (by the steepest descent rule) and the fourth row (having the minimum ratio) as the pivot row, do the pivot operation on a_{41} of the tableau (7.3.18-1) to write

$$\begin{array}{c|cccccc|c|c}
\text{Basic variables} & x_1 & x_2 & s_1 & s_2 & s_3 & s_4 & b_i & b_i/a_{i1} \\
\hline
s_1 & 0 & 6 & 1 & 0 & 0 & -1 & 9 & 9/6 \\
s_2 & 0 & 7/3 & 0 & 1 & 0 & -2/3 & 13/3 & 13/7 \\
s_3 & 0 & 0 & 0 & 0 & 1 & 1 & 0 & \times \\
x_1 & 1 & -2/3 & 0 & 0 & 0 & 1/3 & -2/3 & \times \\
& 0 & -4 & 0 & 0 & 0 & 1 & -2 & -\{-3(-2/3)\}
\end{array}$$

$(7.3.18\text{-}2)$

Now, choosing the second column (having the minimum negative entry among the first $N = 2$ entries in the last row) as the pivot column (by the steepest descent rule) and the first row (having the minimum ratio) as the pivot row, do the pivot operation on a_{12} of the tableau (7.3.18-2) to write

Basic variables
$$\begin{array}{c} \\ x_2 \\ s_2 \\ s_3 \\ x_1 \\ \\ \end{array} \begin{array}{cccccc|c} x_1 & x_2 & s_1 & s_2 & s_3 & s_4 & b_i \\ \hline 0 & 1 & 1/6 & 0 & 0 & -1/6 & 3/2 \\ 0 & 0 & -7/18 & 1 & 0 & -5/18 & 5/6 \\ 0 & 0 & 0 & 0 & 1 & 1 & 0 \\ 1 & 0 & 1/9 & 0 & 0 & 2/9 & 1/3 \\ 0 & 0 & 2/3 & 0 & 0 & 1/3 & 4 \end{array} -\{-3 \cdot 1/3 - 2 \cdot 3/2\}$$

(7.3.18-3)

It is interesting that the entry at the lower-right corner of the tableau is the negative objective function value. Now, we can see no nonzero entry among the first $N = 2$ entries in the last row, which implies that the following optimal result has been reached:

$$[\underset{\text{basic}}{x_1} \ \underset{\text{nonbasic}}{x_2} \ \underset{\text{basic}}{s_1} \ \underset{\text{nonbasic}}{s_2 \ s_3} \ s_4] = [1/3 \ 3/2 \ 0 \ 5/6 \ 0 \ 0] \qquad (7.3.19)$$

where the (current) basic variables $\{x_1, x_2, s_2, s_3\}$ corresponding to single-one columns are set to their corresponding RHS values (in the last column) and those of other (nonbasic) variables $\{s_1, s_4\}$ are set to zero. The value of the objective function at the optimal solution

$$\mathbf{x}^o = \begin{bmatrix} x_1^o \\ x_2^o \end{bmatrix} = \begin{bmatrix} 1/3 \\ 3/2 \end{bmatrix} \qquad (7.3.20)$$

is set to the opposite-signed value of the lower-right corner of the last tableau (7.3.18-3):

$$f(\mathbf{x}^o) = -4 \qquad (7.3.21)$$

as conformed with the solution obtained by 'linprog()' earlier.

(Q) What is implied by the values of slack variables that are $[s_1 \ s_2 \ s_3 \ s_4] = [0 \ 5/6 \ 0 \ 0]$?
(A) It implies that the optimal solution (7.3.20) satisfies the constraints (7.3.16) tightly except for the second one, which is satisfied by a margin of 5/6.

This simplex method has been cast into the following MATLAB function 'simplex()'. To solve the above LP problem, you can run the MATLAB script "test_simplex.m" (listed below), which uses 'simplex()'.

```
»test_simplex
    xo =    1/3         fo =    -4     % Optimal solution
            3/2
```

422 OPTIMIZATION

```
function [xo,fmin,Ab]=simplex(f,A,b,Aeq,beq)
% Uses the simplex method to solve an LP (linear programming) problem:
%    Min f(x) = f.'*x  s.t. A*x <= b and Aeq*x = beq
[M,N] = size(A); % Size of the coefficient matrix for inequality
[Meq,Neq] = size(Aeq); % Size of the coefficient matrix for equality
[M,N] = size(A); % Size of coefficient matrix for inequality
[Meq,Neq] = size(Aeq); % Size of the coefficient matrix for equality
MMeq2=M+Meq*2;   EPS=1e-6;
% Construct the simplex tableau like Eq.(7.3.10)
if Meq==0
  A1=A;  b1=b(:); % No equality constraint
 else
  A1=[A; Aeq; -Aeq]; b1=[b(:); beq; -beq];
  if N~=Neq, error('N must be equal to Neq in simplex'); end
end
Ab = [A1 eye(MMeq2) b1; f.' zeros(1,MMeq2) 0]
M1=MMeq2+1; NM1=N+M1;  % Row/Column sizes of the simplex tableau
% Choose the pivot column with minimum entry in the last row
[AbM1_min,jp]=min(Ab(M1,1:NM)); % Index of the most negative entry
while AbM1_min<-EPS
   indp = find(Ab(1:MMeq2,jp)>0);
   if isempty(indp), disp('Unbounded case!');  end
   ratios=Ab(indp,NM1)./Ab(indp,jp); %indp=indnz(find(ratios>0));
   [Abmin,imin] = min(ratios);
   ip = indp(imin); % Pivot row ip by the minimum ratio rule
   % Pivoting
   Ab(ip,:) = Ab(ip,:)/Ab(ip,jp);
   for i=1:M1
      if i~=ip, Ab(i,:) = Ab(i,:)-Ab(i,jp)*Ab(ip,:); end
   end
   [AbM1_min,jp]=min(Ab(M1,1:NM)); % Index of the most negative entry
end
% Last step to determine the optimal solution
xo = zeros(N,1);
if sum(Ab(1:MMeq2,NM1)<-EPS)>0, disp('Infeasible case!'); end
for j=1:N
   ind1=find(abs(Ab(1:MMeq2,j))>EPS);
   if length(ind1)==1, xo(j,1)=Ab(ind1,NM1); end
end
fmin = -Ab(M1,NM1); % Optimal objective function value
```

```
%test_simplex.m
f=[-3; -2]; % Objective function coefficients to be minimized
A=[3 4; 2 1]; b=[7;3]; % Inequality constraint coefficients and RHS
Aeq=[-3 2]; beq=[2]; % Equality constraint coefficients and RHS
[xo,fmin,Ab_final] = simplex(f,A,b,Aeq,beq)
```

7.3.4 Mixed Integer Linear Programming (MILP)

An MILP problem is a constrained minimization problem with linear objective and constraint functions such as

$$\text{Min } f(\mathbf{x}) = \mathbf{f}^T \mathbf{x} \tag{7.3.22a}$$

$$\text{s.t. } A\mathbf{x} \leq \mathbf{b}, A_{eq}\mathbf{x} = \mathbf{b}_{eq}, \text{ and } \mathbf{l} \leq \mathbf{x} \leq \mathbf{u} \tag{7.3.22b}$$

where some variables x(intcon) are required to be integer-valued. MILP problems can be solved by using the MATLAB built-in function 'intlinprog()' as

`'[xo,fo]=intlinprog(f,intcon,A,b,Aeq,Beq,l,u,options)'.`

where 'intcon' is a set of the indices of integer variables. This MATLAB function produces the solution $\mathbf{x}_o$ (in column vector form) and the minimized value of the objective function $f(\mathbf{x}_o)$ as its first and second output arguments xo and fo.

The usage of 'intlinprog()' is exemplified by the following MATLAB script "test_ILP.m", which uses the function for solving a MILP problem described as

$$\text{Min } f(\mathbf{x}) = \mathbf{f}^T \mathbf{x} = [-2 \ -1] \begin{bmatrix} x_1 \\ x_2 \end{bmatrix} = -2x_1 - x_2 \tag{7.3.23a}$$

$$\text{s.t. } A\mathbf{x} = \begin{bmatrix} 3 & -1 \\ 1 & 5 \end{bmatrix} \begin{bmatrix} x_1 \\ x_2 \end{bmatrix} \leq \begin{bmatrix} 6 \\ 16 \end{bmatrix} = \mathbf{b} \text{ and } \mathbf{l} = \begin{bmatrix} 0 \\ 0 \end{bmatrix} \leq \mathbf{x} = \begin{bmatrix} x_1 \\ x_2 \end{bmatrix} \leq \begin{bmatrix} 10 \\ 10 \end{bmatrix} = \mathbf{u} \tag{7.3.23b}$$

where x_1 and x_2 must be integer-valued (by the integral constraints).
Running the script "test_ILP.m" (listed below) yields the following results as shown in Figure 7.17:

```
»test_ILP
    xo =   2.0000        fo =  -6
           2.0000
```

(Q) Referring to Figure 7.17, can you get the ILP solution simply by rounding up the LP solution?

```
%test_ILP.m
% to solve the MILP problem (7.3.23).
% Min f*x=-3*x(1)-2*x(2) s.t. Ax<=b, Aeq=beq, and l<=x<=u
x0=[0 0]; % Initial point
f=[-2 -1]; % Coefficient vector of the objective function
intcon=[1 2]; % x(intcon)=x([1 2]) must be integer-valued
A=[3 -1; 1 5]; b=[6; 16]; % Inequality constraint Ax<=b
Aeq=[]; beq=[]; % No equality constraint Aeq*x=beq
l=[0 0]; u=[10 10]; % Lower/upper bound l<=x<=u
[xo,fo]=intlinprog(f,intcon,A,b,Aeq,beq,l,u)
```

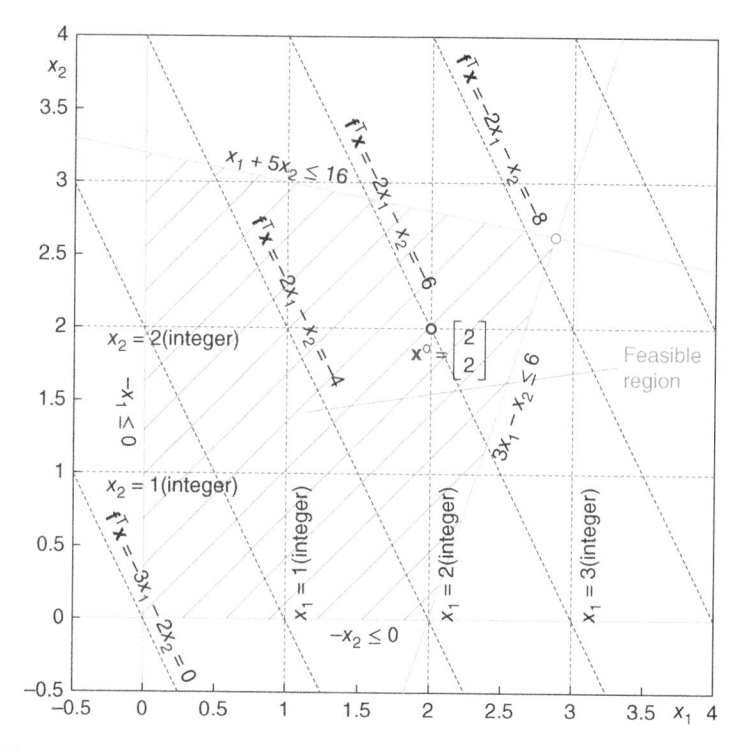

Figure 7.17 The objective function, constraints, and solution of an ILP problem.

There are two techniques that can be used to deal with ILP problems, the *branch-and-bound (BB) method* and the cutting plane method. First, to get the feel of the BB method, let us use it to solve the above ILP problem:

0. Relax (ignore temporarily) the integral constraints to turn the integer linear programming (ILP) problem into an LP one (*LP relaxation*):

$$\text{Min } f(\mathbf{x}) = \mathbf{f}^T \mathbf{x} = [-2 \ -1] \begin{bmatrix} x_1 \\ x_2 \end{bmatrix} = -2x_1 - x_2 \quad (7.3.24a)$$

$$\text{s.t. } A\mathbf{x} = \begin{bmatrix} 3 & -1 \\ 1 & 5 \end{bmatrix} \begin{bmatrix} x_1 \\ x_2 \end{bmatrix} \leq \begin{bmatrix} 6 \\ 16 \end{bmatrix} = \mathbf{b} \quad \text{and} \quad (7.3.24b)$$

$$\mathbf{l} = \begin{bmatrix} 0 \\ 0 \end{bmatrix} \leq \mathbf{x} = \begin{bmatrix} x_1 \\ x_2 \end{bmatrix} \leq \begin{bmatrix} 10 \\ 10 \end{bmatrix} = \mathbf{u}$$

This can be solved by using the LP solver '`linprog()`' as

```
f=[-2 -1]; % Coefficient vector of the objective function
A=[3 -1; 1 5]; b=[6; 16]; Aeq=[]; beq=[]; % Inequality/Equality
l=[0 0]; u=[10 10]; % Lower/upper bound l<=x<=u
[xo1,fo1]=linprog(f,A,b,Aeq,beq,l,u) % -> [23/8; 21/8], f=-67/8
```

to yield the solution

$$\mathbf{x}_1^o = \begin{bmatrix} 23/8 \\ 21/8 \end{bmatrix} \text{ with } f(\mathbf{x}_1^o) = -67/8 \quad (7.3.25)$$

1. First, to trim $x_2 = 21/8$ (which is more fractional than x_1 in the sense that its fractional part is closer to 0.5 than that of $x_1 = 23/8$) of $\mathbf{x}_1^o$ into an integer, we branch the relaxed LP problem into two subproblems (that we call SP11 and SP12, respectively), each with additional constraints $x_2 \leq 2 = \text{floor}(2.625)$ and $x_2 \geq 3 = \text{ceil}(2.625)$ and solve them

```
[xo11,fo11]=linprog(f,A,b,[],[],[0;0],[10;2]) %->[8/3;2], f=-22/3
[xo12,fo12]=linprog(f,A,b,[],[],[0;3],[10;10]) %-> [1;3], f=-5
```

to get

$$\mathbf{x}_{11}^o = \begin{bmatrix} 8/3 \\ 2 \end{bmatrix} \text{ with } f(\mathbf{x}_{11}^o) = -\frac{22}{3}, \ \mathbf{x}_{12}^o = \begin{bmatrix} 1 \\ 3 \end{bmatrix} \text{ with } f(\mathbf{x}_{12}^o) = -5$$
(7.3.26-1)

Here, $\mathbf{x}_{12}^o = \begin{bmatrix} 1 & 3 \end{bmatrix}^T$ is registered as a candidate for optimal solution since it satisfies the integral constraints. But $\mathbf{x}_{11}^o = \begin{bmatrix} 8/3 & 2 \end{bmatrix}^T$ does not satisfy the integral constraint on x_1, which forces us to branch on x_1.

   1-1. To trim $x_1 = 8/3$ of $\mathbf{x}_{11}^o$ into an integer, we branch on x_1 to make two subproblems (that we call SP111 and SP112, respectively), each with additional constraints $x_1 \leq 2 = \text{floor}(8/3)$ and $x_1 \geq 3 = \text{ceil}(8/3)$ and solve them

```
        [xo111,fo111]=linprog(f,A,b,Aeq,beq,[0;0],[2;2])
                %-> [2;2], f=-6
        [xo112,fo112]=linprog(f,A,b,Aeq,beq,[3;0],[10;2])
                %->[3;400/167], f=-8.39
```

to get

$$\mathbf{x}_{111}^o = \begin{bmatrix} 2 \\ 2 \end{bmatrix} \text{ with } f(\mathbf{x}_{111}^o) = -6 \quad (7.3.26-2)$$

$$\mathbf{x}_{112}^o = \begin{bmatrix} 3 \\ 400/167 \end{bmatrix} \text{ with } f(\mathbf{x}_{112}^o) = -\frac{1402}{167}$$

Here, $\mathbf{x}_{111}^o = \begin{bmatrix} 2 & 2 \end{bmatrix}^T$ is registered as a candidate for optimal solution since it satisfies the integral constraints, while $\mathbf{x}_{112}^o = \begin{bmatrix} 3 & 400/167 \end{bmatrix}^T$, violating the (first) inequality constraint (7.3.24b), should be terminated due to infeasibility.

2. Now that all the branches are terminated with feasible or infeasible solution nodes, we choose the best feasible solution (with the least objective function value), which is

$$\mathbf{x}_{111}^o = \begin{bmatrix} 2 \\ 2 \end{bmatrix} \text{ with } f(\mathbf{x}_{111}^o) = -6 \qquad (7.3.26\text{-}3)$$

This process of the BB (branch-and-bound) method is depicted by the search (enumeration) tree in Figure 7.18 and has been cast into the following MATLAB function 'branch_and_bound()'. To solve the above ILP problem, run the following MATLAB script "test_BB.m", which uses 'branch_and_bound()':

```
»test_BB
     xo =      2              fo =    -6   % Optimal solution
              2
```

```
function [xo,fo,icase] = branch_and_bound0(f,A,b,Aeq,beq,l,u)
%Copyleft: Won Y. Yang, wyyang53@hanmail.net, CAU for academic use only
[xo,fo,icase] = linprog(f,A,b,Aeq,beq,l,u);
% Find the index of the most fractional variable of LP optimal solution
if icase<1, xo=NaN; fo=1e10; return; end
[rmax,imf] = max(abs(xo-round(xo))); % Which variable to branch on?
EPS=1e-6;
if rmax>EPS
  l1=l; u1=u; u1(imf)=floor(xo(imf));
  l2=l; u2=u; l2(imf)=ceil(xo(imf));
  [xo1,fos(1),icases(1)]=branch_and_bound0(f,A,b,Aeq,beq,l1,u1);
  [xo2,fos(2),icases(2)]=branch_and_bound0(f,A,b,Aeq,beq,l2,u2);
  icase=max(icases);
  if icases(1)==1
    if icases(2)==1, [fo,imin]=min(fos);
                     if imin==1, xo=xo1;  else xo=xo2; end
       else   xo=xo1; fo=fos(1);
    end
   else
    if icases(2)==1, xo=xo2; fo=fos(2);  else xo=NaN; fo=1e10; end
  end
end
```

```
%test_BB.m
f=[-2; -1]; % Objective function coefficients to be minimized
A=[3 -1; 1 5]; b=[6; 16]; % Inequality constraint Ax<=b
[xo,fmin]=branch_and_bound(f,A,b,[],[])
```

```
function [xo,fo] = branch_and_bound(f,A,b,Aeq,beq,l,u)
% Uses the simplex method to solve an ILP (linear programming) problem:
%    Min f(x) = f.'*x   s.t. A*x <= b and Aeq*x = beq
% for the optimal solution xo with the optimized objective ftn value fo
Nx=length(f); % Number of objective function coefficients and variables
if nargin<6, l=zeros(Nx,1); u=1e10*ones(Nx,1); end
[xo,fo,icase]=branch_and_bound0(f,A,b,Aeq,beq,l,u);
```

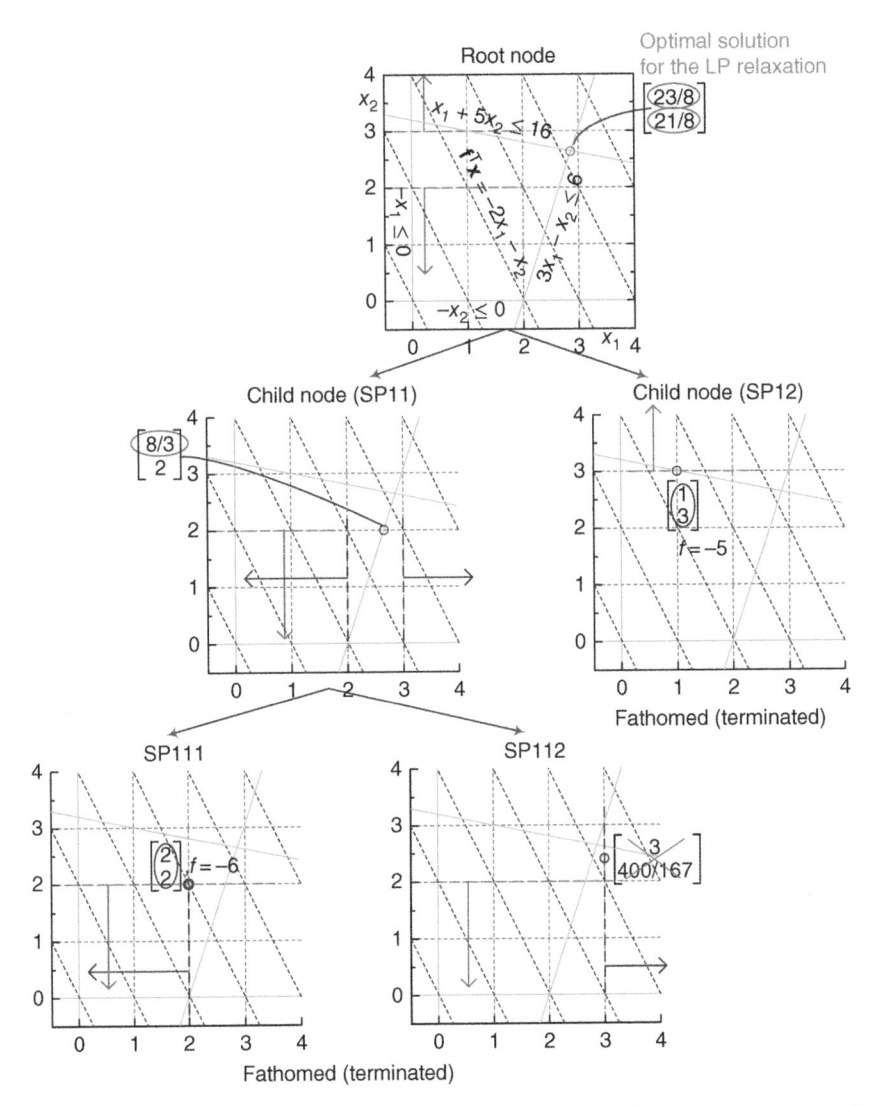

Figure 7.18 Branch-and-bound (BB) search (enumeration) tree for finding the solution of ILP problem (7.3.23).

428 OPTIMIZATION

Now, to get the feel of the *(Gomory's) cutting plane method*, let us use it to solve the above ILP problem (Eq. (7.3.24)):

0. Like the BB method, the cutting plane method also begins with the LP relaxation (without caring about integral constraints), which yields the optimal simplex tableau as follows:

```
f=[-2 -1]; % Coefficient vector of the objective function
A=[3 -1; 1 5]; b=[6; 16]; % Inequality constraint Ax<=b
Aeq=[]; beq=[]; % No equality constraint Aeq*x=beq
[xo1,fo1,Ab]=simplex(f,A,b,Aeq,beq) % -> [23/8; 21/8] with f=-67/8
```

$$\begin{array}{cccc|c} x_1 & x_2 & s_1 & s_2 & b_i \\ \hline 3 & -1 & 1 & 0 & 6 \\ 1 & 5 & 0 & 1 & 16 \\ -2 & -1 & 0 & 0 & 0 \end{array} \xrightarrow{\text{Simplex method}} \begin{array}{cccc|c} x_1 & x_2 & s_1 & s_2 & b_i \\ \hline 1 & 0 & 5/16 & 1/16 & 23/8 \\ 0 & 1 & -1/16 & 3/16 & 21/8 \\ 0 & 0 & 9/16 & 5/16 & 67/8 \end{array}$$

(7.3.27)

This implies that the LP solution, i.e. the optimal solution for the LP relaxation is

$$\mathbf{x}_1^o = \begin{bmatrix} x_{1,1}^o \\ x_{2,1}^o \end{bmatrix} = \begin{bmatrix} 23/8 \\ 21/8 \end{bmatrix} \quad \text{with } f(\mathbf{x}_1^o) = -\frac{67}{8} \qquad (7.3.28)$$

1. Let us impose the integral constraint on x_1 and x_2 one by one. Which one first? Since $x_{2,1}^o = 21/8$ is more fractional than $x_{1,1}^o = 23/8$ in the sense that the distance of 21/8 from 20/8 (the closest midpoints between integers) is shorter than that of 23/8 from 20/8, we first try to fix x_2 by taking the positive fractional part of the constraint involving x_2 (in the second row of the simplex tableau (7.3.27), called a *source row*), as

From the 2nd row of simplex tableau (7.3.27)

$$0x_1 + 1x_2 - \frac{1}{16}s_1 + \frac{3}{16}s_2 = \frac{21}{8} \rightarrow x_2 + \left(-1 + \frac{15}{16}\right)s_1 + \frac{3}{16}s_2 = 2 + \frac{5}{8}$$

$$\xrightarrow{\text{Integer}} x_2 - s_1 - 2 = \frac{5}{8} - \frac{15}{16}s_1 - \frac{3}{16}s_2 \le 0 \rightarrow -\frac{15}{16}s_1 - \frac{3}{16}s_2 \le -\frac{5}{8}$$

(7.3.29a)

writing this inequality (called a *Gomory cut*) with a new slack variable s_3 as

$$-\frac{15}{16}s_1 - \frac{3}{16}s_2 + s_3 = -\frac{5}{8} \qquad (7.3.29b)$$

and add it to the simplex tableau as

$$\begin{bmatrix} x_1 & x_2 & s_1 & s_2 & b_i \\ 1 & 0 & 5/16 & 1/16 & 23/8 \\ 0 & 1 & -1/16 & 3/16 & 21/8 \\ \hline 0 & 0 & 9/16 & 5/16 & 67/8 \end{bmatrix} \xrightarrow[(9.3.25b)]{\text{add Gomory cut}} \begin{bmatrix} x_1 & x_2 & s_1 & s_2 & s_3 & b_i \\ 1 & 0 & 5/16 & 1/16 & 0 & 23/8 \\ 0 & 1 & -1/16 & 3/16 & 0 & 21/8 \\ \hline 0 & 0 & -15/16 & -3/16 & 1 & -5/8 \\ \hline 0 & 0 & 9/16 & 5/16 & 0 & 67/8 \end{bmatrix}$$

$$\text{Ratio} \quad \left|\frac{9/16}{-15/16} = \frac{9}{-15}\right| < \left|\frac{5/16}{-3/16} = \frac{5}{-3}\right|$$

(7.3.30)

2. From the added constraint row $M = 3$ (determined as the pivot row), choose the entry $A(M, j) = -15/16$ with the minimum absolute ratio $\text{Min}\{|A(M+1, j)/A(M, j)|, j = 1:N\}$ ($N = 5$) as the pivot element by the *minimum ratio rule*. Pivot the tableau on the pivot element, which yields

$$\begin{bmatrix} x_1 & x_2 & s_1 & s_2 & s_3 & b_i \\ 1 & 0 & 5/16 & 1/16 & 0 & 23/8 \\ 0 & 1 & -1/16 & 3/16 & 0 & 21/8 \\ 0 & 0 & -15/16 & -3/16 & 1 & -5/8 \\ 0 & 0 & 9/16 & 5/16 & 0 & 67/8 \end{bmatrix} \xrightarrow{\text{Pivoting}} \begin{bmatrix} x_1 & x_2 & s_1 & s_2 & s_3 & b_i \\ 1 & 0 & 0 & 0 & 1/3 & 8/3 \\ 0 & 1 & 0 & 1/5 & -1/15 & 8/3 \\ 0 & 0 & 1 & 1/5 & -16/5 & 2/3 \\ 0 & 0 & 0 & 1/5 & 3/5 & 8 \end{bmatrix}$$

(7.3.31)

(Q) Has the objective function value been improved by adding the new constraint (called a Gomory cut)? Why is that?

3. If all the RHS entries (in the rightest column) are nonnegative integers except for $A(M+1, N+1)$ (which is the negative objective function value), go to the final step where the current basic variables (corresponding the single-one columns) are set to the corresponding RHS values as the optimal feasible solution and stop. If there is any negative entry in the RHS column, pivot the tableau on the pivot element which is in the same row with that entry and the column determined by the minimum ratio rule. Repeat the pivot operations until no negative entry appears in the RHS column. Since all the RHS entries are nonnegative (where $A(M+1, N+1)$ at the lower-right corner does not matter), we go to the next step, which is to impose the integral constraint on another noninteger basic variable as done in Step 1.

4. Which is the most fractional among the $M = 3$ entries in the RHS column? They are all equal in the closeness to the midpoints between integers, and hence, we choose the first one as the source row to make a new Gomory cut:

430 OPTIMIZATION

From the 1st row of simplex tableau (7.3.31)

$$1x_1 + \frac{1}{3}s_3 = \frac{8}{3} = 2 + \frac{2}{3} \xrightarrow{\text{Integer}} x_1 - 2 = \frac{2}{3} - \frac{1}{3}s_3 \leq 0 \to -\frac{1}{3}s_3 \leq -\frac{2}{3}$$

$$\to -\frac{1}{3}s_3 + \overset{\text{A new slack variable}}{s_4} = -\frac{2}{3} \qquad (7.3.32)$$

Then, add it to the simplex tableau, choose the pivot element by the minimum ratio rule, and pivot the tableau as

$$\begin{array}{cccccc} x_1 & x_2 & s_1 & s_2 & s_3 & s_4 & b_i \end{array}$$
$$\begin{bmatrix} 1 & 0 & 0 & 0 & 1/3 & 0 & 8/3 \\ 0 & 1 & 0 & 1/5 & -1/15 & 0 & 8/3 \\ 0 & 0 & 1 & 1/5 & -16/15 & 0 & 2/3 \\ 0 & 0 & 0 & 0 & -1/3 & 1 & -2/3 \\ \hline 0 & 0 & 0 & 1/5 & 3/5 & 0 & 8 \end{bmatrix} \xrightarrow{\text{Pivoting}} \begin{array}{ccccccc} x_1 & x_2 & s_1 & s_2 & s_3 & s_4 & b_i \end{array}$$
$$\begin{bmatrix} 1 & 0 & 0 & 0 & 0 & 1 & 2 \\ 0 & 1 & 0 & 1/5 & 0 & -1/5 & 14/5 \\ 0 & 0 & 1 & 1/5 & 0 & -16/5 & 14/5 \\ 0 & 0 & 0 & 0 & 1 & -3 & 2 \\ 0 & 0 & 0 & 1/5 & 0 & 9/5 & 34/5 \end{bmatrix}$$

Ratio $\left|\frac{3/5}{-1/3}\right| = \frac{9}{5}$ (Only one nonzero ratio) $\qquad (7.3.33)$

where this pivoting operation can be performed by using a MATLAB function 'pivoting' as follows:

```
»A=[1 0 0 0 1/3 0 8/3; 0 1 0 1/5 -1/15 0 8/3; 0 0 1 1/5 -16/15 0 2/3;
    0 0 0 0 -1/3 1 -2/3; 0 0 0 1/5 3/5 0 8]; pivoting(A,4,5)
ans =    1       0       0       0       0       1       2
         0       1       0       1/5     0       -1/5    14/5
         0       0       1       1/5     0       -16/5   14/5
         0       0       0       0       1       -3      2
         0       0       0       1/5     0       9/5     34/5
```

```
function Ap=pivoting(A,ip,jp)
[M,N]=size(A); % Size of coefficient matrix for inequality
Ap(ip,:)=A(ip,:)/A(ip,jp);
for i=1:M
   if i~=ip, Ap(i,:)=A(i,:)-A(i,jp)*Ap(ip,:); end
end
```

5. All the RHS entries are nonnegative (except for $A(M+1, N+1)$ with $M = 4$), but some of them are still fractional. That is why we need to keep imposing the integral constraint on them. Taking the second row as the source row, make a new Gomory cut:

From the 2nd row of simplex tableau (7.3.33)

$$1 \cdot x_2 + \frac{1}{5}s_2 + \left(-1 + \frac{4}{5}\right)s_4 = 2 + \frac{4}{5} \xrightarrow{\text{Integer}} x_2 - s_4 - 2 = \frac{4}{5} - \frac{1}{5}s_2 - \frac{4}{5}s_4 \leq 0$$

$$\to -\frac{1}{5}s_2 - \frac{4}{5}s_4 + \overset{\text{A new slack variable}}{s_5} = -\frac{4}{5} \qquad (7.3.34)$$

MATLAB BUILT-IN FUNCTIONS FOR OPTIMIZATION

Then add it to the simplex tableau, choose the pivot element by the minimum ratio rule, and pivot the tableau as

$$\begin{array}{ccccccc} x_1 & x_2 & s_1 & s_2 & s_3 & s_4 & s_5 & b_i \end{array}$$

$$\begin{bmatrix} 1 & 0 & 0 & 0 & 0 & 1 & 0 & 2 \\ 0 & 1 & 0 & 1/5 & 0 & -1/5 & 0 & 14/5 \\ 0 & 0 & 1 & 1/5 & 0 & -16/5 & 0 & 14/5 \\ 0 & 0 & 0 & 0 & 1 & -3 & 0 & 2 \\ 0 & 0 & 0 & -1/5 & 0 & -4/5 & 1 & -4/5 \\ \hline 0 & 0 & 0 & 1/5 & 0 & 9/5 & 0 & 34/5 \end{bmatrix} \xrightarrow{\text{Pivoting}} \begin{bmatrix} 1 & 0 & 0 & 0 & 0 & 1 & 0 & 2 \\ 0 & 1 & 0 & 0 & 0 & -1 & 1 & 2 \\ 0 & 0 & 1 & 0 & 0 & -4 & 1 & 2 \\ 0 & 0 & 0 & 0 & 1 & -3 & 0 & 2 \\ 0 & 0 & 0 & 1 & 0 & 4 & -5 & 4 \\ \hline 0 & 0 & 0 & 0 & 0 & 1 & 0 & 6 \end{bmatrix}$$

Ratio $\left|\frac{1/5}{-1/5}\right| = 1 < \left|\frac{9/5}{-4/5}\right| = \frac{9}{5}$ (7.3.35)

```
function [xo,fmin,Ab] = cutting_plane(f,A,b,Aeq,beq)
% Uses the cutting plane method to solve an ILP problem:
%   Min f(x) = f.'*x   s.t. A*x <= b and Aeq*x = beq
% for the optimal solution xo with optimized objective ftn value fo
%Copyleft: Won Y. Yang, wyyang53@hanmail.net, CAU for academic use only
Nx=length(f); % Number of objective function coefficients and variables
[xo,fmin,Ab] = simplex(f,A,b,Aeq,beq);
[M1,N1]=size(Ab); M=M1-1; N=N1-1; EPS=1e-6;
indfs=find(sum(abs(Ab(1:M,1:N)-round(Ab(1:M,1:N))),2)>EPS);
[rmax,im] = max(abs(Ab(indfs,N1)-round(Ab(indfs,N1))));
is=indfs(im); % Index of the most fractional RHS entry for source row
while ~isempty(is)&rmax>EPS
   Ab1=Ab;
   Ab1(:,end+1)=Ab(:,end);
   Ab1(:,end-1)=zeros(size(Ab,1),1);
   Ab1(end+1,:)=[Ab(end,1:end-1) 0 Ab(end,end)];
   cut_constraint = floor(Ab(is,:))-Ab(is,:);
   Ab1(end-1,:)=[cut_constraint(1,1:end-1) 1 cut_constraint(1,end)];
   Ab=Ab1  % Tableau with a new Gomory cut added
   [M1,N1]=size(Ab);
   M=M1-1; N=N1-1;
   ip=M; % Pivot row
   indnz = find(abs(Ab(ip,1:N))>EPS&abs(Ab(M1,1:N))>EPS);
   if isempty(indnz), break; end
   [ratio_min,imin] = min(abs(Ab(M1,indnz)./Ab(ip,indnz))); %
   jp=indnz(imin); % Pivot column
   Ab=pivoting(Ab,M,jp) % Tableau updated by pivoting
   indfs=find(sum(abs(Ab(1:M,1:N)-round(Ab(1:M,1:N))),2)>EPS);
   [rmax,im]=max(abs(Ab(indfs,N1)-round(Ab(indfs,N1))));
   is=indfs(im); % Index of the most fractional RHS entry for source row
end
% Last step to determine the optimal solution
for j=1:Nx
   ind1=find(abs(Ab(1:M,j))>EPS);
   if length(ind1)==1, xo(j,1)=Ab(ind1,N1); end
end
fmin = -Ab(M1,N1); % Optimal objective function value
```

Now, all the RHS entries except for $A(M+1, N+1)$ are nonnegative integral numbers so that we can go to the final step.

6. We set the current basic variables (corresponding the single-one columns) to their corresponding RHS values as the optimal feasible solution:

$$[\underbrace{x_1 \; x_2 \; s_1 \; s_2 \; s_3}_{\text{Basic variables}} \; \underbrace{s_4 \; s_5}_{\text{Nonbasic variables}}] = [2 \; 2 \; 2 \; 4 \; 2 \; 0 \; 0] \quad (7.3.35)$$

with the optimized (minimized) objective function value $f_o = -6$ (the opposite-signed entry at the lower-right corner). This conforms with the solution (7.3.26-3) that was obtained by using the BB method and is depicted in Figure 7.16.

This cutting plane method has been cast into the MATLAB function 'cutting_plane()' listed above. To use 'cutting_plane()' for solving the above ILP problem, run the following MATLAB script "test_cutting_plane.m":

```
»test_cutting_plane
    xo =    2              fo =   -6   % Optimal solution
            2
```

```
%test_cutting_plane.m
% Use 'cutting_plane()' to solve the ILP problem (7.3.22).
f=[-2; -1]; % Coefficient vector of the objective function to minimize
A=[3 -1; 1 5]; b=[6; 16]; % Inequality constraint Ax<=b
[xo,fmin,Ab] = cutting_plane(f,A,b,[],[])
```

Before ending this section, let us see a binary integer linear programming (BILP) problem, which is a constrained minimization problem with linear objective and constraint functions such as

$$\text{Min} f(\mathbf{x}) = \mathbf{f}^T \mathbf{x} \quad (7.3.37a)$$

$$\text{s.t. } A\mathbf{x} \le \mathbf{b}, \; A_{eq}\mathbf{x} = \mathbf{b}_{eq}, \text{ and } \mathbf{l} \le \mathbf{x} \le \mathbf{u} \quad (7.3.37b)$$

where some variables x(intcon) are restricted to be binary-valued, i.e. either 0 or 1. BILP problems can be solved by using the MATLAB built-in function 'intlinprog()' with the binary variables bounded in [0, 1]. For example, let us consider a BILP problem described as

```
%test_BILP.m
% to solve the BILP problem (7.3.38).
f=[-3 -2]; % Coefficient vector of the objective function (7.3.38a)
A=[1 4; 4 1]; b=[12; 13]; % Inequality constraint Ax<=b  (7.3.38b)
Aeq=[]; beq=[]; % No equality constraint Aeq*x=beq
intcon=[1]; % x(1) must be binary-valued
l=[0 0]; u=[1 10]; % Lower/upper bound l<=x<=u
[xo,fo]=intlinprog(f,intcon,A,b,Aeq,beq,l,u)
```

$$\text{Min } f(\mathbf{x}) = \mathbf{f}^T\mathbf{x} = [-3 \; -2]\begin{bmatrix}x_1\\x_2\end{bmatrix} = -3x_1 - 2x_2 \qquad (7.3.38a)$$

$$\text{s.t. } A\mathbf{x} = \begin{bmatrix}1 & 4\\4 & 1\end{bmatrix}\begin{bmatrix}x_1\\x_2\end{bmatrix} \leq \begin{bmatrix}12\\13\end{bmatrix} = \mathbf{b} \text{ and } \mathbf{l} = \begin{bmatrix}0\\0\end{bmatrix} \leq \mathbf{x} = \begin{bmatrix}x_1\\x_2\end{bmatrix} \leq \begin{bmatrix}10\\10\end{bmatrix} = \mathbf{u}$$

$$(7.3.38b)$$

where x_1 must be binary-valued and x_2 can take any (real) value in [0, 10].

Running the above script "test_BILP.m" yields the following result:

```
»test_BILP
   xo =    1.0000            fo =   -8.5000
           2.7500
```

7.4 NEURAL NETWORK[K-1]

Figure 7.19 shows the block diagram of a neural network (NN), which consists of three layers (i.e. the input layer, a hidden layer, and the output layer) and employs a supervised learning scheme to update the weights (w_{mn}s) based on the errors ($e_m = d_m - y_m$) between the desired (correct) outputs (d_ms) and the actual outputs (y_ms).

In the NN, each node, say, the mth node (neuron) at the hidden layer receives the inputs $\{x_1, x_2, x_3\}$ from the previous layer, which is the input layer, performs the following jobs on them:

- computes their weighted sum $v_m^{(1)}$ with weights ($w_{mn}^{(1)}$s) possibly including some bias $b^{(1)}$,
- trims it by some activation function $\varphi(\cdot)$:

$$y_m^{(1)} = \varphi(v_m^{(1)}) = \varphi\left(\sum_{n=1}^{M_0} w_{mn}^{(1)} x_n + b^{(1)}\right) \qquad (7.4.1)$$

where $M_0 = M_i$ is the number of nodes at the input layer.
- and transmits it to the next layer, which is the output layer.

Also, each node, say, the mth node (neuron) at the output layer receives the outputs ($y_n^{(m-1)}$s) from the previous layer, which is the hidden layer, performs the same jobs on them, but possibly with different activation function $\varphi(\cdot)$, to generate the overall outputs ($y_m = y_m^{(2)}$s):

$$y_m = y_m^{(2)} = \varphi(v_m^{(2)}) = \varphi\left(\sum_{n=1}^{M_1} w_{mn}^{(2)} y_n^{(1)} + b^{(2)}\right) \qquad (7.4.2)$$

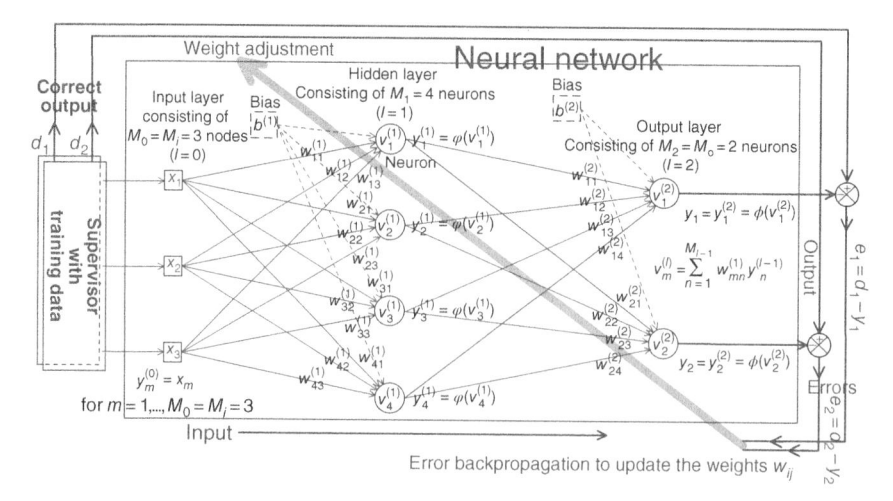

Figure 7.19 A neural network (NN) with supervised learning scheme.

where M_1 is the number of nodes at the hidden layer. Then, the supervisor uses its data matrix (having many inputs in its rows and their corresponding correct outputs in the last entry in each row) to train the NN by the '*back-propagation algorithm*', i.e. updates the weights (starting from the last ones, i.e. $w_{mn}^{(2)}$s) in such a way that the following objective function can be reduced:

$$J = \sum_{m=1}^{M_o} J_m = \frac{1}{2} \sum_{m=1}^{M_o} (d_m - y_m)^2 \ (M_o: \text{the number of nodes at the output layer})$$

(7.4.3)

This is the sum of errors ($e_m = d_m - y_m$) between the desired (correct) outputs (d_ms) and the actual outputs (y_ms). Here, we let the activation function at the hidden layer be the sigmoid function

$$y = \sigma(v) = \frac{1}{1+e^{-v}} \qquad (7.4.4)$$

with the derivative

$$\frac{dy}{dv} = \sigma'(v) = \frac{-(-e^{-v})}{(1+e^{-v})^2} = \frac{1}{1+e^{-v}} \frac{e^{-v}}{1+e^{-v}} = \sigma(v)(1-\sigma(v)) \qquad (7.4.5)$$

and the activation function at the output layer be the *normalized exponential function* or '*softmax*':

$$y_m = \sigma_s(v_m) = \frac{e^{v_m}}{\sum_{i=1}^{M_o} e^{v_i}} \qquad (7.4.6)$$

with the derivative

$$\frac{dy_m}{dv_m} = \sigma_s'(v_m) \stackrel{(7.4.6)}{=} \frac{e^{v_m}\sum_{i=1}^{M_o} e^{v_i} - e^{v_m}(e^{v_m})}{\left(\sum_{i=1}^{M_o} e^{v_i}\right)^2} = \frac{e^{v_m}}{\sum_{i=1}^{M_o} e^{v_i}} \frac{\left(\sum_{i=1}^{M_o} e^{v_i} - e^{v_m}\right)}{\sum_{i=1}^{M_o} e^{v_i}}$$

$$= \sigma_s(v_m)(1 - \sigma_s(v_m)) \tag{7.4.7}$$

Also let the bias be zero for simplicity. Then, the weights are updated[K-1] as follows:

- The weights ($w_{mn}^{(l)}$'s for $l = 1, 2$) are initialized with some values.
- The training data consisting of the input $\{x_n, n = 1:M_i\}$ is entered to yield the output $\{y_m, m = 1:M_o\}$ so that the errors between y_ms and the corresponding desired (correct) output d_ms can be computed:

$$e_m = d_m - y_m = d_m - \sigma_s\left(v_m^{(2)}\right) = d_m - \sigma_s\left(\sum_{n=1}^{M_1} w_{mn}^{(2)} y_n^{(1)}\right) \tag{7.4.8}$$

- The weights ($w_{mn}^{(2)}$'s) from the hidden layer to the output layer are updated in the direction of the negative gradient of the objective function as

$$w_{mn}^{(2)} \leftarrow w_{mn}^{(2)} - \alpha \frac{\partial J}{\partial w_{mn}^{(2)}} \stackrel{(7.4.10)}{=} w_{mn}^{(2)} + \alpha \delta_m^{(2)} y_n^{(1)} \tag{7.4.9a}$$

with

$$\delta_m^{(2)} = (d_m - y_m) y_m^{(2)}\left(1 - y_m^{(2)}\right) = e_m y_m(1 - y_m) \tag{7.4.9b}$$

where α is the learning rate ($0 < \alpha \leq 1$) and

$$\frac{\partial J}{\partial w_{mn}^{(2)}} \stackrel{(7.4.3)}{=} \frac{\partial J_m}{\partial w_{mn}^{(2)}} = \frac{\partial J_m}{\partial y_m^{(2)}} \frac{\partial y_m^{(2)}}{\partial v_m^{(2)}} \frac{\partial v_m^{(2)}}{\partial w_{mn}^{(2)}} \stackrel{(7.4.3)}{\underset{(7.4.2)}{=}} -(d_m - y_m)\sigma_s'(v_m^{(2)}) y_n^{(1)}$$

$$\stackrel{(7.4.5)}{=} -(d_m - y_m)\sigma_s(v_m^{(2)})(1 - \sigma_s(v_m^{(2)})) y_n^{(1)}$$

$$\stackrel{(7.4.6)}{=} -(d_m - y_m) y_m(1 - y_m) y_n^{(1)}$$

$$= -\delta_m^{(2)} y_n^{(1)} \tag{7.4.10}$$

- Then the weights ($w_{mn}^{(1)}$'s) from the input layer to the hidden layer are updated in the direction of the negative gradient of the objective function as

$$w_{mn}^{(1)} \leftarrow w_{mn}^{(1)} - \alpha \frac{\partial J}{\partial w_{mn}^{(1)}} \stackrel{(7.4.10)}{=} w_{mn}^{(1)} + \alpha \delta_m^{(1)} y_n^{(0)} = w_{mn}^{(1)} + \alpha \delta_m^{(1)} x_n \tag{7.4.11a}$$

with

$$\delta_m^{(1)} = \sum_{k=1}^{M_2} w_{km}^{(2)} \delta_k^{(2)} y_m^{(1)} \left(1 - y_m^{(1)}\right) \qquad (7.4.11b)$$

where

$$\frac{\partial J}{\partial w_{mn}^{(l)}} = \frac{\partial J}{\partial y_m^{(l)}} \frac{\partial y_m^{(l)}}{\partial v_m^{(l)}} \frac{\partial v_m^{(l)}}{\partial w_{mn}^{(l)}} \stackrel{(7.4.3)}{=} \sum_{k=1}^{M_{l+1}} \frac{\partial J_k}{\partial v_k^{(l+1)}} \frac{\partial v_k^{(l+1)}}{\partial y_m^{(l)}} \frac{\partial y_m^{(l)}}{\partial v_m^{(l)}} \frac{\partial v_m^{(l)}}{\partial w_{mn}^{(l)}}$$

$$\stackrel{(7.4.1)}{\underset{(7.4.10)}{=}} \sum_{k=1}^{M_{l+1}} -\delta_k^{(l+1)} w_{km}^{(l+1)} \, \sigma'(v)\big|_{v=v_m^{(l)}} y_n^{(l-1)}$$

$$\stackrel{(7.4.5)}{=} -\sum_{k=1}^{M_{l+1}} w_{km}^{(l+1)} \delta_k^{(l+1)} \sigma(v)(1-\sigma(v))\big|_{v=v_m^{(l)}} y_n^{(l-1)}$$

$$\stackrel{(7.4.4)}{=} -\sum_{k=1}^{M_{l+1}} w_{km}^{(l+1)} \delta_k^{(l+1)} y_m^{(l)}(1-y_m^{(l)}) y_n^{(l-1)} = -\delta_m^{(l)} y_n^{(l-1)} \qquad (7.4.12)$$

$$y_n^{(0)} = x_n \text{ (input)}, \quad y_n^{(2)} = y_n \text{ (output)} \qquad (7.4.13a)$$

$M_0 = M_i, M_1, M_2 = M_o$ (the numbers of nodes at each layer) (7.4.13b)

```
function [Ws,er]=train_NN_multiclass(Ws,X,D,alpha)
% Error backpropagation for training a NN (neural network).
% Input: Ws = Cell of weighting matrices
%        X(Nx x Ny x N) = 3D Training input data
%        D(NxMo) = Correct outputs for the training input data X
%        alpha = Learning rate
% Output: Ws = Cell of the updated weighting matrices {W1,W2,..}
%        er = Sum of squared errors (d-y)^2
if nargin<4, alpha=0.9; end
[Nx,Ny,Mo]=size(X); %Size of training data and No of output nodes (classes)
Mi=Nx*Ny; % Number of input layer nodes
L=length(Ws); % Number of layers excluding the input layer
for k=1:Mo
   x=reshape(X(:,:,k),Mi,1); % an input vector of size Mix1
   [y,ys]=NN_multiclass(Ws,x,alpha);
   % Backpropagation
   d=D(k,:).'; % the corresponding desired output for the input x
   e=d-y;  % Error(s)
   er=sum(e.^2); % Sum of squared errors
   delta=y.*(1-y).*e; % Mox1 Eq.(7.4.9b)
   for l=L:-1:1
      yl=ys{l};  % M_(l-1)x1
      dW=alpha*delta*yl.'; % M_lxM_(l-1)   Eq.(7.4.8,11a)
      delta=Ws{l}.'*delta.*yl.*(1-yl); % M_(l-1)x1   Eq.(7.4.11b)
      Ws{l}=Ws{l}+dW;
   end
end
```

```
function [y,ys]=NN_multiclass(Ws,x,alpha)
% To implement a Neural Network with the cell Ws of weight matrices
% Input:  Ws     = Cell of weight matrices
%         x      = An input data
%         alpha  = Learning rate
% Output: y      = Last output for the input x
%         ys     = Cell of the outputs at each layer
%Copyleft: Won Y. Yang, wyyang53@hanmail.net, CAU for academic use only
L=length(Ws); % Number of layers excluding the input layer
ys{1}=x;
for l=1:L
   v=Ws{l}*ys{l};
   if l<L, ys{l+1}=sigmoid(v); % Eq.(7.4.1)
     else ys{l+1}=softmax(v); % Eq.(7.4.2)
   end
end
y=ys{L+1}; % The output (at the last layer)
```

```
function y=sigmoid(x)
y = 1./(1+exp(-x)); % Eq.(7.4.4)
```

```
function y=softmax(x)
ex = exp(x);
y = ex/sum(ex); % Eq.(7.4.6)
```

The above MATLAB function 'train_NN_multiclass()' performs the training process of a neural network to update its weights for classification, which is implemented by the MATLAB function 'NN_multiclass()'.

```
%test_NN_multiclass.m
X=zeros(5,4,3);
Mi=size(X,1)*size(X,2); % Number of input layer nodes
M1=2*Mi; % Number of (the 1st) hidden layer nodes
Mo=size(X,3); % Number of output layer nodes
X0 = [0 0 1 0   1 1 1 0   1 1 1 0;
      0 1 1 0   0 0 0 1   0 0 0 1;
      0 0 1 0   0 1 1 0   0 1 1 0;
      0 0 1 0   1 0 0 0   0 0 0 1;
      0 1 1 0   1 1 1 1   1 1 1 0];
for k=1:3, X(:,:,k) = X0(:,(k-1)*4+[1:4]); end
D = [1 0 0;  % Correct classification 1
     0 1 0;  % Correct classification 2
     0 0 1]; % Correct classification 3
% To train the NN for classification,
W1=2*rand(M1,Mi)-1; W2=2*rand(Mo,M1)-1;
Ws={W1,W2}; % Initialize the weights
alpha=0.9; % Learning rate
N=1000; % Number of epochs for training
for n=1:N % train
   [Ws,ers(n)]=train_NN_multiclass(Ws,X,D,alpha);
end
```

```
nn=0:N-1;
subplot(221), plot(nn,ers)
% Make some minor modifications to generate a real data
X1 = [0 0 1 1  1 1 1 1  1 1 1 1;
      0 0 1 0  0 0 0 1  0 0 0 1;
      0 0 1 0  0 1 1 0  0 1 1 0;
      0 0 1 0  1 0 0 0  0 0 0 1;
      0 1 1 1  1 1 1 1  1 1 1 1];
for k=1:3
   X(:,:,k)=X0(:,(k-1)*4+[1:4]);
end
% To exercise the trained NN for classifying the real data,
L=length(Ws); % Number of layers excluding the input layer
for k=1:Mo
   x = reshape(X(:,:,k),Mi,1); % an Mix1 input vector
   y = NN_multiclass(Ws,x,alpha); yy(:,k)=y;
end
% To check if the NN works fine,
[ymax,imy]=max(yy); % makes classifications
[dmax,imd]=max(D); % as a reference
[imy; imd] % compares the classifications with the true ones
yy % Raw output data for reference
```

The above MATLAB script "test_NN_multiclass.m" uses 'train_NN_multiclass()' to train a two-layer NN for the recognition of three 5-by-4 pixel images as 1, 2, or 3. We can run it to get Figure 7.20 (showing the training performance that the recognition error decreases as the iteration goes on) and the following result:

```
»test_NN_multiclass
    ans =    1    2    3   % Classifications done
             1    2    3   % True classifications
```

This shows that the trained NN has successfully classified the three (contaminated) images X1 into one of {1, 2, and 3} based on the output node with maximum value.

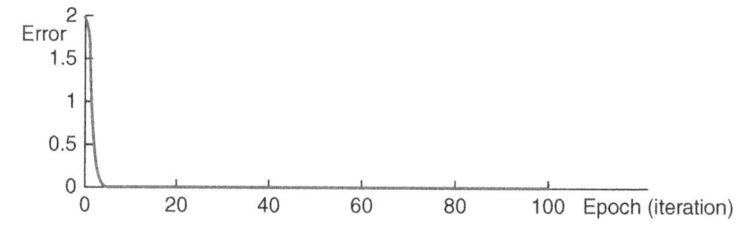

Figure 7.20 Error reducing as the training process goes on.

7.5 ADAPTIVE FILTER[Y-3]

Figure 7.21 shows an Mth-order finite impulse response (FIR) filter, whose output is a linear combination of sequentially delayed inputs as follows:

$$y[n+1] = \mathbf{b}^T\mathbf{x}[n] = \sum_{m=0}^{M} b_m x[n-m] \quad (7.5.1)$$

where $\mathbf{x}[n] = [x[n] \ \ x[n-1] \ \cdots \ x[n-M]]^T$ is the sequential input vector and $\mathbf{b} = [b_0 \ b_1 \ \cdots \ b_M]^T$ is the (weight) coefficient vector.

As an iterative rule of updating the coefficient vector b so that the filter output $y[n]$ can get close to a desired signal $d[n]$, we often use the *steepest descent method* driving the parameter in the direction of negative gradient of the objective function

$$J = \frac{1}{2}\varepsilon\{|e[n+1]|^2\} = \frac{1}{2}\varepsilon\{(d[n+1]-y[n+1])^2\}$$

$$= \frac{1}{2}\varepsilon\{(d[n+1]-\mathbf{b}^T\mathbf{x}[n])^2\} = \frac{1}{2}(D - \mathbf{b}^T\mathbf{p} - \mathbf{p}^T\mathbf{b} + \mathbf{b}^T R\mathbf{b}) \quad (7.5.2)$$

with $\begin{cases} D = \varepsilon\{|d[n+1]|^2\} \\ \mathbf{p} = \varepsilon\{d[n+1]\mathbf{x}[n]\} \\ R = \varepsilon\{\mathbf{x}[n]\mathbf{x}^T[n]\} \quad (R^T = R, \text{ i.e. symmetric}) \end{cases}$

as

$$\mathbf{b}[n+1] = \mathbf{b}[n] - \mu\left.\frac{\partial J}{\partial \mathbf{b}}\right|_{\mathbf{b}=\mathbf{b}[n]} \stackrel{(7.5.2)}{\underset{(C.2),(C.6)}{=}} \mathbf{b}[n] - \mu(R\mathbf{b}[n] - \mathbf{p})$$

$$= (I - \mu R)\mathbf{b}[n] + \mu\mathbf{p} \quad (7.5.3)$$

where ε and μ denote the probabilistic expectation and the *adaptation step-size*, sometimes referred to as the *learning factor*, used to adjust the speed of updating

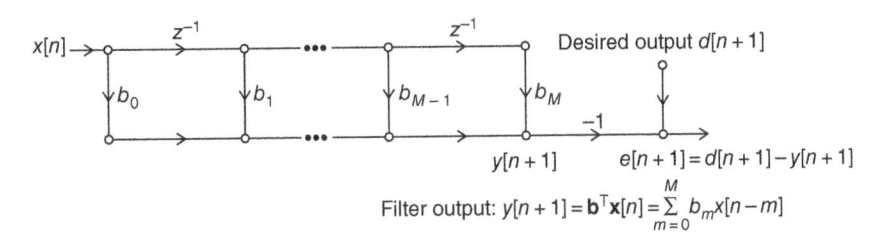

Figure 7.21 An FIR filter.

440 OPTIMIZATION

the parameter estimate, respectively. Note that R and $\mathbf{p}$ are called the input correlation matrix and the crosscorrelation vector (between the input and the desired output), respectively.

If the iteration starts with $\mathbf{b}[0] = 0$, the parameter (estimate) $\mathbf{b}[n]$ updated by Eq. (7.5.3) will converge to the optimal solution

$$\mathbf{b}[n+1] = \sum_{m=0}^{n} (I - \mu R)^m \mu \mathbf{p} \to \mu [I - (I - \mu R)]^{-1} \mathbf{p} = R^{-1} \mathbf{p} = \mathbf{b}^o \quad (7.5.4)$$

on the assumption that R is positive definite and additionally, its (positive) eigenvalue λ_i's and the (adaptive) step-size μ satisfy

$$|1 - \mu \lambda_i| < 1 \quad \forall \ i = 0, \ldots, M \quad \text{or} \quad 0 < \mu < \frac{2}{\text{Max}\{\lambda_i\}} \quad (7.5.5)$$

This convergence condition can be verified by substituting the eigendecomposition of $R = M \Lambda M^{-1}$ (Λ: diagonal matrix having the eigenvalue λ_i's of R on its diagonal, M: modal matrix having an orthogonal set of eigenvectors corresponding to λ_i's of R as its columns) into Eq. (7.5.3):

$$\mathbf{b}[n+1] = \sum_{m=0}^{n} (I - \mu R)^m \mu \mathbf{p} = [I - (I - \mu R)^{n+1}][I - (I - \mu R)]^{-1} \mu \mathbf{p}$$

$$= [I - (I - \mu M \Lambda M^{-1})^{n+1}] M \Lambda^{-1} M^{-1} \mathbf{p}$$

$$= M[I - (I - \mu \Lambda)^{n+1}] \Lambda^{-1} M^{-1} \mathbf{p} \quad (7.5.6)$$

Another iterative rule of updating the FIR filter coefficient vector $\mathbf{b}$ toward the minimum of the objective function (7.5.2) is the *Newton method* using not only the (negative) gradient (first-order partial derivative vector) but also the Hessian (second-order partial derivative matrix) of the objective function w.r.t. $\mathbf{b}$ as

$$\mathbf{b}[n+1] = \mathbf{b}[n] - \mu \left[\frac{\partial^2 J}{\partial \mathbf{b}^2}\right]^{-1} \frac{\partial J}{\partial \mathbf{b}}\bigg|_{\mathbf{b}=\mathbf{b}[n]} \overset{(7.5.2)}{\underset{(C.2,C.6,C.8)}{=}} \mathbf{b}[n] - \mu R^{-1}(R\mathbf{b}[n] - \mathbf{p})$$

$$= (1 - \mu) \mathbf{b}[n] + \mu \mathbf{b}^o \quad (7.5.7)$$

If the iteration starts with $\mathbf{b}[0] = 0$ and (adaptive) step-size constant μ such that $0 < \mu < 2$, this parameter estimate will converge to the optimal solution

$$\mathbf{b}[n+1] = \sum_{m=0}^{n} (1 - \mu)^m \mu \mathbf{b}^o \to \frac{\mu}{1 - (1 - \mu)} \mathbf{b}^o = \mathbf{b}^o \quad (7.5.8)$$

Here, setting $\mu = 1$ may speed up the convergence of the Newton algorithm (7.5.7) so that it theoretically takes just a single iteration to reach the optimal solution from an arbitrary initial guess. Compared with the steepest descent method, the Newton method requires more computation (for finding the inverse of Hessian matrix) per iteration while it usually takes fewer iterations.

However, in the case where the statistical properties of the input signal and desired signal, such as the input autocorrelation matrix R and the crosscorrelation vector $\mathbf{p}$ between the input $\mathbf{x}$ and desired signal d, vary with time, we can use the current value $e^2[n+1]$ itself, instead of the expectation $\varepsilon\{e^2[n+1]\}$, of squared error as the objective function to be minimized:

$$\hat{J} = \frac{1}{2}e^2[n+1] = \frac{1}{2}(d[n+1] - y[n+1])^2 = \frac{1}{2}(d[n+1] - \mathbf{b}^T\mathbf{x}[n])^2 \quad (7.5.9)$$

The *Least Mean Square* (*LMS method*) updates the parameter $\mathbf{b}$ in the direction of negative gradient of this objective function (7.5.2) as

$$\mathbf{b}[n+1] = \mathbf{b}[n] - \mu \left.\frac{\partial \hat{J}}{\partial \mathbf{b}}\right|_{\mathbf{b}=\mathbf{b}[n]} \stackrel{(7.5.9)}{\underset{\text{chain rule}}{=}} \mathbf{b}[n] - \mu e[n+1]\frac{\partial e[n+1]}{\partial \mathbf{b}[n]}$$

$$\stackrel{(7.5.1)}{\underset{(C.2)}{=}} \mathbf{b}[n] + \mu e[n+1]\mathbf{x}[n] \quad (7.5.10)$$

Example 7.7. Adaptive FIR Filter Acting as a One-Step-Ahead Predictor.
Consider a second-order discrete-time system whose input-output relationship is described by the following difference equation (see Figure 7.22a):

$$x[n+1] = 0.7x[n] - 0.1x[n-1] + w[n+1] \quad (E7.7.1)$$

where the input $w[n]$ of this system is a zero-mean uniform white noise with variance $\sigma^2 = 1/4$.

Let us design a first-order FIR filter that makes the one-step-ahead prediction of the output $x[n+1]$ of the system based on the past two output samples $x[n]$ and $x[n-1]$, where the input–output relationship of the adaptive filter is described by the following difference equation:

$$y[n+1] = b_0 x[n] + b_1 x[n-1] \quad (E7.7.2)$$

The following MATLAB script "nm07e07.m" first computes the input autocorrelation matrix R and the crosscorrelation vector $\mathbf{p}$ between the desired signal $d[n+1] = x[n+1]$ and the input $\mathbf{x}[n] = [x[n]\,x[n-1]]^T$ and the desired signal as

$$R = \varepsilon\left\{\begin{bmatrix} x[n] \\ x[n-1] \end{bmatrix} [x[n]\ x[n-1]]\right\}$$

442 OPTIMIZATION

$x[n] = 0.7x[n-1] - 0.1x[n-2] + w[n]$

(a) A second-order system / A one-step-ahead adaptive predictor

Error signal $e[n] = d[n] - y[n]$

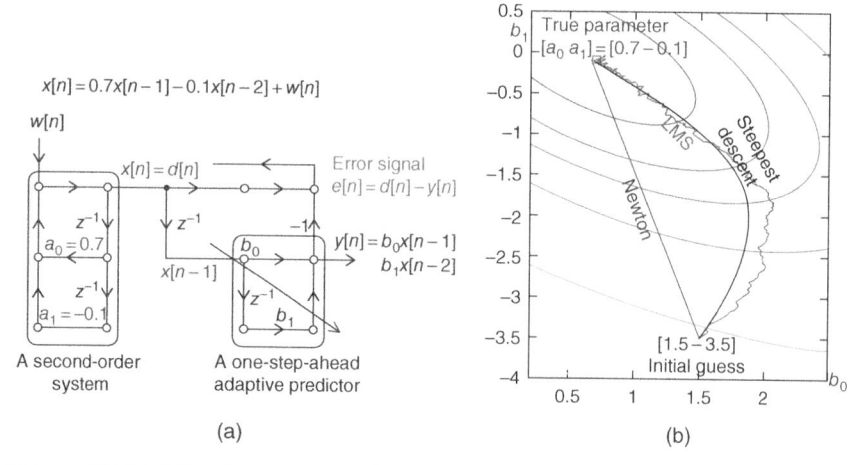

$x[n] = d[n]$, $a_0 = 0.7$, $a_1 = -0.1$

$y[n] = b_0 x[n-1] + b_1 x[n-2]$

(b) True parameter $[a_0\ a_1] = [0.7\ -0.1]$; Initial guess $[1.5\ -3.5]$; curves labeled LMS, Steepest descent, Newton.

Figure 7.22 Updating the parameter of an adaptive predictor using the steepest descent method, the Newton method, and the LMS method. (a) One-step-ahead adaptive predictor. (b) An example of parameter updating by the steepest descent method, the Newton method, and the LMS method.

```
%nm07e07.m
Nx=5000; % Number of data samples for computing R and p
a0=0.7; a1=-0.1; sigma=0.2; noise_amp=sigma*sqrt(12);
noise = noise_amp*(rand(1,Nx)-0.5); % Uniform noise
x(1)=noise(1); x(2)=a0*x(1)+noise(2);
sx2=x(1)^2+x(2)^2; sx12=x(2)*x(1); sx13=0;
for n=2:Nx-1
   x(n+1) = a0*x(n)+a1*x(n-1) + noise(n+1); % Eq.(E7.7.1)
   sx2=sx2+x(n+1)^2; sx12=sx12+x(n+1)*x(n); sx13=sx13+x(n+1)*x(n-1);
end
sx2=sx2/Nx; sx12=sx12/(Nx-1); sx13=sx13/(Nx-2);
R=[sx2 sx12; sx12 sx2]; p=[sx12; sx13]; % Eq.(E7.7.3) and (E7.7.4)
disp('Optimal coefficient vector'), bo=R\p  % Eq.(7.5.4)
b0=[1.5; -3.5]; % Initial guess of the optimal filter coefficient vector
% LMS method
mu=0.05; bn=b0; nf=4000;
for n=2:nf
   x(n+1) = a0*x(n)+a1*x(n-1) + noise_amp*(rand-0.5); % Eq.(E7.7.1)
   b = bn; e = x(n+1) - x([n n-1])*b; % Error signal
   bn = b + mu*e*x([n n-1])'; % Eq.(7.5.10)
   plot([b(1) bn(1)],[b(2) bn(2)],'r'), hold on
end
% Steepest descent method
mu=0.05; bn=b0;
for n=1:nf
   b = bn; bn = b - mu*(R*b-p); % Eq.(7.5.3)
   plot([b(1) bn(1)],[b(2) bn(2)],'k'), hold on
end
% Newton method
mu=1; bn=b0;
for n=1:2
   b = bn; bn = b - mu*R\(R*b-p); % Eq.(7.5.7)
   plot([b(1) bn(1)],[b(2) bn(2)],'b')
end
```

$$R = \varepsilon\left\{\begin{bmatrix} x[n] \\ x[n-1] \end{bmatrix}[x[n]\ x[n-1]]\right\}$$

$$= \begin{bmatrix} \frac{1}{N_x}\sum_{n=0}^{N_x-1} x^2[n] & \frac{1}{N_x-1}\sum_{n=1}^{N_x-1} x[n]x[n-1] \\ \frac{1}{N_x-1}\sum_{n=1}^{N_x-1} x[n]x[n-1] & \frac{1}{N_x}\sum_{n=0}^{N_x-1} x^2[n] \end{bmatrix} \quad (E7.7.3)$$

$$\mathbf{p} = \varepsilon\left\{d[n+1]\begin{bmatrix} x[n] \\ x[n-1] \end{bmatrix}\right\} = \varepsilon\left\{x[n+1]\begin{bmatrix} x[n] \\ x[n-1] \end{bmatrix}\right\}$$

$$= \begin{bmatrix} \frac{1}{N_x-1}\sum_{n=1}^{N_x-1} x[n]x[n-1] \\ \frac{1}{N_x-2}\sum_{n=1}^{N_x-2} x[n+1]x[n-1] \end{bmatrix} \quad (E7.7.4)$$

and then applies the LMS method (7.5.10), the steepest descent method (7.5.3), and the Newton method (7.5.7) to update the predictor coefficient vector $\mathbf{b} = [b_0\ b_1]^T$ starting with an initial guess $\mathbf{b} = [1.5\ -3.5]^T$. The simulation results are depicted in Figure 7.22b, which shows the trade-off between the computational complexity and the convergence speed.

7.6 RECURSIVE LEAST SQUARE ESTIMATION (RLSE)[Y-3]

As an objective function to be minimized for the 'optimal' filter coefficient vector in the FIR filter shown in Figure 7.21, consider the following accumulative sum of squared errors:

$$J(\mathbf{b}, n+1) = \frac{1}{2}\sum_{m=0}^{n} e^2[m+1] = \frac{1}{2}\mathbf{e}^T[m+1]\mathbf{e}[m+1]$$

$$= \frac{1}{2}[\mathbf{d}[m+1] - X[m]\mathbf{b}]^T[\mathbf{d}[m+1] - X[m]\mathbf{b}] \quad (7.6.1)$$

where

$$\mathbf{e}[m+1] = \begin{bmatrix} e[m+1] \\ e[m] \\ \vdots \\ e[0] \end{bmatrix}$$

$$= \begin{bmatrix} d[m+1] \\ d[m] \\ \vdots \\ d[0] \end{bmatrix} - \begin{bmatrix} x[m] & x[m-1] & \cdots & x[m-M] \\ x[m-1] & x[m-2] & \cdots & x[m-M-1] \\ \vdots & \vdots & \ddots & \vdots \\ x[0] & x[-1] & \cdots & x[-M] \end{bmatrix}\begin{bmatrix} b_0 \\ b_1 \\ \vdots \\ b_M \end{bmatrix}$$

$$= \mathbf{d}[m+1] - X[m]\mathbf{b} \quad (7.6.2a)$$

$$\mathbf{b} = \begin{bmatrix} b_0 \\ b_1 \\ \vdots \\ b_M \end{bmatrix} \quad (7.6.2b)$$

$$\mathbf{x}[m] = \begin{bmatrix} x[m] \\ x[m-1] \\ \vdots \\ x[m-M] \end{bmatrix} \quad (7.6.2c)$$

$$X[m] = \begin{bmatrix} \mathbf{x}^T[m] \\ \mathbf{x}^T[m-1] \\ \vdots \\ \mathbf{x}^T[0] \end{bmatrix} = \begin{bmatrix} x[m] & x[m-1] & \cdots & x[m-M] \\ x[m-1] & x[m-2] & \cdots & x[m-M-1] \\ \vdots & \vdots & \ddots & \vdots \\ x[0] & x[-1] & \cdots & x[-M] \end{bmatrix} \quad (7.6.2d)$$

We set the (first) derivative of the objective function (7.6.1) w.r.t. $\mathbf{b}$ to zero to write the normal equation that should be satisfied by the optimal value $\mathbf{b}^o$ of filter coefficient vector:

$$\frac{\partial}{\partial \mathbf{b}} J(\mathbf{b}, n+1) = \left[\frac{\partial \mathbf{e}[n+1]}{\partial \mathbf{b}} \right]^T \mathbf{e}[n+1] \stackrel{(7.6.2a)}{\underset{(C.2)}{=}} -X^T[n]\mathbf{e}[n+1]$$

$$\stackrel{(7.6.2a)}{=} -X^T[n](\mathbf{d}[n+1] - X[n]\mathbf{b}) = \mathbf{0}$$

$$; [X^T[n]X[n]]\mathbf{b} = X[n]\mathbf{d}[n+1] \quad (7.6.3)$$

Then, we might solve this to get the current optimal solution

$$\mathbf{b}[n+1] = [X^T[n]X[n]]^{-1}X[n]\mathbf{d}[n+1] = P[n]\sum_{m=0}^{n}\mathbf{x}[m]d[m+1] \quad (7.6.4)$$

with

$$P[n] = [X^T[n]X[n]]^{-1}_{(M+1)\times(M+1)} = \left[\sum_{m=0}^{n}\mathbf{x}[m]\mathbf{x}^T[m]\right]^{-1} \quad (7.6.5)$$

Referring to Section 2.1.4, we still can improve this solution by rearranging it in such a recursive form that $\mathbf{b}[n+1]$ can be obtained by updating $\mathbf{b}[n]$ with considerably reduced computation. This story begins by writing the inverse of $P[n]$ (Eq. (7.6.5)) in recursive form as

$$P^{-1}[n] = \sum_{m=0}^{n}\mathbf{x}[m]\mathbf{x}^T[m] = \sum_{m=0}^{n-1}\mathbf{x}[m]\mathbf{x}^T[m] + \mathbf{x}[n]\mathbf{x}^T[n]$$

$$= P^{-1}[n-1] + \mathbf{x}[n]\mathbf{x}^T[n] \quad (7.6.6)$$

and writing Eq. (7.6.4) with $n+1$ replaced by n as

$$\mathbf{b}[n] = P[n-1]\sum_{m=0}^{n-1}\mathbf{x}[m]d[m+1] \quad (7.6.7)$$

$$\rightarrow \sum_{m=0}^{n-1}\mathbf{x}[m]d[m+1] = P^{-1}[n-1]\mathbf{b}[n] \stackrel{(7.6.6)}{=} P^{-1}[n]\mathbf{b}[n] - \mathbf{x}[n]\mathbf{x}^T[n]\mathbf{b}[n]$$

$$(7.6.8)$$

Now, we can use Eq. (7.6.8) to rewrite Eq. (7.6.4) as

$$\mathbf{b}[n+1] \stackrel{(7.6.4)}{=} P[n] \left(\sum_{m=0}^{n-1} \mathbf{x}[m]d[m+1] + \mathbf{x}[n]d[n+1] \right)$$

$$\stackrel{(7.6.8)}{=} P[n](P^{-1}[n]\mathbf{b}[n] - \mathbf{x}[n]\mathbf{x}^T[n]\mathbf{b}[n] + \mathbf{x}[n]d[n+1])$$

$$; \mathbf{b}[n+1] = \mathbf{b}[n] + P[n]\mathbf{x}[n](d[n+1] - \mathbf{x}^T[n]\mathbf{b}[n])$$

$$= \mathbf{b}[n] + P[n]\mathbf{x}[n]e[n+1] \qquad (7.6.9)$$

where the P matrix, which is the inverse of Eq. (7.6.6), can also be computed recursively as

$$P[n+1] \stackrel{(7.6.5)}{=} \left[\sum_{m=0}^{n+1} \mathbf{x}[m]\mathbf{x}^T[m] \right]^{-1}$$

$$= \left[\sum_{m=0}^{n} \mathbf{x}[m]\mathbf{x}^T[m] + \mathbf{x}[n+1]\mathbf{x}^T[n+1] \right]^{-1}$$

$$\stackrel{(7.6.6)}{=} [P^{-1}[n] + \mathbf{x}[n+1]\mathbf{x}^T[n+1]]^{-1}$$

$$\stackrel{\substack{(B.14.1)\\ \text{Matrix inversion lemma}}}{=} P[n] - P[n]\mathbf{x}[n+1]$$

$$\times [1 + \mathbf{x}^T[n+1]P[n]\mathbf{x}[n+1]]^{-1}\mathbf{x}^T[n+1]P[n] \qquad (7.6.10)$$

This *recursive least squares estimation (RLSE)* algorithm can be summarized in two ways as follows:

RLSE (Recursive Least Square Estimation) algorithm

$$P[n] \stackrel{(7.6.10)}{=} P[n-1] - P[n-1]\mathbf{x}[n](1 + \mathbf{x}^T[n]P[n-1]\mathbf{x}[n])^{-1}\mathbf{x}^T[n]$$

$$P[n-1] \text{ with } P[-1] = P_0 > 0 \qquad (7.6.11\text{a})$$

$$K[n] \stackrel{(7.6.9)}{=} P[n]\mathbf{x}[n] \qquad (7.6.11\text{b})$$

$$\mathbf{b}[n+1] \stackrel{(7.6.9)}{=} \mathbf{b}[n] + K[n](d[n+1] - \mathbf{x}^T[n]\mathbf{b}[n]) \qquad (7.6.11\text{c})$$

$$\text{with } \mathbf{b}[0] = \mathbf{b}_0$$

$$K[n] \stackrel{(7.6.10)}{=} P[n-1]\mathbf{x}[n](1 + \mathbf{x}^T[n]P[n-1]\mathbf{x}[n])^{-1} \qquad (7.6.12a)$$
$$\text{with } P[-1] = P_0 > 0 \text{ (positive definite)}$$

$$\mathbf{b}[n+1] \stackrel{(7.6.9)}{=} \mathbf{b}[n] + K[n](d[n+1] - \mathbf{x}^T[n]\mathbf{b}[n]) \text{ with } \mathbf{b}[0] = \mathbf{b}_0 \qquad (7.6.12b)$$

$$P[n] \stackrel{(7.6.10)}{=} [I - K[n]\mathbf{x}^T[n]]P[n-1] \text{ with } P[-1] = P_0 > 0 \text{ (positive definite)} \qquad (7.6.12c)$$

Example 7.8. LMS Algorithm vs. RLSE Algorithm.

To compare the performances of the LMS and RLS algorithms for the parameter estimation problem dealt with in Example 7.7, the following MATLAB script "nm07e08.m" has been made and run to yield the mean square errors (MSEs) of the two algorithms:

```
MSE_LMS =    0.0942,     MSE_RLS =    0.0411
```

and Figure 7.23. Figure 7.23a shows how the parameter estimates generated by the two algorithms go toward the true parameter values. Figure 7.23b shows the error signal of the two algorithms, implying that the RLS algorithm converges faster than LMS algorithm in return for an increased computational load. Table 7.3 lists several MATLAB functions that can be used for constrained/unconstrained minimizations.

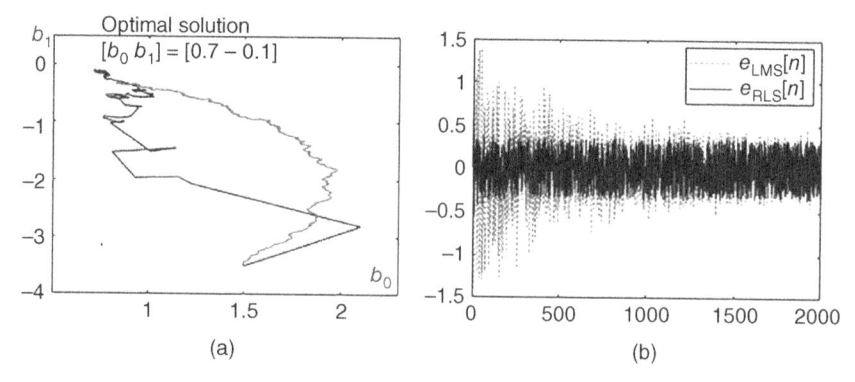

Figure 7.23 LMS and RLS adaptive parameter estimation. (a) Parameter estimates being updated. (b) Error signals of the two algorithms.

Table 7.3 The names of the MATLAB built-in minimization functions.

Minimization methods	Unconstrained minimization			Constrained minimization				
	Bracketing	Nongradient-based	Gradient-based	Linear	Nonlinear	Linear LS	Nonlinear LS	Minimax
MATLAB functions	fminbnd	fminsearch	fminunc	linprog	fmincon	lsqnonlin	lsqlin	minimax

```
%nm07e08.m
% RLS (Recursive Least Square) vs. LMS (Least Mean Square)
clear, clf
Nx=5000; % Number of data samples for computing R and p
a0=0.7; a1=-0.1; % True parameters
sigma=0.2; noise_amp=sigma*sqrt(12); % Amplitude of uniform noise
b0=[1.5; -3.5]; % Initial guess of the optimal filter coefficient vector
subplot(121)
plot(a0,a1,'ms'), hold on % plot the true parameters
% LMS method and RLS method
mu=0.05; % Adaptation step-size or Learning factor
KC=0; % Set as 0/1 to use Eqs.(7.6.11)/(7.6.12)
alpha=1; % Forgetting factor
bn_LMS=b0; bn_RLS=b0; % Initial guess of adaptive filter coefficients
P=10*eye(2);
nf=4000; e2_LMS=0; e2_RLS=0; % Number of iterations
x=rand(1,2)-0.5; % Initialize the process
for n=2:nf
   x(n+1) = a0*x(n)+a1*x(n-1) + noise_amp*(rand-0.5); % Eq.(E7.7.1)
   dn=x(n+1); xnT=x([n n-1]); xn=xnT.';
   b_LMS=bn_LMS; % update the parameter estimate
   e_LMS(n) = dn - xnT*b_LMS; % LMS error
   bn_LMS = b_LMS + mu*e_LMS(n)*xn; % Eq.(7.5.10)
   plot([b_LMS(1) bn_LMS(1)],[b_LMS(2) bn_LMS(2)],'r')
   b_RLS=bn_RLS;
   if KC==0
     P = (P-P*xn*xnT*P/(alpha+xnT*P*xn))/alpha; % Eq.(7.6.11a)
     K = P*xn; % Eq.(7.6.11b)
    else
     K = P*xn/(alpha+xnT*P*xn); % Eq.(7.6.12a)
     P = (P-K*xnT*P)/alpha;    % Eq.(7.6.12c)
   end
   e_RLS(n) = dn - xnT*b_RLS; % RLS error
   bn_RLS = b_RLS + K*e_RLS(n); % Eq.(7.6.12b)
   plot([b_RLS(1) bn_RLS(1)],[b_RLS(2) bn_RLS(2)],'b')
   if n>40, % After settling-down
      e2_LMS=e2_LMS+e_LMS(n)^2; e2_RLS=e2_RLS+e_RLS(n)^2;
   end % Sum of squared estimation error after settling-down
end
subplot(122)
nn=0:nf-1; % Time vector
plot(nn,e_LMS,'r', nn,e_RLS,'b')
legend('e_{LMS}[n]','e_{RLS}[n]')
[[a0 a1].' bn_LMS bn_RLS] % True parameter and LMS/RLS estimates
MSE_LMS=e2_LMS/(nf-40) % Mean square error of LMS
MSE_RLS=e2_RLS/(nf-40) % Mean square error of RLS
```

PROBLEMS

7.1 Modification of Golden Search Method

In fact, the golden search method explained in Section 7.1 requires only one function evaluation per iteration, since one point of a new interval coincides with a point of the previous interval so that only one trial point is updated. In spite of this fact, the MATLAB function 'opt_gs()' implementing the method performs the function evaluations twice per iteration. An improvement may be initiated by modifying the declaration type as

```
[xo,fo]=opt_gs1(f,a,e,fe,r1,b,r,TolX,TolFun,k)
```

so that anyone could use the new function as in the following program, where its input argument list contains another point (e) as well as the new end point (b) of the next interval, its function value (fe), and a parameter (r1) specifying if the point is the left one or the right one. Based on this idea, how do you revise the function 'opt_gs()' to cut down the number of function evaluations?

```
%nm07p01.m  to perform the revised golden search method
f701=@(x)x.*(x-2);
a=0; b=3; r=(sqrt(5)-1)/2;
TolX=1e-4; TolFun=1e-4; MaxIter=100;
h=b-a; rh= r*h;
c=b-rh; d=a+rh;
fc=f701(c);   fd=f701(d);
if fc<fd, [xo,fo]=opt_gs1(f701,a,c,fc,1-r,d,r,TolX,TolFun,MaxIter)
  else [xo,fo]=opt_gs1(f701,c,d,fd,r,b,r,TolX,TolFun,MaxIter)
end
```

7.2 Nelder-Mead, Steepest Descent, Newton, SA, GA and fminunc(), fminsearch()

Consider a two-variable objective function

$$f(\mathbf{x}) = x_1^4 - 12x_1^2 - 4x_1 + x_2^4 - 16x_2^2 - 5x_2 - 20\cos(x_1 - 2.5)\cos(x_2 - 2.9) \quad (P7.2.1)$$

whose gradient vector function is

$$\mathbf{g}(\mathbf{x}) = \nabla f(\mathbf{x}) = \begin{bmatrix} 4x_1^3 - 24x_1 - 4 + 20\sin(x_1 - 2.5)\cos(x_2 - 2.9) \\ 4x_2^3 - 32x_2 - 5 + 20\cos(x_1 - 2.5)\sin(x_2 - 2.9) \end{bmatrix} \quad (P7.2.2)$$

You have the MATLAB functions 'f07p02()' and 'g07p02()' defining the objective function $f(\mathbf{x})$ and its gradient function $\mathbf{g}(\mathbf{x})$, respectively, later. You also have a part of the MATLAB program which plots a mesh/contour-type graphs for $f(\mathbf{x})$. Note that this gradient function has nine zeros as listed in Table P7.2.1.

Table P7.2.1 Extrema (maxima/minima) and saddle points of the function (P7.2.1).

Points	Signs of $\partial^2 f/\partial x_i^2$		Points	Signs of $\partial^2 f/\partial x_i^2$	
(1) [0.6965 −0.1423]	−, −	M	(6) [−1.6926 −0.1183]		
(2) [2.5463 −0.1896]			(7) [−2.6573 −2.8219]	+, +	m
(3) [2.5209 2.9027]	+, +	G	(8) [−0.3227 −2.4257]		
(4) [−0.3865 2.9049]			(9) [2.5216 −2.8946]	+, +	m
(5) [−2.6964 2.9031]					

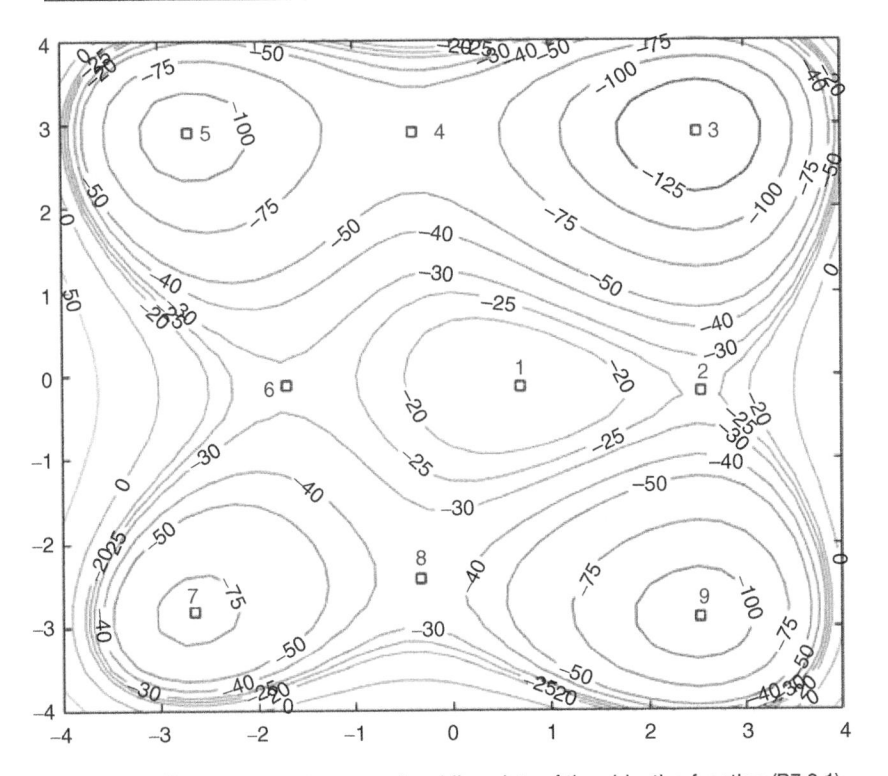

Figure P7.2 The contour, extrema, and saddle points of the objective function (P7.2.1).

(a) From the graphs (like Figure P7.2) which you get by running the (unfinished) program, determine the characteristic of each of the nine points, i.e. whether it is a local maximum(M)/minimum(m), the global minimum(G) or a saddle point(S) which is a minimum with respect to (w.r.t.) one variable and a maximum w.r.t. another variable. Support your judgment by telling the signs of the second derivatives of $f(\mathbf{x})$ w.r.t. x_1 and x_2.

$$\partial^2 f/\partial x_1^2 = 12x_1^2 - 24 + 20\cos(x_1 - 2.5)\cos(x_2 - 2.9)$$
$$\partial^2 f/\partial x_2^2 = 12x_2^2 - 32 + 20\cos(x_1 - 2.5)\cos(x_2 - 2.9) \tag{P7.2.3}$$

```
%nm07p02.m to minimize an objective ftn f(x) by the Newton method
f='f07p02'; g='g07p02';
l=[-4 -4]; u=[4 4];
x1=l(1):.25:u(1); x2=l(2):.25:u(2); [X1,X2]=meshgrid(x1,x2);
for m=1:length(x1)
   for n=1:length(x2), F(n,m)=feval(f,[x1(m) x2(n)]); end
end
figure(1), clf, mesh(X1,X2,F)
figure(2), clf,
v=[-125 -100 -75 -50 -40 -30 -25 -20 0 50];
C=contour(x1,x2,F,v); clabel(C,h)  % Contour with labels
... ... ... ...
```

```
function y=f07p02(x)
y=x(1)^4-12*x(1)^2-4*x(1)+x(2)^4-16*x(2)^2-5*x(2) ...
    -20*cos(x(1)-2.5)*cos(x(2)-2.9);
```

```
function [df,d2f]=g07p02(x) % the 1st/2nd derivatives
df(1)=4*x(1)^3-24*x(1)-4+20*sin(x(1)-2.5)*cos(x(2)-2.9); %Eq.(P7.2.2)
df(2)=4*x(2)^3-32*x(2)-5+20*cos(x(1)-2.5)*sin(x(2)-2.9); %Eq.(P7.2.2)
d2f(1)=12*x(1)^2-24+20*cos(x(1)-2.5)*cos(x(2)-2.9);   % Eq.(P7.2.3)
d2f(2)=12*x(2)^2-32+20*cos(x(1)-2.5)*cos(x(2)-2.9);   % Eq.(P7.2.3)
```

(b) Apply the Nelder-Mead method, the steepest descent method, the Newton method, the simulated annealing (SA), genetic algorithm (GA), and the MATLAB built-in functions 'fminunc()', 'fminsearch()' to minimize the objective function (P7.2.1) and fill in Table P7.2.2 with the number and character of the point reached by each method.

(c) Overall, the point reached by each minimization algorithm depends on the starting point, i.e. the initial value of the independent variable as well as the chatacteristic of the algorithm. Fill in the blanks in the following sentences. Most algorithms succeed to find the global minimum if only they start from the initial point (,),(,),(,) or

Table P7.2.2 Points reached by the several optimization functions.

Initial point x_0	Reached point						
	Nelder	steepest	Newton	fminunc	fminsearch	SA	GA
(0, 0)	(5)/m						
(1, 0)		(3)/G					
(1, 1)			(9)/m				
(0, 1)				(3)/G			
(−1, 1)					(5)/m		
(−1, 0)						≈(3)/G	
(−1, −1)							(3)/G
(0, −1)	(9)/m						
(1, −1)		(9)/m					
(2, 2)			(3)/G				
(−2, −2)			(7)/m				

(,). An algorithm most possibly goes to the closest local minimum (5) if launched from (,) or (,) and it may go to the closest local minimum (7) if launched from (,) or (,). If launched from (,), it may go to one of the two closest local minima (7) and (9) and if launched from (,), it most possibly goes to the closest local minimum (9). But, the global optimization techniques SA and GA seem to work fine almost regardless of the starting point, although not always.

7.3 Minimization of an Objective Function Having Many Local Minima/Maxima
Consider the problem of minimizing the following objective function

$$\text{Min} f(x) = \sin(1/x)/((x - 0.2)^2 + 0.1) \quad \text{(P7.3.1)}$$

which is depicted in Figure P7.3. The graph shows that this function has infinitely many local minima/maxima around $x = 0$ and the global minimum about $x = 0.2$.

(a) Find the solution by using the MATLAB built-in function 'fminbnd()'. Is it plausible?

(b) With nine different values of the initial guess $x_0 = 0.1, 0.2, \ldots, 0.9$, use the four MATLAB functions 'opt_Nelder()', 'opt_steep()', 'fminunc()', and 'fminsearch()' to solve the problem. Among those 36 tryouts, how many times have you got the right solution?

(c) With the values of the parameters set to $l = 0$, $u = 1$, $q = 1$, $\varepsilon_f = 10^{-9}$, $k_{max} = 1000$ and the initial guess $x_0 = 0.1, 0.2, \ldots, 0.9$, use the SA (simulated annealing) function 'sim_anl()' to solve the problem. You can test the performance of the function and your luck by running the function four times for the same problem and finding the probability of getting the right solution.

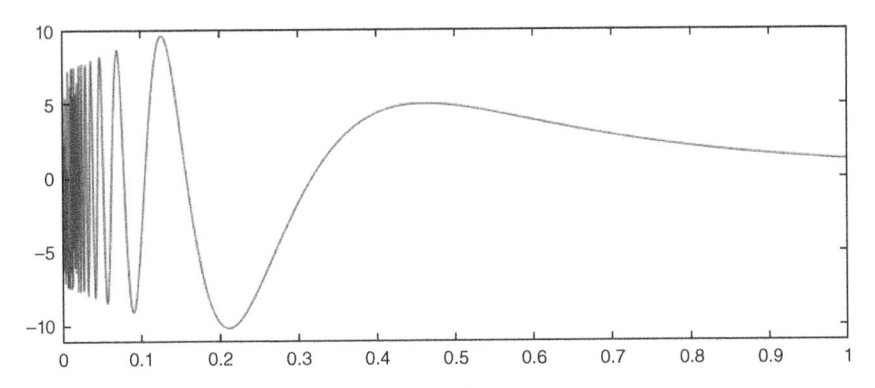

Figure P7.3 The graph of $f(x) = \sin(1/x)/\{(x - 0.2)^2 + 0.1\}$ having many local minima/maxima.

(d) With the values of the parameters set to $l = 0$, $u = 1$, $N_p = 30$, $N_b = 12$, $P_c = 0.5$, $P_m = 0.01$, $\eta = 1$, $k_{max} = 1000$ and the initial guess $x_0 = 0.1$, 0.2, ..., 0.9, use the genetic algorithm (GA) function 'genetic()' to solve the problem. As in (b), you can run the function four times for the same problem and find the probability of getting the right solution in order to test the performance of the function and your luck.

7.4 Constrained Optimization with Lagrange Multiplier Method
Consider the problem of minimizing

$$\text{Min } f(\mathbf{x}) = x_1^2 + 0.5x_1 + 3x_1x_2 + 5x_2^2 \tag{P7.4.1}$$

subject to the constraints

$$3x_1 + 2x_2 \leq -2$$
$$15x_1 - 3x_2 \leq 1 \tag{P7.4.2}$$

(a) Use the Lagrange multiplier method to solve this problem.
(b) Use the MATLAB built-in function 'fmincon()' to find the solution to the problem (P7.4.1).

```
f_=@(x1,x2)x1.^2+0.5*x1+3*x1.*x2+5*x2.^2;
x=linspace(-1,1);  y=linspace(-1,1); [X,Y]=meshgrid(x,y);
Z=f_(X,Y); contour(X,Y,Z); hold on % to plot Figure P7.4
ezplot('3*x+2*y+2'), ezplot('15*x-3*y-1')
```

7.5 Constrained Optimization with Penalty Method
Consider the problem of minimizing a nonlinear objective function

$$\text{Min}_\mathbf{x} f(\mathbf{x}) = -3x_1 - 2x_2 + M(3x_1 - 2x_2 + 2)^2 \quad (M: \text{a large positive number}) \tag{P7.5.1a}$$

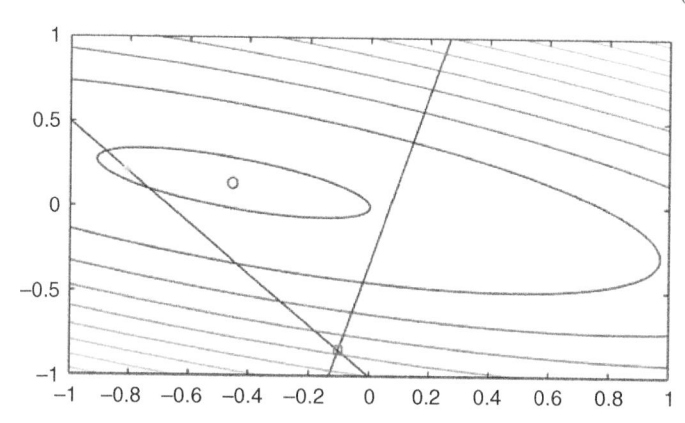

Figure P7.4 The contour for the objective function (P7.4.1) and lines showing the constraints.

subject to the constraints

$$\begin{bmatrix} 3 & 4 \\ -2 & -1 \end{bmatrix} \begin{bmatrix} x_1 \\ x_2 \end{bmatrix} \begin{matrix} \leq 7 \\ \geq -3 \end{matrix} \text{ and } \mathbf{l} = \begin{bmatrix} 0 \\ 0 \end{bmatrix} \leq \mathbf{x} = \begin{bmatrix} x_1 \\ x_2 \end{bmatrix} \leq \begin{bmatrix} 10 \\ 10 \end{bmatrix} = \mathbf{u} \quad \text{(P7.5.1b)}$$

(a) With the two values of the weighting factor $M = 20$ and $10\,000$ in the objective function (P7.5.1a), apply the MATLAB built-in function 'fmincon()' to find the solutions to the above constrained minimization problem. In order to do this job, you might have to make the variable parameter M passed to the objective function (defined in an M-file) either through 'fmincon()' or directly by declaring the parameter as global both in the main program and in the M-file defining (P7.5.1a). In case you are going to have the parameter passed through 'fmincon()' to the objective function, you should have the parameter included in the input argument list of the objective function as

```
function f=f7p05M(x,M)
f= -3*x(1)-2*x(2)+M*(3*x(1)-2*x(2)+2).^2;
```

Additionally, you should give empty matrices ([]) as the ninth input argument (for a nonlinear inequality/equality constraint function 'nonlcon') as well as the tenth one (for 'options') and the value of M as the 11th one of the function 'fmincon()'.

```
xo=fmincon('f7p05M',x0,A,b,[],[],l,u,[],[],M)
```

For reference, type 'help fmincon' into the MATLAB command window.

(b) Noting that the third (squared) term of the objective function (P7.5.1a) has its minimum value of zero for $3x_1 - 2x_2 + 2 = 0$ and thus, it actually represents the penalty (Section 7.2.2) imposed for not satisfying the equality constraint

$$3x_1 - 2x_2 + 2 = 0 \quad \text{(P7.5.2)}$$

tell which of the solutions obtained in (a) is more likely to satisfy this constraint and support your answer by comparing the values of the left-hand side of this equality for the two solutions.

(c) Removing the third term from the objective function and splitting the equality constraint into two reversed inequality constraints, we can modify the problem as follows:

$$\text{Min}_{\mathbf{x}} f(\mathbf{x}) = -3x_1 - 2x_2 \quad \text{(P7.5.3a)}$$

subject to the constraints

$$\begin{bmatrix} 3 & 4 \\ -2 & -1 \\ 3 & -2 \\ 3 & -2 \end{bmatrix} \begin{bmatrix} x_1 \\ x_2 \end{bmatrix} \begin{matrix} \leq 7 \\ \geq -3 \\ \leq -2 \\ \geq -2 \end{matrix} \text{ and } \mathbf{l} = \begin{bmatrix} 0 \\ 0 \end{bmatrix} \leq \mathbf{x} = \begin{bmatrix} x_1 \\ x_2 \end{bmatrix} \leq \begin{bmatrix} 10 \\ 10 \end{bmatrix} = \mathbf{u} \quad \text{(P7.5.3b)}$$

Noting that this fits the linear programming, use the function 'linprog()' to solve this problem.

(d) Treating the equality constraint separately from the inequality constraints, we can modify the problem as follows:

$$\text{Min}_x f(\mathbf{x}) = -3x_1 - 2x_2 \tag{P7.5.4a}$$

subject to the constraints

$$\begin{bmatrix} 3 & -2 \\ 3 & 4 \\ -2 & -1 \end{bmatrix} \begin{bmatrix} x_1 \\ x_2 \end{bmatrix} \begin{matrix} =-2 \\ \le 7 \\ \ge -3 \end{matrix} \quad \text{and} \quad \mathbf{l} = \begin{bmatrix} 0 \\ 0 \end{bmatrix} \le \mathbf{x} = \begin{bmatrix} x_1 \\ x_2 \end{bmatrix} \le \begin{bmatrix} 10 \\ 10 \end{bmatrix} = \mathbf{u} \tag{P7.5.4b}$$

Apply the two functions 'linprog()' and 'fmincon()' to solve this problem and see if the solutions agree with the solution obtained in (c).

(cf) Note that, in comparison with the function 'fmincon()', which can solve a general nonlinear optimization problem, the function 'linprog()' is made solely for dealing with a class of optimization problems having a linear objective function with linear constraints.

7.6 Nonnegative Constrained LS and Constrained Optimization
Consider the problem of minimizing a nonlinear objective function

$$\text{Min}_x \|C\mathbf{x} - \mathbf{d}\|^2 = [C\mathbf{x} - \mathbf{d}]^T [C\mathbf{x} - \mathbf{d}] \tag{P7.6.1a}$$

subject to the constraints

$$\mathbf{x} = \begin{bmatrix} x_1 \\ x_2 \end{bmatrix} \ge \begin{bmatrix} 0 \\ 0 \end{bmatrix} = \mathbf{l} \tag{P7.6.1b}$$

where

$$C = \begin{bmatrix} 1 & 2 \\ 3 & 4 \\ 5 & 1 \end{bmatrix}, \quad \mathbf{d} = \begin{bmatrix} 5.1 \\ 10.8 \\ 6.8 \end{bmatrix} \tag{P7.6.1c}$$

(a) Noting that this problem has no other constraints than the lower bound, apply the constrained linear least squares function 'lsqlin()' to find the solution.

(b) Noting that the lower bounds for all the variables are zeros, apply the MATLAB built-in function 'lsqnonneg()' to find the solution.

(c) Use the general purpose constrained optimization function 'fmincon()' to find the solution.

7.7 Constrained Optimization Problems
Solve the following constrained optimization problems by using the MATLAB built-in function 'fmincon()'.

(a) $\text{Min}_x x_1^3 - 5x_1^2 + 6x_1 + x_2^2 - 2x_2 + x_3$ \hfill (P7.7.1a) subject to the constraints

$$x_1^2 + x_2^2 - x_3 \leq 0$$
$$x_1^2 + x_2^2 + x_3^2 \geq 6 \quad \text{and} \quad \mathbf{x} = \begin{bmatrix} x_1 \\ x_2 \\ x_3 \end{bmatrix} \geq \begin{bmatrix} 0 \\ 0 \\ 0 \end{bmatrix} = \mathbf{1} \quad \text{(P7.7.1b)}$$
$$x_3 \leq 5$$

Try the function 'fmincon()' with the initial guesses listed in Table P7.7.

(b1) $\quad \text{Max}_\mathbf{x} x_1 x_2 x_3$ \hfill (P7.7.2a)

subject to the constraints

$$x_1 x_2 + x_2 x_3 + x_3 x_1 = 3 \quad \text{and} \quad \mathbf{x} = \begin{bmatrix} x_1 \\ x_2 \\ x_3 \end{bmatrix} \geq \begin{bmatrix} 0 \\ 0 \\ 0 \end{bmatrix} \quad \text{(P7.7.2b)}$$

Try the function 'fmincon()' with the initial guesses listed in Table P7.7.

(b2) $\quad \text{Min}_\mathbf{x} \ x_1 x_2 x_3$ \hfill (P7.7.3)

subject to the constraints (P7.7.2b).
Try the function 'fmincon()' with the initial guesses listed in Table P7.7.

(c1) $\quad \text{Max}_\mathbf{x} \ x_1 x_2 + x_2 x_3 + x_3 x_1$ \hfill (P7.7.4a)

subject to the constraints

$$x_1 + x_2 + x_3 = 3 \quad \text{and} \quad \mathbf{x} = \begin{bmatrix} x_1 \\ x_2 \\ x_3 \end{bmatrix} \geq \begin{bmatrix} 0 \\ 0 \\ 0 \end{bmatrix} \quad \text{(P7.7.4b)}$$

Try the function 'fmincon()' with the initial guesses listed in Table P7.7.

(c2) $\quad \text{Min}_\mathbf{x} \ x_1 x_2 + x_2 x_3 + x_3 x_1$ \hfill (P7.7.5)

subject to the constraints (P7.7.4b).
Try the function 'fmincon()' with the initial guesses listed in Table P7.7.

(d) $\quad \text{Min}_\mathbf{x} \dfrac{10\,000}{x_1 x_2^2}$ \hfill (P7.7.6a)

subject to the constraints

$$x_1^2 + x_2^2 = 100 \quad \text{and} \quad \mathbf{x} = \begin{bmatrix} x_1 \\ x_2 \end{bmatrix} > \begin{bmatrix} 0 \\ 0 \end{bmatrix} \quad \text{(P7.7.6b)}$$

Try the function 'fmincon()' with the initial guesses listed in Table P7.7.

(e) Does the function work well with all the initial guesses? If not, does it matter whether the starting point is inside the admissible region?

Table P7.7 The results of using 'fmincon()' with different initial guess.

	Initial guess x_0	Lower bound	$x^o, f(x^o)$	Remark (warning ?)
(a)	[0 0 0]	0		No feasible solution (w)
	[1 1 5]	0		Not a minimum
	[0 0 5]	0		Minimum
	[1 0 2]	0	[1.29 0.57 2], 2.74	
(b1)	[0 0 0]	0	[0 0 0], 0	
	[10 10 10]	0		Maximum (good)
(b2)	[0 0 0]	0		No feasible solution (w)
	[10 10 10]	0		Not a minimum, but the max
	[0.1 0.1 3]	0		One of many minima (w)
(c1)	[0 0 0]	0		
	[0.1 0.1 0.1]	0	[1 1 1], 3	Maximum (good)
	[0 1 2]	0		
(c2)	[0 0 0]	0	[1 1 1], 3	Not a minimum, but the max
	[0.1 0.1 0.1]	0		
	[0 1 2]	0		One of many minima
(d)	[1.0 0.5]	0		Weird (warning)
	[0.2 0.3]	0	[10.25 0], ∞	
	[2 5]	0	[5.77 8.17], 25.98	
	[100 10]	0		Minimum

(cf) Note that, in order to solve the maximization problem by 'fmincon()', we have to reverse the sign of the objective function. Note also that the objective functions (P7.7.3) and (P7.7.5a) have infinitely many minima having the value $f(x) = 0$ in the admissible region satisfying the constraints.

(cf) One might be disappointed with the reliability of the MATLAB optimization functions to see that they may fail to find the optimal solution depending on the initial guess. But how can a human work be perfect in this world? It implies the difficulty of nonlinear constrained optimization problems and can never impair the celebrity and reliability of MATLAB. Actually, it demonstrates the importance of studying some numerical stuff in addition to just getting used to the various MATLAB commands and functions.

Here is a tip for the usage of 'fmincon()': it might be better to use with an initial guess which is not at the origin, but in the admissible region satisfying the constraints, even though it does not guarantee the success of the function. It might also be helpful to apply the function with several values of the initial guess and then choose the best result.

7.8 Constrained Optimization and Penalty Method

Consider again the constrained minimization problem with the objective function (E7.3.1a) and the constraints (E7.3.1b).

$$\text{Min } f(\mathbf{x}) = \{(x_1 + 1.5)^2 + 5(x_2 - 1.7)^2\}\{(x_1 - 1.4)^2 + 0.6(x_2 - 0.5)^2\} \quad \text{(P7.8.1a)}$$

$$\text{s.t. } \mathbf{g}(\mathbf{x}) = \begin{bmatrix} -x_1 \\ -x_2 \\ 3x_1 - x_1 x_2 + 4x_2 - 7 \\ 2x_1 + x_2 - 3 \\ 3x_1 - 4x_2^2 - 4x_2 \end{bmatrix} \leq \begin{bmatrix} 0 \\ 0 \\ 0 \\ 0 \\ 0 \end{bmatrix} \quad \text{(P7.8.1b)}$$

In Example 7.3, we made the MATLAB script "nm0722.m" to solve the problem and defined the objective function (E7.3.2a) having the penalized constraint terms as the MATLAB function 'f722p()':

$$\text{Min } l(\mathbf{x}) = \{(x_1 + 1.5)^2 + 5(x_2 - 1.7)^2\}\{(x_1 - 1.4)^2 + 0.6(x_2 - 0.5)^2\}$$
$$+ \sum_{l=1}^{5} v_l \psi_l(g_l(\mathbf{x})) \quad \text{(P7.8.2a)}$$

where

$$\psi_l(g_l(\mathbf{x})) = \begin{cases} 0 & \text{if } g_l(\mathbf{x}) \leq 0 \text{ (constraint satisfied)} \\ \exp(e_l g_l(\mathbf{x})) & \text{if } g_l(\mathbf{x}) > 0 \text{ (constraint violated)} \end{cases}, \quad \text{(P7.8.2b)}$$

$$e_l = 1 \quad \forall l = 1, \cdots, 5$$

(a) What is the weighting coefficient vector v in the function 'f722p()'? Do the points reached by the functions 'fminsearch()'/ 'opt_steep()'/ 'fminunc()' satisfy all the constraints so that they are in the admissible region? If not, specify the constraint(s) violated by the points.

(b) Suppose the fourth constraint was violated by the point in (a). Then, how would you modify the weighting coefficient vector v so that the violated constraint can be paid more respect? Choose one of the following two vectors

(i) v=[1 1 1 1/3 1]   (ii) v=[1 1 1 3 1]

and modify the function 'f722p()' with this coefficient vector. Then, run the script "nm0722.m", fill in the 28 blanks of Table P7.8 with the results, and see if the fourth constraint is still violated by the points reached by the optimization functions.

Table P7.8 The results of penalty methods depending on the initial guess and weighting factor.

v		Starting point $x_0 = [0.4\ 0.5]$					$x_0 = [0.2\ 4]$				
		Nelder	fminsearch	steep	fminunc	fmincon	Nelder	fminsearch	steep	fminunc	fmincon
1	x^o	1.21		1.34			1.34		1.34		
		0.58		0.62			0.62		0.62		
1	f^o	0.53		0.17			0.17		0.17		
1		−1.21	−1.34	−1.34	−1.38	−1.21	−1.34	−1.34	−1.34	1.26	0.00
		−0.58	−0.62	−0.62	−0.63	−0.58	−0.62	−0.62	−0.62	−1.70	−1.59
1/3	c^o	−1.76	−1.34	−1.34	−1.19	−1.76	−1.34	−1.34	−1.33	−1.84	−0.65
		−0.00				0.00	0.29			−3.82	−1.41
1		−0.00	−0.00	−0.00	−0.00	0.00	−0.00	−0.00	−0.00	−22.1	−16.4
1	x^o	1.21	1.21	1.12	1.18	—	1.21	1.21	1.15	−1.26	—
		0.58	0.58	0.76	0.64		0.58	0.58	0.71	1.70	
1	f^o	0.53	0.53	1.36	0.79	—	0.53	0.53	1.08	0.46	—
1		−1.21	−1.21	−1.12	−1.18		−1.21	−1.21	−1.15	1.26	
		−0.58	−0.58	−0.76	−0.64		−0.58	−0.58	−0.71	−1.70	
3	c^o	−1.76	−1.76	−1.44	−1.65	—	−1.76	−1.76	−1.54	−1.84	—
		−0.00					−0.00			−3.82	
1		−0.00	−0.00	−2.04	−0.70		−0.00	−0.00	−1.39	−22.1	

(c) Instead of the penalty method, apply the intrinsically constrained optimization function 'fmincon()' with the initial guesses $x_0 = [0.4\ 0.5]$ and [0.2 4] to solve the problem described by Eq. (E7.3.1) or (P7.8.1) and fill in Table P7.8 with the results concerning the reached point and the corresponding values of the objective/constraint functions.

(d) Based on the results listed in Table P7.8, circle the right word in each of the parentheses in the following sentences:

- For penalty methods, the nongradient-based minimization functions like 'Nelder()'/'fminsearch()' may work (better, worse) than the gradient-based minimization functions like 'opt_steep()'/'fminunc()'.
- If some constraint is violated, you had better (increase, decrease) the corresponding weight coefficient.

(cf) Besides, unconstrained optimization with the penalized constraints in the objective function sometimes works better than the constrained optimization function 'fmincon()'.

7.9 Constrained Optimization Using 'fmincon()' and 'ga()'

Consider the following constrained minimization problem:

$$\text{Min } f(\mathbf{x}) = \{(x_1 + 1.5)^2 + 5(x_2 - 1.7)^2\}\{(x_1 - 1.4)^2 + 0.6(x_2 - 0.5)^2\} \quad \text{(P7.9.1a)}$$

$$\text{s.t. } \begin{array}{l} x_1 x_2 + x_1 - x_2 \le 2.5 \\ x_1 x_2 + 0.5 x_1 \ge 2 \end{array}, \quad 0 \le x_1 \le 2.5, \text{ and } 0 \le x_2 \le 2.5 \quad \text{(P7.9.1b)}$$

Complete and run the following MATLAB script "nm07p09.m" to use two MATLAB functions 'fmincon()' and 'ga()' for finding two solutions. Are the two solutions the same in terms of the minimized function value? If not, which one is better? If you want to improve the solution obtained using 'ga()', decrease the values of some optional parameters such as {'ConstraintTolerance', 'FunctionTolerance'} by using 'optimoptions()'. Note that to see the optional parameters of 'ga()' with their default values, type 'optimoptions('ga')' at the MATLAB prompt.

```
%nm07p09.m
f=@(x)((x(1)+1.5)^2+5*(x(?)-1.7)^2)*((x(?)-1.4)^2+.6*(x(2)-.5)^2);
% We use fmincon().
A=[]; B=[]; Aeq=[]; Beq=[]; l=[0; 0]; u=[2.5; 2.5];
x0=[1; 1]; % Starting point (Initial guess)
[xo,fo]=fmincon(f,x0,A,B,Aeq,Beq,l,u,@nm07p09_constr)
[c,ceq]=nm07p09_constr(xo) % Are the constraints satisfied by xo?
% Using GA for the constrained minimization
options=optimoptions(@ga,'MutationFcn',@mutationadaptfeasible);
Nx=numel(x0);
[xo_GA,fo_GA]=ga(f,Nx,A,B,Aeq,Beq,l,u,@nm07p09_constr,options)
[c,ceq]=nm07p09_constr(xo_GA) % Are the constraints satisfied by xo_GA?
% To plot the objective function, constraints, and solution(s)
f_=@(x,y)((x+1.5).^2+5*(y-1.7).^2).*((x-1.4).^2+.6*(y-0.5).^2);
ezplot('x*y+x-y-2.5',[l(1) u(1) l(2) u(2)])
hold on, ezplot('2-x*y-0.5*x',[l(1) u(1) l(2) u(2)])
x=linspace(l(1),u(1)); y=linspace(l(2),u(2));
[X,Y]=meshgrid(x,y);
Z=f_(X,Y);
contour(X,Y,Z,0.1*1.5.^[1:20])
plot(xo(1),xo(2),'mo'), axis([l(1) u(1) l(2) u(2)])
```

```
function [c,ceq]=nm07p09_constr(x)
c=[x(?)*x(2)+x(1)-x(?)-2.5; % Inequality constraints
   2-x(1)*x(?)-0.5*x(1)]; % desired to be nonnegative
ceq=[]; % Equality constraints (desired to be zero) if any
```

7.10 **A Constrained Optimization on Location**

A company has three factories that are located at the points $(-16,4)$, $(6,5)$, and $(3,-9)$, respectively, in the $x_1 x_2$-plane (Figure P7.10), and the numbers of deliveries to those factories are 5, 6, and 10 per month, respectively. The company has a plan to build a new warehouse in its site bounded by

$$|x_1 - 1| + |x_2 - 1| \leq 2 \qquad (P7.10.1)$$

```
function [C,Ceq]=fp_warehouse_c(x)
C=sum(abs(x-[1 1]))-2; % Inequality constraint to be nonpositive
Ceq=[]; % No equality constraint
```

```
%nm07p10.m to solve the warehouse location problem
fp_warehouse=@(x)[5 6 ??]* ...
      [norm(x-[-16 ?]); norm(x-[? 5]); norm(x-[3 -?])];
x0=[1 1]; A=[]; b=[]; Aeq=[]; beq=[]; l=[]; u=[];
xo=fmincon(fp_warehouse,x0,A,b,Aeq,beq,l,u,'fp_warehouse_c')
```

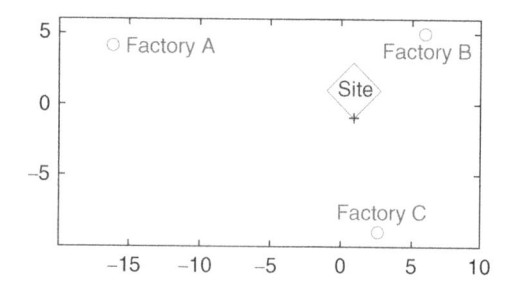

Figure P7.10 The site of a new warehouse and the locations factories.

and is trying to minimize the monthly mileage of delivery trucks in determining the location of a new warehouse on the assumption that the distance between two points represents the driving distance.

(a) What is the objective function that must be defined in the above script "nm07p10.m"?

(b) What is the statement defining the inequality constraint (P7.10.1)?

(c) Complete and run the script "nm07p10.m" to get the optimum location of the new warehouse.

7.11 A Constrained Optimization on Ray Refraction

A light ray follows the path that takes the shortest time when it travels in the space. We want to find the three angles θ_1, θ_2, and θ_3 (measured between the array and the normal to the material surface) of a ray traveling from $P = (0,0)$ to $Q = (L, -(d_1 + d_2 + d_3))$ through a transparent material of thickness d_2 and index of refraction n as depicted in Figure P7.11. Note the following:

- Since the speed of light in the transparent material is $v = c/n$ (c: the speed of light in the free space), the traveling time to be minimized can be expressed as

$$\text{Min } f(\boldsymbol{\theta}, \mathbf{d}, n, L) = \frac{d_1}{c \cos \theta_1} + \frac{n d_2}{c \cos \theta_2} + \frac{d_3}{c \cos \theta_3} \qquad (P7.11.1)$$

- The sum of the three horizontal distances traveled by the light ray must be L:

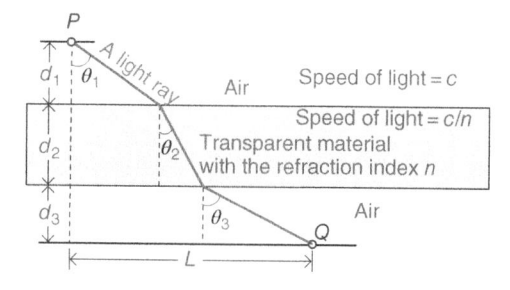

Figure P7.11 Refraction of a light ray at an air-glass interface.

$$g(\boldsymbol{\theta}, \mathbf{d}, n, L) = \sum_{i=1}^{3} c_i \cos \theta_i - L = 0 \qquad (P7.11.2)$$

- The horizontal distance L and the index of refraction n are additionally included in the input argument lists of both the objective function $f(\theta,\mathbf{d},n,L)$ and the constraint function $g(\theta,\mathbf{d},n,L)$ regardless of whether or not they are used in each function. It is because the objective function and the constraint function of the MATLAB function 'fmincon()' must have the same input arguments.

(a) Compose and run a script "nm07p11a.m", which solves the above-constrained minimization problem to find the three angles θ_1, θ_2, and θ_3 for $n = 1.52$, $d_1 = d_2 = d_3 = 1$ cm, and different values of $L = 0.6 : 0.3 : 6$ and plots $\sin(\theta_1)/\sin(\theta_2)$ and $\sin(\theta_3)/\sin(\theta_2)$ vs. L.

(b) Compose and run a script "nm07p11b.m", which finds the three angles θ_1, θ_2, and θ_3 for $L = 3$ cm, $d_1 = d_2 = d_3 = 1$ cm, and different values of $n = 1 : 0.01 : 1.6$ and plots $\sin(\theta_1)/\sin(\theta_2)$ and $\sin(\theta_3)/\sin(\theta_2)$ vs. n.

```
%nm07p11a.m
fp_light_ref=@(th,n,d,L)sum([d(1) n*d(2) d(3)]./cos(??));
options=optimoptions('fmincon','display','off');
n=1.52; d=[1 1 1]; Ls=0.1*[2:20]*sum(d);
for i=1:length(Ls)
   L=Ls(i); th0=[1 1 1];
   l=zeros(size(th0)); u=pi/2*ones(size(th0));
   th(i,:)=fmincon(??_light_ref,th?,[],[],[],[], ...
                   ?,?,'fp_light_ref_c',???????,n,d,L);
end
plot(Ls,sin(th(:,1))./sin(th(:,2)),'b', ...
     Ls,sin(th(:,3))./sin(th(:,2)),'r:')
```

```
function [C,Ceq]=fp_light_ref_c(th,n,d,L)
C=[]; % No inequality constraint
Ceq=sum(d.*tan(th))-L;
```

7.12 Linear Programming Method

Consider the problem of maximizing a linear objective function

$$\text{Max } f(\mathbf{x}) = \mathbf{f}^T \mathbf{x} = \begin{bmatrix} 3 & 2 & -1 \end{bmatrix} \begin{bmatrix} x_1 & x_2 & x_3 \end{bmatrix}^T \qquad (P7.12.1a)$$

subject to the constraints

$$A\mathbf{x} = \begin{bmatrix} 1 & 0 & -2 \\ -3 & -4 & 0 \\ 2 & 1 & 0 \end{bmatrix} \begin{bmatrix} x_1 \\ x_2 \\ x_3 \end{bmatrix} \begin{matrix} = \\ \geq \\ \leq \end{matrix} \begin{bmatrix} 0 \\ -11 \\ 7 \end{bmatrix} = \mathbf{b} \quad \text{and} \qquad (P7.12.1b)$$

$$\mathbf{l} = \begin{bmatrix} 0 \\ 0 \\ 0 \end{bmatrix} \leq \mathbf{x} = \begin{bmatrix} x_1 \\ x_2 \\ x_3 \end{bmatrix} \leq \begin{bmatrix} 10 \\ 10 \\ 10 \end{bmatrix} = \mathbf{u}$$

(a) Jessica has been puzzled by this problem, which is not a minimization but a maximization. How do you suggest her to solve it? Compose a MATLAB script, named "nm07p12a.m", which uses the MATLAB built-in functions 'linprog()' and 'fmincon()' to solve this problem, and run it to get the solutions.

(b) If x_1 and x_3 must be integers, how will you modify so that it can use the MATLAB built-in function 'intlinprog()' to solve the MILP problem? Run the modified script to get the solution. Is the solution same as that of the LP problem (obtained in (a)) with the values of x_1 and x_3 replaced by their closest integers?

7.13 Neural Network

In the supervised learning scheme of the neural network introduced in Section 7.4, consider the following objective function, called the *cross entropy function*, in place of Eq. (7.6.3):

$$J = \sum_{m=1}^{M_o} J_m = -\sum_{m=1}^{M_o} \{d_m \ln y_m + (1-d_m) \ln (1-y_m)\} \quad \text{(P7.13.1)}$$

(a) To compare the two objective functions (7.4.3) and (P7.13.1), plot the graphs of the following two functions vs. $-e = y - d$ for $0.005 \le y \le 0.995$ when $d = 0$ and $d = 1$ as shown in Figure P7.13.

$$J_a = \frac{1}{2}(d-y)^2 = \begin{cases} y^2/2 = (y-d)^2/2 & \text{when } d = 0 \\ (1-y)^2/2 = (d-y)^2/2 & \text{when } d = 1 \end{cases} \quad \text{(P7.13.2a)}$$

Cross entropy function
$$J_b = -d \ln y - (1-d) \ln(1-y)$$
$$= \begin{cases} -\ln(1-y) = -\ln(1-(y-d)) & \text{when } d = 0 \\ -\ln y = -\ln(1-(y-d)) & \text{when } d = 1 \end{cases} \quad \text{(P7.13.2b)}$$

Which one of the two functions is more sensitive to the error $-e = y - d$?

(b) The derivative of the cross entropy function J_b (Eq. (P7.13.2b)) w.r.t. y is

$$\frac{\partial J_b}{\partial y} = -\frac{d}{y} + \frac{1-d}{1-y} = -\frac{d-y}{y(1-y)} \quad \text{(P7.13.3)}$$

so that the gradient of the objective function (P7.13.1) w.r.t. the weights at the last layer L is

```
%nm07p13a.m
yy=0.005:0.001:0.995; % y-Domain
Ja=@(y,d) (d-y).^2/2;
Jb=@(y,d) -d*log(y)-(1-d)*log(1-y);
for i=1:2
   if i<2, d=0; else d=1; end
   subplot(221), plot(yy-d,Ja(yy,d)), axis([-1 1 0 0.5]); hold on
   subplot(222), plot(yy-d,Jb(yy,d)), axis([-1 1 0 5]); hold on
end
```

$$\frac{\partial J}{\partial w_{mn}^{(L)}} = \frac{\partial J_m}{\partial w_{mn}^{(L)}} = \frac{\partial J_m}{\partial y_m^{(L)}} \frac{\partial y_m^{(L)}}{\partial v_m^{(L)}} \frac{\partial v_m^{(L)}}{\partial w_{mn}^{(L)}}$$

$$\overset{(P7.13.3),(7.4.3)}{\underset{(7.4.2),(7.4.4)}{=}} -\left(\frac{d_m}{\sigma(v)} - \frac{1-d_m}{1-\sigma(v)}\right)\sigma'(v)|_{v=v_m^{(L)}} y_n^{(L-1)}$$

$$\overset{(7.4.5)}{=} -\frac{d_m - \sigma(v)}{\sigma(v)(1-\sigma(v))} \sigma(v)(1 - \sigma(v))|_{v=v_m^{(L)}} y_n^{(L-1)}$$

$$= -(d_m - \sigma(v))|_{v=v_m^{(L)}} y_n^{(L-1)}$$

$$\overset{(7.4.4)}{=} -(d_m - y_m^{(L)}) y_n^{(L-1)} = -\delta_m^{(L)} y_n^{(L-1)} \qquad (P7.13.4)$$

where

$$\delta_m^{(L)} = d_m - y_m^{(L)} = e_m \qquad (P7.13.5)$$

Noting that Eq. (P7.13.5) replaces Eq. (7.4.9b) and everything else remains the same, modify the MATLAB function 'train_NN_multiclass()' so that it can train the NN with the objective function (7.4.3) or (P7.13.1) depending on whether the value of the fifth input argument, say, KC is 0 or 1. Then, run the MATLAB script "test_NN_multiclass.m" with KC = 1 to get the error graph like Figure 7.20.

7.14 Deep Neural Network (DNN) for Deep Learning
According to Ref. [K-1], deep learning can be realized by the deep neural network (DNN), that is the multilayer NN containing two or more hidden layers where the rectified linear unit (ReLU) function (similar to the ramp function)

$$\varphi(x) = \text{Max}\{x, 0\} = \begin{cases} x & \text{for } x > 0 \\ 0 & \text{for } x \leq 0 \end{cases} \qquad (P7.14.1)$$

is often used in place of the sigmoid function as the activation function to overcome the vanishing gradient problem of the sigmoid function. Note that the derivative of the ReLU function is like the unit step function:

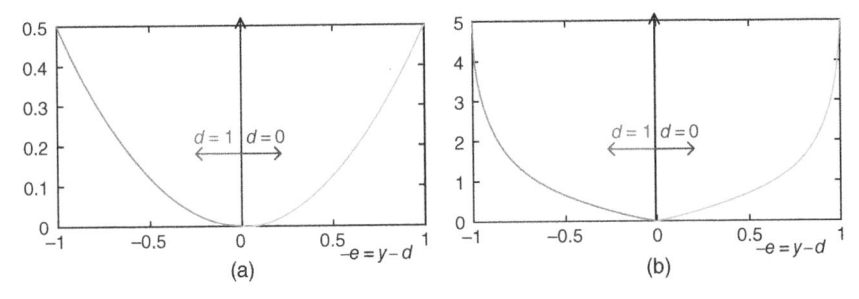

Figure P7.13 Two objective functions for training a neural network. (a) $J_a = (d-y)2/2$, i.e. Eq. (P7.13.2a). (b) $J_b = -d \ln y - (1-d) \ln(1-y)$, i.e. Eq. (P7.13.2b)

464 OPTIMIZATION

$$\varphi'(x) = \begin{cases} 1 & \text{for } x > 0 \\ 0 & \text{for } x \leq 0 \end{cases} \quad \text{(P7.14.2)}$$

Complete the two MATLAB functions 'DNN_multiclass()' and 'train_DNN_multiclass()' listed below so that they can implement the DNN algorithm. Then, run the following script "test_DNN_multiclass.m" to see if the algorithm works fine.

Note that in the script "test_DNN_multiclass.m" (listed below), two bits of the real image data chosen randomly using the MATLAB function 'randperm()' are contaminated, that is into 0/1 if they are 1/0, respectively, by using the MATLAB function 'xor()'. Are the five contaminated image data classified as the corresponding true characters by the trained DNN?

```
%test_DNN_multiclass.m
X=zeros(5,5,5);
Mi=size(X,1)*size(X,2); % Number of input layer nodes
M1=20; M2=20; M3=20; % Numbers of hidden layer nodes
Mo=size(X,3); % Number of output layer nodes
X0 = [0 1 1 0 0   1 1 1 1 0   1 1 1 1 0   0 0 0 1 0   1 1 1 1 1;
      0 0 1 0 0   0 0 0 0 1   0 0 0 0 1   0 0 1 1 0   1 0 0 0 0;
      0 0 1 0 0   0 1 1 1 0   0 1 1 1 0   0 1 0 1 0   1 1 1 1 0;
      0 0 1 0 0   1 0 0 0 0   0 0 0 0 1   1 1 1 1 1   0 0 0 0 1;
      0 1 1 1 0   1 1 1 1 1   1 1 1 1 0   0 0 0 1 0   1 1 1 1 0];
for k=1:5, X(:,:,k)=X0(:,(k-1)*5+[1:5]); end
D = [1 0 0 0 0; % Correct classification
     0 1 0 0 0;
     0 0 1 0 0;
     0 0 0 1 0;
     0 0 0 0 1];
% To initialize the weights for the layers
W1=2*rand(M1,Mi)-1; W2=2*rand(M2,M1)-1;
W3=2*rand(M3,M2)-1; W4=2*rand(Mo,M3)-1;
Ws={W1,W2,W3,W4}; % Initialize the weights
alpha=0.01; % Learning rate
fa='ReLU'; %'sigmoid'; % choose the activation function
% To train the DNN
N=1e4; % Number of epochs
for n=1:N
   [Ws,er(n)]=train_DNN_multiclass(Ws,X,D,alpha,fa);
end
% To exercise the trained NN for classifying the real data,
Nc=2; % Number of bits (among Mi bits) to be contaminated.
for k=1:Mo
   x   = reshape(X(:,:,k),Mi,1); % an input vector of size Mix1
   idx=randperm(Mi,Nc); x(idx)=xor(x(idx),ones(Nc,1)); % contami-
nated
   y=DNN_multiclass(Ws,x,alpha,fa);
   yy(:,k)=y;
end
yy % Raw output data for reference
```

```
function [y,ys,vs]=DNN_multiclass(Ws,x,alpha,fa)
% To implement a Neural Network with the cell Ws of weight matrices
% Input: Ws    = Cell of weighting matrices {W1,W2,...}
%        x     = An input data
%        alpha = Learning rate
%        fa    = Activation function
% Output: y    = Last output for the input x
%         ys   = Cell of the outputs at each layer
%         vs   = Cell of the weighted sums at each layer
L=length(Ws); % Number of layers excluding the input layer
ys{1}=x;
for l=1:L
   v=Ws{l}*ys{l}; vs{l}=v;
   if l<L, ys{l+1}=feval(??,v); % Eq.(P7.14.2)
    else ys{l+1}=???????(v); % Eq.(7.4.2)
   end
end
y=ys{L+1}; % The output (at the last layer)
```

```
function [Ws,er]=train_DNN_multiclass(Ws,X,D,alpha,fa)
% Error backpropagation for training a DNN (deep neural network).
% Input:  Ws = Cell of weighting matrices {W1,W2,...}
%         X(Nx x Ny x N) = 3D Training input data X
%         D(NxMo) = Correct outputs for the training input data
%         alpha   = Learning rate
% Output: Ws = Cell of updated weighting matrices {W1,W2,...}
%         er = Sum of squared errors sum((d-y)^2)
if nargin<7, KC=0; end
if nargin<6, ratio=0; end
if nargin<5, fa='ReLU'; end % Activation function
if nargin<4, alpha=0.01; end
[Nx,Ny,Mo]=size(X); % Size of training data and No of output nodes (classes)
Mi=Nx*Ny; % Number of input layer nodes
L=length(Ws); % Number of layers excluding the input layer
for k=1:Mo
   x=reshape(X(:,:,k),Mi,1); % an Mix1 input vector
   [y,ys,vs]=DNN_multiclass(Ws,x,alpha,fa);
   d=D(k,:).'; % the corresponding desired output for the input x
   e=d-y; er=sum(e.^2); % Error(s)
   delta=e; % Mox1 to minimize the cross entropy
   for l=L:-1:1
      yl=ys{l}; % M_(l-1)x1
      dW=alpha*delta*yl.'; % M_lxM_(l-1)  Eq.(7.4.9a,11a)
      if l>1
        dfa=(vs{l-1}?0); % Eq.(P7.14.2): Derivative of activation ftn
        delta=Ws{l}.'*delta.*???; % M_(l-1)x1  Eq.(7.4.11b)
      end
      Ws{l}=Ws{l}+dW;
   end
end
```

```
function y=ReLU(x)
y = max(x,?); % Eq.(P7.14.2)
```

8

MATRICES AND EIGENVALUES

CHAPTER OUTLINE

8.1	Eigenvalues and Eigenvectors	468
8.2	Similarity Transformation and Diagonalization	469
8.3	Power Method	475
	8.3.1 Scaled Power Method	475
	8.3.2 Inverse Power Method	476
	8.3.3 Shifted Inverse Power Method	477
8.4	Jacobi Method	478
8.5	Gram-Schmidt Orthonormalization and *QR* Decomposition	481
8.6	Physical Meaning of Eigenvalues/Eigenvectors	485
8.7	Differential Equations with Eigenvectors	489
8.8	DoA Estimation with Eigenvectors[Y-3]	493
	Problems	499

In this chapter, we will look at the eigenvalue or characteristic value λ and its corresponding eigenvector or characteristic vector **v** of a matrix.

Applied Numerical Methods Using MATLAB®, Second Edition. Won Y. Yang, Jaekwon Kim, Kyung W. Park, Donghyun Baek, Sungjoon Lim, Jingon Joung, Suhyun Park, Han L. Lee, Woo June Choi, and Taeho Im.
© 2020 John Wiley & Sons, Inc. Published 2020 by John Wiley & Sons, Inc.
Companion website: www.wiley.com/go/yang/appliednumericalmethods

8.1 EIGENVALUES AND EIGENVECTORS

The *eigenvalue* or *characteristic value* and its corresponding *eigenvector* or *characteristic vector* of an $N \times N$ matrix A are defined as a scalar λ and a nonzero vector $\mathbf{v}$ satisfying

$$A\mathbf{v} = \lambda \mathbf{v} \Leftrightarrow (A - \lambda I)\mathbf{v} = 0 \; (\mathbf{v} \neq 0) \quad (8.1.1)$$

where $(\lambda, \mathbf{v})$ is called an *eigenpair*, and there are N eigenpairs for the $N \times N$ matrix A.

How do we get them? Noting that

- in order for the aforementioned equation to hold for any nonzero vector $\mathbf{v}$, the matrix $[A - \lambda I]$ should be singular, i.e. its determinant should be zero ($|A - \lambda I| = 0$), and
- the determinant of the matrix $[A - \lambda I]$ is a polynomial of degree N in terms of λ,

we first must find the eigenvalue λ_i's by solving the so-called *characteristic equation*

$$|A - \lambda I| = \lambda^N + a_{N-1}\lambda^{N-1} + \cdots + a_1\lambda + a_0 = 0 \quad (8.1.2)$$

and then substitute the λ_i's, one by one, into Eq. (8.1.1) to solve it for the eigenvector $\mathbf{v}_i$'s. This is, however, not always so simple, especially if some root (eigenvalue) of Eq. (8.1.2) is of multiplicity $k > 1$, since we have to generate k independent eigenvectors satisfying Eq. (8.1.1) for such an eigenvalue. Still, we do not have to worry about this, thanks to the MATLAB built-in function 'eig()' which finds us all the eigenvalues and their corresponding eigenvectors for a given matrix. How do we use it? All we need to do is to define the matrix, say, A, and type a single statement into the MATLAB Command window as follows:

```
»[V,Lamda]=eig(A)  %e=eig(A) just for eigenvalues
```

Let us take a look at the following example.

Example 8.1. Eigenvalues/Eigenvectors of a Matrix.
Let us find the eigenvalues/eigenvectors of the matrix

$$A = \begin{bmatrix} 0 & 1 \\ 0 & -1 \end{bmatrix} \quad (E8.1.1)$$

First, we find its eigenvalues as

$$|A - \lambda I| = \left| \begin{bmatrix} -\lambda & 1 \\ 0 & -1-\lambda \end{bmatrix} \right| = \lambda(\lambda + 1) = 0; \; \lambda_1 = 0, \; \lambda_2 = -1 \quad (E8.1.2)$$

and then, get the corresponding eigenvectors as

$$[A - \lambda_1 I]\mathbf{v}_1 = \begin{bmatrix} 0 & 1 \\ 0 & -1 \end{bmatrix}\begin{bmatrix} v_{11} \\ v_{21} \end{bmatrix} = \begin{bmatrix} v_{21} \\ -v_{21} \end{bmatrix} = \begin{bmatrix} 0 \\ 0 \end{bmatrix};$$

$$v_{21} = 0; \quad \mathbf{v}_1 = \begin{bmatrix} v_{11} \\ v_{21} \end{bmatrix} = \begin{bmatrix} 1 \\ 0 \end{bmatrix} \tag{E8.1.3a}$$

$$[A - \lambda_2 I]\mathbf{v}_2 = \begin{bmatrix} 1 & 1 \\ 0 & 0 \end{bmatrix}\begin{bmatrix} v_{12} \\ v_{22} \end{bmatrix} = \begin{bmatrix} v_{12} + v_{22} \\ 0 \end{bmatrix} = \begin{bmatrix} 0 \\ 0 \end{bmatrix};$$

$$v_{12} = -v_{22}; \quad \mathbf{v}_2 = \begin{bmatrix} v_{12} \\ v_{22} \end{bmatrix} = \begin{bmatrix} 1/\sqrt{2} \\ -1/\sqrt{2} \end{bmatrix} \tag{E8.1.3b}$$

where we have chosen v_{11}, v_{12}, and v_{22}, so that the norms of the eigenvectors become one.

Alternatively, we can use the MATLAB command 'eig(A)' for finding eigenvalues/eigenvectors or 'roots(poly(A))' just for finding eigenvalues as the roots of the characteristic equation as illustrated by the following script "nm08e01.m".

(Q) Explain the meanings of and the relationship among the values of v, L, e, e1, and L1 that are obtained by running the MATLAB script "nm08e01.m".

```
%nm08e01.m
% to get the eigenvalues & eigenvectors of a matrix A.
A=[0 1;0 -1];
[V,L]=eig(A) % V= modal matrix composed of eigenvectors
% L=diagonal matrix with eigenvalues on its diagonal
e=eig(A)
e1=roots(poly(A)) % just for eigenvalues
L1=V^-1*A*V % diagonalize through similarity transformation
    % into a diagonal matrix having the eigenvalues on diagonal.
```

8.2 SIMILARITY TRANSFORMATION AND DIAGONALIZATION

Premultiplying a matrix A by P^{-1} and postmultiplying it by P makes a similarity transformation

$$A \to P^{-1}AP \tag{8.2.1}$$

Remark 8.1 tells us how a similarity transformation affects the eigenvalues/eigenvectors.

Remark 8.1 Effect of Similarity Transformation on Eigenvalues/Eigenvectors.

(1) The eigenvalues are not changed by a similarity transformation.
$$|P^{-1}AP - \lambda I| = |P^{-1}AP - P^{-1}\lambda I P| = |P^{-1}||A - \lambda I||P| = |A - \lambda I| \tag{8.2.2}$$

(2) Substituting $\mathbf{v} = P\mathbf{w}$ into Eq. (8.1.1) yields

$$A\mathbf{v} = \lambda \mathbf{v} \xrightarrow{\mathbf{v}=P\mathbf{w}} AP\mathbf{w} = \lambda P\mathbf{w} = P\lambda \mathbf{w}; \quad [P^{-1}AP]\mathbf{w} = \lambda \mathbf{w}$$

This implies that the matrix $P^{-1}AP$ obtained by a similarity transformation has $\mathbf{w} = P^{-1}\mathbf{v}$ as its eigenvector if $\mathbf{v}$ is an eigenvector of the matrix A.

To understand the diagonalization of a matrix into a diagonal matrix (having its eigenvalues on the main diagonal) through a similarity transformation, we should know the following theorem:

Theorem 8.1 Distinct Eigenvalues and Independent Eigenvectors.
If the eigenvalues of a matrix A are all distinct, i.e. different from each other, then the corresponding eigenvectors are independent of each other, and, consequently, the modal matrix composed of the eigenvectors as columns is nonsingular.

Now, for an $N \times N$ matrix A whose eigenvalues are all distinct, let us put all the equations (8.1.1) for each eigenvalue-eigenvector pair together to write

$$A[\mathbf{v}_1 \ \mathbf{v}_2 \cdots \mathbf{v}_N] = [\mathbf{v}_1 \ \mathbf{v}_2 \cdots \mathbf{v}_N]\begin{bmatrix} \lambda_1 & 0 & \cdot & 0 \\ 0 & \lambda_2 & \cdot & 0 \\ \cdot & \cdot & \cdot & \cdot \\ 0 & 0 & \cdot & \lambda_N \end{bmatrix}; \quad AV = V\Lambda \tag{8.2.3}$$

Then, noting that the modal matrix V is nonsingular and invertible by Theorem 8.1, we can premultiply the aforementioned equation by the inverse modal matrix V^{-1} to get

$$V^{-1}AV = V^{-1}V\Lambda \equiv \Lambda \tag{8.2.4}$$

This implies that the modal matrix composed of the eigenvectors of a matrix A is the similarity transformation matrix that can be used for converting the matrix A into a diagonal matrix having its eigenvalues on the main diagonal. Here is an example to illustrate the diagonalization.

Example 8.2. Diagonalization Using the Modal Matrix.
Consider the matrix given in the previous example:

$$A = \begin{bmatrix} 0 & 1 \\ 0 & -1 \end{bmatrix} \qquad (E8.2.1)$$

We can use the eigenvectors (E8.1.3) (obtained in Example 8.1) to construct the modal matrix as

$$V = [\mathbf{v}_1 \ \mathbf{v}_2] = \begin{bmatrix} 1 & 1/\sqrt{2} \\ 0 & -1/\sqrt{2} \end{bmatrix} \qquad (E8.2.2)$$

and use this matrix to make a similarity transformation of the matrix A as

$$\begin{aligned} V^{-1}AV &= \begin{bmatrix} 1 & 1/\sqrt{2} \\ 0 & -1/\sqrt{2} \end{bmatrix}^{-1} \begin{bmatrix} 0 & 1 \\ 0 & -1 \end{bmatrix} \begin{bmatrix} 1 & 1/\sqrt{2} \\ 0 & -1/\sqrt{2} \end{bmatrix} \\ &= \begin{bmatrix} 1 & 1 \\ 0 & -\sqrt{2} \end{bmatrix} \begin{bmatrix} 0 & -1/\sqrt{2} \\ 0 & 1/\sqrt{2} \end{bmatrix} = \begin{bmatrix} 0 & 0 \\ 0 & -1 \end{bmatrix} \qquad (E8.2.3) \end{aligned}$$

which is a diagonal matrix having the eigenvalues on its main diagonal.

This job can be performed by the last statement of the MATLAB script "nm08e01.m".

This diagonalization technique can be used to decouple an N-dimensional vector differential equation so that it can be as easy to solve as N independent scalar differential equations. Here is an illustration.

Example 8.3. Decoupling of a Vector Differential Equation through Diagonalization.

(a) For the linear time-invariant (LTI) state equation (6.5.3)

$$\begin{bmatrix} x'_1(t) \\ x'_2(t) \end{bmatrix} = \begin{bmatrix} 0 & 1 \\ 0 & -1 \end{bmatrix} \begin{bmatrix} x_1(t) \\ x_2(t) \end{bmatrix} + \begin{bmatrix} 0 \\ 1 \end{bmatrix} u_s(t) \qquad (E8.3.1)$$

with $\begin{bmatrix} x_1(0) \\ x_2(0) \end{bmatrix} = \begin{bmatrix} 1 \\ -1 \end{bmatrix}$ and $u_s(t) = 1 \ \forall t \geq 0$;

$\mathbf{x}'(t) = A\mathbf{x}(t) + Bu(t)$ with the initial state $\mathbf{x}(0)$ and the input $u(t)$
we use the modal matrix obtained as (E8.2.2) to make a substitution of variable

$$\mathbf{x}(t) = V\mathbf{w}(t); \quad \begin{bmatrix} x_1(t) \\ x_2(t) \end{bmatrix} = \begin{bmatrix} 1 & 1/\sqrt{2} \\ 0 & -1/\sqrt{2} \end{bmatrix} \begin{bmatrix} w_1(t) \\ w_2(t) \end{bmatrix} \qquad (E8.3.2)$$

472 MATRICES AND EIGENVALUES

which converts the aforementioned Eq. (E8.3.1) into

$$V\mathbf{w}'(t) = AV\mathbf{w}(t) + B u_s(t) \qquad (E8.3.3)$$

We premultiply (E8.3.3) by V^{-1} to write it in a decoupled form as

$$\mathbf{w}'(t) = V^{-1}AV\mathbf{w}(t) + V^{-1}B u_s(t)$$
$$= \Lambda \mathbf{w}(t) + V^{-1}B u_s(t) \text{ with } \mathbf{w}(0) = V^{-1}\mathbf{x}(0);$$

$$\begin{bmatrix} w_1'(t) \\ w_2'(t) \end{bmatrix} = \begin{bmatrix} 0 & 0 \\ 0 & -1 \end{bmatrix} \begin{bmatrix} w_1(t) \\ w_2(t) \end{bmatrix} + \begin{bmatrix} 1 & 1 \\ 0 & -\sqrt{2} \end{bmatrix} \begin{bmatrix} 0 \\ 1 \end{bmatrix} u_s(t) = \begin{bmatrix} u_s(t) \\ -w_2(t) - \sqrt{2}u_s(t) \end{bmatrix}$$
(E8.3.4)

$$\text{with } \begin{bmatrix} w_1(0) \\ w_2(0) \end{bmatrix} = \begin{bmatrix} 1 & 1 \\ 0 & -\sqrt{2} \end{bmatrix} \begin{bmatrix} 1 \\ -1 \end{bmatrix} = \begin{bmatrix} 0 \\ \sqrt{2} \end{bmatrix}$$

where there is no correlation between the variables $w_1(t)$ and $w_2(t)$. Then, we can solve these two equations separately to obtain

$$w_1'(t) = u_s(t) \text{ with } w_1(0) = 0;$$

$$sW_1(s) - w_1(0) = \frac{1}{s}; \quad W_1(s) = \frac{1}{s^2}; \quad w_1(t) = t\, u_s(t) \qquad (E8.3.5a)$$

$$w_2'(t) = -w_2(t) - \sqrt{2} u_s(t) \text{ with } w_2(0) = \sqrt{2};$$

$$sW_2(s) - w_2(0) = -W_2(s) - \frac{\sqrt{2}}{s};$$

$$W_2(s) = \frac{w_2(0)}{s+1} - \frac{\sqrt{2}}{s(s+1)} = -\frac{\sqrt{2}}{s} + \frac{2\sqrt{2}}{s+1}; w_2(t) = \sqrt{2}(-1 + 2e^{-t})\, u_s(t)$$
(E8.3.5b)

and substitute this into Eq. (E8.3.2) to get

$$\begin{bmatrix} x_1(t) \\ x_2(t) \end{bmatrix} \overset{(E8.3.2)}{=} \begin{bmatrix} 1 & 1/\sqrt{2} \\ 0 & -1/\sqrt{2} \end{bmatrix} \begin{bmatrix} w_1(t) \\ w_2(t) \end{bmatrix}$$

$$\overset{(E8.3.5)}{=} \begin{bmatrix} 1 & 1/\sqrt{2} \\ 0 & -1/\sqrt{2} \end{bmatrix} \begin{bmatrix} t \\ \sqrt{2}(-1 + 2e^{-t}) \end{bmatrix} u_s(t) = \begin{bmatrix} t - 1 + 2e^{-t} \\ 1 - 2e^{-t} \end{bmatrix} u_s(t)$$
(E8.3.6)

This is the same result as Eq. (6.5.10) obtained in Section 6.5.1.

(b) Suppose Eq. (E8.3.1) has no input term and so we can expect only the natural response resulting from the initial state, with no forced response caused by the input.

$$\begin{bmatrix} x_1'(t) \\ x_2'(t) \end{bmatrix} = \begin{bmatrix} 0 & 1 \\ 0 & -1 \end{bmatrix} \begin{bmatrix} x_1(t) \\ x_2(t) \end{bmatrix} \text{ with } \begin{bmatrix} x_1(0) \\ x_2(0) \end{bmatrix} = \begin{bmatrix} 1 \\ 1 \end{bmatrix} \qquad (E8.3.7)$$

SIMILARITY TRANSFORMATION AND DIAGONALIZATION 473

We apply the diagonalization/decoupling method for this equation to get

$$\begin{bmatrix} w_1'(t) \\ w_2'(t) \end{bmatrix} = \begin{bmatrix} \lambda_1 & 0 \\ 0 & \lambda_2 \end{bmatrix}\begin{bmatrix} w_1(t) \\ w_2(t) \end{bmatrix} = \begin{bmatrix} 0 & 0 \\ 0 & -1 \end{bmatrix}\begin{bmatrix} w_1(t) \\ w_2(t) \end{bmatrix}$$

with $\mathbf{w}(0) = V^{-1}\mathbf{x}(0)$; $\begin{bmatrix} w_1(0) \\ w_2(0) \end{bmatrix} = \begin{bmatrix} 1 & 1 \\ 0 & -\sqrt{2} \end{bmatrix}\begin{bmatrix} 1 \\ 1 \end{bmatrix} = \begin{bmatrix} 2 \\ -\sqrt{2} \end{bmatrix}$;

$$\begin{bmatrix} w_1(t) \\ w_2(t) \end{bmatrix} = \begin{bmatrix} w_1(0)e^{\lambda_1 t} \\ w_2(0)e^{\lambda_2 t} \end{bmatrix} = \begin{bmatrix} 2 \\ -\sqrt{2}e^{-t} \end{bmatrix}; \qquad (E8.3.8)$$

$$\mathbf{x}(t) \stackrel{(E8.3.2)}{=} V\mathbf{w}(t) = [\mathbf{v}_1 \ \mathbf{v}_2]\begin{bmatrix} w_1(0)e^{\lambda_1 t} \\ w_2(0)e^{\lambda_2 t} \end{bmatrix} = w_1(0)e^{\lambda_1 t}\mathbf{v}_1 + w_2(0)e^{\lambda_2 t}\mathbf{v}_2$$

$$= \begin{bmatrix} 1 & 1/\sqrt{2} \\ 0 & -1/\sqrt{2} \end{bmatrix}\begin{bmatrix} 2 \\ -\sqrt{2}e^{-t} \end{bmatrix} = \begin{bmatrix} 2 - e^{-t} \\ e^{-t} \end{bmatrix} \qquad (E8.3.9)$$

As time goes by, this solution converges and so the continuous-time system turns out to be stable, thanks to the fact that all the eigenvalues $(0, -1)$ are distinct and not positive.

Example 8.4. Decoupling of a Vector Difference Equation Through Diagonalization. Consider a discrete-time LTI state equation

$$\begin{bmatrix} x_1[n+1] \\ x_2[n+1] \end{bmatrix} = \begin{bmatrix} 0 & 1 \\ 0.2 & 0.1 \end{bmatrix}\begin{bmatrix} x_1[n] \\ x_2[n] \end{bmatrix} + \begin{bmatrix} 0 \\ 2.2361 \end{bmatrix}u_s[n] \qquad (E8.4.1)$$

with $\begin{bmatrix} x_1[0] \\ x_2[0] \end{bmatrix} = \begin{bmatrix} 1 \\ -1 \end{bmatrix}$ and $u_s[n] = 1 \ \forall n \geq 0$

To diagonalize this equation into a form similar to Eq. (E8.3.4), we use MATLAB to find the eigenvalues/eigenvectors and the modal matrix composed of the eigenvectors, and finally, we do the similarity transformation. Then, we get

$$L = \begin{bmatrix} \lambda_1 & 0 \\ 0 & \lambda_2 \end{bmatrix} = \begin{bmatrix} -0.4 & 0 \\ 0 & 0.5 \end{bmatrix}, \quad V = [\mathbf{v}_1 \ \mathbf{v}_2] = \begin{bmatrix} -0.9285 & -0.8944 \\ 0.3714 & -0.4472 \end{bmatrix} \qquad (E8.4.2)$$

$$A_p = V^{-1}AV = \begin{bmatrix} -0.4 & 0 \\ 0 & 0.5 \end{bmatrix} \text{ and } B_p = V^{-1}B = \begin{bmatrix} 2.6759 \\ -2.7778 \end{bmatrix} \qquad (E8.4.3)$$

so that we can write the diagonalized state equation as

$$\begin{bmatrix} w_1[n+1] \\ w_2[n+1] \end{bmatrix} = \begin{bmatrix} -0.4 & 0 \\ 0 & 0.5 \end{bmatrix}\begin{bmatrix} w_1[n] \\ w_2[n] \end{bmatrix} + \begin{bmatrix} 2.6759 \\ -2.7778 \end{bmatrix}u_s[n]$$

$$= \begin{bmatrix} -0.4w_1[n] + 2.6759 \\ 0.5[n] - 2.7778 \end{bmatrix} \qquad (E8.4.4)$$

```
A=[0 1;0.2 0.1]; B=[0; 2.2361]; % Eq.(E8.4.1)
[V,L]= eig(A) % V= modal matrix composed of eigenvectors (E8.4.2)
   % L= diagonal matrix with eigenvalues on its diagonal
Ap=V^-1*A*V %diagonalize through similarity transformation (E8.4.3)
   % into a diagonal matrix having the eigenvalues on the diagonal
Bp=V^-1*B % Eq.(E8.4.3)
```

Without the input term on the RHS of Eq. (E8.4.1), we would have obtained

$$\begin{bmatrix} w_1[n+1] \\ w_2[n+1] \end{bmatrix} = \begin{bmatrix} \lambda_1 & 0 \\ 0 & \lambda_2 \end{bmatrix} \begin{bmatrix} w_1[n] \\ w_2[n] \end{bmatrix} = \begin{bmatrix} \lambda_1^{n+1} w_1[0] \\ \lambda_2^{n+1} w_2[0] \end{bmatrix} \text{ with } \mathbf{w}[0] = V^{-1}\mathbf{x}[0] \quad (E8.4.5)$$

$$\mathbf{x}[n] = V\mathbf{w}[n] = [\mathbf{v}_1 \ \mathbf{v}_2] \begin{bmatrix} w_1[0] \lambda_1^n \\ w_2[0] \lambda_2^n \end{bmatrix} = w_1[0] \lambda_1^n \mathbf{v}_1 + w_2[0] \lambda_2^n \mathbf{v}_2 \quad (E8.4.6)$$

As time goes by, i.e. as n increases, this solution converges and so the discrete-time system turns out to be stable, thanks to the fact that the magnitude of every eigenvalue $(-0.4, 0.5)$ is less than 1.

Remark 8.2 Physical Meaning of Eigenvalues and Eigenvectors.

(1) As illustrated by the aforementioned examples, we can use the modal matrix to decouple a set of differential equations so that they can be solved one by one as a scalar differential equation in terms of a single variable and then put together to make the solution for the original vector differential equation.

(2) Through the aforementioned examples, we can feel the physical significance of the eigenvalues/eigenvectors of the system matrix A in the state equation on its solution. That is, the state of a LTI system described by an N-dimensional continuous-time (differential) state equation has N modes $\{e^{\lambda_i t}; i = 1, \ldots, N\}$, each of which converges/diverges if the sign of the corresponding eigenvalue is negative/positive and proceeds slowly as the magnitude of the eigenvalue is close to zero. In the case of a discrete-time LTI system described by an N-dimensional difference state equation, its state has N modes $\{\lambda_i^n; i = 1, \ldots, N\}$, each of which converges/diverges if the magnitude of the corresponding eigenvalue is less/greater than 1 and proceeds slowly as the magnitude of the eigenvalue is close to one. To summarize, the convergence property of a state $\mathbf{x}$ or the stability of a LTI system is determined by the eigenvalues of the system matrix A. As illustrated by (E8.3.9) and (E8.4.6), the corresponding eigenvector determines the direction in which each mode proceeds in the N-dimensional state space.

8.3 POWER METHOD

In this section, we will introduce the scaled power method, the inverse power method, and the shifted inverse power method to find the eigenvalues of a given matrix.

8.3.1 Scaled Power Method

This method is used to find the eigenvalue of largest magnitude and is summarized in the following box.

SCALED POWER METHOD

Suppose all the eigenvalues of an $N \times N$ matrix A are distinct with the magnitudes

$$|\lambda_1| > |\lambda_2| \geq |\lambda_3| \geq \cdots \geq |\lambda_N|$$

Then, the dominant eigenvalue λ_1 with the largest magnitude and its corresponding eigenvector $\mathbf{v}_1$ can be obtained by starting with an initial vector $\mathbf{x}_0$ that has some nonzero component in the direction of $\mathbf{v}_1$ and by repeating the following procedure:

Divide the previous vector $\mathbf{x}_k$ by its largest component (in absolute value) for normalization (scaling) and premultiply the normalized vector by the matrix A.

$$\mathbf{x}_{k+1} = A \frac{\mathbf{x}_k}{\|\mathbf{x}_k\|_\infty} \to \lambda_1 \mathbf{v}_1 \quad \text{with} \quad \|\mathbf{x}\|_\infty = \text{Max}\{|x_n|\} \quad (8.3.1)$$

Proof. According to Theorem 8.1, the eigenvectors $\{\mathbf{v}_n, n = 1{:}N\}$ of an $N \times N$ matrix A whose eigenvalues are distinct are independent and so can constitute a basis for an N-dimensional linear space. Consequently, any initial vector $\mathbf{x}_0$ can be expressed as a linear combination of the eigenvectors:

$$\mathbf{x}_0 = \alpha_1 \mathbf{v}_1 + \alpha_2 \mathbf{v}_2 + \cdots + \alpha_N \mathbf{v}_N \quad (8.3.2)$$

Noting that $A\mathbf{v}_n = \lambda_n \mathbf{v}_n$, we premultiply both sides of this equation by A to get

$$A\mathbf{x}_0 = \alpha_1 \lambda_1 \mathbf{v}_1 + \alpha_2 \lambda_2 \mathbf{v}_2 + \cdots + \alpha_N \lambda_N \mathbf{v}_N$$

$$= \lambda_1 \left(\alpha_1 \mathbf{v}_1 + \alpha_2 \frac{\lambda_2}{\lambda_1} \mathbf{v}_2 + \cdots + \alpha_N \frac{\lambda_N}{\lambda_1} \mathbf{v}_N \right)$$

and repeat this multiplication over and over again to obtain

$$\mathbf{x}_k = A^k \mathbf{x}_0 = \lambda_1^k \left\{ \alpha_1 \mathbf{v}_1 + \alpha_2 \left(\frac{\lambda_2}{\lambda_1}\right)^k \mathbf{v}_2 + \cdots + \alpha_N \left(\frac{\lambda_N}{\lambda_1}\right)^k \mathbf{v}_N \right\} \rightarrow \lambda_1^k \alpha_1 \mathbf{v}_1 \quad (8.3.3)$$

which will converge to an eigenvector $\mathbf{v}_1$ as long as $\alpha_1 \neq 0$. Since we keep scaling before multiplying at every iteration, the largest component of the limit vector of the sequence generated by Eq. (8.3.1) must be λ_1.

$$\mathbf{x}_{k+1} = A \frac{\mathbf{x}_k}{\|\mathbf{x}_k\|_\infty} \rightarrow A \frac{\mathbf{v}_1}{\|\mathbf{v}_1\|_\infty} \stackrel{(8.1.1)}{=} \lambda_1 \frac{\mathbf{v}_1}{\|\mathbf{v}_1\|_\infty} \quad (8.3.4)$$

Note that the scaling prevents the overflow or underflow that would result from $|\lambda_1| > 1$ or $|\lambda_1| < 1$. ∎

Remark 8.3 Convergence of Power Method.

(1) In the light of Eq. (8.3.3), the convergence speed of the power method depends on how small the magnitude ratio $(|\lambda_2|/|\lambda_1|)$ of the second largest eigenvalue λ_2 over the largest eigenvalue λ_1 is.

(2) We often use $\mathbf{x}_0 = \begin{bmatrix} 1 & 1 & \cdots & 1 \end{bmatrix}$ as the initial vector. Note that if it has no component in the direction of the eigenvector ($\mathbf{v}_1$) corresponding to the dominant eigenvalue λ_1, i.e. $\alpha_1 = \mathbf{x}_0 \cdot \mathbf{v}_1 / \|\mathbf{v}_1\| = 0$ in Eq. (8.3.2), the iteration of the scaled power method leads to the limit showing the second largest magnitude eigenvalue λ_2 and its corresponding eigenvector $\mathbf{v}_1$. But, if there is more than one largest (dominant) eigenvalue of equal magnitude, it does not converge to either of them.

8.3.2 Inverse Power Method

The objective of this method is to find the (uniquely) smallest (magnitude) eigenvalue λ_N by applying the scaled power method to the inverse matrix A^{-1} and taking the inverse of the component of the limit. It works only in cases where the matrix A is nonsingular and so has no zero eigenvalue. Its idea is based on the equation

$$A\mathbf{v} = \lambda \mathbf{v} \rightarrow A^{-1}\mathbf{v} = \lambda^{-1}\mathbf{v} \quad (8.3.5)$$

obtained from multiplying both sides of Eq. (8.1.1) by $\lambda^{-1}A^{-1}$. This implies that the inverse matrix A^{-1} has the eigenvalues that are the reciprocals of the eigenvalues of the original matrix A, still having the same eigenvectors.

$$\lambda_N = \frac{1}{\text{the largest eigenvalue of } A^{-1}} \quad (8.3.6)$$

8.3.3 Shifted Inverse Power Method

In order to develop a method for finding the eigenvalue that is not necessarily of the largest or smallest magnitude, we subtract $s\mathbf{v}$ (s: a number that does not happen to equal any eigenvalue) from both sides of Eq. (8.1.1) to write

$$A\mathbf{v} = \lambda\mathbf{v} \rightarrow [A - sI]\mathbf{v} = (\lambda - s)\mathbf{v} \qquad (8.3.7)$$

Since this implies that $(\lambda - s)$ is the eigenvalue of $[A - sI]$, we apply the inverse power method for $[A - sI]$ to get its smallest magnitude eigenvalue $(\lambda_k - s)$ with min $\{|\lambda_i - s|, i = 1:N\}$ and add s to it to obtain the eigenvalue of the original matrix A which is closest to the number s.

$$\lambda_s = \frac{1}{\text{the largest eigenvalue of } [A - sI]^{-1}} + s \qquad (8.3.8)$$

Theorem 8.2 Gerschgorin's Disk Theorem[B-1].
Every eigenvalue of a square matrix A belongs to at least one of the disks (in the complex plane) with center a_{mm} (one of the diagonal entries of A) and radius

$r_m = \sum_{n \neq m} a_{mn}$ (the sum of all the entries in the row except the diagonal entry)

Moreover, each of the disks contains at least one eigenvalue of the matrix A.

The prospect of this method is supported by Gerschgorin's disk theorem, which is summarized in the aforementioned box. But, this method is not applicable to the matrix that has more than one eigenvalue of the same magnitude.

```
function [lambda,v,k]=eig_power(A,x,EPS,MaxIter)
% The power method to find the largest eigenvalue (lambda) and
%   the corresponding eigenvector (v) of a matrix A.
if nargin<4, MaxIter=100; end % Maximum number of iterations
if nargin<3, EPS=1e-8; end % Threshold on the difference
N=size(A,2);
if nargin<2, x=[1:N]; end % Initial vector
x=x(:);  xm=max(abs(x));
for k=1:MaxIter
   x0=x;  x0m=xm;
   x = A*x/xm; % Eq.(8.3.1)
   [xm,m]=max(abs(x)); % norm(x,inf);
   if norm(abs(x)-abs(x0))<EPS&abs(xm-x0m)<EPS, break; end
end
v=x/norm(x);
lambda=v'*A*v/(v'*v); % Rayleigh quotient - largest eigenvalue
if k==MaxIter, disp('Warning: you may have to increase MaxIter'); end
```

```
%nm08s03.m
% Use the power method to find the largest/smallest/medium eigenvalue
A=[2 0 1; 0 -2 0; 1 0 2];
x=[1 2 3]'; %x= [1 1 1]'; % with different initial vector
EPS=1e-8; MaxIter=100;
% the largest eigenvalue and its corresponding eigenvector
[lambda_max,v]=eig_power(A,x,EPS,MaxIter)
% the smallest eigenvalue and its corresponding eigenvector
[lambda,v]=eig_power(A^-1,x,EPS,MaxIter);
lambda_min=1/lambda,  v  % Eq.(8.3.6)
% Eigenvalue nearest to a number -3 and its corresponding eigenvector
s=-3; AsI=(A-s*eye(size(A)))^-1; % Eq.(8.3.7)
[lambda,v]=eig_power(AsI,x,EPS,MaxIter);
lambda=1/lambda+s  % Eq.(8.3.8)
fprintf('Eigenvalue closest to %4.2f=%8.4f\nwith eigenvector',s,lambda)
v
[V,LAMBDA]=eig(A) %modal matrix composed of eigenvectors
```

The power method introduced in Section 8.3.1 has been cast into the function 'eig_power()' listed previously. We can run the above MATLAB script "nm08s03.m" to perform the power method, the inverse power method, and the shifted inverse power method for finding the eigenvalues and the corresponding eigenvectors of a matrix, {the largest one 3, the smallest (magnitude) one 1, and the one (−2) nearest to −3}, and compare the results with that of the MATLAB built-in function 'eig()' for cross-check.

```
»nm08s03
  lambda_max =     3.0000
  lambda_min =     1.0000
  lambda     =    -2.0000
```

8.4 JACOBI METHOD

This method finds us all the eigenvalues of a real symmetric matrix. Its idea is based on the following theorem.

Theorem 8.3 Symmetric Diagonalization Theorem.
All of the eigenvalues of an $N \times N$ symmetric matrix A are of real value, and its eigenvectors form an orthonormal basis of an N-dimensional linear space. Consequently, we can make an orthonormal modal matrix V composed of the eigenvectors such that $V^T V = I$; $V^{-1} = V^T$ and use the modal matrix to make the similarity transformation of A, which yields a diagonal matrix having the eigenvalues on its main diagonal:

$$V^T A V = V^{-1} A V = \Lambda \qquad (8.4.1)$$

Now, in order to understand the Jacobi method, we define the *pq*-rotation matrix as

$$R_{pq}(\theta) = \begin{bmatrix} 1 & 0 & \cdot & 0 & \cdot & 0 & \cdot & 0 \\ 0 & 1 & \cdot & 0 & \cdot & 0 & \cdot & 0 \\ \cdot & \cdot & \cdot & & & & & \\ 0 & 0 & \cdot & \cos\theta & \cdot & -\sin\theta & \cdot & 0 \\ \cdot & \cdot & & \cdot & & \cdot & & \cdot \\ 0 & 0 & \cdot & \sin\theta & \cdot & \cos\theta & \cdot & 0 \\ \cdot & \cdot & & & & & & \cdot \\ 0 & 0 & \cdot & 0 & \cdot & 0 & \cdot & 1 \end{bmatrix} \begin{matrix} \\ \\ \\ p\text{th row} \\ \\ q\text{th row} \\ \\ \end{matrix} \quad (8.4.2)$$

with the pth column and qth column indicated.

Since this is an orthonormal matrix whose row/column vectors are orthogonal and normalized

$$R_{pq}^T R_{pq} = I; \quad R_{pq}^T = R_{pq}^{-1} \quad (8.4.3)$$

premultiplying/postmultiplying a matrix A by R_{pq}^T / R_{pq} makes a similarity transformation

$$A_{(1)} = R_{pq}^T A R_{pq} \quad (8.4.4)$$

Noting that the similarity transformation does not change the eigenvalues (Remark 8.1), any matrix resulting from repeating the same operations successively

$$A_{(k+1)} = R_{(k)}^T A_{(k)} R_{(k)} = R_{(k)}^T R_{(k-1)}^T \cdots R^T A R \cdots R_{(k-1)} R_{(k)} \quad (8.4.5)$$

has the same eigenvalues. Moreover, if it is a diagonal matrix, it will have all the eigenvalues on its main diagonal, and the matrix multiplied on the right of the matrix A is the modal matrix V

$$V = R \cdots R_{(k-1)} R_{(k)} \quad (8.4.6)$$

as manifested by matching this equation with Eq. (8.4.1).

What is left for us to think about is how to make this matrix (8.4.5) diagonal. Noting that the similarity transformation (8.4.4) changes only the pth rows/columns and the qth rows/columns as

$$v_{pq} = v_{qp} = a_{qp}(c^2 - s^2) + (a_{qq} - a_{pp})sc = a_{qp}\cos 2\theta + \frac{1}{2}(a_{qq} - a_{pp})\sin 2\theta \quad (8.4.7a)$$

$$v_{pn} = v_{np} = a_{pn}c + a_{qn}s \quad \text{for the } p\text{th row/column with } n \neq p, q \quad (8.4.7b)$$

$$v_{qn} = v_{nq} = -a_{pn}s + a_{qn}c \quad \text{for the } q\text{th row/column with } n \neq p, q \quad (8.4.7c)$$

$$v_{pp} = a_{pp}c^2 + a_{qq}s^2 + 2a_{pq}sc = a_{pp}c^2 + a_{qq}s^2 + a_{pq}\sin 2\theta \quad (8.4.7d)$$

$$v_{qq} = a_{pp}s^2 + a_{qq}c^2 - 2a_{pq}sc = a_{pp}s^2 + a_{qq}c^2 - a_{pq}\sin 2\theta \quad (8.4.7e)$$

$$(c = \cos\theta, s = \sin\theta)$$

```
function [LAMBDA,V,ermsg]=eig_Jacobi(A,EPS,MaxIter)
%Jacobi method finds the eigenvalues/eigenvectors of symmetric matrix A
if nargin<3, MaxIter=100; end
if nargin<2, EPS=1e-8; end
N=size(A,2);
LAMBDA=[]; V=[];
for m=1:N
   if norm(A(m:N,m)-A(m,m:N)')>EPS
      error('asymmetric matrix!');
   end
end
V=eye(N); sq2r=1/sqrt(2);
for k=1:MaxIter
   for m=1:N-1
      [Am(m),Q(m)]= max(abs(A(m,m+1:N)));
   end
   [Amm,p]= max(Am);  q=p+Q(p);
   if Amm<EPS*sum(abs(diag(LAMBDA)))   break;   end
   if abs(A(p,p)-A(q,q))<EPS
      s2=1; s=sq2r; c=s;
      cc=c*c; ss=s*s;
    else
      t2= 2*A(p,q)/(A(p,p)-A(q,q));   %Eq.(8.4.9a)
      c2= 1/sqrt(1+t2*t2); s2=t2*c2;  %Eq.(8.4.9b,c)
      c= sqrt((1+c2)/2); s=s2/2/c;    %Eq.(8.4.9d,e)
      cc=c*c; ss=s*s;
   end
   LAMBDA=A;
   LAMBDA(p,:)= A(p,:)*c +A(q,:)*s;    %Eq.(8.4.7b)
   LAMBDA(:,p)= LAMBDA(p,:)';
   LAMBDA(q,:)=-A(p,:)*s +A(q,:)*c;    %Eq.(8.4.7c)
   LAMBDA(:,q)= LAMBDA(q,:)';
   LAMBDA(p,q)= 0;   LAMBDA(q,p)= 0;   %Eq.(8.4.7a)
   LAMBDA(p,p)= A(p,p)*cc +A(q,q)*ss +A(p,q)*s2; %Eq.(8.4.7d)
   LAMBDA(q,q)= A(p,p)*ss +A(q,q)*cc -A(p,q)*s2; %Eq.(8.4.7e)
   A=LAMBDA;
   V(:,[p q])= V(:,[p q])*[c -s;s c];
end
LAMBDA= diag(diag(LAMBDA)); %for purification
```

```
%nm08s04.m
% applies the Jacobi method
% to find all the eigenvalues/eigenvectors of a symmetric matrix A.
A= [2 0 1;0 -2 0;1 0 2];
EPS=1e-8;  MaxIter=100;
[L,V]= eig_Jacobi(A,EPS,MaxIter)
disp('Using eig()')
[V,LAMBDA]= eig(A) %modal matrix composed of eigenvectors
```

we make the (p, q) entry v_{pq} and the (q, p) entry v_{qp} zero

$$v_{pq} = v_{qp} = 0 \qquad (8.4.8)$$

by choosing the angle θ of the rotation matrix $R_{pq}(\theta)$ in such a way that

$$\tan 2\theta = \frac{\sin 2\theta}{\cos 2\theta} = \frac{2a_{pq}}{a_{pp} - a_{qq}}$$

$$\cos 2\theta = \frac{1}{\sec 2\theta} = \frac{1}{\sqrt{1 + \tan^2 2\theta}}, \quad \sin 2\theta = \tan 2\theta \cos 2\theta \qquad (8.4.9)$$

$$\cos \theta = \sqrt{\cos^2 \theta} = \sqrt{(1 + \cos 2\theta)/2}, \quad \sin \theta = \frac{\sin 2\theta}{2\cos \theta}$$

and computing the other associated entries according to Eqs. (8.4.7b)–(8.4.7e).

There are a couple of things to note. First, in order to make the matrix closer to a diagonal one at each iteration, we should identify the row number and the column number of the largest off-diagonal entry as p and q, respectively, and zero-out the (p, q) entry. Second, we can hope that the magnitudes of the other entries in the pth, qth row/column affected by this transformation process do not get larger, since Eqs. (8.4.7b) and (8.4.7c) implies

$$v_{pn}^2 + v_{qn}^2 = (a_{pn}c + a_{qn}s)^2 + (-a_{pn}s + a_{qn}c)^2 = a_{pn}^2 + a_{qn}^2 \qquad (8.4.10)$$

This so-called Jacobi method has been cast into the function 'eig_Jacobi()'. The MATLAB script "nm0841.m" uses it to find the eigenvalues/eigenvectors of a matrix and compares the result with that of using the MATLAB built-in function 'eig()' for cross-check. The result we may expect is as follows. Interested readers are welcomed to run the script "nm08s04.m".

$$A = \begin{bmatrix} 2 & 0 & 1 \\ 0 & -2 & 0 \\ 1 & 0 & 2 \end{bmatrix} \rightarrow R_{13}^T A R_{13} = \begin{bmatrix} 3 & 0 & 0 \\ 0 & -2 & 0 \\ 0 & 0 & 1 \end{bmatrix} = \Lambda$$

$$\text{with } R_{13} = \begin{bmatrix} 1/\sqrt{2} & 0 & -1/\sqrt{2} \\ 0 & 1 & 0 \\ 1/\sqrt{2} & 0 & 1/\sqrt{2} \end{bmatrix} = V$$

8.5 GRAM-SCHMIDT ORTHONORMALIZATION AND QR DECOMPOSITION

Given a linearly independent set of vectors $\{v_n, n = 1:N\}$, the Gram-Schmidt process generates a set of orthonormal vectors $\{q_n, n = 1:N\}$ that spans the linear space formed by v_n's. Let us define the projection operator of a vector v onto another vector q by

$$\Pr_q\{v\} = \frac{\langle v, q \rangle}{\langle q, q \rangle} q = \frac{v^H q}{q^H q} q \qquad (8.5.1)$$

where the superscript 'H' denotes the Hermitian or conjugate transpose of a matrix or a vector. The (classical) Gram-Schmidt process for a set of linearly independent $M \geq N$-dimensional vectors $\{\mathbf{v}_1, \mathbf{v}_2, \ldots, \mathbf{v}_N\}$ works to make a basis consisting of N orthonormal vectors for the column (vector) space of the matrix $V = [\mathbf{v}_1 \ \mathbf{v}_2 \ \ldots \ \mathbf{v}_N]$ as follows:

$$\mathbf{q}_1 = \mathbf{v}_1/\|\mathbf{v}_1\|$$
$$\mathbf{q}_2 = \mathbf{v}_2 - \mathrm{Pr}_{\mathbf{q}_1}\{\mathbf{v}_2\} \to \mathbf{q}_2 = \mathbf{q}_2/\|\mathbf{q}_2\|$$
$$\mathbf{q}_3 = \mathbf{v}_3 - \mathrm{Pr}_{\mathbf{q}_1}\{\mathbf{v}_3\} - \mathrm{Pr}_{\mathbf{q}_2}\{\mathbf{v}_3\} \to \mathbf{q}_3 = \mathbf{q}_3/\|\mathbf{q}_3\|$$
$$\vdots$$

---

for $n = 1:N$

$$\mathbf{q}_n = \mathbf{v}_n - \sum_{k=1}^{n-1} \mathrm{Pr}_{\mathbf{q}_k}\{\mathbf{v}_n\} \to \mathbf{q}_n = \mathbf{q}_n/\|\mathbf{q}_n\| \quad (8.5.2)$$

end

$$Q = \begin{bmatrix} \mathbf{q}_1 & \mathbf{q}_2 & \cdots & \mathbf{q}_N \end{bmatrix} : \text{a set of orthonormal vectors}$$

This Gram-Schmidt process can be cast into the following function '[Q,R] = Gram_Schmidt(A)', which takes an $M \times N$ matrix A consisting of linearly independent $M \geq N$-dimensional vectors and generates a unitary $M \times N$ matrix Q consisting of N orthonormal column vectors (such that $Q^H Q = I$) together with an upper triangular matrix R such that $QR = A$. Thus, the Gram-Schmidt process can be used to find the QR decomposition.

The modified Gram-Schmidt process which is numerically stabilized toward smaller error in finite-precision arithmetic is as follows:

```
function [Q,R]=Gram_Schmidt(A)
% Gram-Schmidt ortho-normalization of the columns of MxN matrix A
% on the assumption that the columns of A are linearly independent.
% Q = MxN matrix whose columns form an ortho-normal basis
%     for the column space of A.
% R = an invertible upper triangular matrix R so that A = Q*R.
% Warning: For a more stable algorithm, use [Q,R]=qr(A,0).
[M,N]=size(A);
v1=A(:,1); R(1,1)=norm(v1); Q(:,1)=v1/R(1,1);
for n=2:min(M,N)
   %qn = vn - Q*((Q'*Q)\Q'*vn);
   % where Q*(Q'*Q)^-1*Q' is the orthonormal projection operator
   %   onto the column space of Q
   vn=A(:,n);
   R(1:n-1,n)=Q(:,1:n-1)'*vn;
   qn = vn-Q(:,1:n-1)*R(:,n); % Eq.(8.5.2) with Projection Q*(Q'*vn)
   R(n,n)=norm(qn);  Q(:,n)=qn/R(n,n);
end
```

GRAM-SCHMIDT ORTHONORMALIZATION AND QR DECOMPOSITION   483

for $n = 1 : N$
$$\mathbf{q}_n^{(0)} = \mathbf{v}_n$$
for $k = 1 : n-1$
$$\mathbf{q}_n^{(k)} = \mathbf{q}_n^{(k-1)} - \text{Pr}_{\mathbf{q}_k}\{\mathbf{q}_n^{(k-1)}\} \quad (8.5.3)$$
end
$$\mathbf{q}_n = \mathbf{q}_n^{(n-1)}/\|\mathbf{q}_n^{(n-1)}\|$$
end
$$Q = [\mathbf{q}_1 \ \mathbf{q}_2 \ \cdots \ \mathbf{q}_N]$$

This process can be cast into the following function '[Q,R]=Gram_Schmidt_mod(A)', which does the same job as 'Gram_Schmidt()', but with less numerical error.

```
function [Q,R]=Gram_Schmidt_mod(A)
% Modified Gram-Schmidt algorithm and QR Decomposition
[M,N]=size(A); N=min(M,N);
for n=1:N
  qn=A(:,n);
  for k=1:n-1
    R(k,n)=Q(:,k)'*qn;
    qn = qn - Q(:,k)*R(k,n); % Eq. (8.5.3)
  end
  R(n,n)=norm(qn); Q(:,n)=qn/R(n,n);
end
```

You can find some other QR decomposition methods in Problems 8.4 and 8.6.

Example 8.5. Gram-Schmidt Orthogonalization.
Use the Gram-Schmidt process to find an orthonormal basis for a vector space R^3 spanned by

$$\mathbf{v}_1 = \begin{bmatrix} 1 \\ 1 \\ 1 \end{bmatrix}, \quad \mathbf{v}_2 = \begin{bmatrix} 1 \\ 2 \\ 2 \end{bmatrix}, \quad \text{and} \quad \mathbf{v}_3 = \begin{bmatrix} 1 \\ 1 \\ 0 \end{bmatrix} \quad (E8.5.1)$$

Following the steps of Eq. (8.5.2), we get

$$\mathbf{q}_1 = \frac{1}{\|\mathbf{v}_1\|}\mathbf{v}_1 = \frac{1}{\sqrt{3}}\begin{bmatrix} 1 \\ 1 \\ 1 \end{bmatrix} \quad (E8.5.2)$$

$$\mathbf{q}_2 = \mathbf{v}_2 - \text{proj}_{\mathbf{q}_1}\{\mathbf{v}_2\} \stackrel{(8.5.2)}{=} \mathbf{v}_2 - (\mathbf{v}_2 \cdot \mathbf{q}_1)\frac{\mathbf{q}_1}{\|\mathbf{q}_1\|^2}$$

$$= \begin{bmatrix} 1 \\ 2 \\ 2 \end{bmatrix} - \left(\begin{bmatrix} 1 \\ 2 \\ 2 \end{bmatrix} \cdot \begin{bmatrix} 1 \\ 1 \\ 1 \end{bmatrix}\right)\frac{1}{3}\begin{bmatrix} 1 \\ 1 \\ 1 \end{bmatrix} = \begin{bmatrix} -2/3 \\ 1/3 \\ 1/3 \end{bmatrix}$$

$$\xrightarrow{\text{normalize}} \mathbf{q}_2 = \frac{\mathbf{q}_2}{\|\mathbf{q}_2\|} = \frac{1}{\sqrt{6}/3}\begin{bmatrix} -2/3 \\ 1/3 \\ 1/3 \end{bmatrix} = \frac{1}{\sqrt{6}}\begin{bmatrix} -2 \\ 1 \\ 1 \end{bmatrix} \quad (E8.5.3)$$

$$\mathbf{q}_3 = \mathbf{v}_3 - \text{proj}_{\mathbf{q}_1}\{\mathbf{v}_3\} - \text{proj}_{\mathbf{q}_2}\{\mathbf{v}_3\}$$

$$\stackrel{(8.5.2)}{=} \mathbf{v}_3 - (\mathbf{v}_3 \cdot \mathbf{q}_1)\frac{\mathbf{q}_1}{\|\mathbf{q}_1\|^2} - (\mathbf{v}_3 \cdot \mathbf{q}_2)\frac{\mathbf{q}_2}{\|\mathbf{q}_2\|^2}$$

$$= \begin{bmatrix} 1 \\ 1 \\ 0 \end{bmatrix} - \left(\begin{bmatrix} 1 \\ 1 \\ 0 \end{bmatrix} \cdot \begin{bmatrix} 1 \\ 1 \\ 1 \end{bmatrix}\right)\frac{1}{3}\begin{bmatrix} 1 \\ 1 \\ 1 \end{bmatrix} - \left(\begin{bmatrix} 1 \\ 1 \\ 0 \end{bmatrix} \cdot \begin{bmatrix} -2 \\ 1 \\ 1 \end{bmatrix}\right)\frac{1}{6}\begin{bmatrix} -2 \\ 1 \\ 1 \end{bmatrix}$$

$$= \begin{bmatrix} 1 \\ 1 \\ 0 \end{bmatrix} - \begin{bmatrix} 2/3 \\ 2/3 \\ 2/3 \end{bmatrix} + \begin{bmatrix} -2/6 \\ 1/6 \\ 1/6 \end{bmatrix} = \begin{bmatrix} 0 \\ 1/2 \\ -1/2 \end{bmatrix}$$

$$\xrightarrow{\text{normalize}} \mathbf{q}_3 = \frac{\mathbf{q}_3}{\|\mathbf{q}_3\|} = \frac{1}{\sqrt{2}/2}\begin{bmatrix} 0 \\ 1/2 \\ -1/2 \end{bmatrix} = \frac{1}{\sqrt{2}}\begin{bmatrix} 0 \\ 1 \\ -1 \end{bmatrix} \quad (E8.5.4)$$

These computations can be done by typing the following statements at the MATLAB prompt:

```
»A=[1 1 1; 1 2 1; 1 2 0];   [Q,R]=Gram_Schmidt(A);   Q
```

This will yield

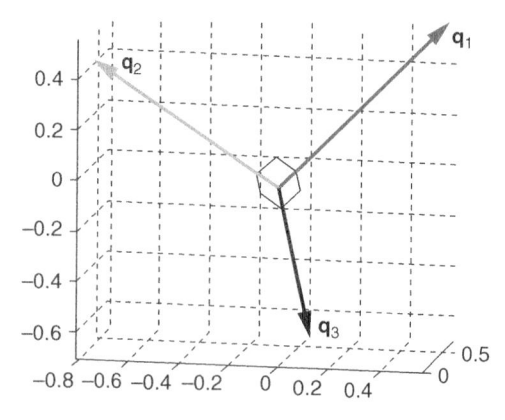

```
Q =    0.5774   -0.8165   -0.0000
       0.5774    0.4082    0.7071
       0.5774    0.4082   -0.7071    % Eqs.(E8.5.2),(E8.5.3),(E8.5.4)
R =    1.7321    2.8868    1.1547
            0    0.8165   -0.4082
            0         0    0.7071
```

Figure 8.1 Three basis vectors obtained using the Gram-Schmidt orthogonalization in Example 8.5.

The three column vectors constituting Q are a set of orthonormal basis vectors for the 3D space R^3 spanned by the three vectors $\mathbf{v}_1$, $\mathbf{v}_2$, and $\mathbf{v}_3$, as depicted in Figure 8.1. If Figure 8.1 does not make you sure of that the column vector $\mathbf{q}_n$'s of $Q = [\mathbf{q}_1\ \mathbf{q}_2\ \mathbf{q}_3]$ are mutually orthogonal, how about checking if the orthonormality condition (B.7.1) holds or aligning the viewing angle of the graph with $\mathbf{q}_3$ by running the following MATLAB statements?

```
»Q'*Q % Is Q orthonormal by satisfying Eq.(B.7.1)?
»hold on, grid on;
S=0; % No scaling
quiver3(0,0,0,Q(1,1),Q(2,1),Q(3,1),S,'Color','r','Linewidth',2);
quiver3(0,0,0,Q(1,2),Q(2,2),Q(3,2),S,'Color','g','Linewidth',2);
quiver3(0,0,0,Q(1,3),Q(2,3),Q(3,3),S,'Color','b','Linewidth',2);
axis('equal'), xlabel('x'), ylabel('y'), zlabel('z')
shg; pause
% To align the viewing angle with q3
view(Q(:,3)) % Now, what do you see?
```

8.6 PHYSICAL MEANING OF EIGENVALUES/EIGENVECTORS

According to Theorem 8.3 (Symmetric Diagonalization Theorem) introduced in the previous section, the eigenvectors $\{\mathbf{v}_n, n = 1:N\}$ of an $N \times N$ symmetric matrix A constitute an orthonormal basis for an N-dimensional linear space.

$$V^T V = I; \quad \mathbf{v}_m^T \mathbf{v}_n = \delta_{mn} = \begin{cases} 1 & \text{for } m = n \\ 0 & \text{for } m \neq n \end{cases} \quad (8.6.1)$$

Consequently, any N-dimensional vector $\mathbf{x}$ can be expressed as a linear combination of these eigenvectors.

$$\mathbf{x} = \alpha_1 \mathbf{v}_1 + \alpha_2 \mathbf{v}_2 + \cdots + \alpha_N \mathbf{v}_N = \sum_{n=1}^{N} \alpha_n \mathbf{v}_n \quad (8.6.2)$$

Thus, the eigenvectors are called the principal axes of matrix A and the squared norm of an eigenvector is the sum of the squares of the components (α_n's) along the principal axis.

$$\|\mathbf{x}\|^2 = \mathbf{x}^T \mathbf{x} = \left(\sum_{m=1}^{N} \alpha_m \mathbf{v}_m\right)^T \left(\sum_{n=1}^{N} \alpha_n \mathbf{v}_n\right) = \sum_{m=1}^{N} \sum_{n=1}^{N} \alpha_m \alpha_n \mathbf{v}_m^T \mathbf{v}_n = \sum_{n=1}^{N} \alpha_n^2 \quad (8.6.3)$$

Premultiplying Eq. (8.6.2) by the matrix A and using Eq. (8.1.1) yields

$$A\mathbf{x} = \lambda_1 \alpha_1 \mathbf{v}_1 + \lambda_2 \alpha_2 \mathbf{v}_2 + \cdots + \lambda_N \alpha_N \mathbf{v}_N = \sum_{n=1}^{N} \lambda_n \alpha_n \mathbf{v}_n \quad (8.6.4)$$

This shows that premultiplying a vector **x** by matrix A has the same effect as multiplying each principal component α_n of **x** along the direction of eigenvector $\mathbf{v}_n$ by the associated eigenvalue λ_n. Therefore, the solution of a homogeneous discrete-time state equation

$$\mathbf{x}(k+1) = A\mathbf{x}(k) \text{ with } \mathbf{x}(0) = \sum_{n=1}^{N} \alpha_n \mathbf{v}_n = V\boldsymbol{\alpha} \qquad (8.6.5)$$

can be written as

$$\mathbf{x}(k) = \sum_{n=1}^{N} \lambda_n^k \alpha_n \mathbf{v}_n \qquad (8.6.6)$$

which was illustrated by Eq. (E8.4.6) in Example 8.4. On the other hand, as illustrated by (E8.3.9) in Example 8.3(b), the solution of a homogeneous continuous-time state equation

$$\mathbf{x}'(t) = A\mathbf{x}(t) \text{ with } \mathbf{x}(0) = \sum_{n=1}^{N} \alpha_n \mathbf{v}_n = V\boldsymbol{\alpha} \qquad (8.6.7)$$

can be written as

$$\mathbf{x}(t) = \sum_{n=1}^{N} e^{\lambda_n t} \alpha_n \mathbf{v}_n \qquad (8.6.8)$$

Equations (8.6.6) and (8.6.8) imply that the eigenvalues of the system matrix characterize the principal modes of the system described by the state equations. That is, the eigenvalues determine not only whether the system is stable or not, i.e. the system state converges to an equilibrium state or diverges, but also how fast the system state proceeds along the direction of each eigenvector. More specifically, in the case of a discrete-time system, the absolute values of all the eigenvalues must be less than 1 for stability and the smaller the absolute value of an eigenvalue (<1) is, the faster the corresponding mode converges. In the case of a continuous-time system, the real parts of all the eigenvalues must be negative for stability, and the smaller a negative eigenvalue is, the faster the corresponding mode converges. The difference among the eigenvalues determines how stiff the system is (see Section 6.5.4). This meaning of eigenvalues/eigenvectors is very important in dynamic systems.

Now, in order to figure out the meaning of eigenvalues/eigenvectors in static systems, we define the mean vector and the covariance matrix of the vectors $\{\mathbf{x}^{(1)}, \mathbf{x}^{(2)}, \ldots, \mathbf{x}^{(K)}\}$ representing K points in a two-dimensional space called the $x_1 x_2$-plane as

$$\mathbf{m}_x = \frac{1}{K} \sum_{k=1}^{K} \mathbf{x}^{(k)}, \quad C_x = \frac{1}{K} \sum_{k=1}^{K} [\mathbf{x}^{(k)} - \mathbf{m}_x][\mathbf{x}^{(k)} - \mathbf{m}_x]^T \qquad (8.6.9)$$

where the mean vector represents the center of the points, and the covariance matrix describes how dispersedly the points are distributed. Let us think about the geometrical meaning of diagonalizing the covariance matrix C_x. As a simple example, suppose we have four points

$$\mathbf{x}^{(1)} = \begin{bmatrix} 0 \\ -1 \end{bmatrix}, \quad \mathbf{x}^{(2)} = \begin{bmatrix} -1 \\ 0 \end{bmatrix}, \quad \mathbf{x}^{(3)} = \begin{bmatrix} 2 \\ 3 \end{bmatrix}, \quad \mathbf{x}^{(4)} = \begin{bmatrix} 3 \\ 2 \end{bmatrix} \quad (8.6.10)$$

for which the mean vector $\mathbf{m}_x$, the covariance matrix C_x and its modal matrix are

$$\mathbf{m}_x = \begin{bmatrix} 1 \\ 1 \end{bmatrix}, \quad C_x = \begin{bmatrix} 2.5 & 2 \\ 2 & 2.5 \end{bmatrix}, \quad V = [\,\mathbf{v}_1 \ \mathbf{v}_2\,] = \frac{1}{\sqrt{2}} \begin{bmatrix} -1 & 1 \\ 1 & 1 \end{bmatrix} \quad (8.6.11)$$

Then, we can diagonalize the covariance matrix as

$$V^T C_x V = \frac{1}{\sqrt{2}} \begin{bmatrix} -1 & 1 \\ 1 & 1 \end{bmatrix} \begin{bmatrix} 2.5 & 2 \\ 2 & 2.5 \end{bmatrix} \frac{1}{\sqrt{2}} \begin{bmatrix} -1 & 1 \\ 1 & 1 \end{bmatrix} = \begin{bmatrix} 0.5 & 0 \\ 0 & 4.5 \end{bmatrix} = \begin{bmatrix} \lambda_1 & 0 \\ 0 & \lambda_2 \end{bmatrix} = \Lambda \quad (8.6.12)$$

which has the eigenvalues on its main diagonal. On the other hand, if we transform the four point vectors by using the modal matrix as

$$\mathbf{y} = V^T(\mathbf{x} - \mathbf{m}_x) \quad (8.6.13)$$

then the new four point vectors are

$$\mathbf{y}^{(1)} = \begin{bmatrix} 1/\sqrt{2} \\ -3/\sqrt{2} \end{bmatrix}, \quad \mathbf{y}^{(2)} = \begin{bmatrix} -1/\sqrt{2} \\ -3/\sqrt{2} \end{bmatrix}, \quad \mathbf{y}^{(3)} = \begin{bmatrix} -1/\sqrt{2} \\ 3/\sqrt{2} \end{bmatrix}, \quad \mathbf{y}^{(4)} = \begin{bmatrix} 1/\sqrt{2} \\ 3/\sqrt{2} \end{bmatrix} \quad (8.6.14)$$

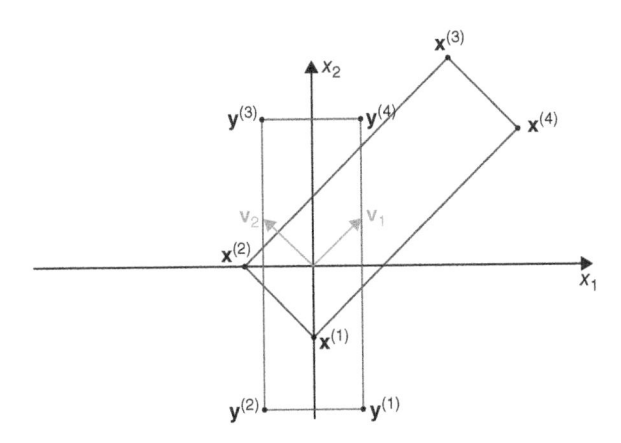

Figure 8.2 Eigenvectors and eigenvalues of a covariance matrix.

for which the mean vector $\mathbf{m}_x$ and the covariance matrix C_x are

$$\mathbf{m}_y = V^T(\mathbf{m}_x - \mathbf{m}_x) = \begin{bmatrix} 0 \\ 0 \end{bmatrix}, \quad C_y = V^T C_x V = \begin{bmatrix} 0.5 & 0 \\ 0 & 4.5 \end{bmatrix} = \Lambda \quad (8.6.15)$$

The original four points and the new points corresponding to them are depicted in Figure 8.2, which shows that the eigenvectors of the covariance matrix for a set of point vectors represents the principal axes of the distribution, and its eigenvalues are related with the lengths of the distribution along the principal axes. The difference among the eigenvalues determines how oblong the overall shape of the distribution is.

Before closing this section, we may think about the meaning of the determinant of a matrix composed of two 2-dimensional vectors and three 3-dimensional vectors. Let us consider a 2×2 matrix composed of two 2-dimensional vectors $\mathbf{x}^{(1)}$ and $\mathbf{x}^{(2)}$.

$$X = [\mathbf{x}^{(1)} \mathbf{x}^{(2)}] = \begin{bmatrix} x_{11} & x_{12} \\ x_{21} & x_{22} \end{bmatrix} \quad (8.6.16)$$

Conclusively, the absolute value of the determinant of this matrix

$$\det(X) = |X| = x_{11}x_{22} - x_{12}x_{21} \quad (8.6.17)$$

equals the area of the parallelogram having the two vectors as its two neighboring sides. In order to certify this fact, let us make a clockwise rotation of the two vectors by the phase angle of $\mathbf{x}^{(2)}$:

$$-\theta_1 = -\tan^{-1}\left(\frac{x_{21}}{x_{11}}\right) \quad (8.6.18)$$

so that the new vector $\mathbf{y}^{(1)}$ corresponding to $\mathbf{x}^{(1)}$ becomes aligned with the x_1-axis (see Figure 8.3). For this purpose, we multiply our matrix X by the rotation matrix defined by Eq. (8.4.2)

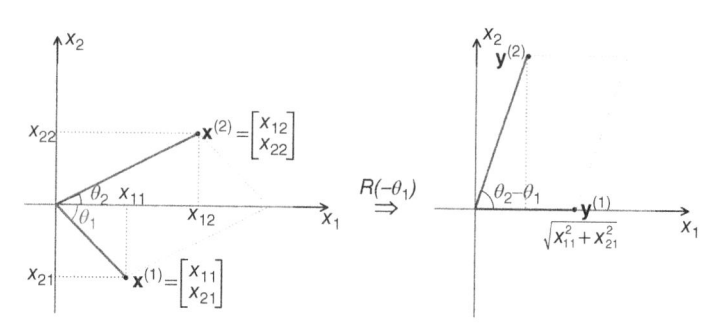

Figure 8.3 Eigenvectors and eigenvalues of a covariance matrix.

$$R(-\theta_1) = \begin{bmatrix} \cos\theta_1 & -\sin(-\theta_1) \\ \sin(-\theta_1) & \cos\theta_1 \end{bmatrix} = \frac{1}{\sqrt{x_{11}^2 + x_{21}^2}} \begin{bmatrix} x_{11} & x_{21} \\ -x_{21} & x_{11} \end{bmatrix} \quad (8.6.19)$$

to get

$$Y = R(-\theta_1)X = \frac{1}{\sqrt{x_{11}^2 + x_{21}^2}} \begin{bmatrix} x_{11} & x_{21} \\ -x_{21} & x_{11} \end{bmatrix} \begin{bmatrix} x_{11} & x_{12} \\ x_{21} & x_{22} \end{bmatrix}; \quad (8.6.20a)$$

$$[\mathbf{y}^{(1)} \; \mathbf{y}^{(2)}] = \frac{1}{\sqrt{x_{11}^2 + x_{21}^2}} \begin{bmatrix} x_{11}^2 + x_{21}^2 & x_{11}x_{12} + x_{21}x_{22} \\ 0 & -x_{12}x_{21} + x_{11}x_{22} \end{bmatrix} \quad (8.6.20b)$$

The parallelograms having the original vectors and the new vectors as their two neighboring sides are depicted in Figure 8.3, where the areas of the parallelograms turn out to be equal to the absolute values of the determinants of the matrices X and Y as

the area of the parallelograms

= the length of the bottom side × the height of the parallelogram

= (the x_1-component of $\mathbf{y}^{(1)}$) × (the x_2-component of $\mathbf{y}^{(2)}$) = $y_{11}y_{22}$ = $\det(Y)$

$$= \frac{x_{11}^2 + x_{21}^2}{\sqrt{x_{11}^2 + x_{21}^2}} \times \frac{-x_{12}x_{21} + x_{11}x_{22}}{\sqrt{x_{11}^2 + x_{21}^2}} \equiv \det(X) \quad (8.6.21)$$

On extension of this result into a three-dimensional situation, the absolute value of the determinant of a 3×3 matrix composed of three 3-dimensional vectors $\mathbf{x}^{(1)}$, $\mathbf{x}^{(2)}$, and $\mathbf{x}^{(3)}$ equals the volume of the parallelepiped having the three vectors as its three edges.

$$\det(X) = |X| = |\mathbf{x}^{(1)} \; \mathbf{x}^{(2)} \; \mathbf{x}^{(3)}| = \begin{vmatrix} x_{11} & x_{12} & x_{13} \\ x_{21} & x_{22} & x_{23} \\ x_{31} & x_{32} & x_{33} \end{vmatrix} \equiv \mathbf{x}^{(1)} \times \mathbf{x}^{(2)} \cdot \mathbf{x}^{(3)} \quad (8.6.22)$$

8.7 DIFFERENTIAL EQUATIONS WITH EIGENVECTORS

In this section, we consider a system of ordinary differential equations that can be formulated as an eigenvalue problem.

The following example will show that Eq. (8.6.8) can be used to find the solution of a system of ordinary homogeneous differential equations.

490　MATRICES AND EIGENVALUES

Example 8.6. Solution of a System of Differential Equation Using Eigenvalues/Eigenvectors.

Consider the following system of homogeneous first-order differential equations

$$\begin{bmatrix} x_1'(t) \\ x_2'(t) \end{bmatrix} = A \begin{bmatrix} x_1(t) \\ x_2(t) \end{bmatrix} = \begin{bmatrix} -4 & -2 \\ 1 & -1 \end{bmatrix} \begin{bmatrix} x_1(t) \\ x_2(t) \end{bmatrix} \text{ with } \begin{bmatrix} x_1(0) \\ x_2(0) \end{bmatrix} = \begin{bmatrix} 1 \\ 0 \end{bmatrix} \quad (E8.6.1)$$

We find the eigenvalues/eigenvectors of the system matrix A:

$$\begin{bmatrix} \lambda_1 \\ \lambda_2 \end{bmatrix} = \begin{bmatrix} -3 \\ -2 \end{bmatrix}, \quad V = [\mathbf{v}_1 \mathbf{v}_2] = \begin{bmatrix} -0.8944 & 0.7071 \\ 0.4472 & -0.7071 \end{bmatrix} \quad (E8.6.2)$$

and use Eq. (8.6.7) to get

$$\boldsymbol{\alpha} \stackrel{(8.6.7)}{=} V^{-1}\mathbf{x}(0) \stackrel{(E8.6.2b)}{=} \begin{bmatrix} -2.2361 \\ -1.4142 \end{bmatrix} \quad (E8.6.3)$$

Then, we substitute these into Eq. (8.6.8) to get the solution

$$\mathbf{x}(t) \stackrel{(8.6.8)}{=} \sum_{n=1}^{N} e^{\lambda_n t} \alpha_n \mathbf{v}_n = V \begin{bmatrix} e^{\lambda_1 t} & 0 \\ 0 & e^{\lambda_2 t} \end{bmatrix} \boldsymbol{\alpha} \quad (E8.6.4)$$

This solution process has been implemented in the following MATLAB script "nm08e06.m".

```
%nm08e06.m
% uses the eigenvalues/eigenvectors for solving the DE (E8.6.1).
syms t s
A=[-4 -2; 1 -1]; % System matrix
df=@(t,x)A*x; % Differential equation (E8.6.1)
t0=0; tf=5; x0=[1; 0]; % Initial/Final times
x0=[1; 0]; % Initial state vector
[t45,x45]=ode45(df,[t0 tf],x0);
[V,L]=eig(A) % Modal matrix composed of eigenvectors
a=V\x0; % Eq.(8.6.7)
x=V*diag(exp(diag(L)*t))*a % Eq.(8.6.8)
xL=ilaplace((s*eye(2)-A)\x0)
t=t45; xE=eval(x.');
plot(t45,x45, t45,xE,':')
```

Now, let us consider the following system of second-order differential equations (DEs):

$$\begin{bmatrix} x_1''(t) \\ x_2''(t) \end{bmatrix} = -\begin{bmatrix} (K_1+K_2)/M_1 & -K_2/M_1 \\ -K_2/M_2 & K_2/M_2 \end{bmatrix} \begin{bmatrix} x_1(t) \\ x_2(t) \end{bmatrix} \text{ with } \begin{bmatrix} x_1(0) \\ x_2(0) \end{bmatrix} \text{ and } \begin{bmatrix} x_1'(0) \\ x_2'(0) \end{bmatrix}$$

$$; \mathbf{x}''(t) = -A\mathbf{x}(t) \text{ with } \mathbf{x}(0) \text{ and } \mathbf{x}'(0) \quad (8.7.1)$$

Let the eigenpairs (eigenvalue-eigenvectors) of the matrix A be $(\lambda_n = \omega_n^2, \mathbf{v}_n)$ with

$$A\mathbf{v}_n = \omega_n^2 \mathbf{v}_n \tag{8.7.2}$$

Noting that the solution of Eq. (8.6.7) can be written as Eq. (8.6.8) in terms of the eigenvectors of the system matrix, we write the solution of Eq. (8.7.1) as

$$\mathbf{x}(t) = \sum_{n=1}^{2} w_n(t)\mathbf{v}_n = [\mathbf{v}_1 \;\; \mathbf{v}_2]\begin{bmatrix} w_1(t) \\ w_2(t) \end{bmatrix} = V\mathbf{w}(t) \tag{8.7.3}$$

and substitute this into Eq. (8.7.1) to have

$$\sum_{n=1}^{2} w_n''(t)\mathbf{v}_n = -A \sum_{n=1}^{2} w_n(t)\mathbf{v}_n \overset{(8.7.2)}{=} -\sum_{n=1}^{2} w_n(t)\omega_n^2 \mathbf{v}_n; \tag{8.7.4}$$

$$w_n''(t) = -\omega_n^2 w_n(t) \quad \text{for } n = 1, 2 \tag{8.7.5}$$

The solution of this equation is

$$w_n(t) = w_n(0)\cos(\omega_n t) + \frac{w_n'(0)}{\omega_n}\sin(\omega_n t) \quad \text{with } \omega_n = \sqrt{\lambda_n} \text{ for } n = 1, 2 \tag{8.7.6}$$

where the initial value of $\mathbf{w}(t) = [w_1(t) \;\; w_2(t)]^T$ can be obtained via Eq. (8.7.3) from that of $\mathbf{x}(t)$ as

$$\mathbf{w}(0) \overset{(8.7.3)}{=} V^{-1}\mathbf{x}(0), \quad \mathbf{w}'(0) = V^{-1}\mathbf{x}'(0) \tag{8.7.7}$$

Finally, we substitute Eq. (8.7.6) into Eq. (8.7.3) to obtain the solution of Eq. (8.7.1).

Example 8.7. Solution of a System of Differential Equation Using Eigenvalues/Eigenvectors.
Consider the following system of second-order differential equations describing the displacements $x_1(t)$ and $x_2(t)$ of the two masses M_1 and M_2 in the undamped mass–spring system depicted in Figure 8.4 where $M_1 = 1$ kg, $M_2 = 2$ kg, $K_1 = 4$ N/m, and $K_2 = 1$ N/m:

$$\begin{bmatrix} x_1''(t) \\ x_2''(t) \end{bmatrix} = -A \begin{bmatrix} x_1(t) \\ x_2(t) \end{bmatrix} = -\begin{bmatrix} 5 & -1 \\ -0.5 & 0.5 \end{bmatrix}\begin{bmatrix} x_1(t) \\ x_2(t) \end{bmatrix} \tag{E8.7.1a}$$

with

$$\begin{bmatrix} x_1(0) \\ x_2(0) \end{bmatrix} = \begin{bmatrix} 1 \\ -0.5 \end{bmatrix} \quad \text{and} \quad \begin{bmatrix} x_1'(0) \\ x_2'(0) \end{bmatrix} = \begin{bmatrix} 0 \\ 0 \end{bmatrix} \tag{E8.7.1b}$$

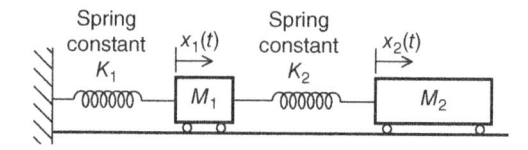

Figure 8.4 An undamped mass–spring system.

(a) First, we find the eigenvalues/eigenvectors of the system matrix A:

$$\begin{bmatrix} \lambda_1 \\ \lambda_2 \end{bmatrix} = \begin{bmatrix} 5.1085 \\ 0.3915 \end{bmatrix}, \quad V = [\mathbf{v}_1 \ \mathbf{v}_2] = \begin{bmatrix} 0.9942 & 0.2121 \\ -0.1079 & 0.9773 \end{bmatrix}, \quad \text{(E8.7.2)}$$

$$\text{and} \quad \begin{bmatrix} \omega_1 \\ \omega_2 \end{bmatrix} \stackrel{(8.7.6b)}{=} \begin{bmatrix} \sqrt{\lambda_1} \\ \sqrt{\lambda_2} \end{bmatrix} = \begin{bmatrix} 2.2602 \\ 0.6257 \end{bmatrix}$$

and use Eq. (8.7.7) to get

$$\mathbf{w}(0) \stackrel{(8.7.7a)}{=} V^{-1}\mathbf{x}(0) \stackrel{(E8.7.2b)}{=} \begin{bmatrix} 1.8094 \\ -0.3914 \end{bmatrix} \text{ and } \mathbf{w}'(0) \stackrel{(8.7.7b)}{=} V^{-1}\mathbf{x}'(0) = \begin{bmatrix} 0 \\ 0 \end{bmatrix}$$
(E8.7.3)

Then, we substitute these into Eq. (8.7.6) to get $\{w_1(t), w_2(t)\}$, which can be substituted into Eq. (8.7.3) to yield the solution $\{x_1(t), x_2(t)\}$.

```
%nm08e07.m
% uses the eigenvalues/eigenvectors for solving a DE.
M1=1; M2=2; K1=4; K2=1;
A=[(K1+K2)/M1 -K2/M1; -K2/M2 K2/M2]; NA=size(A,2);
df=@(t,x)[zeros(NA) eye(NA); -A zeros(NA)]*x(:); % Eq.(E8.7.4)
t0=0; tf=10; % Initial and final times
x0=[1; -0.5; 0; 0]; % Initial state (conditions)
% Use the numerical DE solver ode45()
[t45,x45]=ode45(df,[t0 tf],x0);
% Use the eigenvectors
[V,L]=eig(A); % Modal matrix composed of eigenvectors
w0=V^-1*x0(1:NA);   w10=V^-1*x0(NA+1:end); % Eq.(8.7.7)
omega=sqrt(diag(L)); % Eq.(8.7.6b)
for n=1:NA % Eq.(8.7.6a)
   w45(n,:)=[w0(n) w10(n)]* ...
              [cos(omega(n)*t45) sin(omega(n)/omega(n)*t45)].';
end
xE45=V*w45; % Eq.(8.7.3)
subplot(311), plot(t45,w45(1,:), t45,w45(2,:),'r:')
legend('w_1 (t)','w_2 (t)')
subplot(312), plot(t45,x45(:,1),'g', t45,xE45(1,:),'r:')
legend('x_1(t) - ode45','x_1(t) - Eigenvalue')
subplot(313), plot(t45,x45(:,2),'g', t45,xE45(2,:),'r:')
legend('x_2(t) - ode45','x_2(t) - Eigenvalue')
```

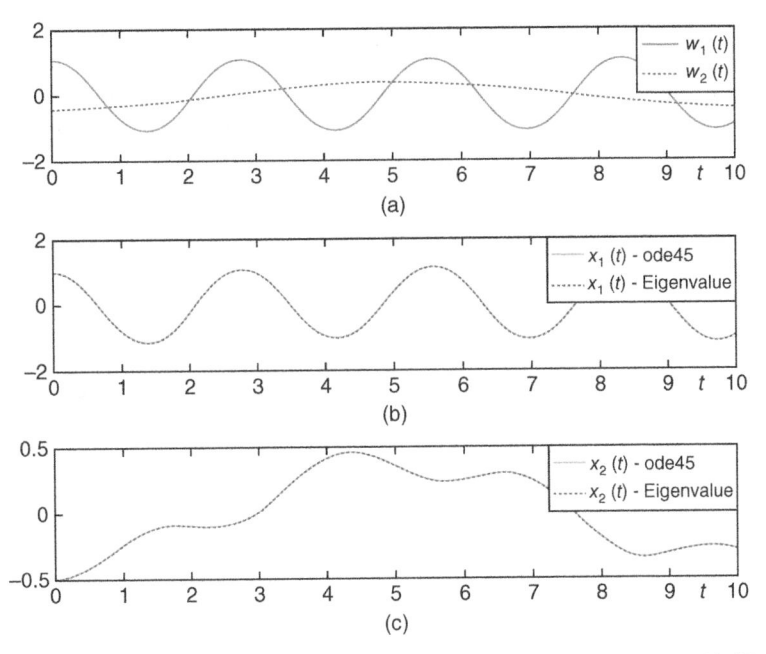

Figure 8.5 Solutions for Example 8.6 (results obtained by running the script "nm08e06.m"). (a) $w_1(t)$ and $w_2(t)$. (b) $x_1(t)$. (c) $x_2(t)$.

(b) To check the validity of the solution obtained in (a), we write the DE (E8.7.1) in state-space form:

$$\begin{bmatrix} \mathbf{x}'(t) \\ \mathbf{x}''(t) \end{bmatrix} = \begin{bmatrix} O & I \\ -A & O \end{bmatrix} \begin{bmatrix} \mathbf{x}(t) \\ \mathbf{x}'(t) \end{bmatrix};$$

$$\begin{bmatrix} x_1'(t) \\ x_2'(t) \\ x_1''(t) \\ x_2''(t) \end{bmatrix} = \begin{bmatrix} 0 & 0 & 1 & 0 \\ 0 & 0 & 0 & 1 \\ -5 & 1 & 0 & 0 \\ 0.5 & -0.5 & 0 & 0 \end{bmatrix} \begin{bmatrix} x_1(t) \\ x_2(t) \\ x_1'(t) \\ x_2'(t) \end{bmatrix} \text{ with } \begin{bmatrix} x_1(0) \\ x_2(0) \\ x_1'(0) \\ x_2'(0) \end{bmatrix} = \begin{bmatrix} 1 \\ -0.5 \\ 0 \\ 0 \end{bmatrix} \quad \text{(E8.7.4)}$$

and then use the numerical DE solver 'ode45()', as implemented in the above MATLAB script "nm08e07.m". We can run it to get the solutions as shown in Figure 8.5.

8.8 DOA ESTIMATION WITH EIGENVECTORS[Y-3]

Figure 8.6 shows a UCA (uniform circular array) having $N_e = 4$ antenna elements equally spaced on a circle of radius R, that is impinged on by $N_s = 2$ signals $s_1(n)$ and $s_2(n)$ with DoAs ($\phi_1 = 60°$, θ_1) and ($\phi_2 = 210°$, θ_2), respectively.

Note that the *degree of arrival*, denoted by DoA (ϕ, θ), of an incident signal is a pair of azimuth and zenith angles where the azimuth angle ϕ and zenith angle θ are the angles of the incident signal with the x_1-axis and x_3-axis. The output signals measured by each element of the UCA at time n can be written as

$$\overset{y(n)}{\begin{bmatrix} y_1(n) \\ y_2(n) \\ \cdot \\ y_{N_e}(n) \end{bmatrix}} = \overset{A:N_e \times N_s \text{ matrix}}{\begin{bmatrix} e^{j\beta R \sin\theta_1 \cos(\phi_1 - 2\pi 0/N_e)} & e^{j\beta R \sin\theta_{Ns} \cos(\phi_{Ns} - 2\pi 0/N_e)} \\ e^{j\beta R \sin\theta_1 \cos(\phi_1 - 2\pi 1/N_e)} & e^{j\beta R \sin\theta_{Ns} \cos(\phi_{Ns} - 2\pi 1/N_e)} \\ \cdot & \cdot \\ e^{j\beta R \sin\theta_1 \cos(\phi_1 - 2\pi(N_e-1)/N_e)} & e^{j\beta R \sin\theta_{Ns} \cos(\phi_{Ns} - 2\pi(N_e-1)/N_e)} \end{bmatrix}} \overset{s(n)}{\begin{bmatrix} s_1(n) \\ s_{N_s}(n) \end{bmatrix}}$$

$$+ \overset{w(n)}{\begin{bmatrix} w_1(n) \\ w_2(n) \\ \cdot \\ w_{N_e}(n) \end{bmatrix}}; \qquad (8.8.1)$$

$$\mathbf{y}(n) = [\mathbf{a}(\phi_1, \theta_1) \mathbf{a}(\phi_{Ns}, \theta_{Ns})] \mathbf{s}(n) + \mathbf{w}(n) \qquad (8.8.2)$$

with $\mathbf{a}(\phi, \theta) = \begin{bmatrix} e^{j\beta R \sin\theta \cos(\phi - 2\pi 0/N_e)} \\ e^{j\beta R \sin\theta \cos(\phi - 2\pi 1/N_e)} \\ \vdots \\ e^{j\beta R \sin\theta \cos(\phi - 2\pi(N_e-1)/N_e)} \end{bmatrix}$ Steering vector

where $w_m(n)$ denotes a complex noise with normal (or Gaussian) distribution $N(0, 0.5)$ of mean 0 and variance 0.5 (see Section 1.1.8 for 'normal' distribution). For simplicity, we suppose that the two signals are

$$s_1(n) = \sin\left(\frac{\pi}{8}n\right) \quad s_2(n) = \cos\left(\frac{\pi}{12}n\right) \qquad (8.8.3\text{a,b})$$

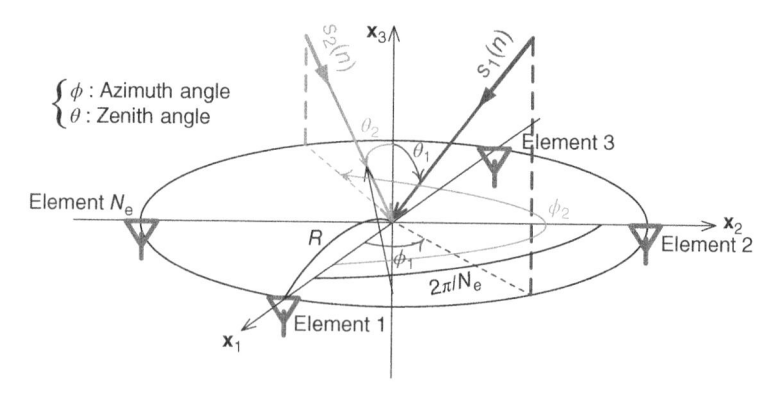

Figure 8.6 An $N_e = 4$-element UCA receiving two signals $s_1(n)$ and $s_2(n)$.

the zenith angles of all the received signals are $\theta = 90°$, the wave propagation factor is $\beta = 2\pi/\lambda = 1$ (λ: wavelength), and the radius of the UCA is $R = 1$.

To estimate the DoAs, we construct the signal correlation matrix R_y from $\{y(n), n = 1: N_d = 200\}$ and use the MATLAB command '[V,Λ]=eig(R_y)' to find its eigenvectors and eigenvalues as

$$R_y = \begin{bmatrix} 1.3728 & 0.87 - j0.3237 & 0.1237 - j0.0103 & 0.213 - j0.0463 \\ 0.87 + j0.3237 & 1.4144 & 0.3024 + j0.0456 & 0.2842 + j0.0811 \\ 0.1237 + j0.0103 & 0.3024 - j0.0456 & 1.6276 & 1.0259 + j0.3658 \\ 0.213 + j0.0463 & 0.2842 - j0.0811 & 1.0259 - j0.3658 & 1.6152 \end{bmatrix}$$

$$\frac{YY^{*T}}{200} =$$

$$\text{with } Y = \begin{bmatrix} y_1(1) & \cdots & y_1(200) \\ y_2(1) & \cdots & y_2(200) \\ y_3(1) & \cdots & y_3(200) \\ y_4(1) & \cdots & y_4(200) \end{bmatrix} \quad (8.8.4)$$

$$\rightarrow V = \begin{bmatrix} -0.1044 - j0.7008 & -0.0251 - j0.0441 & -0.6035 - j0.058 & 0.3575 - j0.008 \\ -0.1354 + j0.6837 & -0.0834 + j0.0774 & -0.5268 - j0.2475 & 0.378 + j0.1399 \\ 0.0723 - j0.0241 & -0.6465 - j0.2482 & 0.3686 + j0.1976 & 0.556 + j0.1749 \\ -0.0802 & 0.7106 & 0.3446 & 0.6082 \end{bmatrix},$$

$$\Lambda = \begin{bmatrix} 0.4614 & 0 & 0 & 0 \\ 0 & 0.5249 & 0 & 0 \\ 0 & 0 & 2.0644 & 0 \\ 0 & 0 & 0 & 2.9793 \end{bmatrix} \quad (8.8.5)$$

Even if the specific values of $y(n)$, R_y, and its eigenvalues/eigenvectors vary with the noise $w(n)$, (the last) two eigenvalues $\lambda_3 \approx 2$ and $\lambda_4 \approx 3$ are conspicuously larger than the other ones $\lambda_1 \approx 0.45$ and $\lambda_2 \approx 0.5$ so that their corresponding eigenvectors v_3 and v_4 can be regarded as spanning the signal subspace, while the eigenvectors v_1 and v_2 corresponding to the two smaller eigenvalues λ_1 and λ_2 can be regarded as spanning the noise subspace. Note the following:

- One way to estimate the azimuth angles of the two signals is to find the local maxima of the spatial spectrum based on the signal subspace

$$P_s(\phi) = \mathbf{a}^{*T}(\phi) SS^{*T} \mathbf{a}(\phi) \text{ with } S = [V(:,3) \ V(:,4)] \quad (8.8.6)$$

which yields the azimuth angle estimates around the true values 60° and 210° (Figure 8.7a1). An alternative is to find the local maxima of the spatial spectrum based on the noise subspace

$$P_n(\phi) = \frac{1}{\mathbf{a}^{*T}(\phi) NN^{*T} \mathbf{a}(\phi)} \text{ with } N = [V(:,1) \ V(:,2)] \quad (8.8.7)$$

MATRICES AND EIGENVALUES

```
%DoA_estimation_with_eigenvector.m
% Use the eigenvectors to estimate the DoA (Degree of Arrival)
b=1; % Wave propagation factor 2*pi/wavelength of the signal to detect
R=1; % Radius of UCA
Ne=4; Ns=2; % Number of antenna elements and Number of signals
% True DoAs (azimuth angles) in degree/radian
phs_deg = [60 210]; phs = phs_deg*pi/180;
Nd=200; % Number of data used for DoA estimation
% Steering function for azimuth angle ph
a = @(ph,th)exp(j*b*R*sin(th)*cos(ph-[0:Ne-1]'*2*pi/Ne));
for m=1:Ns
   A(:,m)=a(phs(m),pi/2); % LxM steering matrix with zenith angle of 90°
end
for n=1:Nd % Make a noisy received signal with some directivity
   s(:,n) = [sin(pi/8*n); cos(pi/12*n)]; % Received signal
   y(:,n) = A*s(:,n)+0.5*(randn(Ne,1)+j*randn(Ne,1)); % Eq.(8.8.1)
end
% UCA output signal correlation matrix
Ry=y*y'/Nd; % Eq.(8.8.4)
% Eigendecomposition of the signal correlation matrix Ry
[V,Lam]=eig(Ry) % Eq.(8.8.5)
% Partition the modal matrix V into Noise/Signal subspace matrices N,S
N = V(:,1:Ne-Ns);    % Noise sub-eigenspace matrix
S = V(:,Ne-Ns+1:Ne); % Signal sub-eigenspace matrix
phph = 0:359; % Whole coverage of azimuth angle
for i=1:length(phph) % To construct the spatial spectra
   phi_rad = phph(i)*pi/180; % degree-to-radian conversion
   a_phi = a(phi_rad,pi/2); % with zenith angle of 90°
   Ps(i)=real(a_phi'*S*S'*a_phi);   % Eq.(8.8.6)
   Pn(i)=1/real(a_phi'*N*N'*a_phi); % Eq.(8.8.7)
end
% Plot the spatial spectra of signal and noise
spsm=max(Ps);  spnm=max(Pn);
subplot(221), plot(phph,Ps) % plot the spatial spectrum based on S
hold on, for m=1:Ns, plot(phs_deg(m)*[1 1],[0 spsm],'r:'); end
subplot(222), plot(phph,Pn) % plot the spatial spectrum based on N
hold on, for m=1:Ns, plot(phs_deg(m)*[1 1],[0 spnm],'r:'); end
% DoA estimation based on the spatial spectrum of noise
[lmaxima,indices]=localmax(Pn,[],false); % Find the local maxima
ph_est = phph(indices) % Estimates of azimuth angles
```

```
%try_beamforming.m
for m=1:Ns % NexNs steering matrix with DoA estimates for beamforming
   Ae(:,m) = a(ph_est(m)*pi/180,pi/2); % Eq.(8.8.8b)
end
W=pinv(Ae); % (Ae'*Ae)^-1*Ae' : NsxNe beamformer gain matrix
sh=real(W*y); % Eq.(8.8.8a): Reconstructed signal by beamforming
% To plot the reconstructed signals with the estimated DoA
nn=[0:Nd-1]; % Time vector
subplot(223), plot(nn,s(1,:), nn,sh(1,:),'r:')
legend('Transmitted signal s1','Reconstructed signal s1')
subplot(224), plot(nn,s(2,:), nn,sh(2,:),'r:')
legend('Transmitted signal s2','Reconstructed signal s2')
```

(Q2) Are the reconstructed signals close to their corresponding transmitted ones, respectively? as shown in Figs. 8.7(b1) and (b2)?

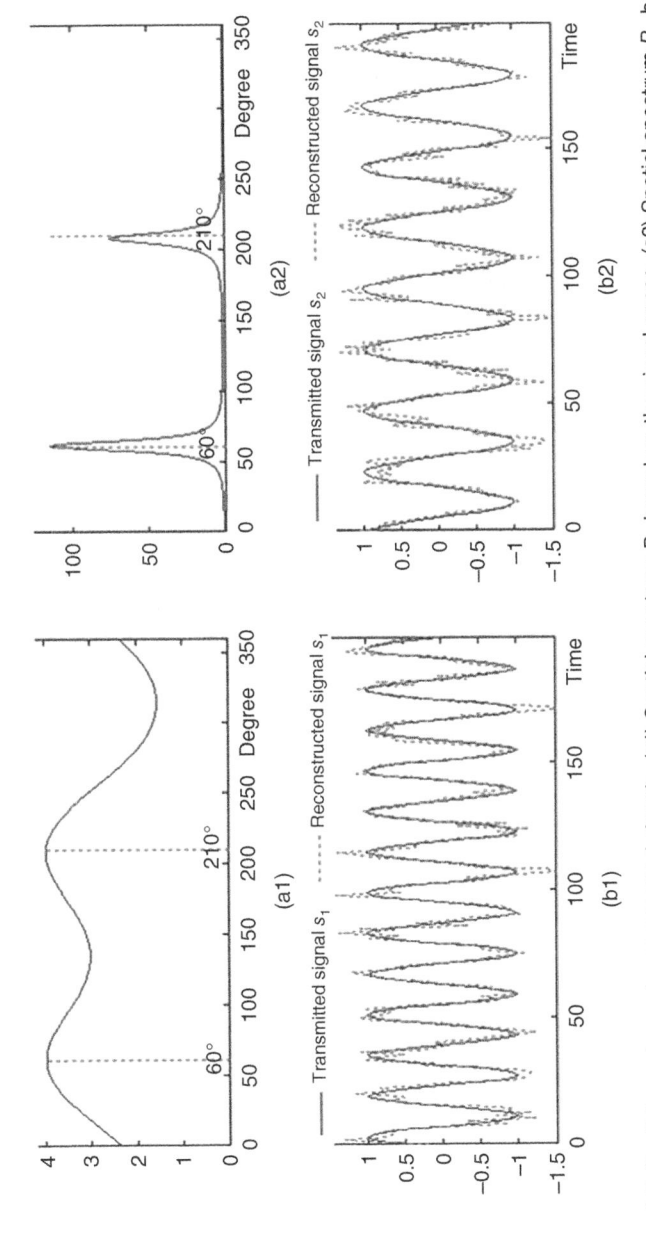

Figure 8.7 Spatial spectra and reconstructed signals. (a1) Spatial spectrum P_s based on the signal space. (a2) Spatial spectrum P_n based on the noise space. (b1) Transmitted signal s_1 and reconstructed signal. (b2) Transmitted signal s_2 and reconstructed signal.

which also yields the azimuth angle estimate $\hat{\phi}$'s around the true values 60° and 210° (Figure 8.7a2). This estimation method is referred to as the *MUSIC* (*Multiple SIgnal Classification*) algorithm. Here is a question:

(Q1) Which one of these two spatial spectra P_s and P_n looks better in terms of selectivity?

To carry out this process, we can run the above MATLAB script "DoA_estimation_with_ eigenvector.m", which yields Figure 8.7a1,a2.

- To reconstruct the received signals s_1 and s_2 from the UCA output signal vector **y**, beamforming with the estimated DoA is needed, and it can be done by premultiplying the UCA output signal vector **y** by the left pseudo inverse (see Remark 1.1) of the (estimated) steering matrix $A_e(\hat{\phi})$ as

$$\hat{\mathbf{s}}(n) = W(\hat{\phi})\mathbf{y}(n) = \overset{N_s \times N_c \text{ matrix}}{[A_e(\hat{\phi})^{*T} A_e(\hat{\phi})]^{-1} A_e(\hat{\phi})^{*T}} \mathbf{y}(n) \qquad (8.8.8a)$$

$$\text{with } A_e(\hat{\phi}) = \overset{N_c \times N_s \text{ matrix}}{[\mathbf{a}(\hat{\phi}_1) \mathbf{a}(\hat{\phi}_2)]} : \text{estimated steering matrix} \qquad (8.8.8b)$$

Together with the transmitted signals $s_1(n)$ and $s_2(n)$, their reconstructions $\hat{s}_1(n)$ and $\hat{s}_2(n)$ obtained through beamforming like this are shown in Figure 8.7b1,b2, respectively. To simulate this receive beamforming process, we can run the above MATLAB script "try_beamforming.m", which yields Figure 8.7b1,b2.

PROBLEMS

8.1 Symmetric Tridiagonal Toeplitz Matrix
Consider the following $N \times N$ symmetric tridiagonal Toeplitz matrix as

$$\begin{bmatrix} a & b & 0 & \cdots & 0 & 0 \\ b & a & b & \cdots & 0 & 0 \\ 0 & b & a & \cdots & 0 & 0 \\ \vdots & \vdots & \vdots & \cdots & \vdots & \vdots \\ 0 & 0 & 0 & \cdots & a & b \\ 0 & 0 & 0 & \cdots & b & a \end{bmatrix} \qquad (P8.1.1)$$

(a) Verify that the eigenvalues and eigenvectors of this matrix are as follows, with $N = 3$ for convenience.

$$\lambda_n = a + 2b \cos\left(\frac{n\pi}{N+1}\right) \quad \text{for } n = 1, \ldots, N \qquad (P8.1.2)$$

$$\mathbf{v}_n = \sqrt{\frac{2}{N+1}} \left[\sin\left(\frac{n\pi}{N+1}\right) \ \sin\left(\frac{2n\pi}{N+1}\right) \ \cdots \ \sin\left(\frac{Nn\pi}{N+1}\right) \right]^T \qquad (P8.1.3)$$

(b) Letting $N = 3$, $a = 2$, and $b = 1$, find the eigenvalues/eigenvectors of the aforementioned matrix by using (P8.1.2,3) and by using the MATLAB function 'eig_Jacobi()' or 'eig()' for cross-check.

8.2 Circulant Matrix
Consider the following $N \times N$ circulant matrix as

$$\begin{bmatrix} h(0) & h(N-1) & h(N-2) & \cdots & h(1) \\ h(1) & h(0) & h(N-1) & \cdots & h(2) \\ h(2) & h(1) & h(0) & \cdots & h(3) \\ \vdots & \vdots & \vdots & \cdots & \vdots \\ h(N-1) & h(N-2) & h(N-3) & \cdots & h(0) \end{bmatrix} \qquad (P8.2.1)$$

(a) Verify that the eigenvalues and eigenvectors of this matrix are as follows, with $N = 4$ for convenience.

$$\lambda_n = h(0) + h(1)e^{-j2\pi n/N} + h(2)e^{-j2\pi 2n/N} + \cdots$$
$$+ h(N-1)e^{-j2\pi(N-1)n/N} \qquad (P8.2.2)$$

$$\mathbf{v}_n = \begin{bmatrix} 1 & e^{j2\pi n/N} & e^{j2\pi 2n/N} & \cdots & e^{j2\pi(N-1)n/N} \end{bmatrix}^T \qquad (P8.2.3)$$

for $n = 0$ to $N - 1$

(b) Letting $N = 4$, $h(0) = 2$, $h(3) = h(1) = 1$, and $h(2) = 0$, find the eigenvalues/eigenvectors of the aforementioned matrix by using (P8.2.2,3) and by using the MATLAB function 'eig_Jacobi()' or 'eig()'. Do they agree? Do they satisfy Eq. (8.1.1)?

8.3 *Companion Matrix* and Its Characteristic Equation
Consider the following forms of companion matrix:

$$A_{c1} = \begin{bmatrix} -a_1 & -a_2 & -a_3 & -a_4 & -a_5 \\ 1 & 0 & 0 & 0 & 0 \\ 0 & 1 & 0 & 0 & 0 \\ 0 & 0 & 1 & 0 & 0 \\ 0 & 0 & 0 & 1 & 0 \end{bmatrix} \text{ or } A_{c2} = \begin{bmatrix} 0 & 0 & 0 & 0 & -a_5 \\ 1 & 0 & 0 & 0 & -a_4 \\ 0 & 1 & 0 & 0 & -a_3 \\ 0 & 0 & 1 & 0 & -a_2 \\ 0 & 0 & 0 & 1 & -a_1 \end{bmatrix} \quad (P8.3.1)$$

The characteristic equations of these matrices (that are used to find the eigenvalues) can be obtained by using Eq. (8.1.2) as

$$|xI - A_{c1}| = 0 \text{ or } |xI - A_{c2}| = 0:$$
$$A_5(x) = x^5 + a_1 x^4 + \cdots + a_4 x + a_5 = 0 \quad (P8.3.2)$$

Running the following MATLAB statements, find the MATLAB built-in function which returns the characteristic equation of a given matrix:

```
a=[1  -3  -23  51  94 -120]; % A 5th-degree polynomial
Ac1=[-a(2:6); eye(4) zeros(4,1)]; % The companion matrix
Ac1=compan(a) % Alternatively
syms x; det(x*eye(size(Ac1))-Ac1) % Eq.(P8.3.2)
char_eq=poly(Ac1) % Characteristic equation of matrix Ac1
Lam=roots(char_eq).' % Eigenvalues of matrix A
Lam=eig(Ac1).' % Eigenvalues of matrix A
```

Noting that the matrices given in Eq. (P8.3.1) are called the (*Frobenius*) *companion matrices* of the monic polynomial (with unit leading coefficient), show that the characteristic equation of another companion matrix A_{c2} is also the same as Eq. (P8.3.2).

8.4 Solving a Vector Differential Equation by Decoupling – Diagonalization of the System Matrix
Consider the following two-dimensional vector differential equation (state equation) as

$$\begin{bmatrix} x_1'(t) \\ x_2'(t) \end{bmatrix} = \begin{bmatrix} 0 & 1 \\ -2 & -3 \end{bmatrix} \begin{bmatrix} x_1(t) \\ x_2(t) \end{bmatrix} + \begin{bmatrix} 0 \\ 1 \end{bmatrix} u_s(t) \quad (P8.4.1)$$

with $\begin{bmatrix} x_1(0) \\ x_2(0) \end{bmatrix} = \begin{bmatrix} 1 \\ 0 \end{bmatrix}$ and $u_s(t) = 1 \quad \forall \ t \geq 0$,

which was solved by using Laplace transform in Problem 6.2. In this problem, we solve it again by the decoupling method through diagonalization of the system matrix.

(a) Show that the eigenvalues and eigenvectors of the system matrix are as follows.

$$\lambda_1 = -1, \lambda_2 = -2; \quad \mathbf{v}_1 = \begin{bmatrix} 1 \\ -1 \end{bmatrix}, \mathbf{v}_2 = \begin{bmatrix} 1 \\ -2 \end{bmatrix} \quad \text{(P8.4.2)}$$

(b) Show that the diagonalization of the aforementioned vector differential equation using the modal matrix $V = [\mathbf{v}_1 \ \mathbf{v}_2]$ yields the following equation:

$$\begin{bmatrix} w_1'(t) \\ w_2'(t) \end{bmatrix} = \begin{bmatrix} -1 & 0 \\ 0 & -2 \end{bmatrix} \begin{bmatrix} w_1(t) \\ w_2(t) \end{bmatrix} + \begin{bmatrix} 1 \\ -1 \end{bmatrix} u_s(t)$$

$$= \begin{bmatrix} -w_1(t) + u_s(t) \\ -2w_2(t) - u_s(t) \end{bmatrix} \text{ with } \begin{bmatrix} w_1(0) \\ w_2(0) \end{bmatrix} = \begin{bmatrix} 2 \\ -1 \end{bmatrix} \quad \text{(P8.4.3)}$$

(c) Show that these equations can be solved individually by using the Laplace transform to yield the following solution, which is the same as Eq. (P6.2.2) obtained in Problem 6.2(b).

$$W_1(s) = \frac{1}{s} + \frac{1}{s+1}; \quad w_1(t) = (1 + e^{-t}) u_s(t) \quad \text{(P8.4.4a)}$$

$$W_2(s) = \frac{-1/2}{s} - \frac{1/2}{s+2}; \quad w_2(t) = -\frac{1}{2}(1 + e^{-2t}) u_s(t) \quad \text{(P8.4.4b)}$$

$$; \begin{bmatrix} x_1(t) \\ x_2(t) \end{bmatrix} = \begin{bmatrix} 1/2 + e^{-t} - (1/2)e^{-2t} \\ -e^{-t} + e^{-2t} \end{bmatrix} u_s(t) \quad \text{(P8.4.5)}$$

8.5 Householder Method and QR Factorization

This method can zero out several elements in a column vector at each iteration and make any $N \times N$ matrix a (lower) triangular matrix in $(N-1)$ iterations.

(a) Householder reflection

Referring Figure P8.5 shows that the transformation matrix by which we can multiply a vector $\mathbf{x}$ to generate another vector $\mathbf{y}$ having the same norm is

$$H = I - 2\mathbf{w}\mathbf{w}^T \quad \text{(P8.5.1)}$$

with $\mathbf{w} = \dfrac{\mathbf{x} - \mathbf{y}}{\|\mathbf{x} - \mathbf{y}\|_2} = \dfrac{1}{c}(\mathbf{x} - \mathbf{y}), c = \|\mathbf{x} - \mathbf{y}\|_2, \|\mathbf{x}\| = \|\mathbf{y}\|$

and that this is an orthonormal symmetric matrix such that $H^T H = HH = I$; $H^{-1} = H$.

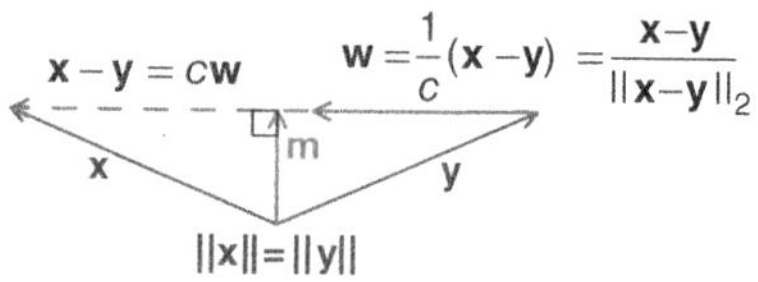

MATRICES AND EIGENVALUES

Figure P8.5 Householder reflection.

Note the following facts.

(i) $$y = x - (x - y) \stackrel{(P8.4.1)}{=} x - cw \quad (P8.5.2a)$$

(ii) $$w^T w \stackrel{(P8.4.1)}{=} 1 \text{ and } \|x\| = \|y\| \quad (P8.5.2b)$$

(iii) $$m = (x + y)/2 = x - (c/2)w \quad (P8.5.2c)$$

(iv) The mean vector m of x and y is orthogonal to the difference vector $w = (x - y)/c$.

Thus, we have

$$w^T(x - (c/2)w) = 0; \quad w^T x - (c/2)w^T w = w^T x - (c/2) = 0 \quad (P8.5.3)$$

This gives an expression for $c = \|x - y\|_2$ as

$$c = \|x - y\|_2 = 2w^T x \quad (P8.5.4)$$

We can substitute this into (P8.5.2a) to get the desired result:

$$y = x - cw = x - 2\,ww^T x = [I - 2ww^T]\,x \equiv Hx \quad (P8.5.5)$$

On the other hand, the Householder transform matrix is an orthogonal matrix since

$$H^T H = HH = [I - 2ww^T][I - 2ww^T]$$
$$= I - 4ww^T + 4ww^T ww^T = I - 4ww^T + 4ww^T = I \quad (P8.5.6)$$

(b) Householder transform

To show that the Householder matrix can be used to zero out some part of a vector, let us find the kth Householder matrix H_k transforming any vector

$$x = \begin{bmatrix} x_1 & \cdots & x_{k-1} & x_k & x_{k+1} & \cdots & x_N \end{bmatrix} \quad (P8.5.7)$$

into

$$y = \begin{bmatrix} x_1 & \cdots & x_{k-1} & -g_k & 0 & \cdots & 0 \end{bmatrix} \quad (P8.5.8)$$

where g_k is fixed in such a way that the norms of these two vectors are the same:

$$g_k = \sqrt{\sum_{n=k}^{N} x_n^2} \quad (P8.5.9)$$

First, we find the difference vector of unit norm as

$$\mathbf{w}_k = \frac{1}{c}(\mathbf{x} - \mathbf{y}) = \frac{1}{c}\begin{bmatrix} 0 & \cdots & 0 & x_k + g_k & x_{k+1} & \cdots & x_N \end{bmatrix} \quad \text{(P8.5.10)}$$

$$\text{with } c = \|\mathbf{x} - \mathbf{y}\|_2 = \sqrt{(x_k + g_k)^2 + x_{k+1}^2 + \cdots + x_N^2} \quad \text{(P8.5.11)}$$

Then, one more thing we should do is to substitute this difference vector into Eq. (P8.5.1).

$$H_k = [I - 2\mathbf{w}_k \mathbf{w}_k^T] \quad \text{(P8.5.12)}$$

Complete the following MATLAB function 'Householder()' by permuting the statements and try it with $k = 1, 2, 3$, and 4 for a five-dimensional vector generated by the MATLAB command rand(5,1) to check if it works fine.

»x=rand(5,1), for k=1:4, householder(x,k)*x, end

```
function H=Householder(x,k)
%Householder transform to zero out tail part starting from k+1
H=eye(N)-2*w*w'; % Householder matrix
N=length(x);
w=zeros(N,1);
w(k)=(x(k)+g)/c;  w(k+1:N)=x(k+1:N)/c;  % Eq.(P8.5.10)
tmp=sum(x(k+1:N).^2);
c=sqrt((x(k)+g)^2 +tmp); % Eq.(P8.5.11)
g=sqrt(x(k)^2+tmp); % Eq.(P8.5.9)
```

(c) *QR* factorization using Householder transform

We can use Householder transform to zero out the part under the main diagonal of each column of an $N \times N$ matrix A successively and then make it a lower triangular matrix R in $(N - 1)$ iterations. The necessary operations are collectively written as

$$H_{N-1} H_{N-2} \cdots H_1 A = R \quad \text{(P8.5.13)}$$

which implies that

$$A = [H_{N-1} H_{N-2} \cdots H_1]^{-1} R = H_1^{-1} \cdots H_{N-2}^{-1} H_{N-1}^{-1} R$$
$$= H_1 \cdots H_{N-2} H_{N-1} R = QR \quad \text{(P8.5.14)}$$

where the product of all the Householder matrices

$$Q = H_1 \cdots H_{N-2} H_{N-1} \quad \text{(P8.5.15)}$$

turns out to be not only symmetric but also orthogonal like each H_k:

$$Q^T Q = [H_1 \cdots H_{N-2} H_{N-1}]^T H_1 \cdots H_{N-2} H_{N-1}$$
$$= H_{N-1}^T H_{N-2}^T \cdots H_1^T H_1 \cdots H_{N-2} H_{N-1} = I$$

This suggests a *QR* factorization method that has been cast into the following MATLAB function 'qr_my()'. You can try it for a nonsingular 3×3 matrix generated by the MATLAB command rand(3) and compare the result with that of the MATLAB built-in function 'qr()'.

```
function [Q,R]=qr_my(A)
%QR factorization
N= size(A,1); R=A; Q=eye(N);
for k=1:N-1
   H=Householder(R(:,k),k);
   R=H*R; % Eq.(P8.5.13)
   Q=Q*H; % Eq.(P8.5.15)
end
```

8.6 Hessenberg Form Using Householder Transform

We can make use of Householder transform (introduced in Problem 8.4) to zero out the elements below the lower subdiagonal of a matrix so that it becomes an upper Hessenberg form, which is almost upper-triangular matrix. Complete the aforementioned MATLAB function 'Hessenberg()' by filling in the second input argument of the function 'Householder()' and try it (together with the MATLAB built-in function 'hess') for a 5×5 matrix generated by the MATLAB command rand(5) to check if it works.

```
function [Hs,HH]=Hessenberg(A)
%Transform into an almost upper triangular matrix
% having only zeros below lower subdiagonal
N=size(A,1);
Hs=A; HH=eye(N); %HH*A*HH'=Hs
for k=1:N-2
   H=Householder(Hs(:,k),???);
   Hs=H*Hs*H; HH=H*HH;
end
```

8.7 *QR* Factorization of Hessenberg Form Using the Givens Rotation

We can make use of the Givens rotation to get the *QR* factorization of Hessenberg form by the procedure implemented in the aforementioned MATLAB function 'qr_Hessenberg()', where each element on the lower subdiagonal is zeroed out at each iteration. Generate a 4×4 random matrix A by the MATLAB command 'rand(4)', transform it into a Hessenberg form Hs by using the function 'Hessenberg()', and try this function 'qr_Hessenberg()' for the matrix of Hessenberg form. Check the validity by seeing if norm(Hs-Q*R) $\approx$ 0 or not.

```
function [Q,R]=qr_Hessenberg(Hs)
%QR factorization of Hessenberg form by Givens rotation
N=size(Hs,1);
Q=eye(N); R=Hs;
for k=1:N-1
   x=R(k,k); y=R(k+1,k); r=sqrt(x*x+y*y); c=x/r; s=-y/r;
   R0=R; Q0=Q;
   R(k,:)=c*R0(k,:)-s*R0(k+1,:);
   R(k+1,:)=s*R0(k,:)+c*R0(k+1,:);
   Q(:,k)=c*Q0(:,k)-s*Q0(:,k+1);
   Q(:,k+1)=s*Q0(:,k)+c*Q0(:,k+1);
end
```

8.8 Diagonalization by Using *QR* Factorization to Find Eigenvalues

You will see that a real symmetric matrix A can be diagonalized into a diagonal matrix having the eigenvalues on its diagonal if we repeat the similarity transformation by using the orthogonal matrix *Q* obtained from the *QR* factorization. For this purpose, take the following steps:

(a) Make the following MATLAB function 'eig_QR()' that uses the MATLAB built-in function 'qr()' and then apply it to a 4 × 4 random symmetric matrix A generated by the following MATLAB statements.

```
»A=rand(4); A=A+A';
```

(b) Make the following function 'eig_QR_Hs()' that transforms a given matrix into a Hessenberg form by using the function 'Hessenberg()' (appeared in Problem 8.5) and then, repetitively makes the *QR* factorization by using the function 'qr_Hessenberg()' (appeared in Problem 8.7) and the similarity transformation by the orthogonal matrix Q until the matrix becomes diagonal. Apply it to the 4 × 4 random symmetric matrix A generated in (a) and compare the result with those obtained in (a) and by using the MATLAB built-in function 'eig()' for cross-check.

```
function [eigs,A]=eig_QR(A,kmax)
%Find eigenvalues by using QR factorization
if nargin<2, kmax=200; end
for k=1:kmax
   [Q,R]=qr(A); %A=Q*R; R=Q'*A=Q^-1*A
   A=R*Q; %A=Q^-1*A*Q
end
eigs=diag(A);
```

```
function [eigs,A]=eig_QR_Hs(A,kmax)
%Find eigenvalues by using QR factorization via Hessenberg
if nargin<2, kmax=200; end
Hs=Hessenberg(A);
for k=1:kmax
   [Q,R]=qr_Hessenberg(Hs); %Hs=Q*R; R=Q'*Hs=Q^-1*Hs
   Hs=R*Q; %Hs=Q^-1*Hs*Q
end
eigs=diag(Hs);
```

8.9 Differential/Difference Equation, State Equation, and Eigenvalue
As mentioned in Section 6.5.3, a high-order scalar differential equation such as

$$x^{(3)}(t) + a_2 x^{(2)}(t) + a_1 x'(t) + a_0 x(t) = u(t) \tag{P8.9.1}$$

can be transformed into a first-order vector differential equation, called a state equation, as

$$\begin{bmatrix} x_1'(t) \\ x_2'(t) \\ x_3'(t) \end{bmatrix} = \begin{bmatrix} 0 & 1 & 0 \\ 0 & 0 & 1 \\ -a_0 & -a_1 & -a_2 \end{bmatrix} \begin{bmatrix} x_1(t) \\ x_2(t) \\ x_3(t) \end{bmatrix} + \begin{bmatrix} 0 \\ 0 \\ 1 \end{bmatrix} u(t) \tag{P8.9.2a}$$

$$x(t) = \begin{bmatrix} 1 & 0 & 0 \end{bmatrix} \begin{bmatrix} x_1(t) \\ x_2(t) \\ x_3(t) \end{bmatrix} \tag{P8.9.2b}$$

The characteristic equation of the differential equation (P8.9.1) is

$$s^3 + a_2 s^2 + a_1 s + a_0 = 0 \tag{P8.9.3}$$

and its roots are called the characteristic roots.

(a) What is the relationship between these characteristic roots and the eigenvalues of the system matrix A of the previously mentioned state equation (P8.9.2)? To answer this question, write the equation $|\lambda I - A| = 0$ to solve for the eigenvalues of A, and show that it is equivalent to Eq. (P8.9.3). To extend your experience or just for practice, you can try the symbolic computation of MATLAB by running the following program "nm08p09a.m".

```
%nm08p09a.m
syms a0 a1 a2 s
A=[0 1 0;0 0 1;-a0 -a1 -a2]; % Eq.(P8.9.2a)
det(s*eye(3)-A) %characteristic polynomial
ch_eq=charpoly(A) %or, equivalently
```

(b) Let the input $u(t)$ in the state equation (P8.9.2) be dependent on the state as

$$u(t) = Kx(t) = [K_0 x_1(t) \quad K_1 x_2(t) \quad K_2 x_3(t)] \tag{P8.9.4}$$

Then, the state equation can be written as

$$\begin{bmatrix} x_1'(t) \\ x_2'(t) \\ x_3'(t) \end{bmatrix} = \begin{bmatrix} 0 & 1 & 0 \\ 0 & 0 & 1 \\ K_0 - a_0 & K_1 - a_1 & K_2 - a_2 \end{bmatrix} \begin{bmatrix} x_1(t) \\ x_2(t) \\ x_3(t) \end{bmatrix} \tag{P8.9.5}$$

If the parameters of the original system matrix are $a_0 = 1$, $a_1 = -2$, and $a_2 = 3$, what are the values of the gain matrix $K = [K_0 \ K_1 \ K_2]$ you will fix so that the virtual system matrix in the state equation (P8.9.5) has the eigenvalues of $\lambda = -1, -2$, and -3? Note that the characteristic equation of the system whose behavior is described by the state equation (P8.9.5) is

$$s^3 + (a_2 - K_2)s^2 + (a_1 - K_1)s + a_0 - K_0 = 0 \quad \text{(P8.9.6)}$$

and the equation having the roots of $\lambda = -1, -2$, and -3 is

$$(s + 1)(s + 2)(s + 3) = s^3 + 6s^2 + 11s + 6 = 0 \quad \text{(P8.9.7)}$$

8.10 A Homogeneous Differential Equation – An Eigenvalue Equation

Consider the following homogeneous DE (differential equation):

$$\mathbf{x}'(t) = A\mathbf{x}(t); \quad \begin{bmatrix} x_1'(t) \\ x_2'(t) \end{bmatrix} = \begin{bmatrix} -2 & 1 \\ -2 & 0 \end{bmatrix} \begin{bmatrix} x_1(t) \\ x_2(t) \end{bmatrix} \text{ with } \begin{bmatrix} x_1(0) \\ x_2(0) \end{bmatrix} = \begin{bmatrix} 1 \\ -1 \end{bmatrix} \quad \text{(P8.10.1)}$$

Since the eigenvalues of the system matrix A are $-1 + j$ and $-1 - j$, the solution can be written as

$$\mathbf{x}(t) \stackrel{(8.6.8)}{=} \sum_{n=1}^{N} e^{\lambda_n t} \alpha_n \mathbf{v}_n = [\alpha_1 \mathbf{v}_1 \cdots \alpha_N \mathbf{v}_N] \begin{bmatrix} e^{\lambda_1 t} \\ \vdots \\ e^{\lambda_N t} \end{bmatrix} \quad \text{(P8.10.2)}$$

where $\mathbf{v}_n$'s are the eigenvectors and α_n's can be found from the initial conditions by using Eq. (8.6.7) as

$$\begin{bmatrix} \alpha_1 \\ \alpha_2 \end{bmatrix} \stackrel{(8.6.7)}{=} V^{-1}\mathbf{x}(0) \quad \text{(P8.10.3)}$$

Complete the following program "nm08p10.m" whose objective is to solve the DE (P8.10.1) for the time interval [0,5] in two ways, i.e., by using the ODE solver 'ode45()' (Section 6.5.1) and by using the eigenvalue method (Section 8.7) and plot the two solutions. Run the completed program to obtain the solution graphs for $x_1(t)$ and $x_2(t)$.

```
%nm08p10.m
A=[-2 1; -2 0]; x0=[1;-1];
dx=@(t,x)A*x;
t0=0; tf=5; % Initial/Final times
[t4,x4]=ode45(dx,[t0 tf],x0);
[V,L]=eig(A); % Modal matrix composed of eigenvectors
NA=numel(x0);
for n=1:NA, lambda(n,1)=L(n,n); end
a=V???*x0; % Eq.(P8.10.3)
xE=real(?.*repmat(?.',NA,1)*exp(??????*t4.')); % Eq.(P8.10.2)
for n=1:NA
   subplot(2,1,n), plot(t4,x4(:,n), t4,xE(n,:),'r:')
end
```

9

PARTIAL DIFFERENTIAL EQUATIONS

CHAPTER OUTLINE

9.1 Elliptic PDE	510
9.2 Parabolic PDE	515
9.2.1 The Explicit Forward Euler Method	515
9.2.2 The Implicit Backward Euler Method	516
9.2.3 The Crank-Nicholson Method	518
9.2.4 Using the MATLAB function 'pdepe()'	520
9.2.5 Two-Dimensional Parabolic PDEs	523
9.3 Hyperbolic PDES	526
9.3.1 The Explicit Central Difference Method	526
9.3.2 Two-Dimensional Hyperbolic PDEs	529
9.4 Finite Element Method (FEM) for Solving PDE	532
9.5 GUI of MATLAB for Solving PDES – PDEtool	543
9.5.1 Basic PDEs Solvable by PDEtool	543
9.5.2 The Usage of PDEtool	545
9.5.3 Examples of Using PDEtool to Solve PDEs	549
Problems	559

Applied Numerical Methods Using MATLAB®, Second Edition. Won Y. Yang, Jaekwon Kim, Kyung W. Park, Donghyun Baek, Sungjoon Lim, Jingon Joung, Suhyun Park, Han L. Lee, Woo June Choi, and Taeho Im.
© 2020 John Wiley & Sons, Inc. Published 2020 by John Wiley & Sons, Inc.
Companion website: www.wiley.com/go/yang/appliednumericalmethods

What is a partial differential equation (PDE)? It is a class of differential equations involving more than one independent variable. In this chapter, we consider a general second-order PDE in two independent variables x and y, which is written as

$$A(x,y)\frac{\partial^2 u}{\partial x^2} + B(x,y)\frac{\partial^2 u}{\partial x \partial y} + C(x,y)\frac{\partial^2 u}{\partial y^2} = f\left(x,y,u,\frac{\partial u}{\partial x},\frac{\partial u}{\partial y}\right) \quad (9.0.1)$$

for $x_0 \leq x \leq x_f$ and $y_0 \leq y \leq y_f$

subject to the boundary conditions (BCs) given by

$$u(x,y_0) = b_{y_0}(x),\ u(x,y_f) = b_{y_f}(x),\ u(x_0,y) = b_{x_0}(y),\ \text{and}\ u(x_f,y) = b_{x_f}(y) \quad (9.0.2)$$

These PDEs are classified into three groups:

elliptic PDE: if $B^2 - 4AC < 0$
parabolic PDE: if $B^2 - 4AC = 0$
hyperbolic PDE: if $B^2 - 4AC > 0$

These three types of PDE are associated with equilibrium states, diffusion states, and oscillating system, respectively. We will study some numerical methods for solving these PDEs, since their analytical solutions are usually difficult to find.

9.1 ELLIPTIC PDE

As an example, we will deal with a special type of elliptic equation called *Helmholtz's equation*, which is written as

$$\nabla^2 u(x,y) + g(x,y)u(x,y) = \frac{\partial^2 u(x,y)}{\partial x^2} + \frac{\partial^2 u(x,y)}{\partial y^2} + g(x,y)u(x,y) = f(x,y) \quad (9.1.1)$$

over a domain $D = \{(x,y)|x_0 \leq x \leq x_f,\ y_0 \leq y \leq y_f\}$ with some *Dirichlet* (fixed) BCs of

$$u(x_0,y) = b_{x0}(y),\ u(x_f,y) = b_{xf}(y),\ u(x,y_0) = b_{y0}(x),\ \text{and}\ u(x,y_f) = b_{yf}(x) \quad (9.1.2)$$

(cf) Eq. (9.1.1) is called *Poisson's equation* if $g(x,y) = 0$, and it is called *Laplace's equation* if $g(x,y) = 0$ and $f(x,y) = 0$.

To use the difference method as a numerical approach, we divide the domain into M_x sections, each of length $\Delta x = (x_f - x_0)/M_x$ along the x-axis and into M_y sections (Figure 9.1), each of length $\Delta y = (y_f - y_0)/M_y$ along the y-axis,

ELLIPTIC PDE 511

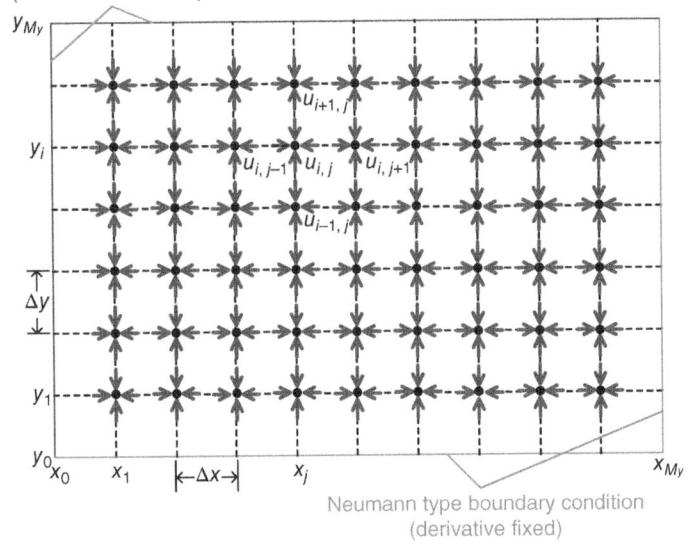

Figure 9.1 Grid for an elliptic PDE with Dirichlet/Neumann boundary conditions.

respectively, and then replace the second derivatives by the three-point central difference approximation:

$$\left.\frac{\partial^2 u(x,y)}{\partial x^2}\right|_{x_j, y_i} \cong \frac{u_{i,j+1} - 2u_{i,j} + u_{i,j-1}}{\Delta x^2} \quad \text{with } x_j = x_0 + j\Delta x, y_i = x_0 + i\Delta y$$
(9.1.3a)

$$\left.\frac{\partial^2 u(x,y)}{\partial y^2}\right|_{x_j, y_i} \cong \frac{u_{i+1,j} - 2u_{i,j} + u_{i-1,j}}{\Delta y^2} \quad \text{with } u_{i,j} = u(x_j, y_i) \quad (9.1.3b)$$

so that, for every interior point (x_j, y_i) with $1 \le i \le M_y - 1$ and $1 \le j \le M_x - 1$, we obtain the difference equation

$$\frac{u_{i,j+1} - 2u_{i,j} + u_{i,j-1}}{\Delta x^2} + \frac{u_{i+1,j} - 2u_{i,j} + u_{i-1,j}}{\Delta y^2} + g_{i,j} u_{i,j} = f_{i,j} \quad (9.1.4)$$

where

$$u_{i,j} = u(x_j, y_i), \quad f_{i,j} = f(x_j, y_i), \quad \text{and} \quad g_{i,j} = g(x_j, y_i)$$

These equations can somehow be arranged into a set of simultaneous equations in the $(M_y - 1)(M_x - 1)$ variables $\{u_{1,1}, u_{1,2}, \ldots, u_{1,M_x-1}, u_{2,1}, \ldots, u_{2,M_x-1}, \ldots, u_{M_y-1,1}, u_{M_y-1,2}, \ldots, u_{M_y-1,M_x-1}\}$, but it seems to be messy to work

```
function [u,x,y]=pde_poisson(f,g,bx0,bxf,by0,byf,D,Mx,My,tol,MaxIter)
%  To solve u_xx +u_yy +g(x,y)u = f(x,y)
%   over the region D=[x0,xf,y0,yf]={(x,y)|x0<=x<=xf, y0<=y<=yf}
%  subject to the BCs (boundary conditions):
%   u(x0,y)=bx0(y), u(xf,y)=bxf(y)
%   u(x,y0)=by0(x), u(x,yf)=byf(x)
%  D = [x0 xf y0 yf] : Domain
%    = [x0 xf y0 yf; 0 1 1 0] : Dirichlet/Neumann/Neumann/Dirichlet BCs
%  Mx = Number of subintervals along x-axis
%  My = Number of subintervals along y-axis
%  tol   : Error tolerance
%  MaxIter: Maximum number of iterations
if nargin<11, MaxIter=500; end
if nargin<10, tol=1e-8; end
if nargin<9
   if nargin==8, My=Mx; else My=50; end
end
if nargin<8, Mx=50; end
x0=D(1); xf=D(2); y0=D(3); yf=D(4);
dx=(xf-x0)/Mx;   x=x0+[0:Mx]*dx;
dy=(yf-y0)/My;   y=y0+[0:My]'*dy;
Mx1=Mx+1; My1=My+1;
% Boundary conditions (9.1.5b)
for m=1:My1
   u(m,[1 Mx1])=[bx0(y(m)) bxf(y(m))]; % Left/Right side
end
for n=1:Mx1
   u([1 My1],n)=[by0(x(n)); byf(x(n))]; % Bottom/Top
end
% Initialize as the average of boundary values
sum_of_bv=sum(sum([u(2:My,[1 Mx1]) u([1 My1],2:Mx)']));
u(2:My,2:Mx)=sum_of_bv/(2*(Mx+My-2));
for i=1:My
   for j=1:Mx
      F(i,j)=f(x(j),y(i)); G(i,j)=g(x(j),y(i));
   end
end
dx2=dx*dx; dy2=dy*dy;   dxy2=2*(dx2+dy2);
rx=dx2/dxy2; ry=dy2/dxy2;  rxy=rx*dy2;
for itr=1:MaxIter
   for j=2:Mx
      for i=2:My
         u(i,j)= ry*(u(i,j+1)+u(i,j-1)) +rx*(u(i+1,j)+u(i-1,j))...
                 +rxy*(G(i,j)*u(i,j)-F(i,j)); % Eq.(9.1.5a)
      end
   end
   if itr>1&max(max(abs(u-u0)))<tol, break; end
   u0=u;
end
```

with, and we may be really in trouble as M_x and M_y become large. A simpler way is to use the iterative methods introduced in Section 2.5. In order to do so, we first need to shape the equations and the BCs into the following form:

$$u_{i,j} = r_y(u_{i,j+1} + u_{i,j-1}) + r_x(u_{i+1,j} + u_{i-1,j}) + r_{xy}(g_{i,j}u_{i,j} - f_{i,j}) \quad (9.1.5a)$$

$$u_{i,0} = b_{x_0}(y_i), \quad u_{i,M_x} = b_{x_f}(y_i), \quad u_{0,j} = b_{y_0}(x_j), \quad \text{and} \quad u_{M_y,j} = b_{y_f}(x_j) \quad (9.1.5b)$$

where

$$\frac{\Delta y^2}{2(\Delta x^2 + \Delta y^2)} = r_y, \quad \frac{\Delta x^2}{2(\Delta x^2 + \Delta y^2)} = r_x, \quad \text{and} \quad \frac{\Delta x^2 \Delta y^2}{2(\Delta x^2 + \Delta y^2)} = r_{xy} \quad (9.1.6)$$

How do we initialize this algorithm? If we have no priori knowledge about the solution, it is reasonable to take the average value of the boundary values as the initial values of $u_{i,j}$.

This process to solve Eq. (9.1.1) with BC (9.1.2) has been cast into the above MATLAB function 'pde_poisson()'.

Example 9.1. Laplace's Equation – Steady-state Temperature Distribution over a Plate.

Consider Laplace's equation

$$\nabla^2 u(x, y) = \frac{\partial^2 u(x, y)}{\partial x^2} + \frac{\partial^2 u(x, y)}{\partial y^2} = 0 \text{ for } 0 \leq x \leq \pi \text{ and } 0 \leq y \leq \pi \quad (E9.1.1)$$

subject to the BCs

$$u(0, y) = 0 \quad (E9.1.2a)$$

$$u(\pi, y) = 0 \quad (E9.1.2b)$$

$$u(x, 0) = \sin(2x) \quad (E9.1.2c)$$

$$u(x, \pi) = 0 \quad (E9.1.2d)$$

where the unknown 2D function $u(x,y)$ may describe the temperature distribution over a square plate having each side 4 units long.

We made the script "nm09e01.m", which uses the MATLAB function 'pde_poisson()' to solve the Laplace's equation given previously, and run it to obtain the solution graph shown in Figure 9.2.

(Q) Does the solution graph satisfy the BCs (E9.1.2a)-(E9.1.2d)?

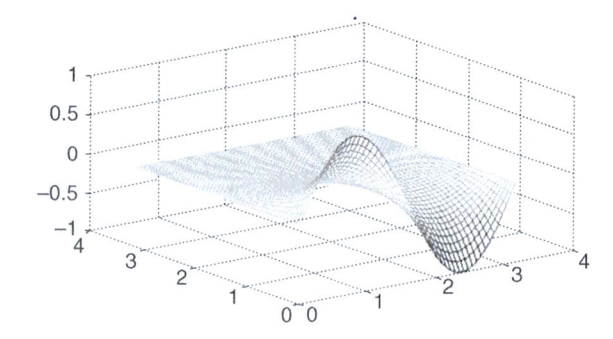

Figure 9.2 Temperature distribution over a rectangular plate for Example 9.1.

```
%nm09e01.m
f=@(x,y)0; g=@(x,y)0;   % Eq.(E9.1.1) w.r.t. Eq.(9.1.1)
% (Rectangular) Domain and BC types
x0=0; xf=pi; y0=0; yf=pi;  % 0/1 for Dirichlet/Neumann conditions
D=[x0 xf y0 yf; 0 0 0 0];
% BCs
bx0=@(y)0; bxf=@(y)0;   % Eq.(E9.1.2a,b)
by0=@(x)sin(2*x); byf=@(x)0;  % Eq.(E9.1.2c,d)
% Numbers of grid points along x-/y-axes, Error tolerance,
Mx=50; My=50; tol=1e-5; MaxIter=600;
[un,xx,yy]=pde_poisson(f,g,bx0,bxf,by0,byf,D,Mx,My,tol,MaxIter);
mesh(xx,yy,un) % Plot the solution graph
```

Now, let us consider the so-called Neumann BCs described as

$$\frac{\partial u(x, y)}{\partial x}\bigg|_{x=x_0} = b'_{x_0}(y) \quad \text{for } x = x_0 \text{ (the left-side boundary)} \quad (9.1.7)$$

Replacing the first derivative on the left-side boundary ($x = x_0$) by its 3-point central difference approximation (5.1.8),

$$\frac{u_{i,1} - u_{i,-1}}{2\Delta x} \approx b'_{x_0}(y_i); \quad u_{i,-1} \approx u_{i,1} - 2b'_{x_0}(y_i)\Delta x \quad \text{for } i = 1, 2, \ldots, M_y - 1$$
(9.1.8)

and then substituting this constraint into Eq. (9.1.5a) at the boundary points, we have

$$u_{i,0} = r_y(u_{i,1} + u_{i,-1}) + r_x(u_{i+1,0} + u_{i-1,0}) + r_{xy}(g_{i,0}u_{i,0} - f_{i,0})$$

$$= r_y(u_{i,1} + u_{i,1} - 2b'_{x_0}(y_i)\Delta x) + r_x(u_{i+1,0} + u_{i-1,0}) r_{xy}(g_{i,0}u_{i,0} - f_{i,0})$$

$$= 2r_y u_{i,1} + r_x(u_{i+1,0} + u_{i-1,0}) + r_{xy}(g_{i,0}u_{i,0} - f_{i,0} - 2b'_{x_0}(y_i)/\Delta x) \quad (9.1.9)$$
$$\text{for } i = 1, 2, \ldots, M_y - 1$$

If the BC on the lower side boundary ($y = y_0$) is also of Neumann type, then we need to write similar equations for $j = 1, 2, \ldots, M_x - 1$

$$u_{0,j} = r_y(u_{0,j+1} + u_{0,j-1}) + 2r_x u_{1,j} + r_{xy}(g_{0,j}u_{0,j} - f_{0,j} - 2b'_{y_0}(x_j)/\Delta y) \quad (9.1.10)$$

and additionally for the left-lower corner point (x_0, y_0),

$$u_{0,0} = 2(r_y u_{0,1} + r_x u_{1,0}) + r_{xy}(g_{0,0} u_{0,0} - f_{0,0} - 2(b'_{x_0}(y_0)/\Delta x + 2b'_{y_0}(x_0)/\Delta y)) \quad (9.1.11)$$

9.2 PARABOLIC PDE

An example of a parabolic PDE is a one-dimensional heat equation describing the temperature distribution $u(x,t)$ (x: position, t: time) as

$$a^2 \frac{\partial^2 u(x,t)}{\partial x^2} = \frac{\partial u(x,t)}{\partial t} \quad \text{for} \quad 0 \le x \le x_f, \ 0 \le t \le t_f \quad (9.2.1)$$

In order for this equation to be solvable, the BCs $u(0,t) = b_0(t)$ and $u(x_f,t) = b_{xf}(t)$ as well as the IC (initial condition) $u(x,0) = i_0(x)$ should be provided.

9.2.1 The Explicit Forward Euler Method

To apply the finite difference method, we divide the spatial domain $[0,x_f]$ into M sections, each of length $\Delta x = x_f/M$ and the time domain $[0,t_f]$ into N segments, each of duration $\Delta t = t_f/N$, and then replace the second partial derivative on the LHS and the first partial derivative on the RHS of Eq. (9.2.1) by the central difference approximation (5.3.1) and the forward difference approximation (5.1.4), respectively, so that we have

$$a^2 \frac{u_{i+1}^k - 2u_i^k + u_{i-1}^k}{\Delta x^2} = \frac{u_i^{k+1} - u_i^k}{\Delta t} \quad (9.2.2)$$

This can be made into the following algorithm, called the *explicit forward Euler method*, which is to be solved iteratively:

$$u_i^{k+1} = r(u_{i+1}^k + u_{i-1}^k) + (1 - 2r)u_i^k \quad \text{with} \quad r = a^2 \frac{\Delta t}{\Delta x^2} \quad \text{for} \quad i = 1, 2, \ldots, M-1 \quad (9.2.3)$$

To find the stability condition of this algorithm, we substitute a trial solution

$$u_i^k = \lambda^k e^{ji\pi/P} \quad (P: \text{any nonzero integer}) \quad (9.2.4)$$

into Eq. (9.2.3) to get

$$\lambda = r(e^{j\pi/P} + e^{-j\pi/P}) + (1 - 2r) = 1 - 2r\left(1 - \cos\left(\frac{\pi}{P}\right)\right) \quad (9.2.5)$$

Since we must have $|\lambda| \leq 1$ for nondivergence, the stability condition turns out to be

$$r \overset{(9.2.3)}{=} a^2 \frac{\Delta t}{\Delta x^2} \leq \frac{1}{2} \quad (9.2.6)$$

This implies that as we decrease the spatial interval Δx for better accuracy, we must also decrease the time step Δt at the cost of more computations in order not to lose the stability.

The explicit forward Euler method has been cast into the following MATLAB function 'pde_heat_exp()'.

```
function [u,x,t]=pde_heat_exp(a,xf,tf,it0,bx0,bxf,M,N)
% To solve a^2*u_xx = u_t  for 0<=x<=xf, 0<=t<=tf
%   IC : u(x,0)=it0(x)
%   BCs: u(0,t)=bx0(t), u(xf,t)=bxf(t)
% M = Number of subintervals along x-axis
% N = Number of subintervals along t-axis
dx=xf/M;   x=[0:M]'*dx;
dt=tf/N;   t=[0:N]*dt;
for i=1:M+1, u(i,1)= it0(x(i)); end
for n=1:N+1, u([1 M+1],n)=[bx0(t(n)); bxf(t(n))]; end
r=a^2*dt/dx/dx,  r1=1-2*r;
for k=1:N
   for i=2:M
      u(i,k+1)= r*(u(i+1,k)+u(i-1,k)) +r1*u(i,k); % Eq.(9.2.3)
   end
end
```

9.2.2 The Implicit Backward Euler Method

In this section, we consider another algorithm called the implicit backward Euler method that comes out from substituting the backward difference approximation (5.1.6) for the first partial derivative on the RHS of Eq. (9.2.1) as

$$a^2 \frac{u_{i+1}^k - 2u_i^k + u_{i-1}^k}{\Delta x^2} = \frac{u_i^k - u_i^{k-1}}{\Delta t} \quad (9.2.7)$$

$$; -r u_{i-1}^k + (1 + 2r)u_i^k - r u_{i+1}^k = u_i^{k-1} \quad \text{with } r = a^2 \frac{\Delta t}{\Delta x^2} \quad (9.2.8)$$

$$\text{for } i = 1, 2, \ldots, M-1$$

If the values of u_0^k and u_M^k at both end points are given from the Dirichlet type of BC, then these equations will be cast into a system of simultaneous equations:

$$\begin{bmatrix} 1+2r & -r & 0 & \cdots & 0 & 0 \\ -r & 1+2r & -r & \cdots & 0 & 0 \\ 0 & -r & 1+2r & \cdots & 0 & 0 \\ \vdots & \vdots & \vdots & \cdots & \vdots & \vdots \\ 0 & 0 & 0 & \cdots & 1+2r & -r \\ 0 & 0 & 0 & \cdots & -r & 1+2r \end{bmatrix} \begin{bmatrix} u_1^k \\ u_2^k \\ u_3^k \\ \vdots \\ u_{M-2}^k \\ u_{M-1}^k \end{bmatrix} = \begin{bmatrix} u_1^{k-1} + ru_0^k \\ u_2^{k-1} \\ u_3^{k-1} \\ \vdots \\ u_{M-2}^{k-1} \\ u_{M-1}^{k-1} + ru_M^k \end{bmatrix} \quad (9.2.9)$$

How about the case where the values of $\partial u/\partial x|_{x=0} = b'_0(t)$ at one end are given? In that case, we approximate this Neumann type of BC by

$$\frac{u_1^k - u_{-1}^k}{2\Delta x} = b'_0(k) \tag{9.2.10}$$

and mix it up with one more equation associated with the unknown variable u_0^k:

$$-ru_{-1}^k + (1+2r)u_0^k - ru_1^k = u_0^{k-1} \tag{9.2.11}$$

to get

$$(1+2r)u_0^k - 2ru_1^k = u_0^{k-1} - 2rb'_0(k)\Delta x \tag{9.2.12}$$

We augment Eq. (9.2.9) with this to write

$$\begin{bmatrix} 1+2r & -2r & 0 & 0 & \cdots & 0 & 0 \\ -r & 1+2r & -r & 0 & \cdots & 0 & 0 \\ 0 & -r & 1+2r & -r & \cdots & 0 & 0 \\ 0 & 0 & -r & 1+2r & \cdots & 0 & 0 \\ \vdots & \vdots & \vdots & \vdots & \cdots & \vdots & \vdots \\ 0 & 0 & 0 & \cdots & \cdots & 1+2r & -r \\ 0 & 0 & 0 & \cdots & \cdots & -r & 1+2r \end{bmatrix} \begin{bmatrix} u_0^k \\ u_1^k \\ u_2^k \\ u_3^k \\ \vdots \\ u_{M-2}^k \\ u_{M-1}^k \end{bmatrix}$$

$$= \begin{bmatrix} u_0^{k-1} - 2rb'_0(k)\Delta x \\ u_1^{k-1} \\ u_2^{k-1} \\ u_3^{k-1} \\ \vdots \\ u_{M-2}^{k-1} \\ u_{M-1}^{k-1} + ru_M^k \end{bmatrix} \tag{9.2.13}$$

Equations such as Eq. (9.2.9) or (9.2.13) are really nice in the sense that they can be solved very efficiently by exploiting their tridiagonal structures and are guaranteed to be stable owing to their diagonal dominancy. The unconditional stability of Eq. (9.2.9) can be shown by substituting Eq. (9.2.4) into Eq. (9.2.2.8):

$$-re^{-j\pi/P} + (1+2r) - re^{j\pi/P} = \frac{1}{\lambda}; \quad \lambda = \frac{1}{1+2r(1-\cos(\pi/P))}; \quad |\lambda| \leq 1 \quad (9.2.14)$$

The following function 'pde_heat_imp()' implements this algorithm to solve the PDE (9.2.1) with the ordinary (Dirichlet type of) BC via Eq. (9.2.9).

```
function [u,x,t]=pde_heat_imp(a,xf,tf,it0,bx0,bxf,M,N)
% To solve a^2*u_xx = u_t  for 0<=x<=xf, 0<=t<=tf
%  IC : u(x,0)=it0(x)
%  BCs: u(0,t)=bx0(t), u(xf,t)=bxf(t)
%  M = Number of subintervals along x-axis
%  N = Number of subintervals along t-axis
dx=xf/M;  x=[0:M]'*dx;   dt=tf/N;  t=[0:N]*dt;
for i=1:M+1, u(i,1)=it0(x(i)); end
for n=1:N+1, u([1 M+1],n)=[bx0(t(n)); bxf(t(n))]; end
r=a^2*dt/dx/dx;  r2=1+2*r;
for i=1:M-1
   A(i,i)=r2; % Eq.(9.2.9)
   if i>1, A(i-1,i)= -r; A(i,i-1)= -r; end
end
for k=2:N+1
   b= [r*u(1,k); zeros(M-3,1); r*u(M+1,k)] +u(2:M,k-1); % Eq.(9.2.9)
   u(2:M,k)= trid(A,b);
end
```

9.2.3 The Crank-Nicholson Method

Here, let us go back to see Eq. (9.2.2.7) and try to improve the implicit backward Euler method. The difference approximation on the LHS is taken at time point k, while the difference approximation on the RHS is taken at the midpoint between time k and $k-1$, if we regard it as the central difference approximation with time step $\Delta t/2$. Doesn't this seem to be inconsistent? How about taking the difference approximation of both sides at the same time point, say, the midpoint between $k+1$ and k, for balance? In order to do so, we take the average of the central difference approximations of the LHS at the midpoint between $k+1$ and k, yielding

$$\frac{a^2}{2}\left(\frac{u_{i+1}^{k+1} - 2u_i^{k+1} + u_{i-1}^{k+1}}{\Delta x^2} + \frac{u_{i+1}^k - 2u_i^k + u_{i-1}^k}{\Delta x^2}\right) = \frac{u_i^{k+1} - u_i^k}{\Delta t} \quad (9.2.15)$$

which leads to the so-called *Crank-Nicholson method*:

$$-ru_{i+1}^{k+1} + 2(1+r)u_i^{k+1} - ru_{i-1}^{k+1} = ru_{i+1}^k + 2(1-r)u_i^k + ru_{i-1}^k \quad \text{with } r = a^2\frac{\Delta t}{\Delta x^2}$$
(9.2.16)

With the Dirichlet/Neumann type of BCs on x_0 and x_M, respectively, this can be made into the following tridiagonal system of equations.

$$\begin{bmatrix} 2(1+r) & -r & 0 & \cdots & 0 & 0 \\ -r & 2(1+r) & -r & \cdots & 0 & 0 \\ 0 & -r & 2(1+r) & \cdots & 0 & 0 \\ \vdots & \vdots & \vdots & \cdots & \vdots & \vdots \\ 0 & 0 & 0 & \cdots & 2(1+r) & -r \\ 0 & 0 & 0 & \cdots & -2r & 2(1+r) \end{bmatrix} \begin{bmatrix} u_1^{k+1} \\ u_2^{k+1} \\ u_3^{k+1} \\ \vdots \\ u_{M-1}^{k+1} \\ u_M^{k+1} \end{bmatrix}$$

$$= \begin{bmatrix} 2(1-r) & r & 0 & \cdots & 0 & 0 \\ r & 2(1-r) & r & \cdots & 0 & 0 \\ 0 & r & 2(1-r) & \cdots & 0 & 0 \\ \vdots & \vdots & \vdots & \cdots & \vdots & \vdots \\ 0 & 0 & 0 & \cdots & 2(1-r) & r \\ 0 & 0 & 0 & \cdots & 2r & 2(1-r) \end{bmatrix} \begin{bmatrix} u_1^k \\ u_2^k \\ u_3^k \\ \vdots \\ u_{M-1}^k \\ u_M^k \end{bmatrix}$$

$$+ \begin{bmatrix} r(u_0^{k+1} + u_0^k) \\ 0 \\ 0 \\ \vdots \\ 0 \\ 2r(b_M'(k+1) + b_M'(k)) \end{bmatrix} \quad (9.2.17)$$

This system of equations can also be solved very efficiently, and its unconditional stability can be shown by substituting Eq. (9.2.4) into Eq. (9.2.16):

$$2\lambda\{1 + r(1 - \cos(\pi/P))\} = 2\{1 - r(1 - \cos(\pi/P))\};$$

$$\lambda = \frac{1 - r(1 - \cos(\pi/P))}{1 + r(1 - \cos(\pi/P))}; \quad |\lambda| \le 1 \quad (9.2.18)$$

This algorithm has been cast into the following MATLAB function 'pde_heat_CN()'.

```
function [u,x,t]=pde_heat_CN(a,xf,tf,it0,bx0,bxf,M,N)
%solve a^2*u_xx = u_t   for 0<=x<=xf, 0<=t<=tf
% IC : u(x,0)=it0(x)
% BCs: u(0,t)=bx0(t), u(xf,t)=bxf(t)
% M = Number of subintervals along x-axis
% N = Number of subintervals along t-axis
dx=xf/M;  x=[0:M]'*dx;
dt=tf/N;  t=[0:N]*dt;
for i=1:M+1, u(i,1)= it0(x(i)); end
for n=1:N+1, u([1 M+1],n)= [bx0(t(n)); bxf(t(n))]; end
r=a^2*dt/dx/dx;
r1=2*(1-r); r2=2*(1+r);
for i=1:M-1
   A(i,i)=r1; % Eq.(9.2.17)
   if i>1
     A(i-1,i)= -r; A(i,i-1)= -r;
   end
end
for k=2:N+1
   b=[r*u(1,k); zeros(M-3,1); r*u(M+1,k)] ...
        +r*(u(1:M-1,k-1)+u(3:M+1,k-1)) +r2*u(2:M,k-1);
   u(2:M,k)=trid(A,b); % Eq.(9.2.17)
end
```

9.2.4 Using the MATLAB function 'pdepe()'

The MATLAB function 'pdepe()' can be used to solve a 1D parabolic or elliptic PDE subject to some ICs and BCs:

$$x^{-m}\frac{\partial}{\partial x}\left(x^m f\left(x,t,u,\frac{\partial u}{\partial x}\right)\right) + s\left(x,t,u,\frac{\partial u}{\partial x}\right) = c\left(x,t,u,\frac{\partial u}{\partial x}\right)\frac{\partial u}{\partial t} \quad (9.2.19)$$
$$\text{for} \quad x_0 \leq x \leq x_f, t_0 \leq t \leq t_f$$

Note that m can be 0, 1, or 2, corresponding to slab, cylindrical, or spherical symmetry, and if $m > 0$, x_0 must be nonnegative. In order for this equation to be solvable, the BCs

$$p_0(x,t,u) + q_0(x,t)f(x,t,u,\partial u/\partial x) = 0 \,\forall\, t \quad \text{for } x = x_0$$
$$p_f(x,t,u) + q_f(x,t)f(x,t,u,\partial u/\partial x) = 0 \,\forall\, t \quad \text{for } x = x_f$$

as well as the IC $u(x,t_0) = i_0(x)$ should be provided.

The syntax of 'pdepe()' is

```
sol = pdepe(m,pdefun,icfun,bcfun,xmesh,tspan,options)
```

where its arguments are as follows:

m	=	Parameter to be set as 0/1/2 depending on the symmetry of slab/cylinder/sphere.
Pdefun	=	A handle to a function returning the components $c, f,$ and s of the PDE (9.2.19) with the syntax of

`[c,f,s]=pdefun(x,t,u,dudx)`

Note that $c, f,$ and s are column vectors of the dimension of the system of PDEs where c stores the diagonal elements of the matrix $c(x, t, u, \partial u/\partial x)$.

icfun	=	A handle to a function $i_0(x)$ returning the column vector of ICs at x with the syntax of

`u = icfun(x)`

bcfun	=	A handle to a function that defines the BCs with the syntax of

`[p0,q0,pf,qf]=bcfun(x0,u0,xf,uf,t)`

xmesh	=	A vector $[x_0 \ x_1 \ \cdots \ x_n]$ (of length greater than 2) specifying the points at which a numerical solution is requested for every value in tspan.
tspan	=	A vector $[t_0 \ t_1 \ \cdots \ t_f]$ (of length greater than 2) specifying the points at which a solution is requested for every value in xmesh.
options	=	A structure of optional parameters which can be built as the output argument of 'odeset()' like

`options = odeset('RelTol',1e-4,'AbsTol',[1e-4 1e-4]);`

sol	=	A 3D array where sol(j,k,i) contains the ith component of the solution at (tspan(j),xspan(k)).

Example 9.2. 1D (One-dimensional) Parabolic PDE – Heat Flow Equation.

Consider the following parabolic PDE

$$a^2 \frac{\partial^2 u(x,t)}{\partial x^2} = \frac{\partial u(x,t)}{\partial t} \quad (a=1) \quad \text{for } 0 \leq x \leq 1 \text{ and } 0 \leq t \leq 0.1 \quad \text{(E9.2.1)}$$

subject to the IC and BCs

$$u(x,0) = \sin \pi x, \quad u(0,t) = 0, \quad u(1,t) = 0 \quad \text{(E9.2.2a,b,c)}$$

We made the following MATLAB function 'nm09e02()' to use the functions 'pde_heat_ana()', 'pde_heat_exp()', 'pde_heat_imp()', 'pde_heat_CN()', and 'pdepe()' for solving this problem (with $a = 1$) and run it to obtain the results shown in Figure 9.3. Note that, with the spatial interval $\Delta x = x_f/M = 1/20$ and the time step $\Delta t = t_f/N = 0.1/100 = 0.001$, we have

$$r = a^2 \frac{\Delta t}{\Delta x^2} = 1 \frac{0.001}{(1/20)^2} = 0.4 \quad \text{(E9.2.3)}$$

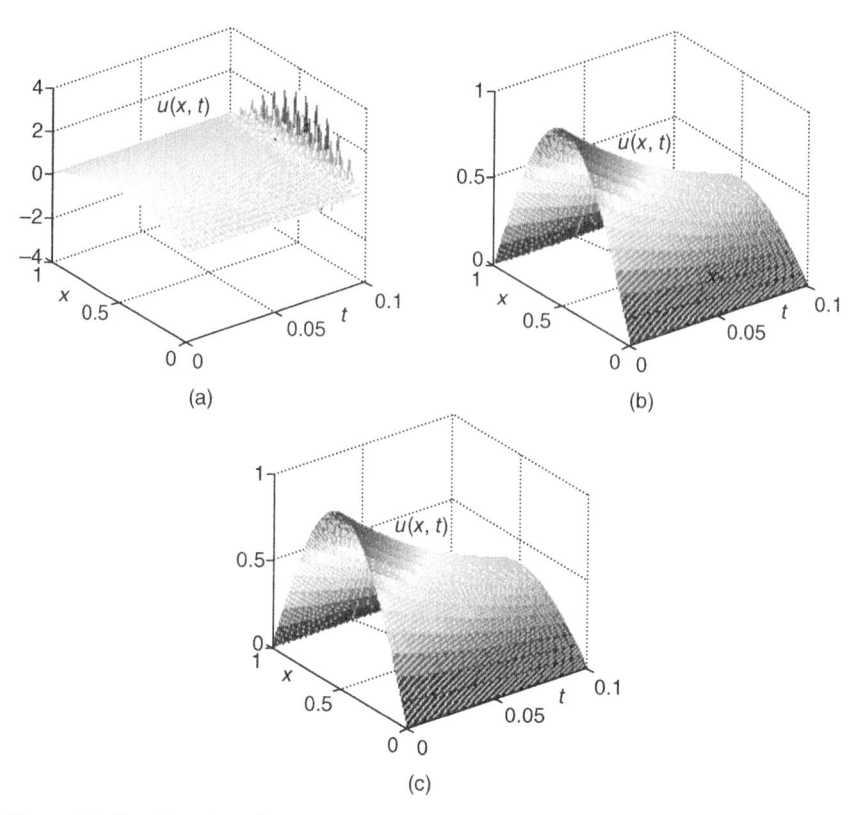

Figure 9.3 Results of various algorithms for a 1D parabolic PDE – heat equation for Example 9.2. (a) The explicit method. (b) The Crank-Nicholson method. (c) Using pdepe().

which satisfies the stability condition ($r \leq 1/2$) (9.2.6), and all the three methods lead to fair results with a relative error of about 0.013. But, if we decrease the spatial interval to $\Delta x = 1/25$ for better resolution so that $r = 0.625$, violating the stability condition, then the explicit forward Euler method ('pde_heat_exp()') blows up because of instability as shown in Figure 9.3a, while the implicit backward Euler method ('pde_heat_imp()') and the Crank-Nicholson method ('pde_heat_CN()') work quite well as shown in Figure 9.3b,c. Now, with the spatial interval $\Delta x = 1/25$ and the time step $\Delta t = 0.1/120$, the explicit method as well as the other ones work well with a relative error less than 0.001 in return for somewhat (30%) more computations, despite that $r = 0.5208$ does not strictly satisfy the stability condition. This implies that the condition ($r \leq 1/2$) for stability of the explicit forward Euler method is not a necessary one, but only a sufficient one. Besides, if it converges, its accuracy may be better than that of the implicit backward Euler method, but generally no better than that of the Crank-Nicholson method.

```
function [u_pdepe,x,t]=nm09e02
a=1; % The parameter of Eq.(E9.2.1)
it0=@(x)sin(pi*x); % IC (E9.2.2a)
bx0=@(t)0; bxf=@(t)0; % BC (E9.2.2b,c)
xf=1; M=25; tf=0.1; N=100; % r=0.625
[u_exp,x,t]=pde_heat_exp(a,xf,tf,it0,bx0,bxf,M,N); % Explicit method
[u_imp,x,t]=pde_heat_imp(a,xf,tf,it0,bx0,bxf,M,N); % Implicit method
[u_CN,x,t]=pde_heat_CN(a,xf,tf,it0,bx0,bxf,M,N); % Crank-Nicholson method
subplot(131), mesh(t,x,u_exp),
xlabel('t'), ylabel('x'), title('Numerical solution using pde heat exp()')
subplot(132), mesh(t,x,u_CN),
xlabel('t'), ylabel('x'), title('Numerical solution using pde heat CN()')
% Using pdepe() ..........................................................
m=0; sol=pdepe(m,@pdex1pde,@pdex1ic,@pdex1bc,x,t);
u_pdepe=sol(:,:,1); % Extract the first solution component as u.
% Analytical solution
uo=@(x,t)sin(pi*x)*exp(-pi*pi*t);
Uo=uo(x,t); aUo=abs(Uo)+eps; % Values of true analytical solution
Nterm=10;
[u_an,A]=pde_heat_ana(a,xf,tf,it0,M,N,Nterm);
MN=M*N; err_an= norm((u_an-Uo)./aUo)/MN
err_exp= norm((u_exp-Uo)./aUo)/MN
err_imp= norm((u_imp-Uo)./aUo)/MN
err_CN=norm((u_CN-Uo)./aUo)/MN
err_pdepe=norm((u_pdepe.'-Uo)./aUo)/MN
err_an= norm((u_an-Uo)./aUo)/MN
subplot(133), mesh(t,x,u_pdepe.') % surf(t,x,u)
xlabel('t'), ylabel('x'), title('Numerical solution using pdepe()')
% ........................................................................
function [c,f,s] = pdex1pde(x,t,u,DuDx)
% defines the PDE by setting the parameters
c = 1; f = DuDx; s = 0; % c=1/a;
% ........................................................................
function u0 = pdex1ic(x)
u0 = sin(pi*x); % Initial condition (E9.2.2a)
% ........................................................................
function [pl,ql,pr,qr] = pdex1bc(xl,ul,xr,ur,t)
pl = ul; ql = 0; % Left boundary condition
pr = ur; qr = 0; % Right boundary condition
```

9.2.5 Two-Dimensional Parabolic PDEs

Another example of a parabolic PDE is a 2D *heat equation* describing the temperature distribution $u(x,y,t)$ ((x,y): position, t: time) as

$$a^2 \left(\frac{\partial^2 u(x,y,t)}{\partial x^2} + \frac{\partial^2 u(x,y,t)}{\partial y^2} \right) = \frac{\partial u(x,y,t)}{\partial t} \quad (9.2.21)$$

for $x_0 \leq x \leq x_f, y_0 \leq y \leq y_f, t_0 \leq t \leq t_f$

In order for this equation to be solvable, we should be provided with the BCs

$$u(x_0,y,t) = b_{x_0}(y,t), \quad u(x_f,y,t) = b_{x_f}(y,t),$$
$$u(x,y_0,t) = b_{y_0}(x,t), \quad \text{and} \quad u(x,y_f,t) = b_{y_f}(x,t)$$

as well as the IC $u(x,y,t_0) = i_{t_0}(x,y)$.

We replace the first-order time derivative on the RHS by the 3-point central difference at the midpoint $(t_{k+1} + t_k)/2$ just as with the Crank-Nicholson method. We also replace one of the second-order derivatives, u_{xx} and u_{yy}, by the 3-point central difference approximation at time t_k and the other at time t_{k+1}, yielding

```
function [u,x,y,t]=pde_heat2_ADI(a,D,tf,it0,bxyt,Mx,My,N)
%To solve a^2*(u_xx+u_yy)=u_t for D(1)<=x<=D(2), D(3)<=y<=D(4), 0<=t<=tf
%  IC: u(x,y,0)  = it0(x,y)
%  BC: u(x,y,t) = bxyt(x,y,t) for (x,y)cB
% Mx/My = Number of subintervals along x/y-axis
% N = Number of subintervals along t-axis
dx=(D(2)-D(1))/Mx;   x=D(1)+[0:Mx]*dx;
dy=(D(4)-D(3))/My;   y=D(3)+[0:My]'*dy;
dt=tf/N;  t=[0:N]*dt;
% Initialization
for j=1:Mx+1,  for i=1:My+1, u(i,j)= it0(x(j),y(i)); end, end
rx= a^2*dt/(dx*dx); rx1=1+2*rx; rx2=1-2*rx;
ry= a^2*dt/(dy*dy); ry1=1+2*ry; ry2=1-2*ry;
for j=1:Mx-1 % Eq.(9.2.24a)
   Ay(j,j)= ry1; if j>1, Ay(j-1,j)= -ry; Ay(j,j-1)= -ry; end
end
for i=1:My-1 % Eq.(9.2.24b)
   Ax(i,i)= rx1; if i>1, Ax(i-1,i)= -rx; Ax(i,i-1)= -rx; end
end
for k=1:N
  u_1=u;   t=k*dt;
  for i=1:My+1 % BCs
     u(i,[1 Mx+1])= feval(bxyt,x([1 Mx+1]),y(i),t);
  end
  for j=1:Mx+1
     u([1 My+1],j)= feval(bxyt,x(j),y([1;My+1]),t);
  end
  if mod(k,2)==0
    for i=2:My
       jj=2:Mx;
       bx= [ry*u(i,1) zeros(1,Mx-3) ry*u(i,My+1)] ...
                +rx*(u_1(i-1,jj)+u_1(i+1,jj)) +rx2*u_1(i,jj);
       u(i,jj)= trid(Ay,bx')'; % Eq.(9.2.24a)
    end
   else
    for j=2:Mx
       ii=2:My;
       by= [rx*u(1,j); zeros(My-3,1); rx*u(Mx+1,j)] ...
                +ry*(u_1(ii,j-1)+u_1(ii,j+1)) +ry2*u_1(ii,j);
       u(ii,j)= trid(Ax,by); % Eq.(9.2.24b)
    end
  end
end
```

$$a^2 \left(\frac{u_{i,j+1}^k - 2u_{i,j}^k + u_{i,j-1}^k}{\Delta x^2} + \frac{u_{i+1,j}^{k+1} - 2u_{i,j}^{k+1} + u_{i-1,j}^{k+1}}{\Delta y^2} \right) = \frac{u_{i,j}^{k+1} - u_{i,j}^k}{\Delta t} \quad (9.2.22)$$

PARABOLIC PDE **525**

This seems to be attractive since it can be formulated into a tridiagonal system of equations in $u_{i+1,j}^{k+1}$, $u_{i,j}^{k+1}$, and $u_{i-1,j}^{k+1}$. But, why do we treat u_{xx} and u_{yy} with discrimination, i.e. evaluate one at time t_k and the other at time t_{k+1} in a fixed manner? In an alternate manner, we write the difference equation for the next time point t_{k+1} as

$$a^2 \left(\frac{u_{i,j+1}^{k+1} - 2u_{i,j}^{k+1} + u_{i,j-1}^{k+1}}{\Delta x^2} + \frac{u_{i+1,j}^{k} - 2u_{i,j}^{k} + u_{i-1,j}^{k}}{\Delta y^2} \right) = \frac{u_{i,j}^{k+2} - u_{i,j}^{k+1}}{\Delta t} \quad (9.2.23)$$

This solution process, proposed by Peaceman and Rachford[P-1] and referred to as the *alternating direction implicit* (*ADI*) method, can be put into the following algorithm:

$$-r_y(u_{i-1,j}^{k+1} + u_{i+1,j}^{k+1}) + (1 + 2r_y)u_{i,j}^{k+1} = r_x(u_{i,j-1}^{k} + u_{i,j+1}^{k}) + (1 - 2r_x)u_{i,j}^{k} \quad (9.2.24a)$$

$$\text{for } j = 1, 2, \ldots, M_x - 1$$

$$-r_x(u_{i,j-1}^{k+2} + u_{i,j+1}^{k+2}) + (1 + 2r_x)u_{i,j}^{k+2} = r_y(u_{i-1,j}^{k+1} + u_{i+1,j}^{k+1}) + (1 - 2r_y)u_{i,j}^{k+1}$$

$$\text{for } i = 1, 2, \ldots, M_y - 1 \quad (9.2.24b)$$

with

$$r_x = a^2 \Delta t / \Delta x^2, \quad r_y = a^2 \Delta t / \Delta y^2, \quad \Delta x = (x_f - x_0)/M_x,$$
$$\Delta y = (y_f - y_0)/M_y, \quad \Delta t = (t_f - t_0)/N$$

This algorithm for solving a 2D heat equation (9.2.21) has been cast into the above MATLAB function 'pde_heat2_ADI()'.

Example 9.3. A Parabolic PDE – 2D Temperature Diffusion over a Plate. Consider a 2D parabolic PDE

$$10^{-4} \left(\frac{\partial^2 u(x,y,t)}{\partial x^2} + \frac{\partial^2 u(x,y,t)}{\partial y^2} \right) = \frac{\partial u(x,y,t)}{\partial t} \quad (E9.3.1)$$

$$\text{for } 0 \le x \le 4, \ 0 \le y \le 4, \text{ and } 0 \le t \le 5000$$

subject to the IC and BCs:

$$u(x,y,0) = 0 \quad \text{for all } (x,y) \quad (E9.3.2a)$$

$$u(x,y,t) = e^y \cos x - e^x \cos y \text{ for } x = 0, \ x = 4, \ y = 0, \ y = 4, \text{ and } \forall\, t \quad (E9.3.2b)$$

The following MATLAB script "nm09e03.m", which uses the function 'pde_heat2_ADI()' to solve this equation, has been run to yield the solution graph at the final time as shown in Figure 9.4.

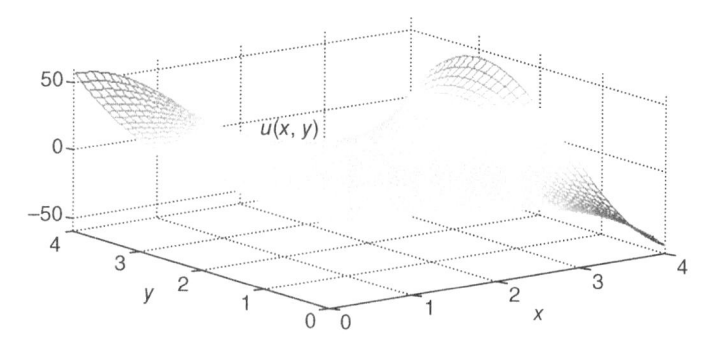

Figure 9.4 Temperature distribution over a plate for Example 9.3.

```
%nm09e03.m
a=1e-2;
it0=@(x,y)0; % IC (E9.3.2a)
bxyt=@(x,y,t)exp(y)*cos(x)-exp(x)*cos(y); % BC (E9.3.2b)
D=[0 4 0 4]; % Solution region
t0=0; tf=5000; Mx=40; My=40; N=50;
[u,x,y,t]=pde_heat2_ADI(a,D,tf,it0,bxyt,Mx,My,N);
mesh(x,y,u)
```

9.3 HYPERBOLIC PDES

An example of a hyperbolic PDE is a 1D *wave equation* for the amplitude function $u(x,t)$ (x: position, t: time) as

$$a^2 \frac{\partial^2 u(x,t)}{\partial x^2} = \frac{\partial^2 u(x,t)}{\partial t^2} \quad \text{for } 0 \le x \le x_f \text{ and } 0 \le t \le t_f \quad (9.3.1)$$

In order for this equation to be solvable, the BCs $u(0, t) = b_{x0}(t)$ and $u(x_f, t) = b_{xf}(t)$ as well as the ICs $u(x,0) = i_0(x)$ and $\partial u/\partial t|_{t=0}(x,0) = i'_0(x)$ should be provided.

9.3.1 The Explicit Central Difference Method

In the same way with the parabolic PDEs, we replace the second derivatives on both sides of Eq. (9.3.1) by their 3-point central difference approximations (5.3.1) as

$$a^2 \frac{u_{i+1}^k - 2u_i^k + u_{i-1}^k}{\Delta x^2} = \frac{u_i^{k+1} - 2u_i^k + u_i^{k-1}}{\Delta t^2} \quad \text{with} \quad \Delta x = \frac{x_f}{M} \text{ and } \Delta t = \frac{t_f}{N} \quad (9.3.2)$$

which leads to the explicit central difference method:

$$u_i^{k+1} = r(u_{i+1}^k + u_{i-1}^k) + 2(1-r)u_i^k - u_i^{k-1} \text{ with } r = a^2 \frac{\Delta t^2}{\Delta x^2} \quad (9.3.3)$$

Since $u_i^{-1} = u(x_i, -\Delta t)$ is not given, we cannot get u_i^1 directly from this (9.3.3) with $k = 0$:

$$u_i^1 = r(u_{i+1}^0 + u_{i-1}^0) + 2(1-r)u_i^0 - u_i^{-1} \quad (9.3.4)$$

Therefore, we approximate the IC on the derivative by the central difference as

$$\frac{u_i^1 - u_i^{-1}}{2\Delta t} = i_0'(x_i) \quad (9.3.5)$$

and make use of this to remove u_i^{-1} from Eq. (9.3.3):

$$u_i^1 = r(u_{i+1}^0 + u_{i-1}^0) + 2(1-r)u_i^0 - (u_i^1 - 2i_0'(x_i)\Delta t);$$

$$u_i^1 = \frac{1}{2}r(u_{i+1}^0 + u_{i-1}^0) + (1-r)u_i^0 + i_0'(x_i)\Delta t \quad (9.3.6)$$

We use Eq. (9.3.6) together with the ICs to get u_i^1 and then go on with Eq. (9.3.3) for $k = 1, 2, \ldots$. Here, note the following:

- We must have $r \leq 1$ to guarantee the stability.
- The accuracy of the solution gets better as r becomes larger so that Δx decreases.

It is therefore reasonable to select $r = 1$.

```
function [u,x,t]=pde_wave(a,xf,tf,it0,i1t0,bx0,bxf,M,N)
% To solve a^2*u_xx = u_tt   for 0<=x<=xf, 0<=t<=tf
%   IC: u(x,0)=it0(x),  u_t(x,0)=i1t0(x)
%   BC: u(0,t)=bx0(t),  u(xf,t)=bxf(t)
% M = Number of subintervals along x-axis
% N = Number of subintervals along t-axis
dx=xf/M;   x=[0:M]'*dx;
dt=tf/N;   t=[0:N]*dt;
for i=1:M+1, u(i,1)= it0(x(i)); end
for k=1:N+1, u([1 M+1],k)=[bx0(t(k)); bxf(t(k))]; end
r=(a*dt/dx)^2;  r1=r/2;  r2=2*(1-r);
u(2:M,2)= r1*u(1:M-1,1) +(1-r)*u(2:M,1) +r1*u(3:M+1,1) ...
             +dt*i1t0(x(2:M)); % Eq.(9.3.6)
for k=3:N+1
   u(2:M,k)= r*u(1:M-1,k-1) +r2*u(2:M,k-1) +r*u(3:M+1,k-1)...
             -u(2:M,k-2); % Eq.(9.3.3)
end
```

528 PARTIAL DIFFERENTIAL EQUATIONS

The stability condition can be obtained by substituting Eq. (9.2.4) into Eq. (9.3.3) and applying the Jury test[P-3]:

$$\lambda = 2r\cos\left(\frac{\pi}{P}\right) + 2(1-r) - \lambda^{-1}; \quad \lambda^2 + 2\left\{r\left(1-\cos\left(\frac{\pi}{P}\right)\right) - 1\right\}\lambda + 1 = 0$$

We need the solution of this equation to be inside the unit circle for stability, which requires

$$r \leq \frac{1}{1-\cos(\pi/P)}; \quad r \stackrel{(9.3.3)}{=} a^2\frac{\Delta t^2}{\Delta x^2} \leq 1 \qquad (9.3.7)$$

This process to solve a 1D wave equation like Eq. (9.3.1) has been cast into the above MATLAB function 'pde_wave()'.

Example 9.4. A Hyperbolic PDE – 1D Wave (Vibration) on a String Fixed at Both Ends.
Consider a 1D hyperbolic PDE

$$\frac{\partial^2 u(x,t)}{\partial x^2} = \frac{\partial^2 u(x,t)}{\partial t^2} \quad \text{for} \quad 0 \leq x \leq 1 \quad \text{and} \quad 0 \leq t \leq 2 \qquad \text{(E9.4.1)}$$

subject to the BCs and ICs

$$u(0,t) = 0, \cdots u(1,t) = 0 \quad \text{for all} \quad t \qquad \text{(E9.4.2a,b)}$$

$$u(x,t)|_{t=0} = x(1-x), \quad \left.\frac{\partial u}{\partial t}(x,t)\right|_{t=0} = 0 \quad \text{for} \quad t = 0 \qquad \text{(E9.4.3a,b)}$$

We can compose the following MATLAB script "nm09e04.m", which uses the MATLAB function 'pde_wave()' to solve this equation for $u(x,t)$ and shows a dynamic (motion) picture showing how $u(x,t)$ changes with time $t = n\Delta t$ as illustrated in Figure 9.5.

```
%nm09e04.m
a=1;
% BCs and ICs
bx0t=@(t)0;  bxft=@(t)0;      % Eq.(E9.4.2a,b)
it0=@(x)x.*(1-x); i1t0=@(x)0; % Eq.(E9.4.3a,b)
xf=1; M=20;  tf=2; N=50;
% To solve the wave equation (E9.4.1)
[un,x,t]=pde_wave(a,xf,tf,it0,i1t0,bx0t,bxft,M,N);
figure(1), clf
mesh(t,x,un)
figure(2), clf
for n=1:N % Motion picture
   plot(x,un(:,n)), axis([0 xf -0.3 0.3]), pause(0.2)
end
```

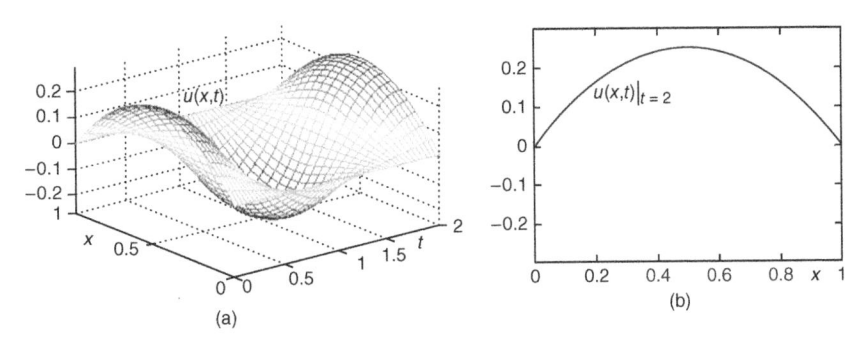

Figure 9.5 Solution to the 1D hyperbolic PDE for Example 9.4. (a) The solution $u(x,t)$. (b) A snapshot of the solution at $t = t_f$.

9.3.2 Two-Dimensional Hyperbolic PDEs

In this section, we consider a 2D wave equation for the amplitude function $u(x,y,t)$ ((x,y): position, t: time) as

$$a^2 \left(\frac{\partial^2 u(x,y,t)}{\partial x^2} + \frac{\partial^2 u(x,y,t)}{\partial y^2} \right) = \frac{\partial^2 u(x,t)}{\partial t^2} \quad (9.3.8)$$

for $0 \le x \le x_f, 0 \le y \le y_f,$ and $0 \le t \le t_f$

In order for this equation to be solvable, we should be provided with the BCs

$$u(0,y,t) = b_{x_0}(y,t), \quad u(x_f,y,t) = b_{x_f}(y,t) \quad (9.3.9a,b)$$

$$u(x,0,t) = b_{y_0}(x,t), \quad u(x,y_f,t) = b_{y_f}(x,t) \quad (9.3.9c,d)$$

as well as the ICs

$$u(x,y,0) = i_0(x,y) \text{ and } \left.\frac{\partial u}{\partial t}\right|_{t=0}(x,y,0) = i'_0(x,y) \quad (9.3.10)$$

In the same way with the 1D case, we replace the second derivatives on both sides by their 3-point central difference approximation as

$$a^2 \left(\frac{u_{i,j+1}^k - 2u_{i,j}^k + u_{i,j-1}^k}{\Delta x^2} + \frac{u_{i+1,j}^k - 2u_{i,j}^k + u_{i-1,j}^k}{\Delta y^2} \right) = \frac{u_{i,j}^{k+1} - 2u_{i,j}^k + u_{i,j}^{k-1}}{\Delta t^2}$$

(9.3.11)

with $\Delta x = \dfrac{x_f}{M_x}, \Delta y = \dfrac{y_f}{N_y},$ and $\Delta t = \dfrac{T}{N}$

which leads to the explicit central difference method:

$$u_{i,j}^{k+1} = r_x(u_{i,j+1}^k + u_{i,j-1}^k) + 2(1 - r_x - r_y)u_{i,j}^k + r_y(u_{i+1,j}^k + u_{i-1,j}^k) - u_{i,j}^{k-1} \quad (9.3.12)$$

$$\text{with} \quad r_x = a^2 \frac{\Delta t^2}{\Delta x^2} \quad \text{and} \quad r_y = a^2 \frac{\Delta t^2}{\Delta y^2}$$

Since $u_{i,j}^{-1} = u(x_j, y_i, -\Delta t)$ is not given, we cannot get $u_{i,j}^1$ directly from this (9.3.12) with $k = 0$:

```
function [u,x,y,t]=pde_wave2(a,D,tf,it0,i1t0,bxyt,Mx,My,N)
%To solve a^2*(u_xx+u_yy)=u_tt for D(1)<=x<=D(2),D(3)<=y<=D(4),0<=t<=tf
%  IC: u(x,y,0) = it0(x,y),   u_t(x,y,0) = i1t0(x,y)
%  BC: u(x,y,t) = bxyt(x,y,t) for (x,y)cBoundary
%  Mx/My = Number of subintervals along x/y-axis
%  N = Number of subintervals along the t-axis
dx=(D(2)-D(1))/Mx;  x=D(1)+[0:Mx]*dx;
dy=(D(4)-D(3))/My;  y=D(3)+[0:My]'*dy;
dt=tf/N;  t=[0:N]*dt;
% Initialization
u=zeros(My+1,Mx+1);  ut=zeros(My+1,Mx+1);
for j=2:Mx
   for i=2:My,  u(i,j)=it0(x(j),y(i));  ut(i,j)=i1t0(x(j),y(i));  end
end
adt2= a^2*dt*dt;  rx= adt2/(dx*dx);  ry=adt2/(dy*dy);
rxy1= 1-rx-ry;  rxy2= rxy1*2;
u_1=u;
for k=0:N
  t=k*dt;
  for i=1:My+1 % BC
    u(i,[1 Mx+1])= [bxyt(x(1),y(i),t) bxyt(x(Mx+1),y(i),t)];
  end
  for j=1:Mx+1
    u([1 My+1],j)= [bxyt(x(j),y(1),t); bxyt(x(j),y(My+1),t)];
  end
  if k==0
    for i=2:My
      for j=2:Mx  % Eq.(9.3.15)
        u(i,j)=0.5*(rx*(u_1(i,j-1)+u_1(i,j+1))...
              +ry*(u_1(i-1,j)+u_1(i+1,j)))+rxy1*u(i,j)+dt*ut(i,j);
      end
    end
   else
    for i=2:My
      for j=2:Mx  % Eq.(9.3.12)
        u(i,j)= rx*(u_1(i,j-1)+u_1(i,j+1))...
              +ry*(u_1(i-1,j)+u_1(i+1,j)) +rxy2*u(i,j) -u_2(i,j);
      end
    end
  end
  u_2=u_1;  u_1=u; % update the buffer memory
  mesh(x,y,u), axis([0 2 0 2 -0.1 0.1]), pause
end
```

$$u_{i,j}^1 = r_x(u_{i,j+1}^0 + u_{i,j-1}^0) + 2(1 - r_x - r_y)u_{i,j}^0 + r_y(u_{i+1,j}^0 + u_{i-1,j}^0) - u_{i,j}^{-1} \quad (9.3.13)$$

Therefore, we approximate the IC on the derivative by the central difference as

$$\frac{u_{i,j}^1 - u_{i,j}^{-1}}{2\Delta t} = i_0'(x_j, y_i) \quad (9.3.14)$$

and make use of this to remove $u_{i,j}^{-1}$ from Eq. (9.3.13) to have

$$u_{i,j}^1 = \frac{1}{2}\{r_x(u_{i,j+1}^0 + u_{i,j-1}^0) + r_y(u_{i+1,j}^0 + u_{i-1,j}^0)\} + (1 - r_x - r_y)u_{i,j}^0 + i_0'(x_j, y_i)\Delta t \quad (9.3.15)$$

We use this equation together with the ICs to get $u_{i,j}^1$ and then go on using Eq. (9.3.12) for $k = 1, 2, \ldots$. A sufficient condition for stability [S-1, Section 9.6] is

$$r = \frac{4a^2 \Delta t^2}{\Delta x^2 + \Delta y^2} \leq 1 \quad (9.3.16)$$

This process to solve the 2D wave equation (9.3.8) has been cast into the above MATLAB function 'pde_wave2()'.

Example 9.5. A Hyperbolic PDE – 2D Wave (Vibration) over a Square Membrane. Consider a 2D hyperbolic PDE

$$\frac{1}{4}\left(\frac{\partial^2 u(x, y, t)}{\partial x^2} + \frac{\partial^2 u(x, y, t)}{\partial y^2}\right) = \frac{\partial^2 u(x, y, t)}{\partial t^2} \quad (E9.5.1)$$

for $0 \leq x \leq 2, 0 \leq y \leq 2$, and $0 \leq t \leq 2$

subject to the zero BCs and the ICs:

$$u(0, y, t) = 0 \quad (E9.5.2a)$$
$$u(2, y, t) = 0 \quad (E9.5.2b)$$
$$u(x, 0, t) = 0 \quad (E9.5.2c)$$
$$u(x, 2, t) = 0 \quad (E9.5.2d)$$
$$u(x, y, t)|_{t=0} = 0.1 \sin(\pi x) \sin(\pi y/2) \quad (E9.5.3a)$$
$$\partial u(x, y, t)/\partial t|_{t=0} = 0 \quad (E9.5.3b)$$

We made the following MATLAB script "nm09e05.m" to use the function 'pde_wave2()' for solving the PDE and run it to see a motion picture showing how $u(x, t)$ changes with time $t = n\Delta t$ as illustrated in Figure 9.6. Note that we can be sure of stability since we have

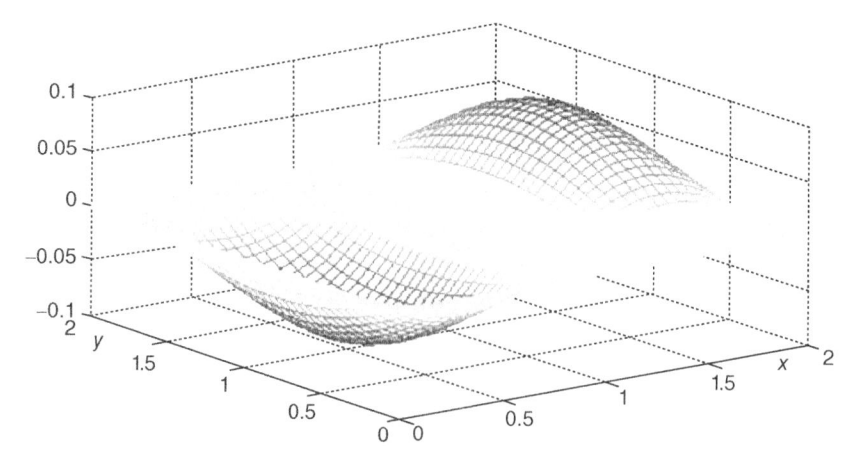

Figure 9.6 Solution to the 2D hyperbolic PDE – vibration of a square membrane (Example 9.5).

$$r \stackrel{(9.3.16)}{=} \frac{4a^2 \Delta t^2}{\Delta x^2 + \Delta y^2} = \frac{4(1/4)(2/20)^2}{(2/20)^2 + (2/20)^2} = \frac{1}{2} \leq 1 \qquad (E9.5.4)$$

```
%nm09e05.m
bxyt=@(x,y,t)0; % BC (E9.5.2)
it0=@(x,y)0.1*sin(pi*x)*sin(pi*y/2); % IC (E9.5.3a)
i1t0=@(x,y)0; % IC (E9.5.3b)
a=1/2; D=[0 2 0 2]; tf=2; Mx=40; My=40; N=40;
[u,x,y,t]=pde_wave2(a,D,tf,it0,i1t0,bxyt,Mx,My,N);
```

9.4 FINITE ELEMENT METHOD (FEM) FOR SOLVING PDE

The FEM method is another procedure used in finding approximate numerical solutions to BVPs/PDEs. It can handle irregular boundaries in the same way as regular boundaries[R-1, S-2, Z-1]. It consists of the following steps to solve the elliptic PDE

$$\frac{\partial^2 u(x,y)}{\partial x^2} + \frac{\partial^2 u(x,y)}{\partial y^2} + g(x,y)u(x,y) = f(x,y) \qquad (9.4.1)$$

for the domain D enclosed by the boundary B on which the BC is given as

$$u(x,y) = b(x,y) \text{ on the boundary } B \qquad (9.4.2)$$

FINITE ELEMENT METHOD (FEM) FOR SOLVING PDE **533**

1. Discretize the 2D domain D into, say, N_s subregions $\{S_1, S_2, \ldots, S_{N_s}\}$ such as triangular elements, neither necessarily of the same size nor necessarily covering the entire domain completely and exactly.
2. Specify the positions of N_n nodes and number them starting from the boundary nodes, say, $n = 1, \ldots, N_b$ and then, the interior nodes, say, $n = N_b + 1, \ldots, N_n$.
3. Define the basis/shape/interpolation functions

$$\varphi_n(x, y) = \{\phi_{n,s}, \text{ for } s = 1, \cdots, N_s\} \forall (x, y) \in D \quad (9.4.3a)$$

$$\phi_{n,s}(x, y) = p_{n,s}(1) + p_{n,s}(2)x + p_{n,s}(3)y \quad \text{for each subregion } S_s \quad (9.4.3b)$$

collectively for all subregions $s = 1, \ldots, N_s$ and for each node $n = 1, \ldots, N_n$ so that ϕ_n is one (1) only at node n, and zero (0) at all other nodes. Then, the approximate solution of the PDE is a linear combination of basis functions $\phi_n(x,y)$ as

$$u(x, y) = \mathbf{c}^T \boldsymbol{\phi}(x, y) = \sum_{n=1}^{N_n} c_n \phi_n(x, y)$$

$$= \sum_{n=1}^{N_b} c_n \phi_n + \sum_{n=N_b+1}^{N_n} c_n \phi_n = \mathbf{c}_1^T \boldsymbol{\phi}_1 + \mathbf{c}_2^T \boldsymbol{\phi}_2 \quad (9.4.4)$$

where

$$\boldsymbol{\phi}_1 = [\phi_1 \; \phi_2 \; \cdot \; \phi_{N_b}]^T, \quad \mathbf{c}_1 = [c_1 \; c_2 \; \cdot \; c_{N_b}]^T \quad (9.4.5a)$$

$$\boldsymbol{\phi}_2 = [\phi_{N_b+1} \; \phi_{N_b+2} \; \cdot \; \phi_{N_n}]^T, \quad \mathbf{c}_2 = [c_{N_b+1} \; c_{N_b+2} \; \cdot \; c_{N_n}]^T \quad (9.4.5b)$$

For each subregion $s = 1, \ldots, N_s$, this solution can be written as

$$\phi_s(x, y) = \sum_{n=1}^{N_b} c_n \phi_{n,s}(x, y) = \sum_{n=1}^{N_n} c_n (p_{n,s}(1) + p_{n,s}(2)x + p_{n,s}(3)y) \quad (9.4.6)$$

4. Set the values of the boundary node coefficients in $\mathbf{c}_1$ to the boundary values according to the BC.
5. Determine the values of the interior node coefficients in $\mathbf{c}_2$ by solving the system of equations

$$A_2 \mathbf{c}_2 = \mathbf{d} \quad (9.4.7)$$

where

$$A_1 = \sum_{s=1}^{N_s} \left\{ \left[\frac{\partial}{\partial x}\boldsymbol{\phi}_{2,s}\right]\left[\frac{\partial}{\partial x}\boldsymbol{\phi}_{1,s}\right]^T + \left[\frac{\partial}{\partial y}\boldsymbol{\phi}_{2,s}\right]\left[\frac{\partial}{\partial y}\boldsymbol{\phi}_{1,s}\right]^T \right. \\ \left. - g(x_s, y_s)\boldsymbol{\phi}_{2,s}\boldsymbol{\phi}_{1,s}^T \right\} \Delta S_s \quad (9.4.8)$$

$$\boldsymbol{\phi}_{1,s} = [\phi_{1,s} \; \phi_{2,s} \; \cdots \; \phi_{N_b,s}]^T$$

$$\frac{\partial}{\partial x}\boldsymbol{\phi}_{1,s} = [p_{1,s}(2) \; p_{2,s}(2) \; \cdots \; p_{N_b,s}(2)]^T$$

$$\frac{\partial}{\partial y}\boldsymbol{\phi}_{1,s} = [p_{1,s}(3) \; p_{2,s}(3) \; \cdots \; p_{N_b,s}(3)]^T$$

$$A_2 = \sum_{s=1}^{N_s} \left\{ \left[\frac{\partial}{\partial x}\boldsymbol{\phi}_{2,s}\right]\left[\frac{\partial}{\partial x}\boldsymbol{\phi}_{2,s}\right]^T + \left[\frac{\partial}{\partial y}\boldsymbol{\phi}_{2,s}\right]\left[\frac{\partial}{\partial y}\boldsymbol{\phi}_{2,s}\right]^T \right. \\ \left. - g(x_s, y_s)\boldsymbol{\phi}_{2,s}\boldsymbol{\phi}_{2,s}^T \right\} \Delta S_s \quad (9.4.9)$$

$$\boldsymbol{\phi}_{2,s} = [\phi_{N_b+1,s} \; \phi_{N_b+2,s} \; \cdots \; \phi_{Nn,s}]^T$$

$$\frac{\partial}{\partial x}\boldsymbol{\phi}_{2,s} = [p_{N_b+1,s}(2) \; \phi_{N_b+2,s}(2) \; \cdots \; \phi_{Nn,s}(2)]^T$$

$$\frac{\partial}{\partial x}\boldsymbol{\phi}_{2,s} = [p_{N_b+1,s}(3) \; \phi_{N_b+2,s}(3) \; \cdots \; \phi_{Nn,s}(3)]^T$$

$$\mathbf{d} = -A_1 \mathbf{c}_1 - \sum_{s=1}^{Ns} f(x_s, y_s)\boldsymbol{\phi}_{2,s}\Delta S \quad (9.4.10)$$

(x_s, y_s): the centroid (gravity center) of the sth subregion S_s

The FEM is based on the variational principle that a solution to Eq. (9.4.1) can be obtained by minimizing the functional

$$I = \iint_R \left\{ \left(\frac{\partial}{\partial x}u(x,y)\right)^2 + \left(\frac{\partial}{\partial y}u(x,y)\right)^2 \right. \\ \left. - g(x,y)u^2(x,y) + 2f(x,y)u(x,y) \right\} dx\,dy \quad (9.4.11)$$

which, with $u(x,y) = \mathbf{c}^T\boldsymbol{\phi}(x,y)$, can be written as

$$I = \iint_R \left\{ \mathbf{c}^T\frac{\partial}{\partial x}\boldsymbol{\phi}\frac{\partial}{\partial x}\boldsymbol{\phi}^T\mathbf{c} + \mathbf{c}^T\frac{\partial}{\partial y}\boldsymbol{\phi}\frac{\partial}{\partial y}\boldsymbol{\phi}^T\mathbf{c} \right. \\ \left. - g(x,y)\mathbf{c}^T\boldsymbol{\phi}\boldsymbol{\phi}^T\mathbf{c} + 2f(x,y)\mathbf{c}^T\boldsymbol{\phi} \right\} dx\,dy \quad (9.4.12)$$

The condition for this functional to be minimized with respect to **c** is

$$\frac{d}{d\mathbf{c}_2} I = \iint_R \left\{ \frac{\partial}{\partial x}\boldsymbol{\phi}_2 \frac{\partial}{\partial x}\boldsymbol{\phi}^T \mathbf{c} + \frac{\partial}{\partial y}\boldsymbol{\phi}_2 \frac{\partial}{\partial y}\boldsymbol{\phi}^T \mathbf{c} \right.$$

$$\left. - g(x,y)\boldsymbol{\phi}_2 \boldsymbol{\phi}^T \mathbf{c} + f(x,y)\boldsymbol{\phi}_2 \right\} dx\, dy = 0 \quad (9.4.13)$$

$$\approx A_1 \mathbf{c}_1 + A_2 \mathbf{c}_2 + \sum_{s=1}^{N_s} f(x_s, y_s)\boldsymbol{\phi}_{2,s} \Delta S_s = 0 \quad (9.4.14)$$

See [R-1] for details.

The objectives of the following MATLAB functions 'fem_basis_ftn()' and 'fem_coef()' are to construct the basis function $\phi_{n,s}(x,y)$'s for each node $n = 1, \ldots, N_n$ and each subregion $s = 1, \ldots, N_s$ and to get the coefficient vector **c** of the solution (9.4.4) via Eq. (9.4.7) and the solution polynomial $\phi_s(x,y)$'s via Eq. (9.4.6) for each subregion $s = 1, \ldots, N_s$, respectively.

Before going into a specific example of applying the FEM method to solve a PDE, let us take a look at the basis (shape) function $\phi_n(x,y)$ for each node $n = 1, \ldots, N_n$, which is defined collectively for all the (triangular) subregions so that ϕ_n is one (1) only at node n and zero (0) at all other nodes. We will plot the basis (shape) functions for the region divided into four triangular subregions as shown in Figure 9.7 in two ways. First, we generate the basis functions by using the function 'fem_basis_ftn()' and plot one of them for node $n = 1$ by using the MATLAB functions 'inpolygon()' and 'mesh()', as depicted in Figure 9.8a. Second, without generating the basis functions, we use the MATLAB function 'trimesh()' or 'trisurf()' to plot the shape functions for nodes $n = 2,3,4$, and 5 as depicted in Figure 9.8b–e, each of which is one (1) only at the corresponding node n and zero (0) at all other nodes. Figure 9.8f is the graph of a linear combination of basis functions

$$u(x,y) = \mathbf{c}^T \boldsymbol{\phi}(x,y) = \sum_{n=1}^{Nn} c_n \phi_n(x,y) \quad (9.4.15)$$

having the given value c_n at each node n. This can be plotted by using the MATLAB function 'trimesh()' as

```
»trimesh(S,N(:,1),N(:,2),c)
```

where the first input argument s has the node numbers for each subregion, the second/third input argument N has the x/y-coordinates for each node, and the fourth input argument c has the function values at each node as follows.

```
function p=fem_basis_ftn(N,S)
% p(i,s,1:3): Coefficients of each basis function phi_i
%                for s-th subregion(triangle)
% N(n,1:2)   : x & y coordinates of the n-th node
% S(s,1:3)   : Node numbers of the s-th subregion(triangle)
N_n= size(N,1); % Total number of nodes
N_s= size(S,1); % Total number of subregions(triangles)
for n=1:N_n
  for s=1:N_s
    for i=1:3
       A(i,1:3)=[1 N(S(s,i),1:2)];
       b(i)=(S(s,i)==n); % The nth basis ftn is 1 only at node n.
    end
    pnt=A\b';
    for i=1:3, p(n,s,i)=pnt(i); end
  end
end
```

```
function [U,c]=fem_coef(f,g,p,c,N,S,N_i)
% p(i,s,1:3): Coefficients of basis ftn phi_i for the s-th subregion
% c=[ .1 1 . 0 0 .] with value for boundary and 0 for interior nodes
% N(n,1:2)   : x & y coordinates of the n-th node
% S(s,1:3)   : Node numbers of the s-th subregion(triangle)
% N_i        : Number of the interior nodes
% U(s,1:3)   : Coefficients of p1+p2(s)x+p3(s)y for each subregion
N_n=size(N,1); % Total number of nodes = N_b+N_i
N_s=size(S,1); % Total number of subregions(triangles)
d=zeros(N_i,1);
N_b=N_n-N_i;
for i=N_b+1:N_n
  for n=1:N_n
    for s=1:N_s
      xy=(N(S(s,1),:)+N(S(s,2),:)+N(S(s,3),:))/3; % Gravity center
      % phi_i,x*phi_n,x + phi_i,y*phi_n,y - g(x,y)*phi_i*phi_n
      p_vctr=[p([i n],s,1) p([i n],s,2) p([i n],s,3)];
      tmpg(s)=sum(p(i,s,2:3).*p(n,s,2:3))...
             -g(xy(1),xy(2))*p_vctr(1,:)*[1 xy]'*p_vctr(2,:)*[1 xy]';
      dS(s)=det([N(S(s,1),:) 1; N(S(s,2),:) 1;N(S(s,3),:)])/2;
       % Area of triangular subregion
      if n==1, tmpf(s)= -f(xy(1),xy(2))*p_vctr(1,:)*[1 xy]'; end
    end
    A12(i-N_b,n)=tmpg*abs(dS)'; % Eqs.(9.4.8),(9.4.9)
  end
  d(i-N_b)=tmpf*abs(dS)'; % Eq.(9.4.10)
end
d=d-A12(1:N_i,1:N_b)*c(1:N_b)'; % Eq.(9.4.10)
c(N_b+1:N_n)=A12(1:N_i,N_b+1:N_n)\d; % Eq.(9.4.7)
for s=1:N_s
  for j=1:3, U(s,j)= c*p(:,s,j); end % Eq.(9.4.6)
end
```

FINITE ELEMENT METHOD (FEM) FOR SOLVING PDE

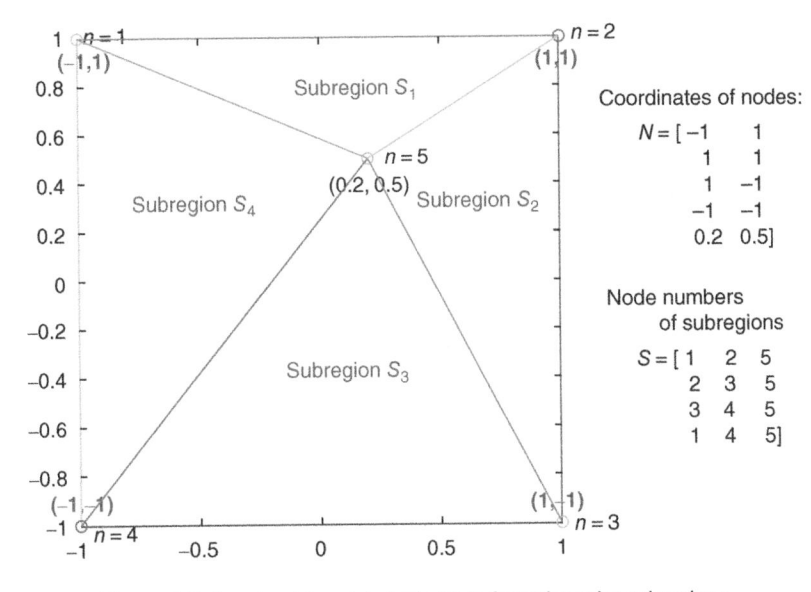

Figure 9.7 A region (domain) divided into four triangular subregions.

Coordinates of nodes:

$$N = \begin{bmatrix} -1 & 1 \\ 1 & 1 \\ 1 & -1 \\ -1 & -1 \\ 0.2 & 0.5 \end{bmatrix}$$

Node numbers of subregions

$$S = \begin{bmatrix} 1 & 2 & 5 \\ 2 & 3 & 5 \\ 3 & 4 & 5 \\ 1 & 4 & 5 \end{bmatrix}$$

$$S = \begin{bmatrix} 1 & 2 & 5 \\ 2 & 3 & 5 \\ 3 & 4 & 5 \\ 1 & 4 & 5 \end{bmatrix}, \quad N = \begin{bmatrix} -1 & 1 \\ 1 & 1 \\ 1 & -1 \\ -1 & -1 \\ 0 & 0 \end{bmatrix}, \quad c = \begin{bmatrix} 0 \\ 1 \\ 2 \\ 3 \\ 0 \end{bmatrix} \quad (9.4.16)$$

To this end, we make the following script "show_FEM_basis.m" and run it to get Figures 9.7 and 9.8 together with the coefficients of each basis function as

$$p(:,:,1) = \begin{bmatrix} -3/10 & 0 & 0 & -1/8 \\ -7/10 & -7/16 & 0 & 0 \\ 0 & 3/16 & 1/10 & 0 \\ 0 & 0 & 7/30 & 7/24 \\ 2 & 5/4 & 2/3 & 5/6 \end{bmatrix},$$

$$p(:,:,2) = \begin{bmatrix} -1/2 & 0 & 0 & -5/8 \\ 1/2 & 15/16 & 0 & 0 \\ 0 & 5/16 & 1/2 & 0 \\ 0 & 0 & -1/2 & -5/24 \\ 0 & -5/4 & 0 & 5/6 \end{bmatrix}, \quad (9.4.17)$$

538 PARTIAL DIFFERENTIAL EQUATIONS

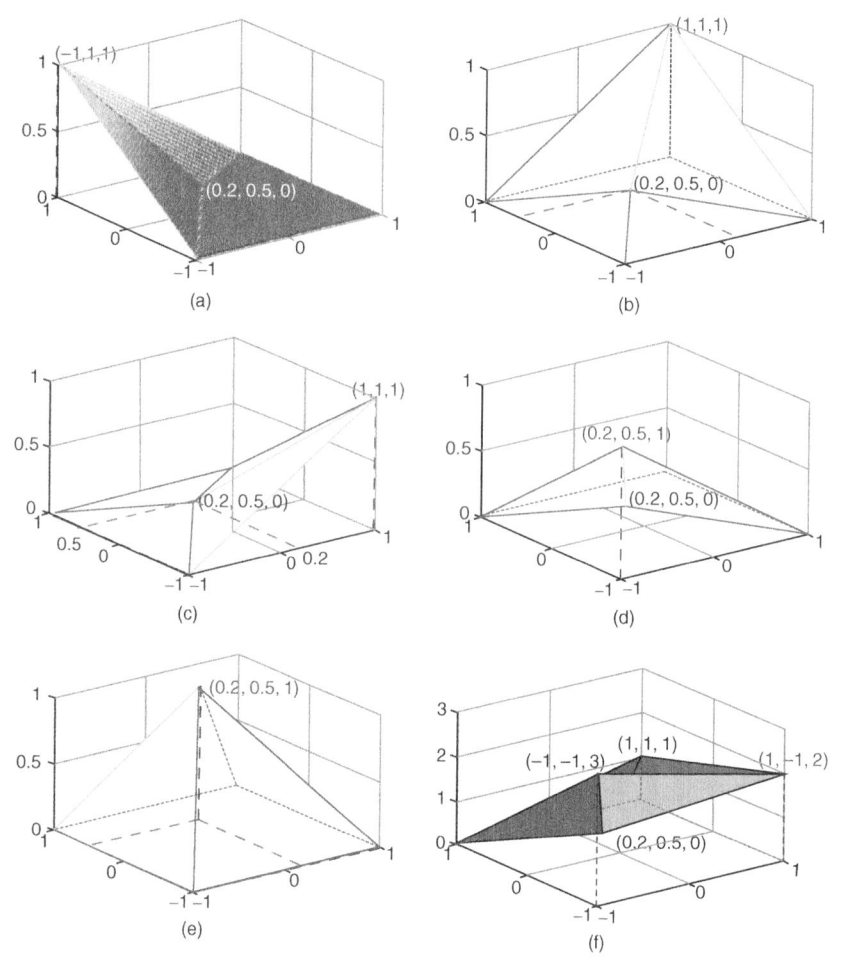

Figure 9.8 The basis (shape) functions for nodes in Figure 9.7 and a composite function. (a) $\phi_1(x,y)$. (b) $\phi_2(x,y)$. (c) $\phi_3(x,y)$. (d) $\phi_4(x,y)$. (e) $\phi_5(x,y)$. (f) $\phi_2(x,y) + 2\phi_3(x,y) + 3\phi_4(x,y)$.

$$p(:,:,3) = \begin{bmatrix} 4/5 & 0 & 0 & 1/2 \\ 6/5 & 1/2 & 0 & 0 \\ 0 & -1/2 & -2/5 & 0 \\ 0 & 0 & -4/15 & -1/2 \\ -2 & 0 & 2/3 & 0 \end{bmatrix}$$

The meaning of this N_n (the number of nodes: 5) $\times N_s$ (the number of subregions: 4) $\times 3$ array p is that, say, the second rows of the three subarrays constitute the coefficient vectors of the basis function for node 2 as

```
%show_FEM_basis.m
clear
N=[-1 1;1 1;1 -1;-1 -1;0.2 0.5]; % List of nodes in Fig.9.7
N_n= size(N,1); % Number of nodes
S=[1 2 5;2 3 5;3 4 5;1 4 5]; % List of subregions in Fig.9.7
N_s=size(S,1); % Number of subregions

figure(1), clf
for s=1:N_s
  nodes=[S(s,:) S(s,1)];
  for i=1:3
    plot([N(nodes(i),1) N(nodes(i+1),1)], ...
         [N(nodes(i),2) N(nodes(i+1),2)]), hold on
  end
end

% Basis/shape function
p=fem_basis_ftn(N,S);
x0=-1; xf=1; y0=-1; yf=1; % Graphic region
figure(2), clf
Mx=50; My=50;
dx=(xf-x0)/Mx; dy=(yf-y0)/My;
xi=x0+[0:Mx]*dx; yi=y0+[0:My]*dy;

i_ns=[1 2 3 4 5]; % List of node numbers whose basis function to plot
for itr=1:5
  i_n=i_ns(itr);
  if itr==1
    for i=1:length(xi)
      for j=1:length(yi)
        Z(j,i)=0;
        for s=1:N_s
          if inpolygon(xi(i),yi(j), N(S(s,:),1),N(S(s,:),2))>0
          Z(j,i)=p(i_n,s,1)+p(i_n,s,2)*xi(i)+p(i_n,s,3)*yi(j);
          break;
          end
        end
      end
    end
    subplot(321), mesh(xi,yi,Z) % Basis function for node 1
  else
    c1=zeros(size(c)); c1(i_n)=1;
    subplot(320+itr)
      trimesh(S,N(:,1),N(:,2),c1) % Basis function for node 2-5
  end
end
c=[0 1 2 3 0]; % Values for all nodes

subplot(326)
trimesh(S,N(:,1),N(:,2),c) % Fig.9.8(f): a composite function
```

$$\phi_2(x,y) = \begin{cases} -7/10 + (1/2)x + (6/5)y & \text{for subregion } S_1 \\ -7/16 + (15/16)x + (1/2)y & \text{for subregion } S_2 \\ 0 + 0 \cdot x + 0 \cdot y & \text{for subregion } S_3 \\ 0 + 0 \cdot x + 0 \cdot y & \text{for subregion } S_4 \end{cases} \quad (9.4.18)$$

which turns out to be one (1) only at node 2, i.e. (1,1) and zero (0) at all other nodes and on the subregions that do not have node 2 as their vertex, as depicted in Figure 9.8b.

With the script "show_FEM_basis.m" saved on your computer, type the following commands at the MATLAB prompt and see the graphical/textual output.

```
»show_FEM_basis, p
```

Now, let us see the following example.

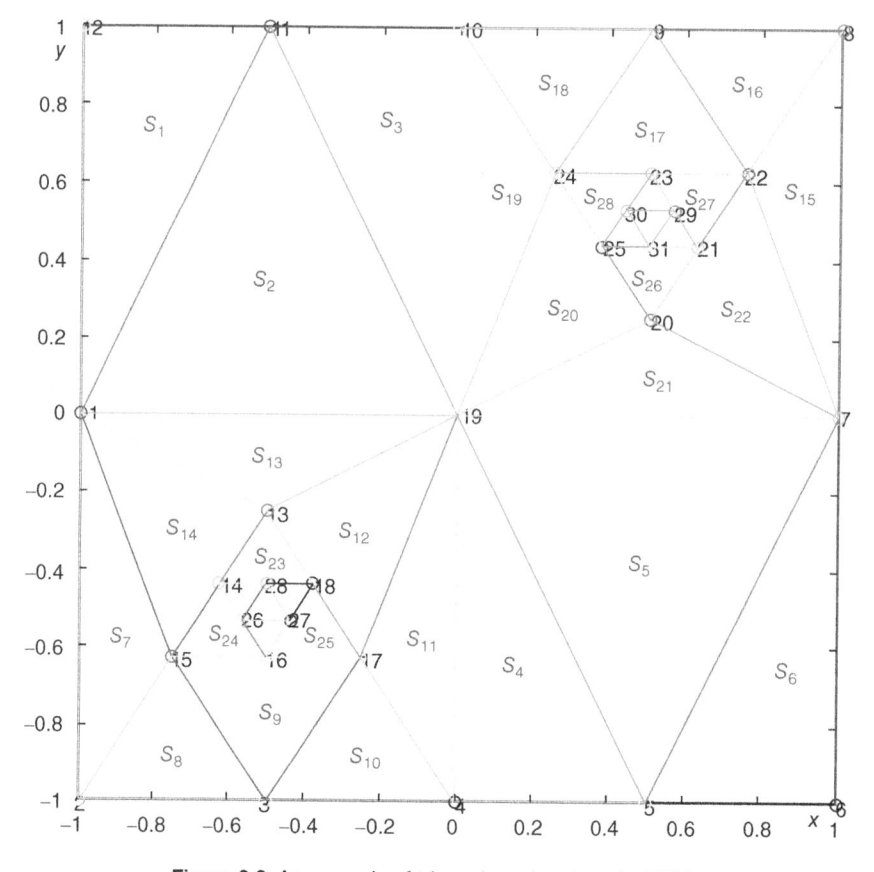

Figure 9.9 An example of triangular subregions for FEM.

FINITE ELEMENT METHOD (FEM) FOR SOLVING PDE **541**

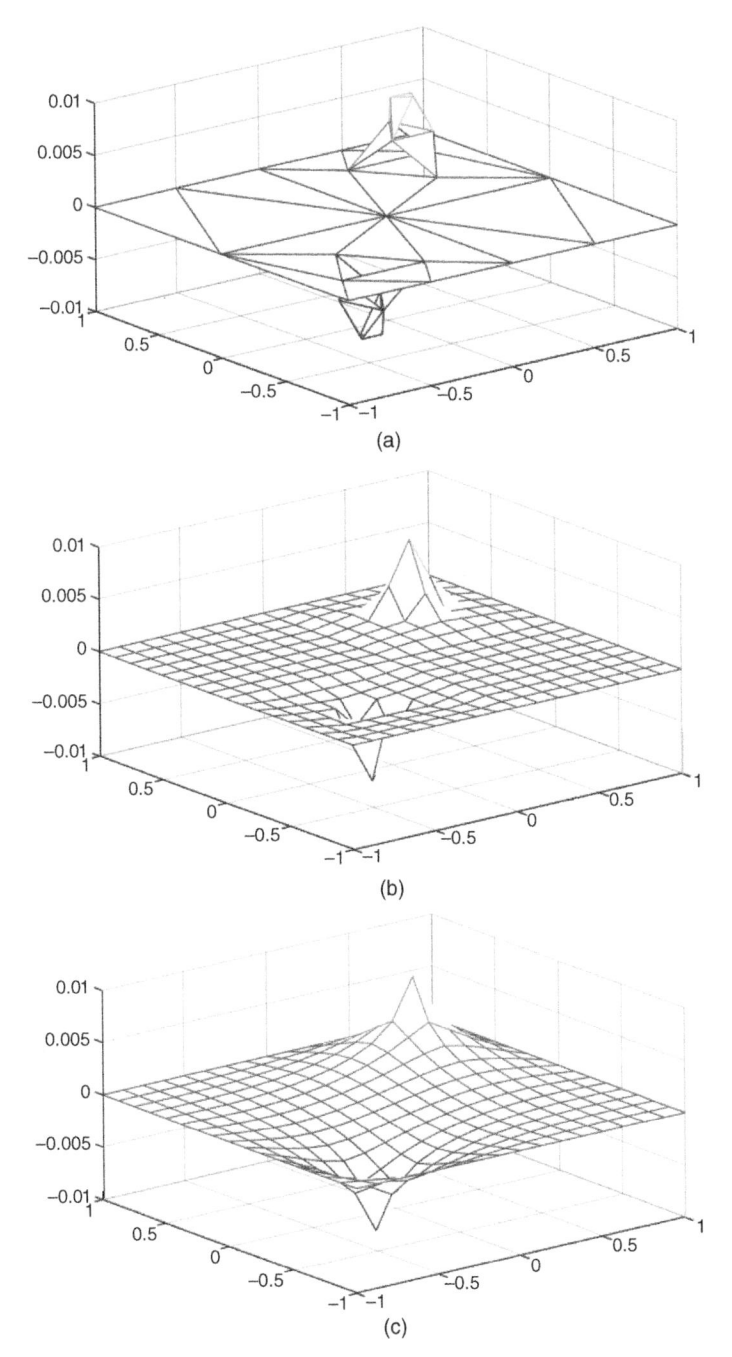

Figure 9.10 FEM solutions for Example 9.6. (a) 31-point FEM (finite element method) solution plotted by using trimesh(). (b) 31-point FEM (finite element method) solution plotted by using mesh(). (c) 16 × 15-point FDM (finite difference method) solution.

```
%do_fem.m
% for Example 9.6
clear
p1=[-0.5 -0.5]; p2=[0.5 0.5]; % Points at which +/- charges are located
N=[-1 0;-1 -1;-1/2 -1;0 -1;1/2 -1; 1 -1;1 0;1 1;1/2 1; 0 1;
    -1/2 1;-1 1; -1/2 -1/4; -5/8 -7/16;-3/4 -5/8;-1/2 -5/8;
    -1/4 -5/8;-3/8 -7/16; 0  0; 1/2 1/4;5/8 7/16;3/4 5/8;
    1/2 5/8;1/4 5/8;3/8 7/16;-9/16 -17/32;-7/16 -17/32;
    -1/2 -7/16;9/16 17/32;7/16 17/32;1/2 7/16]; % Nodes
N_b=12; % Number of boundary nodes
S=[1 11 12;1 11 19;10 11 19;4 5 19;5 7 19; 5 6 7;1 2 15; 2 3 15;
    3 15 17;3 4 17;4 17 19;13 17 19;1 13 19;1 13 15;7 8 22;8 9 22;
    9 22 24;9 10 24; 10 19 24; 19 20 24;7 19 20; 7 20 22;13 14 18;
    14 15 16;16 17 18;20 21 25;21 22 23;23 24 25;14 26 28;
    16 26 27;18 27 28; 21 29 31;23 29 30;25 30 31;
    26 27 28; 29 30 31]; % Triangular subregions

f=@(x,y)(norm([x y]-p1)<0.01)-(norm([x y]-p2)<0.01); % Eq.(E9.6.2)
g=@(x,y)0;

N_n=size(N,1); % Total number of nodes
N_i=N_n-N_b; % Number of interior nodes
c=zeros(1,N_n); % Boundary value or 0 for boundary/interior nodes
p=fem_basis_ftn(N,S);
[U,c]=fem_coef(f,g,p,c,N,S,N_i);
% Output through the triangular mesh-type graph
figure(1), clf, trimesh(S,N(:,1),N(:,2),c)

% Output through the rectangular mesh-type graph
N_s=size(S,1); % Total number of subregions(triangles)
x0=-1; xf=1; y0=-1; yf=1;
Mx=16; dx=(xf-x0)/Mx; xi=x0+[0:Mx]*dx;
My=16; dy=(yf-y0)/My; yi=y0+[0:My]*dy;
for i=1:length(xi)
    for j=1:length(yi)
        for s=1:N_s % Which subregion the point belongs to
            if inpolygon(xi(i),yi(j), N(S(s,:),1),N(S(s,:),2))>0
                Z(i,j)=U(s,:)*[1 xi(i) yi(j)]'; % Eq.(9.4.5b)
                break;
            end
        end
    end
end
figure(2), clf, mesh(xi,yi,Z)

% For comparison
bx0=@(y)0; bxf=@(y)0; by0=@(x)0; byf=@(x)0;
[U,x,y]=pde_poisson(f,g,bx0,bxf,by0,byf,[x0 xf y0 yf],Mx,My);
figure(3)
clf, mesh(x,y,U)
```

Example 9.6. Laplace's Equation – Electric Potential over a Plate with Point Charge.

Consider the following Laplace's equation

$$\nabla^2 u(x, y) = \frac{\partial^2 u(x,y)}{\partial x^2} + \frac{\partial^2 u(x,y)}{\partial y^2} = f(x,y) \text{ for } -1 \le x \le +1 \text{ and } -1 \le x \le +1 \quad \text{(E9.6.1)}$$

where

$$f(x,y) = \begin{cases} -1 & \text{for } (x,y) = (0.5, 0.5) \\ +1 & \text{for } (x,y) = (-0.5, -0.5) \\ 0 & \text{elsewhere} \end{cases} \quad \text{(E9.6.2)}$$

and the BC is $u(x,y) = 0$ for all boundaries of the rectangular domain.

To solve this equation by using the FEM, we locate 12 boundary points and 19 interior points, number them and divide the domain into 36 triangular subregions as depicted in Figure 9.9. Note that we have made the size of the subregions small and their density high around the points $p_1 = (+0.5,+0.5)$ and $p_2 = (-0.5,-0.5)$, since they are only two points at which the value of the RHS of Eq. (9.6.1) is not zero, and consequently, the value of the solution $u(x,y)$ is expected to change sensitively around them.

We made the above MATLAB program "do_fem.m" to use the functions 'fem_basis_ftn()' and 'fem_coef()' for solving this equation. For comparison, we have added the statements to solve the same equation by using the function 'pde_poisson()' (Section 9.1). The results obtained by running this program are depicted in Figure 9.10a-c.

9.5 GUI OF MATLAB FOR SOLVING PDES – PDEtool

In this section, we will see what problems can be solved by using the GUI (graphical user interface) tool of MATLAB for PDEs and then apply the tool to solve the elliptic/parabolic/hyperbolic equations dealt with in Examples 9.1/9.3/9.5 and 9.6.

9.5.1 Basic PDEs Solvable by PDEtool

Basically, the PDE toolbox can be used for the following kinds of PDE.

1. Elliptic PDE

$$-\nabla \cdot (c \nabla u) + a u = f \text{ over a domain } \Omega \quad (9.5.1)$$

with some BCs like

$$h\,u = r \text{ (Dirichlet condition)} \tag{9.5.2}$$

$$\text{or } \mathbf{n} \cdot c\,\nabla u + q\,u = g \text{ (generalized Neumann condition)}$$

on the boundary $\partial\Omega$, where $\mathbf{n}$ is the outward unit normal vector to the boundary.

Note that, in case u is a scalar-valued function on a rectangular domain as depicted in Figure 9.1, Eq. (9.5.1) becomes

$$-c\left(\frac{\partial^2 u(x,y)}{\partial x^2} + \frac{\partial^2 u(x,y)}{\partial y^2}\right) + a\,u(x,y) = f(x,y) \tag{9.5.3}$$

and if the BC for the left-side boundary segment is of Neumann type like Eq. (9.1.7), Eq. (9.5.2) can be written as

$$-\mathbf{i} \cdot c\left(\frac{\partial u(x,y)}{\partial x}\mathbf{i} + \frac{\partial u(x,y)}{\partial y}\mathbf{j}\right) + q\,u(x,y) = -c\frac{\partial u(x,y)}{\partial x} + q\,u(x,y)$$
$$= g(x,y) \tag{9.5.4}$$

since the outward unit normal vector to the left-side boundary is the unit vector along the x-axis.

2. Parabolic PDE

$$-\nabla \cdot (c\,\nabla u) + a\,u + d\frac{\partial u}{\partial t} = f \tag{9.5.5}$$

over a domain Ω and for a time range $0 \le t \le T$

with BCs like Eq. (9.5.2) and additionally, the IC $u(t_0)$.

3. Hyperbolic PDE

$$-\nabla \cdot (c\nabla u) + au + d\frac{\partial^2 u}{\partial t^2} = f \tag{9.5.6}$$

over a domain Ω and for a time range $0 \le t \le T$

with BCs like Eq. (9.5.2) and additionally, the ICs $u(t_0)/u'(t_0)$.

4. Eigenmode PDE

$$-\nabla \cdot (c\nabla u) + au = \lambda\,du \tag{9.5.7}$$

over a domain Ω and for an unknown eigenvalue λ

with some BCs like Eq. (9.5.2).

The PDE toolbox can also deal with a system of PDEs like

$$-\nabla \cdot (c_{11}\nabla u_1) - \nabla \cdot (c_{12}\nabla u_2) + a_{11}u_1 + a_{12}u_2 = f_1$$
$$-\nabla \cdot (c_{21}\nabla u_1) - \nabla \cdot (c_{22}\nabla u_2) + a_{21}u_1 + a_{22}u_2 = f_2 \quad \text{over a domain } \Omega$$
(9.5.8)

with Dirichlet BCs like

$$\begin{bmatrix} h_{11} & h_{12} \\ h_{21} & h_{22} \end{bmatrix} \begin{bmatrix} u_1 \\ u_2 \end{bmatrix} = \begin{bmatrix} r_1 \\ r_2 \end{bmatrix} \quad (9.5.9)$$

or generalized Neumann BCs like

$$\mathbf{n} \cdot (c_{11}\nabla u_1) + \mathbf{n} \cdot (c_{12}\nabla u_2) + q_{11}u_1 + q_{12}u_2 = g_1$$
$$\mathbf{n} \cdot (c_{21}\nabla u_1) + \mathbf{n} \cdot (c_{22}\nabla u_2) + q_{21}u_1 + q_{22}u_2 = g_2 \quad (9.5.10)$$

or mixed BCs, where

$$c = \begin{bmatrix} c_{11} & c_{12} \\ c_{21} & c_{22} \end{bmatrix}, \quad a = \begin{bmatrix} a_{11} & a_{12} \\ a_{21} & a_{22} \end{bmatrix}, \quad f = \begin{bmatrix} f_1 \\ f_2 \end{bmatrix}, \quad u = \begin{bmatrix} u_1 \\ u_2 \end{bmatrix}$$

$$h = \begin{bmatrix} h_{11} & h_{12} \\ h_{21} & h_{22} \end{bmatrix}, \quad r = \begin{bmatrix} r_1 \\ r_2 \end{bmatrix}, \quad q = \begin{bmatrix} q_{11} & q_{12} \\ q_{21} & q_{22} \end{bmatrix}, \quad g = \begin{bmatrix} g_1 \\ g_2 \end{bmatrix}$$

9.5.2 The Usage of PDEtool

The PDEtool in MATLAB solves PDEs by using the FEM (finite element method). We should take the following steps to use it.

0. Type 'pdetool' at the MATLAB prompt to have the PDE toolbox window on the screen as depicted in Figure 9.11. You can toggle on/off the grid by clicking 'Grid' in the Options pull-down menu (Figure 9.12a). You can also adjust the ranges of the *x*-axis and the *y*-axis in the box window opened by clicking 'Axes_Limits' in the Options pull-down menu. If you want the rectangles to be aligned with the grid lines, click 'Snap(-to-grid)' in the Options pull-down menu (Figure 9.12a). If you want to have the *x*-axis and *y*-axis of equal scale so that a circle/square may not look like an ellipse/rectangle, click 'Axes_Equal' in the Options pull-down menu. You can choose the type of PDE problem you want to solve in the submenu popped out by clicking 'Application' in the Options pull-down menu (Figure 9.12a).

    (cf) In order to be able to specify the BC for a boundary segment by clicking it, the segment must be inside in the graphic region of PDEtool.

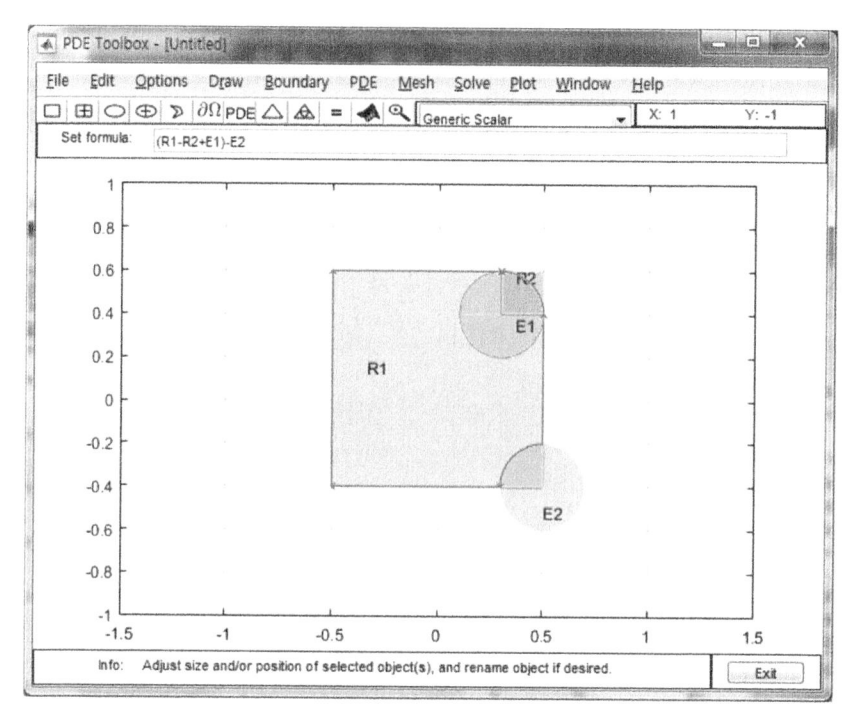

Figure 9.11 The GUI (graphical user interface) window of the MATLAB PDEtool.

1. In *Draw* mode, you can create the 2D geometry of domain Ω by using the constructive solid geometry (CSG) paradigm, which enables us to make a set of solid objects such as rectangles, circles/ellipses, and polygons. In order to do so, click the object which you want to draw in the Draw pull-down menu (Figure 9.12b) or click the button with the corresponding icon (□, ⊞, ○, ⊕, ⊃) in the toolbar just below the top menu bar (Figure 9.11). Then, you can click-and-drag to create/move the object of any size at any position as you like. Once an object has been drawn, it can be selected by clicking on it. Note that the selected object becomes surrounded by a black solid line and can be deleted by pressing Delete or ^R(Ctrl-R) key. The created object is automatically labeled, but it can be relabeled and resized (numerically) through the Object dialog box opened by double-clicking the object and even rotated (numerically) through the box opened by clicking 'Rotate' in the Draw pull-down menu. After creating and positioning the objects, you can make a CSG model by editing the set formula appropriately in the set formula field of the second line below the top menu bar to take the union (by default), the intersection, and/or the set difference of the objects to form the shape of

GUI OF MATLAB FOR SOLVING PDES – PDEtool 547

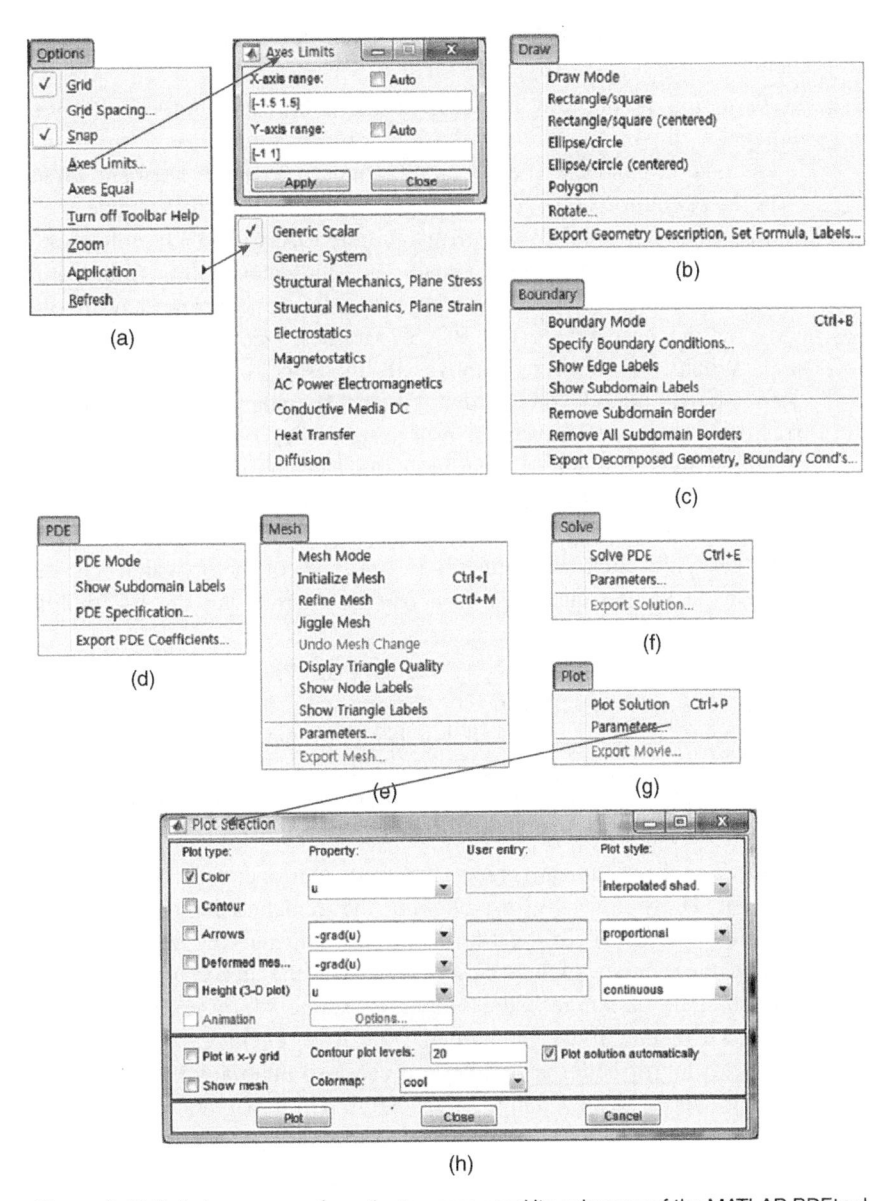

Figure 9.12 Pull-down menus from the top menu and its submenu of the MATLAB PDEtool. (a) Options pull-down menu. (b) Draw pull-down menu. (c) Boundary pull-down menu. (d) PDE pull-down menu. (e) Mesh pull-down menu. (f) Solve pull-down menu. (g) Plot pull-down menu. (h) The dialog box popped up by clicking Parameters in the Plot pull-down menu.

the domain Ω (Figure 9.11). If you want to see the overall shape of the domain you created, click 'Boundary_mode' in the Boundary pull-down menu.

2. In *Boundary* mode, you can remove the subdomain borders that are induced by the intersections of the solid objects but are not between different materials and also specify the BC for each boundary segment. First, click the $\partial\Omega$ button in the toolbar (Figure 9.11) or 'Boundary_mode(^B)' in the Boundary pull-down menu (Figure 9.12c), which will make the boundary segments appear with red/blue/green colors (indicating Dirichlet(default)/Neumann/mixed type of BC) and arrows toward its end (for the case where the BC is parameterized along the boundary). When you want to remove all the subdomain borders, click 'Remove_All_Subdomain_Borders' in the Boundary pull-down menu. You can set the parameters h, r or g, q in Eq. (9.5.2) to a constant or a function of x and y specifying the BC, through the dialog box opened by double-clicking each boundary segment. In case you want to specify/change the BC for multiple segments at a time, you had better use shift-click the segments to select all of them (which will be colored black) and click again on one of them to get the BC dialog box.

3. In *PDE* mode, you can specify the type of PDE (Elliptic/Parabolic/Hyperbolic/Eigenmode) and its parameters. In order to do so, open the PDE specification dialog box by clicking the PDE button in the toolbar or 'PDE_Specification' in the PDE pull-down menu (Figure 9.12d), check the type of PDE, and set its parameters in Eqs. (9.5.1)/(9.5.5)/(9.5.6)/(9.5.7).

4. In *Mesh* mode, you can create the triangular mesh for the domain drawn in Draw mode by just clicking the Δ button in the toolbar or 'Initialize_Mesh(^I)' in the Mesh pull-down menu (Figure 9.12e). To improve the accuracy of the solution, you can refine successively the mesh by clicking the button in the toolbar or 'Refine_Mesh(^M)' in the Mesh pull-down menu. You can jiggle the mesh by clicking 'Jiggle_Mesh' in expectation of better accuracy. You can also undo any refinement by clicking 'Undo_Mesh_Change' in the Mesh pull-down menu.

5. In *Solve* mode, you can solve the PDE and plot the result by just clicking the = button in the toolbar or 'Solve_PDE(^E)' in the Solve pull-down (Figure 9.12f). But, in the case of parabolic or hyperbolic PDE, you must click 'Parameters' in the Solve pull-down menu to set such parameters as the maximum number of triangles, the maximum number of refinements, the triangle selection method, the time range, the ICs, ... before solving the PDE.

6. In *Plot* mode, you can change the plot option in the Plot selection dialog box opened by clicking the ⬥ button in the toolbar or 'Parameters' in the Plot pull-down menu (Figure 9.12g). In the Plot selection dialog box (Figure 9.12h), you can set the plot type to, say, Color/Height(3D) and the plot style to, say, interpolated shading and continuous(interpolated) height. If you want the mesh to be shown in the solution graph, check the box of Show_mesh (in the second line from the bottom of the dialog box). In case you want to plot the graph of a known function, change the option(s) of the Property into 'user_entry', type in the MATLAB expression describing the function, and click the Plot button. You can save the plot parameters as the current default by clicking the Done button. You can also change the color map in the second line from the bottom of the dialog box.

(cf) We can extract the parameters involved in the domain geometry by clicking 'Export ..' in the Draw pull-down menu, the parameters specifying the boundary by clicking 'Export ..' in the Boundary pull-down menu, the parameters specifying the PDE by clicking 'Export ..' in the PDE pull-down menu, the parameters specifying the mesh by clicking 'Export ..' in the Mesh pull-down menu, the parameters related to the solution by clicking 'Export ..' in the Solve pull-down menu, and the parameters related to the graph by clicking 'Export ..' in the Plot pull-down menu. Whenever you want to save what you have worked in PDEtool, select File>Save in the top menubar.

(cf) Visit the website "http://www.mathworks.com/access/helpdesk/help/helpdesk.html" for more details.

9.5.3 Examples of Using PDEtool to Solve PDEs

In this section, we will make use of PDEtool to solve some PDE problems that were dealt with in the previous sections.

Example 9.7. Laplace's Equation – Steady-state Temperature Distribution over a Plate.
Consider the Laplace's equation (Example 9.1)

$$\nabla^2 u(x,y) = \frac{\partial^2 u(x,y)}{\partial x^2} + \frac{\partial^2 u(x,y)}{\partial y^2} = 0 \quad \text{for } 0 \leq x \leq 4 \text{ and } 0 \leq y \leq 4 \quad \text{(E9.7.1)}$$

with the following BCs:

$$u(0,y) = e^y - \cos y, u(4,y) = e^y \cos 4 - e^4 \cos y \quad \text{(E9.7.2)}$$

$$u(x,0) = \cos x - e^x, u(x,4) = e^4 \cos x - e^x \cos 4 \quad \text{(E9.7.3)}$$

The procedure for using PDEtool to solve this problem is as follows:

0. Type 'pdetool' at the MATLAB prompt to have the PDE toolbox window on the screen. Then, adjust the ranges of the *x*-axis and *y*-axis to [0 5] and [0 5], respectively, in the dialog box opened by clicking 'Axes_Limits' in the Options pull-down menu. You can also click 'Axes_Equal' in the Options pull-down menu to have the *x*-axis and *y*-axis of equal scale so that a circle/square may not look like an ellipse/rectangle.
1. Click the ☐ button in the toolbar and click-and-drag on the graphic region to create a rectangle of domain. Then, in the Object dialog box opened by double-clicking the rectangle, set the Left/Bottom/Width/Height to 0/0/4/4. In this case, you need not construct a CSG model by editing the set formula because the domain consists of a single object: the rectangle.
2. Click the $\partial \Omega$ button in the toolbar and double-click each boundary segment to specify the BC as Eqs. (E9.7.2) and (E9.7.3) in the BC dialog box (see Figure 9.13a).
3. Open the PDE specification dialog box by clicking the PDE button in the toolbar, check the box on the left of Elliptic as the type of PDE, and set its parameters in Eq. (E9.7.1) as depicted in Figure 9.13b.
4. Click the Δ button in the toolbar to divide the domain into a number of triangular subdomains to get the triangular mesh as depicted in Figure 9.13c. You can click the ⚠ button in the toolbar to refine the mesh successively for better accuracy.
5. Click the = button in the toolbar to plot the solution in the form of 2D graph with the value of *u*(*x*,*y*) shown in color.
6. If you want to plot the solution in the form of a 3D graph with the value of *u*(*x*,*y*) shown in height as well as color, check the box before Height on the far-left side of the Plot selection dialog box opened by clicking the 🐟 button in the toolbar. If you want the mesh shown in the solution plot as Figure 9.13d, check the box before Show_mesh on the far-left and low side and click the Plot button at the bottom of the Plot selection dialog box (Figure 9.12h). You can compare the result with that of Example 9.3 depicted in Figure 9.4.
7. If you have the true analytical solution

$$u(x, y) = e^y \cos x - e^x \cos y \qquad (E9.7.4)$$

and you want to plot the difference between the PDEtool (FEM) solution and the true analytical solution, change the entry 'u' into 'user entry' in the Color/Contour row and the Height row of the Property column and write 'u-(exp(y).*cos(x)-exp(x).*cos(y))' into the corresponding fields in the User_entry column of the Plot selection dialog box opened by clicking the 🐟 button in the toolbar and click the Plot button at the bottom of the dialog box.

GUI OF MATLAB FOR SOLVING PDES – PDEtool 551

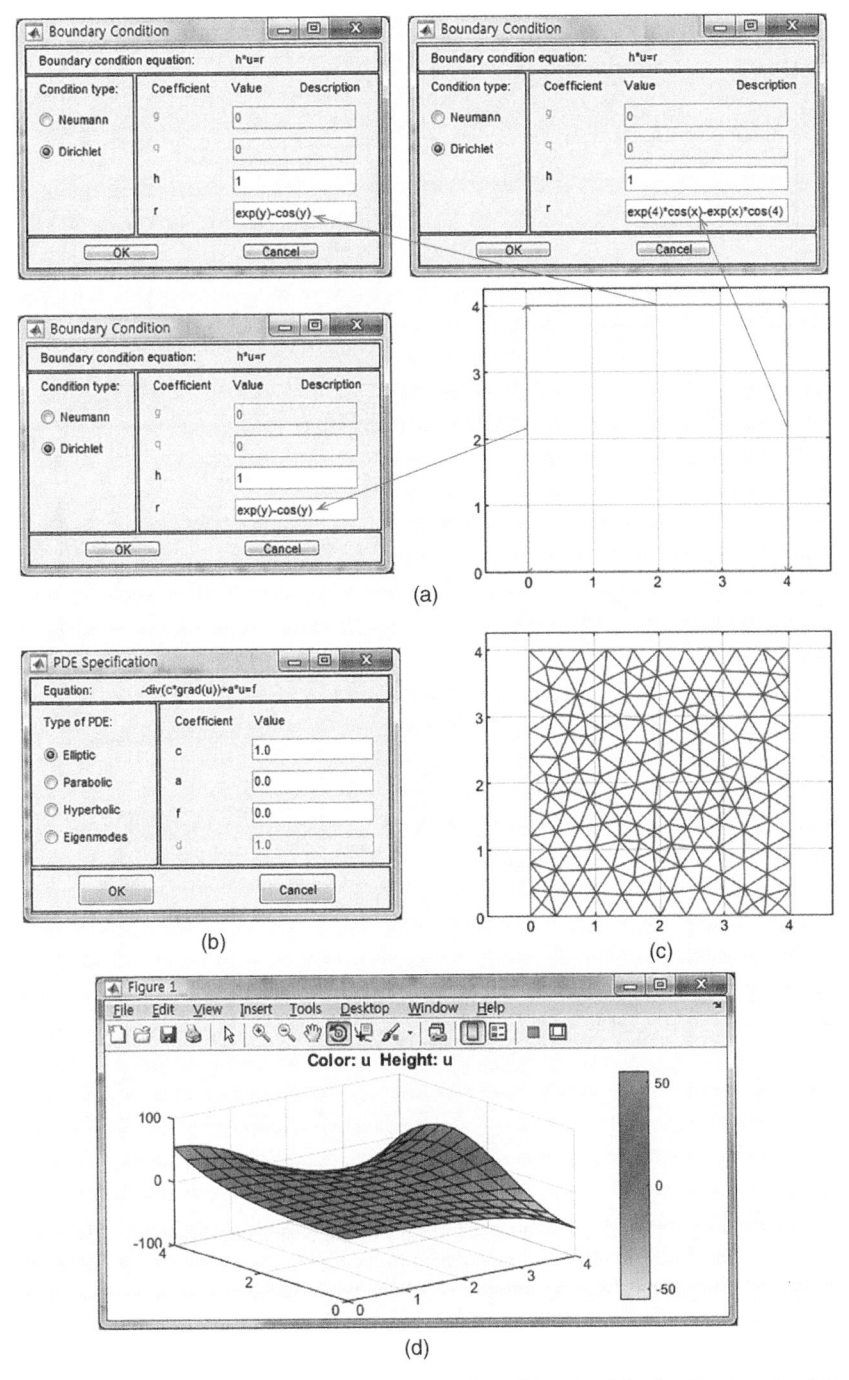

Figure 9.13 Using PDEtool Example 9.1/9.7. (a) Specifying the BCs for the domain of the PDE. (b) PDE Specification. (c) Initialize_Mesh. (d) The mesh plot of solution.

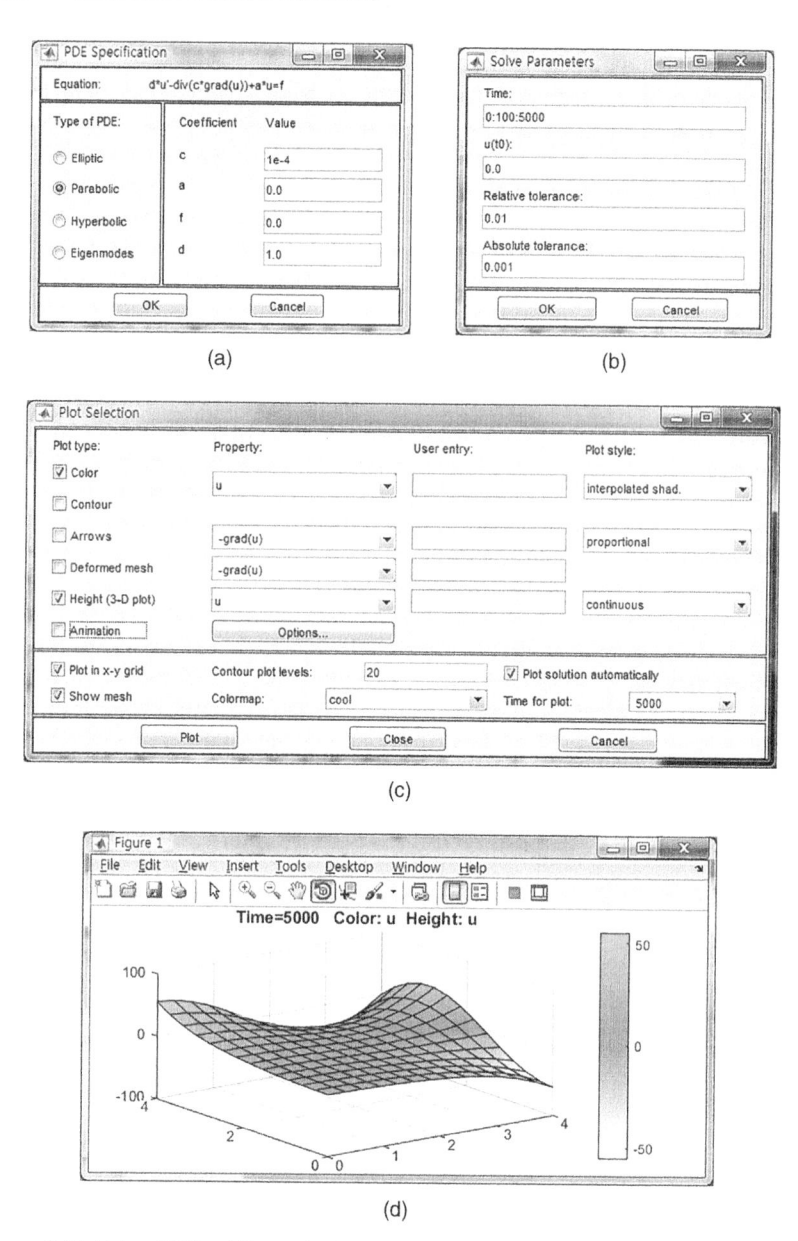

Figure 9.14 Using PDEtool Example 9.3/9.8. (a) PDE Specification dialog box. (b) Solve Parameters dialog box. (c) Plot Selection dialog box popped up by clicking Parameters in the Plot pull-down menu. (d) The mesh plot of solution at $t = 5000$.

Example 9.8. A Parabolic PDE – 2D Temperature Diffusion over a Plate.
Consider a 2D (two-dimensional) parabolic PDE

$$10^{-4}\left(\frac{\partial^2 u(x,y,t)}{\partial x^2} + \frac{\partial^2 u(x,y,t)}{\partial y^2}\right) = \frac{\partial u(x,y,t)}{\partial t} \quad \text{(E9.8.1)}$$

for $0 \le x \le 4$, $0 \le y \le 4$, and $0 \le t \le 5000$

with the ICs and BCs

$$u(x,y,0) = 0 \quad \text{for } t = 0 \quad \text{(E9.8.2a)}$$

$$u(x,y,t) = e^y \cos x - e^x \cos y \quad \text{for } x=0, x=4, y=0, \text{ and } y=4 \quad \text{(E9.8.2b)}$$

The procedure for using the PDEtool to solve this problem is as follows:

(0-2) Do exactly the same things as Steps 0-2 for the case of an elliptic PDE in Example 9.7.

(3) Open the PDE specification dialog box by clicking the PDE button, check the box on the left of 'Parabolic' as the type of PDE, and set its parameters in Eq. (E9.8.1) as depicted in Figure 9.14a.

(4) Exactly as in Step 4 (for the case of elliptic PDE) in Example 9.7, click the Δ button to get the triangular mesh. You can click the ⟁ button to refine the mesh successively for better accuracy.

(5) Unlike the case of an elliptic PDE, you must click 'Parameters' in the Solve pull-down menu (Figure 9.12c) to set the time range, say, as 0 : 100 : 5000 and the ICs as Eq. (E9.8.2a) before clicking the = button to solve the PDE. (See Figure 9.14b.)

(6) As in Step 6 of Example 9.7, you can check the box before Height in the Plot selection dialog box opened by clicking the ⬢ button, check the box before Show_mesh and click the Plot button. If you want to plot the solution graph at a time other than the final time, select the time for plot from

$$\{0, 100, 200, \ldots, 5000\}$$

in the far-right field of the Plot selection dialog box and click the Plot button again. If you want to see a movie-like dynamic picture of the solution graph, check the box before Animation, click Options right after Animation, fill in the fields of animation rate in fps, i.e. the number of frames per second and the number of repeats in the Animation Options dialog box, click the OK button and then click the Plot button in the Plot selection dialog box.

(cf) If the dynamic picture is too oblong, you can scale up/down the solution by changing the Property of the Height row from 'u' into 'user entry' and filling in the corresponding field of User_entry with, say, 'u/25' in the Plot selection dialog box.

According to your selection, you will see a movie-like dynamic picture or the (final) solution graph like Figure 9.14d, which is the steady-state solution for Eq. (E9.8.1) with $\partial u(x, y, t)/\partial t = 0$, virtually the same as the elliptic PDE (E9.7.1) whose solution is depicted in Figure 9.13d.

Before closing this example, let us have an experience of exporting the values of some parameters. For example, we extract the mesh data {p,e,t} by clicking 'Export_Mesh' in the Mesh pull-down menu and then clicking the OK button in the Export dialog box. Among the mesh data, the matrix p contains the x-/y-coordinates in the first and second rows, respectively. We also extract the solution u by clicking 'Export_Solution' in the Solve pull-down menu and then clicking the OK button in the Export dialog box. Now, we can estimate how far the graphical/numerical solution deviates from the true steady-state solution $u(x,y) = e^y \cos x - e^x \cos y$ by running the following MATLAB statements.

```
»x=p(1,:)'; y=p(2,:)'; %x,y-coordinates of nodes in column vector
»err=exp(y).*cos(x)-exp(x).*cos(y)-u(:,end); %deviation from true sol
»err_max=max(abs(err)) %maximum absolute error
```

Note that the dimension of the solution matrix u is 177×51 and the solution at the final stage is stored in its last column `u(:,end)`, where 177 is the number of nodes in the triangular mesh, and $51 = 5000/100 + 1$ is the number of frames or time stages.

Example 9.9. A Hyperbolic PDE – 2D Wave (Vibration) over a Suare Membrane. Consider a 2D hyperbolic PDE

$$\frac{1}{4}\left(\frac{\partial^2 u(x, y, t)}{\partial x^2} + \frac{\partial^2 u(x, y, t)}{\partial y^2}\right) = \frac{\partial u^2(x, y, t)}{\partial t^2} \quad (E9.9.1)$$

for $0 \leq x \leq 2, 0 \leq y \leq 2$, and $0 \leq t \leq 2$

with the zero BCs and the ICs

$$u(0, y, t) = 0, \quad u(2, y, t) = 0, \quad u(x, 0, t) = 0, \quad u(x, 2, t) = 0 \quad (E9.9.2)$$
$$u(x, y, 0) = 0.1 \sin(\pi x) \sin(\pi y/2), \quad \partial u/\partial t(x, y, 0) = 0 \text{ for } t = 0 \quad (E9.9.3)$$

The procedure for using the PDEtool to solve this problem is as follows:

(0-2) Do the same things as steps 0-2 for the case of elliptic PDE in Example 9.7, except for the following:

- Set the ranges of the x-axis and the y-axis to [0 3] and [0 3].
- Set the Left/Bottom/Width/Height to 0/0/2/2 in the Object dialog box opened by double-clicking the rectangle.
- Set the BC to zero as specified by Eqs. (E9.9.2) in the BC dialog box opened by clicking the $\partial \Omega$ button in the toolbar, shift-clicking the four boundary segments, and double-clicking one of the boundary segments.

GUI OF MATLAB FOR SOLVING PDES – PDEtool 555

(a)

(b)

(c)

(d)

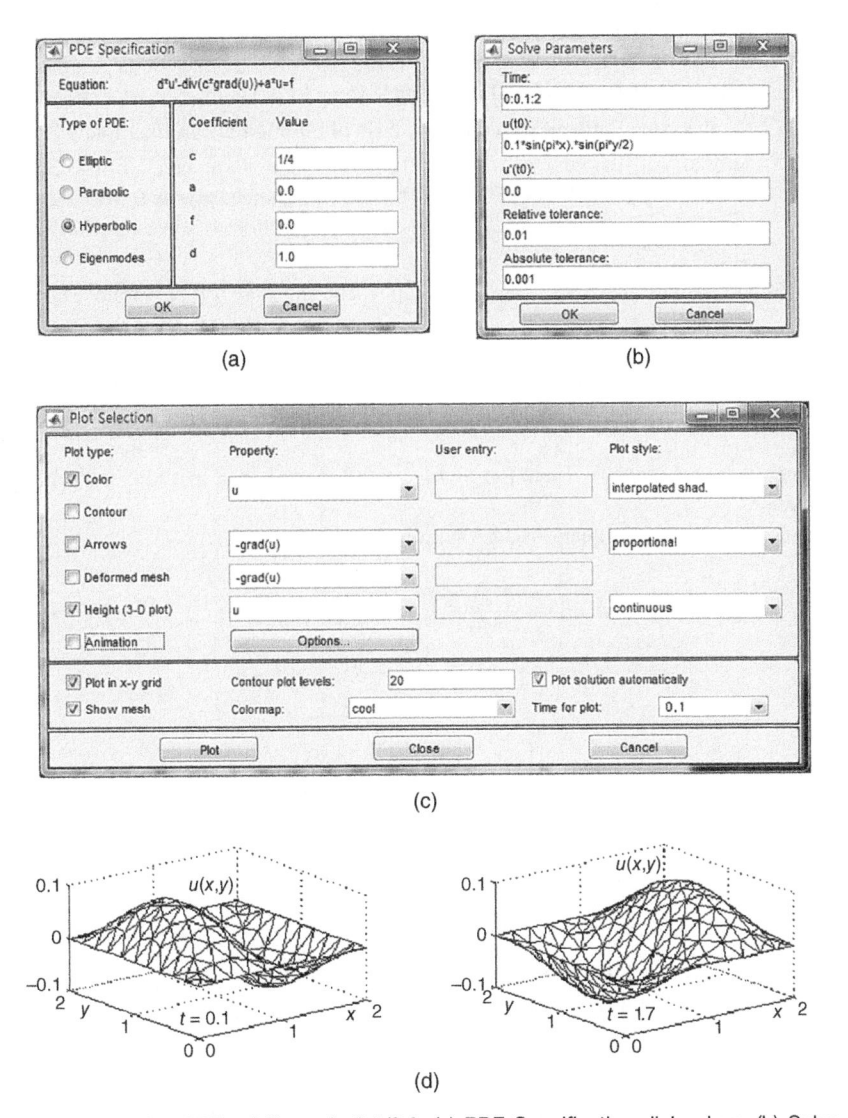

Figure 9.15 Using PDEtool Example 9.5/9.9. (a) PDE Specification dialog box. (b) Solve Parameters dialog box. (c) Plot Selection dialog box popped up by clicking Parameters in the Plot pull-down menu. (d) The mesh plots of solution at $t = 0.1$ and 1.7.

556 PARTIAL DIFFERENTIAL EQUATIONS

(3) Open the PDE specification dialog box by clicking the PDE button, check the box on the left of 'Hyperbolic' as the type of PDE, and set its parameters in Eq. (E9.9.1) as depicted in Figure 9.15a.

(4) Do the same thing as Step 4 for the case of elliptic PDE in Example 9.8.

(5) Similarly to the case of a parabolic PDE, click 'Parameters' in the Solve pull-down menu (Figure 9.12c) to set the time range, say, as 0 : 0.1 : 2 and the ICs as Eq. (E9.9.3) before clicking the = button to solve the PDE (see Figure 9.15b.)

(6) Do almost the same thing as Step 6 for the case of parabolic PDE in Example 9.8.

Finally, you could see the solution graphs like Figure 9.15d, that are similar to Figure 9.6.

Example 9.10. Laplace's Equation – Electric Potential over a Plate with Point Charge.
Consider the Laplace's equation (dealt with in Example 9.6)

$$\nabla^2 u(x, y) = \frac{\partial^2 u(x, y)}{\partial x^2} + \frac{\partial^2 u(x, y)}{\partial y^2} = f(x, y) \tag{E9.10.1}$$

$$\text{for } -1 \leq x \leq +1, -1 \leq y \leq +1$$

where

$$f(x, y) = \begin{cases} -1 & \text{for } (x, y) = (0.5, 0.5) \\ +1 & \text{for } (x, y) = (-0.5, -0.5) \\ 0 & \text{elsewhere} \end{cases} \tag{E9.10.2}$$

and the BC is $u(x,y) = 0$ for all boundaries of the rectangular domain.
The procedure for using the PDEtool to solve this problem is as follows:

(0-2) Do the same thing as Step 0–2 for the case of elliptic PDE in Example 9.7, except for the following.
- Set the Left/Bottom/Width/Height to −1/−1/2/2 in the Object dialog box opened by double-clicking the rectangle.
- Set the BC to zero in the BC dialog box opened by clicking the $\partial\Omega$ button in the toolbar, shift-clicking the four boundary segments and double-clicking one of the boundary segments.

(3) Open the PDE specification dialog box by clicking the PDE button, check the box on the left of 'Elliptic' as the type of PDE, and set its parameters in Eq. (E9.10.1,2) as depicted in Figure 9.16a.

(4) Click the △ button to initialize the triangular mesh.

(5) Click the button to open the Plot selection dialog box, check the box before 'Height', and check the box before 'Show_mesh' in the dialog box.

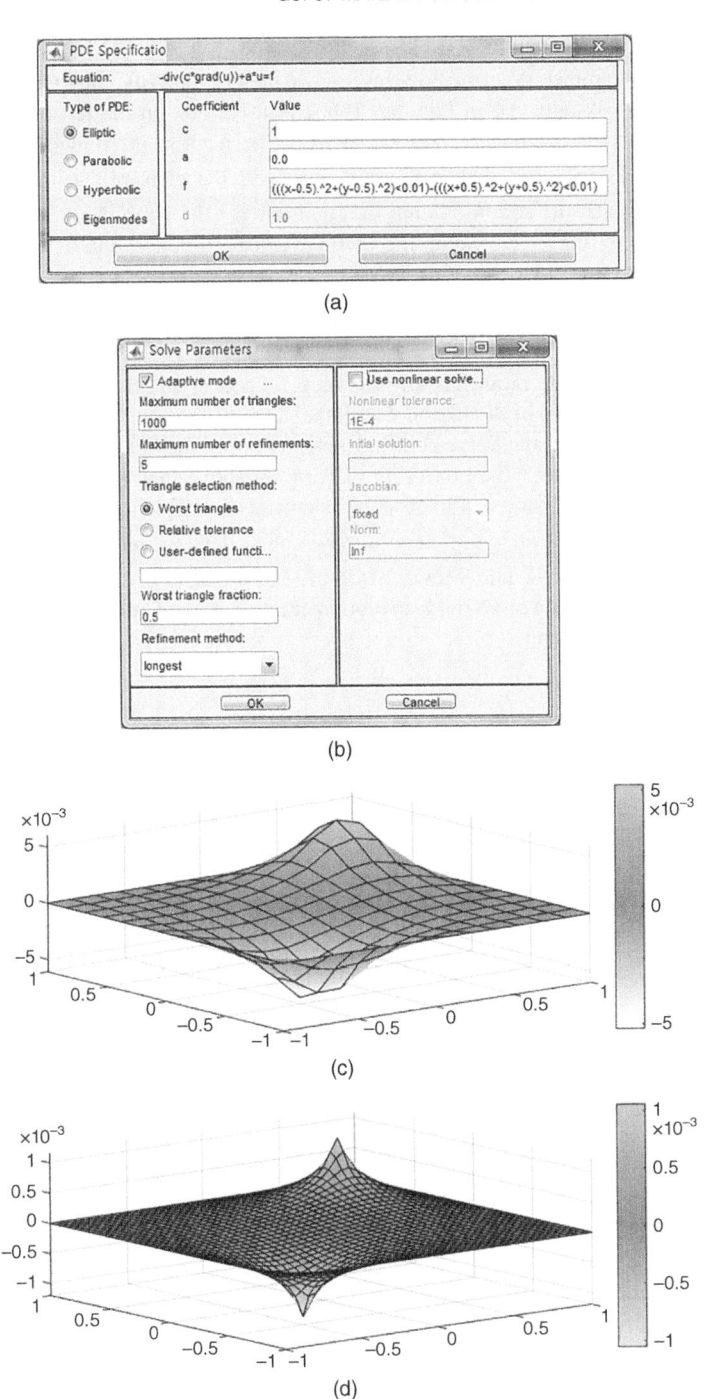

Figure 9.16 Using PDEtool Example 9.6/9.10. (a) PDE Specification dialog box. (b) Solve Parameters dialog box. (c) The mesh plot of solution with Initialize Mesh. (d) The mesh plot of solution with Adaptive Mesh.

(6) Click the Plot button to get the solution graph as depicted in Figure 9.16c.
(7) Click 'Parameters' in the Solve pull-down menu to open the 'Solve Parameters' dialog box depicted in Figure 9.16b, check the box on the left of 'Adaptive mode' and click the OK button to activate the adaptive mesh mode.
(8) Click the = button to get a solution graph with the adaptive mesh.
(9) Noting that the solution is not the right one for the point charge distribution given by (E9.10.2), reopen the PDE specification dialog box by clicking the PDE button and rewrite f as follows:

```
f (((x-0.5).^2+(y-0.5).^2)<0.00064)-(((x+0.5).^2+(y+0.5).^2)<0.00064)
```

(10) Noting that the mesh has already been refined in an adaptive way to yield smaller meshes in the region where the slope of the solution is steeper, click 'Parameters' in the Solve pull-down menu to open the 'Solve Parameters' dialog box, uncheck the box on the left of 'Adaptive mode' and click the OK button in the dialog box in order to inactivate the adaptive mesh mode.
(11) Click the = button to get the solution graph as depicted in Figure 9.16d.
(12) You can click 'Refine_Mesh(^M)' in the Mesh pull-down menu and click the = button to get a more refined solution graph (with higher resolution) as many times as you want.

PROBLEMS

9.1 Elliptic PDEs – Poisson Equations
Use the function 'pde_poisson()' (in Section 9.1) to solve the following PDEs and plot the solutions by using the MATLAB function 'mesh()'.

(a)
$$\nabla^2 u(x,y) = \frac{\partial^2 u(x,y)}{\partial x^2} + \frac{\partial^2 u(x,y)}{\partial y^2} = x+y \quad \text{for } 0 \le x,y \le 1 \quad \text{(P9.1.1)}$$

with the BCs

$$u(0,y) = y^2, \quad u(1,y) = 1, \quad u(x,0) = x^2, \quad u(x,1) = 1 \quad \text{(P9.1.2)}$$

Divide the solution region (domain) into $M_x \times M_y = 5 \times 10$ sections.

(b)
$$\frac{\partial^2 u(x,y)}{\partial x^2} + \frac{\partial^2 u(x,y)}{\partial y^2} - 12.5\pi^2 u(x,y) = -25\pi^2 \cos\left(\frac{5\pi}{2}x\right)\cos\left(\frac{5\pi}{2}y\right)$$
$$\text{for } 0 \le x, y \le 0.4 \quad \text{(P9.1.3)}$$

with the BCs

$$u(0,y) = \cos\left(\frac{5\pi}{2}y\right), \quad u(0.4,y) = -\cos\left(\frac{5\pi}{2}y\right) \quad \text{(P9.1.4)}$$

$$u(x,0) = \cos\left(\frac{5\pi}{2}x\right), \quad u(x,0.4) = -\cos\left(\frac{5\pi}{2}x\right) \quad \text{(P9.1.5)}$$

Divide the solution region into $M_x \times M_y = 40 \times 40$ sections.

(c)
$$\frac{\partial^2 u(x,y)}{\partial x^2} + \frac{\partial^2 u(x,y)}{\partial y^2} + 4\pi(x^2+y^2)u(x,y) = 4\pi \cos(\pi(x^2+y^2)) \quad \text{(P9.1.6)}$$
$$\text{for } 0 \le x, y \le 1$$

with the BCs

$$u(0,y) = \sin(\pi y^2), \, u(1,y) = \sin(\pi y^2 + 1) \quad \text{(P9.1.7)}$$

$$u(x,0) = \sin(\pi x^2), \, u(x,1) = \sin(\pi x^2 + 1) \quad \text{(P9.1.8)}$$

Divide the solution region into $M_x \times M_y = 40 \times 40$ sections.

(d)
$$\frac{\partial^2 u(x,y)}{\partial x^2} + \frac{\partial^2 u(x,y)}{\partial y^2} = 10\,e^{2x+y} \quad \text{for } 0 \le x \le 1, 0 \le y \le 2 \quad \text{(P9.1.9)}$$

with the BCs

$$u(0,y) = 2\,e^y, \quad u(1,y) = 2\,e^{2x+y},$$
$$u(x,0) = 2\,e^{2x}, \quad u(x,2) = 2\,e^{2x+2} \quad \text{(P9.1.10)}$$

Divide the solution region into $M_x \times M_y = 20 \times 40$ sections.

(e)
$$\frac{\partial^2 u(x,y)}{\partial x^2} + \frac{\partial^2 u(x,y)}{\partial y^2} = 0 \quad \text{for } 0 \le x \le 1, 0 \le y \le \pi/2 \quad \text{(P9.1.11)}$$

with the BCs

$$u(0,y) = 4\cos(3y), \quad u(1,y) = 4e^{-3}\cos(3y),$$
$$u(x,0) = 4e^{-3x}, \quad u(x,\pi/2) = 0 \quad \text{(P9.1.12)}$$

Divide the solution region into $M_x \times M_y = 20 \times 20$ sections.

9.2 More General PDE Having Nonunity Coefficients

Consider the following PDE having nonunity coefficients.

$$A\frac{\partial^2 u(x,y)}{\partial x^2} + B\frac{\partial^2 u(x,y)}{\partial x \partial y} + C\frac{\partial^2 u(x,y)}{\partial y^2} + g(x,y)u(x,y) = f(x,y) \quad \text{(P9.2.1)}$$

Modify the function 'pde_poisson()' so that it can solve this kind of PDEs and declare it as

```
function[u,x,y]=pde_poisson_abc(ABC,f,g,bx0,bxf,...,Mx,My,tol,imax)
```

where the first input argument ABC is supposed to carry the vector containing three coefficients A, B, and C. Use the function to solve the following PDEs and plot the solutions by using the MATLAB function 'mesh()'.

(a)
$$\frac{\partial^2 u(x,y)}{\partial x^2} + 2\frac{\partial^2 u(x,y)}{\partial y^2} = 10 \quad \text{for } 0 \le x, y \le 1 \quad \text{(P9.2.2)}$$

with the BCs
$$u(0,y) = y^2, \quad u(1,y) = (y+2)^2,$$
$$u(x,0) = 4x^2, \quad u(1,y) = (2x+1)^2 \quad \text{(P9.2.3)}$$

Divide the solution region (domain) into $M_x \times M_y = 20 \times 40$ sections.

(b)
$$\frac{\partial^2 u(x,y)}{\partial x^2} + 3\frac{\partial^2 u(x,y)}{\partial x \partial y} + 2\frac{\partial^2 u(x,y)}{\partial y^2} = 0 \quad \text{for } 0 \le x, y \le 1 \quad \text{(P9.2.4)}$$

with the BCs

$$u(0,y) = e^y + \cos y, \quad u(1,y) = e^{y-1} + \cos(y-2) \quad \text{(P9.2.5)}$$
$$u(x,0) = e^{-x} + \cos(-2x), \quad u(x,1) = e^{1-x} + \cos(1-2x) \quad \text{(P9.2.6)}$$

Divide the solution region into $M_x \times M_y = 40 \times 40$ sections.

(c)
$$\frac{\partial^2 u(x,y)}{\partial x^2} + 3\frac{\partial^2 u(x,y)}{\partial x \, \partial y} + 2\frac{\partial^2 u(x,y)}{\partial y^2} = x \sin y \quad \text{(P9.2.7)}$$
for $0 \le x \le 2, 0 \le y \le \pi$

with the BCs

$$u(0,y) = (3/4)\cos y, \quad u(2,y) = -\sin y + (3/4)\cos y \quad \text{(P9.2.8)}$$
$$u(x,0) = 3/4, \quad u(x,\pi) = -3/4 \quad \text{(P9.2.9)}$$

Divide the solution region into $M_x \times M_y = 20 \times 40$ sections.

(d)
$$4\frac{\partial^2 u(x,y)}{\partial x^2} - 4\frac{\partial^2 u(x,y)}{\partial x \, \partial y} + \frac{\partial^2 u(x,y)}{\partial y^2} = 0 \quad \text{for } 0 \le x,y \le 1 \quad \text{(P9.2.10)}$$

with the BCs

$$u(0,y) = y\,e^{2y}, \quad u(1,y) = (1+y)e^{1+2y},$$
$$u(x,0) = x\,e^x, \quad u(x,1) = (x+1)e^{x+2} \quad \text{(P9.2.11)}$$

Divide the solution region into $M_x \times M_y = 40 \times 40$ sections.

```
function [u,x,y]=pde_poisson_Neuman(f,g,bx0,bxf,by0,byf,x0,xf,
y0,yf,...)
.. .. .. .. .. ..
Neum=zeros(1,4); % Not Neumann, but Dirichlet condition by default
if length(x0)>1, Neum(1)=x0(2); x0=x0(1); end
if length(xf)>1, Neum(2)=xf(2); xf=xf(1); end
if length(y0)>1, Neum(3)=y0(2); y0=y0(1); end
if length(yf)>1, Neum(4)=yf(2); yf=yf(1); end
.. .. .. .. .. .. ..
dx_2=dx*dx; dy_2=dy*dy; dxy2=2*(dx_2+dy_2);
rx=dx_2/dxy2; ry=dy_2/dxy2; rxy=rx*dy_2; rx=rx;
dx2=dx*2; dy2=dy*2; rydx=ry*dx2; rxdy=rx*dy2;
u(1:My1,1:Mx1)=zeros(My1,Mx1);
sum_of_bv= 0; num= 0;
if Neum(1)==0 %Dirichlet BC
   for m=1:My1, u(m,1)= bx0(y(m)); end %side a
  else %Neumann BC
   for m=1:My1, duxa(m)= bx0(y(m)); end %du/dx(x0,y)
end
if Neum(2)==0 %Dirichlet BC
   .. .. .. .. .. .. .. .. .. ..
end
if Neum(3)==0 %Dirichlet BC
   n1=1; nM1=Mx1;
   if Neum(1)==0, u(1,1)=(u(1,1)+by0(x(1)))/2; n1=2; end
   if Neum(2)==0, u(1,Mx1)=(u(1,Mx1)+by0(x(Mx1)))/2; nM1=Mx;
     end
   for n=n1:nM1, u(1,n)= by0(x(n)); end %side c
  else %Neumann BC
   for n=1:Mx1, duyc(n)= by0(x(n)); end %du/dy(x,y0)
end
```

```
   if Neum(4)==0 %Dirichlet BC
   .. .. .. .. .. .. .. .. .. .. ..
   end
   for itr=1:imax
     if Neum(1) %Neumann BC
       for i=2:My
         u(i,1)= 2*ry*u(i,2) +rx*(u(i+1,1)+u(i-1,1)) ...
                +rxy*(G(i,1)*u(i,1)-F(i,1))-rydx*duxa(i);
               %(9.1.9)
       end
       if Neum(3), u(1,1)= 2*(ry*u(1,2) +rx*u(2,1)) ...
                +rxy*(G(1,1)*u(1,1)-F(1,1))-rydx*duxa(1)-rxdy*duyc(1);
               %(9.1.11)
       end
       if Neum(4), u(My1,1)= 2*(ry*u(My1,2) +rx*u(My,1)) ...
                +rxy*(G(My1,1)*u(My1,1)-F(My1,1))
                +rxdy*duyd(1)-rydx*duxa(My1);
       end
     end
     if Neum(2) %Neumann BC
     .. .. .. .. .. .. .. .. .. .. ..
     end
     if Neum(3) %Neumann BC
       for j=2:Mx
         u(1,j)= 2*rx*u(2,j)+ry*(u(1,j+1)+u(1,j-1)) ...
                +rxy*(G(1,j)*u(1,j)-F(1,j))-rxdy*duyc(j);
               %(9.1.10)
       end
     end
     if Neum(4) %Neumann BC
     .. .. .. .. .. .. .. .. .. .. ..
     end
     .. .. .. .. .. .. .. .. .. .. ..
   end
```

9.3 Elliptic PDEs with Neumann BC

Consider the PDE (E9.1.1) (dealt with in Example 9.1)

$$\frac{\partial^2 u(x,y)}{\partial x^2} + \frac{\partial^2 u(x,y)}{\partial y^2} = 0 \quad \text{for } 0 \leq x, y \leq 4 \quad \text{(P9.3.1)}$$

with different BCs of Neumann type, which was discussed in Section 9.1. Modify the function 'pde_poisson()' so that it can deal with the Neumann BC and declare it as

function [u,x,y]=pde_poisson_Neuman(f,g,bx0,bxf,by0,byf,x0,xf,y0,yf,...)

where the 3rd/4th/5th/6th input arguments are supposed to carry the functions of

$u(x_0,y)/u(x_f,y)/u(x,y_0)/u(x,y_f)$ or

$\partial u(x,y)/\partial x|_{x=x_0}/\partial u(x,y)/\partial x|_{x=x_f}/\partial u(x,y)/\partial y|_{y=y_0}/\partial u(x,y)/\partial y|_{y=y_f}$

and the 7th/8th/9th/10th input arguments are to carry $x_0/x_f/y_0/y_f$ or $[x_0\ 1]/[x_f\ 1]/[y_0\ 1]/[y_f\ 1]$ depending on whether each BC is of Dirichlet or Neumann type. Use it to solve the PDE with the following BCs and plot the solutions by using the MATLAB command 'mesh()'. Divide the solution region (domain) into $M_x \times M_y = 20 \times 20$ sections.

(cf) You may refer to the related part of the script in the previous page.

(a) $\partial u(x,y)/\partial x|_{x=0} = -\cos y$, $u(4,y) = e^y \cos 4 - e^4 \cos y$ (P9.3.2)

$\partial u(x,y)/\partial y|_{y=0} = \cos x$, $u(x,4) = e^4 \cos x - e^x \cos 4$ (P9.3.3)

(b) $u(0,y) = e^y - \cos y$, $\partial u(x,y)/\partial x|_{x=4} = -e^y \sin 4 - e^4 \cos y$ (P9.3.4)

$u(x,0) = \cos x - e^x$, $\partial u(x,y)/\partial y|_{y=4} = e^4 \cos x + e^x \sin 4$ (P9.3.5)

(c) $\partial u(x,y)/\partial x|_{x=0} = -\cos y$, $u(4,y) = e^y \cos 4 - e^4 \cos y$ (P9.3.6)

$u(x,0) = \cos x - e^x$, $\partial u(x,y)/\partial y|_{y=4} = e^4 \cos x + e^x \sin 4$ (P9.3.7)

(d) $u(0,y) = e^y - \cos y$, $\partial u(x,y)/\partial x|_{x=4} = -e^y \sin 4 - e^4 \cos y$ (P9.3.8)

$\partial u(x,y)/\partial y|_{y=0} = \cos x$, $u(x,4) = e^4 \cos x - e^x \cos 4$ (P9.3.9)

(e) $\partial u(x,y)/\partial x|_{x=0} = -\cos y$, $\partial u(x,y)/\partial x|_{x=4} = -e^y \sin 4 - e^4 \cos y$ (P9.3.10)

$\partial u(x,y)/\partial y|_{y=0} = \cos x$, $u(x,4) = e^4 \cos x - e^x \cos 4$ (P9.3.11)

(f) $\partial u(x,y)/\partial x|_{x=0} = -\cos y$, $u(4,y) = e^y \cos 4 - e^4 \cos y$ (P9.3.12)

$\partial u(x,y)/\partial y|_{y=0} = \cos x$, $\partial u(x,y)/\partial y|_{y=4} = e^4 \cos x + e^x \sin 4$ (P9.3.13)

(g) $u(0,y) = e^y - \cos y$, $\partial u(x,y)/\partial x|_{x=4} = -e^y \sin 4 - e^4 \cos y$ (P9.3.14)

$\partial u(x,y)/\partial y|_{y=0} = \cos x$, $\partial u(x,y)/\partial y|_{y=4} = e^4 \cos x + e^x \sin 4$ (P9.3.15)

(h) $\partial u(x,y)/\partial x|_{x=0} = -\cos y$, $\partial u(x,y)/\partial x|_{x=4} = -e^y \sin 4 - e^4 \cos y$ (P9.3.16)

$\partial u(x,y)/\partial y|_{y=0} = \cos x$, $\partial u(x,y)/\partial y|_{y=4} = e^4 \cos x + e^x \sin 4$ (P9.3.17)

9.4 Parabolic PDEs – Heat Equations

Modify the script "solve_heat.m" (in Section 9.2.3) so that it can solve the following PDEs by using the explicit forward Euler method, the implicit backward Euler method, and the Crank-Nicholson method.

(a) $$\frac{\partial^2 u(x,t)}{\partial x^2} = \frac{\partial u(x,t)}{\partial t} \quad \text{for } 0 \leq x \leq 1, 0 \leq t \leq 0.1 \quad (P9.4.1)$$

with the IC/BCs

$$u(x,0) = x^4, \cdots u(0,t) = 0, \cdots u(1,t) = 1 \quad (P9.4.2)$$

(i) With the solution region divided into $M \times N = 10 \times 20$ sections, does the explicit forward Euler method converge? What is the value of $r = A\Delta t/(\Delta x)^2$?

(ii) If you increase M and N to make $M \times N = 20 \times 40$ for better accuracy, does the explicit forward Euler method still converge? What is the value of $r = A\Delta t/(\Delta x)^2$?

(iii) What is the number N of subintervals along the t-axis that we should choose in order to keep the same value of r for $M = 20$? With that value of r, does the explicit forward Euler method converge?

(b)
$$10^{-5}\frac{\partial^2 u(x,t)}{\partial x^2} = \frac{\partial u(x,t)}{\partial t} \quad \text{for } 0 \le x \le 1, 0 \le t \le 6000 \quad \text{(P9.4.3)}$$

with the IC/BCs

$$u(x,0) = 2x + \sin(2\pi x), \quad u(0,t) = 0, u(1,t) = 2 \quad \text{(P9.4.4)}$$

(i) With the solution region divided into $M \times N = 20 \times 40$ sections, does the explicit forward Euler method converge? What is the value of $r = A\Delta t/(\Delta x)^2$? Does the numerical stability condition (9.2.6) seem to be so demanding?

(ii) If you increase M and N to make $M \times N = 40 \times 160$ for better accuracy, does the explicit forward Euler method still converge? What is the value of $r = A\Delta t/(\Delta x)^2$? Does the numerical stability condition (9.2.6) seem to be so demanding?

(iii) With the solution region divided into $M \times N = 40 \times 200$ sections, does the explicit forward Euler method converge? What is the value of $r = A\Delta t/(\Delta x)^2$?

(c)
$$2\frac{\partial^2 u(x,t)}{\partial x^2} = \frac{\partial u(x,t)}{\partial t} \quad \text{for } 0 \le x \le \pi, 0 \le t \le 0.2 \quad \text{(P9.4.5)}$$

with the IC/BCs

$$u(x,0) = \sin(2x), \quad u(0,t) = 0, \quad u(\pi,t) = 0 \quad \text{(P9.4.6)}$$

(i) By substituting
$$u(x,t) = \sin(2x)e^{-8t} \quad \text{(P9.4.7)}$$
into Eq. (P9.4.5), verify that this is a solution to the PDE.

(ii) With the solution region divided into $M \times N = 40 \times 100$ sections, does the explicit forward Euler method converge? What is the value of $r = A\Delta t/(\Delta x)^2$?

(iii) If you increase N (the number of subintervals along the t-axis) to 125 for improving the numerical stability, does the explicit forward Euler method converge? What is the value of $r = A\Delta t/(\Delta x)^2$? Use the MATLAB statements in the following box to find the maximum absolute errors of the numerical solutions obtained by the three methods. Which method yields the smallest error?

```
uo=@(x,t)sin(2*x)*exp(-8*t); % True analytical sol
Uo= uo(x,t);
err= max(max(abs(u1-Uo)))
```

(iv) If you increase N to 200, what is the value of $r = A\Delta t/(\Delta x)^2$? Find the maximum absolute errors of the numerical solutions obtained by the three methods as in (iii). Which method yields the smallest error?

(d)
$$\frac{\partial^2 u(x,t)}{\partial x^2} = \frac{\partial u(x,t)}{\partial t} \quad \text{for } 0 \le x \le 1, 0 \le t \le 0.1 \quad \text{(P9.4.8)}$$

with the IC/BCs

$$u(x,0) = \sin(\pi x) + \sin(3\pi x), \quad u(0,t) = 0, u(1,t) = 0 \quad \text{(P9.4.9)}$$

(i) By substituting

$$u(x,t) = \sin(\pi x)e^{-\pi^2 t} + \sin(3\pi x)e^{-(3\pi)^2 t} \quad \text{(P9.4.10)}$$

into Eq. (P9.4.5), verify that this is a solution to the PDE.

(ii) With the solution region divided into $M \times N = 25 \times 80$ sections, does the explicit forward Euler method converge? What is the value of $r = A\Delta t/(\Delta x)^2$?

(iii) If you increase N (the number of subintervals along the t-axis) to 100 for improving the numerical stability, does the explicit forward Euler method converge? What is the value of $r = A\Delta t/(\Delta x)^2$? Find the maximum absolute errors of the numerical solutions obtained by the three methods as in (c)(iii).

(iv) If you increase N to 200, what is the value of $r = A\Delta t/(\Delta x)^2$? Find the maximum absolute errors of the numerical solutions obtained by the three methods as in (c)(iii). Which one gained the accuracy the most of the three methods through increasing N?

9.5 Parabolic PDEs with Neumann BCs

Let us modify the functions 'pde_heat_exp()', 'pde_heat_imp()', and 'pde_heat_cn()' (in Section 9.2) so that they can accommodate the heat equation (9.2.1) with Neumann BCs:

$$\partial u(x,t)/\partial x|_{x=x_0} = b_{x_0}(t), \quad \partial u(x,t)/\partial x|_{x=x_f} = b_{x_f}(t) \quad \text{(P9.5.1)}$$

(a) Consider the explicit forward Euler algorithm described by Eq. (9.2.3)

$$u_i^{k+1} = r(u_{i+1}^k + u_{i-1}^k) + (1 - 2r)u_i^k \text{ with } r = A\frac{\Delta t}{\Delta x^2} \quad (P9.5.2)$$

$$\text{for } i = 1, 2, \ldots, M - 1$$

In case of Dirichlet BC, we do not need to get u_0^{k+1} and u_M^{k+1} because they are already given. But, in case of Neumann BC, we must get them by using this equation for $i = 0$ and M as

$$u_0^{k+1} = r(u_1^k + u_{-1}^k) + (1 - 2r)u_0^k \quad (P9.5.3a)$$

$$u_M^{k+1} = r(u_{M+1}^k + u_{M-1}^k) + (1 - 2r)u_M^k \quad (P9.5.3b)$$

and the BCs approximated as

$$\frac{u_1^k - u_{-1}^k}{2\Delta x} = b_0'(k); \quad u_{-1}^k = u_1^k - 2b_0'(k)\Delta x \quad (P9.5.4a)$$

$$\frac{u_{M+1}^k - u_{M-1}^k}{2\Delta x} = b_M'(k); \quad u_{M+1}^k = u_{M-1}^k + 2b_M'(k)\Delta x \quad (P9.5.4b)$$

Substituting Eqs. (P9.5.4a) and (P9.5.4b) into (P9.5.3) yields

$$u_0^{k+1} = 2r(u_1^k - b_0'(k)\Delta x) + (1 - 2r)u_0^k \quad (P9.5.5a)$$

$$u_M^{k+1} = 2r(u_{M-1}^k + b_M'(k)\Delta x) + (1 - 2r)u_M^k \quad (P9.5.5b)$$

Modify the function 'pde_heat_exp()' so that it can use this scheme to deal with the Neumann BCs for solving the heat equation and declare it as

```
function [u,x,t]=pde_heat_exp_Neuman(a,xfn,T,it0,bx0,bxf,M,N)
```

where the 2nd input argument xfn and the 5th, 6th input arguments bx0, bxf are supposed to carry [xf 0 1] and $b_{x_0}'(t)$, $b_{x_f}'(t)$, respectively, if the BC at x_0/x_f is of Dirichlet/Neumann type and they are also supposed to carry [xf 1 1] and $b_{x_0}'(t)$, $b_{x_f}'(t)$, respectively, if both of the BCs at x_0/x_f are of Neumann type.

(b) Consider the implicit backward Euler algorithm described by Eq. (9.2.13), which deals with the Neumann BC at the one end for solving the heat equation (9.2.1). With reference to Eq. (9.2.13), modify the function 'pde_heat_imp()' so that it can solve the heat equation with the Neumann BCs at two end points x_0 and x_f and declare it as

```
function [u,x,t]=pde_heat_imp_Neuman(a,xfn,T,it0,bx0,bxf,M,N)
```

(c) Consider the Crank-Nicholson algorithm described by Eq. (9.2.17), which deals with the Neumann BC at the one end for solving the heat equation (9.2.1). With reference to Eq. (9.2.17), modify the function 'pde_heat_cn()' so that it can solve the heat equation with the Neumann BCs at two end points x_0 and x_f and declare it as

```
function [u,x,t]=pde_heat_cn_Neuman(a,xfn,T,it0,bx0,bxf,M,N)
```

(d) Solve the following heat equation with three different BCs by using the three modified functions in (a), (b), and (c) with M=20, N=100 and find the maximum absolute errors of the three solutions as in Problem 9.4(c)(iii).

$$\frac{\partial^2 u(x,t)}{\partial x^2} = \frac{\partial u(x,t)}{\partial t} \quad \text{for } 0 \le x \le 1, 0 \le t \le 0.1 \quad \text{(P9.5.6)}$$

with the IC/BCs

(i) $u(x,0) = \sin(\pi x)$, $\partial u(x,t)/\partial x|_{x=0} = \pi\, e^{-\pi^2 t}$, $u(x,t)|_{x=1} = 0$ (P9.5.7)

(ii) $u(x,0) = \sin(\pi x)$, $u(x,t)|_{x=0} = 0$, $\partial u(x,t)/\partial x|_{x=1} = -\pi\, e^{-\pi^2 t}$ (P9.5.8)

(iii) $u(x,0) = \sin(\pi x)$, $\partial u(x,t)/\partial x|_{x=0} = \pi\, e^{-\pi^2 t}$,

$$\partial u(x,t)/\partial x|_{x=1} = -\pi\, e^{-\pi^2 t} \quad \text{(P9.5.9)}$$

Note that the true analytical solution is

$$u(x,t) = \sin(\pi x) e^{-\pi^2 t} \quad \text{(P9.5.10)}$$

9.6 Hyperbolic PDEs – Wave Equations

Modify the script "solve_wave.m" (in Section 9.3) so that it can solve the following PDEs by using the explicit central difference method.

(a)
$$4\frac{\partial^2 u(x,t)}{\partial x^2} = \frac{\partial^2 u(x,t)}{\partial t^2} \quad \text{for } 0 \le x \le 1, 0 \le t \le 1 \quad \text{(P9.6.1)}$$

with the ICs/BCs

$$u(x,0) = 0, \partial u(x,t)/\partial t|_{t=0} = 5\sin(\pi x), \quad u(0,t) = 0, u(1,t) = 0 \quad \text{(P9.6.2)}$$

Note that the true analytical solution is

$$u(x,t) = \frac{2.5}{\pi} \sin(\pi x) \sin(2\pi t) \quad \text{(P9.6.3)}$$

(i) With the solution region divided into $M \times N = 20 \times 50$ sections, what is the value of $r = A(\Delta t)^2/(\Delta x)^2$? Use the MATLAB statements in Problem 9.4(c)(iii) to find the maximum absolute error of the solution obtained by using the function 'pde_wave()'.

(ii) With the solution region divided into $M \times N = 40 \times 100$ sections, what is the value of r? Find the maximum absolute error of the numerical solution.

(iii) If we increase M (the number of subintervals along the x-axis) to 50 for better accuracy, what is the value of r? Find the maximum absolute error of the numerical solution and determine whether it has been improved.

(iv) If we increase the number M to 52, what is the value of r? Can we expect better accuracy in the light of the numerical stability condition (9.3.7)? Find the maximum absolute error of the numerical solution and determine whether it has been improved or not.

(v) What do you think the best value of r is?

(b)
$$6.25 \frac{\partial^2 u(x,t)}{\partial x^2} = \frac{\partial^2 u(x,t)}{\partial t^2} \quad \text{for } 0 \leq x \leq \pi, 0 \leq t \leq 0.4\pi \quad \text{(P9.6.4)}$$

with the ICs/BCs

$$u(x,0) = \sin(2x), \quad \partial u(x,t)/\partial t|_{t=0} = 0, \quad u(0,t) = 0, u(1,t) = 0 \quad \text{(P9.6.5)}$$

Note that the true analytical solution is

$$u(x,t) = \sin(2x)\cos(5t) \quad \text{(P9.6.6)}$$

(i) With the solution region divided into $M \times N = 50 \times 50$ sections, what is the value of $r = A(\Delta t)^2/(\Delta x)^2$? Find the maximum absolute error of the solution obtained by using the function 'pde_wave()'.

(ii) With the solution region divided into $M \times N = 50 \times 49$ sections, what is the value of r? Find the maximum absolute error of the numerical solution.

(iii) If we increase N (the number of subintervals along the t-axis) to 51 for better accuracy, what is the value of r? Find the maximum absolute error of the numerical solution.

(iv) What do you think the best value of r is?

(c)
$$\frac{\partial^2 u(x,t)}{\partial x^2} = \frac{\partial^2 u(x,t)}{\partial t^2} \quad \text{for } 0 \leq x \leq 10 \text{ and } 0 \leq t \leq 10 \quad \text{(P9.6.7)}$$

with the ICs/BCs

$$u(x,0) = \begin{cases} (x-2)(3-x) & \text{for } 2 \leq x \leq 3 \\ 0 & \text{elsewhere} \end{cases} \quad \text{(P9.6.8)}$$

$$\partial u(x,t)/\partial t|_{t=0} = 0, \quad u(0,t) = 0, u(10,t) = 0 \quad \text{(P9.6.9)}$$

(i) With the solution region divided into $M \times N = 100 \times 100$ sections, what is the value of $r = A(\Delta t)^2/(\Delta x)^2$?

(ii) Noting that the initial condition (P9.6.8) can be implemented by the MATLAB statement as

```
»it0=@(x)(x-2).*(3-x).*(2<x&x<3);   % A function handle
```

solve the PDE (P9.6.7) in the same way as in the script "nm09e04.m" and make a dynamic picture out of the numerical solution, with the current time printed on screen. Estimate the time when one of the two separated pulses propagating leftward is reflected and reversed. How about the time when the two separated pulses are reunited?

```
a=1;
it0=@(x)(x-2).*(3-x).*(2<x&x<3);  i1t0=@(x)0;  % ICs (P9.6.8)
bx0t=@(t)0;  bxft=@(t)0;  % BCs (P9.6.9)
xf=10; M=100;  T=10; N=100;  dt=T/N;
[u,x,t]=pde_wave(a,xf,T,it0,i1t0,bx0t,bxft,M,N);
............................
```

9.7 FEM (Finite Element Method)

In expectation of better accuracy/resolution, modify the script "do_fem.m" (in Section 9.6) by appending the following lines

```
;-17/32 -31/64; -1/2 -17/32;-15/32 -31/64
;17/32 31/64; 1/2 17/32; 15/32 31/64
```

to the last part of the Node array N and replacing the last line of the subregion array S with

```
26 32 33; 27 33 34; 28 32 34; 29 35 36;
30 36 37; 31 35 37; 32 33 34; 35 36 37
```

This is equivalent to refining the triangular mesh in the subregions nearest to the point charges at (0.5,0.5) and (−0.5,−0.5) as depicted in Figure P9.7. Plot the new solution obtained by running the modified script. You may have to change a statement of the script as follows.

```
f=@(x,y)(norm([x y]-p1)<1e-3) -(norm([x y]-p2)<1e-3);
```

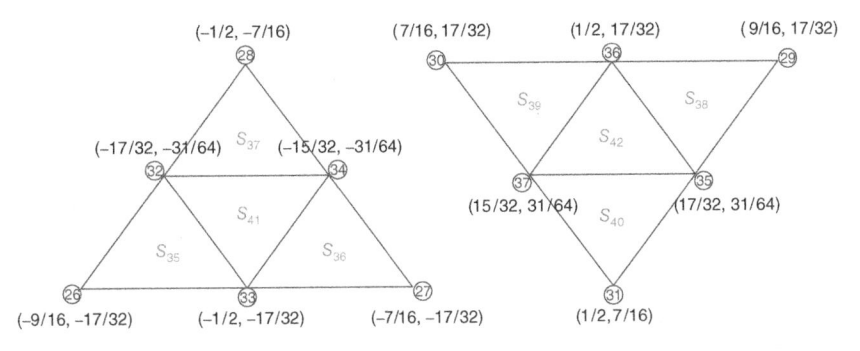

Figure P9.7 Refined triangular meshes for Problem 9.7.

9.8 PDEtool – GUI (Graphical User Interface) of MATLAB for Solving PDEs

Table P9.8.1 The maximum absolute error and the number of nodes.

	The maximum absolute error	The number of nodes
pde_poisson()	1.9256	41 × 41
PDEtool with initialize mesh	0.1914	177
PDEtool with refine mesh		
PDEtool with 2nd refine mesh		
PDEtool with adaptive mesh		
PDEtool with 2nd adaptive mesh		

(a) Consider the PDE

$$4\frac{\partial^2 u(x,y)}{\partial x^2} - 4\frac{\partial^2 u(x,y)}{\partial x\, \partial y} + \frac{\partial^2 u(x,y)}{\partial y^2} = 0 \quad \text{for } 0 \le x, y \le 1 \quad (P9.8.1)$$

with the BCs

$$u(0,y) = y\, e^{2y}, \quad u(1,y) = (1+y)e^{1+2y},$$
$$u(x,0) = x\, e^x, \quad u(x,1) = (x+1)e^{x+2} \quad (P9.8.2)$$

Noting that the field of coefficient c should be filled in as

⊙ Elliptic ║ c [4 -2 -2 1] or [4 -2 1]

in the PDE specification dialog box and the true analytical solution is

$$u(x,y) = (x+y)e^{x+2y} \quad (P9.8.3)$$

use the PDEtool to solve this PDE and fill in Table P9.8.1 with the maximum absolute error and the number of nodes together with those of Problem 9.2(d) for comparison.

You can refer to Example 9.8 for the procedure to get the numerical value of the maximum absolute error. Notice that the number of nodes is the number of columns of p, which is obtained by clicking 'Export_Mesh' in the Mesh pull-down menu and then, clicking the OK button in the Export dialog box. You can also refer to Example 9.10 for the usage of 'Adaptive Mesh', but in this case, you only have to check the box on the left of 'Adaptive mode', click the OK button in the 'Solve Parameters' dialog box opened by clicking 'Parameters' in the Solve pull-down menu and then, the mesh is adaptively refined every time you click the = button in the toolbar to get the solution. With the box on the left of 'Adaptive mode' unchecked in the 'Solve Parameters' dialog box, the mesh is nonadaptively refined every time you click 'Refine Mesh' in the Mesh pull-down menu. You can restore the previous mesh by clicking 'Undo Mesh Change' in the Mesh pull-down menu.

(b) Consider the PDE

$$\frac{\partial^2 u(x,y)}{\partial x^2} + \frac{\partial^2 u(x,y)}{\partial y^2} = 0 \quad \text{for } 0 \leq x, y \leq 4 \tag{P9.8.4}$$

with the Dirichlet/Neumann BCs

$$u(0,y) = e^y - \cos y, \quad \partial u(x,y)/\partial x|_{x=4} = -e^y \sin 4 - e^4 \cos y \tag{P9.8.5}$$

$$\partial u(x,y)/\partial y|_{y=0} = \cos x, \quad \partial u(x,y)/\partial y|_{y=4} = e^4 \cos x + e^x \sin 4 \tag{P9.8.6}$$

Noting that the true analytical solution is

$$u(x,y) = e^y \cos x - e^x \cos y \tag{P9.8.7}$$

use the PDEtool to solve this PDE and fill in Table P9.8.2 with the maximum absolute error and the number of nodes together with those of Problem 9.3(g) for comparison.

Table P9.8.2 The maximum absolute error and the number of nodes.

	The maximum absolute error	The number of nodes
pde_poisson()	0.2005	21×21
PDEtool with initialize mesh	0.5702	177
PDEtool with refine mesh		
PDEtool with 2nd refine mesh		
PDEtool with adaptive mesh		
PDEtool with 2nd adaptive mesh		

(c) Consider the PDE

$$2\frac{\partial^2 u(x,t)}{\partial x^2} = \frac{\partial u(x,t)}{\partial t} \quad \text{for } 0 \leq x \leq \pi, 0 \leq t \leq 0.2 \tag{P9.8.8}$$

with the IC and BCs

$$u(x,0) = \sin(2x), \quad u(0,t) = 0, \quad u(\pi,t) = 0 \tag{P9.8.9}$$

Noting that the true analytical solution is

$$u(x,t) = \sin(2x)e^{-8t} \tag{P9.8.10}$$

use the PDEtool to solve this PDE and fill in Table P9.8.3. with the maximum absolute error and the number of nodes together with those obtained with the MATLAB function 'pde_heat_CN()' in Problem 9.4(c) for comparison. In order to do this job, take the following steps:

Table P9.8.3 The maximum absolute error and the number of nodes.

	The maximum absolute error	The number of nodes
pde_poisson()	7.5462×10^{-4}	41×101
PDEtool with initialize mesh		
PDEtool with refine mesh		
PDEtool with 2nd refine mesh		

(1) Click the ▭ button in the toolbar and click-and-drag on the graphic region to create a rectangular domain. Then, double-click the rectangle to open the Object dialog box and set the Left/Bottom/Width/Height to 0/0/pi/0.01 to make a long rectangular domain.

(cf) Even if the PDEtool is originally designed to deal with only 2D PDEs, we can use it to solve 1D PDEs like (P9.8.8) by proceeding in this way.

(2) Click the $\partial\Omega$ button in the toolbar, double-click the upper/lower boundary segments to set the homogeneous Neumann BC ($g = 0$, $q = 0$) and double-click the left/right boundary segments to set the Dirichlet BC ($h = 1, r = 0$) as given by Eq. (P9.8.9).

(3) Open the PDE specification dialog box by clicking the PDE button, check the box on the left of 'Parabolic' as the type of PDE, and set its parameters in Eq. (9.5.5) as $c = 2$, $a = 0$, $f = 0$, and $d = 1$, which corresponds to Eq. (P9.8.8).

(4) Click 'Parameters' in the Solve pull-down menu to set the time range, say, as 0 : 0.002 : 0.2 and the initial conditions as Eq. (P9.8.9).

(5) In the Plot selection dialog box opened by clicking the 🖉 button, check the box before Height and click the Plot button. If you want to plot the solution graph at a time other than the final time, select the time for plot from {0,0.002,0.004, ... ,0.2} in the far-right field of the Plot selection dialog box and click the Plot button again.

(6) If you want to see a movie-like dynamic picture of the solution graph, check the box before Animation and then click the Plot button in the Plot selection dialog box.

(7) Click 'Export_Mesh' in the Mesh pull-down menu and then, click the OK button in the Export dialog box to extract the mesh data {p,e,t}. Also click 'Export_Solution' in the Solve pull-down menu and then, click the OK button in the Export dialog box to extract the solution u. Now, you can estimate how far the graphical/numerical solution deviates from the true solution (P9.8.10) by typing the following statements at the MATLAB prompt:

```
»x=p(1,:)'; y=p(2,: )'; % x,y-coordinates of nodes in columns
»tt=0:0.01:0.2; % time vector in row
»err=sin(2*x)*exp(-8*tt)-u; %deviation from true sol.(P9.8.10)
»err_max=max(abs(err)) % maximum absolute error
```

(d) Consider the PDE

$$\frac{\partial^2 u(x,t)}{\partial x^2} = \frac{\partial^2 u(x,t)}{\partial t^2} \quad \text{for } 0 \le x \le 10, 0 \le t \le 10 \quad \text{(P9.8.11)}$$

with the ICs/BCs

$$u(x,0) = \begin{cases} (x-2)(3-x) & \text{for } 2 \le x \le 3 \\ 0 & \text{elsewhere} \end{cases}, \quad \left.\frac{\partial u(x,t)}{\partial t}\right|_{t=0} = 0 \quad \text{(P9.8.12)}$$

$$u(0,t) = 0, \quad u(10,t) = 0 \quad \text{(P9.8.13)}$$

Use the PDEtool to make a dynamic picture out of the solution for this PDE and see if the result is about the same as that obtained in Problem 9.6(c) in terms of the time when one of the two separated pulses propagating leftward is reflected and reversed and the time when the two separated pulses are reunited.

(cf) Even if the PDEtool is originally designed to solve only 2D PDEs, we can solve 1D PDE like (P9.8.11) by proceeding as follows

(0) In the PDE toolbox window, adjust the ranges of the x-axis and the y-axis to [−0.5 10.5] and [−0.01 +0.01], respectively, in the box opened by clicking 'Axes_Limits' in the Options pull-down menu.

(1) Click the ☐ button in the toolbar and click-and-drag on the graphic region to create a long rectangle of domain ranging from $x_0 = 0$ to $x_f = 10$. Then, double-click the rectangle to open the Object dialog box and set the Left/Bottom/Width/Height to 0/−0.01/10/0.02.

(2) Click the ∂Ω button in the toolbar, double-click the upper/lower boundary segments to set the homogeneous Neumann BC ($g = 0$, $q = 0$) and double-click the left/right boundary segments to set the Dirichlet BC ($h = 1$, $r = 0$) as given by Eq. (P9.8.13).

(3) Open the PDE specification dialog box by clicking the PDE button, check the box on the left of 'Hyperbolic' as the type of PDE, and set its parameters in Eq. (P9.8.11) as $c = 1$, $a = 0$, $f = 0$, and $d = 1$ (see Figure 9.15(a).)

(4) Click 'Parameters' in the Solve pull-down menu to set the time range to, say, as 0 : 0.2 : 10, the BC as (P9.8.13) and the initial conditions as (P9.8.12) (see Figure 9.15(b) and Problem 9.6(c)(ii).)

(5) In the Plot selection dialog box opened by clicking the ⚑ button, check the box before 'Height' and the box before 'Animation' and then click the Plot button in the Plot selection dialog box to see a movie-like dynamic picture of the solution graph.

(6) If you want to have better resolution in the solution graph, click Mesh in the top menu bar, click 'Refine Mesh' in the Mesh pull-down menu. Then, select Plot in the top menu bar or type CTRL+P(^P) on the keyboard and click 'Plot_Solution' in the Plot pull-down menu to see a smoother animation graph.

(7) In order to estimate the time when one of the two separated pulses propagating leftward is reflected and reversed and the time when the two separated pulses are reunited, count the flickering frame numbers, noting that one flickering corresponds to 0.2 seconds according to the time range set in step (4).

(8) If you want to save the PDEtool program, click File in the top menu bar, click 'Save_As' in the File pull-down menu and input the file name of your choice.

APPENDIX A

MEAN VALUE THEOREM

Theorem A.1 Mean Value Theorem.
Let a function f(x) be continuous on the interval [a,b] and differentiable over (a,b). Then, there exists at least one point ξ between a and b at which

$$f'(\xi) = \frac{f(b) - f(a)}{b - a}; \quad f(b) = f(a) + f'(\xi)(b - a) \quad (A.1)$$

In other words, the curve of a continuous function f(x) has the same slope as the straight line connecting the two end points (a,f(a)) and (b,f(b)) of the curve at some point $\xi \varepsilon [a,b]$, as in Figure A.1.

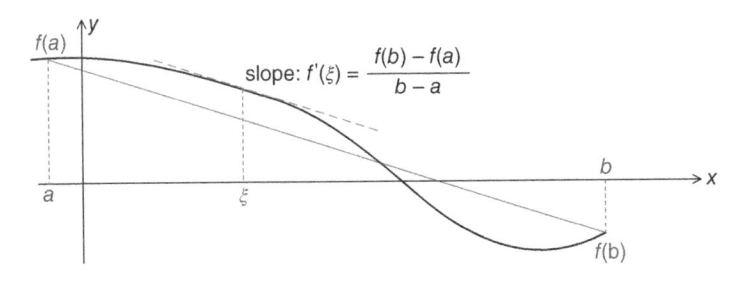

Figure A.1 Mean value theorem.

Theorem A.2 Taylor Series Theorem.

If a function f(x) is continuous and its derivatives up to order (K+1) are also continuous on an open interval D containing some point a, then the value of the function f(x) at any point x∈D can be represented by

$$f(x) = \sum_{k=0}^{K} \frac{f^{(k)}(a)}{k!}(x-a)^k + R_{K+1}(x) \qquad (A.2)$$

where the first term of the RHS is called the Nth-degree Taylor polynomial and the second term called the remainder (error) term is

$$R_{K+1}(x) = \frac{f^{(K+1)}(\xi)}{(K+1)!}(x-a)^{K+1} \quad \text{for some } \xi \text{ between } a \text{ and } x \qquad (A.3)$$

Moreover, if the function f(x) has continuous derivatives of all orders on D, then the aforementioned representation becomes

$$f(x) = \sum_{k=0}^{\infty} \frac{f^{(k)}(a)}{k!}(x-a)^k \qquad (A.4)$$

which is called the (infinite) Taylor series expansion of f(x) about a.

APPENDIX B

MATRIX OPERATIONS/PROPERTIES

CHAPTER OUTLINE

B.1	Addition and Subtraction	578
B.2	Multiplication	578
B.3	Determinant	578
B.4	Eigenvalues and Eigenvectors of a Matrix[1]	579
B.5	Inverse Matrix	580
B.6	Symmetric/Hermitian Matrix	580
B.7	Orthogonal/Unitary Matrix	581
B.8	Permutation Matrix	581
B.9	Rank	581
B.10	Row Space and Null Space	581
B.11	Row Echelon Form	582
B.12	Positive Definiteness	582
B.13	Scalar (Dot) Product and Vector (Cross) Product	583
B.14	Matrix Inversion Lemma	584

Applied Numerical Methods Using MATLAB®, Second Edition. Won Y. Yang, Jaekwon Kim, Kyung W. Park, Donghyun Baek, Sungjoon Lim, Jingon Joung, Suhyun Park, Han L. Lee, Woo June Choi, and Taeho Im.
© 2020 John Wiley & Sons, Inc. Published 2020 by John Wiley & Sons, Inc.
Companion website: www.wiley.com/go/yang/appliednumericalmethods

578 MATRIX OPERATIONS/PROPERTIES

B.1 ADDITION AND SUBTRACTION

$$A \pm B = \begin{bmatrix} a_{11} & a_{12} & \cdots & a_{1N} \\ a_{21} & a_{22} & \cdots & a_{2N} \\ \vdots & \vdots & & \vdots \\ a_{M1} & a_{M2} & \cdots & a_{MN} \end{bmatrix} \pm \begin{bmatrix} b_{11} & b_{12} & \cdots & b_{1N} \\ b_{21} & b_{22} & \cdots & b_{2N} \\ \vdots & \vdots & & \vdots \\ b_{M1} & b_{M2} & \cdots & b_{MN} \end{bmatrix} = \begin{bmatrix} c_{11} & c_{12} & \cdots & c_{1N} \\ c_{21} & c_{22} & \cdots & c_{2N} \\ \vdots & \vdots & & \vdots \\ c_{M1} & c_{M2} & \cdots & c_{MN} \end{bmatrix} = C$$

(B.1.1)

$$\text{with } a_{mn} \pm b_{mn} = c_{mn} \tag{B.1.2}$$

B.2 MULTIPLICATION

$$AB = \begin{bmatrix} a_{11} & a_{12} & \cdots & a_{1K} \\ a_{21} & a_{22} & \cdots & a_{2K} \\ \vdots & \vdots & & \vdots \\ a_{M1} & a_{M2} & \cdots & a_{MK} \end{bmatrix} \begin{bmatrix} b_{11} & b_{12} & \cdots & b_{1N} \\ b_{21} & b_{22} & \cdots & b_{2N} \\ \vdots & \vdots & & \vdots \\ b_{K1} & b_{K2} & \cdots & b_{KN} \end{bmatrix} = \begin{bmatrix} c_{11} & c_{12} & \cdots & c_{1N} \\ c_{21} & c_{22} & \cdots & c_{2N} \\ \vdots & \vdots & & \vdots \\ c_{M1} & c_{M2} & \cdots & c_{MN} \end{bmatrix} = C \quad (B.2.1)$$

$$\text{with } c_{mn} = \sum_{k=1}^{K} a_{mk} b_{kn} \tag{B.2.2}$$

(cf) For this multiplication to be done, the number of columns of A must equal the number of rows of B.

(cf) The commutative law does not hold for the matrix multiplication, i.e. $AB \neq BA$.

B.3 DETERMINANT

The *determinant* of a $K \times K$ (square) matrix $A = [a_{mn}]$ is defined by

$$\det(A) = |A| = \sum_{k=0}^{K} a_{kn}(-1)^{k+n} M_{kn} \text{ or } \sum_{k=0}^{K} a_{mk}(-1)^{m+k} M_{mk} \tag{B.3.1}$$

for any fixed $1 \leq n \leq K$ or $1 \leq m \leq K$

where the minor M_{kn} is the determinant of the $(K-1) \times (K-1)$ (minor) matrix formed by removing the kth row and the nth column from A and $A_{kn} = (-1)^{k+n} M_{kn}$ is called the cofactor of a_{kn}.

Especially, the determinants of a 2×2 matrix $A_{2 \times 2}$ and a 3×3 matrix $A_{3 \times 3}$ are

$$\det(A_{2 \times 2}) = \begin{vmatrix} a_{11} & a_{12} \\ a_{21} & a_{22} \end{vmatrix} = \sum_{k=0}^{2} a_{kn}(-1)^{k+n} M_{kn} = a_{11}a_{22} - a_{12}a_{21} \qquad (B.3.2)$$

$$\det(A_{3 \times 3}) = \begin{vmatrix} a_{11} & a_{12} & a_{13} \\ a_{21} & a_{22} & a_{23} \\ a_{31} & a_{32} & a_{33} \end{vmatrix} = a_{11}\begin{vmatrix} a_{22} & a_{23} \\ a_{32} & a_{33} \end{vmatrix} - a_{12}\begin{vmatrix} a_{21} & a_{23} \\ a_{31} & a_{33} \end{vmatrix} + a_{13}\begin{vmatrix} a_{21} & a_{22} \\ a_{31} & a_{32} \end{vmatrix}$$

$$= a_{11}(a_{22}a_{33} - a_{23}a_{32}) - a_{12}(a_{21}a_{33} - a_{23}a_{31})$$
$$+ a_{13}(a_{21}a_{32} - a_{22}a_{31}) \qquad (B.3.3)$$

Note the following properties:

- If the determinant of a matrix is zero, the matrix is singular.
- The determinant of a matrix equals the product of the eigenvalues of a matrix.
- If A is upper/lower triangular having only zeros below/above the diagonal in each column, its determinant is the product of the diagonal elements.
- $\det(A^T) = \det(A)$; $\det(AB) = \det(A) \cdot \det(B)$; $\det(A^{-1}) = 1/\det(A)$.

B.4 EIGENVALUES AND EIGENVECTORS OF A MATRIX[1]

The *eigenvalue* or *characteristic value* and its corresponding *eigenvector* or *characteristic vector* of an $N \times N$ matrix A are defined to be a scalar λ and a nonzero vector $\mathbf{v}$ satisfying

$$A\mathbf{v} = \lambda \mathbf{v} \quad \Leftrightarrow \quad (A - \lambda I)\mathbf{v} = \mathbf{0} \text{ (with } \mathbf{v} \neq \mathbf{0}) \qquad (B.4.1)$$

where I denotes the $N \times N$ identity matrix having ones on its diagonal and zeros elsewhere. Note that $(\lambda, \mathbf{v})$ is called an *eigenpair*, and there are N eigenpairs for an $N \times N$ matrix A.

The eigenvalues of a matrix can be computed as the roots of the characteristic equation

$$|A - \lambda I| = 0 \qquad (B.4.2)$$

and the eigenvector corresponding to an eigenvalue λ_i can be obtained by substituting λ_i into Eq. (B.4.1) and solve it for $\mathbf{v}$.

[1] See the website at http://www.sosmath.com/index.html.

Note the following properties:

- If A is symmetric, all the eigenvalues are real-valued.
- If A is symmetric and positive definite, all the eigenvalues are real and positive.
- If $\mathbf{v}$ is an eigenvector of A, so is $c\mathbf{v}$ for any nonzero scalar c.

B.5 INVERSE MATRIX

The *inverse* matrix of a $K \times K$ (square) matrix $A = [a_{mn}]$ is denoted by A^{-1} and defined to be a matrix which is premultiplied/postmultiplied by A to form an identity matrix, i.e. it satisfies

$$AA^{-1} = A^{-1}A = I \qquad (B.5.1)$$

An element of the inverse matrix $A^{-1} = [\alpha_{mn}]$ can be computed as

$$\alpha_{mn} = \frac{1}{\det(A)}(-1)^{m+n}M_{nm} \qquad (B.5.2)$$

where M_{nm} is the minor of a_{nm} and $(-1)^{n+m}M_{nm}$ is the cofactor of a_{nm}.
Note that a square matrix A is invertible/nonsingular if and only if

- no eigenvalue of A is zero, or equivalently,
- the rows (and the columns) of A are linearly independent, or equivalently,
- the determinant of A is nonzero.

B.6 SYMMETRIC/HERMITIAN MATRIX

A square matrix A is said to be *symmetric*, if it is equal to its transpose, i.e.

$$A^T \equiv A \qquad (B.6.1)$$

A complex-valued matrix is said to be *Hermitian* if it is equal to its complex conjugate transpose, i.e.

$$A \equiv A^{*T} \text{ where the superscript } * \text{ means the conjugate.} \qquad (B.6.2)$$

Note the following properties of a symmetric/Hermitian matrix:

- All the eigenvalues are real.
- If all the eigenvalues are distinct, the eigenvectors can form an orthogonal/unitary matrix U.

B.7 ORTHOGONAL/UNITARY MATRIX

A nonsingular (square) matrix A is said to be *orthogonal* if its transpose is equal to its inverse, i.e.

$$A^{\mathrm{T}}A \equiv I; \quad A^{\mathrm{T}} \equiv A^{-1} \tag{B.7.1}$$

A complex-valued (square) matrix is said to be *unitary* if its conjugate transpose is equal to its inverse, i.e.

$$A^{*\mathrm{T}}A \equiv I; \quad A^{*\mathrm{T}} \equiv A^{-1} \tag{B.7.2}$$

Note the following properties of an orthogonal/unitary matrix.

- The magnitude (absolute value) of every eigenvalue is one.
- The product of two orthogonal matrices is also orthogonal; $(AB)^{*\mathrm{T}}(AB) = B^{*\mathrm{T}}(A^{*\mathrm{T}}A)B \equiv I$.

B.8 PERMUTATION MATRIX

A matrix P having only one nonzero element of value 1 in each row and column is called a *permutation matrix* and has the following properties.

- Premultiplication/postmultiplication of a matrix A by a permutation matrix P, i.e. PA or AP yields the row/column change of the matrix A, respectively.
- A permutation matrix A is orthogonal, i.e. $A^{\mathrm{T}}A \equiv I$.

B.9 RANK

The *rank* of an $M \times N$ matrix is the number of linearly independent rows/columns, and if it equals $\min(M,N)$, then the matrix is said to be of *maximal* or *full rank*; otherwise, the matrix is said to be *rank-deficient* or to have *rank-deficiency*.

B.10 ROW SPACE AND NULL SPACE

The *row space* of an $M \times N$ matrix A, denoted by $\mathscr{R}(A)$, is the space spanned by the row vectors, i.e. the set of all possible linear combinations of row vectors of A that can be expressed by $A^{\mathrm{T}}\boldsymbol{\alpha}$ with an M-dimensional column vector $\boldsymbol{\alpha}$. On the other hand, the *null space* of the matrix A, denoted by $\mathscr{N}(A)$, is the space orthogonal (perpendicular) to the row space, i.e. the set of all possible linear combinations of the N-dimensional vectors satisfying $A\mathbf{x} = \mathbf{0}$.

B.11 ROW ECHELON FORM

A matrix is said to be of *row echelon form* if

- each nonzero row having at least one nonzero element has a 1 as its first nonzero element,
- the leading 1 in a row is in a column to the right of the leading 1 in the upper row, and
- all-zero rows are below the rows that have at least one nonzero element.

A matrix is said to be of *reduced row echelon form* if it satisfies the aforementioned conditions, and additionally, each column containing a leading 1 has no other nonzero elements.

Any matrix, singular or rectangular, can be transformed into this form through the Gaussian elimination procedure, i.e. a series of *elementary row operations*, or equivalently, by using the MATLAB built-in function "rref()". For example, we have

$$A = \begin{bmatrix} 0 & 0 & 1 & 3 \\ 2 & 4 & 0 & -8 \\ 1 & 2 & 1 & -1 \end{bmatrix} \xrightarrow[\text{change}]{\text{Row}} \begin{bmatrix} 2 & 4 & 0 & -8 \\ 0 & 0 & 1 & 3 \\ 1 & 2 & 1 & -1 \end{bmatrix} \xrightarrow[\text{Row subtraction}]{\text{Row division}} \begin{bmatrix} 1 & 2 & 0 & -4 \\ 0 & 0 & 1 & 3 \\ 0 & 0 & 1 & 3 \end{bmatrix}$$

$$\xrightarrow[\text{subtraction}]{\text{Row}} \begin{bmatrix} 1 & 2 & 0 & -4 \\ 0 & 0 & 1 & 3 \\ 0 & 0 & 0 & 0 \end{bmatrix} = \text{rref}(A)$$

Once this form is obtained, it is easy to compute the rank, the *determinant* and the inverse of the matrix, if only the matrix is invertible.

B.12 POSITIVE DEFINITENESS

A square matrix A is said to be *positive definite* if

$$\mathbf{x}^{*T} A \mathbf{x} > 0 \text{ for any nonzero vector } \mathbf{x} \qquad \text{(B.12.1)}$$

A square matrix A is said to be *positive semi-definite* if

$$\mathbf{x}^{*T} A \mathbf{x} \geq 0 \text{ for any nonzero vector } \mathbf{x} \qquad \text{(B.12.2)}$$

Note the following properties of a positive definite matrix A:

- A is nonsingular and all of its eigenvalues are positive, and
- the inverse of A is also positive definite.

There are similar definitions for negative definiteness and negative semi-definiteness.

Note the following property, which can be used to determine if a matrix is positive (semi-)definite or not. A square matrix is positive definite if and only if

(i) every diagonal element is positive and
(ii) every leading principal minor has positive determinant.

On the other hand, a square matrix is positive semi-definite if and only if

(i) every diagonal element is nonnegative and
(ii) every principal minor has nonnegative determinant.

Note also that the principal minors are the submatrices taking the diagonal elements from the diagonal of the matrix A and, say for a 3×3 matrix, the principal minors are

$$a_{11}, a_{22}, a_{33}, \begin{bmatrix} a_{11} & a_{12} \\ a_{21} & a_{22} \end{bmatrix}, \begin{bmatrix} a_{22} & a_{23} \\ a_{32} & a_{33} \end{bmatrix}, \begin{bmatrix} a_{11} & a_{13} \\ a_{31} & a_{33} \end{bmatrix}, \begin{bmatrix} a_{11} & a_{12} & a_{13} \\ a_{21} & a_{22} & a_{23} \\ a_{31} & a_{32} & a_{33} \end{bmatrix}$$

among which the leading ones are

$$a11, \begin{bmatrix} a_{11} & a_{12} \\ a_{21} & a_{22} \end{bmatrix}, \begin{bmatrix} a_{11} & a_{12} & a_{13} \\ a_{21} & a_{22} & a_{23} \\ a_{31} & a_{32} & a_{33} \end{bmatrix}$$

B.13 SCALAR (DOT) PRODUCT AND VECTOR (CROSS) PRODUCT

A scalar (or inner or dot) product of two N-dimensional vectors $\mathbf{x}$ and $\mathbf{y}$ is denoted by $\mathbf{x} \cdot \mathbf{y}$ and is defined by

$$\mathbf{x} \cdot \mathbf{y} = \sum_{n=1}^{N} x_n y_n = \mathbf{x}^T \mathbf{y} \qquad (B.13.1)$$

A vector (or outer or cross) product of two 3-dimensional column vectors $\mathbf{x} = [x_1 \ x_2 \ x_3]^T$ and $\mathbf{y} = [y_1 \ y_2 \ y_3]^T$ is denoted by $\mathbf{x} \times \mathbf{y}$ and is defined by

$$\mathbf{x} \times \mathbf{y} = \begin{bmatrix} x_2 y_3 - x_3 y_2 \\ x_3 y_1 - x_1 y_3 \\ x_1 y_2 - x_2 y_1 \end{bmatrix} \tag{B.13.2}$$

B.14 MATRIX INVERSION LEMMA

Matrix Inversion Lemma. Let A, C, and $[C^{-1} + DA^{-1}B]$ be well-defined with nonsingularity as well as compatible dimensions. Then, we have

$$[A + BCD]^{-1} = A^{-1} - A^{-1}B[C^{-1} + DA^{-1}B]^{-1}DA^{-1} \tag{B.14.1}$$

Proof. We will show that postmultiplying Eq. (B.14.1) by $[A + BCD]$ yields an identity matrix.

$$[A^{-1} - A^{-1}B[C^{-1} + DA^{-1}B]^{-1}DA^{-1}][A + BCD]$$
$$= I + A^{-1}BCD - A^{-1}B[C^{-1} + DA^{-1}B]^{-1}D - A^{-1}B[C^{-1} + DA^{-1}B]^{-1}DA^{-1}BCD$$
$$= I + A^{-1}BCD - A^{-1}B[C^{-1} + DA^{-1}B]^{-1}C^{-1}CD - A^{-1}B[C^{-1} + DA^{-1}B]^{-1}DA^{-1}BCD$$
$$= I + A^{-1}BCD - A^{-1}B[C^{-1} + DA^{-1}B]^{-1}[C^{-1} + DA^{-1}B]CD$$
$$= I + A^{-1}BCD - A^{-1}BCD \equiv I$$

∎

APPENDIX C

DIFFERENTIATION W.R.T. A VECTOR

The first derivative of a scalar-valued function $f(\mathbf{x})$ w.r.t. the vector $\mathbf{x} = [x_1 \ x_2]^T$ is called the *gradient* of $f(\mathbf{x})$ and defined as

$$\nabla f(\mathbf{x}) = \frac{d}{d\mathbf{x}} f(\mathbf{x}) = \begin{bmatrix} \partial f / \partial x_1 \\ \partial f / \partial x_2 \end{bmatrix} \quad \text{(C.1)}$$

Based on this definition, we can write the following equation:

$$\frac{\partial}{\partial \mathbf{x}} \mathbf{x}^T \mathbf{y} = \frac{\partial}{\partial \mathbf{x}} \mathbf{y}^T \mathbf{x} = \frac{\partial}{\partial \mathbf{x}} (x_1 y_1 + x_2 y_2) \stackrel{(B.1)}{=} \begin{bmatrix} y_1 \\ y_2 \end{bmatrix} = \mathbf{y} \quad \text{(C.2)}$$

$$\frac{\partial}{\partial \mathbf{x}} \mathbf{x}^T \mathbf{x} = \frac{\partial}{\partial \mathbf{x}} (x_1^2 + x_2^2) = 2 \begin{bmatrix} x_1 \\ x_2 \end{bmatrix} = 2\mathbf{x} \quad \text{(C.3)}$$

Also with an $M \times N$ matrix $A = [a_{mn}]$, we have

$$\frac{\partial}{\partial \mathbf{x}} \mathbf{x}^T A \mathbf{y} = \frac{\partial}{\partial \mathbf{x}} \mathbf{y}^T A^T \mathbf{x} = A\, \mathbf{y} \quad \text{(C.4a)}$$

$$\frac{\partial}{\partial \mathbf{x}} \mathbf{y}^T A \mathbf{x} = \frac{\partial}{\partial \mathbf{x}} \mathbf{x}^T A^T \mathbf{y} = A^T \mathbf{y} \quad \text{(C.4b)}$$

Applied Numerical Methods Using MATLAB®, Second Edition. Won Y. Yang, Jaekwon Kim, Kyung W. Park, Donghyun Baek, Sungjoon Lim, Jingon Joung, Suhyun Park, Han L. Lee, Woo June Choi, and Taeho Im.
© 2020 John Wiley & Sons, Inc. Published 2020 by John Wiley & Sons, Inc.
Companion website: www.wiley.com/go/yang/appliednumericalmethods

where

$$\mathbf{x}^T A \mathbf{y} = \sum_{m=1}^{M} \sum_{n=1}^{N} a_{mn} x_m y_n \xrightarrow{\text{if } M=N=2} a_{11} x_1 y_1 + a_{12} x_1 y_2 + a_{21} x_2 y_1 + a_{22} x_2 y_2 \tag{C.5}$$

Especially for a symmetric (and square) matrix A (with $M = N$), we have

$$\frac{\partial}{\partial \mathbf{x}} \mathbf{x}^T A \mathbf{x} = (A + A^T)\mathbf{x} \xrightarrow{\text{if } A \text{ is symmetric}} 2A\mathbf{x} \tag{C.6}$$

The second derivative of a scalar function $f(\mathbf{x})$ w.r.t. the vector $\mathbf{x} = [x_1 \ \ x_2]^T$ is called the *Hessian* (*matrix*) of $f(\mathbf{x})$ and is defined as

$$H(\mathbf{x}) = \nabla^2 f(\mathbf{x}) = \frac{d^2}{d\mathbf{x}^2} f(\mathbf{x}) = \begin{bmatrix} \partial^2 f / \partial x_1^2 & \partial^2 f / \partial x_1 \partial x_2 \\ \partial^2 f / \partial x_2 \partial x_1 & \partial^2 f / \partial x_2^2 \end{bmatrix} \tag{C.7}$$

Based on this definition, we can write the following equation:

$$\frac{d^2}{d\mathbf{x}^2} \mathbf{x}^T A \mathbf{x} = A + A^T \xrightarrow{\text{if } A \text{ is symmetric}} 2A \tag{C.8}$$

The first derivative of a vector-valued function $\mathbf{f}(\mathbf{x})$ w.r.t. the vector $\mathbf{x} = [x_1 \ \ x_2]^T$ is called the *Jacobian* (*matrix*) of $\mathbf{f}(\mathbf{x})$ and is defined as

$$J(\mathbf{x}) = \frac{d}{d\mathbf{x}} \mathbf{f}(\mathbf{x}) = \begin{bmatrix} \partial f_1 / \partial x_1 & \partial f_1 / \partial x_2 \\ \partial f_2 / \partial x_1 & \partial f_2 / \partial x_2 \end{bmatrix} \tag{C.9}$$

Note that the Hessian of a scalar function $f(\mathbf{x})$ of a vector $\mathbf{x}$ is equal to the *Jacobian* of the gradient of $f(\mathbf{x})$:

$$H\{f(\mathbf{x})\} = J\{\nabla f(\mathbf{x})\} \tag{C.10}$$

APPENDIX D

LAPLACE TRANSFORM

Table D.1 Laplace transforms of basic functions.

$x(t)$	$X(s)$	$x(t)$	$X(s)$	$x(t)$	$X(s)$
(1) $\delta(t)$	1	(5) $e^{-at} u_s(t)$	$\dfrac{1}{s+a}$	(9) $e^{-at} \sin \omega t \; u_s(t)$	$\dfrac{\omega}{(s+a)^2 + \omega^2}$
(2) $\delta(t - t_1)$	$e^{-t_1 s}$	(6) $t^m e^{-at} u_s(t)$	$\dfrac{m!}{(s+a)^{m+1}}$	(10) $e^{-at} \cos \omega t \; u_s(t)$	$\dfrac{s+a}{(s+a)^2 + \omega^2}$
(3) $u_s(t)$	$\dfrac{1}{s}$	(7) $\sin \omega t \; u_s(t)$	$\dfrac{\omega}{s^2 + \omega^2}$		
(4) $t^m u_s(t)$	$\dfrac{m!}{s^{m+1}}$	(8) $\cos \omega t \; u_s(t)$	$\dfrac{s}{s^2 + \omega^2}$		

Applied Numerical Methods Using MATLAB®, Second Edition. Won Y. Yang, Jaekwon Kim, Kyung W. Park, Donghyun Baek, Sungjoon Lim, Jingon Joung, Suhyun Park, Han L. Lee, Woo June Choi, and Taeho Im.
© 2020 John Wiley & Sons, Inc. Published 2020 by John Wiley & Sons, Inc.
Companion website: www.wiley.com/go/yang/appliednumericalmethods

Table D.2 Properties of Laplace transform.

(0) Definition	$X(s) = \mathscr{L}\{x(t)\} = \int_0^\infty x(t)e^{-st}\,dt$
(1) Linearity	$\alpha\,x(t) + \beta\,x(t) \to \alpha\,X(s) + \beta\,Y(s)$
(2) Time shifting	$x(t - t_1)u_s(t - t_1),\ t_1 > 0\ \to e^{-st_1}\left\{X(s) + \int_{-t_1}^0 x(\tau)e^{-s\tau}\,d\tau\right\}$
(3) Frequency shifting	$e^{s_1 t}x(t) \to X(s - s_1)$
(4) Real convolution	$g(t) * x(t) \to G(s)X(s)$
(5) Time derivative	$x'(t) \to sX(s) - x(0)$
(6) Time integral	$\int_{-\infty}^t x(\tau)\,d\tau \to \dfrac{1}{s}X(s) + \dfrac{1}{s}\int_{-\infty}^0 x(\tau)\,d\tau$
(7) Complex derivative	$t\,x(t) \to -\dfrac{d}{ds}X(s)$
(8) Complex convolution	$x(t)y(t) \to \dfrac{1}{2\pi j}\int_{\sigma_0-\infty}^{\sigma_0+\infty} X(v)Y(s - v)\,dv$
(9) Initial value theorem	$x(0) \to \lim\limits_{s \to \infty} sX(s)$
(10) Final value theorem	$x(\infty) \to \lim\limits_{s \to 0} sX(s)$

APPENDIX E

FOURIER TRANSFORM

Table E.1 Definition and properties of the CTFT (continuous-time Fourier transform.)

(0) Definition	$X(j\omega) = \mathcal{F}\{x(t)\} = \int_{-\infty}^{\infty} x(t)\, e^{-j\omega t}\, dt$
(1) Linearity	$a\, x(t) + b\, y(t) \stackrel{\mathcal{F}}{\leftrightarrow} a\, X(\omega) + b\, Y(\omega)$
(2) Time reversal	$x(-t) \stackrel{\mathcal{F}}{\leftrightarrow} X(-\omega)$
(3) Symmetry for real-valued functions	Real-valued $x(t) = x_e(t) + x_o(t) \stackrel{\mathcal{F}}{\leftrightarrow} X(\omega) = X^*(-\omega)$ Real-valued and even $x_e(t) \stackrel{\mathcal{F}}{\leftrightarrow} X_e(\omega) = \text{Re}\{X(\omega)\}$ Real-valued and odd $x_o(t) \stackrel{\mathcal{F}}{\leftrightarrow} X_o(\omega) = j\text{Im}\{X(\omega)\}$
(4) Conjugation	$x^*(t) \stackrel{\mathcal{F}}{\leftrightarrow} X^*(-\omega)$
(5) Time shifting (real translation)	$x(t - t_1) \stackrel{\mathcal{F}}{\leftrightarrow} X(\omega)\, e^{-j\omega t_1} = X(\omega)\angle -t_1\omega$
(6) Frequency shifting (complex translation)	$x(t)\, e^{j\omega_1 t} \stackrel{\mathcal{F}}{\leftrightarrow} X(\omega - \omega_1)$
(7) Duality	$g(t) \stackrel{\mathcal{F}}{\leftrightarrow} f(\omega) \Leftrightarrow f(t) \stackrel{\mathcal{F}}{\leftrightarrow} 2\pi g(-\omega)$

Applied Numerical Methods Using MATLAB®, Second Edition. Won Y. Yang, Jaekwon Kim, Kyung W. Park, Donghyun Baek, Sungjoon Lim, Jingon Joung, Suhyun Park, Han L. Lee, Woo June Choi, and Taeho Im.
© 2020 John Wiley & Sons, Inc. Published 2020 by John Wiley & Sons, Inc.
Companion website: www.wiley.com/go/yang/appliednumericalmethods

Table E.1 (*Continued*)

(8) Real convolution	$y(t) = x(t) * g(t) \overset{\mathcal{F}}{\leftrightarrow} Y(\omega) = X(\omega)\, G(\omega)$
(9) Complex convolution (modulation)	$y(t) = x(t)\, m(t) \overset{\mathcal{F}}{\leftrightarrow} Y(\omega) = \frac{1}{2\pi} X(\omega) * M(\omega)$
(10) Time differentiation	$\frac{dx(t)}{dt} \overset{\mathcal{F}}{\leftrightarrow} j\omega X(\omega)$
(11) Time integration	$\int_{-\infty}^{t} x(\tau)\, d\tau \overset{\mathcal{F}}{\leftrightarrow} \pi X(0)\delta(\omega) + \frac{1}{j\omega} X(\omega)$
(12) Scaling	$x(at) \overset{\mathcal{F}}{\leftrightarrow} \frac{1}{|a|} X\left(\frac{\omega}{a}\right)$
(13) Time multiplication-Frequency differentiation	$t\, x(t) \overset{\mathcal{F}}{\leftrightarrow} j\frac{dX(\omega)}{d\omega}$
(14) Parseval's relation	$\int_{-\infty}^{\infty} |x(t)|^2 dt = \frac{1}{2\pi} \int_{-\infty}^{\infty} |X(\omega)|^2 d\omega$

Table E.2 Properties of DTFT (discrete-time Fourier transform.)

(0) Definition	$X(\Omega) = \sum_{n=-\infty}^{\infty} x[n] e^{-j\Omega n}$
(1) Linearity	$\alpha\, x[n] + \beta\, x[n] \to \alpha\, X(\Omega) + \beta\, Y(\Omega)$
(2) Symmetry	$x[n] = x_e[n] + x_o[n]$: real $\to$ $X(\Omega) \equiv X^*(-\Omega)$
	$x_e[n]$: real and even $\to$ $X_e(\Omega) = \mathrm{Re}\,\{X(\Omega)\}$
	$x_o[n]$: real and odd $\to$ $X_o(\Omega) = j\,\mathrm{Im}\,\{X(\Omega)\}$
	$x[-n] \to X(-\Omega)$
(3) Time shifting	$x[n - n_1] \to e^{-j\Omega n_1} X(\Omega)$
(4) Frequency shifting	$e^{j\Omega_1 n} x[n] \to X(\Omega - \Omega_1)$
(5) Real convolution	$g[n] * x[n] = \sum_{n=-\infty}^{\infty} g[m] x[n-m] \to G(\Omega) X(\Omega)$
(6) Complex derivative	$n\, x[n] \to j\frac{d}{d\Omega} X(\Omega)$
(7) Complex convolution	$x[n] y[n] \to \frac{1}{2\pi} X(\Omega) * Y(\Omega)$ (periodic/circular convolution)
(8) Scaling	$\begin{cases} x[n/M] & \text{if } n = mM\,(m : \text{an integer}) \\ 0, & \text{otherwise} \end{cases} \to X(M\Omega)$
(9) Parseval's relation	$\sum_{n=-\infty}^{\infty} |x[n]|^2 = \frac{1}{2\pi} \int_{2\pi} |x(\Omega)|^2 d\Omega$

APPENDIX F

USEFUL FORMULAS

Formulas for summation of finite number of terms

$$\sum_{n=0}^{N} a^n = \frac{1 - a^{N+1}}{1 - a} \quad \text{(F.1)} \qquad \sum_{n=0}^{N} na^n = a\frac{1 - (N+1)a^N + Na^{N+1}}{(1-a)^2} \quad \text{(F.2)}$$

$$\sum_{n=0}^{N} n = \frac{N(N+1)}{2} \quad \text{(F.3)} \qquad \sum_{n=0}^{N} n^2 = \frac{N(N+1)(2N+1)}{6} \quad \text{(F.4)}$$

$$\sum_{n=0}^{N} n(n+1) = \frac{N(N+1)(N+2)}{3} \quad \text{(F.5)}$$

$$(a+b)^N = \sum_{n=0}^{N} NC_n a^{N-n} b^n \quad \text{with} \quad NC_n = NC_{N-n} = \frac{NP_n}{n!} = \frac{N!}{(N-n)!n!} \quad \text{(F.6)}$$

Formulas for summation of infinite number of terms

$$\sum_{n=0}^{\infty} x^n = \frac{1}{1-x}, \quad |x| < 1 \quad \text{(F.7)} \qquad \sum_{n=0}^{\infty} nx^n = \frac{x}{(1-x)^2}, \quad |x| < 1 \quad \text{(F.8)}$$

$$\sum_{n=0}^{\infty} n^k x^n = \lim_{a \to 0} (-1)^k \frac{\partial^k}{\partial a^k} \left\{ \frac{x}{x - e^{-a}} \right\}, \quad |x| < 1 \quad \text{(F.9)}$$

Applied Numerical Methods Using MATLAB®, Second Edition. Won Y. Yang, Jaekwon Kim, Kyung W. Park, Donghyun Baek, Sungjoon Lim, Jingon Joung, Suhyun Park, Han L. Lee, Woo June Choi, and Taeho Im.
© 2020 John Wiley & Sons, Inc. Published 2020 by John Wiley & Sons, Inc.
Companion website: www.wiley.com/go/yang/appliednumericalmethods

$$\sum_{n=0}^{\infty} \frac{(-1)^n}{2n+1} = 1 - \frac{1}{3} + \frac{1}{5} - \frac{1}{7} + \cdots = \frac{1}{4}\pi \tag{F.10}$$

$$\sum_{n=0}^{\infty} \frac{1}{n^2} = 1 + \frac{1}{2^2} + \frac{1}{3^2} + \frac{1}{4^2} + \cdots = \frac{1}{6}\pi^2 \tag{F.11}$$

$$e^x = \sum_{n=0}^{\infty} \frac{1}{n!}x^n = 1 + \frac{1}{1!}x + \frac{1}{2!}x^2 + \frac{1}{3!}x^3 + \cdots \tag{F.12}$$

$$a^x = \sum_{n=0}^{\infty} \frac{(\ln a)^n}{n!}x^n = 1 + \frac{\ln a}{1!}x + \frac{(\ln a)^2}{2!}x^2 + \frac{(\ln a)^3}{3!}x^3 + \cdots \tag{F.13}$$

$$\ln(1 \pm x) = -\sum_{n=1}^{\infty} (\pm 1)^n \frac{1}{n}x^n = \pm x - \frac{1}{2}x^2 \pm \frac{1}{3}x^3 - \cdots, \quad |x| < 1 \tag{F.14}$$

$$\sin x = \sum_{n=0}^{\infty} \frac{(-1)^n}{(2n+1)!}x^{2n+1} = x - \frac{1}{3!}x^3 + \frac{1}{5!}x^5 - \frac{1}{7!}x^7 + \cdots \tag{F.15}$$

$$\cos x = \sum_{n=0}^{\infty} \frac{(-1)^n}{(2n)!}x^{2n} = 1 - \frac{1}{2!}x^2 + \frac{1}{4!}x^4 - \frac{1}{6!}x^6 + \cdots \tag{F.16}$$

$$\tan x = x + \frac{1}{3}x^3 + \frac{2}{15}x^5 + \cdots, \quad |x| < \frac{\pi}{2} \tag{F.17}$$

$$\tan^{-1} x = \sum_{n=0}^{\infty} \frac{(-1)^n}{2n+1}x^{2n+1} = x - \frac{1}{3}x^3 + \frac{1}{5}x^5 - \frac{1}{7}x^7 + \cdots \tag{F.18}$$

Trigonometric formulas

$$\sin(A \pm B) = \sin A \cos B \pm \cos A \sin B \tag{F.19}$$

$$\cos(A \pm B) = \cos A \cos B \mp \sin A \sin B \tag{F.20}$$

$$\tan(A \pm B) = \frac{\tan A \pm \tan B}{1 \mp \tan A \tan B} \tag{F.21}$$

$$\sin A \sin B = \frac{1}{2}\{\cos(A - B) - \cos(A + B)\} \tag{F.22}$$

$$\sin A \cos B = \frac{1}{2}\{\sin(A + B) + \sin(A - B)\} \tag{F.23}$$

$$\cos A \sin B = \frac{1}{2}\{\sin(A + B) - \sin(A - B)\} \tag{F.24}$$

$$\cos A \cos B = \frac{1}{2}\{\cos(A + B) + \cos(A - B)\} \tag{F.25}$$

$$\sin A + \sin B = 2 \sin\left(\frac{A+B}{2}\right) \cos\left(\frac{A-B}{2}\right) \tag{F.26}$$

$$\cos A + \cos B = 2 \cos\left(\frac{A+B}{2}\right) \cos\left(\frac{A-B}{2}\right) \tag{F.27}$$

$$a \cos A - b \sin A = \sqrt{a^2 + b^2} \cos(A + \theta), \quad \theta = \tan^{-1}\left(\frac{b}{a}\right) \tag{F.28}$$

$$a \sin A + b \cos A = \sqrt{a^2 + b^2} \sin(A + \theta), \quad \theta = \tan^{-1}\left(\frac{b}{a}\right) \tag{F.29}$$

$$\sin^2 A = \frac{1}{2}(1 - \cos 2A) \tag{F.30}$$

$$\cos^2 A = \frac{1}{2}(1 + \cos 2A) \tag{F.31}$$

$$\sin^3 A = \frac{1}{4}(3 \sin A - \sin 3A) \tag{F.32}$$

$$\cos^3 A = \frac{1}{4}(3 \cos A + \cos 3A) \tag{F.33}$$

$$\sin 2A = 2 \sin A \cos A \tag{F.34}$$

$$\sin 3A = 3 \sin A - 4 \sin^3 A \tag{F.35}$$

$$\cos 2A = \cos^2 A - \sin^2 A = 1 - 2\sin^2 A = 2\cos^2 A - 1 \tag{F.36}$$

USEFUL FORMULAS

$$\cos 3A = 4\cos^3 A - 3\sin A \qquad \text{(F.37)}$$

$$\frac{a}{\sin A} = \frac{b}{\sin B} = \frac{c}{\sin C} \qquad \text{(F.38)}$$

$$a^2 = b^2 + c^2 - 2bc\cos A \qquad \text{(F.39a)}$$

$$b^2 = c^2 + a^2 - 2ca\cos B \qquad \text{(F.39b)}$$

$$c^2 = a^2 + b^2 - 2ab\cos C \qquad \text{(F.39c)}$$

$$e^{\pm j\theta} = \cos\theta \pm j\sin\theta \qquad \text{(F.40)}$$

$$\sin\theta = \frac{1}{j2}(e^{j\theta} - e^{-j\theta}) \qquad \text{(F.41a)}$$

$$\cos\theta = \frac{1}{2}(e^{j\theta} + e^{-j\theta}) \qquad \text{(F.41b)}$$

$$\tan\theta = \frac{1}{j}\frac{e^{j\theta} - e^{-j\theta}}{e^{j\theta} + e^{-j\theta}} \qquad \text{(F.41c)}$$

APPENDIX G

SYMBOLIC COMPUTATION

CHAPTER OUTLINE

G.1 How to Declare Symbolic Variables and Handle Symbolic
  Expressions 595
G.2 Calculus 597
  G.2.1 Symbolic Summation 597
  G.2.2 Limits 597
  G.2.3 Differentiation 598
  G.2.4 Integration 598
  G.2.5 Taylor Series Expansion 599
G.3 Linear Algebra 600
G.4 Solving Algebraic Equations 601
G.5 Solving Differential Equations 601

G.1 HOW TO DECLARE SYMBOLIC VARIABLES AND HANDLE SYMBOLIC EXPRESSIONS

To declare any variable(s) as a symbolic variable, you should use the command '`sym`' or '`syms`' as follows:

Applied Numerical Methods Using MATLAB®, Second Edition. Won Y. Yang, Jaekwon Kim, Kyung W. Park, Donghyun Baek, Sungjoon Lim, Jingon Joung, Suhyun Park, Han L. Lee, Woo June Choi, and Taeho Im.
© 2020 John Wiley & Sons, Inc. Published 2020 by John Wiley & Sons, Inc.
Companion website: www.wiley.com/go/yang/appliednumericalmethods

SYMBOLIC COMPUTATION

```
»a=sym('a'); t=sym('t'); x=sym('x');
»syms a x y t % or, equivalently and more efficiently
```

Once the variables have been declared as symbolic, they can be used in expressions and as arguments to many functions without being evaluated as numeric:

```
»f=x^2/(1+tan(x)^2);
»ezplot(f,[-pi pi])
»simplify(cos(x)^2+sin(x)^2) % simplify an expression
    ans = 1
»simplify(cos(x)^2-sin(x)^2) % simplify an expression
    ans = cos(2*x)
»simplify(2*sin(x)*cos(x)) % simple expression
    ans = sin(2*x)
»simplify(sin(x)*cos(y)+cos(x)*sin(y)) % simple expression
    ans = sin(x+y)
»eq1=expand((x+y)^3-(x+y)^2) % expand
    eq1 = x^3 + 3*x^2*y - x^2 + 3*x*y^2 - 2*x*y + y^3 - y^2
»collect(eq1,y) % collect similar terms in descending order w.r.t. y
    ans = y^3 + (3*x-1)*y^2 + (3*x^2-2*x)*y + x^3 - x^2
»factor(eq1) % factorize
    ans = [ x + y - 1, x + y, x + y]
»horner(eq1) % nested multiplication form
    ans = x*(x*(x + 3*y - 1) - 2*y + 3*y^2) - y^2 + y^3
»pretty(ans) % pretty form
                              2        2      3
    x (x (x + 3 y - 1) - 2 y + 3 y ) - y  + y
```

If you need to substitute numeric values or other expressions for some symbolic variables in an expression, you can use the function 'subs()' as follows:

```
»subs(eq1,x,0) % substitute numeric value
    ans= y^3 - y^2
»subs(eq1,{x,y},{0,x-1}) % substitute numeric and symbolic values
    ans= (x-1)^3 - (x-1)^2
```

The command 'sym' allows you to declare symbolic real variables by using the 'real' option as follows:

```
»x=sym('x','real'); y=sym('y','real');
»syms x y real % or, equivalently
»z=x+i*y; % declare z as a symbolic complex variable
»conj(z) % complex conjugate
    ans = x - y*1i
»angle(z) % phase angle of z
    ans = atan2(y, x)
```

The command 'sym' can be used to convert numeric values into their symbolic expressions.

```
»sym(1/2)+0.2
    ans= 7/10 % symbolic expression
```

On the other hand, the functions 'double()' and 'eval()' convert symbolic expressions into their numeric (double-precision floating-point) values. The vpa command finds the variable-precision arithmetic (VPA) expression (as a symbolic representation) of a numeric or symbolic expression with d significant decimal digits, where d is the current setting of DIGITS that can be set by the 'digits' command. Note that the output of the function 'vpa()' is a symbolic expression even if it may look like a numeric value. Let us see some examples.

```
»f=sym('exp(i*pi/4)') % may not work in higher MATLAB versions
   f = 2^(1/2)*(1/2 + i/2)
»double(f) % or eval(f)
   ans = 0.7071 + 0.7071i % numeric value
»vpa(ans,2) % variable-precision arithmetic
   ans = 0.71 + 0.71*i % symbolic expression with 2 significant digits
```

G.2 CALCULUS

G.2.1 Symbolic Summation

We can use the MATLAB function 'symsum()' to obtain the sum of an indefinite/definite series as follows:

```
»syms x n N % declare x,n,N as symbolic variables
»simplify(symsum(n,0,N))
   ans= (N*(N+1))/2    % ∑_{n=0}^{N} n = N(N+1)/2
»simplify(symsum(n^2,0,N))
   ans= (N*(N+1)*(2*N+1))/6    % ∑_{n=0}^{N} n² = N(N+1)(2N+1)/6
»symsum(1/n^2,1,inf)
   ans= pi^2/6    % ∑_{n=0}^{∞} 1/n² = π²/6
»symsum(x^n,n,0,inf)
   ans = piecewise(1<=x, Inf, x in Dom::Interval(-1,1), -1/(x-1))
   % ∑_{n=0}^{∞} xⁿ = ∞ for |x|≥1 and 1/(1-x) for |x|<1
»symsum(x^n/factorial(n),n,0,Inf)
   ans = exp(x)
```

G.2.2 Limits

We can use the MATLAB function 'limit()' to get the (two-sided) limit and the right/left-sided limits of a function as the following.

```
»syms h n x
»limit(sin(x)/x,x,0)    % lim_{x→0} sin x / x = 1
   ans = 1
»limit(x/abs(x),x,0,'right')    % lim_{x→0⁺} x/|x| = 1
   ans = 1
»limit(x/abs(x),x,0,'left')    % lim_{x→0⁻} x/|x| = -1
   ans = -1
»limit(x/abs(x),x,0)    % lim_{x→0} x/|x| =?
   ans = NaN    %Not a Number
```

```
»limit((cos(x+h)-cos(x))/h,h,0)  % lim_{h→0} (cos(x+h)-cos(x))/h = d/dx cos x = -sin x
   ans = -sin(x)
»limit((1+x/n)^n,n,inf)  % lim_{n→∞} (1 + x/n)^n = e^x
   ans = exp(x)
```

G.2.3 Differentiation

The function 'diff()' differentiates a symbolic expression w.r.t. the variable given as one of its second and third input arguments or its free variable that might be determined by using 'findsym()'.

```
»syms a b x n t
»diff(x^n)
   ans = n*x^(n-1)
»f=exp(a*x)*cos(b*t)
»diff(f)    % equivalently, diff(f,x): the 1st derivative w.r.t. x
   ans = a*exp(a*x)*cos(b*t)    % d/dx f = d/dx e^{ax} cos(bt) = ae^{ax} cos(bt)
»diff(f,t)    % the 1st derivative w.r.t. t
   ans = -b*exp(a*x)*sin(b*t)    % d/dt f = d/dt e^{ax} cos(bt) = -be^{ax} sin(bt)
»diff(f,2)    % equivalently, diff(f,x,2): the 2nd derivative w.r.t. x
   ans = a^2*exp(a*x)*cos(b*t)    % d^2/dx^2 f = a^2 e^{ax} cos(bt)
»diff(f,t,2)    % the 2nd derivative w.r.t. t
   ans = -b^2*exp(a*x)*cos(b*t)    % d^2/dt^2 f = -e^{ax} cos(bt)b^2
»g=[cos(x)*cos(t)  cos(x)*sin(t)];
»jacob_g=jacobian(g,[x t])
   jacob_g= [ -cos(t)*sin(x),  -cos(x)*sin(t)]
            [ -sin(t)*sin(x),   cos(t)*cos(x)]
```

Note that the MATLAB function 'jacobian()' finds the jacobian defined by (C.9), i.e. the derivative of a vector function $[g_1\ g_2]^T$ w.r.t. a vector variable $[x\ t]^T$ as

$$J = \begin{bmatrix} \partial g_1/\partial x & \partial g_1/\partial t \\ \partial g_2/\partial x & \partial g_2/\partial t \end{bmatrix} \tag{G.1}$$

G.2.4 Integration

The MATLAB function 'int()' returns the indefinite/definite integral (antiderivative) of a function or an expression w.r.t. the variable given as its second input argument or its free variable, which might be determined by using 'findsym()'.

```
»syms a x y t
»int(x^n)
   ans = piecewise(n==-1, log(x), n~=-1, x^(n+1)/(n+1))   % ∫ x^n dx = 1/(n+1) x^{n+1}
»int(1/(1+x^2))
   ans = atan(x)    % ∫ 1/(1+x^2) dx = tan^{-1} x
»int(a^x)   %equivalently diff(f,x,2)
   ans = a^x/log(a)    % ∫ a^x dx = 1/(log a) a^x
»int(sin(a*t),0,pi)   %equivalently int(sin(a*t),t,0,pi)
   -(2*sin((pi*a)/2)^2)/a   % ∫_0^π sin(at) dt = -1/a cos(at)|_0^π = -1/a cos(aπ) + 1/a  (F.30)= 2sin^2(aπ/2)/a
```

```
»int(exp(-(x-a)^2),a,inf) %equivalently int(exp(-(x-a)^2),x,0,inf)
  ans = pi^(1/2)/2    % $\int_0^\infty e^{-(x-a)^2}dx = \frac{1}{2}\sqrt{\pi}$
```

G.2.5 Taylor Series Expansion

We can use the MATLAB function 'taylor()' to find the Taylor series expansion of a function or an expression w.r.t. the variable given as its second or third input argument or its free variable that might be determined by using the function 'findsym()'.

One may put 'help taylor' into the MATLAB command window to see its usage, which is restated in the following. Let us try applying it.

```
»syms x t; N=3;
»Tx0=taylor(exp(-x),'Order',N+1)  %f(x) ≅ $\sum_{n=0}^{N} \frac{1}{n!} f^{(n)}(0) x^n$
  Tx0 = - x^3/6 + x^2/2 - x + 1
»sym2poly(Tx0) % extract the coefficients of Taylor series polynomial
  ans =  -0.1667    0.5000   -1.0000    1.0000
```

- `taylor(f)` gives the fifth-order Maclaurin series expansion of f about $x = 0$.
- `taylor(f,x,'expansionpoint',a)` the fifth-order Taylor series expansion of f w.r.t. x about $x = a$.
- `taylor(sin(x),'order',N)` the Nth-order Taylor series expansion of f about a.

(cf) The target function f given directly as the first input argument must be a legitimate function defined as a symbolic expression like `exp(x)` (with x declared as a symbolic variable).

(cf) Before using the command 'taylor()' with the target function defined as a symbolic expression, one should declare the arguments of the function as symbols by running the statement like 'syms x t'.

(cf) If the independent variable is not given as the second input argument, the default independent variable will be one closest (alphabetically) to 'x' among the (symbolic or literal) variables contained as arguments of the function f.

(cf) You should use the MATLAB command 'sym2poly()' if you want to extract the coefficients from the Taylor series expansion obtained as a symbolic expression.

```
»xo=1; Tx1=taylor(exp(-x),'expansionpoint',xo,'order',N+1)
            %f(x) ≅ $\sum_{n=0}^{N} \frac{1}{n!} f^{(n)}(x_o)(x-x_o)^n$
  Tx1 = exp(-1)-exp(-1)*(x-1)+(exp(-1)*(x-1)^2)/2-(exp(-1)*(x-1)^3)/6
»f=exp(-x)*sin(t);
»Tt=taylor(f,t,'Order',N+1)  %f(x) ≅ $\sum_{n=0}^{N} \frac{1}{n!} f^{(n)}(0) t^n$
  Tt = t*exp(-x) - (t^3*exp(-x))/6
```

G.3 LINEAR ALGEBRA

Several MATLAB commands and functions can be used to manipulate the vectors or matrices consisting of symbolic expressions as well as those consisting of numerics.

```
»syms a11 a12 a21 a22
»A=[a11 a12; a21 a22];
»det(A)
   ans = a11*a22-a12*a21
»AI=A^-1
   AI = [  a22/(a11*a22-a12*a21), -a12/(a11*a22-a12*a21)]
        [ -a21/(a11*a22-a12*a21),  a11/(a11*a22-a12*a21)]
»A*AI
   ans = [a11*a22/(a11*a22-a12*a21)-a12*a21/(a11*a22-a12*a21),        0]
        [          0, a11*a22/(a11*a22-a12*a21)-a12*a21/(a11*a22-a12*a21)]
»simplify(ans) % simplify an expression
   ans = [ 1,  0]
        [ 0,  1]
»syms x t;
»G=[cos(t) sin(t); -sin(t) cos(t)] % Givens transformation matrix
   G = [  cos(t),  sin(t)]
       [ -sin(t),  cos(t)]
»det(G), simplify(ans)
   ans = cos(t)^2+sin(t)^2
   ans = 1
»G2=G^2, simplify(G2)
   G2 = [  cos(t)^2-sin(t)^2,   2*cos(t)*sin(t)]
        [ -2*cos(t)*sin(t), cos(t)^2-sin(t)^2]
   ans = [  cos(2*t),  sin(2*t)]
        [ -sin(2*t),  cos(2*t)]
»GTG=G.'*G, simplify(GTG)
   GTG = [ cos(t)^2+sin(t)^2,                    0]
         [                 0,  cos(t)^2+sin(t)^2]
   ans = [ 1, 0]
        [ 0, 1]
»simplify(G^-1) % inv(G) for the inverse of Givens matrix
   G = [ cos(t), -sin(t)]
       [ sin(t),  cos(t)]
»syms b c
»A=[0 1; -c -b];
»[V,E]=eig(A), pretty([V E])
  /                2                    2                                              \
  |  b    sqrt(b -4c)        b    sqrt(b -4c)                                          |
  | - + --------------    - - --------------                       2                   |
  |  2        2              b   2      2           b    b    sqrt(b -4c)|             |
  | ------------------ - -, ------------------ - -, - - ------------------,      0|    |
  |         c                        c             c    2         2                    |
  |                                                                                    |
  |                                                                     2              |
  |                                                              sqrt(b -4c)    b      |
  |            1,                      1,                   0, ------------- - -       |
  \                                                                     2         2   /
```

Besides, other MATLAB functions such as 'jordan(A)' and 'svd(A)' can be used to get the Jordan canonical form together with the corresponding similarity transformation matrix and the singular value decomposition of a symbolic matrix.

G.4 SOLVING ALGEBRAIC EQUATIONS

We can use the backslash(\) operator to solve a set of linear equations written in a matrix-vector form.

```
»syms R11 R12 R21 R22 b1 b2
»R=[R11 R12; R21 R22]; b=[b1; b2];
»x=R\b
    x= -(R12*b2 - R22*b1)/(R11*R22 - R12*R21)
       (R11*b2 - R21*b1)/(R11*R22 - R12*R21)
```

We can also use the MATLAB function 'solve()' to solve symbolic algebraic equations.

```
»syms a b c x
»fx=a*x^2+b*x+c;
»solve(fx)  % formula for roots of 2nd-order polynomial eq
    -(b + (b^2 - 4*a*c)^(1/2))/(2*a)
    -(b - (b^2 - 4*a*c)^(1/2))/(2*a)
»syms x1 x2 b1 b2
»fx1=x1+x2-b1; fx2=x1+2*x2-b2; % a set of simultaneous algebraic eqs.
»[x1o,x2o]=solve(fx1,fx2)  % solve the set of simultaneous eqs.
    x1o = 2*b1 - b2
    x2o = b2 - b1
```

G.5 SOLVING DIFFERENTIAL EQUATIONS

We can use the MATLAB function 'dsolve()' to solve symbolic differential equations (DEs).

```
»syms a b c x
»xo=dsolve('Dx+a*x=0')  % DE w/o initial condition
    xo= C4*exp(-a*t)  % a solution with undetermined constant
»xo=dsolve('Dx+a*x=0','x(0)=2')  % a d.e. with initial condition
    xo= 2*exp(-a*t)  % a solution with undetermined constant
»xo=dsolve('Dx=1+x^2')  % Differential eq. w/o initial condition
    xo= tan(C7 + t)  % a solution with undetermined constant
»xo=dsolve('Dx=1+x^2','x(0)=1')  % with the initial condition
    xo= tan(t + pi/4)  % a solution with determined constant
»yo=dsolve('D2u=-u','t')  % 2nd-order d.e. without initial condition
    yo= C12*cos(t) - C13*sin(t)
»xo=dsolve('D2u=-u','u(0)=1,Du(0)=0')  % with the initial condition
    xo= cos(t))
```

```
»yo=dsolve('(Dy)^2+y^2=1','y(0)=0','x')  %1st-order nonlinear DE (NLDE)
    yo= [  sin(x)]    % different shape depending on the MATLAB version
        [ -sin(x)]
»yo=dsolve('D2y=cos(2*x)-y','y(0)=1,Dy(0)=0','x')  % a 2nd-order NLDE
»simplify(yo) % a 2nd-order NLDE
    1 - (8*sin(x/2)^4)/3   % yo = 4/3*cos(x)-2/3*cos(x)^2+1/3
»S=dsolve('Df=3*f+4*g','Dg=-4*f+3*g');
»f=S.f, g=S.g
    f = C43*cos(4*t)*exp(3*t) + C42*sin(4*t)*exp(3*t)
    g = C42*cos(4*t)*exp(3*t) - C43*sin(4*t)*exp(3*t)
»[f,g]=dsolve('Df=3*f+4*g,Dg=-4*f+3*g','f(0)=0,g(0)=1')
    f = sin(4*t)*exp(3*t)
    g = cos(4*t)*exp(3*t)
```

APPENDIX H

SPARSE MATRICES

A matrix is said to be sparse if it has a large portion of zero elements. MATLAB has some built-in functions/routines, which enables us to exploit the sparsity of a matrix for computational efficiency.

- The MATLAB function 'sparse()' can be used to convert a (regular) matrix into a sparse form by squeezing out any zero elements and to generate a sparse matrix having the elements of a vector given together with the row/column index vectors. On the other hand, the MATLAB function 'full()' can be used to convert a matrix of sparse form into a regular one.

```
»row_index=[1 1 2 3 4]; col_index=[1 2 2 3 4]; elements=[1 2 3 4 5];
»m=4; n=4; As=sparse(row_index,col_index,elements,m,n)
    As=  (1,1)       1
         (1,2)       2
         (2,2)       3
         (3,3)       4
         (4,4)       5
»Af=full(As)
    Af=  1      2      0      0
         0      3      0      0
         0      0      4      0
         0      0      0      5
```

Applied Numerical Methods Using MATLAB®, Second Edition. Won Y. Yang, Jaekwon Kim, Kyung W. Park, Donghyun Baek, Sungjoon Lim, Jingon Joung, Suhyun Park, Han L. Lee, Woo June Choi, and Taeho Im.
© 2020 John Wiley & Sons, Inc. Published 2020 by John Wiley & Sons, Inc.
Companion website: www.wiley.com/go/yang/appliednumericalmethods

604 SPARSE MATRICES

We can use the MATLAB function '`sprandn(m,n,nzd)`' to generate an m×n sparse matrix having the given nonzero density `nzd`. Let us see how efficient the operations can be on the matrices in sparse forms.

```
»As=sprandn(10,10,0.2); % A sparse matrix and
»Af=full(As); % Its full version
»AsA=As*As; % in sparse forms (50 flops)
»AfA=Af*Af; % in full(regular) forms (2000 flops)
»b=ones(10,1);
»x=As\b; % in sparse forms (160 flops)
»x=Af\b; % in full(regular) forms (592 flops)
»inv(As); % in sparse forms (207 flops)
»inv(Af); % in full(regular) forms (592 flops)
»[L,U,P]=lu(As); % in sparse forms (53 flops)
»f[L,U,P]=lu(Af); % in full(regular) forms (92 flops)
```

Additionally, the MATLAB function '`speye(n)`' is used to generate an n×n identity matrix and the MATLAB function '`spy(n)`' is used to visualize the sparsity pattern. The computational efficiency of LU factorization can be upgraded if one preorders the sparse matrix by the symmetric minimum degree permutation, which has been cast into the MATLAB function '`symmmd()`'.

Interest readers are welcomed to run the following script "do_sparse.m" to figure out the functions of several sparsity-related MATLAB functions.

```
%do_sparse.m
clear, clf
%create a sparse mxn random matrix
m=4; n=5; A1=sprandn(m,n,.2)
%create a sparse symmetric nxn random matrix with nonzero density nzd
nzd=0.2; A2=sprandsym(n,nzd)
%create a sparse symmetric random nxn matrix with condition number r
r=0.1; A3=sprandsym(n,nzd,r)
%a sparse symmetric random nxn matrix with the set of eigenvalues eigs
eigs=[0.1 0.2 .3 .4 .5]; A4=sprandsym(n,nzd,eigs)
eig(A4)
tic, A1A=A1*A1', time_sparse=toc
A1f=full(A1); tic, A1Af=A1f*A1f'; time_full=toc
spy(A1A), full(A1A), A1Af
sparse(A1Af)
n=10; A5=sprandsym(n,nzd)
tic, [L,U,P]=lu(A5); time_lu=toc
tic, [L,U,P]=lu(full(A5)); time_full=toc
mdo=symamd(A5); %symmetric minimum degree permutation
tic, [L,U,P]=lu(A5(mdo,mdo)); time_md=toc
```

APPENDIX I

MATLAB

On a Windows-based computer with MATLAB® software installed, you can double-click on the MATLAB icon (like the one on the right side) to start MATLAB. Once MATLAB has been started, you will see the MATLAB desktop like Figure I.1, which has the MATLAB Command window in the lower middle where a blinking cursor appears to the right of the MATLAB prompt '»' waiting for you to type in MATLAB commands/statements. It also contains other windows each labeled Current Folder, Editor, and Workspace (showing the contents of MATLAB memory).

How do we work with the MATLAB Command window?

- By clicking on 'New'/'Open' in the groups of functionalities under HOME tab and then selecting an item in their dropdown menus, you can create/edit any file with the MATLAB editor.
- By typing any MATLAB commands/statements in the MATLAB Command window, you can use various powerful mathematic/graphic functions of MATLAB.
- If you have an M-file which contains a series of commands/statements for performing a target procedure, you can type in the filename (without the extension '.m') to run it.

Applied Numerical Methods Using MATLAB®, Second Edition. Won Y. Yang, Jaekwon Kim, Kyung W. Park, Donghyun Baek, Sungjoon Lim, Jingon Joung, Suhyun Park, Han L. Lee, Woo June Choi, and Taeho Im.
© 2020 John Wiley & Sons, Inc. Published 2020 by John Wiley & Sons, Inc.
Companion website: www.wiley.com/go/yang/appliednumericalmethods

- By clicking on 'Set Path' in the right part of the toolstrip, you can make the MATLAB search path list include or exclude the paths containing the files you want to or not to be run.
- By typing a search term like a keyword or a function name into the Search Documentation box at the top-right part, you can search help documentation for what you want to know about.
- To undock, i.e. separate the Editor window from the MATLAB desktop, click on the Editor Actions button (down-arrow-shaped) in the top-right part of the Editor window and select 'Undock' in its drop-down menu. To dock the Editor window back to the desktop, click on the Editor Actions button and select 'Dock Editor' in its drop-down menu.

How do we edit an existing M-file (script or function) that will be called a program here?

1. With the program (you want to edit), say, named "nm03p17.m", loaded into the MATLAB Editor/Debugger window, set breakpoint(s) at any statement(s) which you think are suspicious to be the source(s) of error, by clicking the pertinent statement line of the program with the left mouse button and pressing the F12 key or clicking 'Set/Clear' in the 'Breakpoints' pull-down menu. Then you will see a small red disk in front of every statement at which you set the breakpoint.

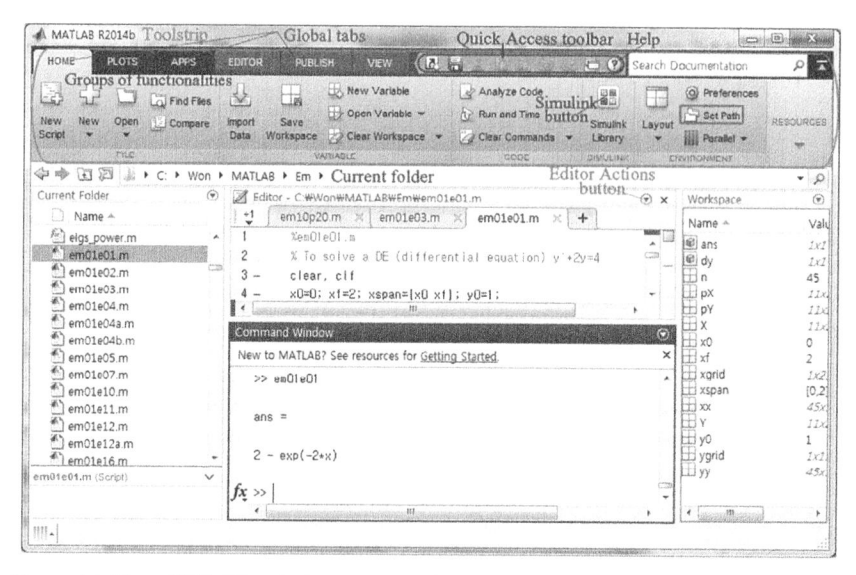

Figure I.1 The MATLAB desktop with Command/Editor/Current folder/Workspace windows.

2. Going to the MATLAB Command window, type in the name of the file containing the main program or click Run/Continue button (in the tool bar) to try running the program. Then, go back to the Editor/Debugger window and you will see the cursor blinking just after a green arrow between the red disk and the first statement line at which you set the breakpoint (see Figure I.2).
3. Determining which variable to look into, go to the Command window and type in the variable name(s) (just after the prompt 'K»') or whatever statement you want to run for debugging.
4. If you want to proceed to the next statement line in the program, go back to the Editor/Debugger window and press F10 (single_step) key or the F11 (step_in) key to dig into a called function. If you want to jump to the next breakpoint, press F5 or click 'Run (Continue)'. If you want to run the program till just before a statement, move the cursor to the line and click 'Run to Cursor'.
5. If you have figured out what is wrong, edit the pertinent part of the program, save the edited program, go to the Command window, and type in the name of the file containing the main program or click Run button to try running the program for test. If the result seems to reflect that the program still has a bug, go back to step 1 and restart the whole procedure.

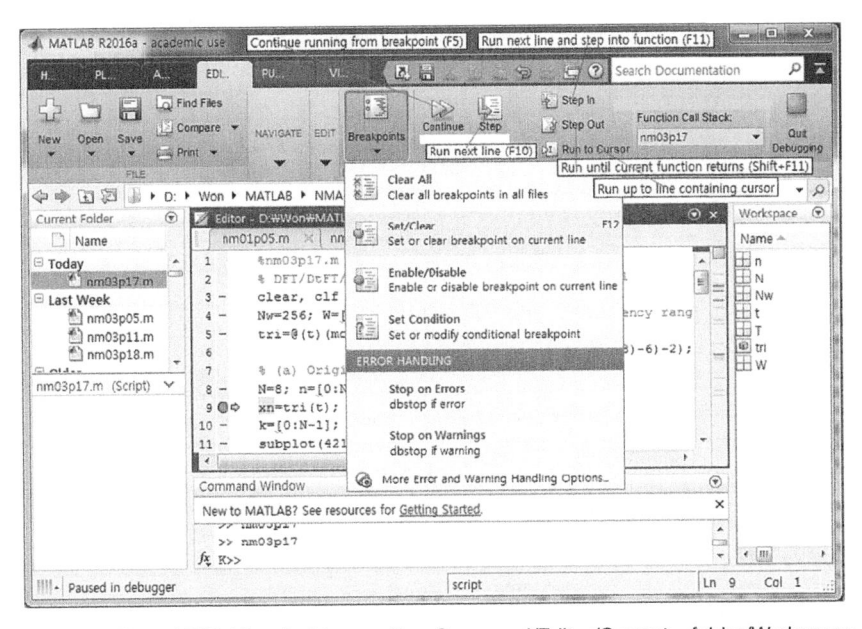

Figure I.2 The MATLAB desktop with Command/Editor/Current folder/Workspace windows.

Table I.1 Some frequently used MATLAB commands.

	General commands
break	to exit from a for or while loop
fprintf	fprintf('\n x(%d)=%6.4f \a',ind,x(ind))
keyboard	stop execution till the user types any key
return	terminate a routine and go back to the calling routine
load *** x y	read the values of x and y from the MATLAB file ***.mat
load x.dat	read the value(s) of x from the ASCII file x.dat
save *** x y	save the values of x and y into the MATLAB file ***.mat
save x.dat x	save the value(s) of x into the ASCII file x.dat
clear	remove all or some variables/functions from memory

	Two-dimensional graphic commands
bar(x,y),plot(x,y),stairs(x,y) stem(x,y),loglog(x,y) semilogx(x,y),semilogy(x,y)	plot the values of y vs. x in a bar/continuous/stairs/discrete/xy-log/x-log/y-log graph
plot(y) (y: read-valued)	plot the values of vector/array over the index
plot(y) (y: complex-valued)	plot the imaginary part vs. the real part: plot(real(y),imag(y))
bar(y, $s_1s_2s_3$)	The string of three characters $s_1s_2s_3$, given as one of the input
plot(y, $s_1s_2s_3$)	arguments to these graphic commands specifies the color, the
stairs(y, $s_1s_2s_3$)	symbol, and the line types:
stem(y, $s_1s_2s_3$)	s_1 (color): y(ellow), m(agenta), c(yan), r(ed), g(reen), b(lue),
loglog(y, $s_1s_2s_3$)	w(hite), (blac)k
semilogx(y, $s_1s_2s_3$)	s_2 (symbol): .(point), o,x,+,*, s(quare: □), d(iamond:◊), v(∇),
semilogy(y, $s_1s_2s_3$)	ˆ(Δ), <(◁), >(▷), p(entagram:★), h(exagram)
plot(y1, $s_1s_2s_3$, y2, $s_1s_2s_3$)	s_3 (line symbol): -(solid, default), :(dotted), -.(dashdot), --(dashed)
	(ex) `plot(x,'b+:')` plots x(n) with the + symbols on a blue dotted line
polar(theta,r)	plot the graph in polar form with the phase theta and magnitude r

	Auxiliary graphic commands
axis([xmin xmax ymin ymax])	specify the ranges of graph on horizontal/vertical axes
clf(clear figure)	clear the existent graph(s)
grid on/off	draw the grid lines
hold on/off	keep/remove the existent graph(s)
subplot(ijk)	divide the screen into i x j sections and use the kth one
text(x,y,plot(y,'***')	print the string '***' in the position (x,y) on the graph
title('**'), xlabel('**'), ylabel('**')	print the string '**' into the top/low/left side of graph

	Three-dimensional graphic commands
mesh(X,Y,Z)	connect the points of height Z at points (X,Y) where X,Y and Z are the matrices of the same dimension
mesh(x,y,Z)	connect the points of height Z(j,i) at points specified by the two vectors (x(i),y(j))
mesh(Z),surf(),plot3(),contour()	connect the points of height Z(j,i) at points specified by (i,j)

It should be noted that

1. the index of an array in MATLAB starts from 1, not 0, and
2. a dot(.) must be put before an operator to make a term-wise (element-by-element) operation.

If you have further interest in MATLAB, visit the homepage of MathWorks (to find many useful resources) and/or read some references like [B-2], [B-3], [C-2], [F-1], [H-1], [K-1], [K-2], [K-3], [L-1], [M-1], [M-2], [N-1], [O-1], [P-1], [P-2], [P-3], [R-1], [R-2], [S-1], [S-2], [S-3], [W-1], [W-2], [W-3], [W-4], [W-5], [W-6], [W-7], [W-8], [Y-1], [Y-2], [Y-3], [Z-1].

Table I.1 shows some frequently used MATLAB commands.

Table I.2 shows various the graphic line specifications used in the plot( ) command. Some of mathematical functions and special reserved constants/variables defined in MATLAB are listed in Table I.3.

Table I.2 Graphic line specifications used in the `plot()` command.

Line type	Point type (marker symbol)			Color	
—— solid line	.: dot	+: plus	*: asterisk	r: red	m: magenta
.... dotted line	^: △	>: >	°: circle	g: green	y: yellow
--- dashed line	p:*	v: ▽	x: x-mark	b: blue	c: cyan (sky blue)
-.- dash-dot	d: ◊	<: <	s: square (□)	k: black	

Table I.3 Functions and variables in MATLAB.

Function	Remark	Function	Remark
cos(x)		exp(x)	Exponential function
sin(x)		log(x)	Natural logarithm
tan(x)		log10(x)	Common logarithm
acos(x)	$\cos^{-1}(x)$	abs(x)	Absolute value
asin(x)	$\sin^{-1}(x)$	angle(x)	Phase of a complex number (rad)
atan(x)	$-\pi/2 \leq \tan^{-1}(x) \leq \pi/2$	sqrt(x)	Square root
atan2(y,x)	$-\pi \leq \tan^{-1}(y/x) \leq \pi$	real(x)	Real part
cosh(x)	$(e^x + e^{-x})/2$	imag(x)	Imaginary part
sinh(x)	$(e^x - e^{-x})/2$	conj(x)	Complex conjugate
tanh(x)	$(e^x - e^{-x})/(e^x + e^{-x})$	round(x)	The nearest integer (roundoff)
acosh(x)	$\cos h^{-1}(x)$	fix(x)	The nearest integer toward 0
asinh(x)	$\sin h^{-1}(x)$	floor(x)	The greatest integer $\leq x$

(*continued overleaf*)

Table I.3 (*Continued*)

Function	Remark	Function	Remark
atanh(x)	$\tanh^{-1}(x)$	ceil(x)	The smallest integer $\geq x$
max	Maximum and its index	sign(x)	1(positive)/0/−1(negative)
min	Minimum and its index	mod(y,x)	Remainder of y/x
sum	sum	rem(y,x)	Remainder of y/x
prod	product	eval(f)	Evaluate an expression
norm	norm	feval(f,a)	Function evaluation
sort	Sort in the ascending order	polyval	Value of a polynomial function
clock	Present time	poly	Polynomial with given roots
find	Index of element(s)	roots	roots of polynomial
tic	Start a stopwatch timer	toc	Read the stopwatch timer (elapsed time from tic)
date	Present date		

Reserved variables with special meaning			
i, j	$\sqrt{-1}$	pi	π
eps	Machine epsilon	Inf, inf	Largest number (∞)
realmax, realmin	Largest/smallest positive number	NaN	Not_a_Number (undetermined)
end	The end of for-loop or if, while, case statement or an array index	break	Exit while/for loop
nargin	Number of input arguments	nargout	# of output arguments
varargin	Variable input argument list	varargout	Variable output argument list

REFERENCES

[B-1] Bell, Howard E., "Gerschgorin's theorem and the zeros of polynomials," *Am. Math. Mon.* **72**, 292–295, 1965.

[B-2] Burden, Richard L. and J. Douglas Faires, *Numerical Analysis*, 7th ed., Brooks/Cole Thomson, CA, 2001.

[B-3] Burstein, Leonid, *Primary MATLAB® for Life Sciences: Guide for Beginners*, Bentham Science, 2013.

[C-1] Canale, Raymond and Steven C. Chapra, *Numerical Methods for Engineers: With Software and Programming Applications*, McGraw Hill, 2002.

[C-2] Cheng, David K., *Field and Wave Electromagnetics*, 2nd ed., Addison Wesley, 1989.

[F-1] Fausett, Laurene V., *Applied Numerical Analysis using MATLAB*, Prentice Hall, Upper Saddle River, New Jersey, 1999.

[H-1] Hamming, Richard W., *Numerical Methods for Scientists and Engineers*, 2nd ed., McGraw-Hill, New York, 1973.

[K-1] Kim, Phil, *MATLAB Deep Learning with Machine Learning, Neural Networks and Artificial Intelligence*, Apress, 2017.

[K-2] Kreyszig, Erwin, *Advanced Engineering Mathematics*, 8th ed., John Wiley, New York, 1999.

[K-3] Kreyszig, Erwin, *Introductory Functional Analysis with Applications*, John Wiley, New York, 1978.

[L-1] Lindfield, George R. and John E.T. Penny, *Numerical Methods using MATLAB*, 8th ed., Prentice Hall, Inc., Upper Saddle River, New Jersey, 2000.

[M-1] Mathews, John H. and Kurtis D. Fink, *Numerical Methods Using MATLAB*, Prentice Hall, Inc., Upper Saddle River, New Jersey, 1999.

[M-2] Maron, Melvin J., *Numerical Analysis*, Macmillan Publishing Co., Inc., New York, 1982

[N-1] Nakamura, Shoichiro, *Numerical Anaylsis and Graphic Visualization with MATLAB*, 2nd ed., Prentice Hall, Inc., New Jersey, 2002

[O-1] Oppenheim, Alan V. and Ronald W. Schafer, *Discrete-time Signal Processing*, Prentice Hall, Inc., Upper Saddle River, New Jersey, 1989.

Applied Numerical Methods Using MATLAB®, Second Edition. Won Y. Yang, Jaekwon Kim, Kyung W. Park, Donghyun Baek, Sungjoon Lim, Jingon Joung, Suhyun Park, Han L. Lee, Woo June Choi, and Taeho Im.
© 2020 John Wiley & Sons, Inc. Published 2020 by John Wiley & Sons, Inc.
Companion website: www.wiley.com/go/yang/appliednumericalmethods

[P-1] Peaceman, D. W. and H. H. Rachford, Jr., "The numerical solution of parabolic and elliptic differential equations," *J. Soc. Ind. Appl. Math.* **3**, 28–41, 1955.

[P-2] Pham, Duc T. and D. Karaboga, *Intelligent Optimization Techniques*, Springer-Verlag, London, 1998

[P-3] Phillips, Charles L. and Troy Nagle, *Digital Control System Analysis and Design*, Prentice Hall, Inc., Upper Saddle River, New Jersey, 2002.

[R-1] Rao, Singiresu S., *The Finite Element Method in Engineering*, 3rd ed., Butterworth Heinemann, Boston, MA, 1999.

[R-2] Recktenwald, Gerald W., *Numerical Methods with MATLAB*, Prentice Hall, Inc., Upper Saddle River, New Jersey, 2000.

[S-1] Schilling, Robert J. and Sandra L. Harris, *Applied Numerical Methods for Engineers using MATLAB and C*, Brooks/Cole Publishing Company, Pacific Grove, CA, 2000.

[S-2] Silvester, Peter P. and Ronald L. Ferrari, *Finite Elements for Electrical Engineers*, 3rd ed., Cambridge University Press, Cambridge, UK, 1996.

[S-3] Stoer, J. and R. Bulirsch, *Introduction to Numerical Analysis*, Springer-Verlag, New York, 1980.

[W-1] Website https://en.wikipedia.org/wiki/Bes_sel_function: Bessel Function.

[W-2] Website https://en.wikipedia.org/wiki/Dis_crete-time_Fourier_transform: DtFT (Discrete- Time Fourier Transform).

[W-3] Website https://en.wikipedia.org/wiki/ Her_mite_polynomials: Hermit polynomial.

[W-4] Website https://en.wikipedia.org/wiki/_Poi sson_distribution: Poisson distribution.

[W-5] Website https://en.wikipedia.org/wiki/Tay_lor_series: Taylor series.

[W-6] Webpage http://mathfaculty.fullerton.edu/mathews/n2003/NumericalUndergrad-Mod.html: about Numerical Methods.

[W-7] Website http://mathworld.wolfram.com/L HospitalsRule.html: L'Hospital's rule.

[W-8] Website http://web.eecs.utk.edu/~roberts/ECE315/PresentationSlides/Chapter6.pdf: CtFT (Continuous-Time Fourier Transform).

[Y-1] Yang, Won Y. et al., *Circuit Systems with MATLAB and PSpice*, John Wiley, New York, 2010.

[Y-2] Yang, Won Y. et al., *Electronic Circuits with MATLAB, PSpice, and Smith Chart*, John Wiley, New York, 2019.

[Y-3] Yang, Won Y. et al., *MATLAB/Simulink for Digital Signal Processing*, Hongreung Science, Seoul, Korea, 2012.

[Z-1] Zienkiewicz, Olek C. and Robert L. Taylor, *The Finite Element Method*, 4th ed., Vol. **1**, McGraw-Hill, London, 1989.

INDEX

A

absolute error, 37–38
acceleration of Aitken, 228
Adams-Bashforth-Moulton (ABM) method, 314–318
adaptive filter, 439–443
adaptive input argument, 51
adaptive quadrature, *see* quadrature
alternating direction implicit (ADI) method, 527
animation, 352, 555, 574
apostrophe, 17

B

back-propagation algorithm, 434
backslash, 21–22, 65–66, 82, 127
backward difference approximation, 247, 254
backward substitution, 87–90, 104–105
Bairstow's method, 212–214
basic variable, 418–419, 432
basis function (of FEM), 535–540
beamforming, 496, 498
bilinear interpolation, 155–157
bisection method, 201–203
BJT circuit, 219–222
Boltzmann, 390
boundary condition, 333, 356–360, 510–512
  $\sim$ for a cubic spline, 147
  Dirichlet $\sim$, 510, 544, 545
  Neumann $\sim$, 514, 544
Boundary mode, 548
boundary node, 533
boundary value problem (BVP), 306, 333–340
bracketing method, 205, 207
branch-and-bound (BB) method, 425–427
Bulirsch-Stoer, 178–179

C

cantilever beam, 346
case, 29
catastrophic cancellation, 37
central difference approximation, 247, 249, 254
cftool (curve fitting tool), 165–166
characteristic equation, 468, 579
characteristic value, *see* eigenvalue
characteristic vector, *see* eigenvector
Chebyshev coefficient polynomial, 139–140
Chebyshev node, 137–139, 177
Chebyshev polynomial, 140–141
chemical reactor, 345
Cholesky decomposition (factorization), 105
circuit, 219–222, 232–237
circulant matrix, 499
conjugate gradient method, 387–389
  Fletcher-Reeves (FR) $\sim$, 388
  Polak-Ribiere (PR) $\sim$, 388
constrained linear least squares, *see* least squares
constrained optimization, 376, 399–409, 413–433
constructive solid geometry (CSG), 546
contour, 12, 410
corrector, 314, 315

Applied Numerical Methods Using MATLAB®, Second Edition. Won Y. Yang, Jaekwon Kim, Kyung W. Park, Donghyun Baek, Sungjoon Lim, Jingon Joung, Suhyun Park, Han L. Lee, Woo June Choi, and Taeho Im.
© 2020 John Wiley & Sons, Inc. Published 2020 by John Wiley & Sons, Inc.
Companion website: www.wiley.com/go/yang/appliednumericalmethods

correlation matrix, 495–496
covariance matrix, 486–488
Crank-Nicholson method, 518–520
cross entropy function, 462
CtFS (Continuous-time Fourier series), 270
CtFT (Continuous-time Fourier transform), 74–76, 589–590
cubic spline, 146–153
curve fitting, 130, 158–166, 182–185
cutting plane method, 428–432

D

damped Newton method, 211–212, 244
dat-file, 3–4
data file, 3–4
debugging, 625
decoupling, 471–474
deep neural network (DNN), *see* neural network
departing (leaving) variable, 420
determinant, 578
DFT (Discrete Fourier Transform), 167–171
  recursive computation of ∼, 195
diagonalization, 469–474
  symmetric ∼ theorem, *see* symmetric
difference approximation
  ∼ of the first derivative, 246–248
  ∼ of the second or higher derivative, 253–257
differential equation, 306–340, 601
Dirichlet boundary condition, *see* boundary condition
discrete Fourier transform, *see* DFT
discrete-time Fourier Transform, *see* DtFT
discretization of LTI state equation, *see* state equation
distinct eigenvalues, 470
divided difference, 133–135
DoA (degree of arrival) estimation, *see* eigenvector
double integral/integration, 278–281, 303
Draw mode, 546
DtFT (Discrete-time Fourier Transform), 590

E

eigenmode PDE, 544
eigenpair, 468
eigenvalue, 468–469, 579
  physical meaning of ∼, 485–489
  ∼problem, 366–372
eigenvector, 468–469, 579

differential equation with ∼s, 489–493
DoA (degree of arrival) estimation with ∼s, 493–498
physical meaning of ∼, 485–489
electric potential, 543, 556
electronic circuit, 219–222, 232–237
element-by-element operation, 15, 18
elliptic PDE, 510–515, 532, 543
entering (incoming) variable, 419–420
eps, 15, 34
error, 35–36
  absolute ∼, 37–38
  ∼ analysis of difference approximation, 246, 249–252
  ∼ analysis of midpoint rule, 261
  ∼ analysis of Simpson's rule, 262
  ∼ analysis of trapezoidal rule, 262
  ∼ estimate, 263
  ∼ magnification, 36
  ∼ propagation, 38
  relative ∼, 37–38
errorbar(), 163
Euler's method, 306–309
explicit central difference method, 526–528
explicit forward Euler method, 515–516
exponent field, 31–32

F

false position (regular falsi), 203–205
FFT (Fast Fourier Transform), 167–168
finite difference method (FDM), 336–340
finite element method (FEM), 532–543
fixed-point
  ∼ iteration, 198–201, 223–227
  ∼ theorem, 198
Fletcher-Reeves (FR) conjugate gradient method, *see* conjugate gradient method
forward difference approximation, 246, 254
forward substitution, 104–105
Fourier transform
  continuous-time ∼, *see* CtFT
  discrete-time ∼, *see* DtFT
full pivoting, 95, 119

G

GA, *see* genetic algorithm
Gauss(ian) elimination, 86–95
Gauss quadrature, *see* quadrature
Gauss-Chebyshev, 277–278
Gauss-Hermite, 275–276, 294
Gauss-Jordan elimination, 97–99

Gauss-Laguerre integration, 277, 295
Gauss-Legendre integration, 272–275
Gauss-Seidel iteration, 111–116
Gaussian (normal) distribution, 26–27
genetic algorithm (GA), 393–399
Gerschgorin's disk theorem, 477
golden search method, 376–378, 448
Gomory cut, 428–430
gradient, 383–384, 585
graphic, 6–13, 608

H

Hamming method, 316–317
heat equation, 515, 523, 525
heat flow equation, 521
Helmholtz's equation, 510
Hermite interpolating polynomial, 153–154
Hermite polynomial, 72, 275
Hermitian, 580
Hessenberg form, 504–505
Hessian, 385, 386, 404, 586
Heun's method, 309–310
hidden bit, 32–35, 69
Hilbert matrix, 96–97
histogram, 10–11, 26
Householder, 501–504
hyperbolic PDE, 526–532

I

IDFT, 167
IEEE 64-bit floating-point number, 31–35
ill-condition, 95–96
impedance matching, 238–242
implicit backward Euler method, 516–518, 522
improper integral, 289–294
inactive inequality, 400
inconsistency, 91, 93, 95
interpolation
    $\sim$ by Chebyshev polynomial, 137–142
    $\sim$ by Lagrange polynomial, 130–131
    $\sim$ by Newton polynomial, 132–136
    two-dimensional (2D) $\sim$, 155–158
    $\sim$ using DFS, 172–174
interpolation function, *see* basis function
intersection of two circles, 243–244
inverse matrix, 100, 580
inverse power method, *see* power method
    shifted $\sim$, *see* power method
IVP (initial value problem), 327

J

Jacobi iteration, 108–111
Jacobi method, 478–481
Jacobian, 210, 586, 598

K

keyboard input, 3, 6

L

Lagrange coefficient polynomial, 130–131
Lagrange multiplier method, 80, 399–405
Lagrange polynomial, 130–131
Laguerre polynomial, 277
Laplace transform, 322, 588
Laplace's equation, 510, 513, 543, 549, 556
largest number in MATLAB, 15, 30–31, 34
learning factor, 439, 447
least mean square (LMS), 441
least square (LS), 82–83, 159
    constrained linear $\sim$, 415
    $\sim$ error (LSE), 78
    nonlinear $\sim$ (NLLS), 411–412
    nonnegative $\sim$ (NLS), 416, 454
    recursive $\sim$ estimation (RLSE), 83–86, 443–447
    weighted $\sim$ (WLS), 159
Legendre polynomial, 273–274
length of arc/curve, 298
limit, 597–598
linear programming (LP), 416–433
    mixed integer $\sim$, 423–433
logical operator, 28
loop, 29–30
    $\sim$ iteration, 29, 43–44
Lorenz equation, 345
loss of significance, 36–39
LP, *see* linear programming
    $\sim$ relaxation, 424
LSE, *see* least square
LU decomposition (factorization), 100–104

M

mantissa field, 31–32
mat-file, 3–4
mathematical functions, 13–14
matrix, 16–25, 578–584
matrix inversion lemma, 84, 445, 584
mean value theorem, 575
mesh, 12–13, 54, 608
Mesh mode, 548
midpoint rule, 259–262

N

minimum ratio rule, 419, 422
minimum-norm solution, 79–82
mixed boundary condition, 339, 357–360
mixed integer linear programming (MILP), *see* linear programming
modal matrix, 470
mode, 328, 474–486
modification formula, 315, 317
mu (μ)-law, 57–58
mu-inverse (μ^{-1}) law, 57–58, 391–393

N

negligible addition, 35
Nelder-Mead Algorithm, 380–382
nested
~ computing, 42–43, 71–72, 134
Neumann boundary condition, *see* boundary condition
neural network (NN), 433–438
deep ~ (DNN), 463–465
Newton (-Raphson) method, 205–208, 385–386
Newton polynomial, 132–137
nonbasic variables, 419, 432
nonlinear BVP, 339, 363–365, 372
nonlinear least squares (NLLS), *see* least squares
nonnegative least squares (NLS), *see* least squares
normal (Gaussian) distribution, 26–27
normalized exponential function, 434
normalized range, 33–34
null space, 79–81, 125, 581
numerical differentiation, 246–258
numerical integration, 259–271

O

orthogonal, 581
orthogonalization, 483–484
orthonormal, 478, 479, 481–482
overdetermined, 82–83, 106–107
overflow, 35, 39, 70

P

Pade approximation, 142–146, 177
parabolic PDE, 510, 515–526
two-dimensional ~, 423–526
parallelepiped, 489
parallelogram, 488–489
parameter passing through VARARGIN, 50–51

parameter sharing via GLOBAL, 49–50
partial differential equation (PDE), 510–543
partial pivoting, 88–95, 117–118
scaled ~, 93
PDE mode, 548
PDEtool, 543–558
penalty function method, 406–409
permutation, 100, 581
persistent excitation, 186
plot, *see* graphic
Plot mode, 549
Poisson's equation, 510
Polak-Ribiere (PR) method, *See* conjugate gradient method
polynomial
~ approximation, 137
~ wiggle, 137
positive definite, 582
power method, 475–478
inverse ~, 476
scaled ~, 475–476
shifted inverse ~, 477
predictor, 309, 312–317
product
cross (outer, vector) ~, 584
dot (inner, scalar) ~, 583
inner (dot, scalar) ~, 583
outer (cross, vector) ~, 584
scalar (dot, inner) ~, 583
vector (cross, outer) ~, 584
projection operator, 80, 481
pseudo (generalized) inverse, 80, 82, 124–125
PWL (piecewise linear) function, 281–282

Q

QR decomposition (factorization), 105–107, 481–483
quadratic approximation method, 378–380
quadratic interpolation, 175
quadratically convergent, 206
quadrature, 259
adaptive ~, 268–271
Gauss ~, 272–278
quantization error, 35, 69, 70, 249
quenching factor, 390, 391

R

rank, 581
rational interpolation, 178–179
rectified linear unit (ReLU), 463

recursive, 45, 132, 195, 227, 265
~ (self-calling) routine, 45, 227
recursive least square estimation (RLSE), *see* least squares
redundancy, 90, 93, 95
regula falsi, *see* false position
relational operators, 28
relaxation, 116
reserved constants/variables, 15
Richardson's extrapolation, 248
RLSE, *see* least squares
robot path planning, 180–181
Romberg integration, 265–267
rotation, 488
~ matrix, 478–481
round-off error, 35, 39
row echelon form, 582
row space, 581
row switching, 92
Runge phenomenon, 137
Runge-Kutta (RK4), 310–311

S

saddle point, 347
satellite orbit problem, 216–218
scaled partial pivoting, *see* partial pivoting
Schroder method, 230
secant method, 208–209
self-calling routine, *see* recursive routine
set path, 2
shape function, *see* basis function
shifted inverse power method, *see* power method
shooting method, 333–336
shooting position, 356
similarity transformation, 469–474
simplex method, 417–422
simplex tableau, 417, 419
Simpson's rule, 259–264
simulated annealing (SA), 389–393
singular value decomposition (SVD), 106–108, 124–127
slack variable, 400, 419, 421
smallest positive number in MATLAB, 15, 30, 33
softmax function, *see* normalized exponential function,
Solve mode, 548
SOR (successive over-relaxation), 116
source row, 428
sparse, 603–604

stability, 515–516, 518–519, 522, 528, 531
state equation, 320–328
  discretization of LTI, 324–327
  high-order differential equation to ~, 327–328
steepest descent, 383–384
  ~ rule, 420
Steffensen method, 228–229
step-size, 211, 246, 249–252, 387
  adaptation ~, 439, 440
  ~ dilemma, 250
stiff, 328–332, 347–349, 486
Sturm-Liouville (BVP) equation, 370–372
surface area of revolutionary object, 299–300
SVD, *see* singular value decomposition
symbolic, 257, 291, 595–602
  ~ algebraic equation, 601
  ~ computation, 595–602
  ~ differential equation, 601
  ~ differentiation, 598
  ~ integration, 598
  ~ linear algebra, 600
  priority of ~ variables, 215
  ~ solution, 215
  ~ variable, 595–596
symmetric, 580
  ~ diagonalization theorem, 402

T

Taylor series theorem, 576, 599
term-wise operation, *see* element-by-element operation
Toeplitz matrix, 499
trapezoidal rule, 259–260, 263–265
tridiagonal, 120
truncation error, 35, 246, 248–250

U

unbounded case, 419
unconstrained optimization, 376, 376–399, 409–413
underdetermined, 79–82
underflow, 35, 70
uniform probabilistic distribution, 25
unitary, 581
un-normalized range, 33–34

V

Van der Pol equation, 331–333, 344
Van der Waals Isotherms, 242–243

vector
    differentiation w.r.t. a ~, 585–586
    ~ operation, 43–44
volume, 279
    ~ of revolutionary 3D object, 300

W

wave equation, 526, 528–531
    2D ~, 529–531
weight least square (WLS), *see* least square

INDEX FOR MATLAB FUNCTIONS

(cf) A/C/E/P/R/S/T stand for Appendix/Chapter/Example/Problems/Remark/Section/Table, respectively.

Name	Place	Description
ABMc()	S6.4.1	Predictor/Corrector coefficients in Adams-Bashforth-Moulton ODE solver
adapt_smpsn()	S5.8	integration by the adaptive Simpson method
adc1()	P1.10	AD conversion
adc2()	P1.10	AD conversion
axis()	S1.1.4	specify axis limits or appearance
backslash(\)	S1.1.7/R1.2	left matrix division
backsubst()	S2.4.1	backward substitution for lower-triangular matrix equation
bar()/barh()	S1.1.4	a vertical/horizontal bar chart
bisct()	S4.2	bisection method to solve a nonlinear equation
bisct_r()	P4.4	bisection method to solve a nonlinear equation (self-calling)
BJT_DC_analysis_exp0()	S4.9	DC analysis of a BJT circuit
BJT2_complementary0()	S4.9	Analysis of a complementary BJT circuit
branch_and_bound()	S7.3.4	use the branch-and-bound method to solve a MILP problem
break	S1.1.9	terminate execution of a `for` loop or `while` loop
bvp2_eig()	P6.12	solve an eigenvalue BVP2

Applied Numerical Methods Using MATLAB®, Second Edition. Won Y. Yang, Jaekwon Kim, Kyung W. Park, Donghyun Baek, Sungjoon Lim, Jingon Joung, Suhyun Park, Han L. Lee, Woo June Choi, and Taeho Im.
© 2020 John Wiley & Sons, Inc. Published 2020 by John Wiley & Sons, Inc.
Companion website: www.wiley.com/go/yang/appliednumericalmethods

620 INDEX FOR MATLAB FUNCTIONS

bvp2_fdf()	S6.6.2	FDM (Finite difference method) for a BVP
bvp2_fdfp()	P6.7	FDM for a BVP with initial derivative fixed
bvp2_shoot()	S6.6.1	Shooting method for a BVP (boundary value problem)
bvp2_shootp()	P6.7	Shooting method for a BVP with initial derivative fixed
bvp2_fdf()	S6.6.2	FDM (Finite difference method) for a BVP
bvp2_fdfp()	P6.7	FDM for a BVP with initial derivative fixed
bvp2m_shoot()	P6.8	Shooting method for BVP with mixed boundary condition I
bvp2m_fdfp()	P6.8	FDM for a BVP with mixed boundary condition I
bvp2mm_shoot()	P6.9	Shooting method for BVP with mixed boundary condition II
bvp2mm_fdf()	P6.9	FDM for a BVP with mixed boundary condition II
bvp4c()	S6.6.2, P6.11~13	Finite difference method for a BVP with initial derivative fixed
c2d_steq()	S6.5.2	continuous-time state equation to discrete-time one
ceil()	S1.1.5/T1.3	round toward infinity
Cheby()	S3.3	Chebyshev polynomial approximation
chol()	S2.4.2	Cholesky factorization
clear	S1.1.2	remove items from workspace, freeing up system memory
clf	S1.1.4	clear current figure window
compare_DFT_FFT	S3.9.1	compare DFT with FFT
cond()	S2.2.2	condition number
contour()	S1.1.5	2-D contour plot of a scalar-valued function of 2-D variable
conv()	S1.1.6	convolution of two sequences or multiplication of two polynomials
cspline()	S3.5	cubic spline interpolation
CTFS_exp()	S5.11	exponential CtFS (Continuous-time Fourier Series) coefs
CtFT1()	P1.28	CtFT (Continuous-time Fourier Transform)
curve_fit()	P3.11, p157	weighted least square (WLS) curve fitting
cutting_plane()	S7.3.4	use the cutting plane method to solve a MILP problem
dblquad()	S5.10	2-D (double) integral

diag()	S1.1.7/R1.3	construct a diagonal matrix or get diagonals of a matrix
difapx()	S5.3	difference approximation for numerical derivatives
diff()	AG2.3, S5.4	differences between neighboring elements in an array
disp()	S1.1.3	display text or array onto the (monitor) screen
DNN_multiclass()	P7.14	implement a deep neural network (DNN)
do_Cheby	S3.3	approximate by Chebyshev polynomial
do_condition	S2.2.2	condition numbers for ill-conditioned matrices
do_csplines	S3.5	interpolate by cubic splines
do_FFT	S3.9.1	do FFT (Fast Fourier Transform)
do_Gauss	S2.2.2	do Gauss elimination
do_interp2	S3.7	do 2D interpolation
do_Lagranp	S3.1	do Lagrange polynomial interpolation
do_lagnewch	S3.3	try Lagrange/Newton/Chebyshev polynomial
do_lu_dcmp	S2.4.1	do LU decomposition (factorization)
do_MBK	P6.4	simulate a mass-damper-spring system
do_Newtonp	S3.2	do Newton polynomial interpolation
do_Newtonp1	S3.2	do Newton polynomial interpolation
do_Pade	S3.4	do Pade (rational polynomial) approximation
do_polyfit	S3.8.2	do polynomial curve fitting
do_RDFT	P3.21	do recursive DFT
do_quiver	P6.0	use quiver() to plot the gradient vectors
do_rlse	S2.1.4	do recursive least-squares estimation
do_wlse	S3.8.2	do weighted least-squares curve fitting
DoA_estimation_with_eigenvector	S8.8	use eigenvectors to estimate the DoA (degree of arrival)
double()	AG.1	convert to double-precision
draw_MBK	P6.4	simulate a mass-damper-spring system
dsolve()	S6.5.1, AG.5	symbolic differential equation solver
eig()	S8.1	eigenvalues and eigenvectors of a matrix
eig_Jacobi()	S8.4	find the eigenvalues/eigenvectors of a symmetric matrix
eig_power()	S8.3	find the largest eigenvalue & the corresponding eigenvector
eig_QR()	P8.8	find eigenvalues using QR factorization
eig_QR_Hs()	P8.8	find eigenvalues using QR factorization via Hessenberg

else	S1.1.9	for conditional execution of statements
elseif	S1.1.9	for conditional execution of statements
end	S1.1.9	terminate `for/while/witch/try/if` statements or last index
error_DE_sol()	P6.5	evaluate the error of solution of differential eq.
error_DE2_sols()	P6.10	evaluate the error of solution(s) of 2nd-order differential eq.
eval()	S1.1.5/T1.3	evaluate a string containing a literal/symbolic expression
eye()	S1.1.7	identity matrix (having 1/0 on/off its diagonal)
ezplot()	S1.3.6	easy plot
falsp()	S4.3	false position method to solve a nonlinear equation
fem_basis_ftn()	S9.4	coefficients of each basis function for subregions
fem_coef()	S9.4	coefficients for subregions
feval()	S1.1.6	evaluation of a function defined by inline() or in an M-file
find()	P1.10	find indices of nonzero (true) elements
find_CTFS_PWL()	S5.11	find the CtFS coefficients of a PWL function
findsym()	S4.8	find symbolic variables in a symbolic expression
fix()	S1.1.6/T1.3	round towards zero
fixpt()	S4.1	fixed-point iteration to solve a nonlinear equation
fliplr()	S1.1.7	flip the elements of a matrix left-right
flipud()	S1.1.7	flip the elements of a matrix up-down
floor()	S1.1.6/T1.3	round to −infinity
fminbnd()	S7.1.2	unconstrained minimization of one-variable function
fmincon()	S7.3.2	constrained minimization
fminimax()	S7.3.2	minimize the maximum of vector/matrix-valued function
fminsearch()	S7.2.2, S7.3.1	unconstrained nonlinear minimization (Nelder-Mead)
fminunc()	S7.2.2, S7.3.1	unconstrained nonlinear minimization (gradient-based)
for	S1.1.9	repeat statements a specific number of times
format	S1.1.3	control display format for numbers
forsubst()	S2.4.1	forward substitution for lower-triangular matrix equation
fprintf()	S1.1.3, P1.2	write formatted data to screen or file

INDEX FOR MATLAB FUNCTIONS 623

fsolve()	S4.6,4.9, E4.3	solve nonlinear equations by a least squares (LS) method
Gauseid()	S2.5.2	Gauss-Seidel method to solve a system of linear equations
Gauss()	S2.2.2	Gauss elimination to solve a system of linear equations
Gauss_legendre()	S5.9.1	Gauss-Legendre integration
Gausslp()	S5.9.1	grid points of Gauss-Legendre integration formula
Gausshp()	S5.9.2	grid points of Gauss-Hermite integration formula
genetic()	S7.1.8	optimization by the genetic algorithm (GA)
ginput()	S1.1.4	input the x- & y-coordinates of point(s) clicked by mouse
global	S1.3.5	declare global variables
gradient()	P6.0	numerical gradient
grid on/off	S1.1.4	grid lines for 2-D or 3-D graphs
gtext()	S1.1.4	mouse placement of text in a 2-D graph
help	S1.1	display help comments for MATLAB routines
Hermit()	S3.6	Hermite polynomial interpolation
Hermitp()	S5.9.2	Hermite polynomial
Hermits()	S3.6	multiple Hermite polynomial interpolations
Hessenberg()	P8.6	transform a matrix into almost upper-triangular one
hist()	S1.1.4, 1.1.8	plot a histogram
hold on/off	S1.1.4	hold on/off current graph in the figure
Housholder()	P8.5	Householder matrix to zero-out the tail part of a vector
ICtFT1()	P1.28	ICtFT (Inverse Continuous-time Fourier Transform)
iD_NMOS_at_vDS_vGS()	P4.12	drain current of an NMOS
if	S1.1.9	for conditional execution of statements
ilaplace()		inverse Laplace transform
imp_match_1stub0()	P4.14	single-stub impedance matching
imp_match_1stub1()	P4.14	single-stub impedance matching
inline()	S1.1.6	define a function inside the program
inpolygon()	S9.4	is the point inside an polygonal region?
input()	S1.1.3	request and get user input
int()	S5.8, AG2.4	numerical/symbolic integration
interp1()	S3.5	1-D interpolation
interp2()	S3.7	2-D interpolation
intlinprog()	S7.3.4	solve an integer linear programming (ILP) problem
intrp1()	P3.10	1-D interpolation
intrp2()	S3.7	2-D interpolation

624 INDEX FOR MATLAB FUNCTIONS

interpolate_by_DFS	S3.9.3	interpolation using DFS
int2s()	S5.10, P5.15	2-D (double) integral
inv()	S2.1.1	the inverse of a matrix
isempty()	P1.1, P1.10	is it empty (no value)?
isnumeric()	S1.3.7	has it a numeric value?
jacob()	S4.6	Jacobian matrix of a given function
jacob1()	P5.3	Jacobian matrix of a given function
jacobi()	S2.5.1	Jacobi iteration to solve a equation
Jkb()	P1.23	1st kind of k-th order Bessel function
Lagranp()	S3.1	Lagrange polynomial interpolation
Lgndrp()	S5.9.1	Legendre polynomial
length()	S1.1.7	the length of a vector (sequence) or a matrix
limit()	AG2.2	limit of a symbolic expression
lin_eq()	S2.1.3	solve linear equation(s)
linprog()	S7.3.3	solve a linear programming (LP) problem
load	S1.1.2,4	read variable(s) from file
loglog()	S1.1.4	plot data as logarithmic scales for the x-axis and y-axis
lookfor	S1.1	search for string in the first comment line in all M-files
lscov()	S3.8.1	weighted least-squares with known (error) covariance
lsqcurvefit()	P3.11, S7.3.1	weighted nonlinear least-squares curve fitting
lsqlin()	S7.3.1	solve a linear least squares (LLS) problem
lsqnonlin()	S7.3.1	solve a non-linear least squares (NLLS) problem
lsqnonneg()	S7.3.2	find a non-negative least squares (NNLS) solution
lu()	S2.4.1	LU decomposition (factorization)
lu_dcmp()	S2.4.1	LU decomposition (factorization)
max()	S1.1.7	find the maximum element(s) of an array
mesh()	S1.1.5, S3.7	plot a mesh-type graph of f(x,y)
meshgrid()	S1.1.5, S3.7	grid points for plotting a mesh-type graph
min()	S1.1.7	find the minimum element(s) of an array
mkpp()	P1.13	make a piece-wise polynomial
mod()	S1.1.6/T1.3	remainder after division
mulaw()	P1.9	μ-law
mu_inv()	S7.1.7	μ^{-1}-law
multiply_matrix()	P1.14	matrix multiplication
Newton()	S4.4	Newton method to solve a nonlinear equation
Newtonp()	S3.2	Newton polynomial interpolation
Newtons()	S4.6	Newton method to solve a system of nonlinear equation

nm07e07	S7.5	apply the LMS/steepest descent/Newton methods for ...
nm07e08	S7.6	apply the RLS/LMS methods for parameter estimation
NN_multiclass()	S7.4	perform a NN(neural network) based multi-classification
norm()	P1.15	norm of vector/matrix
numel()	S1.1.7	number of elements in a given array
ode_ABM()	S6.4.1	solve a state equation by Adams-Bashforth-Moulton solver
ode_Euler()	S6.1	solve a state equation by Euler's method
ode_Ham()	S6.4.2	solve a state equation by Hamming ODE solver
ode_Heun()	S6.2	solve a state equation by Heun's method
ode_RK4()	S6.3	solve a state equation by Runge-Kutta method
ode23()/ode45()/ode113()	S6.4.3	ODE solver
ode15s()/ode23s()/ ode23t()/ode23tb()	S6.5.4	solve (stiff) ODEs
ones()	S1.1.7	constructs an array of ones
opt_conjg()	S7.1.6	optimization by Conjugate gradient method
opt_gs()	S7.1.1	optimization by Golden search
opt_quad()	S7.1.2	optimization by quadratic approximation
opt_Nelder()	S7.1.3	optimization by Nelder-Mead method
opt_steep()	S7.1.4	optimization by steepest descent
Padeap()	S3.4	Pade approximation
pdetool	S9.5	start the PDE toolbox GUI (graphical user interface)
pinv()	S1.1.7/R1.1	pseudo-inverse (generalized inverse)
pivoting()	S7.3.4	pivoting at a given pivot element
plot()	S1.1.4	linear 2-D plot
plot3()	S1.1.5	linear 3-D plot
pde_heat_exp()	S9.2.1	explicit forward Euler method for parabolic PDE (heat eq)
pde_heat_imp()	S9.2.2	implicit backward Euler method for parabolic PDE (heat eq)
pde_heat_CN()	S9.2.3	Crank-Nicholson method for parabolic PDE (heat eq)
pde_heat2_ADI()	S9.2.5	ADI method for parabolic PDE (2-D heat equation)
pde_poisson()	S9.1	central difference method for elliptic PDE (Poisson's eq)
pde_wave()	S9.3.1	central difference method for hyperbolic PDE (wave eq)
pde_wave2()	S9.3.2	central difference method for hyperbolic PDE (2-D wave eq)

Function	Section	Description
pdepe()	S9.2.4	solve a 1D parabolic or elliptic PDE
polar()	S1.1.4	plot polar coordinates in a Cartesian plane with polar grid
poly_der()	P1.13	derivative of polynomial
polyder()	P1.13	derivative of polynomial
polyfit()	P3.13	polynomial curve fitting
polyfits()	S3.8.2	polynomial curve fitting
polyint()	P1.13	integral of polynomial
polyval()	S1.1.6	evaluate a polynomial
ppval()	P1.13	evaluate a set of piece-wise polynomials
pretty()	AG	print symbolic expression like in type-set form
prod()	S1.1.7/T1.3	product of array elements
qr()	S2.4.2	QR factorization
qr_Hessenberg()	P8.7	QR factorization of Hessenberg form by Givens rotation
quad()	P1.8, S5.8	numerical integration
quadl()	S5.8	numerical integration
quiver()	P6.0	plot gradient vectors
quiver3()	P6.0	plot normal vectors on a surface
rand()	S1.1.8	uniform random number generator
randn()	S1.1.8	Gaussian random number generator
rational_interpolation()	P3.6	rational polynomial interpolation
repetition()	P1.17	repetition of a matrix
repmat()	S1.1.7	repetition of a matrix
reshape()	S1.1.7	a matrix into one with given numbers of row/columns
residue()	P1.13	partial fraction expansion of Laplace-transformed function
residuez()	P1.13	partial fraction expansion of z-transformed rational function
rlse_online()	S2.1.4	on-line Recursive Least-Squares Estimation
Rmbrg()	S5.7	Integration by Romberg method
robot_path	P3.9	determine a path of robot using cubic splines
roots()	P1.13	roots of a polynomial equation
roots_Bairstow()	S4.7	roots of a polynomial equation (using Bairstow's method)
round()	S1.1.6/T1.3	round to nearest integer
rot90()	S1.1.7	rotate a matrix by 90 degrees
save	S1.1.2	save variable(s) into a file
secant()	S4.5	secant method to solve a nonlinear equation
semilogx()	S1.1.4	plot data as logarithmic scales for the x-axis

INDEX FOR MATLAB FUNCTIONS 627

Function	Reference	Description
semilogy()	S1.1.4	plot data as logarithmic scales for the y-axis
sigmoid()	S7.4	sigmoid function defined (Eq. (7.4.4))
size()	S1.1.7	the numbers of rows/columns/ ... of a 1-D/2-D/3-D array
sim_anl()	S7.1.7	optimization by simulated annealing (SA)
simplex()	S7.3.3	implement the simplex method for linear programming (LP)
simplify()	AG	simplify a symbolic expression
Smpsns()	S5.6	integration by Simpson rule
Smpsns_fxy()	S5.10, P5.16	2D integration of a function f(x,y) along y
softmax()	S7.4	softmax or 'normalized exponential function' (Eq. (7.4.6))
solve()	P3.1, S4.8, AG.4	solve a (set of) symbolic algebraic equation(s)
sort()	S1.1.4	arranges the elements of an array in ascending order
spline()	S3.5	cubic spline
sprintf()	S1.1.4	make formatted data to a string
stairs()	S1.1.4	stair-step plot of zero-hold signal of sampled data systems
stem()	S1.1.4	plot discrete sequence data
Stfns()	P4.6	Steffensen Method to solve a nonlinear equation
subplot()	S1.1.4	divide the current figure into rectangular panes
subs()	AG.1	substitute
sum()	S1.1.7/R1.3	sum of elements of an array
surf()	P6.0	plot a surface-type graph of f(x,y)
surfnorm()	P6.0	generate vectors normal to a surface
svd()	S2.4.2	$\underline{s}$ingular $\underline{v}$alue $\underline{d}$ecomposition
switch	S1.1.9	switch among several cases
syms	AG.1	declare symbolic variable(s)
sym2poly()	S5.3, AG.2.5	extract the coefficients of symbolic polynomial expression
taylor()	S5.3, AG.2.5	Taylor series expansion
test_DNN_multiclass()	P7.14	test 'train_DNN_multiclass()'/'DNN_multiclass()'
test_NN_multiclass()	S7.4	test 'train_NN_multiclass()'/'NN_multiclass()'
text()	S1.1.4	add a text at the specified location on the graph
title()	S1.1.4	add a title string to current axes
train_DNN_multiclass	P7.14	train a (multi-layer) DNN for multi-class classification

train_NN_multiclass()	S7.4	train a (possibly multi-layer) NN for multi-class classification
trid()	S6.6.2	solve a tri-diagonal system of linear equations
trimesh()	S9.4	plot a triangular-mesh-type graph
trpzds()	S5.6	Integration by trapezoidal rule
try_beamforming	S8.8	beamforming toward estimated DoA to get the arriving signal
varargin()	S1.3.6	<u>var</u>iable length <u>in</u>put <u>arg</u>ument list
view()	S1.1.5, P1.4	3-D graph viewpoint specification
vpa()	AG.1	evaluate double array by <u>v</u>ariable <u>p</u>recision <u>a</u>rithmetic
while	S1.1.9	repeat statements an indefinite number of times
windowing()	P3.18	multiply a sequence by the specified window sequence
xlabel()/ylabel()	S1.1.4	label the x-axis/y-axis
zeros()	S1.1.7	make an array of zeros
zeroing()	P1.17	cross out every (kM-m)th element to zero

INDEX FOR TABLES

Table Number	Place	Description
Table 1.1	S1.1.3	Conversion type specifiers & special characters in fprintf()
Table 1.2	S1.1.4	Graphic line specifications used in the plot() command
Table 1.3	S1.1.6	Functions and variables inside MATLAB
Table 1.4	S1.1.9	Relational operators and logical operators
Table 2.1	S2.4.1	Residual error and the number of floating-point operations of various solutions
Table 3.1	S3.2	Divided difference table
Table 3.2	S3.2	Divided differences
Table 3.3	S3.3	Chebyshev coefficient polynomials
Table 3.4	S3.5	Boundary conditions for a cubic spline
Table 3.5	S3.8.3	Linearization of nonlinear functions by parameter/data transformation
Table 5.1.1	S5.2	The forward difference approximation (5.1.4) for the 1st derivative and its error depending on the step-size
Table 5.1.2	S5.2	The central difference approximation (5.1.8) for the 1st derivative and its error depending on the step-size
Table 5.2	S5.3	The difference approximation formulas for the 1st and 2nd derivatives
Table 5.3	S5.7	Romberg table
Table 6.1	S6.1	A numerical solution of DE (6.1.1) obtained by the Euler's method
Table 6.2	S6.4.3	Results of applying several routines for solving a simple DE
Table 7.1	S7.1.7	Results of running several unconstrained optimization routines with various initial values
Table 7.2	S7.3.1	Results of running several unconstrained optimization routines with various initial values
Table 7.3	S7.6	The names of the MATLAB built-in minimization routines

(cf) A: Appendix, P: Problem, S: Section, T: Table, R: Remark

Applied Numerical Methods Using MATLAB®, Second Edition. Won Y. Yang, Jaekwon Kim, Kyung W. Park, Donghyun Baek, Sungjoon Lim, Jingon Joung, Suhyun Park, Han L. Lee, Woo June Choi, and Taeho Im.
© 2020 John Wiley & Sons, Inc. Published 2020 by John Wiley & Sons, Inc.
Companion website: www.wiley.com/go/yang/appliednumericalmethods

Printed and bound by CPI Group (UK) Ltd, Croydon, CR0 4YY